D1460418

The Mitchell Beazley Joy of Knowledge Library

History and Culture 2

Scientiam non dedit natura semina scientiae nobis dedit

"Nature has given us not knowledge itself, but the seeds thereof."

Seneca

The Joy of Knowledge Encyclopaedia is affectionately dedicated to the memory of John Beazley 1932–1977, Book Designer, Publisher and Co-Founder of the publishing house of Mitchell Beazley Limited, by all his many friends and colleagues in the company.

The Joy of Knowledge Library

General Editor: James Mitchell
With an overall preface by Lord Butler, Master of Trinity College, University of Cambridge

Science and The Universe	Introduced by Sir Alan Cottrell, Master of Jesus College, University of Cambridge; and Sir Bernard Lovell, Professor of Radio Astronomy, University of Manchester
The Physical Earth	Introduced by Dr William Nierenberg, Director, Scripps Institution of Oceanography, University of California
The Natural World	Introduced by Dr Gwynne Vevers, Assistant Director of Science, the Zoological Society of London
Man and Society	Introduced by Dr Alex Comfort, Professor of Pathology, Irvine Medical School, University of California, Santa Barbara
History and Culture 1 **History and Culture** 2	Introduced by Dr Christopher Hill, Master of Balliol College, University of Oxford
Man and Machines	Introduced by Sir Jack Callard, former Chairman of Imperial Chemical Industries Limited
The Modern World	Introduced by Dr Michael Wise, Professor of Geography, London School of Economics and Political Science

Fact Index A–K

Fact Index L–Z

The Mitchell Beazley Joy of Knowledge Library

History and Culture 2

MITCHELL BEAZLEY

The Joy of Knowledge Library

ISBN 0 85533 110 0

Typesetting by Filmtype Services Limited, England
Photoprint Plates Ltd, Rayleigh, Essex, England

Printed in England by Balding + Mansell

Major contributors and advisers to The Joy of Knowledge Library

Fabian Acker CEng, MIEE, MIMarE; Professor Leslie Alcock; Professor H.C. Allen MC; Leonard Amey OBE; Neil Ardley BSc; Professor H.R.V. Arnstein DSc, PhD, FIBiol; Russell Ash BA(Dunelm), FRAI; Norman Ashford PhD, CEng, MICE, MASCE, MCIT; Professor Robert Ashton; B.W. Atkinson BSc, PhD; Anthony Atmore BA; Professor Philip S. Bagwell BSc(Econ), PhD; Peter Ball MA; Edwin Banks MIOP; Professor Michael Banton; Dulan Barber; Harry Barrett; Professor J.P. Barron MA, DPhil, FSA; Professor W.G. Beasley FBA; Alan Bender PhD, MSc, DIC, ARCS; Lionel Bender BSc; Israel Berkovitch PhD, FRIC, MIChemE; David Berry MA; M.L. Bierbrier PhD; A.T.E. Binsted FBBI (Dipl); David Black; Maurice E.F. Block BA, PhD(Cantab); Richard H. Bomback BSc (London), FRPS; Basil Booth BSc (Hons), PhD, FGS, FRGS; J. Harry Bowen MA(Cantab), PhD(London); Mary Briggs MPS, FLS; John Brodrick BSc(Econ); J.M. Bruce ISO, MA, FRHistS, MRAeS; Professor D.A. Bullough MA, FSA, FRHistS; Tony Buzan BA(Hons) UBC; Dr Alan R. Cane; Dr J.G. de Casparis; Dr Jeremy Catto MA; Denis Chamberlain; E.W. Chanter MA; Professor Colin Cherry D Sc(Eng), MIEE; A.H. Christie MA, FRAI, FRAS; Dr Anthony W. Clare MPhil(London), MB, BCh, MRCPI, MRCPsych; Professor Aidan Clarke MA, PhD, FTCD; Sonia Cole; John R. Collis MA, PhD; Professor Gordon Connell-Smith BA, PhD, FRHistS; Dr A.H. Cook FRS; Professor A.H. Cook FRS; J.A.L. Cooke MA, DPhil; R.W. Cooke BSc, CEng, MICE; B.K. Cooper; Penelope J. Corfield MA; Robin Cormack MA, PhD, FSA; Nona Coxhead; Patricia Crone BA, PhD; Geoffrey P. Crow BSc(Eng), MICE, MIMunE, MInstHE, DIPTE; J.G. Crowther; Professor R.B. Cundall FRIC; Noel Currer-Briggs MA, FSG; Christopher Cviic BA(Zagreb), BSc(Econ, London); Gordon Daniels BSc(Econ, London), DPhil(Oxon); George Darby BA; G.J. Darwin; Dr David Delvin; Robin Denselow BA; Professor Bernard L. Diamond; John Dickson; Paul Dinnage MA; M.L. Dockrill BSc(Econ), MA, PhD; Patricia Dodd BA; James Dowdall; Anne Dowson MA(Cantab); Peter M. Driver BSc, PhD, MIBiol; Rev Professor C.W. Dugmore DD; Herbert L. Edlin BSc, Dip in Forestry; Pamela Egan MA(Oxon); Major S.R. Elliot CD, BComm; Professor H.J. Eysenck PhD, DSc; Dr Peter Fenwick BA, MB, BChir, DPM, MRCPsych; Jim Flegg BSc, PhD, ARCS, MBOU; Andrew M. Fleming MA; Professor Antony Flew MA(Oxon), DLitt (Keele); Wyn K. Ford FRHistS; Paul Freeman DSc(London); G.S.P. Freeman-Grenville DPhil, FSA, FRAS, G.E. Fussell DLitt, FRHistS; Kenneth W. Gatland FRAS, FBIS; Norman Gelb BA; John Gilbert BA(Hons, London); Professor A.C. Gimson; John Glaves-Smith BA; David Glen; Professor S.J. Goldsack BSc, PhD, FINSTP, FBCS; Richard Gombrich MA, DPhil; A.F. Gomm; Professor A. Goodwin MA; William Gould BA(Wales); Professor J.R. Gray; Christopher Green PhD; Bill Gunston; Professor A. Rupert Hall LittD; Richard Halsey BA(Hons, UEA); Lynette K. Hamblin BSc; Norman Hammond; Peter Harbison MA, DPhil; Professor Thomas G. Harding PhD; Professor D.W. Harkness; Richard Harris; Dr Randall P. Harrison; Cyril Hart MA, PhD, FRICS, FIFor; Anthony P. Harvey; Nigel Hawkes BA(Oxon); F.P. Heath; Peter Hebblethwaite MA (Oxon), LicTheol; Frances Mary Heidensohn BA; Dr Alan Hill MC, FRCP; Robert Hillenbrand MA, DPhil; Catherine Hills PhD; Professor F.H. Hinsley; Dr Richard Hitchcock; Dorothy Hollingsworth OBE, BSc, FRIC, FIBiol, FIFST, SRD; H.P. Hope BSc(Hons, Agric); Antony Hopkins CBE, FRCM, LRAM, FRSA; Brian Hook; Peter Howell BPhil, MA(Oxon); Brigadier K. Hunt; Peter Hurst BDS, FDS, LDS, RSCEd, MSc(London); Anthony Hyman MA, PhD; Professor R.S. Illingworth MD, FRCP, DPH, DCH; Oliver Impey MA, DPhil; D.E.G. Irvine PhD; L.M. Irvine BSc; E.W. Ives BA, PhD; Anne Jamieson cand mag(Copenhagen), MSc (London); Michael A. Janson BSc; G.H. Jenkins PhD; Professor P.A. Jewell BSc (Agric), MA, PhD. FIBiol; Hugh Johnson; Commander I.E. Johnston RN; I.P. Jolliffe BSc, MSc, PhD, ComplCE, FGS; Dr D.E.H. Jones ARCS, FCS; R.H. Jones PhD, BSc, CEng, MICE, FGS, MASCE, Hugh Kay; Dr Janet Kear; Sam Keen; D.R.C. Kempe BSc, DPhil, FGS; Alan Kendall MA (Cantab); Michael Kenward; John R. King BSc(Eng), DIC, CEng, MIProdE; D.G. King-Hele FRS; Professor J.F. Kirkaldy DSc; Malcolm Kitch; Michael Kitson MA; B.C. Lamb BSc, PhD; Nick Landon; Major J.C. Larminie QDG,Retd; Diana Leat BSc(Econ), PhD; Roger Lewin BSc, PhD, Harold K. Lipset; Norman Longmate MA(Oxon); John Lowry; Kenneth E. Lowther MA; Diana Lucas BA(Hons); Keith Lye BA, FRGS; Dr Peter Lyon; Dr Martin McCauley; Sean McConville BSc; D.F.M. McGregor BSc, PhD(Edin); Jean Macqueen PhD; William Baird MacQuitty MA(Hons), FRGS, FRPS; Professor Rev F.X. Martin OSA; Jonathan Martin MA; Rev Cannon E.L. Mascall DD; Christopher Maynard MSc, DTh; Professor A.J. Meadows; Dr T.B. Millar; John Miller MA, PhD; J.S.G. Miller MA, DPhil, BM, BCh; Alaric Millington BSc, DipEd, FIMA; Rosalind Mitchison MA, FRHistS; Peter L. Moldon; Patrick Moore OBE; Robin Mowat MA, DPhil; J. Michael Mullin BSc; Alistair Munroe BSc, ARCS; Professor Jacob Needleman; John Newman MA, FSA; Professor Donald M. Nicol MA PhD; Gerald Norris; Professor F.S. Northedge PhD; Caroline E. Oakman BA(Hons. Chinese); S. O'Connell MA(Cantab), MInstP; Dr Robert Orr; Michael Overman; Di Owen BSc; A.R.D. Pagden MA, FRHistS; Professor E.J. Pagel PhD; Liam de Paor MA; Carol Parker BA(Econ), MA (Internat. Aff.); Derek Parker; Julia Parker DFAstrolS; Dr Stanley Parker; Dr Colin Murray Parkes MD, FRC(Psych), DPM; Professor Geoffrey Parrinder MA, PhD, DD(London), DLitt(Lancaster); Moira Paterson; Walter C. Patterson MSc; Sir John H. Peel KCVO, MA, DM, FRCP, FRCS, FRCOG; D.J. Penn; Basil Peters MA. MInstP, FBIS; D.L. Phillips FRCR, MRCOG; B.T. Pickering PhD, DSc; John Picton; Susan Pinkus; Dr C.S. Pitcher MA, DM, FRCPath; Alfred Plaut FRCPsych; A.S. Playfair MRCS, LRCP, DObstRCOG; Dr Antony Polonsky; Joyce Pope BA; B.L. Potter NDA, MRAC, CertEd; Paulette Pratt; Antony Preston Frank J. Pycroft; Margaret Quass; Dr John Reckless; Trevor Reese BA, PhD, FRHistS; M.M. Reese MA (Oxon); Derek A. Reid BSc, PhD; Clyde Reynolds BSc; John Rivers; Peter Roberts; Colin A. Ronan MSc, FRAS; Professor Richard Rose BA(Johns Hopkins), DPhil (Oxon); Harold Rosenthal; T.G. Rosenthal MA(Cantab); Anne Ross MA, MA(Hons, Celtic Studies), PhD, (Archaeol and Celtic Studies, Edin); Georgina Russell MA; Dr Charles Rycroft BA (Cantab), MB(London), FRCPsych; Susan Saunders MSc(Econ); Robert Schell PhD; Anil Seal MA, PhD(Cantab); Michael Sedgwick MA(Oxon); Martin Seymour-Smith BA(Oxon), MA(Oxon); Professor John Shearman; Dr Martin Sherwood; A.C. Simpson BSc; Nigel Sitwell; Dr Alan Sked; Julie and Kenneth Slavin FRGS, FRAI; Professor T.C. Smout; Alec Xavier Snobel BSc(Econ); Terry Snow BA, ATCL; Rodney Steel; Charles S. Steinger MA, PhD; Geoffrey Stern BSc(Econ); Maryanne Stevens BA(Cantab), MA(London); John Stevenson DPhil, MA; J. Sidworthy MA; D. Michael Stoddart BSc, PhD; Bernard Stonehouse DPhil, MA, BSc, MInstBiol; Anthony Storr FRCP, FRCPsych; Richard Storry; Charles Stuart-Jervis; Professor John Taylor; John W.R. Taylor FRHistS, MRAeS. FSLAET; R.B. Taylor BSc(Hons, Microbiol); J. David Thomas MA, PhD; D. Thompson BSc(Econ); Harvey Tilker PhD; Don Tills PhD, MPhil, MIBiol, FIMLS; Jon Tinker; M. Tregear MA; R.W. Trender; David Trump MA, PhD, FSA; M.F. Tuke PhD; Christopher Tunney MA; Laurence Urdang Associates (authentication and fact check); Sally Walters BSc; Christopher Wardle; Dr D. Washbrook; David Watkins; George Watkins MSc; J.W.N. Watkins; Anthony J. Watts; Dr Geoff Watts; Melvyn Westlake; Anthony White MA(Oxon), MAPhil(Columbia); Dr Ruth D. Whitehouse; P.J.S. Whitmore MBE, PhD; Professor G.R. Wilkinson; Rev H.A. Williams CR; Christopher Wilson BA; Professor David M. Wilson; John B. Wilson BSc, PhD, FGS, FLS; Philip Windsor BA, DPhil(Oxon), Roy Wolfe BSc(Econ), MSc; Donald Wood MA PhD, Dr David Woodings MA, MRCP, MRCPath; Bernard Yallop PhD, BSc, ARCS, FRAS Professor John Yudkin MA, MD, PhD(Cantab), FRIC, FIBiol, FRCP.

The General Editor wishes particularly to thank the following for all their support:
Nicolas Bentley
Bill Borchard
Adrianne Bowles
Yves Boisseau
Irv Braun
Theo Bremer
the late Dr Jacob Bronowski
Sir Humphrey Browne
Barry and Helen Cayne
Peter Chubb
William Clark
Sanford and Dorothy Cobb
Alex and Jane Comfort
Jack and Sharlie Davison
Manfred Denneler
Stephen Elliott
Stephen Feldman
Orsola Fenghi
Professor Richard Gregory
Dr Leo van Grunsven
Jan van Gulden
Graham Hearn
the late Raimund von Hofmansthal
Dr Antonio Houaiss
the late Sir Julian Huxley
Alan Isaacs
Julie Lansdowne
Professor Peter Lasko
Andrew Leithead
Richard Levin
Oscar Lewenstein
The Rt Hon Selwyn Lloyd
Warren Lynch
Simon macLachlan
George Manina
Stuart Marks
Bruce Marshall
Francis Mildner
Bill and Christine Mitchell
Janice Mitchell
Patrick Moore
Mari Pijnenborg
the late Donna Dorita de Sa Putch
Tony Ruth
Dr Jonas Salk
Stanley Schindler
Guy Schoeller
Tony Schulte
Dr E. F. Schumacher
Christopher Scott
Anthony Storr
Hannu Tarmio
Ludovico Terzi
Ion Trewin
Egil Tveteras
Russ Voisin
Nat Wartels
Hiroshi Watanabe
Adrian Webster
Jeremy Westwood
Harry Williams
the dedicated staff of MB Encyclopaedias who created this *Library* and of MB Multimedia who made the IVR Artwork Bank.

History and Culture 2/Contents

Keystone

Lord Butler, Master of Trinity College, Cambridge, knocks on the great door of the college during his installation ceremony on October 7, 1965

Preface

I do not think any other group of publishers could be credited with producing so comprehensive and modern an encyclopaedia as this. It is quite original in form and content. A fine team of writers has been enlisted to provide the contents. No library or place of reference would be complete without this modern encyclopaedia, which should also be a treasure in private hands.

The production of an encyclopaedia is often an example that a particular literary, scientific and philosophic civilization is thriving and groping towards further knowledge. This was certainly so when Diderot published his famous encyclopaedia in the eighteenth century. Since science and technology were then not so far developed, his is a very different production from this. It depended to a certain extent on contributions from Rousseau and Voltaire and its publication created a school of adherents known as the encyclopaedists.

In modern times excellent encyclopaedias have been produced, but I think there is none which has the wealth of illustrations which is such a feature of these volumes. I was particularly struck by the section on astronomy, where the illustrations are vivid and unusual. This is only one example of illustrations in the work being, I would almost say, staggering in their originality.

I think it is probable that many responsible schools will have sets, since the publishers have carefully related much of the contents of the encyclopaedia to school and college courses. Parents on occasion feel that it is necessary to supplement school teaching at home, and this encyclopaedia would be invaluable in replying to the queries of adolescents which parents often find awkward to answer. The "two-page-spread" system, where text and explanatory diagrams are integrated into attractive units which relate to one another, makes this encyclopaedia different from others and all the more easy to study.

The whole encyclopaedia will literally be a revelation in the sphere of human and humane knowledge.

Butler

Master of Trinity College,
Cambridge

The Structure of the Library

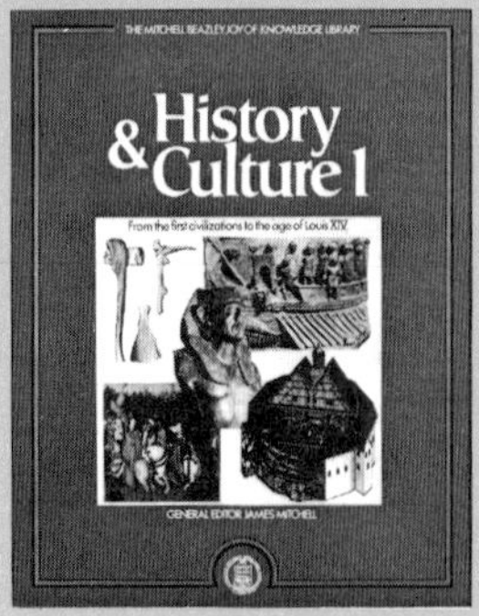

Science and The Universe

The growth of science
Mathematics
Atomic theory
Statics and dynamics
Heat, light and sound
Electricity
Chemistry
Techniques of astronomy
The Solar System
Stars and star maps
Galaxies
Man in space

The Physical Earth

Structure of the Earth
The Earth in perspective
Weather
Seas and oceans
Geology
Earth's resources
Agriculture
Cultivated plants
Flesh, fish and fowl

The Natural World

How life began
Plants
Animals
Insects
Fish
Amphibians and reptiles
Birds
Mammals
Prehistoric animals and plants
Animals and their habitats
Conservation

Man and Society

Evolution of man
How your body works
Illness and health
Mental health
Human development
Man and his gods
Communications
Politics
Law
Work and play
Economics

History and Culture

Volume 1 From the first civilizations to the age of Louis XIV

The art of prehistory
Classical Greece
India, China and Japan
Barbarian invasions
The crusades
Age of exploration
The Renaissance
The English revolution

History and Culture 2 is a book of popular history from the beginning of the eighteenth century to the present day. It is a self-contained book with its own index and its own internal system of cross-references to help you to build up a rounded picture of the background to our times.

It is one volume in Mitchell Beazley's intended ten-volume library of individual books we have entitled *The Joy of Knowledge Library*—a library which, when complete, will form a comprehensive encyclopaedia.

For a new generation brought up with television, words alone are no longer enough—and so we intend to make the *Library* a new sort of pictorial encyclopaedia for a visually oriented age, a new "family bible" of knowledge which will find acceptance in every home.

Seven other colour volumes in the *Library* are planned to be *Man and Society, The Physical Earth, The Natural World, History and Culture 1, Science and The Universe, Man and Machines*, and *The Modern World. The Modern World* will be arranged alphabetically: the other volumes will be organized by topic and will provide a comprehensive store of general knowledge rather than isolated facts.

The last two volumes in the *Library* will provide a different service. Split up for convenience into A-K and L-Z references, these volumes will be a fact index to the whole work. They will provide factual information of all kinds on peoples, places and things through approximately 25,000 mostly short entries listed in alphabetical order. The entries in the A-Z volumes also act as a comprehensive index to the other eight volumes, thus turning the whole *Library* into a rounded *Encyclopaedia*, which is not only a comprehensive guide to general knowledge in volumes 1–7 but which now also provides access to specific information as well in *The Modern World* and the fact index volumes.

Access to knowledge

Whether you are a systematic reader or an unrepentant browser, my aim as General Editor has been to assemble all the facts you really ought to know into a coherent and logical plan that makes it possible to build up a comprehensive general knowledge of the subject.

Depending on your needs or motives as a reader in search of knowledge, you can find things out from *History and Culture 2* in four or more ways: for example, you can simply browse pleasurably about in its pages haphazardly (and that's my way!) or you can browse in a more organized fashion if you use our "See Also" treasure hunt system of connections referring you from spread to spread. Or you can gather specific facts by using the index. Yet again, you can set yourself the solid task of finding out literally everything in the book in logical order by reading it from cover to cover: in this the Contents List (page 6) is there to guide you.

Our basic purpose in organizing the volumes in *The Joy of Knowledge Library* into two elements—the three volumes of A-Z factual information and the seven volumes of general knowledge—was functional. We devised it this way to make it easier to gather the two different sorts of information—simple facts and wider general knowledge, respectively—in appropriate ways.

The functions of an encyclopaedia

An encyclopaedia (the Greek word means "teaching in a circle" or, as we might say, the provision of a *rounded* picture of knowledge) has to perform these two distinct functions for two sorts of users, each seeking information of different sorts.

First, many readers want simple factual answers to straightforward questions such as "Who was William Wilberforce?" They may be intrigued to learn that he was an English philanthropist who lived between 1759 and 1833, was a Member of Parliament for more than 40 years, and was a leading figure in the fight for abolition of the slave trade. Such facts are best supplied by a short entry and in the Library they will be found in the *Fact Index* volumes.

But secondly, for the user looking for in-depth knowledge on a subject or on a series of subjects—such as "What effect did the abolition of the slave trade have on the economies of England and the West Indies?"—short alphabetical entries alone are bitty and disjointed. What do you look up first—"England"? "West Indies"? "Slave trade"?—and do you have to read all the entries or only some? You normally have to look up lots of entries in a purely alphabetical encyclopaedia to get a comprehensive answer to such wide-ranging questions. Yet comprehensive answers are what general knowledge is all about.

A long article or linked series of longer articles, organized

History and Culture

Volume 2 From the Age of Reason to the modern world

Neoclassicism
Colonizing Australasia
World War I
Ireland and independence
Twenties and the depression
World War II
Hollywood

Man and Machines

The growth of technology
Materials and techniques
Power
Machines
Transport
Weapons
Engineering
Communications
Industrial chemistry
Domestic engineering

The Modern World

Flags of the world
Nations of the world
Almanac
Atlas
Gazetteer

Fact Index A-K

The first of two volumes containing 25,000 mostly short factual entries on people, places and things in A-Z order. The Fact Index also acts as an index to the eight colour volumes. In this volume, everything from Aachen to Kyzyl.

Fact Index L-Z

The second of the A-Z volumes that turn the Library into a complete encyclopaedia. Like the first, it acts as an index to the eight colour volumes. In this volume, everything from Ernest Laas to Zyrardow.

by related subjects, is clearly much more helpful to the person wanting such comprehensive answers. That is why we have adopted a logical, so-called *thematic* organization of knowledge, with a clear system of connections relating topics to one another, for teaching general knowledge in *History and Culture 2* and the six other general knowledge volumes in the *Library*.

The spread system

The basic unit of all the general knowledge books is the "spread"—a nickname for the two-page units that comprise the working contents of all these books. The spread is the heart of our approach to explaining things.

Every spread in *History and Culture 2* tells a story—almost always a self-contained story—a story on the English in Ireland, for example (pages 24 to 25) or on African art (pages 60 to 61) or on the rural consequences of industrialization (pages 92 to 93) or on World War II (pages 228 to 229). The spreads all work to the same discipline. which is to tell you all you need to know in two facing pages of text and pictures. The discipline of having to get in all the essential and relevant facts in this comparatively short space actually makes for better results—text that has to get to the point without any waffle, pictures and diagrams that illustrate the essential points in a clear and coherent fashion, captions that really work and explain the point of the pictures.

The spread system is a strict discipline but once you get used to it, I hope you'll ask yourself why you ever thought general knowledge could be communicated in any other way.

The structure of the spread system will also, I hope prove reassuring when you venture out from the things you *do* know about into the unknown areas you don't know, but want to find out about. There are many virtues in being systematic. You will start to feel at home in all sorts of unlikely areas of knowledge with the spread system to guide you. The spreads are, in a sense, the building blocks of knowledge. Like living cells which are the building blocks of plants and animals, they are systematically "programmed" to help you to learn more easily and to remember better. Each spread has a main article of 850 words summarising the subject. The article is illustrated by an average of ten pictures and diagrams, the captions of which both complement *and* supplement the information in the article (so please read the captions, incidentally, or you may miss something!). Each spread, too, has a "key" picture or diagram in the top right-hand corner. The purpose of the key picture is twofold: it summarises the story of the spread visually and it is intended to act as a memory stimulator to help you to recall all the integrated facts and pictures on a subject.

Finally, each spread has a box of connections headed "See Also" and, sometimes, "Read First". These are cross-reference suggestions to other connecting spreads. The "Read Firsts" normally appear only on spreads with particularly complicated subjects and indicate that you might like to learn to swim a little in the elementary principles of a subject before being dropped in the deep end of its complexities.

The "See Alsos" are the treasure hunt features of *The Joy of Knowledge* system and I hope you'll find them helpful and, indeed, fun to use. They are also essential if you want to build up a comprehensive general knowledge. If the spreads are individual living cells, the "See Alsos" are the secret code that tells you how to fit the cells together into an organic whole which is the body of general knowledge.

Level of readership

The level for which we have created *The Joy of Knowledge Library* is intended to be a universal one. Some aspects of knowledge are more complicated than others and so readers will find that the level varies in different parts of the *Library* and indeed in different parts of this volume, *History and Culture 2*. This is quite deliberate: *The Joy of Knowledge Library* is a library for all the family.

Some younger people should be able to enjoy and to absorb most of the pages in this volume on Victorian London, for example, from as young as ten or eleven onwards—but the level has been set primarily for adults and older children who need some basic knowledge to make sense of the pages on twentieth-century sociology or the world's monetary system, for example.

Whatever their level, the greatest and the bestselling popular encyclopaedias of the past have always had one thing in common—simplicity. The ability to make even

Main text Here you will find an 850-word summary of the subject.

Connections "Read Firsts" and "See Alsos" direct you to spreads that supply essential background information about the subject.

Illustrations Cutaway artwork, diagrams, brilliant paintings or photographs that convey essential detail, re-create the reality of art or highlight contemporary living.

Annotation Hard-working labels that identify elements in an illustration or act as keys to descriptions contained in the captions.

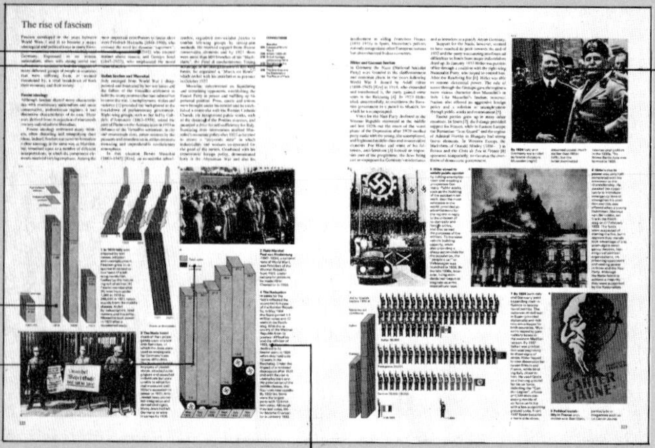

A typical spread Text and pictures are integrated in the presentation of comprehensive general knowledge on the subject.

Captions Detailed information that supplements and complements the main text and describes the scene or object in the illustration.

Key The illustration and caption that sum up the theme of the spread and act as a recall system.

complicated subjects clear, to distil, to extract the simple principles from behind the complicated formulae, the gift of getting to the heart of things: these are the elements that make popular encyclopaedias really useful to the people who read them. I hope we have followed these precepts throughout the *Library*: if so our level will be found to be truly universal.

Philosophy of the Library

The aim of *all* the books—general knowledge and *Fact Index* volumes—in the *Library* is to make knowledge more readily available to everyone, and to make it fun. This is not new in encyclopaedias. The great classics enlightened whole generations of readers with essential information, popularly presented and positively inspired. Equally, some works in the past seem to have been extensions of an educational system that believed that unless knowledge was painfully acquired it couldn't be good for you, would be inevitably superficial, and wouldn't stick. Many of us know in our own lives the boredom and disinterest generated by such an approach at school, and most of us have seen it too in certain types of adult books. Such an approach locks up knowledge instead of liberating it.

The great educators have been the men and women who have enthralled their listeners or readers by the self-evident passion they themselves have felt for their subjects. Their joy is natural and infectious. We remember what they say and cherish it for ever. The philosophy of *The Joy of Knowledge Library* is one that precisely mirrors that enthusiasm. We aim to seduce you with our pictures, absorb you with our text, entertain you with the multitude of facts we have marshalled for your pleasure—yes, *pleasure*. Why not pleasure?

There are three uses of knowledge: education (things you ought to know because they are important); pleasure (things which are intriguing or entertaining in themselves); application (things we can do with our knowledge for the world at large).

As far as education is concerned there are certain elementary facts we need to learn in our schooldays. The *Library*, with its vast store of information, is primarily designed to have an educational function—to inform, to be a constant companion and to guide everyone through school, college and other forms of higher education.

But most facts, except to the student or specialist (and these books are not only for students and specialists, they are for everyone) aren't vital to know at all. You don't *need* to know them. But discovering them can be a source of endless pleasure and delight, nonetheless, like learning the pleasures of food or wine or love or travel. Who wouldn't give a king's ransom to know when man really became man and stopped being an ape? Who wouldn't have loved to have spent a day at the feet of Leonardo or to have met the historical Jesus or to have been there when Stephenson's *Rocket* first moved? The excitement of discovering new things is like meeting new people—it is one of the great pleasures of life.

There is always the chance, too, that some of the things you find out in these pages may inspire you with a lifelong passion to apply your knowledge in an area which really interests you. My friend Patrick Moore, the astronomer, who first suggested we publish the *Library* and wrote much of the astronomy section in this volume on *Science and The Universe*, once told me that he became an astronomer through the thrill he experienced on first reading an encyclopaedia of astronomy called *The Splendour of the Heavens*, published when he was a boy. Revelation is the reward of encyclopaedists. Our job, my job, is to remind you always that the joy of knowledge knows no boundaries and can work untold miracles.

In an age when we are increasingly creators (and less creatures) of our world, the people who *know*, who have a sense of proportion, a sense of balance, above all perhaps a sense of insight (the inner as well as the outer eye) in the application of their knowledge, are the most valuable people on earth. They, and they alone, will have the capacity to save this earth as a happy and a habitable planet for all its creatures. For the true joy of knowledge lies not only in its acquisition and its enjoyment, but in its wise and loving application in the service of the world.

Thus the Latin tag "Scientiam non dedit natura, semina scientiae nobis dedit" on the first page of this book. It translates as "Nature has given us not knowledge itself, but the seeds thereof."

It is, in the end, up to each of us to make the most of what we find in these pages.

General Editor's Introduction

The Structure of this Book

History and Culture 2 is a book about the events that made the world we live in, and about that world itself. It covers the last 300 years—the rest of recorded history is the subject of Volume 1—in such a way that the forces that shaped the world are traced from the great intellectual impetus of the Enlightenment to the current problems of the Third World. Although the story centres on Europe, the changes that occurred in the East and the Americas are not neglected: there are spreads of China, India, Japan, the Pacific, and North and South America.

Although this is a book of world history, with a balanced coverage of events, a disproportionately large amount of space is devoted to Britain and the Commonwealth. The deliberate editorial intention has been to make the coverage of our national heritage so dense that if the relevant spreads were separated from the chronology of world events into which they fall, they would form a complete text in themselves. In all, the coverage of British and Commonwealth history and culture amounts to a third of the two volumes.

Before itemizing the contents of *History and Culture 2* I'm going to assume that you—just like me when I began planning the book—are coming to history more as a "know-nothing" than as a "know-all". Incidentally, knowing nothing can be a great advantage as a reader—or as an editor, as I discovered in the early days of selecting topics for this book. If you admit to knowing nothing, but want to extend your knowledge, you ask awkward questions all the time. I spent much of my time as General Editor of this *Library* asking acknowledged experts awkward questions and refusing to be fobbed off with complicated answers I could not understand. As a result, *History and Culture 2*, like every other book in this *Library*, has been through the sieve of my personal ignorance in its attempt to inform simply and understandably.

In the first of the *Library's* two volumes on history and culture, Dr Christopher Hill discusses the historian's job, the sources he should use and the questions he should ask. His conclusion is that the real nature of the task is to discover what kind of questions ordinary men and women were answering in the past when they did whatever it was they did—stormed the Bastille or fought the Battle of Waterloo or sailed with Drake. Dr Hill also sounds a warning against the view of history presented by official sources, which up until the last few centuries make up the bulk of our surviving evidence on what happened in the past, and points out that it is only recently that we began to have adequate records of the lives of ordinary people. "If we are not merely to repeat historical mythology, the self-selected, self-justifying legends of past ruling classes, we are up against great difficulties," he says. It would be presumptuous of me to pretend that *The Joy of Knowledge Library* volumes on history and culture have surmounted those difficulties, but we have tried to be aware of them as we approach the modern era, and our records of the lives of ordinary people improve, our writers have endeavoured to make use of those records.

Treatment of the subject

Both volumes on history and culture in *The Joy of Knowledge Library* tackle their subject chronologically. The story begins in the first volume with the earlist civilizations of which written records exist and ends in the second with a consideration of the current political situation in Europe. Europe figures largely in our treatment, principally because so many of the modern world's attitudes to life derive from European models, notably Greece and Rome. Taking Europe as the main element in the story also enables us to assess simultaneous events in other civilizations, such as the Chinese.

In both volumes there are frequent pauses in the chronological treatment so that social and economic progress, and the development of science, can be discussed. In a similar way, spreads on the history of the arts fall adjacent to those on the history of the civilizations that bore the most significant artistic fruit. To give a full picture of any given age, nation or trend, works of art have often been used to illustrate historical events.

Time Chart

Finally, both volumes contain the relevant parts of an extensive time chart that begins in 4000 BC. The chart covers milestones in politics, religion and philosophy, music, literature, art and architecture, and science and technology. A special section covers British national events. The chart may be used in three ways. First, readers who

History and Culture 2, like most volumes in *The Joy of Knowledge Library*, tackles its subject topically on a two-page spread basis. Although the spreads are self-contained, you may find some of them easier to understand if you read certain basic spreads first. Those spreads are illustrated here. They are "scene-setters" that will give you an understanding of the major civilizations of the past and of how historians approach their subjects. They will also demonstrate how the material in *History and Culture 2* is organized. The eight spreads are:

wish to get the flavour of a period may do so by reading the introductory paragraphs to each subject. These paragraphs, which are set in bold type, sum up the events covered on each spread of the time chart. Secondly, the chart may be used to follow the development of a particular discipline, such as philosophy, through the ages. Thirdly, the chart may be used to discover what progress was being made in other fields at the time a particular discovery was made or a particular event took place. Thus if you find a reference to the invention of the steam engine in the main body of the book, you may turn to the time chart and discover against what political, philosophical, artistic and scientific background the discovery was set.

The periods of time covered by each spread of the chart shorten as the present day is approached so that proportionately more space is given to recent events.

Plan of the book

Because of our chronological treatment, neither volume of *History and Culture* lends itself to an easy division into topics. But it is possible to select certain cultures whose progress is charted regularly through the book and other major events—such as the two World Wars in this volume —which are treated in depth at the appropriate time in the story. In *History and Culture 2*, these are the subjects that receive such attention.

Britain and the Commonwealth

Of 110 spreads in the two volumes on history and culture devoted solely or largely to the the history of Britain, its empire and the Commonwealth, 68 are in this volume. The stories of England, Scotland, Wales and Ireland are treated in individual spreads inserted chronologically through the main structure of the book. There are spreads on the West Indies, Canada, South Africa, Australia and New Zealand. One spread is devoted to a "family tree" of the rulers of Britain since 828.

Revolution and industrial progress

The 200 years following the American Declaration of Independence established revolution as a commonly used means of drastic political change. Running parallel to political changes were changes in technology that were sufficiently wide-ranging to be described as revolutions. The introduction in 1701 of Jethro Tull's seed drill is generally regarded as the beginning of the mechanization of agriculture; 50 years later, at the beginning of the Industrial Revolution, Britain was self-sufficient. Britain was the first industrial nation in the world and by the first decade of the nineteenth century had entered a period of self-sustaining economic growth that lasted until the 1870s or 1880s, and ensured general prosperity until World War 1.

The revolutions that occurred between 1776 and 1848 had as their philosophical bases the ideas of nationalism and liberalism. In one form or another, these twin ideas were to dominate much of the history of the world up to the present day. *History and Culture 2* traces their development and details the events they inspired or influenced.

India, China and Japan

Volume 1 left the story of India, China and Japan at the point where those three countries had just been, or were about to be, exposed to Western influences. India was set on the path towards becoming a member of the British Empire by the arrival of the East India Company in the first decade of the seventeenth century. The Portuguese established a Jesuit mission in Japan in 1549 and trade with China was begun by the Dutch in 1557. Both Japan and China, however, strongly resisted contact with the West, and agreed to it only reluctantly in the nineteenth century.

The Americas

The development of the United States, from the first settlements in North America through the Declaration of Independence and on to its advent as a major world power, falls into this volume. Parallel with the growth of the United States, the Spanish and Portuguese empires in South America were breaking up. The struggle for independence in Latin America, and the subsequent development of the Latin American countries, are subjects also explored.

Imperialism

The theme of the growth and decline of empires—notably the British Empire—runs through *History and Culture 2*. In addition to spreads on individual colonies, there are a dozen spreads that deal with imperial policy or its results.

The American Revolution

The French Revolution

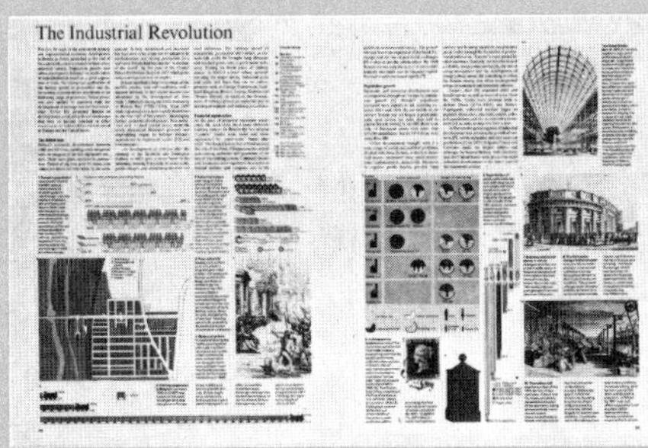

The Industrial Revolution

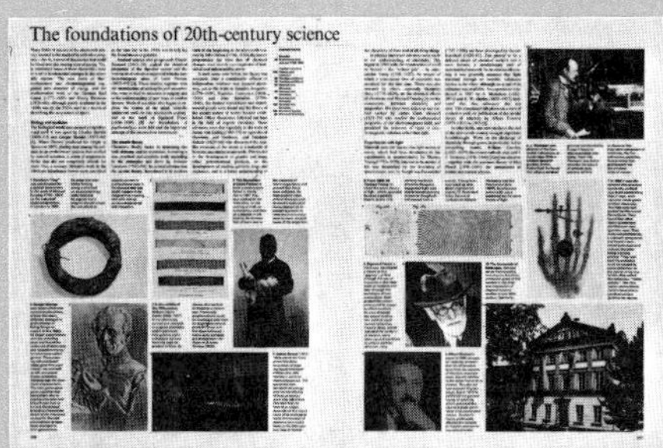

The foundations of 20th-century science

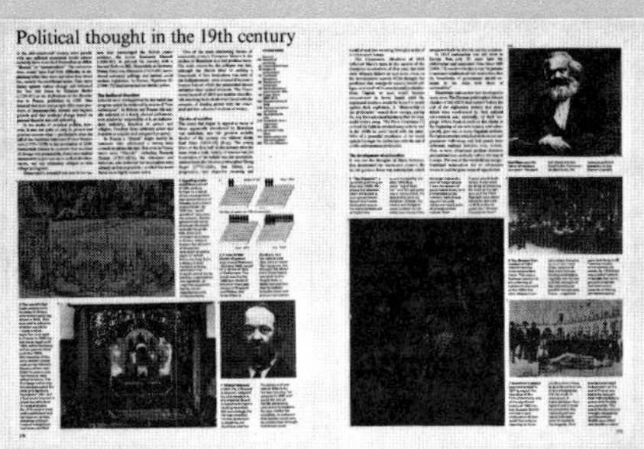

Political thought in the 19th century

British foreign policy since 1914

Decolonization and related topics such as the rise of the Third World, the United Nations, and underdevelopment are also described.

Art
Forty spreads on art include three on the film—its origins, Hollywood, and the cinema as art. The art spreads also include a number devoted to Scottish, Welsh and Irish culture. As in Volume 1, many of the illustrations on history spreads are of works of art, particularly in the period before World War I.

Science and technology
Although there are separate volumes in *The Joy of Knowledge Library* on science and technology, the history of their development properly finds its place in *History and Culture*. This volume takes the story from 1700 to the present day.

Religion
As a force affecting international relations, religion has steadily declined. The development of Christianity is still, however, of fundamental importance to millions of people around the world, and a spread is devoted to that topic.

The twentieth century
The bulk of *History and Culture 2* is given over to an examination of the history of the twentieth century. It has been a century of upheaval, realignment, changing values and tremendous technological progress. So rapid has been the change that new disciplines, such as sociology, have had to be conceived to measure them. The very recording of events has become more complex and historians have tended to specialize, producing political history, economic history, military history and, most recently, social history. The last, as Christopher Hill points out in his introduction to Volume 1, is increasingly becoming a major theme. Film and television have provided source material that is at once richer and more dramatic than anything available in the past and at the same time more difficult to use, both because of its sheer bulk and also because the techniques that can turn documentary into propaganda are more difficult to detect. In setting down the history of contemporary events—if that is not a contradiction in itself—we have tried to be aware of the need to avoid pitfalls as well as of the difficulty of imposing any pattern, or of making any judgement, that will not require almost instant reassessment. How well we have done our job, the reader must judge. If we have done it properly, *History and Culture 2* ought both to inform and enthral you.

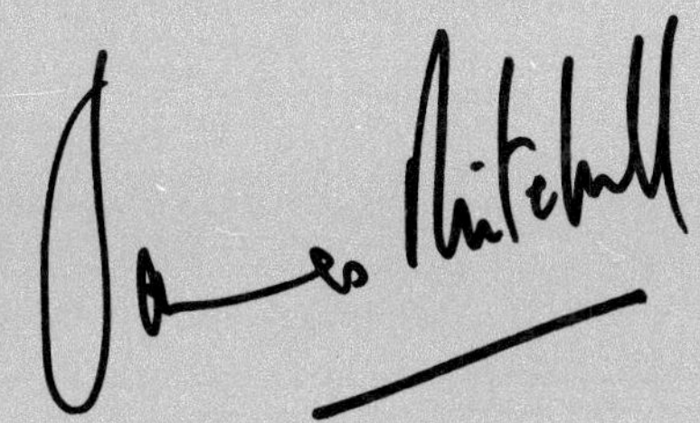

Europe: economy and society, 1700–1800

The most important economic developments of the eighteenth century [2] took place in Britain, France, the Low Countries and parts of Germany; eastern and southern Europe made much less progress. At one extreme Britain was, by the end of the century, well on the way to becoming the first industrial nation, a development given theoretical backing by the work of economists such as Adam Smith (1723–90). At the other extreme countries such as Russia and Italy retained economic systems little different from those of the medieval period. Socially, too, the most striking developments occurred in the countries on the Atlantic seaboard.

Population growth and better harvests

One development common to Europe as a whole in this period was population growth, caused chiefly by a declining mortality rate [5]. In advanced countries, economic expansion also increased the demand for labour, permitted more children to be supported and encouraged earlier marriage.

Population growth was encouraged by, and stimulated, agricultural improvement [4]. Large tracts of land were brought into cultivation and new techniques and crops were harnessed to meet the demand for foodstuffs for a growing population. Yields and animal weights rose during the century, especially in northwestern Europe. Elsewhere, agriculture often remained backward and near to subsistence. Famine continued to afflict many communities in southern and eastern Europe until the nineteenth century, although the adoption of the potato as a subsistence crop helped to increase supplies.

Agricultural practice varied enormously along with different systems of tenure and landholding throughout Europe. In Russia and much of central and eastern Europe, serfdom was still in force. In other areas a landowning peasantry had emerged, ranging from the relatively prosperous peasants of the Low Countries to the poorer ones of Brittany. In the British Isles the poor tenants of southern Ireland, the landless agricultural labourers of the "enclosed" counties of southern England and the semi-feudal "crofters" of Scotland, contrasted with the prosperous tenant farmers and great landowners.

International trade played a vital part in the economy of many states. Trade expanded in the eighteenth century, especially for the Atlantic countries with their easy access to West Indian, American, African and Asiatic markets. Britain and France both made large strides in trade, founding the prosperity of a merchant middle class in cities such as Bristol, Liverpool, Bordeaux and Nantes.

Domestic trade and manufacturing

Internal trade also flourished, aided by improvements in river and road transport [Key]. In many countries agricultural produce, in particular grain, was the major commodity traded, but timber, coal, mineral ores, stone and other products were carried in ever-increasing quantities.

Growing population and increasing wealth in Britain and other European countries stimulated the demand for a wide range of manufactured goods, but this increased output was still achieved by pre-industrial methods [3]. The domestic system and handworking remained the principal means of producing textiles, iron, pottery and a wide

CONNECTIONS

See also
60 The early Industrial Revolution
22 The Enlightenment
18 England under the Hanoverians
20 The agricultural revolution
24 The English in Ireland
68 Pitt, Fox and the call for reform
32 Exploration and science 1750–1850
44 International economy 1700–1800

300 *In other volumes* History and Culture 1
317 Man and Society

1

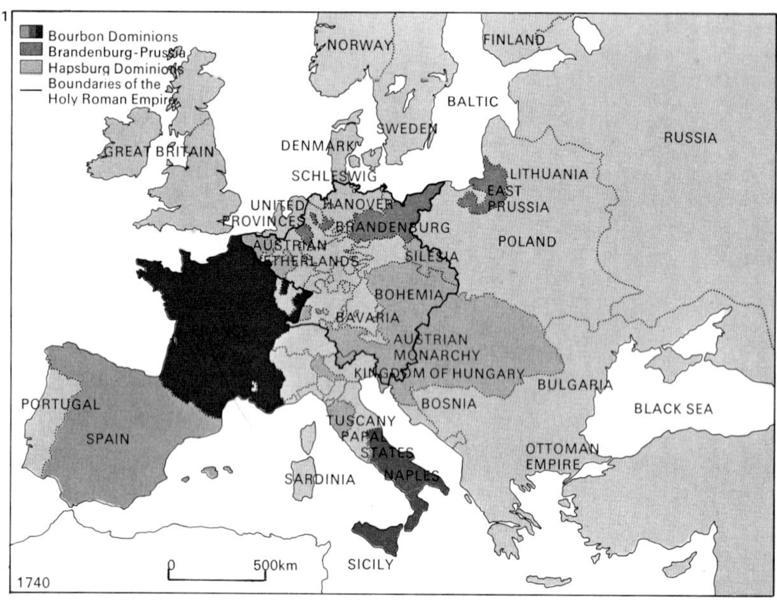

1 Political stability was a characteristic of 18th-century Europe until the French Revolution in 1789. The previous century witnessed the consolidation of a number of powerful nation states, ruled by well-established monarchies (in Britain, France, Prussia, Austria and Russia). Despite changing dynasties and disputed successions, political authority had been established in these countries for a long period. Much of Germany and Italy, however, remained a jumble of petty states and kingdoms separated until the middle of the next century. Poland was another unstable monarchy, subject to encroachment and partition by other powers. Europe's political boundary in the southeast was the Ottoman Empire – in decline but still dominant in the Balkans.

2

1701	1710	1720	1730	1740	1750	1760	1770	1780	1790	1800
100	110	96	95	92	91	99	95	110	114	175

2 Price indices for a range of goods in 18th century Britain suggest that the cost of living decreased until the latter part of the century. Favourable economic conditions and an expansive social environment were the key to political stability. Good harvests made for cheaper food while early industrial enterprises began to reduce the cost of many everyday items such as clothing, furniture and domestic utensils which, together with widely available foodstuffs, comprise the bulk of this index. One result of the overall price stability was a higher birthrate, reduced infant mortality and a longer life expectancy because of better diet. The consequent strain on food supplies, exacerbated by the effect of war with France, and a series of bad harvests led to much higher prices in the 1790s, particularly affecting the new industrial classes.

3

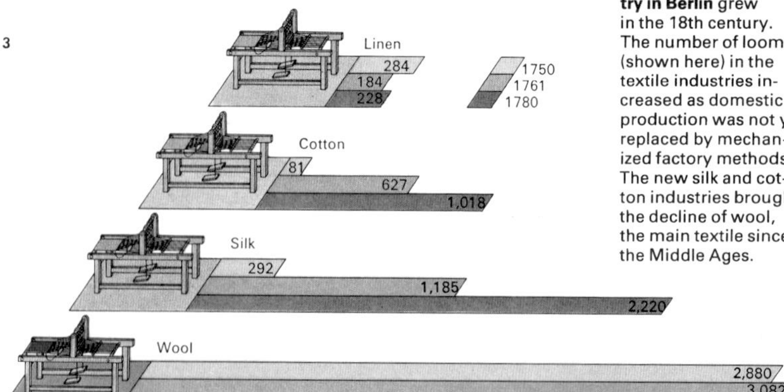

3 The textile industry in Berlin grew in the 18th century. The number of looms (shown here) in the textile industries increased as domestic production was not yet replaced by mechanized factory methods. The new silk and cotton industries brought the decline of wool, the main textile since the Middle Ages.

4

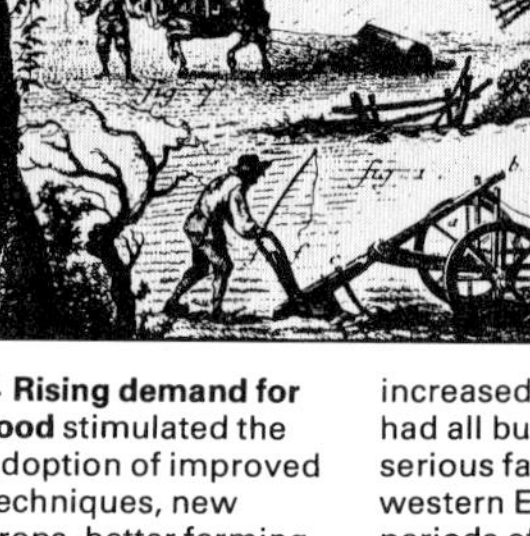

4 Rising demand for food stimulated the adoption of improved techniques, new crops, better farming implements and machinery. By the 1750s increased production had all but eliminated serious famine in western Europe, but periods of scarcity persisted, especially as population grew towards the end of the century. Improvements were propagated by enthusiasts such as Arthur Young (1741–1820) and the French economists.

range of other goods. In Britain, water power was the most widely used source of energy for manufacturing, as steam power was obtaining a foothold only by the end of the century. The units of production were generally small, domestic organization being by far the most widespread, although the expanding coal mines and textile towns of Britain, France and Belgium were beginning to concentrate production. Overall, outside Britain, the picture was one of gradual expansion affecting most European countries.

Minuets and misery

European society remained essentially traditional in the eighteenth century. Most of the population still lived in a hierarchic, rural society based upon agriculture. The village and small market town continued to dominate social environment and culture for most people. In these areas, religion still played an important role in everyday life and helped to cement social bonds. Parishes, manors and guilds remained the typical forms of social organization and in most of Europe social structure was little affected by economic changes. Although serfdom was attacked in some "enlightened states" and was abolished in the Hapsburg dominions by Joseph II (1741–90), in Russia it was increasing. Peasant revolts and periodic riots over various causes did little to shake social structure before the French Revolution.

In contrast, the capital cities such as London, Paris, Vienna and St Petersburg continued to grow in size, adding many magnificent buildings and becoming centres of government and major markets. Merchant cities, such as Liverpool and Bordeaux, also grew rapidly, bringing new standards of comfort to the upper and middle classes [7]. For these classes education and prosperity led to the development of a sophisticated urban culture, shown in the demand for literature and art and in the access to wider means of communication such as newspapers and journals. For the poorer urban populations, living conditions were much harsher, with bad housing, health and low incomes [6]. Crime was rife in most towns and cities and was usually met by savage punishment, including torture and death for trivial offences.

KEY

Improved travel was one of the most important features of 18th-century Europe. Much easier movement by road, river and canal greatly stimulated trade and the first industrial developments. For the wealthy, better communications created the basis for a common European culture.

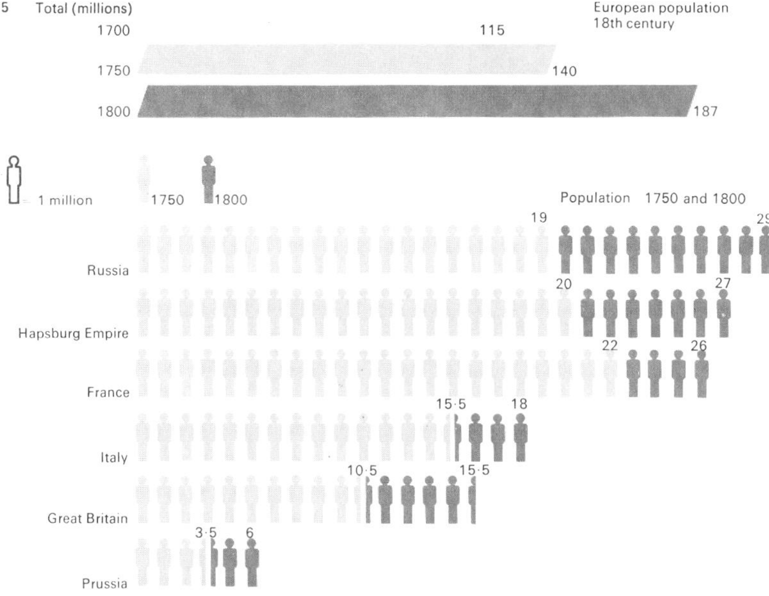

5 The major factor in population growth was a fall in the death-rate. Better diet, an improved climate, the decline of some major diseases and an increase in personal hygiene (aided by increasing availability of cheap soap and washable cotton clothing) contributed to lower mortality. Smallpox inoculation and improved midwifery perhaps also played a part. Increased life expectancy in Western Europe thus led to a population increase, which had previously been checked by epidemics and famine. Even in Eastern Europe and Russia, where the improvements had less impact, the population increase was marked.

6 Harshness and squalor were still the lot of the poor, even in the age of enlightenment. Cheap spirits became readily available through large-scale distilling and caused chronic drunkenness. Food and drink adulteration by crooked traders was common, promoting death and disease in the larger urban slums. Moral tracts condemned the "Gin Lane" conditions of the urban poor.

7 Improved living standards among the upper and middle classes led to the building of elegant town houses in many European cities. New squares, like those in London's West End and Edinburgh's New Town, evinced Britain's commercial wealth. The merchant cities of the Continent, such as Amsterdam and Bordeaux, and the great governmental centres such as Versailles provided the basis for an affluent urban society. Spas and fashionable resorts such as Bath (shown here) and Dresden provided recreation for the rich, who generated a demand for luxury goods and cultural amusements. The rich also enjoyed improvements such as street lighting, parks and water supply.

8 Armies in the 18th century consisted of long-service volunteers or conscripts. Harsh discipline, low pay and bad conditions attracted only the poor and criminals into the ranks, while officers, from the nobility, used influence to gain commissions. Distinctive uniforms were chosen by commanders – warfare was formalized and long-range weapons were inaccurate.

England under the Hanoverians

England of the Hanoverian period, which began on 18 September 1714, with the arrival of Georg Ludwig (1660–1727), Elector of Hanover, at Greenwich to become George I of England [1], was a country generally free from the turbulence of the seventeenth century and the social and intellectual ferment of the nineteenth. Its apparent stability and prosperous complacency were revealed in politics, in religion, in commerce and in letters; the reigns of both George I and George II (reigned 1727–60) exhibited an external, though superficial, calmness that rested on the settlement of old quarrels at home and expanding power, naval and trading, overseas. Samuel Johnson (1709–84) [Key] represented the age, in his rational thought, political conservatism, and dry religious orthodoxy.

The Hanoverian succession

George I knew no English when he came to the throne. Both he and his son were more interested in Hanover than in England. George II was the last English monarch to lead his army into battle in Europe, at Dettingen in 1743. They both spent much of their time out of England, so that the House of Hanover did not become fully naturalized until the reign of George III (reigned 1760–1820).

Yet the royal court remained the centre of social life and of politics, the source of the patronage that was the cement of the political structure. The strife of the Civil Wars and the party rancour of the age of Queen Anne (reigned 1702–14) gave way to the politics of consensus, presided over by the Whig oligarchy established between 1721 and 1742 when Robert Walpole (1676–1745) [3] served as first minister.

The Whig politicians who had backed the Hanoverian succession undertook a long period of effectively one-party rule. The Jacobite uprisings of 1715 and 1745, attempting to restore the Stuarts to the throne, were dismal failures and tainted Toryism with disaffection and rebellion. The Whig landed gentry fused its interests with the court and the great merchant-financiers [2,6] and gained a stranglehold on the House of Commons and government service. Elections [5] remained unruly affairs, but the tiny electorate, the heavy cost of fighting elections (more or less direct bribery of the voters was common practice), and the large number of rotten and "pocket" boroughs kept the House of Commons under the control of the Whig government managers. The most famous and untiring of these was the Duke of Newcastle (1693–1768), who was Secretary of State (1724–48) and first minister (1754–62).

The growth of the empire and the expansion of trade and industry meant that the civil service became more elaborate. The Treasury, the Customs and Excise, the Admiralty and the War Office all expanded, and the dispensers of royal patronage often rewarded the place-hunting "friends" of the government with jobs in these departments.

The growth of the cabinet system

The cabinet began to develop as the main organ of government under the Hanoverians. Walpole, often considered to be the first prime minister (the term is still an unofficial one), bypassed the old Cabinet Councils

CONNECTIONS

See also
16 Europe: economy and society 1700–1800
20 The agricultural revolution
22 The Enlightenment
28 Scotland in the 18th century
34 European literature in the 18th century
70 Georgian art and architecture
68 Pitt, Fox and the call for reform

1 George I, aged 54 on his accession to the English throne, became king by virtue of the Act of Settlement of 1701, which had made his mother, Sophia, Electress of Hanover, heir to Anne's throne. This was done to prevent a return of the Stuart line. George, who had ruled as a despot in Hanover, had little liking for the English people or their liberal constitution and left the day-to-day running of affairs to his ministers.

2 Marine insurance companies, such as Sun-Fire, whose sign is shown here, were important to the development and the protection of 18th-century commerce. Most of these companies were financed as joint-stock enterprises. *Lloyd's List,* with news about merchant shipping, first published in 1734, is the oldest daily paper still published in London. Hanoverian prosperity rested on the profits of Britain's expanding trade empire.

3 Robert Walpole, here talking to Arthur Onslow (1691–1768), Speaker of the Commons, was defeated on his 1733 Excise Bill. He found that patronage alone could not guarantee a majority in the Commons as the substantial minority of "country gentlemen" was not bound to ministers by patronage. The development of the cabinet system was widely feared at this time and did not become fully established until the second half of the century. Walpole was criticized as a secretive, power-seeking man.

4 The War of Jenkins' Ear (1739–41) interrupted a long spell of peace for England. It was declared in response to the merchants' demand for protection at sea against the Spaniards. The ear, here being shown to Walpole by Capt. Jenkins, was allegedly torn off by Spanish coastguards. The war was fought mainly at sea near to the Spanish colonies, and although Britain won no important territorial advantages, the disruption hastened the decline of the Spanish Empire.

of privy councillors and relied on an inner cabinet of four or five of the Crown's principal ministers. They were not yet united by party – the resignation of the first minister rarely entailed that of any other – but the process had begun by which the cabinet and party were to oust the Crown and patronage from the centre of the political stage. The eighteenth century was the golden age of the mixed constitution, with a much-praised balance of King, Lords and Commons in the constitution or limited monarchy established by the "Whig revolution" of 1688–9.

The rise of Methodism

The Church of England was dormant following the bitter sectarian disputes of the previous two centuries. Protestant dissenters and Catholics were excluded from the universities as well as many public offices, unless they paid nominal allegiance to the established Church. Secure in its monopoly, the Church upheld a bland, unquestioning view of the truth of Christianity. Political obeisance to a powerful Whig patron was the way to a bishopric [8].

In such circumstances, it is not surprising that the most important religious movement of the century was the evangelical revival led by John Wesley (1703–91) [9]. His highly organized preaching tours, combined with the founding of "cells" in towns and villages, made Methodism by 1760 the most dynamic body of opinion in the country, and by 1800 there were more than 100,000 Methodists. The Anglican Church was implacably opposed to Wesley and no bishop would ordain for him or his assistants. In 1784 he therefore broke with the Church and began to ordain his ministers himself.

Wesley appealed principally to the poor of a society marked by great inequalities. The Duke of Newcastle had an annual income of more than £50,000 from his estates in 12 counties, whereas a handloom weaver worked for less than a shilling (5p) a day. Nevertheless, although the mass of the people possessed neither the vote nor property, they had basic political rights – freedom from arbitrary arrest, trial by jury, the right to political demonstration – which were denied to most of their European contemporaries.

KEY

Samuel Johnson, essayist, poet, critic and lexicographer, gave the English language its first systematic and formal, if idiosyncratic, setting in his *Dictionary* of 1755. He was gregarious and nocturnal, and drank water and tea rather than wine. He lived comfortably but was never rich. In 1760 George III rewarded him with a civil-list pension of £300 a year. A resolute Tory, Johnson condemned the American rebels and defended the wealth and doctrines of the Church of England. He was renowned for his acerbic wit, recorded by James Boswell (1740–95), in the first great English biography.

5

5 Electoral violence, seen in this painting by William Hogarth (1697–1764), resulted from bribery of the voters, but it became less common as the century progressed. The Septennial Act (1716) greatly reduced the frequency of elections, and in 1761 only four elections were contested for county seats.

6

6 The Stock Exchange provided for the easy reinvestment of funds in new trading ventures or in industry. Since there was little social distinction drawn between wealth acquired in trade and that derived from land, the aristocracy happily contributed to the financial expansion that won for Britain the title "A nation of shopkeepers".

7 Cricket was widely played and was first organized on a county basis in the 18th century, thanks to the patronage of the great landowners. It epitomized the relative social harmony of rural areas, in contrast to the often violent towns, where riots such as the Gordon Riots of 1780 might occur. Lord's cricket ground was opened in 1787 by Thomas Lord for the White Conduit Club, later the Marylebone Cricket Club (MCC).

7

8

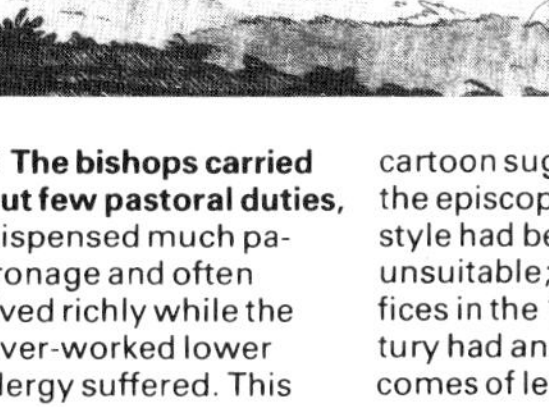

8 The bishops carried out few pastoral duties, dispensed much patronage and often lived richly while the over-worked lower clergy suffered. This cartoon suggests that the episcopal lifestyle had become quite unsuitable; 1,200 benefices in the 18th century had annual incomes of less than £20.

9

9 John Wesley offered the poor a promise of individual salvation, an idea that seemed to be genuinely egalitarian. His appeal to the personal worth of each individual quickened a response, especially in the new industrial towns of Wales and the north, where the Church of England was inactive.

The agricultural revolution

Historians have often described the changes that occurred in British farming during the course of the eighteenth and nineteenth centuries as an agricultural "revolution". The phrase was coined in the nineteenth century by those who saw comparable changes in the mode of production and social relationships on the land to what was happening in industry. More recent research has tended to emphasize the long drawn-out evolution of agricultural change and the varied pattern that it presents over the country as a whole. The earliest books on farming techniques had appeared in the early sixteenth century and enclosure had started in the thirteenth century, accelerating in the sixteenth century.

The results of new techniques

Whether or not the phrase "agricultural revolution" is an exaggeration, changes in British agriculture between, say, 1700 and 1870, were real and substantial. Greatly improved output, new crops, and improved techniques were matched by a number of important social developments.

Many of the changes in agricultural practice made wider use of established ideas. The use of rotation of crops, for example, particularly the utilization of root crops such as the turnip as a part of the rotation and a source of winter feed for animals, was known in the seventeenth century. Selective breeding of livestock was familiar to earlier generations but was impracticable under the open field system of agriculture where animals were herded together in the common field.

Agricultural historians have identified the hundred years before 1750 as one of slack demand for farm products because of an upward trend in harvests and a largely static population. From 1750 onward, however, a discernible rise in population and generally poorer harvests provided a stimulus to investment in agriculture and the application of fresh techniques in order to increase production. A number of pioneers, such as Jethro Tull (1674–1741), Charles "Turnip" Townshend (1674–1738), and Robert Bakewell (1725–95) popularized the new techniques. Probably the most important of these were the use of crop rotation, scientific breeding of animals, and the use of crops such as turnips and lucerne as animal fodder. These improved yields led to heavier and healthier animals – within the century the average weight of sheep sold in London nearly trebled and permitted the wintering of livestock on stored feed. New implements such as improved ploughs [7] and harrows, as well as Tull's revolutionary seed drill, contributed to the improvement of yields [2].

The effects of enclosure

Many of these techniques could not have been applied without reorganization of landholding. Enclosure of land [Key, 4] into self-contained units had been going on for centuries; more than one-third of England was enclosed by 1600, usually by agreement among local landowners. From the mid-eighteenth century, much land was enclosed through private Acts of Parliament. Enclosure involved creating separate holdings out of the medieval common and "open" fields. Generally it eliminated the inefficiency of farming strips in each field and the wasteful system of leaving one field fallow each year, and it also permitted farmers to experiment

CONNECTIONS

See also
66 The early Industrial Revolution
92 The rural consequences of industrialization
18 England under the Hanoverians
16 Europe: economy and society 1700–1800

In other volumes
196 History and Culture 1
28 Man and Machines
158 The Physical Earth

1 Agricultural manuals and tracts helped to spread the use of new techniques among farmers from as early as the 16th century. During the late 18th century agricultural improvement became a fashionable concern: King George III (r. 1760–1820) himself ran a model farm at Windsor.

2 Jethro Tull's famous horse-drawn seed drill, first used in 1701, is regarded as having initiated the mechanization of agriculture. Before this invention, seed was laboriously broadcast by hand, which was a wasteful and uncertain procedure.

A
DISCOURS
OF
HUSBANDRIE
USED IN
Brabant and Flanders:
SHEWING
The wonderful improvement of Land there;
and serving as a pattern for our
practice in this
COMMON-WEALTH.

The Second Edition, Corrected and Inlarged.

LONDON,
Printed by *William Du-Gard*, dwelling in *Suffolk-lane*, near London-stone, *Anno Dom.* 1652.

3 Richard Weston's (1591–1652) *A Discours of the Husbandrie used in Brabant and Flanders* (1645) spread Flemish agricultural ideas among English farmers. It described methods of crop rotation and several techniques by which poor soils were improved.

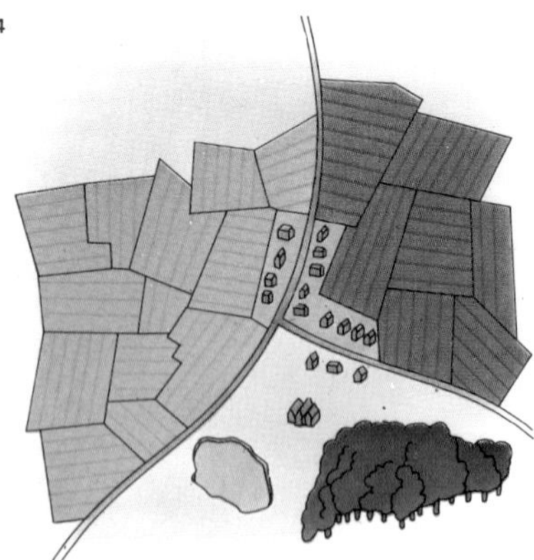

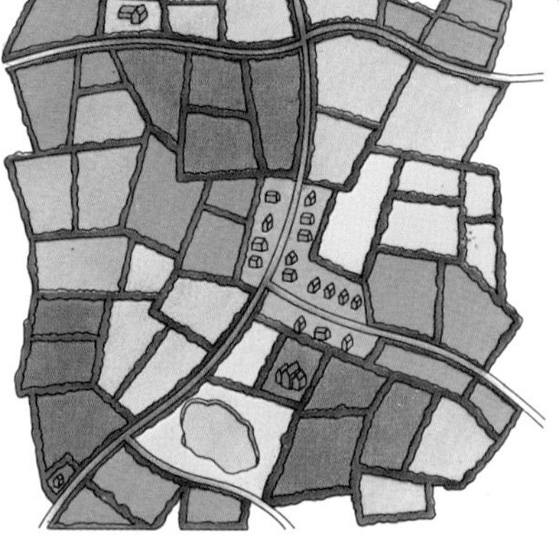

4 The effects of enclosure upon landholding can be seen from the plans of a typical parish. Before enclosure, many villages preserved the medieval layout of large open fields in which each inhabitant held by custom a few strips. The intention was to give everybody a share of good and bad land. But this meant that the land between each strip was wasted and involved unnecessary journeys between different fields. It was impossible to experiment with new techniques, and most people used a simple system of rotation that left one field fallow for a year. Enclosure consolidated holdings and permitted improved agriculture. The progressive farmers supported it, and those with large holdings often provided incentives in order to obtain better tenants at greater rents. After enclosure, all the land was divided up, and hedges were usually planted to mark the new field boundaries, in this way creating what is now the familiar English landscape.

5 The Game Laws restricted the taking of game to men of property. In 1671 freeholders with less than £100 worth of property and leaseholders with less than £150 were prohibited from taking deer, hares, rabbits, pheasants or partridges. As enclosure progressed, the areas in which labourers and smallholders could legitimately take game were still further reduced. Harsher game laws were introduced, including imprisonment, transportation and even death in the case of resistance to gamekeepers. Mantraps (such as the one shown) and spring-guns were used to deter poachers, and gamekeepers were given wide powers. But the 19th century saw a relaxation of these laws.

with new techniques on a consolidated holding. Nearly 3,000 enclosure acts were passed between 1751 and 1810, the largest number being passed during the Napoleonic Wars (1803–15) when, due to trade dislocation, food prices were at their highest [10].

Poverty and prosperity

The enclosures of this period were once believed to have contributed to the pauperization of the agricultural labourer by depriving him of his rights of grazing on the common and rendering his smallholding uneconomic. This view has now changed. Many smallholders remained and the number of families working on the land actually rose between 1750 and 1831. Migration did take place from the land, and many smallholders were pushed into the ranks of wage-labourers, but this was more the effect of population growth than of enclosure. Pauperization of the agricultural labourer arose from chronic rural unemployment and concomitantly depressed wage levels.

On the other hand, owners of large farms tended to prosper. The landed classes of the eighteenth century had ample wealth for housebuilding, the creation of rich collections, and foreign travel. The wealthier of their tenants and professional farmers were also able to build substantial farmhouses. That tireless observer of rural life, William Cobbett (1762–1835), among others, noticed that the social status of farmers had greatly improved by the 1820s, but that it had the effect of making them more distant from their employees. Much contemporary comment satirized the social pretensions of farmers and their families in the early nineteenth century. But it took several decades for the typical pattern of Victorian rural society – which was that of large landowners, tenant farmers, and landless labourers – to emerge.

Expansion of the cultivatable acreage and improved yields provided the food for a growing urban population. Precise production figures are not available, but Britain's ability to feed itself despite the virtual trebling of its population between 1750 and 1850 was not the least remarkable feature of the development of its economy.

KEY

Enclosure was generally completed with little serious disagreement. To obtain an Act of Parliament took time and required the agreement of many of the local landowners. The land also had to be accurately surveyed and the appropriate legal titles established. The allocation of land was usually fair. Each enclosure act appointed commissioners to distribute the land.

6

6 The "Pangborn Hog" was a gigantic prize pig reared on Tidmarsh Farm in Berkshire as a result of systematically controlled and scientific breeding. There were also new strains of other animals by selective breeding. Among the farmers who popularized new breeds of sheep was Robert Bakewell of Dishley, Leicestershire. He was so successful that he managed to double the amount of meat obtainable from each of his sheep.

7

7 Agricultural improvements owed little to new machinery, apart from Tull's seed drill. But many small improvements were carried out on existing implements. An improved plough with a metal blade, for example, was produced in 1703, and wooden ploughs were gradually superseded. The Rotherham plough, shown here, included a metal blade and appeared in 1730.

8 Arthur Young (1741–1820) was a famous propagandist for the techniques of agricultural improvement in Britain. In his books and articles he argued that large-scale farming, using enclosure, the latest techniques and plentiful capital would greatly increase production. His writings provide historians with a rich source of information about the social, political and economic life of the 18th century.

8

9

9 Holkham Gatherings was the name given to a series of agricultural shows organized by Thomas Coke (1752–1842) of Holkham in Norfolk. Coke was a pioneer of agricultural improvement and, like many others, he was a propagandist for the new methods. He experimented with root crops, especially the swede, helped to improve breeds of cattle, sheep and pigs, and was the first to grow wheat instead of rye in western Norfolk.

10 Agricultural output increased during the 18th century as new methods of farming were introduced. Rapid population growth stimulated demand at home; exports also increased. Here the average annual export of corn is shown for each ten-year period. Domestic prices reached a peak at the time of the Napoleonic Wars but fell slightly after 1815. The protective Corn Laws kept prices up, but even when these were repealed, in 1846, farming remained prosperous.

Period	Export
1697–1706	74,100
1707–1716	118,700
1717–1726	133,700
1727–1736	168,200
1737–1746	280,000
1747–1756	448,700

1 sack = 10,000 quarters (56lbs)

The Enlightenment

The seventeenth century saw the emergence of a belief among European philosophers and writers that the truth about the nature of man and his world could be discovered by the use of reason. "In the search for truth in the sciences", the *Discours de la Méthode* (1637) by René Descartes (1596–1650) led to a more sceptical approach to astrology and history. At the same time, critical interpretation of the Scriptures and an interest in comparative religion brought about a weakening of religious orthodoxy.

In England, the execution of Charles I discredited belief in the divine right of kings and the revolution of 1688 established a liberal constitution that served as an ideal in continental Europe for a century. This shift in opinion and attitude was given the name of Enlightenment (*Les Lumières* in French, *Aufklärung* in German).

Scope of the movement

The Enlightenment was a literary and philosophical movement against superstition, ignorance, traditional knowledge and accepted wisdom. Politically it ran from John Locke (1632–1704) to the French Revolution a century later. Lessing (1729–81) and Goethe (1749–1832) were its representatives in Germany, Franklin [3] and Jefferson (1743–1826) in America and Algarotti (1712–64), Alfieri (1749–1803) and Beccaria (1738–94) in Italy. It was little known in Spain and Eastern European countries, but attempts at Westernizing Russia under Peter the Great (1672–1725) and Catherine II (1729–96) owed their inspiration to the French and German Enlightenments.

Scientific and literary inquiry

Although this movement did not give rise to any particular school, its leaders shared a spirit of scientific inquiry, an acceptance of Newtonian physics in the idea that the universe was regulated by discoverable laws, a belief that knowledge should be widened so that all men could exercise their God-given gift of reason and a wish to combat the errors of antiquity that led to superstition and had become the pretext for persecution [4].

Literature centred on humane contemporary topics. There was a tendency towards cosmopolitanism and the use of non-European settings, as in Defoe's *Robinson Crusoe* [6] and Montesquieu's *Lettres persanes* (1721). Satire, whether in verse or prose, became the main genre and vehicle for criticism, bringing with it a strong plea for toleration and liberty of conscience.

In religion, the scientific spirit of the age was reflected in a denial of miracles and revelation and also an abhorrence of "enthusiasm". Enlightened writers (the *philosophes*), while not usually aesthetic, considered on humanitarian grounds that it was absurd to foist one's religion on another individual or nation; differences of creed were seen by Voltaire in *Zaïre* (1732) and Lessing in *Nathan der Weise* (1779) as mere accidents of birth and education. Accounts of voyages turned interest towards the common denominator of all religions: a belief in a God that could serve as a rational hypothesis to account for existence. As a result, there arose the concept of an honest pagan and of deism, which held that, having created a workable universe, God left it running rather in the manner of a clock, which Alexander Pope [2]

CONNECTIONS

See also
32 Exploration and science 1750–1850
34 European literature in the 18th century
72 Origins of romanticism
74 The French revolution
64 The American revolution
28 Scotland in the 18th century
30 Scottish culture to 1850
18 England under the Hanoverians
170 Political thought in the 19th century
172 Masters of sociology

In other volumes
324 History and Culture 1
234 Man and Society
272 Man and Society
214 Man and Machines

1 *De rerum natura* by Lucretius (*c.* 99–55 BC) was a popular text for Enlightenment writers. Lucretius had attacked religion and favoured rational explanation to combat superstitious fears of natural disasters such as volcanoes, earthquakes or plague (depicted here in a 1725 edition of the work).

2 Alexander Pope (1688–1744), although primarily a satirical poet, was best known in Europe for his *Essay on Man* and *The Universal Prayer.* They summarized what his contemporaries felt about the inflated theories of metaphysics and the possibility of a form of deism that would be universally acceptable.

3 Benjamin Franklin (1706–90) – depicted here in an allegory – was an intellectual all-rounder embodying the very spirit of the Enlightenment. He was an artisan, apprentice, journalist, author, founder of a philosophical society and later of an academy that acquired university status, experimental scientist, traveller, and statesman of international repute.

4 The Massacre of St Bartholomew's Day was held up by Voltaire (1694–1778) in *La Henriade* as an abhorrent example of religious fanaticism. This epic, secretly published in 1723, praised Henri IV for being an enlightened king and granting the Edict of Nantes. This religious tolerance had been gradually eroded and the edict was revoked in 1685. The flight of French refugees to Holland, Britain and the German states was an important facet of internationalism in the Enlightenment.

5 Göttingen University was founded in 1737 by the Elector of Hanover, George II of England. The aula is a notable example of 18th-century civic architecture. Academic expansion was one of the practical effects of the Enlightenment.

expounded in his popular *Essay on Man* (1733). Rousseau (1712–78) [7] expounded in *Emile* (1762) a more emotive form of deism while there were others who became overtly atheistic.

The *Encyclopédie* [Key] of Diderot (1713–84) and d'Alembert (1717–83) embodied the central idea of the Enlightenment that knowledge is power. A final monument to human endeavour before the days of specialization, this massive encyclopaedia concentrated attention largely on the skill of the artisan at a time when literature too was dealing more and more with common people, the trader and merchant. Jeremy Bentham's Utilitarian plea for "the greatest happiness of the greatest number" was an offshoot of this interest while the Physiocratic movement in France and Adam Smith's *Wealth of Nations* (1776) also typify the Enlightenment in seeking rational solutions to economic problems. Their plea for commercial liberty (*laissez-faire*) is the counterpart of the philosophers' plea for liberty of conscience.

Socio-political problems provided the most obvious targets for satire, frequently in the form of imaginary journeys, as in Swift's *Gulliver's Travels* (1726). Serious historical writing of the period again shows the desire to reject accepted judgments and to replace them with the kind of critical appraisal of original material undertaken by Bayle and Montesquieu in France, Hume and Gibbon in Britain and Alfieri in Italy.

The arts and politics

In the arts, there was a trend away from emotion towards order and disciplined form. Civic architecture such as the New Town, Edinburgh, the theatre at Bordeaux or the aula (great hall) at Göttingen University [5] clearly embodied Enlightenment ideals. Diderot's *Salons* (1759–79) and Lessing's *Laokoon* (1766) exemplified the new importance given to criticism of the arts.

To ascribe the French Revolution to the direct influence of the Enlightenment is an oversimplification – the leaders of the Enlightenment [8] were optimistic reformers not revolutionaries, opposing all forms of dogmatism and believing in gradual progress towards a better world.

KEY

The *Encyclopédie*, published between 1751 and 1772, had this frontispiece by C. N. Cochin sold as a separate item. It captures the semi-poetic, almost religious fervour with which the *philosophes* pursued enlightenment. In the background the obscurity of Antiquity is ruptured by a burst of light. Reason appears emblazoned in the centre; at her feet the Sciences and Useful Arts bring their tribute. Nor are the Fine Arts forgotten; in the left foreground silversmiths and goldsmiths bring a luxurious contribution of elegantly wrought plate. The emotive quality is reminiscent of rococo church decoration.

6

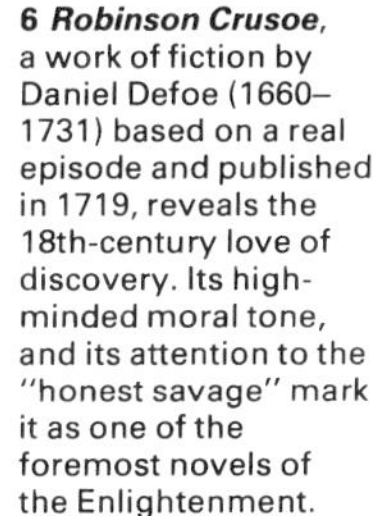

6 *Robinson Crusoe*, a work of fiction by Daniel Defoe (1660–1731) based on a real episode and published in 1719, reveals the 18th-century love of discovery. Its high-minded moral tone, and its attention to the "honest savage" mark it as one of the foremost novels of the Enlightenment.

7

7 The French Revolution claimed Jean Jacques Rousseau as its spiritual progenitor, although his major work, *The Social Contract* (1762), was widely read only after 1789. This pictorial allegory shows Rousseau presiding over the eye of truth, above the tree of liberty and other symbols of the revolution.

8 Madame Geoffrin's salon [A, B] brought together in Paris during the mid-1700s many of the artists and *philosophes* of the time, together with such distinguished visitors as Gibbon, Hume and Horace Walpole. Marie Thérèse Geoffrin (1699–1777), who appears [7] in the painting, was a wealthy bourgeoise and patroness of men of letters. Le Kain [4] addresses a group that includes Buffon [1], Mlle de Lespinasse [2], Mlle Clairon [3], D'Alembert [5], Helvetius [6], Fontenelle [8], Montesquieu [9], Mairan [10], Turgot [11], Diderot [12], Quesnay [13], Saint-Lambert [14], Rousseau [16], Raynal [17], Thomas [18], Marmontel [19], Marivaux [20], Condillac[21] and Réaumur [22]. Presiding over the assembly is a bust of Voltaire [15], who, in a letter to M Lefebvre in 1732, expressed the opinion that such salons fostered a spirit of coterie.

B

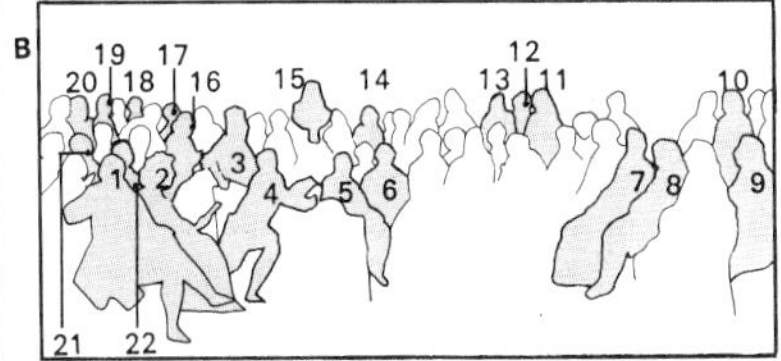

8
A

The English in Ireland

At the end of the Middle Ages, Ireland remained a partially conquered country in which opposed cultures coexisted uneasily. English influence was confined to the colonized counties around Dublin. Beyond that "Pale" were some 90 independent lordships, two-thirds of them ruled by Gaelic dynasts, the rest by gaelicized Anglo-Norman lords. The only significant governmental function was defence, and this was entrusted to the FitzGeralds of Kildare.

The Reformation and anglicization

The weakness of the state was of merely local importance until the Reformation, by altering England's relations with Europe, gave Ireland a new strategic significance. After 1534, an attempt was made to improve international security by extending English control in Ireland. Direct rule was introduced on the basis of imported governors, civil servants [1] and armies, but experience soon revealed that control could not rest on military conquest alone and there followed an associated programme of anglicization.

Both Gaelic and Anglo-Norman lords resisted encroachments on their autonomy, and the situation was complicated by the tenacity with which both natives and settlers adhered to their Catholicism.

The conquest was poorly financed, piecemeal and protracted. It was brought to completion only when the outbreak of war between England and Catholic Spain made Ireland a strategic liability. In the Nine Years War (1595–1603), an Ulster-based confederacy of Gaelic lords led by Hugh O'Neill (*c*. 1540–1616) was defeated, and the incoming James I (reigned 1603–25) became the first English king to rule all Ireland. Thereafter, anglicization proceeded quickly.

The self-exile of the defeated northern leaders made possible the systematic "plantation" of Ulster with English and Scots settlers [4, 7]. The discriminatory enforcement of English property law allowed a widespread public and private expropriation of the Irish to take place elsewhere. A sizeable group of immigrant Protestant landowners gradually developed. Their influence was contested by the older Catholic colonists, who vowed loyalty to the Crown and sought guarantees of their property rights but were rebuffed in the 1630s when Lord Deputy Wentworth (1593–1641), later the Earl of Strafford [5], confiscated a proportion of their land to further a colonizing scheme he was promoting in Connaught.

Protestant conquest

When the Ulster Irish rose in rebellion in 1641, the Catholic colonists joined them. Both were united in fearing that the growing influence of the English Parliament and the Scots would overturn the practical toleration of Catholicism in Ireland. There was little unity of purpose, because the colonists, who possessed one-third of Ireland, had much to lose, while the Irish, joined by returning exiles from Europe, had much to gain. The English Civil Wars created disabling divisions on the English side, but after Charles I (reigned 1625–49) had been executed, Oliver Cromwell (1599–1658) conquered Ireland easily and ruthlessly [6]. A vast expropriation followed, in which no distinction was made between the Gaelic Irish and the

CONNECTIONS

See also

1 Edmund Spenser (1552-99) served as a minor official in Ireland and acquired plantation lands in County Cork where he wrote much of *The Faerie Queene*. His grandson was designated an "Irish papist" by the Cromwellians, deprived of his estate and transplanted to Connaught with many other Catholics.

2 Thomas Lee (*d*. 1601), here fancifully portrayed as an Irish knight dressed for bogland terrain, was one of the many English adventurers who sought their fortunes in 16th-century Ireland. Others included Richard Grenville (*c*. 1542-91), Humphrey Gilbert (*c*. 1539-83) and Walter Raleigh (*c*. 1552-1618).

3 Crannogs, artificial islands of brushwood, peat, logs and stones sometimes surrounded by a timber palisade, mostly date from the early Bronze Age, although Neolithic remains have been found in some. They provided a secure home for the more important families in low-lying and marshy areas. Some were still inhabited in Ulster during the 16th century.

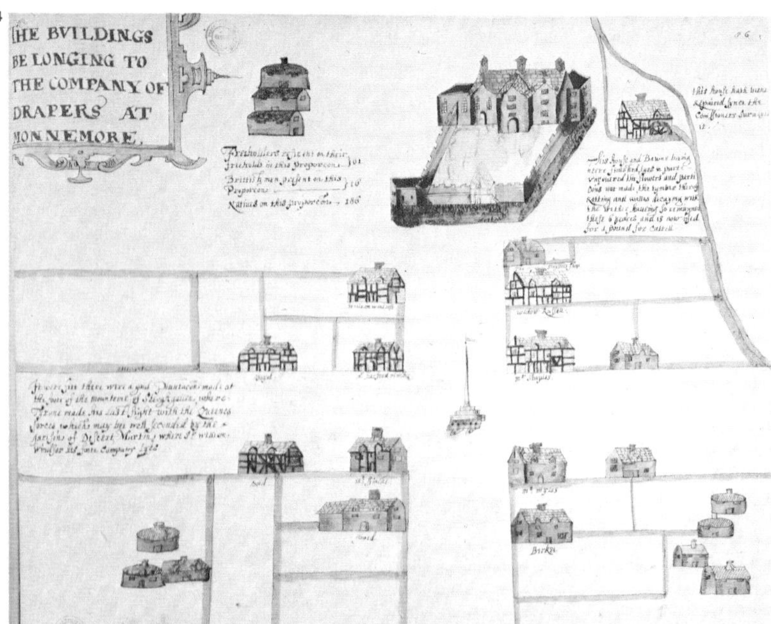

4 The village, in plantation areas, was not only a unit of defence, but a symbol of civilization. In Ulster, as this contemporary map detail shows, they were composed of neat timber-framed houses and cottages that contrasted sharply with the native Irish settlements in which wattle and turf houses clustered together.

5 The independence of the Lord Lieutenant of Ireland, the Earl of Strafford, posed a threat to the power of the English and Scottish parliaments when they joined forces against the Crown in 1640. He was charged with treason, the Irish Parliament readily attested to his misgovernment and the discontents of three kingdoms converged to lead to his execution.

Catholic colonial community. This was not accompanied by systematic settlement however, and existing Protestant settlers benefitted largely.

When Charles II (reigned 1660–85) was restored in 1660, his dependence upon Protestant support ensured that only token modifications of this arrangement were possible. But the fact that his brother and heir, James II (reigned 1685–8), was a Catholic gave hope of redress, and Catholics in Ireland rallied to James's support when he was deposed in 1688, while Protestants transferred their allegiance to William of Orange (reigned 1689–1702) [Key].

The Protestants were confirmed in their ownership of Irish land, and the government confirmed in its power to rule Ireland without reference to the interests of its Catholic population. Protestant supremacy was secured by a system of laws designed to depress Catholics, and particularly the remaining Catholic landholders, rather than to suppress Catholicism. Important changes within the Catholic community followed. As the population increased steadily, and landlords responded to demand by letting their land in ever smaller units, settlers and natives gradually merged into a depressed peasantry. In the towns, by contrast, where economic activity was less affected by penal constraints, a Catholic middle class desirous of its full rights slowly developed.

Union with England

The privileged society of Irish Protestants quickly acquired local interest and ambitions. No longer needing English support to uphold their position, they came to resent English control. Their claims for recognition of Ireland's legislative independence were conceded in 1782.

Circumstances soon challenged the basis of Protestant ascendancy. Indeed, the growth of republican separatism produced an abortive rising in 1798 so that when the government proposed the political union of England and Ireland as the most secure arrangement, Irish Protestants recognized the scheme as the best means of protecting their position in the future [10]. In 1800, the Irish Parliament voted itself out of existence.

KEY

The Battle of the Boyne between James II and William of Orange in 1690 was fought for control of the English throne. The battle was part of a wider European conflict and the armies were international, but it incidentally decided the future of Ireland. William's victory established a Protestant domination that excluded the Catholic descendants of English settlers from the colonial community and led them to assimilate with the Irish.

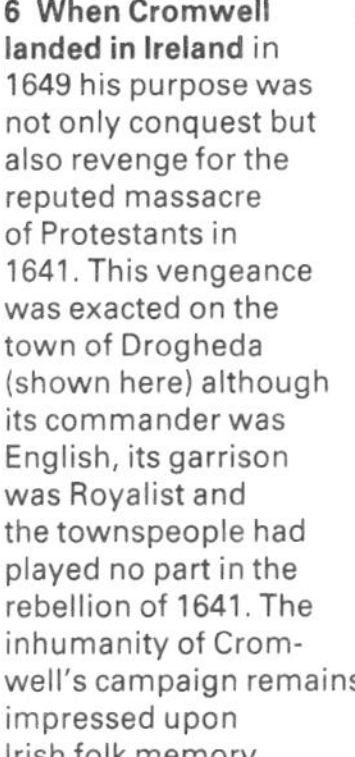

6 **When Cromwell landed in Ireland** in 1649 his purpose was not only conquest but also revenge for the reputed massacre of Protestants in 1641. This vengeance was exacted on the town of Drogheda (shown here) although its commander was English, its garrison was Royalist and the townspeople had played no part in the rebellion of 1641. The inhumanity of Cromwell's campaign remains impressed upon Irish folk memory.

6

8

7

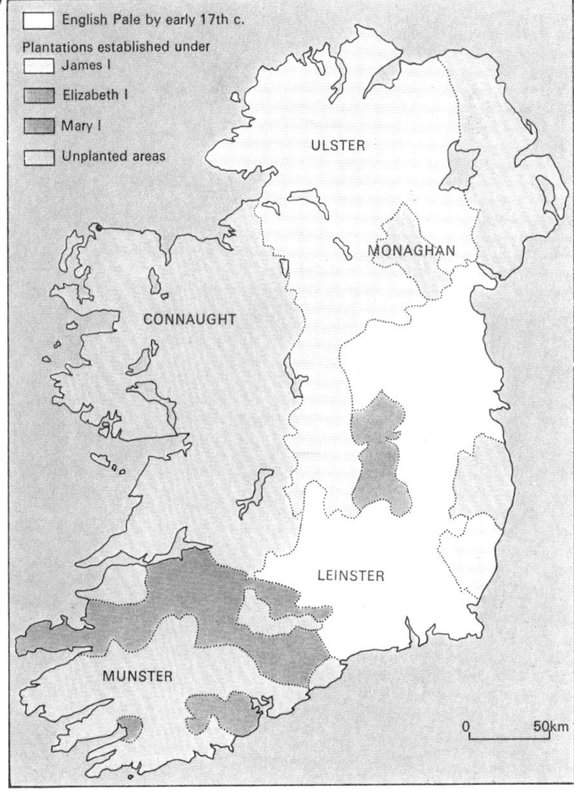

7 **The character of settlement** varied widely. In all planted areas, settlers were interspersed among natives, but in Ulster, particularly in the unofficial, Scottish-based northeastern settlements, they were a fair reflection of society; elsewhere the lower classes were greatly under-represented. In unplanted areas, land ownership changed radically: in 1641 Catholics held 59 per cent of Irish land; by 1703 their share had fallen to 14 per cent.

8 **The impressive classical façade** of the Custom House in Dublin symbolizes the prosperity of the privileged in the late 18th century and suggests the extravagance of their life-style.

9

9 **Edmund Burke** (1729–97), the statesman and philosopher, left Ireland as a young man, although his parliamentary championship of the interests of the American colonists was informed by an Irishman's understanding of their situation.

10

10 **By the Act of Union,** the centuries-old Irish Parliament exchanged its recently won legislative independence for Irish representation at Westminster. Its passage was widely believed to have been procured by bribery. In fact Irish Protestants chose to surrender their power to a protective England of their own accord. Despite Protestant identification with England, the English persistently regarded them as Irish, as this contemporary cartoon suggests.

Irish culture to 1850

Early Celtic art has been described as "the first great contribution by the barbarians to European art". The swirling curves of this so-called La Tène style created by the Celts reached Ireland centuries before Christianity, and are found on metalwork and on such stone sculpture as the Turoe Stone [1].

Early Christian art

This Celtic La Tène style, influenced by Germanic animal ornament and interlacing from the eastern Mediterranean, formed the basic elements of decoration used on objects of Christian metalwork from the golden age of early Christian Ireland, about AD 700. The best examples include the Ardagh chalice [3] and the Tara brooch. For many centuries the monasteries remained the main inspiration and sanctuary of culture, producing beautiful illuminated manuscripts. The *Cathach*, of about 600, is the earliest surviving Irish manuscript. By the mid-seventh century, these became more elaborate, as exemplified by manuscripts such as the *Book of Durrow* [2], and they reached their greatest intricacy with works such as the *Book of Kells*, dated about 800, and regarded as one of the world's most beautiful books. On the margins of their manuscripts, Irish monks wrote lyric poetry in the vernacular of remarkable quality.

The remains of these monasteries, which are still visible, include small stone churches and oratories (such as that at Gallarus [4]), round towers and high crosses. These majestic stone crosses, dating largely from the eighth and ninth centuries, often depict scenes from the Old and New Testaments, serving as open-air picture Bibles. One of the finest crosses is that erected by Muiredach at Monasterboice [Key].

From 1000 to 1660

Although the Viking invasions of the ninth and tenth centuries may have somewhat stunted artistic development in Ireland, the eleventh and twelfth centuries saw a lively cultural revival. In architecture, Romanesque churches include a cathedral at Clonfert [5] and Cormac's Chapel at Cashel, showing an admixture of continental and Norman styles suffused by Scandinavian-inspired animal ornament. Although the manuscripts of that period failed to match the intricacy of the *Book of Kells*, items of metalwork such as the Cross of Cong are reminiscent of the golden age of the Ardagh chalice, while introducing Romanesque and animal ornament as well. Stone high crosses with figures of Christ and bishops or abbots in high relief were carved in the west of Ireland.

The advent of new religious orders such as the Cistercians from 1140 onwards introduced new monasteries in the Gothic style. The Normans, after their arrival in 1169, built fine Gothic cathedrals at Dublin and Kilkenny, and erected strong castles such as Carrickfergus and Trim to defend the lands they had conquered.

But after 1350, Gaelic culture reasserted itself after the effects of the Norman invasion, and with men such as Goffraidh Fionn Ó Dálaigh a vigorous tradition of bardic poetry was created that continued uninterruptedly down to the eighteenth century. Typically Irish Franciscan friaries sprang up in the fifteenth century, at a time when many older churches were refurnished, and

CONNECTIONS

See also
24 The English in Ireland
34 European literature in the 18th century

In other volumes
138 History and Culture 1
140 History and Culture 1
86 History and Culture 1

1 The Turoe Stone is the finest example in Ireland of a decorated Celtic ritual stone, the original purpose of which is unknown. The top part of the stone is ornamented with tendrils and triskeles (a device showing three branches, arms or legs radiating from a common centre). The decoration is executed in an art style that was developed by the Celts on the European mainland about the 4th century BC, reaching Ireland soon after 300 BC.

2 The *Book of Durrow* (*c.* AD 650–700) is the earliest of the great illuminated manuscripts to have survived in Ireland. Full-page decorations of Celtic motifs were innovatory, and the human figure, such as that of St Matthew (seen here), show borrowings from enamelling techniques.

3 The Ardagh chalice was made about AD 700, the golden age of metalwork in Ireland, and is one of its most splendid products. It has 354 separate parts. The body is made of silver and it is profusely decorated with gold panels. It was found by a boy at Ardagh, Co Limerick, in 1868.

4 Stone oratories, such as this one at Gallarus, Co Kerry, were used for prayer by the monks of some of the small Celtic monasteries on the west coast of Ireland from about the 8th to the 12th centuries. The walls are not vertical but slope inwards until they meet at the roof-ridge.

5 The doorway of Clonfert Cathedral, in Co Galway, with its profuse decoration, is perhaps the finest expression of the Romanesque style of architecture and carving practised in Ireland in the 12th century. It stands on the site of a monastery, founded by St Brendan in 558. The great triangular pediment above the door displays 15 sculptured heads. Donkeys', dragons' and horses' heads and grotesque masks carved in the sandstone give the doorway a semi-pagan feeling. The grey limestone frame to the door itself is probably a 15th-century addition.

6 Jonathan Swift was Dean of St Patrick's Cathedral from 1713 until his death in 1745. Much loved by the poor of Dublin, whose cause he espoused, he is best known for his *Gulliver's Travels*. His anonymous *Drapier's Letters* (1724) brought English attention to Irish grievances.

the Irish built tower-houses such as Dun Guaire, which flourished into the seventeenth century. About 1400, much important historical and genealogical material was compiled, such as that in the *Book of Ballymore* and the *Book of Lecan*; the De Burgo genealogy of 1583, has in addition, finely decorated pages. The Limerick mitre and crozier and the Ballylongford cross are fifteenth century works of European standing.

Eighteenth century cultural revival

The wars of the seventeenth century drained Ireland of much of its artistic vigour, but in the eighteenth century a lively culture was sponsored by the landed aristocracy. Large new public buildings such as Gandon's Custom House in Dublin, and splendid domestic structures such as Castletown House [11] and the Casino at Marino, were built in the classical style of the Georgian period. Dublin, with its town houses and squares, became an elegant capital, and Ireland became famous for its silver, glass [7, 8] and bookbinding.

From 1660 onwards, Ireland enjoyed a lively theatrical life, encouraged by men such as George Farquhar (1678–1707) and Richard Brinsley Sheridan (1751–1816). Irish writers began to play an important role in English literature in the eighteenth century. Jonathan Swift's (1667–1745) *Gulliver's Travels* (1726) [6] and Oliver Goldsmith's (*c.* 1730–74) *The Deserted Village* [10] have both become classics of the English language. The visit of George Friedrich Handel (1685–1759) to Dublin where his *Messiah* was first performed in 1742, encouraged Dublin to become a great musical city for the rest of the century. John Field (1782–1837), the inventor of the nocturne, William Wallace (1813–65) of *Maritana* fame, and Michael Balfe (1808–70), well-known to nineteenth-century music lovers for his *The Bohemian Girl*, are among Ireland's best-known composers. At the same time, a lively interest was taken in traditional Irish music, encouraged by Carolan (1670–1738), the blind harpist. Irish tunes collected by Edward Bunting (1773–1843) and Thomas Moore (1779–1852) in *Irish Melodies* (1807–34) were widely influential.

KEY

Muiredach's Cross is a fine example of a Celtic high cross, with its characteristic ring (possibly suggesting a halo of glory) surrounding the intersection of arms and shaft. These standing crosses are among the most striking remains to be seen on many early Christian monastic sites. Most were carved in stone during the 8th, 9th and 10th centuries, and some were intended as memorials to individuals. This cross, erected by Abbot Muiredach at Monasterboice, Co Louth, like many other late crosses of its kind has its surface decorated with a selection of scenes from the Old and New Testaments. These probably helped to explain the Bible in pictorial form to those who could not read.

7

7 Sauce-boats made in silver by Robert Calderwood of Dublin in 1737 are typical of the work of 18th-century Irish craftsmen who excelled in objects such as these, designed for the tables of the rich. Early silverware was quite plain, but as the century progressed, decoration became more intricate and profuse, graduating from rococo swags to the delicately proportioned classical style of Robert Adam.

8

8 This richly decorated wine decanter was made by Penrose in Waterford about 1795. It typifies the outstanding glassware produced there and in other Irish centres during the 18th and early 19th centuries.

9

9 George Berkeley (1685–1753), Bishop of Cloyne, Co Cork, was Ireland's most renowned philosopher. His theory of vision laid the foundations for later research on the subject, and in his treatise "The Querist" (1735–7) he anticipated the economic theories of David Hume (1711–76) and Adam Smith (1723–90). He denied the existence of matter as substance but subscribed to the conventional laws of physics.

10

10 Oliver Goldsmith, poet, novelist and dramatist, was born in County Longford. He is one of the greatest Irish writers of the 18th century and is probably best known for his only novel, *The Vicar of Wakefield* (1766). Critics believe that some of his finest works, such as the play *She Stoops to Conquer* (1773) and his poem *The Deserted Village* (1770), embody scenes from his Irish childhood. He also wrote *A History of England* (1771). Goldsmith spent much of his life in London, where he tried unsuccessfully to practise medicine. The verdict of his friends was that "he wrote like an angel and talked like poor Poll". This did not debar him from Dr Johnson's literary circle, where he was always welcome.

11

11 Castletown House, in Castletown, Co Kildare, is a splendid 18th-century Irish mansion built for himself by William Connolly, Speaker of the Irish House of Commons. The building was designed by the Italian architect Alessandro Galilei. The tall central block joined to its lower wings by an open colonnade inspired the designs of many country houses throughout Ireland at that time. But the Georgian style ebbed after the Act of Union united the English and Irish parliaments in 1801. Although Castletown House is an outstanding example, it was by no means unique because in the 18th century the Irish landed gentry amassed sufficient wealth to build splendid country seats for themselves. Among the foremost architects designing in the Palladian style were Edward Lovett Pearce and Richard Cassells.

Scotland in the 18th century

With the passing of the Act of Union in 1707, Scotland ceased to be an independent country and became part of Great Britain. Sixteen Scottish peers were elected to join the House of Lords (English membership 190) and 45 MPs sat in the House of Commons (English membership 513) at Westminster. Scottish MPs were notoriously pliable to the government's will at Westminster. The effective management of Scottish affairs in Parliament passed to government "managers" – the Dukes of Argyll for much of the century and thereafter usually the lord advocates, of whom Henry Dundas (1742–1811) William Pitt the Younger's confidant, was the most famous and effective.

The Jacobite rebellions

Enemies of the Union were few and far between, except for the Jacobites who supported the exiled House of Stewart's claim to the throne. In 1715, the Earl of Mar (1675–1732), a former supporter of the Union whose political ambitions had been blocked, attempted to raise the country for James Edward, the Old Pretender (1688–1766), James II's son. The rising won most support in the Highlands, but petered out after the inconclusive Battle of Sheriffmuir in November 1715 [1].

In 1714, the Young Pretender, Prince Charles Edward Stewart (1720–88), made a second attempt on the throne. In August he landed in Inverness-shire from a French ship and proclaimed his father King of Scotland and England. Even among the Highland clans only a minority followed him, and despite initial successes in which he took Edinburgh and defeated a Hanoverian army at Prestonpans, Charles was relying upon English Jacobite and French support when he marched on London in November [3]. The Jacobite army reached Derby and after some hesitation, in the face of mounting opposition forces under the Duke of Cumberland (1721–65), turned back northwards.

In April 1746, Cumberland caught up with Charles at Culloden [4] where the Jacobite army, outnumbered and poorly organized, was heavily defeated. The rebellion was finally crushed and Charles forced to flee back to France. The Union remained intact. Even the Jacobites had primarily wanted to regain the British throne for the Stewarts rather than to re-establish Scottish independence. With their defeat there was no further challenge to Westminster government for nearly half a century, and in the aftermath of Culloden the power of the Highlanders was broken for ever.

Economic consequences of the Union

The satisfaction that most Scots felt for the Union in the eighteenth century rested largely on an economic base and the new markets opened up by the Union with England. Although there was little dramatic change in the condition of the country until after 1760, even in the first half of the century food became more plentiful, the cattle trade with England expanded and linen emerged as the first major Scottish industry.

The tobacco trade, too, became an important source of prosperity to Glasgow and the west of Scotland when in the 1740s Scottish merchants secured bulk contracts to supply tobacco to France. The link forged with Virginia remained important until after

CONNECTIONS

See also
18 England under the Hanoverians
30 Scottish culture to 1850
94 The British labour movement to 1868
164 Scotland in the 19th century

In other volumes
298 History and Culture 1
214 Man and Machines

1

1 After the Battle of Sheriffmuir the cause of the Old Pretender was doomed. The rebel prisoners were taken south to be tried by courts in Carlisle and London because it was feared that they would not be punished severely enough under Scottish law. Here one of the two executions of the rebels, that of Lord Derwentwater (1689–1716), is shown.

2

2 Rioting broke out in Glasgow in 1725 following the decision by Westminster to extend the tax on malt to include Scotland as well as England. An angry mob attacked the house of Campbell of Shawfield who had supported the measure in Parliament. It was essentially an anti-government riot and not pro-Jacobite.

3

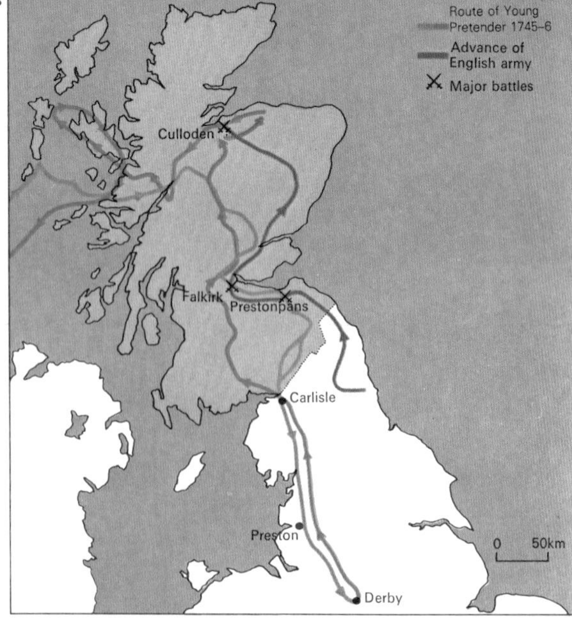

4

3 Prince Charles's march through Scotland and into England was almost unopposed. His army took Edinburgh without a shot, routed a Hanoverian army at Prestonpans, captured Carlisle and marched on London. But after reaching Derby, the Prince retreated northwards on 6 December 1745.

4 The bloody battle at Culloden, near Inverness, on 16 April 1746 was an overwhelming victory for the Hanoverian forces and effectively broke the Jacobite cause forever. In its aftermath many of the defeated clansmen were butchered by the victorious army under the Duke of Cumberland.

the American Revolution (1775–83).

Towards the end of the century economic change became more rapid and far-reaching. In the Lowlands landowners and their tenants began enclosures, turnip husbandry and more intensive forms of farming both for animals and grain. By 1790 this had transformed the Lothians and was beginning to have an impact elsewhere. In the Highlands the "crofting system" was introduced whereby tenants had smallholdings along the shore and spent part of their time fishing or gathering seaweed ('kelp') from which an industrial alkali was manufactured. Everywhere in the countryside population increased rapidly, many migrated to gain employment in the towns or settled round the new industries such as Carron Iron Works (founded 1759) or the cotton mills at New Lanark (founded 1785) [9].

The Scottish Enlightenment

During this period of economic prosperity, Edinburgh flourished exceedingly. Its enterprising town council planned a New Town focused on Princes Street and George Street; by 1800 this was attracting fashionable and middle-class families to its splendid homes in large numbers [Key]. The Scottish universities, especially Edinburgh and Glasgow, gained a worldwide reputation. Thinkers such as the philosopher David Hume (1711–76), economist Adam Smith (1723–90), the poet Robert Burns (1759–96), painters and architects such as Henry Raeburn (1756–1823) and Robert Adam (1728–92) and inventors such as James Watt (1736–1819), created "Scottish Enlightenment" of learning and ingenuity without parallel in the past [6, 8].

Not everyone, however, was convinced that the system was incapable of improvement. In the last decade of the century Scottish radical sympathizers with the French Revolution, especially the Friends of the People and the United Scotsmen began to make demands for a more democratic form of government [7]. Dundas and Pitt suppressed them as they had also suppressed English radical clubs, but in the nineteenth century the radicals' challenge proved more enduring than that of the Jacobites.

KEY

Charlotte Square in Edinburgh was the most splendid of the squares built in the New Town in the 18th century. The planning and construction of the New Town reflects the growing wealth and prosperity of the professional classes in 18th-century Edinburgh. Robert Adam was the main architect of this square and the elegant, classical style for which he is famous dominates the New Town.

5

5 The Royal Bank of Scotland was founded in 1727 largely because the Bank of Scotland's directors were suspected of Jacobitism. Even today the rival banks issue different notes.

6

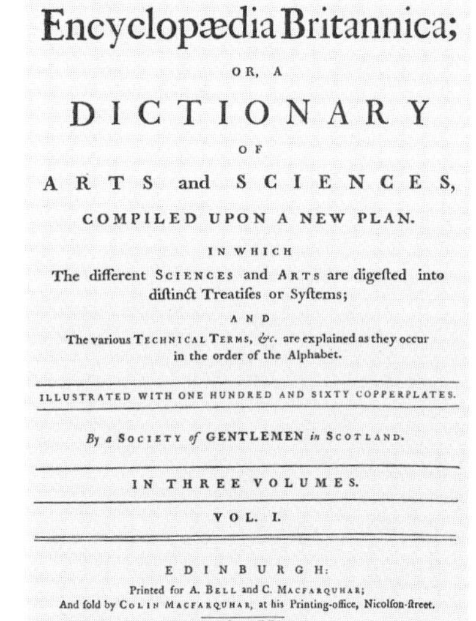

Encyclopædia Britannica;

OR, A

DICTIONARY

OF

ARTS and SCIENCES,

COMPILED UPON A NEW PLAN.

IN WHICH

The different SCIENCES and ARTS are digested into distinct Treatises or Systems;

AND

The various TECHNICAL TERMS, &c. are explained as they occur in the order of the Alphabet.

ILLUSTRATED WITH ONE HUNDRED AND SIXTY COPPERPLATES.

By a SOCIETY of GENTLEMEN in SCOTLAND.

IN THREE VOLUMES.

VOL. I.

EDINBURGH:

Printed for A. BELL and C. MACFARQUHAR;

And sold by COLIN MACFARQUHAR, at his Printing-office, Nicolson-street.

M.DCC.LXXI.

6 The *Encyclopaedia Britannica*, compiled by a "society of gentlemen", first began to come off the Edinburgh presses in weekly numbers in December 1768. The editor was William Smellie (1740–95), a local printer, and the work aimed to provide "a dictionary of the arts and sciences". The three volumes of the first edition were completed in 1771 and an expanded second edition was begun in 1776.

7

7 Thomas Muir (1765–98), an Edinburgh lawyer and leader of the Friends of the People, was tried for sedition before Lord Braxfield in 1793. This was one of a series of trials held in the wake of the French Revolution that were aimed at Jacobin radical societies. After a hearing notorious for the violent bias of Braxfield, Muir was sentenced to transportation to Botany Bay for 14 years.

8

8 The school system in Scotland was the cornerstone of the great flowering of intellectual life in the late 18th century that was **the Scottish Enlightenment**. The ability to read and write was much more widespread than in England, and in many areas the rural parish school (supported by a tax levied on local landowners) educated both rich and poor. In the Highlands and towns, however, the parish system was inadequate and would often be supplemented by the wealthy with private tuition. The universities in the 18th century, in particular Edinburgh, flourished and established an eminent tradition in law, medicine and philosophy.

9

9 The new technology of Richard Arkwright (1732–92), made it possible to spin cotton fibres by water power. Arkwright introduced his new technique into Scotland at New Lanark (shown here) because he wished to undercut English labour costs. In 1799 the mills were sold to Robert Owen (1771–1858). Owen made the community at New Lanark world-famous for his pioneering social reforms: in working conditions, housing and education.

Scottish culture to 1850

The first flowering of Scottish art was Celtic in its forms. The Pictish stones of eastern Scotland [1] from the sixth to the ninth centuries are often enigmatic, with a fondness for displaying snouted monsters and mounted horsemen. More attractive is the almost contemporary art of Dalriada in western Scotland which is Irish in character, and is best represented by the beautiful "high crosses" of Islay and Iona, by jewels such as the Hunterston brooch [2] and by the *Book of Kells* now in Trinity College, Dublin, which was probably illuminated by the monks of Iona around 800. The Celtic tradition went on in the Highlands through the Middle Ages, although it was gradually coarsened and weakened with time.

The development of Scottish architecture

Scottish medieval art survives today – largely in architecture – as Gothic ruins such as the cathedrals of Elgin and St Andrews or the Border abbeys [3]. A grace and sweetness of line often compensates for the relatively modest scale of most of these buildings. From the earlier Anglo-Norman influences, castles and mansions in Scotland in medieval times developed a distinctive national character, which often showed a strong French flavour. Fine medieval castle keeps remain at Borthwick, Bothwell and the Hermitage [4], each overwhelming in their impression of strength rather than comfort.

In the sixteenth and seventeenth centuries secular architecture mellowed. The new royal palaces of Stirling and Falkland showed French renaissance influence of a type unparalleled in England. The Scottish lords preferred to develop a "Scottish baronial" style untouched by classic models but graceful and picturesque with its emphasis on verticals and fantastic roof lines [5]. In the towns, whitewashed houses and crowstepped gables were in fashion.

After 1660 the classical architecture of southern Europe entered Scotland under the polished hand of Sir William Bruce (1630–1710) and his eighteenth-century successors, William Adam (died 1748) and Robert Adam (1728–92) [8]. The last named was the most famous architect Scotland ever produced: under his influence "the whole world went Adamitic", and his interpretation of Etruscan and Roman forms was copied from St Petersburg to Boston, Massachusetts. The eighteenth century was also the age of town and village planning – Edinburgh's gracious New Town is only the best known example of a movement that stretched from Inverary in Argyll to Cupar in Fife. By the early nineteenth century the romantic revival was producing a new kind of "Scottish baronial" country house inspired by the nostalgic writings of Sir Walter Scott (1771–1832), and his home, Abbotsford.

Early writers and the Enlightenment

In Scottish intellectual life the most important figures were of the eighteenth century, although Duns Scotus (*c.* 1265–1308) was an important scholastic philosopher, and in the sixteenth century a number of major figures emerged. The poetry of William Dunbar (*c.* 1460–*c.* 1520) and that of Gavin Douglas (*c.* 1474–1522) was part of a great flowering of Scottish culture. Other eminent scholars of this period include the theologian John Major (1469–1550), the Latinist and histo-

CONNECTIONS

1 This Pictish slab, with a cross on one side and a battle scene on the reverse, dates from the 8th century and stands at Aberlemno in Angus. Its zoomorphic designs are characteristic.

2 The Hunterston brooch is an outstanding example of Celtic craftsmanship of the 8th and 9th centuries. Wrought in silver-gilt and set with amber, it resembles the Tara brooch in Dublin.

3 The Border abbeys date mainly from the 12th century, but many were damaged in English raids and were substantially rebuilt in the 14th and 15th centuries. Indeed, so much of Scottish medieval architecture lies in ruins that it is hard to form an impression of how fine it was. Melrose Abbey (shown here) was a Cistercian foundation under David I (1084–1153), who introduced many new monastic orders to Scotland. The existing remains, of 1450–1505, show a strong French influence and include some fine late Gothic tracery. Among the other abbeys, Kelso includes some unique late Norman work.

4 The formidable Hermitage, one of the most famed of the Border castles, dates mainly from the late 14th and early 15th centuries. The Hermitage illustrates the typically uncompromising style of these fortresses. Only with the union of the crowns in 1603 did the Border castles fall into disuse and disrepair.

5 The Scottish baronial house of the 16th and 17th centuries is one of the most distinctive national contributions to architecture. Craithes, shown here, is typical. Lightly fortified against raids – it has only one door and its windows were protected by grilles – it is nevertheless very graceful, culminating in fanciful turrets and pitched roofs.

rian George Buchanan (1506–82) [7] and the mathematician John Napier (1550–1617), who invented logarithms.

The Scottish Enlightenment of the eighteenth century was primarily a philosophical movement associated with David Hume (1711–76), perhaps the most important and original philosopher ever to write in English. Other important figures were Adam Smith (1723–90), the philosopher and father of political economy, and Adam Ferguson (1723–1816) and John Millar (1735–1801), who laid the foundations of sociology. On the scientific and technical side outstanding contributions were made in the eighteenth century by Joseph Black (1728–99), who formulated a theory of latent heat, James Watt (1736–1819), of steam engine fame, and Thomas Telford (1757–1834), the "Colossus of Roads". Edinburgh, the "Athens of the north", proved a magnet for the philosophical and literary society of the age.

The literary achievement of eighteenth-century Scotland ranged from the work of diarist and biographer James Boswell (1740–95) to that of Robert Burns (1759–96) [Key], whose poetry was the deliberate revival of a national tradition of writing in the vernacular that went back to late medieval poets such as William Dunbar. Burns's lyrics also stood at the beginning of Romanticism, a movement advanced by the forged "translations" of epic Gaelic poetry by James Macpherson (1736–96) and by the novels of Sir Walter Scott [9].

Painting and a portrait

The same cultural upsurge can be seen rather less fruitfully in painting, where the classicism of Allan Ramsay (1713–84) and Henry Raeburn (1756–1823) degenerated into the anecdotal paintings of David Wilkie (1785–1841), which gained considerable popularity in the early nineteenth century.

The enormous vigour of the period 1740–1830 makes it a true golden age of Scottish culture. No better portrait of its closing years is possible than the lively diaries of Henry Cockburn (1779–1854), the Edinburgh lawyer and politician whose writings have again grown in popularity in the late twentieth century.

KEY

Robert Burns has a unique place in the hearts of all Scots, partly because he was what so many people would like to be – romantic, handsome, irresistible to women, and incapable of resisting them. The exceptional quality of his lyrics spring from the perfect match of words and tune: his love songs are moving without being cloying, passionate without being false. Burns is one of the few poets who can express happiness without making it sound more boring than misery. The other side to his genius was expressed in his great patriotism and personal radicalism in castigating hypocrisy and extolling the liberty and dignity of the common man.

6

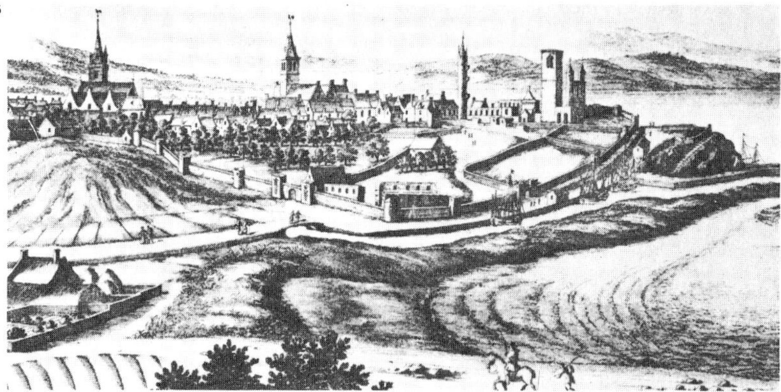

6 The town of St Andrews boasts the first university in Scotland. It dates from 1410, when Bishop Wardlaw (*c.* 1370–1440) established the College of St Mary in the shadow of his cathedral. This followed the Great Schism within the papacy (1378–1417) which divided the Church between two contesting popes and divided Scotland and England between the candidates, cutting off Scottish scholars from attendance at English universities. Thus, before the end of the 15th century, two more colleges had been founded in Scotland – at Glasgow and Aberdeen. This 17th-century illustration shows the town as it was in post-Reformation times. The cathedral, on the right, is in ruins after being attacked by Protestants in 1559. A defensive wall surrounds the burgh and unenclosed land lies beyond.

7

7 George Buchanan gained widespread renown as a Latin poet, historian and political philosopher in 16th-century Europe. He was condemned as a heretic following his attacks on the Franciscans and forced to flee to Bordeaux, where the French essayist Montaigne was one of his pupils. James VI of Scotland and I of England was also taught by Buchanan and spoke of his tutor as "a great master" of languages. Buchanan's distaste of corruption and convention has left him with a mixed reputation, but he is generally regarded as a principal figure of the Scottish Reformation, especially because of his conflicts with the Church and the state.

8

8 Robert Adam's castellated style derives not from Scottish baronial models, but from Italian ones: for example Culzean in Ayrshire, shown here, built 1777–90, exploits with classical aplomb its splendid cliff-top site. The interior has rich plasterwork in the typical Roman and Etruscan style of Adam's work. Among his most important works are Osterley Park (1761–80) and Syon House (1762–9), near London, and parts of Edinburgh New Town and University.

9 Walter Scott did more than anyone to impress on the world the romantic image of Scotland. His many novels and poems reveal his intense love of the past which is saved from sentimentality by his narrative gifts. Throughout Europe, Scott was hailed as the leader of the new romantic attitude in the arts. Although his writings were immensely successful, in particular *Waverley* (1814) and *The Heart of Midlothian* (1818), Scott spent much of his life writing desperately and in the teeth of painful illness, in order to pay off debts acquired by ill choice of publishing partners. His writing speed and prolific output were extraordinary, especially in view of the undiminished quality of his later romances.

9

Exploration and science 1750–1850

The advances in exploration and science made in the eighteenth century were based upon the scientific movement and overseas expansion of Europe in the years before 1700. Scientists and explorers were assimilating and extending the pioneer exploits of seventeenth-century giants such as René Descartes (1596–1650) and Isaac Newton (1642–1727). Greater knowledge of the non-European world excited interest in its huge range of exotic flora and fauna and its geography, and questioned many accepted notions about the universe and its origins.

Growth of technical knowledge

In part the growth of knowledge depended upon technical innovations: improvements in navigational instruments, such as the sextant and the marine chronometer [1], facilitated scientific exploration. Captain James Cook (1728–79) commanded three voyages of exploration between 1768 and 1776 and the precision of his navigation depended on these new techniques. On his expeditions Cook charted large areas of the Pacific, including New Zealand, and added the islands of Tahiti and Samoa to existing maps. He also contributed greatly to knowledge of the variety of cultures and natural history.

Scientific discovery also depended upon improvements in instrumentation. The use of a greatly improved telescope by William Herschel (1738–1822) led to the discovery of the planet Uranus in 1781. More sensitive balances contributed to chemical discoveries that displaced the phlogiston theory of combustion. In mathematics trigonometry and calculus were greatly advanced by men such as Johann Bernouilli (1667–1748) and Joseph Lagrange (1736–1813). In physics the properties of heat were investigated by Joseph Black (1728–99) in Glasgow where he met James Watt, inventor of the first efficient steam engine.

Impact of electricity and chemistry

Electricity was one of the principal fields of the physical sciences to be developed. Luigi Galvani [7] discovered "galvanic" (current) electricity, and Alessandro Volta obtained it from his "voltaic pile" or battery, which quickly made electrolysis possible.

In chemistry many natural elements were isolated for the first time. Joseph Priestley (1733–1804) is usually credited with the discovery of oxygen, while Antoine Lavoisier (1743–94) laid the basis for the modern treatment of chemical reactions. In botany the standard classification of plants was devised by the Swedish botanist Carolus Linnaeus (1707–78) [3], while the French naturalist Georges Buffon (1707–88) in his *Histoire Naturelle* suggested that the earth had evolved through a far longer time than the six thousand years of biblical history. Other pioneers included Charles Lyell (1797–1875), whose *Principles of Geology* [5] emphasized the theory that the processes of geological change were long, slow and uniform, and this was as true of the remote past as of the present. Many more speculative but competent naturalists such as Erasmus Darwin (1731–1802) and Jean Baptiste Lamarck (1744–1829) were already propounding the idea of organic evolution, that is, the descent of the present races of animals and plants from antecedent, dissimilar forms.

Many advances in the physical sciences

CONNECTIONS

See also
22 The Enlightenment
44 International economy 1700–1800
156 The foundations of 20th-century science
124 Colonizing Oceania and Australasia

In other volumes
302 History and Culture 1
130 The Physical Earth
22 The Natural World
32 The Natural World
124 The Natural World
252 The Natural World
112 Man and Society
228 The Natural World

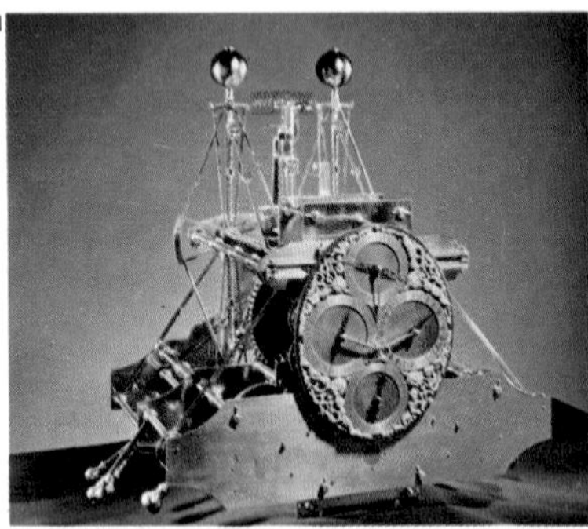

1 The marine chronometer, developed by John Harrison (1693–1776), revolutionized maritime navigation in the 18th century. Latitude had been easy to fix for centuries, but the chronometer made it possible for the first time to establish longitude to within half a degree. Harrison's invention won him a government prize of £20,000.

2 The paintings of animals by George Stubbs (1724–1806) revealed a deep interest in and knowledge of anatomy. A friend of eminent natural scientists, he published a series of anatomical studies. Among them were reconstructions of animals from material brought by Joseph Banks from the Pacific. Stubbs painted this study for William Hamilton.

3 Carolus Linnaeus, the Swedish botanist, founded modern biological classification with his *Systema Naturae* published in 1735.

4 Alexander von Humboldt (1769–1859) made his name as a traveller to many parts of the world and as a geographer, natural historian, ethnographer and oceanographer. His main work was the multi-disciplinary *Kosmos* (1845–62), which attempted to synthesize all knowledge of the universe. He is shown here with his close companion, the Frenchman Aimé Bonpland (1773–1858) (right). He examined wildlife on Mount Chimborazo in the Andes and ascended to a record height in these mountains. At this time vast areas of the world were still unknown. Although the existence of all the continents was known by 1800, there remained large tracts of unexplored land beyond the coastlines of many of them. Settlement was steadily filling the gaps in the Americas but great areas of the Pacific and elsewhere remained untouched. Original accounts of Africa, South America and Arabia were also published during the 19th century.

were directly related to and stimulated by economic activity, trade and industrial processes. Physics played an important part in the rise of the new technologies upon which the Industrial Revolution was based, especially the harnessing of steam power.

The scientific investigations of the Enlightenment very quickly spilled over into other spheres that touched more directly upon politics. The rationalists of the eighteenth century who discovered laws underpinning the physical world were soon joined by writers and thinkers who saw similar laws in politics; thus the scientific movement also gave rise to the "laws of political economy". *Essay on Population* (1798) by Thomas Malthus (1766–1834) produced a mechanical and inevitable theory of demography, similar to that accepted by the classical economists, Adam Smith (1723–90) and David Ricardo (1772–1823). The concept of immutable laws governing political behaviour was adopted by the utilitarian thinkers, such as Jeremy Bentham (1748–1832) and James Mill (1773–1836). Similarly the scientific movement extended to the study of man and by the early nineteenth century produced the first works of sociology and comparative law.

Science and religion

Undoubtedly the most important effect of the scientific movement was its impact upon conventional religious beliefs. Clearly the account of the Creation contained in the Old Testament was not literally true and the world was older than the Bible suggested. Fossil evidence was clear proof of the existence of life on earth at hitherto unthinkably remote periods. Moreover, to some the evidence of geology and of fossils disproved the Old Testament idea of simultaneous creation of all living things. Although these ideas encountered intense hostility and were at first rejected, they provoked profound questioning about the place of revealed religion in the scientific age. *On the Origin of Species by Means of Natural Selection* (1859) by Charles Darwin (1809–82) [Key] detonated an explosion of ideas that had been gradually accumulating during the late eighteenth and early nineteenth centuries.

KEY

Lampooned as an ape, Charles Darwin attracted, with other scientists of his time, the wrath of society when his discoveries and arguments undermined the traditional and biblical accounts of man's creation. The investigation of foreign cultures, and the new scientific approach to the world of nature, threw European institutions and conventional thought into sharper focus and led to questions about their assumptions and legitimacy. As a result, Darwin and others who supported his theory were treated for some time with hostility by their contemporaries who held more orthodox opinions.

5

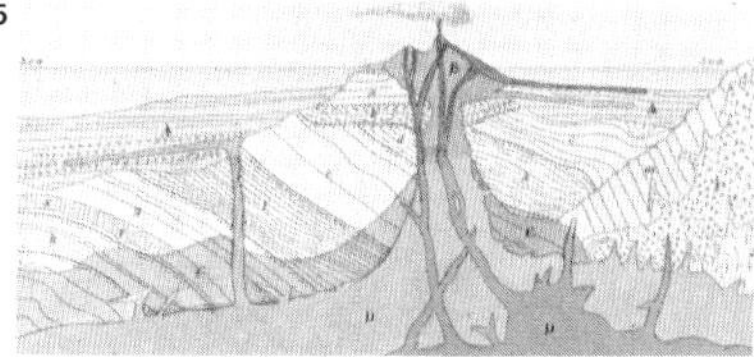

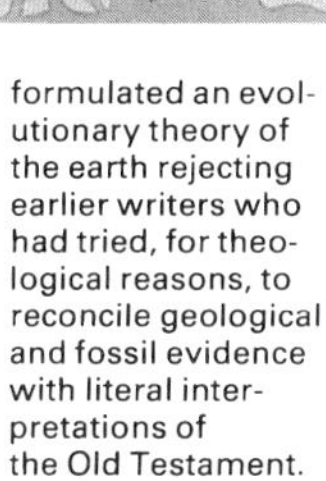

5 The foundations of modern geology were laid down in the 18th century with the study of rock strata and with attempts at geological dating. The frontispiece of Lyell's *Principles of Geology,* (1830–33) is shown here. The work formulated an evolutionary theory of the earth rejecting earlier writers who had tried, for theological reasons, to reconcile geological and fossil evidence with literal interpretations of the Old Testament.

6 A

B

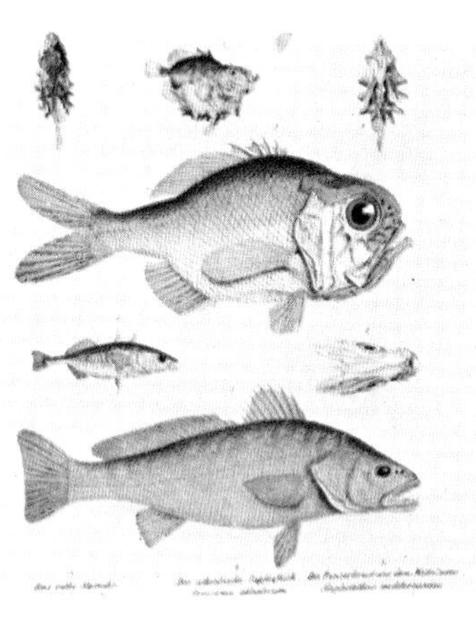

6 Accurate observations of flora and fauna provided one of the most important vehicles of early scientific enquiry. Natural history attracted professional and amateur collectors and precisely illustrated studies proliferated. This illustration of *Phyllanthus niruriodes* [A] is from a collection called *Hindustan Plants.* The fish [B] formed part of a collection that appeared in a German publication by Schinz (1836).

7

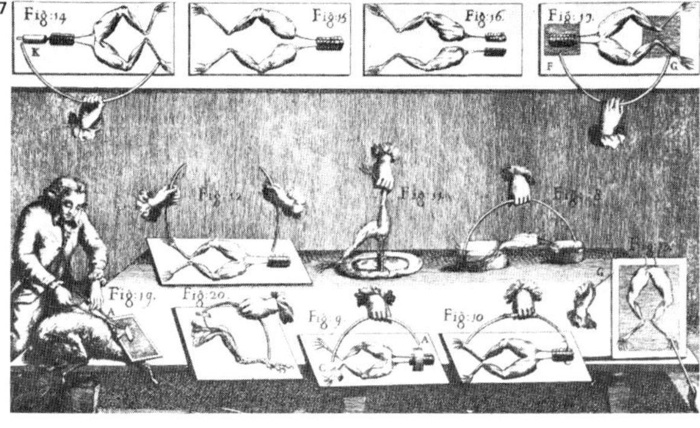

7 In the physical sciences one of the most important fields of development in the 18th century was electricity. Luigi Galvani (1737–98) of Bologna is seen here demonstrating in 1786 the effect of electrical impulses on a frog's nerves. Benjamin Franklin (1706–90), Volta (1745–1827), Ampère (1775–1836) and Ohm (1787–1854) also experimented with electricity.

8

8 The academic study of anatomy became accepted by the 18th century. Its teaching was much advanced by the work of Hermann Boerhaave (1668–1738), and John Hunter (1728–93) and his brother William (1718–83), seen here giving a lecture to the Royal Academy in London. It was not until the advances in chemistry and bacteriology of the 19th century that the causes of diseases were diagnosed accurately and aseptic and antiseptic procedures made surgery less risky.

9

9 The old wives' tale that dairy maids never caught smallpox led to an effective counter to the disease – Jenner's vaccination. Epidemics of the disease were rife and regularly killed or scarred great numbers of people, especially children and young persons. In many countries it had replaced the plague as a major check on population growth. Inoculation using human virus was used among the poor in the 18th century but it was expensive and dangerous with a high risk of serious infection. Edward Jenner (1749–1823) examined the traditional view, saw that people who had caught the relatively harmless cowpox were immune from smallpox and was able to develop a vaccine that carried a minimum risk of ill effects and could be made widely available. Inoculation remained in use in many European countries even after the development of vaccination, but it was finally made illegal in Britain in 1840.

European literature in the 18th century

The rise of the novel in the eighteenth century appears paradoxical in that increasingly scientific age. However, the expanding and newly literate middle class with their "enlightened" attitudes learnt to accept first fiction disguised as fact, then the novel of character and predicament, followed by the picaresque romance, works of sentiment and sensibility, and finally, by the end of the century, the terrors of the Gothic novel.

Early development of the novel

It was Daniel Defoe (1660–1731) who gave the English novel its vigorous start with *Robinson Crusoe* (1719), *Moll Flanders* (1722), *Colonel Jack* (1722) and *Roxana* (1724). He was a realistic narrator, suited to a credulous public, but the novel demanded the elements of social observation and conversational style that were being developed by Richard Steele and Joseph Addison [3], whose essays and sketches created lively characters and stressed the morality of a rational social order. The final step towards the true novel was the addition of a coherent plot to a prose narrative dealing with the interplay of human relations. Samuel Richardson (1689–1761) effected this with *Pamela* [7], which in turn encouraged the corrective burlesque novels such as *Joseph Andrews* (1742) and *Tom Jones* (1749) by Henry Fielding (1707–54).

The comic masterpiece *Don Quixote* by Cervantes (1547–1616) probably had the most influence on the European novel. In France, Alain le Sage (1668–1747) published *Gil Blas* (1715–35), a romance of picaresque adventures, and his Scottish translator, Tobias Smollett (1721–71), was able to develop the novel of diverse characters and incidents – coarse, realistic and episodic. His novels – *Roderick Random* (1748), *Peregrine Pickle* (1751), *Humphrey Clinker* (1771) – all lack the subtlety and irony that mark Laurence Sterne's *Tristram Shandy* (1760–67) and *A Sentimental Journey* (1768). Sterne (1713–68) wrote a fluid, surreal fantasy that lies outside the mainstream development of the novel, but his success was immense and his imitators many, notably Denis Diderot (1713–84) in *Jacques Le Fataliste* (1796). A more central novel, *The Vicar of Wakefield* (1766), is by Oliver Goldsmith (*c.* 1730–74), a delicate stylist who dealt largely with the trials of virtue and domesticity.

With Fanny Burney (1752–1840) and her *Evelina, or, the History of a Young Lady's Entrance Into the World* (1778), a new sensibility was added to the theme of a young girl's adaptation to society.

The change in the novel after 1760

At this point the novel, which had seen occasional outbursts of fancy and imagination such as Horace Walpole's *The Castle of Otranto* (1764) or *Niels Klim* (1741), a Gulliver-like satire by the Dane Ludvig Holberg (1684–1754), turned almost entirely to Gothic sentimentalism and terror. Matthew Lewis's (1775–1818) *The Monk* (1796) and Mrs Radcliffe's *The Mysteries of Udolpho* (1794) were widely acclaimed, and they unleashed both fine novels and a flood of indifferent fiction. In the last decade of the eighteenth century, however, Jane Austen (1775–1817) began writing *Sense and Sensibility* and a series of other novels whose

CONNECTIONS

See also
22 The Enlightenment
72 Origins of romanticism
26 Irish culture to 1850
68 Pitt, Fox and the call for reform
16 Europe: economy and society 1700–1800
60 African Art

In other volumes
316 History and Culture 1
312 History and Culture 1
324 History and Culture 1

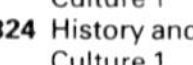

1 The nocturnal wanderings of Nicolas Restif (1734–1806) among the low life of Paris produced a vast output of moralizing, multi-volumed novels and tracts written between 1767 and 1802. Part fact, part fantasy, often tinged with erotic fancy (as illustrated), his work is a unique but diffuse picture of urban life. The novels focus on the vicissitudes of the human heart and the inevitable corruption of country innocence by the fascinating vices of the city.

3 The literary partnership of Richard Steele (1672–1729) and Joseph Addison (1672–1719) set the civilized tone of the Augustan age in their publications, the *Tatler* (1709–11), the *Spectator* (1711–12) and the *Guardian* (1713). Both were among the 48 wits, politicians and men of letters who were members of the Kit-Cat Club. The *Spectator* embodied the clubman's concern with manners, morals and character.

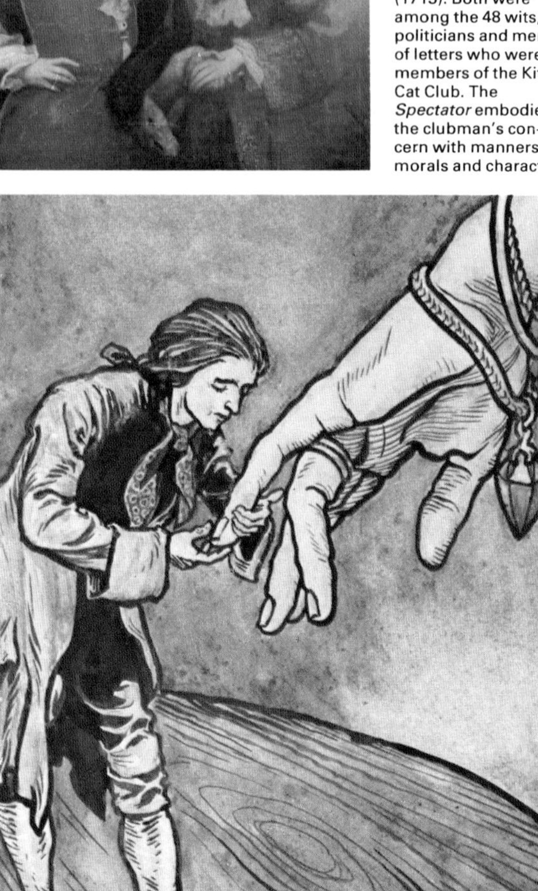

2 *Gulliver's Travels* has outlasted its original political and social targets. This work by Jonathan Swift (1667–1745) has become a landmark in English fiction. Gulliver, the one flesh-and-blood character, moves in a fantasy world that owes nothing to conventional fiction. He is tossed between a world of giants and a world of midgets, meets with nobility and vileness, and is at last returned to his own tragic race, incapable of perfection in Swift's eyes.

balanced insights were a corrective to rampant sensationalism.

The pre-eminence of the English novel did not obscure the brilliance of the French authors who chose this form. The Abbé Prévost (1697–1763) was successful with *Manon Lescaut*, the seventh volume of his dull *Memoirs of a Man of Quality* (1728–31), and he translated all Richardson's work into French. Choderlos de Laclos (1741–1803) published *Les Liaisons Dangereuses* in 1782, a novel that analysed corruption just as *Manon* depicted the degredation caused by passion. Fiction for Voltaire (1694–1778) [4] was only a device to convey ideas: with the pessimistic *Candide* (1759), the optimistic *Zadiq* (1747), through *L'Ingénu* (1767) and *La Princesse de Babylone* (1768) he conveyed his freethinking and rationalism.

The development of the press

Daniel Defoe is at the well-head of that other great manifestation of prose in the eighteenth century, the press. His weekly *Review* (1704–13) created a new class of reader, which was also brilliantly served for a while by the *Tatler* and *Spectator* as well as by the Tory *Examiner.*

The Stamp Act of 1712 ended this brief golden age of journalism, but it led to a new form of weekly paper that by its size escaped duty and also assured the alliance of literature with the press. Such commendable ventures as the *Gentleman's Magazine* (1731), which ran for nearly two centuries, raised journalism above the Grub Street level and away from rival party politics.

The first English daily paper, the *Daily Courant*, appeared in 1702 and flourished with six pages until 1735. The *Daily Universal Register* (1785) became *The Times* and began its unbroken run in 1788, by which time there were 40 London newspapers and countless provincial papers largely supported by advertising revenue.

Inevitably, as newspapers began to see themselves as champions of public opinion, a struggle for freedom of speech was engaged. There were numerous bitter clashes before the controversy over John Wilkes (1727–97) and his weekly anti-government journal, the *North Briton*, was resolved in his favour.

KEY

Eighteenth-century booksellers were entrepreneurs who bought copyright, commissioned authors, dealt with printers and sold books over the counter. Samuel Johnson considered them to be "liberal and enterprising", but the popular view was of monopolists, contemptuous of genius, panders to the vilest taste and employers of worthless hacks.

4

4 Voltaire was a prolific writer (he is here dictating as he dresses) who produced in *Candide* a kaleidoscope of the adventures and misfortunes endured by Candide and Dr Pangloss. Polite irony and cheerful common sense pervade this fine tale, yet beneath the surface the characters are merely instruments of Voltaire's dialectic. *Candide*'s motto, "All is for the best in the best of possible worlds" is ironically subverted by the action of the book; an intractable universe defies human attempts to impose reason on it; the author's ironical response to disillusionment is that "we should cultivate our garden".

5 A

DICTIONARY OF THE ENGLISH LANGUAGE.

B

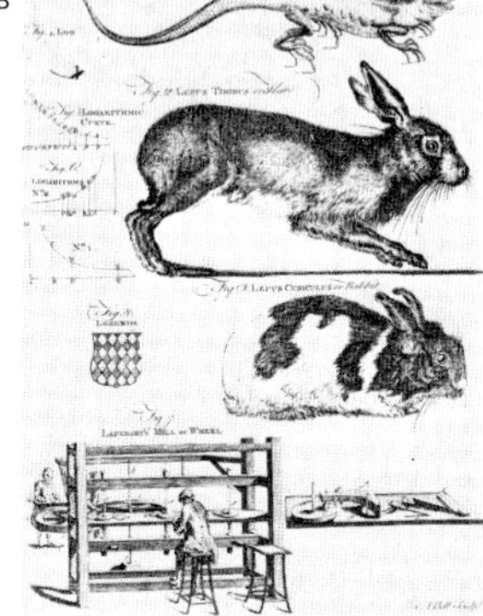

5 The *Dictionary* (1755) of Samuel Johnson (1709–84), commissioned by seven booksellers for £1,575, was a great single-handed feat of scholarship [A]. With his idiosyncratic etymologies, definitions and quotations, Johnson surpassed the struggling French Academy which was engaged on an identical task. The *Encyclopaedia Britannica* was a Scottish dictionary [B] issued in parts by William Smellie from 1768–71.

6

7

6 A mass of cheap occasional papers and social journals was published in France in the later 18th century. This cartoon of a penniless columnist getting his news on the cheap shows how unprofessional a trade journalism was.

7 *Pamela*, the first work of a London printer, Samuel Richardson, titillated the reading public of 1740. The author was a prim, unadventurous man who had nevertheless revealed the ways of the female heart. *Pamela* is a moral tale in which the heroine is a maid – here seen asking her master's blessing – exposed to the designs of the son of the house. Vivid in dialogue and told in a series of confidential letters, its instant fame encouraged Richardson to produce *Clarissa*, another novel in epistolary form. Henry Fielding's parallel, and at times competitive, career yielded more robust novels, including *Jonathan Wild* (1743), *Tom Jones* (1749) and *Amelia* (1751).

The Rococo

The Rococo was perhaps the first purely pleasure-giving style in the history of art. It had no didactic purpose or message, it was not a form of thanks-offering to God, and was not particularly adapted to telling stories or supplying an all-round picture of the world. Its primary aim was to delight.

The spirit of the Rococo

This is not to say that the Rococo could never be a vehicle for serious feeling, as the paintings of Antoine Watteau (1684–1721) demonstrate (although they are admittedly fairly exceptional), or that it was altogether cut off from reality. On the contary, Rococco artists were fascinated by nature, especially plants, small animals and birds, and at least one painter, Jean-Baptiste-Siméon Chardin (1699–1779), was a supreme natural observer. But these considerations apart, the Rococo was essentially playful, decorative and witty, a style in which taste and ingenuity were pursued as ends in themselves. Unsurprisingly, it was not a style for a wide public or for official monuments, rather its sphere was the private apartments and pleasure gardens of a small, sophisticated élite.

In keeping with all this, the Rococo was developed at least as much in the so-called applied arts of furniture, porcelain, silverware and interior decoration as it was in the fine arts of painting, sculpture and architecture. Indeed, the term can hardly be used of architecture at all, except in relation to its surface decoration, as Rococo had no role in the field of structures. The applied arts, on the other hand, were ideally suited to the style and it is noteworthy that they re-entered the mainstream of stylistic developments for the first time since the Middle Ages (when they were not, as they were with porcelain, entirely new).

Historically, Rococo grew out of the Baroque style that preceded it and of which it was in one sense a continuation, although at the same time it represented a reaction against Baroque gravity and pomp. Its heyday coincided almost exactly with the first half of the eighteenth century. It occurred chiefly in France and the German states – in the latter partly under French influence – but it also spread in some degree to Britain, Italy and other countries on the outer ring of Western Europe. The word "rococo" probably derives from *rocaille* ("rockwork"), meaning the shells and bits of rock used in sixteenth-century decorative schemes; this is appropriate because many characteristics of the style, such as irregular S-curves, wavy spikes and asymmetry, are to be found summed up in a certain type of shell [Key].

Rococo interiors

The first Rococo interiors appeared about 1700 in the very citadel of French Baroque –Versailles [1] and its associated royal châteaux. In their various private apartments the style of decoration used, instead of heavy architectural features, a series of tall wooden panels, painted ivory-white and covered with low-relief carving in gilt. Mirrors, especially over the chimney-pieces, formed an important part of the decoration. The effect was rich and even restless, for there was no break in the scheme, but it was in a sense intimate – the setting for a "private" rather than a "public" life.

The style soon spread from Versailles to

CONNECTIONS

See also
16 Europe: economy and society 1700–1800
40 Baroque and classical music

In other volumes
304 History and Culture 1
306 History and Culture 1

1 The Cabinet de la Pendule (1738) forms part of the private apartments created by Louis XV at Versailles. They were deliberately planned to be lighter and more intimate than the state apartments made for Louis XIV and are typical of French Rococo.

2 Watteau, the creator of French Rococo painting, designed a new genre for it, the *fête galante*, in which a kind of game of ideal romantic love is played out by real people clothed in fancy dress. In his most famous painting, "Departure for the Island of Cythera" (1717), part of which is reproduced here, the lovers are shown preparing to leave the island that the Greeks knew as the birthplace of Aphrodite, after a day's pleasure. The style is colourful and the picture charged with the atmosphere of passion and sweet regret.

3 Among the most dazzling of all secular Rococo interiors is the Hall of Mirrors in the Amalienburg, the park pavilion built near Munich by François de Cuviliés in 1734–9 for the Bavarian Elector, Karl Albrecht. Although the style shows French influence, the design is developed with a great freedom and plasticity typical of Germany. Decorative motifs inspired by an overgrown garden climb up the walls between the mirrors and spill out on to the ceiling.

the smart houses of Paris, where it became even more extravagant, and from there it was carried to the francophile German courts. As for the designers, they often came from a north European provincial background, where there were long traditions of craftsmanship, trained in Paris, and either stayed there or moved on to one of the French provincial or German centres of patronage. One who followed this pattern is the Flemish-born François de Cuvilliés (1695–1768) who, after a Parisian training, became court architect to the Elector of Bavaria. His Hall of Mirrors in the Amalienburg, an ornamental villa in the grounds of Schloss Nymphenburg, near Munich, is more fantastic in design than anything in France and is perhaps the finest example of Rococo decoration anywhere [3].

Besides this French-inspired Rococo there was in Germany a more native form of the style, particularly in churches. Unlike French Rococo this emerged by a process of natural evolution from the Baroque, which still flourished in Germany in the eighteenth century. Often indeed a Baroque church, such as Ottobeuren [4], will contain Rococo decoration inside, in a style produced mainly by changing the powerful, dynamic rhythms of the Baroque into the quicker, more playful and more sinuous ones of the Rococo without altering the forms themselves.

Watteau and Boucher

Rococo painting, on the other hand, was created by a single artist of genius, Antoine Watteau. Basing his style on the Flemish Baroque painter Peter Paul Rubens (1577–1640), Watteau invented a new pictorial genre, the *fête galante*, in which the loves of the classical gods are replaced by the fantasy loves of human beings, as in a ballet or opera [2]. As the nineteenth-century critics, the brothers Goncourt, wrote: "Watteau renewed the quality of grace". His truest successor was Chardin, whose subject-matter – still life and scenes of gentle bourgeois domesticity [6] – was real, not ideal; but Watteau also provided the starting-point for the decorative painter, François Boucher (1703–70), the most erotic and the most successful artist of the age [5].

KEY

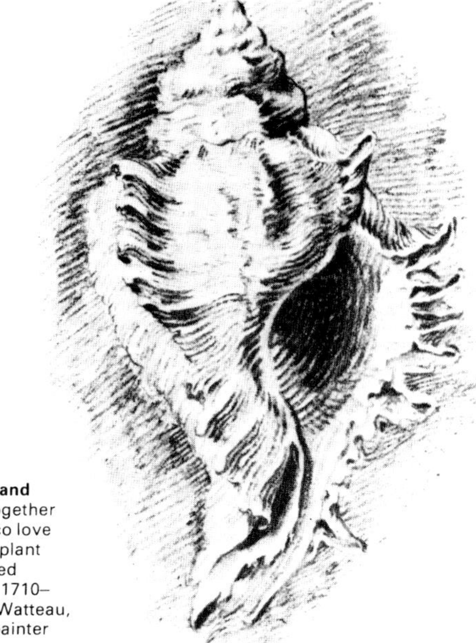

Rococo forms and asymmetry, together with the Rococo love of marine and plant life, are reflected in this shell (*c.* 1710–15), drawn by Watteau, the foremost painter of the Rococo.

4 The Abbey Church of Ottobeuren, Bavaria (1737–67), like many German 18th-century works, is by several hands and combines more than one style. The architecture, late Baroque with regularly grouped columns and a heavy cornice, was completed by Johann Michael Fischer (1692–1766). The Rococo ornament – playful, inventive and brilliant in white, gold and other colours – was executed by the stuccoist, Johann Michael Feuchtmayr (1709–72).

5 Eroticism was a favoured quality of art at the sophisticated, pleasure-loving court of the French king Louis XV. Its most fluent exponent was François Boucher, whose "Cupid a Captive" (detail) was part of a series painted in 1754 for the boudoir of Louis' most celebrated mistress, Madame de Pompadour. The result is a decorative asymmetrical mixture of forms similar in feeling to a figure group in porcelain or a Rococo silver table ornament of the period.

6 Using the themes of the love between a mother and her children and the innocence of childhood, Jean-Baptiste-Siméon Chardin produced, in "Saying Grace" (1740), a remarkably unsentimental picture. He relies on a simple pyramidal composition, balanced yet still with a hint of asymmetry, and restricts his facility with still life to subordinate details, such as the toy drum, the pot in the foreground and the objects on the table. But while he rejects the Rococo spirit of frivolity his style retains a Rococo grace.

Early ballet

Ballet was in its infancy at the beginning of the seventeenth century, having been introduced to the French court from Italy by Catherine de' Medici (1519–89). A ballet then consisted of a number of *entrées* by masked dancers in elaborate costumes, and at the conclusion there was a *grand ballet* in which the king and queen generally took part. The *entrées* for the ballets of Louis XIII (1601–43) were related to a theme, which was explained by speech or song.

Court ballet in Italy and France

Two of the most famed of the court ballets were *La Liberazione di Tirrenio* [1], given in Italy in 1616, and *La Déliverance de Renaud*, performed at the French court a year later. Many of the ballets had political significance; *La Déliverance de Renaud* was typical in that it was intended to reassure the Spanish and Austrian ambassadors that Louis XIII was still ardent towards his queen. For the ballet's presentation seats were erected in tiers along the length of the Grande Salle of the Louvre, with the royal throne at one end facing a stage concealed from the audience by a painted drop curtain. On either side of the hall musicians were concealed in clumps of trees. After a short opening chorus the curtain rose to reveal a grotto at the back of the stage with Monsieur de Luynes, the dancer portraying Renaud, lying on the grass. The king and 12 of his gentlemen represented devils left by another character, Armide, to watch over Renaud. Louis, de Luynes and two other gentlemen came down to the floor of the hall and performed the first *entrée*.

Following further *entrées*, there was a magical transformation on stage in which a mountain revolved to show a garden with three fountains; a nymph, portrayed by a young boy, rose from a fountain and sang; six monsters appeared and attacked two knights before fleeing; a giant cart, on which there were singers and a small wood, entered the hall and a choir of 92 voices sang the triumph of Renaud. Finally the king and his gentlemen came down into the great hall to perform the grand ballet.

During the reign of Louis XIV (1638–1715) [Key] the ballets attained an unparalleled splendour with dances arranged by Pierre Beauchamps, music by Jean Baptiste Lully (1632–87), words by Molière (1622–73) and designs by Jean Berain. These ballets sometimes took place out of doors. *Le Carrousel de Louis XIV* was a celebrated horse ballet given in front of the Louvre in 1662. Two years later *Le Palais d'Alcine* [3] was given in the grounds of the Palace of Versailles, using one of the lakes as a convenient part of the setting.

The Royal Academy of Music

The Royal Academy of Music (now the Paris Opéra) was founded in 1669, and three years later Louis XIV authorized Lully to add to this academy a school to educate pupils in dancing as well as singing, playing the violin and other instruments. However, in creating the Academies, Louis unwittingly helped bring about the decline of ballet at court. With the king's retirement from dancing the amateur gradually gave way to the professional dancer, and in 1681 the first professional women dancers took the stage.

Ballets at the turn of the century still combined dancing, music and singing – in other

CONNECTIONS

See also
274 Classical and modern Ballet
40 Baroque and classical music

In other volumes
318 History and Culture 1
314 History and Culture 1

1 *La Liberazione di Tirrenio* was given at the theatre of the Uffizi Palace in Florence in 1616 to celebrate the wedding of the Duke of Mantua to Catherine de' Medici. The spectators were seated on raised platforms on either side, with the royal host in the foreground. Divinities were lowered on a cloud machine and the dancers descended from the stage to join in the *grand ballet* on the floor of the hall.

2 This horse ballet was performed in the open air in Florence to celebrate a visit by the Duke of Urbino; 42 horse riders and 300 foot soldiers were used in this representation of a battle.

3 The island of Alcina appeared in *Le Palais d'Alcine* performed at Versailles in May 1664 on the third day of a series of entertainments. Alcina is shown riding on a sea monster with nymphs perched on dolphins. On both sides are long rocky "islands" where the orchestra sits. On another "island" in the background is Alcina's enchanted palace which was "destroyed" in a huge fireworks display.

4 *Le Triomphe de l'Amour*, a ballet by Jean Baptiste Lully, was presented at St Germain in 1681 with ladies of the court. When the work was given a little later in public in Paris at the Academy of Music, the part of Mars, danced by Pierre Beauchamps, the king's dancing master, was given to a professional dancer. Mademoiselle Lafontaine, the first *première danseuse*, led three other *danseuses*. The costumes were designed by Jean Berain, who had the men appear in Roman style while the women wore theatricalized adaptations of court dress.

words, operas with the dancing subservient to the singing. It was not until 1717, in John Weaver's *The Loves of Mars and Venus* at the Theatre Royal, Drury Lane, London, that the dancers alone conveyed meaning through movement. In this ballet Weaver made a clear distinction for the first time between the steps given to the man in the *pas de deux* and those given to the woman.

The dancers: teachers and pupils

The eighteenth century saw the rise of many famous *danseuses*, such as Marie Camargo [5], Marie Sallé and Madeleine Guimard, and *danseurs* such as the Vestris, father and son [7], and Pierre Gardel. Both Gaetano and Auguste Vestris appeared in ballets by Jean-Georges Noverre (1727–1810), one of the most influential men in the history of ballet. Noverre put into practice the theories set out in his *Lettres sur la Danse et les Ballets* published in 1760. Reforms he suggested included abolishing the mask (it was actually Maximilien Gardel who implemented this) and replacing the series of conventional dances which then consituted a ballet with the *ballet d'action*, in which dance and story were united.

Noverre's pupil, Jean Dauberval, moved ballet a further step forward in 1789 with the first comedy ballet *La Fille mal gardée*, in which ordinary people replaced the usual collection of conventional shepherds and shepherdesses.

Charles Didelot [8], who trained in turn with Dauberval, Noverre and Auguste Vestris, synthesized the contributions of his three teachers. From Dauberval he took ideas for comedy ballet, from Noverre the *ballet d'action* and from Vestris the pure dancing. Didelot's ballet *Flore et Zéphyr*, produced at the King's Theatre, Haymarket, in 1796, not only put these ideas into practice but also introduced dancing on point (on block dancing shoes). This was probably brought about by an improved flying machine that enabled dancers to fly across or round the stage, because the dancer rose on to the tips of her toes before beginning her flight. The development of dancing on point paved the way for the romantic ballets of the early nineteenth century.

KEY

Louis XIV appeared as the Sun King in *Le Ballet de la Nuit* given in 1653. The ballet's 43 *entrées* symbolically spanned 12 hours, the king appearing in the fourth and last section covering 3am to sunrise. Louis, aged 14, was dressed in the male costume typical of the period – short, scalloped skirt and plumed headdress embellished with the sun and its rays. In ballets during his reign Louis represented exalted figures: Apollo, Neptune and Jupiter. His last appearance was in 1669. Louis' ballet master, Beauchamps, invented the five accepted modern positions of the feet in order to make the dancers appear more elegant.

5

5 Marie-Anne de Cupis de Camargo (1710–70), the first famous ballerina, was painted by Nicolas Lancret in this idealized, pastoral setting. In 1730 she created a sensation by shortening her skirt several centimetres to enable her to perform *entrechats* (the crossing and uncrossing of the legs in a vertical jump). Camargo was renowned for her technical brilliance and vivacity while Marie Sallé was more expressive and graceful. Sallé appeared in her own creation of *Pygmalion* wearing a plain muslin dress draped like a Greek robe – Isadora Duncan 200 years later was the next dancer to adopt this revolutionary costume.

6

6 The basic costume for the male dancer in the mid-18th century was plumed helmet and tunic, with short hooped skirt (*tonnelet*). It was adapted by varying the decoration and type of mantle to show the character. The costume shown here for a faun, possibly designed by P. Lior, is draped and spotted to suggest an animal's skin, with vine leaves in the hair, at the elbows and across the tunic, and a pan pipe nestling in the leaves on the *tonnelet*. The dancer is wearing buskins on his legs. In 1907 Alexandre Benois was to revive this style in his costumes for *Le Pavillon d'Armide*, which brought fame to Vaslav Nijinsky.

7

7 Vestris father and son were both remarkable dancers. Gaetano (1729–1808) was known as the God of Dance and his son Auguste (1760–1842) was one of the greatest dancers in the entire history of ballet. He was renowned for the extraordinary height of his jumps and was the *premier danseur* at the Opéra for 36 years, coming out of retirement in 1835 to partner the young Marie Taglioni in a minuet. He is shown here dancing at Vauxhall. The significance of the geese becomes clear in a quotation from Plutarch: "A stranger at Sparta standing upon one leg said to a Lacedaemonian, 'I do not believe you can do as much.' 'True,' said he, 'but every goose can.'"

8

8 Charles Didelot and Mademoiselle Théodore appeared in *Amphion and Thalia* at the Pantheon, London, in 1791. Thomas Rowlandson's painting shows not only the two dancers, but also the musicians and crowded auditorium of a typical theatre of the time. The French Revolution two years before had been responsible for alterations in costumes which were taken up on the ballet stage. Dresses, now modelled on the tunics of ancient Greece and Rome, were lighter and slightly transparent. The shorter dress of transparent material led to the introduction of tights or *maillots*. Pink tights were worn by Didelot, who also introduced the transparent tunic in *Corisandre* in 1791.

Baroque and classical music

The Baroque era in European music dawned in Italy at the beginning of the seventeenth century with a new expressive style of singing modelled on natural speech and established in the first tentative operas. The genius of Claudio Monteverdi, who was master of music at St Mark's, Venice, for 30 years from 1613, was able to embody both old and new styles in his *Vespers* (1610). The richness of his sacred and secular music (especially his operas) stems largely from his imaginative synthesis of both traditions. In contrast to the older, other-worldly polyphony, an urge to express the feelings came to dominate the music of the Baroque age, a trend that led to excessive ornamentation comparable with some architecture of the period.

The rise of instrumental composition

The Catholic Church was not slow to sanction the attractive new style in its Counter-Reformation drive, and the Oratory of Philip Neri in Rome was the scene of sacred dramatic works by Giacomo Carissimi (giving music the "oratorio"). Alessandro Scarlatti in Naples contributed brilliantly to a Neapolitan opera style and established the *sinfonia avanti l'opera*, the Italian opera overture, an ancestor of the movements of the classical symphony in its quick-slow-quick movement format. The new operatic convention of emphasis on a solo voice over a bass line was also to have large implications for the development of instrumental music, where small groups of solo instruments separated out, as in the divided orchestras of Alessandro Stradella, in the beginnings of the concerto. Guiseppe Torelli, Arcangelo Corelli, Tommaso Albinoni, and Antonio Vivaldi [1] were principals in the development of the instrumental *concerto grosso* (a small instrumental group in contrast and in combination with a larger group) and the solo concerto. The development of violin-making in the great Cremona families of Amati, Stradivari and Guarneri greatly assisted composers to obtain orchestral expressiveness.

In small chamber groupings [2], the new style demanded an harmonic "filling". This led to the Baroque convention of *basso continuo* [7], using a harpsichordist to direct a performance from the keyboard while "filling in" the harmonies as indicated in a shorthand notation in the bass part. Music for these small groups formed the core of the popular trio sonata. *Sonata* meant simply a piece played instrumentally, as opposed to *cantata*, to be sung, and only later acquired its specialized use. A distinction was made between the *sonata da camera* (chamber), movements for dancing, and the *Sonata da chiesa* (church), usually four movements starting with an adagio (slow) movement.

The style spreads through Europe

Eventually Italy's dominant influence in operatic and instrumental styles spread throughout Europe. In Germany, Heinrich Schütz – whose Passion compositions influenced J. S. Bach – Michael Praetorius and Johann Jakob Froberger (1616–67) were leading disciples of the style. In France, Louis XIV's court dictated taste for nearly a century, and was served by an Italian-born ballet and opera composer, Jean-Baptiste Lully. Jean Philippe Rameau maintained the independent French tradition in orchestral music and opera.

CONNECTIONS

See also
106 Music: the Romantic period
120 Opera in the 19th and 20th centuries
72 Origins of romanticism
104 Development of the orchestra

In other volumes
248 History and Culture 1
324 History and Culture 1

1 Antonio Vivaldi (*c.* 1675–1741) was a prolific Venetian composer and master of the concerto – a form that emerged in Italy during the Baroque. He was a virtuoso violinist and his more than 400 works include many for solo violin and for groups of violins with orchestra. He worked as director of a combined music school and orphanage, the Ospedale della Pietà in Venice (1703–40), composing many works for his charges.

2 The collegium musicum, or amateur musical society, was a major support of music in Germany in the 17th century. These institutions reflected the rise in middle-class prosperity and were frequently attached to the universities. They spread to nearby countries and later to America. Many of the societies gave public concerts. The famous Leipzig Collegium was founded by George Philipp Telemann in 1701.

3 King Frederick the Great of Prussia, who reigned from 1740–86, was a typical royal patron of his time. An iron ruler of the old Prussian military class he was also an amateur flute player. His palace in Berlin and summer residence in Potsdam provided employment for a band of notable musicians which included C. P. E. Bach and Johann Joachim Quantz, who was a gifted flautist. Church and aristocratic patronage, which would see their final flowering in the late 1700s, had been the support of music up to that time. There had been many famous associations of composer and patron: Lully, Couperin and Louis XIV; James II, William and Mary and Purcell; Handel and George I; Haydn and Prince Esterhazy.

4 Franz Joseph Haydn (1732–1809) was the major composer of the early classical period and was responsible for giving such structural cohesion and organization to the sonata principle in the symphony and string quartet that accepted forms of these compositions can be said to date from his time. The Hungarian Esterhazy family were his employers, but he also visited London twice, where he was greatly admired.

5 The Mozart family was portrayed by della Croce. The composer Wolfgang Amadeus is at the harpsichord with his sister Maria Anna. His father, Leopold, is prominent with his violin, while the composer's dead mother, Anna Maria, gazes down from a portrait on the wall. Leopold toured the children as musical prodigies and the precocious Wolfgang astonished musical Europe with his compositions and performances.

In England, after the Restoration in 1660, Henry Purcell created vital English music that would be matched only by Handel. In the momentous year 1685 Handel, J. S. Bach [Key] and Domenico Scarlatti were born. Of the three, Scarlatti's influence was slightest, but in his hundreds of harpsichord works one can hear hints of the classical keyboard sonata.

Handel and J. S. Bach were the complementary giants of the High Baroque era. They followed distinct yet not opposed paths. Handel, extrovert and worldly, was at home in the intrigues of the opera house and happy in the massed performances of his English oratorios such as *Messiah* and *Judas Maccabeus*. J. S. Bach was confined to provincial Germany, devotional in his music, an industrious court and municipal composer who always had to fight to maintain the meagre musical forces at his disposal. He summarized rather than innovated and contemporaries associated the name Bach more readily with his son Carl Philipp Emanuel, whose dramatic style was more akin to the natural simplicity and passion advocated by the French philosopher Jean Jacques Rousseau than to the grand, unified and somewhat rigid expressions which large Baroque forms like the fugue had become.

The classical period emerges

The classical period which rose on the foundations of the Enlightenment in response to the overdeveloped, ornamental style of High Baroque and Rococo music, found two masters in Joseph Haydn [4] and Wolfgang Amadeus Mozart [5]. Their classical style distinguished itself from the many-voiced Baroque texture by its greater clarity and simplicity of thematic treatment and chordal accompaniment. The prolific Haydn drew on his peasant origins to maintain a vitality in his music, and his innovations in form and the use of key relationships were to be highly influential. Mozart, drawing on Haydn, the continuing Italian influence and his own genius, proved a brilliant exponent of all the forms bequeathed by Baroque and early classical music. Only Beethoven [8] would be comparable, spanning the classical and Romantic styles with the stamp of genius.

KEY

Johann Sebastian Bach (1685–1750) brought the great contrapuntal forms of Baroque music to their peak in a glorious summary of the musical development of several centuries. A member of a long-established dynasty of musicians, Bach lived most of his life in the eastern German province of Thuringia, working mainly at the courts of Weimar (1708–17) where much of his organ music was composed, and at Cöthen (1717–23), a fruitful time for composition which included his *Brandenburg Concertos*. His last post was at St Thomas's Church in Leipzig (1723–50) for which he wrote his great church works, among them the *St Matthew Passion* and most of his cantatas.

6

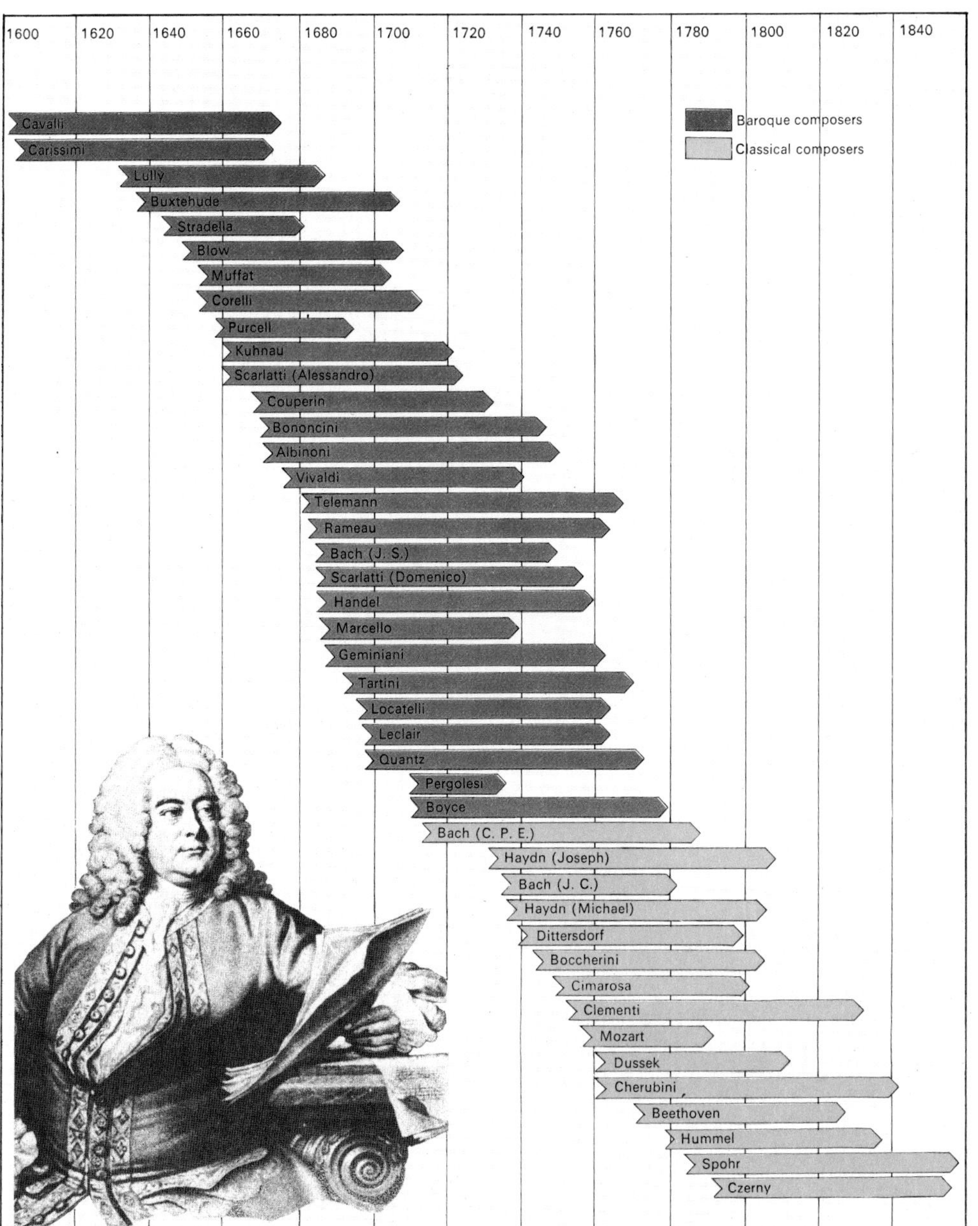

6 The beginning of the Baroque period is generally dated around 1600 by virtue of the rise of Italian opera at that time. The period effectively came to an end in 1750 with the death of J. S. Bach, who recognized the influence of the classical style of his young contemporaries. Handel, as a major Baroque composer, is shown on the left. The classical era centres on the works of Haydn and Mozart.

7

7 A positive organ built by the celebrated organ builder Gottfried Silbermann at Leipzig (1724) is characteristic of the High Baroque period with its taste for flamboyant ornament. In the 17th century organs were substantially improved and had an important role in the *basso continuo* or supporting bass in Baroque music.

8 Ludwig van Beethoven (1770–1827) is the key transitional figure between classical and Romantic styles in European music, between the decorum of the 18th century and the free-ranging expression of the 19th century. In his early work one sees influences from Haydn and Mozart but a dynamic originality is apparent.

8

Neoclassicism

Neoclassicism, a movement of the late eighteenth century, was a reaction against the excesses of the Baroque and the Rococo, but more important, the fruit of a new age of inquiry into ancient Greek and Roman art.

Knowledge of antiquity

Some classical buildings, especially in Rome, had always been familiar as they were above ground. With the passionate interest in the Antique that inspired the Renaissance, pieces of classical sculpture, including several important ones, were dug up in Italy in the fifteenth and sixteenth centuries. But the very vitality and freedom that characterizes Renaissance art in a sense depended on a degree of ignorance of its sources; the few examples that were known fired the imagination in a way that the thousands of pieces that became available later could not have done. More than this, no distinction was yet drawn between Greek and Roman art or between different periods of either. Although the importance of the Greek contribution was known in theory from texts – this was particularly so in painting, of which almost no examples survived – classical Greek sculpture of the fifth and fourth centuries BC was known only through Roman copies.

This situation altered only gradually in the seventeenth century, with the research now done on more scholarly lines. The decisive change came in the eighteenth century, when not only were far more classical statues excavated and a museum of them opened on the Capitol in Rome, but whole unknown ancient arts were rediscovered, such as Greek painting at Herculaneum and Pompeii, near Naples, and Greek vases, then paradoxically thought to be Etruscan, in southern Italy. In short a new science – archaeology – was born. Above all, Greece itself was explored, notably by the British architects James Stuart (1713–88) and Nicholas Revett (1720–1804), and by the end of the century a real perception existed of the differences between Greek and Roman architecture. A further revelation was brought about by the arrival of the Parthenon sculptures in London, imported from Athens by Lord Elgin (1766–1841) in the early nineteenth century, for these showed that the art of the greatest Greek sculptor, Phidias, did not conform to the so-called "classical ideal" as it had been previously understood but consisted instead of something at once more primitive and more naturalistic.

These developments were associated with a new sense of history. Scholars and amateurs (who played a key role in the Neoclassical movement) were now concerned to place things in chronological order and in the context of the culture that produced them. In particular, they were fascinated by what came *first* and was thus the most primitive.

The new conception of the Antique

Out of this in turn was evolved a new conception of the Antique on which Neoclassicism was based. The Antique was now austere, simple, static, with clear, hard outlines and, in sculpture and painting, almost without shadows or foreshortening.

One further piece in the jigsaw remained: the moral element. It was a preoccupation with this that brought the Neoclassicists in France out in full cry against the Rococo. The French reformers were full of admiration for

CONNECTIONS

See also
74 The French Revolution
76 Napoleonic Europe
22 The Enlightenment
72 Origins of romanticism
80 Romantic art: figure painting
102 Romantic art: landscape painting
70 Georgian art and architecture

1

1 This secretaire was made by J. H. Riesener (1734–1806) in 1783 for Marie Antoinette in the "Louis Seize" style. It has Neoclassical lines but a richness of ornament that was typical of the court.

2

2 Collecting antiques was a central part of the Neoclassical movement. Sometimes it reached absurd proportions as with Charles Towneley and his marbles (1782) painted by Johann Zoffany (1734–1810).

3

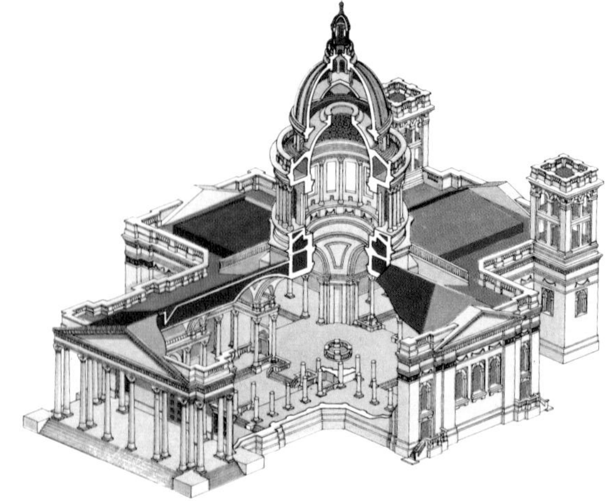

3 Jacques-Germain Soufflot's Church of Ste-Geneviève (now the Panthéon), Paris (built 1757–92), was the first large Neoclassical building. It was also influenced by the dome of St Paul's, London.

4

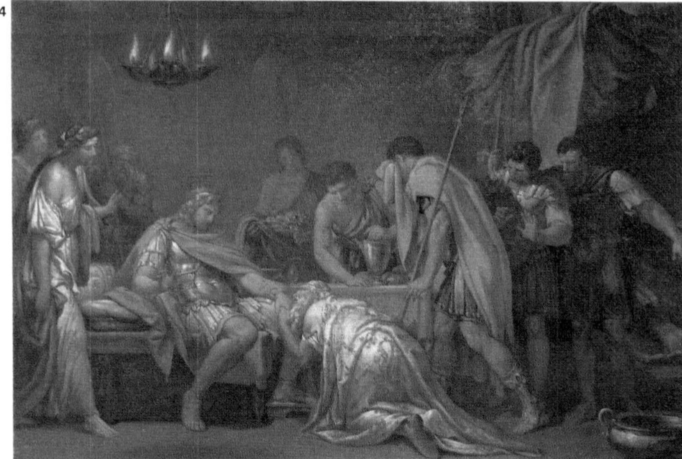

4 The revival of Greek poetry as a subject for art is reflected in "Priam Begging Achilles for the Body of Hector" (c. 1770) by Gavin Hamilton (1723–98).

the Roman ideals of civic virtue and the Greek experiment in democracy. The Rococo began to seem a decadent, frivolous style associated with a decadent, frivolous society. There was a strong move towards bourgeois rectitude, a wave of moral idealism which carried over into the Revolution. The Antique, as it was newly conceived, appeared the perfect visual counterpart of this.

Neoclassicism in France

Consequently from about 1760 onwards there was in all the arts in France, but first in architecture and the applied arts, perhaps last in painting and sculpture, a reaction in favour of straight, simple lines and austere ornament derived from classical and sometimes specifically Greek models.

The first important French building to show this trend was the Panthéon in Paris, although it was more Roman than Greek, designed by Jacques-Germain Soufflot (1713–80) as early as 1755–6 [3]; a fuller, more imaginative brand of Neoclassical architecture was developed some years later by Claude-Nicolas Ledoux (1736–1806). In painting it was less easy to depend on classical precedent because surviving examples were few. The solution was to use as an intermediary the greatest seventeenth-century master of classicism, Poussin: his influence was of vital importance for Jacques-Louis David (1748–1825), the first major Neoclassical painter and the interpreter of both the heroic spirit of the Revolution [6] and the imperious genius of Napoleon. His successor, Jean-Auguste-Dominique Ingres (1780–1867), evolved a purer, more refined form of Neoclassicism [7], harking back to Raphael and Pompeian painting.

Although the grandest Neoclassical paintings and buildings were French, the movement was an international one. By far the greatest Neoclassical sculptor was an Italian, Antonio Canova (1757–1822) [5], but German architects evolved the most scholarly manifestations of the style in the early nineteenth century. Neoclassicism was by no means a merely backward looking or monolithic phase of art history; on the contrary, it reflected the radical questioning of established values in a period of revolution.

KEY

As with the Rococo, some of the most attractive and ingenious expressions of Neoclassicism are to be found in the applied arts. Devotees of Neoclassical taste liked to own ornaments derived (often rather freely) from objects that once had a practical or symbolic function in antiquity. This "Pedestal and Vase in the Etruscan Taste" (*c.* 1795), by the British architect James Wyatt (1747–1813), has a base like a Roman altar, with Greek decoration and a Greek vase on top. While the curling candle-brackets and lighthearted mood recall the Rococo, the chaste lines and figured details are in the purest spirit of Neoclassicism.

5

5 Startling illusionism is combined with Neoclassical control in Antonio Canova's "Maria Christina Monument" (1799–1805): a group of allegorical mourning figures led by Piety (with the urn) appears to move through an open door into the grave. The deceased was an obscure Austrian duchess, but the conception of the tomb goes beyond a particular death to pose questions about the mystery of death in general.

6

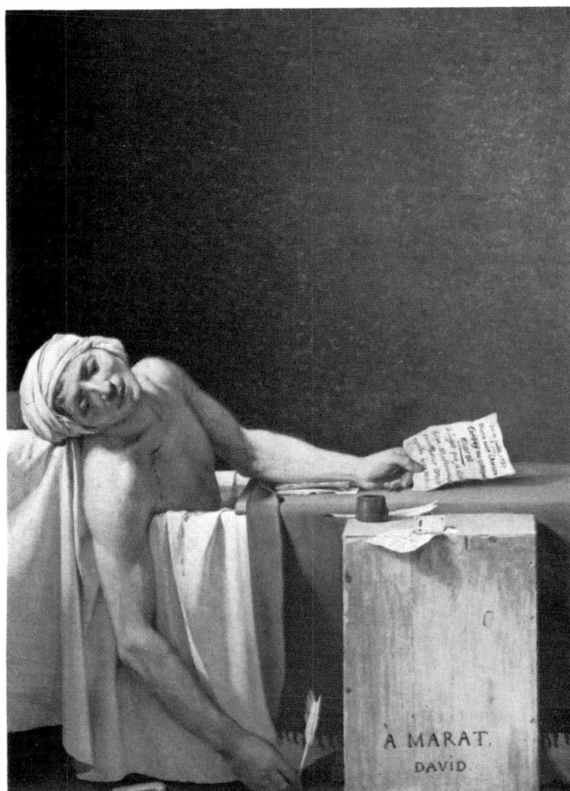

6 Jacques-Louis David's painting, "The Death of Marat" (1793), is a commemorative monument but with everything concentrated on the individual. David depicts Jean Paul Marat, the French rationalist writer and demagogue, stabbed at work in his bath by Charlotte Corday, as a revolutionary martyr. Neoclassical touches include the echo in the pictorial treatment of the figure of the deaths of the classical philosophers, Socrates and Seneca, and the clarity and economy of the painting's design.

7

7 A less brutal, less realistic painter than David, Jean-Auguste-Dominique Ingres interpreted Neoclassicism mainly in terms of line. He created an ideal, dignified world of pure contours and clean colours, yet not a world without feeling – all well exemplified in his "Baigneuse" (1808). He was the father of 19th-century academic art in Europe.

8

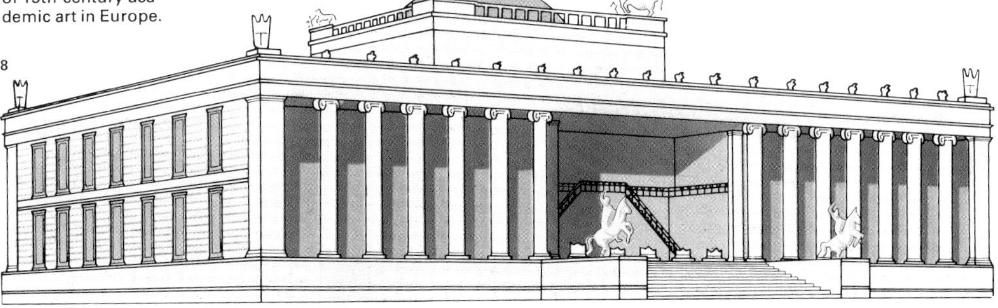

8 The founding of public museums in the major European cities was one of the products of Neoclassicism. The Old Museum, Berlin (1823–30), by Friedrich Schinkel (1781–1841), built to house the art collection of the Prussian state, is an important example of Neoclassical architecture.

International economy 1700–1800

During the eighteenth century trading links between Europe and the rest of the world strengthened into commercial bonds that were vital to the prosperity of several European states and to many of their merchants and workmen. The rise of Great Britain to a dominant position in trade, overtaking the Dutch, Spanish, and French, showed itself in extensive contacts and trading arrangements with the Americas and Asia [5]. There was a general expansion of trade at the same time, particularly in the Atlantic, in which other countries, especially France, shared. Expansion of trade brought with it specialization in shipping and financial business and provided a stimulus to the increased production of manufactured goods in Europe.

Incentives for trade expansion

The expansion of European influence in the years before 1700 had introduced a wide range of precious and tropical products which were the staples of extra-European trade. Gold and silver bullion had provided one major component of the trade of the declining Spanish Empire, but this trade had also included spices, tobacco, coffee, sugar and cocoa. A growing taste in Europe for tropical products gave merchants the incentive to incur the risks of long overseas voyages. Merchants from countries such as Britain and France, which were relative latecomers to colonial trade, made greater efforts in the 1700s to capture part of the trade controlled by the older empires.

Several commercial wars were fought by the other European nations to open up the colonial trade of the declining Spanish Empire to their merchants. In 1715 British merchants obtained permission in the Asiento Treaty to deliver 4,800 slaves annually to the Spanish colonies and for one ship a year to trade with Panama. In all other respects, Spain tried to keep her colonial trade closed. But because Spain was unable to supply all the needs of her colonists, there was much illicit traffic, especially in slaves and in manufactured goods. After the Seven Years War (1756–63) this rigid control was relaxed and Britain and France seized a greater part of the Spanish American trade.

Similarly, British and Dutch ships captured an increasing share of Portuguese trade. In the East, the Dutch had taken over much of the old Portuguese Empire. With stations in Ceylon, Bengal, Malabar and Batavia their influence extended as far as Japan. Organized through the Dutch East India Company based on Batavia in Java, they carried on a lucrative trade in spices, coffee, and silks. By the end of the century, however, they began to feel the effects of competition from the British and French.

French and British power

France had colonies in North America, including Quebec and Louisiana, and in the West Indies, including Guadeloupe and Martinique; she held Senegal in Africa and colonized Madagascar and Mauritius in the Indian Ocean. In India, France maintained important trading positions at Pondicherry and Chandernagore. From these colonies and contacts French trade grew rapidly in the late eighteenth century, especially with the West Indies. After the Seven Years War, however, France lost Quebec and much of her influence in India. Nonetheless her

CONNECTIONS

See also
16 Europe: economy and society 1700–1800
18 England under the Hanoverians
134 The story of the West Indies
136 The story of Canada
46 British colonial policy in the 18th century
32 Exploration and science 1750–1850
66 The early Industrial Revolution

In other volumes
238 History and Culture 1
208 The Physical Earth
216 The Physical Earth

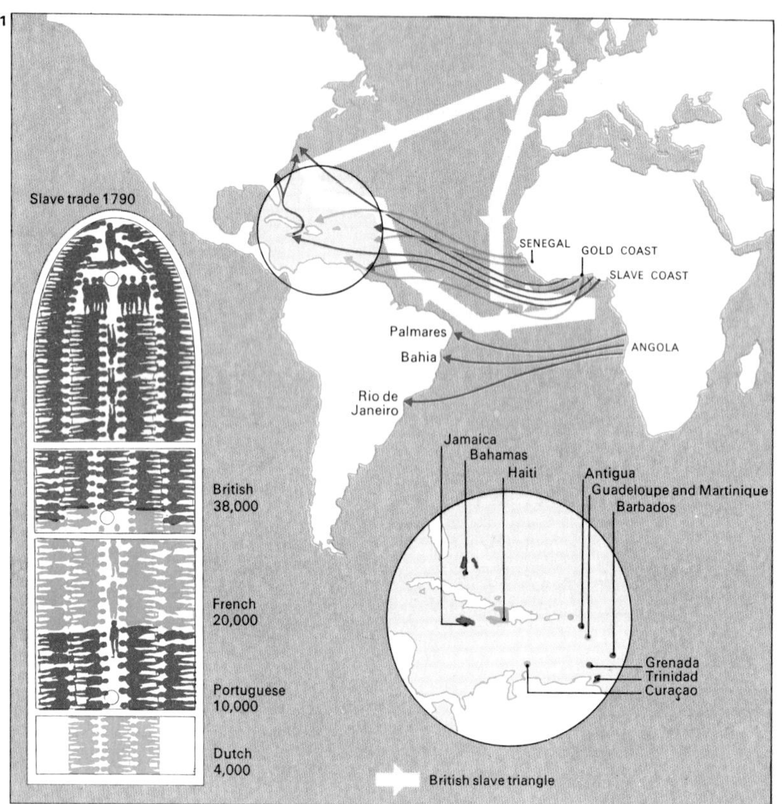

1 The triangular route taken by slave ships from European ports such as Liverpool, Bristol and Bordeaux took them to Africa to collect slaves, across the Atlantic to sell them and back again with cargoes bartered in exchange. The major share of the slave traffic was carried on by Great Britain by the end of the 18th century, supplying plantations in the West Indies and in the Americas.

2 Brest, on the Atlantic coast of France, and other Atlantic ports such as Bristol, Bordeaux and Liverpool grew enormously rich on the profits of the slave trade and colonial traffic.

Sugar production and slavery in St Domingue, Jamaica and Cuba
156,000 789,000
1797
105,000 377,000
1767
24,000 120,000
1720
5,000 tons sugar
50,000 slaves

3 Economic development in the West Indies depended on a steady supply of slave labour from Africa. The slave population and sugar production grew to satisfy the European demand for sugar.

4 Smuggling in the 18th century was a common way of avoiding heavy duties on commodities such as spirits, wines and tobacco. Revenue men waged an intermittent war with smugglers in many parts of Europe, especially in remoter areas.

busy trade with the remainder of her colonies led to the expansion of ports such as Bordeaux, Nantes, and Brest [2]; Marseilles prospered with the Levant trade.

The major rising power was Great Britain, which exerted its superiority over the older empires by the middle of the century and conducted all the trade of its colonies in British-registered ships. In spite of the loss of the North American colonies, trade continued to expand with the Americas both through chartered companies, such as the Hudson's Bay Company [8], and through more open commerce. The vastly profitable but inhuman slave traffic was primarily in British hands by the end of the century, the "Triangular Route" [1] between Britain, Africa, and the Americas providing great profits upon which mercantile cities such as London, Liverpool and Bristol flourished. Second only to the slave traffic in importance was the sugar trade [3], which provided, with other tropical produce, a valuable re-export trade to continental Europe. In the Indian Ocean the British East India Company triumphed over the French and controlled Bombay, Bengal, Madras and most of southern India; but by the end of the century the trade was opened up and the company's monopoly broken.

Stimulus for industrialization

Expansion of trade led to increases in merchant shipping and the development of more versatile credit and financial institutions. In London, firms of shipping insurers, such as Lloyd's, began to supplant the Dutch. Growing trade led to relaxation of the old mercantile ideas of monopolistic companies and favoured the adoption of *laissez-faire* ideas of free trade and open competition.

In many countries industries grew up that refined tropical products and re-exported them either to other European countries or back to the colonies as manufactured goods. Thus sugar-refining, tanning, distilling and many other processes became early examples of industrial activity. The most striking development by the late eighteenth century was the rapid rise in raw cotton exports from the southern states of America to the growing British textile industry.

KEY

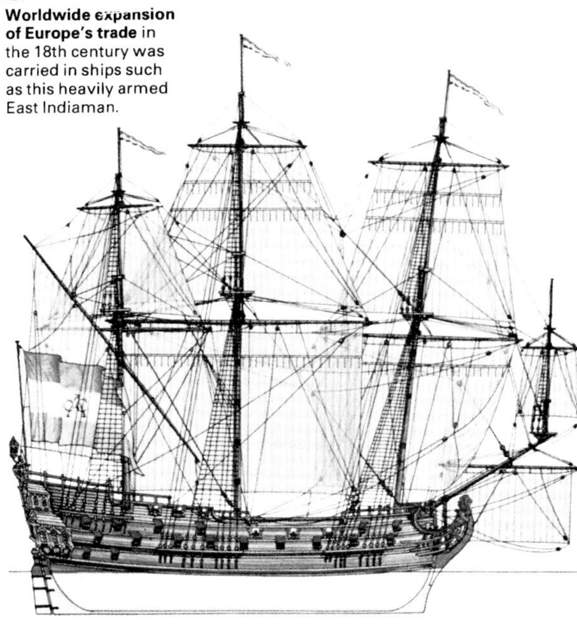

Worldwide expansion of Europe's trade in the 18th century was carried in ships such as this heavily armed East Indiaman.

5

5 Trading horizons widened in the 18th century as European contact with hitherto exotic countries became more regular. Tea plantations in China were developed to meet a taste for tea in Britain and an extensive trade also grew up in silk, porcelain and spices. Although China limited European influence to specified trading centres, these provided a foothold that was extended in the next century.

6

6 Speculation, greed and dishonesty were caricatured by the satirist William Hogarth (1697–1764) in 1720 when the South Sea Company, upon which heavy speculation had centred, collapsed and led to a change of government. Financial and credit institutions in the 18th century were often unstable, partly because governments were apt to draw on them for credit and also because investors panicked easily.

7

7 Spanish financial institutions, established in the 16th century, shared in the great expansion of European trade in the 18th century. This deposit certificate, dated 1759, is from the Real Compañia de Comercio of Barcelona.

8

8 The Hudson's Bay Company, whose Fort Garry, Manitoba, trading post is shown here, was founded by the British in 1670 to open up a lucrative trade in Canada, supplying Europe with furs, timber and salted fish. Simultaneously, the colonists at Albany and New York were extending trade with Indians. France was also active in the Canadian trade but lost its share in 1763 when it was defeated in the Seven Years War and thereafter surrendered control of Canada to Britain.

British colonial policy in the 18th century

British exploration overseas began in earnest in the sixteenth century and the foundations of the Thirteen Colonies in America were laid in the reign of Charles II (1660–85) [1]. Nevertheless the modern British Empire is usually dated from the Treaty of Utrecht (1713). The treaty gave Britain some minor gains in the Caribbean; France yielded its claims to Newfoundland and Nova Scotia; Spain ceded Gibraltar and Minorca to Britain and, most important, transferred to it the *asiento*, giving Britain the right to furnish slaves to the Spanish colonies and to send one ship a year to trade in those colonies.

Anglo-French rivalry

For a century the power of Spain had been on the wane. After 1713 France and Britain were the great rivals and during the next century a series of wars between the two nations, ending in Wellington's victory at Waterloo (1815), raised Britain to the rank of the world's dominant colonial power. In the eighteenth century it was understood that trade was power. The Anglo-French wars were fought to win control of the world's trade, especially the highly profitable trade in slaves and sugar.

The eighteenth-century empire was based on the "triangular trade". British ships carried manufactured goods to Africa, exchanged them for slaves [9], sold those in the West Indies and the southern American colonies, and returned home with cargoes of sugar, tobacco and raw cotton. By 1750 hardly a trading or manufacturing town in England was not connected with the trade. Between 1670 and 1786 more than two million slaves were transported from Africa to British colonies.

The imperial economy was run on the principle that the colonies provided raw materials and accepted in return manufactured goods from the home country. This system was strictly maintained by the Navigation Act, passed in 1660, which required that trade between Britain and the colonies be carried by British-built ships manned by British crews. Produce going from the colonies to foreign parts had first to pass through a British port. This trade with the protected market of the American colonies was important, too, to British industry, which had to seek new outlets after the colonies won independence. It also hindered the growth of colonial industry and thus encouraged the American demand for independence.

Defeat of the French

Before the American Revolution began, Britain, led by William Pitt (1708–78) later Earl of Chatham, won a great victory over France in the Seven Years War (1756–63). The war was fought to gain two aims: supremacy at sea and the capture of French trading posts, and it spread over Europe, India, the West Indies and Canada. Wolfe's victory at Quebec (1759) secured Canada's fish, fur and timber for Britain [2]; Eyre Coote (1726–83) and Robert Clive (1725–74) won victories in India which brought Bengal under the rule of the East India Company.

At the Peace of Paris (1763) the French sugar islands – Guadeloupe and Martinique – were returned to France, and Cuba to Spain, in order to satisfy the West Indian planters who feared lower prices from an increased sugar supply. But the war achieved one great

CONNECTIONS

See also
64 The American Revolution
44 International economy 1700–1800
48 India from the Moguls to 1800
32 Exploration and science 1750–1850
18 England under the Hanoverians
68 Pitt, Fox and the call for reform
124 Colonizing Oceania and Australasia
132 The British Empire in the 19th century
134 The story of the West Indies
136 The story of Canada

1 The presentation of the first pineapple grown in England to Charles II is an exotic example of the change in British eating habits that followed the colonization of the New World. Maize, which has a yield ten times higher than wheat; the potato, which thrives in poor soil and made possible Ireland's population explosion; the kidney bean and pumpkin were all introduced into Britain in the late 16th century, but it was not until the 18th century that their widespread cultivation brought about any significant improvement in the diet of the peasantry. In this period, the colonies were intended to benefit the mother country rather than become independent economic entities.

2 British interests in Canada in the 18th century were mostly in the hands of chartered companies. The largest of these was the Hudson's Bay Company, whose founders were given a royal charter by Charles II in 1670. Canada provided furs, fish and, most important, timber to build warships for the navy. But from a commercial point of view the country was insignificant compared, for example, with the Sugar Islands. From 1762 to 1776, the value of British imports from Grenada alone was eight times that from Canada, whereas exports from Britain to Grenada were double the value of those to Canada.

3 An English fleet attacked Portobelo, a Spanish port on the Colón peninsula of Panama, in 1739. Spain was declining as a colonial power by the mid-17th century and Britain was anxious to ensure that France, the chief rival for control of the Spanish dominions, did not gain control of the most important Spanish colonies. The import of bullion from the New World had mostly stopped by this time; sugar was the valuable crop for which the European nations were struggling for the Caribbean.

4 New Yorkers pulled down the statue of George III, symbolizing their revolt against Britain. But the loss of the American colonies meant merely a reorientation of the empire towards India; and British trade with the USA itself soon grew.

result: it deprived France of huge markets in Canada and India and, by making those places British, provided scope for the industrial expansion of Britain. Without India, the cotton industry of Lancashire would not have had the market to sustain its spectacular growth between 1780 and 1820.

Britain and India

In 1776, the very year that the American nation was born, British control over India was tightened. At the beginning of the century there were no more than 1,500 English in all of India – traders and their families, concentrated in the ports of Madras, Calcutta, Surat and Bombay. They were concerned only with making money and were not interested in penetrating inland nor in the destiny of Britain to rule over a native population. But conflict with the French, who had established themselves at the ports of Pondicherry and Chandernagore, was endemic throughout the first half of the century. In 1754 royal troops were for the first time sent out to assist the private armed forces of the East India Company.

At the end of the Seven Years War Britain had won supremacy in India. But the area was still under the management of the East India Company, which steadily tightened its grip on Bengal, collecting taxes and administering justice. The wealth that it extracted from the Indian peasantry [5], and the severity of its judicial administration, aroused criticism at home and led to the great political scandal of the century, the impeachment of Warren Hastings (1732–1818) [6]. It also produced the India Act of 1784. By that date control of Bengal had, under the governorship of Charles Cornwallis (1738–1805) and Richard Wellesley (1760–1842), been extended over most of the subcontinent. It was in these circumstances that the power of the East India Company was curtailed by making it responsible for the administration of the whole of India jointly with the Crown department, the Board of Control.

France made one last, heroic attempt under Napoleon to regain its position as the world's leading power. This ended with Wellington's victory at Waterloo in 1815.

KEY

The port of Bristol prospered from having almost a monopoly of the lucrative West Indian slave trade during Walpole's administration. With a population of about 50,000, it was, after London, the second largest town in 18th-century Britain.

5

5 Some Britons made huge fortunes out of their service with the East India Company. These "nabobs" lived ostentatiously in India, and at home bought seats in Parliament to influence policy.

6

6 The impeachment of Warren Hastings, for corruption and mistreatment of natives when he was governor of Bengal (1772–85), began in 1788. The trial dragged on for seven years and ended in Hastings' acquittal.

7 Gibraltar, first occupied in 1704, became a British possession by the Treaty of Utrecht (1713). Spain joined France and the USA in the war against England in 1779 and besieged the colony for four years. George III tried to barter Gibraltar for Spanish colonies in America, but by the Peace of Paris (1783) Britain gave up East and West Florida to Spain and retained Gibraltar.

8 Penal settlements were set up in the new colonies of Australia in the late 18th century. The first convicts were sent to New South Wales in 1788, and to Van Dieman's Land (Tasmania), shown here, in 1803. Transportation was generally for life, but it could often allow the convict to establish himself in the colony; at the time there were more than 100 offences in England that could bring the death penalty.

8

7

9

9 Commemorative medallions, such as this one by Josiah Wedgwood (1730–95), were struck to encourage the abolition of the slave trade in the British Empire. This took place in 1807 and resulted partly from an evangelical campaign by William Wilberforce (1759–1833), and partly because British sugar-planters wished to stop the flow of slaves to their French rivals when their own plantations were at their maximum production and could survive without the import of more slaves.

India from the Moguls to 1800

In 1192 Muhammad of Ghur destroyed the Rajput princes at Tarain. In the next ten years the Muslims overran the Ganges plains and founded the Sultanate of Delhi (1206). This marked the beginning of Islam's long dominance over Hindustan (northern India) which survived into the 1700s and created an entity that was distinct in the Islamic world.

The three centuries that followed saw the rise and fall of dynasties – the Turco-Afghans, Khaljis, Tughluks, Sayyids and Lodis – under which the Delhi Empire sometimes expanded as far as the Deccan and the south (as it did under the megalomaniac Alaudin Kalji, or under Muhammad-bin-Tughluk the greatest of the sultans) and sometimes contracted into the narrow regions around Delhi and Agra (as it did under the Lodis) [2].

The regions continued to challenge the centre; invasions from the north (of which Timur's sack of Delhi in 1398 was the most ferocious example) still threatened the empires, and Delhi remained unable to dominate the Deccan.

Two new forces, however, were to alter the old patterns: Mogul rule and European expansion. In 1526 Babur (1483–1530), driven from his mountain stronghold in Firghana, crushed the Lodis at Panipat and established the Moguls in Delhi. In 1498 the Portuguese explorer Vasco da Gama sailed into Calicut, the precursor of other seaborne powers from Europe who in time were to be the heirs of the Mogul.

Mogul Empire

The great Mogul rulers – Babur, Humayun, Sher Shah and, above all, Akbar [3] and Aurungzebe [4] – brought more of India under one empire than any ruler since the time of Ashoka in the third century BC and gave it a more sophisticated administration than it had ever possessed. But the Great Mogor (Mogul), sitting on his peacock throne, surrounded by his great entourage of courtiers, concubines and slaves, his vast numbers of elephants and camels, and backed by his 200,000 cavalry, had built an empire on a fragile base. Mogul rule was the rule of foreigners and it was expensive. Built by conquest, it could not be sustained by force alone. Despite all the trappings of imperial power, the army and the centrally controlled military bureaucrats (the *mansabhadars*), the Moguls were forced to come to terms with their Indian subjects. This meant they had to give reasonable freedom to the magnates who actually controlled the land. They had to tolerate Hinduism and above all they had to avoid meddling too much in local affairs. All this qualified the notion of Mogul power – more overrule than real empire – and helps to explain the decline of the Mogul Empire [6].

The problem of revenue

Like all land empires in pre-modern times, the fundamental difficulty the Mogul rulers faced was how to raise revenues from limited sources, while retaining the co-operation of local notables, the *zomindars*. One obvious device was to extend the empire but, as Aurungzebe discovered, more conquest meant more expense. Another was to try to extract more land revenue from their subjects, but in the end the Moguls lost the co-operation of the local notables, who were squeezed from above and resisted by the

CONNECTIONS

See also
140 India in the 19th century
46 British colonial policy in the 18th century
130 Imperialism in the 19th century
50 Indian art: the Mogul age

In other volumes
118 History and Culture 1

1 The ruins of the fortress and city of Tughlukabad, near Delhi, are an example of the feckless manner in which Indian cities were made and unmade by rulers who sought to commemorate themselves by building new capitals. Constructed between 1321 and 1323, Tughlukabad was soon abandoned by Muhammad Tughluk Shah in favour of Deogir, ostensibly because of bad water.

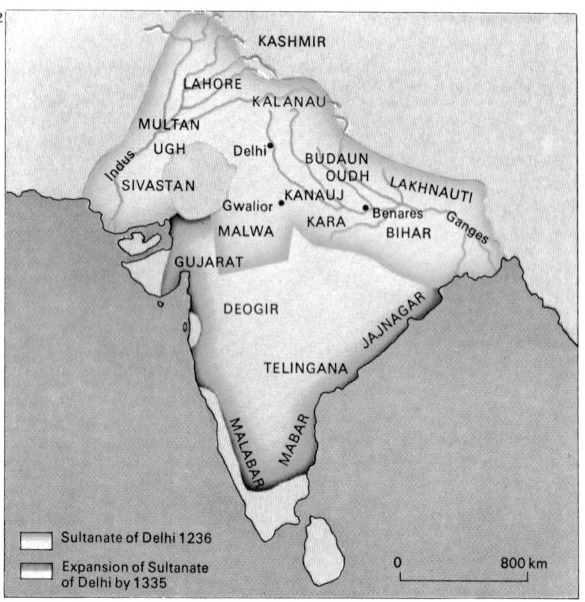

2 The Sultanate of Delhi at the death of Iltumish (1236) and, 100 years later, under Muhammad-bin-Tughluk (*c.* 1325–51), when it reached its greatest extent, are shown on the map. It indicates that Muslim dynasties had dominated northern India for centuries before the Moguls and had made attempts apparently, only temporarily successful, to dominate the Deccan and the south of India.

3 Akbar (*r.* 1556–1605), the most celebrated Mogul emperor, was the almost exact contemporary of Queen Elizabeth I. He vastly expanded. The Mogul's territories, gave them an efficient system of government and tried to encourage new sources of work. In his effort to conciliate his Hindu subjects and to stamp out Muslim bigotry, the emperor tried, but without lasting success, to create for India a new religion reconciling the different beliefs.

4 Under Aurungzebe (*r.* 1658–1707), the Mogul Empire reached its greatest extent: from Kabul to the Cauvery. But the empire he bequeathed did not survive much beyond the five years of his son's reign. The seeds of disruption, always present, were germinated by his religious intolerance; he alienated the Rajputs, the military prop of the empire; at great cost he conquered, but failed to integrate, the Deccan; he neglected Hindustan and fought the Marathas in the south.

5 The Red Fort at Delhi, built by Shah Jehan (1592–1666), was surrounded by massive red sandstone walls 22.8m (75ft) high and enclosed a complex of palaces, gardens, military barracks and other buildings. The fort, like many other masterpieces of this reign, including the Taj Mahal, illustrates the splendour of Mogul rule in India.

peasantry below. These inherent difficulties were made worse by Aurungzebe's religious intolerance, which alienated the Rajput allies and confirmed the Marathas in their hostility. A sequence of ineffectual successors after Aurungzebe reduced the throne at Delhi to the plaything of factions; more invasions from the north culminated in Nader Shah's sack of Delhi in 1739.

Rise of the East India Company

With the collapse of Mogul rule, India was parcelled out among a set of virtually independent "country" powers – Bengal and Oudh under the Nawabs, Mysore under the Muslim dynasty of an adventurer, Hyderabad under the Nizam and the Maratha confederacy marauding throughout India from its western base.

This was the background to the territorial rise of the British-owned East India Company. The company had gone to India to trade not conquer. Throughout the eighteenth century its directors protested that commerce, not dominion, was their aim. But by the mid-eighteenth century the British, long established in their factories at Calcutta, Madras and Bombay [8], had a trade (mainly in Indian cotton goods) that was worth protecting both against the smaller French company and against Indian disruption [7].

When Britain and France fought in Europe, the companies in India could not stay neutral. Faced by a more powerful English company, Joseph François Dupleix (1697–1763) of the French East India Council decided, during the Austrian War of Succession, to improve the odds by using Indian allies against his rival.

The English soon adopted the French tactic. Robert Clive (1725–74) used it to win the Battle of Plassey (1757) and thereby gain Bengal. Dupleix's idea of using Indian resources to pay for European expansion gave Britain the richest province of India and set her firmly on the road to dominion [9].

By the time William Pitt (1759–1806) and Henry Dundas (1742–1811) began to think of an Indian Empire as part of the swing eastwards of British expansion, the East India Company had already become the heir of the Moguls.

KEY

After the Afghan defeat of the Marathas at Panipat (1761) Mahadaji Sindhia (1727–94), a bluff but literate soldier of fortune, built a large empire in Hindustan. Acting nominally as the soldier of the *peshwa* (sovereign), he became the master of the Mogul ruler, Shah Alam, and the "actual sovereign of Hindustan from the Sutlej to Agra, the conqueror of the princes of Rajputana, the commander of an army". But although he modernized the Maratha armies, giving them artillery and recruiting Muslim and Jat soldiers and European officers in his service (one of whom he is entertaining here), his power was fragile because his domains lacked the resources to support his armies.

6

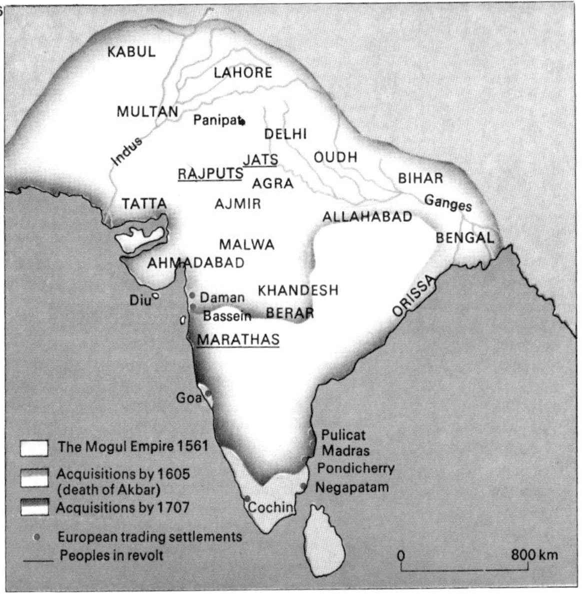

6 The rise of the Mogul Empire and the early signs of its decline are indicated on the map. By 1561 Akbar had created a unified Mogul Empire in the north of India; by 1707 when Aurungzebe died, the Mogul Empire had reached its utmost limits. But the Rajputs, Jats and Marathas were already in revolt and the European seaborne powers, who were to be the heirs of the Moguls, occupied the coast.

7

7 The English settlement at Fort St George in Madras was one of the earliest in India, set up in 1639. The struggle with the French for mastery centred upon the south and the fort was crucial.

8

8 Bombay was ceded by Portugal to the English in 1661 and granted to the East India Company in 1668. Its fort helped to defend English commercial interests located on the west coast of India.

9

9 In 1765 Shah Alam made his grant of the *diwani* – the right to collect land revenue – to Clive, by then the real master of Bengal. The fact that the Mogul, although politically impotent by this time, was still seen as the source of legitimacy in ratifying the company's authority in Bengal, shows the conservatism of Indian politics; it is also indicative of British reluctance to move from trade to dominion that Clive did not demand more power.

10

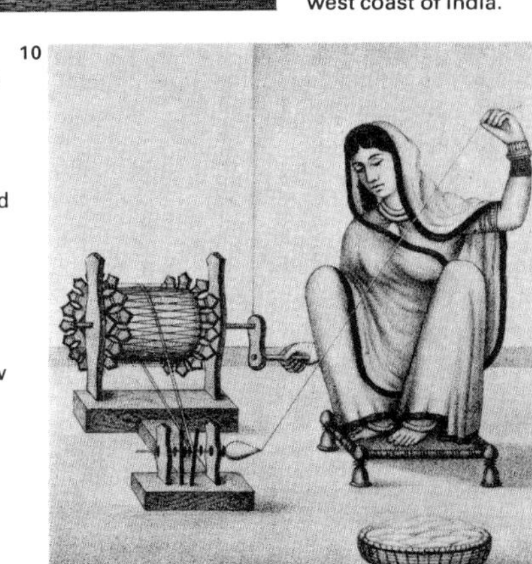

10 The East India Company's fortunes in the late 17th and 18th centuries were built upon the export of hand loom cottons whose thread was spun in villages throughout India, sometimes with the aid of the *gharkha*, or spinning wheel, illustrated here. The 19th century saw the decline of village spinning and weaving, and India became the largest market for Lancashire's cotton goods. In the 20th century Gandhi revived the hand loom.

Indian art: the Mogul age

The Muslim conquests in India in the twelfth century brought about a fundamental change in the art of the subcontinent. The invaders imposed a foreign style – Persian – upon the country; the new royal patronage encouraged the survival of secular art and there was a systematic attack on religious art. Before the invasions the royal court had patronized art from religious establishments. Afterwards there was far more secular art.

The contrasts with Islam

There could hardly be a sharper contrast than that between the religious monuments of the Hindus, Buddhists and Jains, with their heavy forms and exuberant figure sculpture and the light, airy mosques of Islam with arcades, domes and slender towers [3]. These mosques were without sculptured figures but decorated with disciplined floral and geometric ornaments carried out in bright colours, either flat or in low relief. The floors were often covered with carpets and this helped to foster the Indian carpet industry which, as with other crafts, began with a marked Persian influence but eventually developed its own national style. The same architectural idioms that applied to mosques were used in the erection of other buildings. An important technique of surface decoration consisting of patterns of inlaid coloured stone [6] is similar to Italian *pietra dura* and may have been imported from the West.

The independence of some cultures within the area of Muslim influence is shown by Jain painting, which developed as a form of manuscript illumination [1]. This painting is quite distinct from the more naturalistic wall painting at Ajanta that preceded it. The Jains adopted a highly stylized flat and decorative form that made use of bright colours, particularly red, blue and gold. A characteristic trait is the use of wiry, angular figures, each having the head in three-quarter front view with the farther eye protruding beyond the line of the face.

A studio system set up

The establishment of the Mogul dynasty in northern India created conditions under which Muslim art was able to flourish. The emperor Akbar (1542–1605), who was an enlightened monarch, expanded the studio system set up by his father in which Indian painters had worked under the supervision of two Persian painters. This arrangement, together with a prevailing Persian taste, imposed a strong Persian influence on painting at the Mogul court. Under Akbar's patronage, however, a style quickly evolved that was unmistakably Indian, showing a refined naturalism and the use of rich colours to impart vigorous movement [2].

In the early seventeenth century an interest in portraiture and natural history (perhaps under Western influence) was added to heroic and mythological subjects. A sense of serene contemplation took the place of violent motion; it was coupled with a simpler, more subtle sense of composition and deeper psychological insight. This continued into the reign of Shah Jehan (1629–58), [Key and 4] but while the high standard of technique remained, a stiffness of execution inhibited the former supple spontaneity. Under Aurungzebe (1618–1707), Mogul painting declined – a trend that, in general, continued almost uninterruptedly.

CONNECTIONS

See also
48 India from the Moguls to 1800
130 Imperialism in the 19th century

In other volumes
122 History and Culture 1

1 This Jain religious diagram (*yantra*), of which a detail is shown, was painted in Gujarat in western India in 1447. It makes no attempt at realism since its purpose was not primarily aesthetic; such considerations were subordinated to the stimulation of religious ideas through the use of traditional artistic forms. Jainist paintings often took the form of manuscript illuminations made on palm leaves, even after the introduction of paper in the late 14th century.

2 The Annals of Akbar *(Akbar-nama)*, c. 1600, illustrates the festivities that marked the birth of Prince Salim at Fathpur Sikri in 1569. Under Akbar Indian painting absorbed the Persian style and became unquestionably Indian. Attentive to detail, its bright colours and agitated composition produced pictures that were like mosaics but seemed to be in constant movement.

3 The Qutb-Minar at Lalkot, Delhi, was built *c.* 1230 with later additions. As well as being a minaret it also served as a watch-tower and war memorial. It is a monument to early Islamic architecture in India since the minaret is one of the more obvious characteristics of a mosque. In this case the mosque was constructed partly of material from a nearby ruined Hindu temple.

4 This carved jade wine cup in the form of a gourd bears a legend indicating that it was made for the Mogul emperor Shah Jehan in 1657. It was during his reign that Mogul sumptuousness reached its zenith, represented by the Taj Mahal and the peacock throne. The naturalism of the cup is entirely Indian, showing the emancipation of Mogul art from that of Persia.

In the Deccan a style similar to that farther north – but somewhat modified by pre-Islamic elements – continued during the eighteenth century and produced, among other subjects, illustrations corresponding to musical modes (scales). To the northwest, in Rajasthan, a large number of schools developed under the patronage of local courts of which those at Mewar, Bundi, Kotah, Bikaner and Jaipur are probably the best known. Their style was characterized in the seventeenth century by emotional intensity and in the eighteenth century by the vivid use of bright colours and an acute sense of drama. Under similar patronage, but below the Himalayas in the hill states of the Punjab such as Kangra and Guler, several schools arose that exhibited a marked individuality of style. Known as Pahari (meaning "of the hills") paintings, these pictures took as their subjects romantic themes found in Hindi and Sanskrit poems such as the *Gitagovinda* of Jayadeva. This epic dating from the twelfth century recounts the amorous adventures of the Krishna, such as those describing his love for Radha, his favourite *gopis* (milkmaid). The adventures are admirably illustrated in the Pahari miniatures, which convey the delicate passion of the poem by the expressive use of line and subtle colour effects [5].

The decorative arts

Under the Moguls decorative arts flourished as an aspect of the ostentatious luxury of the court and minor rulers. Intricately wrought gold and silver work was produced as well as jewellery (for men and women) set with diamonds, pearls, rubies and other precious stones. Vessels and other objects such as rings and dagger hilts were carved in rock crystal and jade [4] and were also sometimes inlaid with gold and gems. In addition to embroideries, sumptuous textiles were woven – velvets, brocades and silk fabrics.

In southern India, Hindu traditions in decorative art generally held their own in spite of Muslim influence from the north, but they did not remain entirely unaffected. This resulted in a style, best illustrated by some superb ivory carvings [7], that made use of tightly controlled organic motifs combined with figures sometimes in erotic scenes.

KEY

The Mogul emperor Shah Jehan looks at a piece of jewellery in this portrait (*c.* 1618) that reflects most of the characteristics of post-Muslim conquest Indian art. Opulent, tender and fastidious, it glorifies the individual. It records also the name of the artist (Abu'l Hassan), who earlier would have been anonymous. Although some attempt has been made to achieve a likeness, the result is formalized and without character. The floral background, in which plants are treated in a naturalistic but gem-like way, reflects the interest in flower painting that was fashionable during this period.

5

5 Divine love as an ideal for human love was a popular theme in Indian, as well as European, art, as seen in the poetic paintings of Krishna and Radha. In this picture from the Punjab hills their love is depicted with a poignant tenderness. The complete painting follows the Indian convention of showing two or more related events within the same frame; above the lovers' meeting, shown here, there is another picture of their forest walk.

6

6 A mausoleum at Agra, erected in 1628, is constructed of marble inlaid with patterns of coloured stone. It shows the Mogul interest in surface decoration and linear movement.

7

7 Ivory carving has had a long history in India. It was probably most fully developed in the south, as this 18th-century comb from Mysore, showing Gaja-Lakshmi, goddess of fortune and prosperity, magnificently reveals.

8

8 Cotton paintings ("chintz") on an 18th-century coverlet combine Indian technique and European designs, such as engravings of ornamental motifs. Hand-painted at first, they were later block-printed.

China from 1368 to *c.* 1800

The Ming or "Brilliant" dynasty (1368–1644) was founded by a Buddhist peasant who became leader of the rebel bands that overthrew the Mongol rule. He made his capital at Nanking and gave himself the title Hung Wu (reigned 1368–98). Under the new emperor the government reverted to the T'ang system of Confucianism, taxes were reduced and peasants were encouraged to work harder by being allowed to keep the land they reclaimed. The empire expanded, taking in vassal states which included Annam and Siam. Maritime expansion also occurred and under the able command of Admiral Cheng Ho fleets of exceptionally large Chinese vessels made seven expeditions between 1405 and 1433 to places as far away as Sri Lanka, the Persian Gulf and parts of Africa. The ships carried merchandise and Cheng Ho was able to establish trading relations with about 30 ports.

Protection of the homeland

At home the government was determined to defeat further invasions from the north. The Great Wall was repaired, many towns were fortified and Nanking was surrounded by a wall 32km (20 miles) long and 18m (60ft) high. This "closed door" policy of the Ming was similar to that of the early Sung period, which had also established Chinese rule over the Middle Kingdom after a period that had witnessed foreign intervention.

In spite of this attitude the Ming were to see the arrival of Europeans from the south. The Portuguese established a settlement in Macao and were followed by missionaries.

The Ming built numerous palaces and splendid tombs, the most famous being the Imperial Palace in Peking [1] and the tombs of the emperors north of the city [3, 4]. In 1958 that of the Wan-li emperor was opened to reveal three lacquer coffins containing the bodies of the emperor and two of his wives.

The end of the Ming dynasty

In spite of all the elaborate precautions for security, the Brilliant dynasty reached its end when northern barbarians, in this case the Manchus, entered north China. They came at the request of the Ming commander who sought their help to unseat a rebel, Li Tzu-ch'eng, who had made himself master of Peking. Rather than submit to humiliation the last Ming emperor hanged himself and the rebel emperor in his turn was overthrown by the Manchus whose leader, Fu-lin (1638–61), proclaimed himself emperor and set up the Ch'ing or "Pure" dynasty.

For a century and a half the Manchus governed wisely; they provided the country with domestic prosperity and extended its boundaries beyond all previous limits. The emperors ruled as conquerors and skilfully protected themselves against rebellion by instituting a system of banners. These were military and administrative divisions in which the people were systematically registered, taxed and conscripted.

The Manchus were unable either to read or write and accepted the Ming examination procedure for the selection of civil service officials. They were careful, however, to forbid the Chinese to hold office in their native provinces and they also divided responsibility in such a way that the officials were obliged to keep a check on one another. The Manchus were more sophisticated than

CONNECTIONS

See also
54 Chinese art 1368 to the 20th century
84 European empires in the 19th century
138 The expansion of Christianity

In other volumes
224 History and Culture 1
126 History and Culture 1

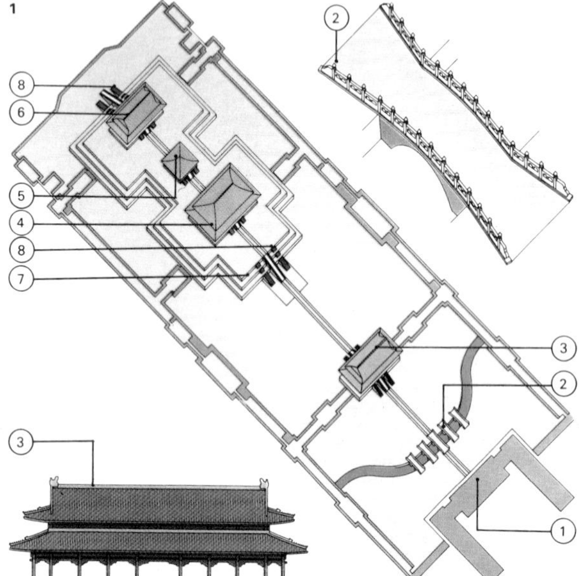

1 The Imperial Palace in Peking was the seat of 22 Ming, and Ch'ing emperors after Peking became the capital in 1421. Begun in 1406, the palace now covers 720,000m² (7,750,000 sq ft). The palace is an assemblage of imperial buildings, the most important [shown here] containing the ceremonial halls. The principal entrance is through the Meridian gatehouse [1] leading to the Golden Water River with its five marble bridges [2]. After the second gatehouse [3] lies the first great hall [4], which holds the imperial throne. The halls [5, 6] stand on the marble, three-tiered Dragon Pavement [7] with triple staircases [8].

2 Chinese astronomers were the most persistent and accurate observers of celestial bodies anywhere before the Renaissance. The importance of the calendar for a primarily agrarian society and the state interest in astrology meant that astronomy was of central importance in their lives. In contrast with Europe, scientific work was not a private concern and astronomers often worked from the Imperial Observatory in Peking. Their observations, like the work of scientists in many other fields, were included in large encyclopedias that were compiled and published at the instigation and sole expense of the state.

3 The tomb of the Ming Wan-li emperor (1573–1620) is situated at the foot of rugged mountains to the northwest of Peking. The complex is approached by an avenue lined with impressive stone sculptures of guardian animals, soldiers and officials. Work on the tombs began in 1584 and took four years.

4 The throne of the Ming Wan-li emperor was placed in his tomb. Ming emperors generally enjoyed an unprecedented, although often abused, degree of power. Ultimately the Ming dynasty fell to the invading Manchus who seized Peking in 1644, but already during Wan-li's reign internal dissent and foreign attacks threatened Ming power.

the Mongols and proved to be neither barbarous nor destructive, but in fact grew to admire Chinese culture. One of their first acts was to request a group of scholars to write a history of the Ming dynasty which was followed by a vast encyclopedia, the *Ssu-k'u ch'uan-shu*, comprising 36,000 volumes, begun in 1772 and completed in 1781.

The influence of the Manchu

Abroad Manchu authority spread to Manchuria, Mongolia, Tibet and Turkestan. During the Ch'ien Lung era their armies entered Burma, Nepal and Annam [6]. As the empire grew in size and wealth foreigners increased their pressure to trade with this huge untapped territory. The Jesuits, who had been accepted during the Ming dynasty, were now followed by other Catholic orders from Europe and before the end of the seventeenth century Franciscans, Dominicans, Augustinians and members of the Society of Foreign Missions in Paris had established themselves in several cities in the interior.

European merchants did not penetrate China as readily as the missionaries, but they persisted in their efforts and eventually limited trading facilities became available for the French, British, Dutch and Portuguese. In 1784 the first of many ships from the United States arrived. By the middle of the eighteenth century British trade, the monopoly of the British East India Company, had outpaced all others [8].

Chinese silk, tea, cotton and porcelain were in endless demand in Europe. However, business could be carried out only through selected groups of Chinese merchants, the Co-Hung. There were no fixed tariffs, a policy that increased corruption amongst officials. The Chinese were forbidden to teach the foreigners their language and foreigners had to submit to Chinese law in Chinese courts, where the Chinese with their belief in group responsibility held the foreign communities liable for the misdemeanour of any of their members.

Such conditions soon became impossible. The Western trading countries therefore sent missions to Peking [5], which began the lengthy process of bringing China out of her diplomatic, cultural and economic isolation.

KEY

This Ming imperial crown, decorated with a phoenix, is indicative of Ming wealth and their patronage of the arts.

5 The first official British mission to China was made in 1792 by Lord Macartney (1737–1806) and came at a time of demands for increased trading rights as the British and other foreign powers pressed for a foothold. Already the British East India Company had a monopoly over the trade in tea. However, the mission was treated by the Ch'ien-lung emperor as purely diplomatic.

6

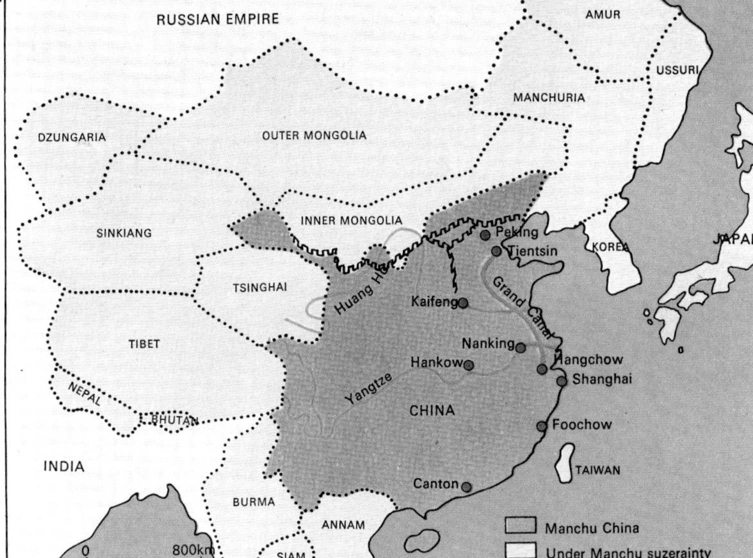

6 The Manchu Empire in 1800 covered a large area of central and South-East Asia. But already European and Russian expansion in search of trade threatened this shaky conglomerate.

7

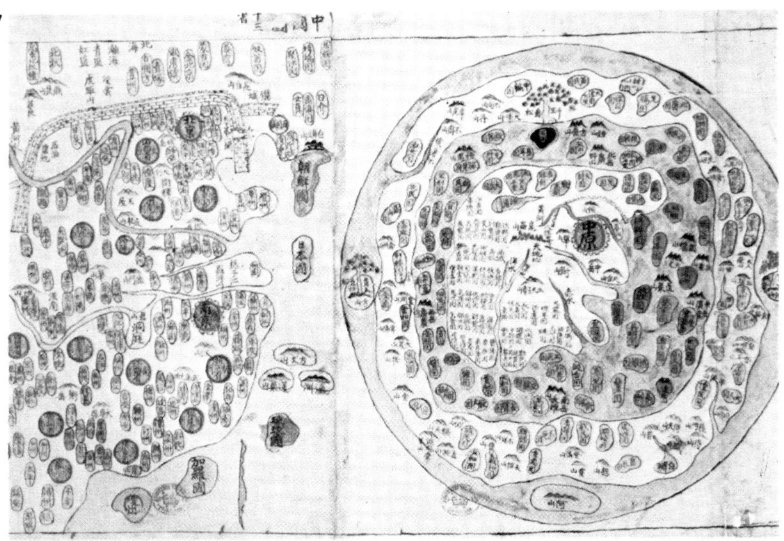

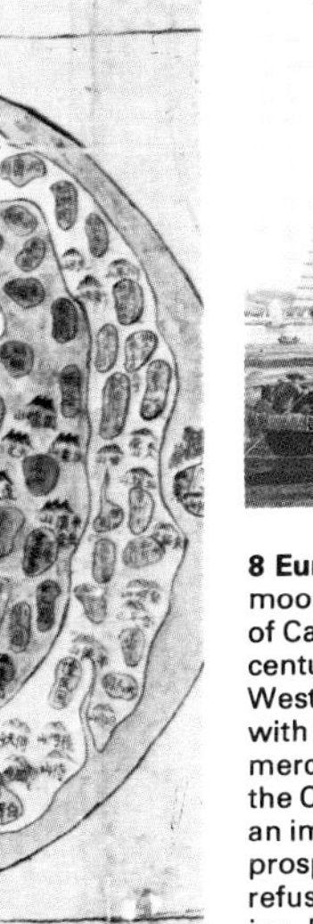

7 This Sinocentric map (*c.* 1800) shows China at the centre of the world, both culturally and geographically, as a "middle kingdom" surrounded by barbarians. China was generally self-sufficient and pursued a policy of aggressive isolationism, believing that little was to be gained by contact with other peoples.

8

8 European ships moored in the port of Canton in the 18th century reflected the West's desire to trade with China. Western merchants regarded the China trade as an immensely rich prospect, but China refused to allow trading. From 1557 the Portuguese had had a monopoly of trade with China from Macao until the Dutch began to trade but attempts by the British East India Company in the early 17th century failed to gain official approval. The Europeans pressed for greater trading concessions, but from the 1750s all foreign trade had to go through Canton and was strictly supervised by the state. British trade in tea grew during the 18th century and in the 1780s they began to smuggle in opium profitably from India.

Chinese art 1368 to the 20th century

With the establishment of the Ming dynasty, Chinese society, as reflected in the arts, underwent a renaissance. The style of the arts of the succeeding three centuries is reminiscent of the T'ang dynasty in its boldness and grandeur. The Imperial Palace (Forbidden City) [7] and the imperial tombs of Peking embody much of this style. The large scale of the buildings and the use of immense space in the layout of courtyards and processional ways show the confidence and flair of the designers.

The arts during the Ming dynasty

The applied arts also flourished, particularly during the earlier centuries of the dynasty. Carved lacquer of beautiful quality was made by masters in small workshops. By contrast the anonymous ceramic industry had grown to considerable production and was by now centralized on the great kiln area of Ching-te-chen, Kiangsi.

Added to the home market the overseas trade in ceramics was rapidly accelerating. The taste for decorated porcelain led to the development of techniques hitherto unexplored. First, the use of underglaze painting in cobalt blue was perfected through various styles [1]; second, by the sixteenth century low-fired coloured glazes painted on top of the primary porcelain glaze and refired produced the increasingly popular polychrome decorated wares. Cloisonné enamels on metal bodies also produced rich effects. This technique, of much earlier origin, originally came from the Near East. The grandly simple jade carving is rare and it shows an affinity in style with early Ming lacquer work. In the later years of the dynasty there was a move towards elaboration exemplified by the use of inlay and onlay of semi-precious stones on both jade and porcelain.

The fine arts of the Ming reflect the tastes and position of the scholar classes. At the beginning of the dynasty a variety of styles of painting were practised. The court attempted to re-establish an academy on traditional lines. Painters such as Tai Chin (1390–1460) [2] served for a time but soon retired and became the centre of a group known as the Che School (centred on Chekiang). Using colour and a broad, wet brushwork they produced marvellously evocative genre paintings of country people and elegant landscape and bamboo compositions. Because of their early association with the fashionable court and its intrigues, the members of this school gained a reputation for slight, showy work.

The Wu School and its successors

The great rivals of the Che School were the Wu School, headed by Shen Chou (1427–1509) [3], a native of Wu Hsi. Painters of this group kept away from court and followed the great masters of Yuan, living as scholar literati and painting in an eclectic scholarly style. Shen Chou himself never took up official life. He was the first painter to use his own poetry as part of his painting and also to use figures in the composition in such a way as to invite the viewer to identify with the figure and so become directly involved with the painting.

Wen Cheng-ming (1470–1559) [4], Shen Chou's pupil, showed a certain decorative quality in his work, which typified a move that eventually caused concern to the purists. These scholars made an analytical and histor-

CONNECTIONS

See also
52 China from 1368 to c. 1800
144 The opening up of China

In other volumes
128 History and Culture 1

1 White porcelain was made in a wide range of styles and decorated in cobalt and copper under glaze, as is this large wine jar from the 14th century, which was made at Ching-te-chen, Kiangsi. Cut-through techniques often included an inner decorated vessel.

2 A native of Chekiang, Tai Chin became a court artist for a short time but he was banished to Hangchow where he painted his major works. He is regarded as the leader of the Che School. This handscroll in ink and colour on paper is called "Fishermen on the river".

3 Shen Chou came from a scholarly family. He first followed the painter Huang Kung-wang but later established his own simple style. This album leaf in ink is called "Poet singing in the Mountain".

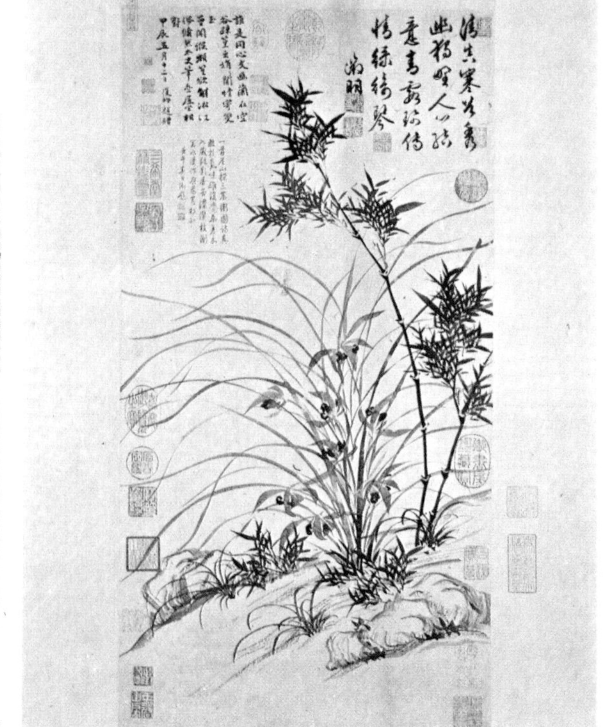

4 Wen Cheng-ming was the head of a notable family of painters from Tsang Chou. His style was both elegant and decorative and included the popular subject shown here of "Bamboo and Epidendrum".

ical study of painting that was intended to rectify what they saw as a trivialization. Their precepts and classification were a great influence on painters and critics of later periods.

The painters of the generation following the purist Tung Ch'i-ch'ang (1555–1636) [8] were mindful of his analysis and the four Wangs (men with the surname Wang, but not all related) are often regarded as being the followers of Tung. The two elder Wangs, Wang Shin-min (1592–1680) and Wang Chien (1598–1677), follow Wen Chengming and the later Wu School. They painted large eclectic landscape compositions but were also part of the general movement towards a high Romantic style that flowered in the eighteenth century.

Contemporary with the elder Wangs were Shih T'ao (1641–1720) and Chu Ta (1626–1705). Shih T'ao was by inclination a scholar-painter who relied a great deal on literary inspiration. He was an innovator in the use of ink, colour and brush. Chu Ta was an instinctive Ch'an (Zen) artist with an elegant brush style.

At this time a generation of Romantic painters appeared. With the establishment of the grand courts of the Ch'ing, K'ang Hsi, Yung Cheng and Chien Lung, Romantic art was served by the European Jesuit painters led by Castiglione (Lang Shi-ning [1698–1768]), whose curious Italianate-Chinese style was true *chinoiserie*.

Chinese art to the present day

In the seventeenth century, the newly wealthy towns of Souchow and Yangchow were the homes of an innovative group of painters known as the Eccentrics, who advertised their wares and painted for money. The eighteenth century saw a strong taste for archaism in all crafts. With the dawn of the nineteenth century both artist and patron seem to have lost confidence: painting moved slowly and hesitantly along the old paths of the seventeenth century.

Twentieth-century Chinese artists have been unsure of their direction and they experimented first with European and Japanese styles. Communism has re-directed the arts to serve political aims, so they have a dominant propaganda function.

KEY

An inventive artist in both techniques and subject-matter, Shih T'ao was a late 17th century recluse who corresponded with his contemporaries and was one of the most influential artists in his own and later generations. His quite distinctive use of the surface texture of the paper, as in this landscape in ink and colour, is still being explored today.

5

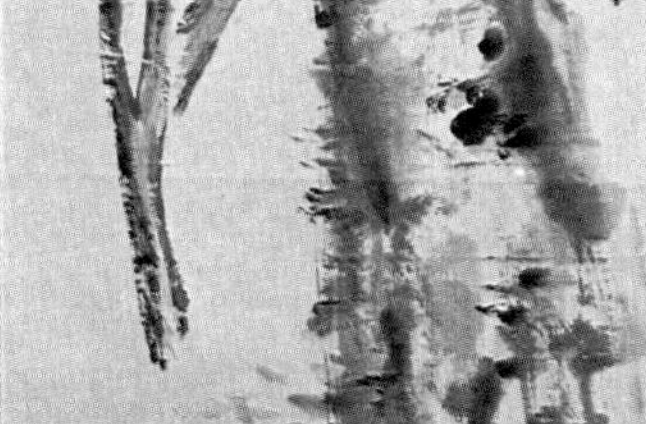

5 Wu Ch'ang-shih (1842–1927) followed the broader ink style of the late 19th century as well as adopting the current subjects – birds and flowers.

6

6 Carved by Chu San-sung, in the 16th century, this bamboo brush pot was designed to hold a scholar's pens and brushes.

7

7 The Forbidden City, palace of the Ming emperors, is enclosed within a large, rectangular, moated, walled compound. The central north-south axis is marked by a series of huge audience halls each flanked by smaller halls and surrounded by large courtyards and tiered walks. The buildings are of gaily painted wood with yellow glazed roofs; the surrounding courts are constructed of white marble.

8

8 This calligraphy by Tung Ch'i-ch'ang (1555–1636) is strong and fine and typifies the Chinese scholar-painter's command of his art form. Tung was a theorist and historian who also painted landscapes and lighter works.

9

9 Fu Pao-shih (1904–65), one of the major Chinese painters of the 20th century, studied in Japan in the 1930s and was highly influential as a teacher at the Nanking Academy of Arts. This is his "Scholar in his study".

Japan 1185–1868

In 1185, following five years of civil war, Minamoto Yoritomo ruled much of Japan from his headquarters at Kamakura. In Kyōto the emperor reigned, but effective power lay in the hands of the Minamoto house which had defeated the previous ruling family, the Fujiwara.

The new Japanese rulers

In 1192 the emperor recognized this new reality. Yoritomo was made shōgun (general), as head of the military class, and his followers were rewarded with further rights in land at the expense of their defeated enemies. Some became military governors of distant provinces, others supervised tax collection on the estates of imperial courtiers. On these foundations Yoritomo created a practical and efficient administration which gradually but inevitably eroded the basis of imperial power.

This new power produced a new spirit and a new culture. Warriors (samurai) rejected the ceremonial of established Buddhist sects and turned to simpler, more austere, forms of religion. Zen temples were built in Kamakura. A new Buddhism was preached and the scriptures were translated into Japanese.

In 1221 this dynamic government defeated an imperial revolt against its authority. Fifty years later it overcame a greater challenge from Asia. In 1274 and 1281 the vast Mongol armies of Kublai Khan [1] invaded Japan but effective defences and the power of the "divine wind" (kamikaze) repulsed these expeditions.

In time jealousy and hatred of the Hōjō grew and in 1333 Ashikaga Takauji joined Emperor Daigo II in destroying their power [3]. The emperor then tried to re-establish his authority but the Ashikaga resented their poor rewards and in 1336 Takauji drove Daigo II from Kyōto, replacing him with a puppet. Two years later he became shōgun.

The long civil war

By 1467 the rising economic and military strength of the great provincial houses was uncontrollable. As the shōguns grew weaker civil war broke out and continued for a hundred years. In this unrest, Europeans first reached Japan and began to influence its domestic politics. In 1543 the Portuguese arrived and soon Jesuits began to spread their missionary activities.

When Oda Nobunaga (1534–82) began the task of unification, religious and secular forces obstructed his advance. In 1571 he destroyed the temples of monastic armies and two years later extinguished the power of the Ashikaga line. In 1582 he was assassinated in Kyōto but his chief commander Toyotomi Hideyoshi (1537–98) continued his campaign. Soon Hideyoshi dominated Japan and threw his armies against Korea. His death ended these adventures and left Tokugawa Ieyasu supreme at home. Ieyasu officially became shōgun in 1603 and sought to establish peace throughout Japan [2].

Like his predecessors Ieyasu re-measured lands and regulated taxes. He relegated his enemies to outer provinces and concentrated his allies in central Japan. Fortifications were controlled and spies recruited. Lords or their families were compelled to reside at his new capital of Edo (later Tokyo). Confucianism became the official ideology to create a comprehensive social philosophy from the

CONNECTIONS

See also
146 Japan: the Meiji Restoration

In other volumes
130 History and Culture 1
224 History and Culture 1
123 History and Culture 1

1 In 1274 Kublai Khan launched a force of almost 30,000 troops against western Japan, but violent storms drove his ships back to their bases. Seven years later a second expedition (shown here) landed in Kyushu. Well-organized resistance, stone defences and gales combined to repel the invaders. There were fears of a further onslaught, but in 1294 Kublai Khan died and his preparations were abandoned.

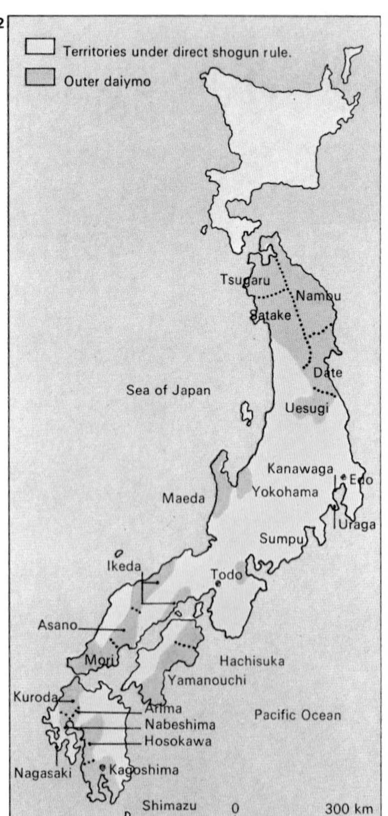

2 Following his victory at Sekigahara (1600), Ieyasu redrew the political map of Japan to reduce possible threats to his supremacy. Eighty-seven families lost all their lands. As a result the territory of Ieyasu and his allies was increased and concentrated in central Japan. Traditional enemies were relegated to outer provinces. With minor adjustments, these boundaries survived until 1868.

3 Ashikaga Takauji (1305–58) supported the Emperor Daigo II in destroying the Kamakura military government. Later he drove the emperor from his capital and made himself shōgun in 1338.

4 Tokugawa Ieyasu (1543–1616) was the son of a middle-rank lord who rose to become political master of Japan. After serving as a successful general under Nobunaga and Hideyoshi, he destroyed his rivals at the battle of Sekigahara. In 1603 he became shōgun and proved to be an astute and determined politician. He limited the power of the emperor and his regime continued until 1868.

samurai code of conduct (bushido) and society was modelled on an agrarian ideal. Warriors were a privileged élite, farmers next in importance; craftsmen and merchants of lowest esteem. Ieyasu's successors developed his policies [5, 7]. They also suppressed Christianity and restricted foreign trade in the interest of political stability. A few Dutch [5] and Chinese merchants resided at Nagasaki but no Japanese could go abroad.

The nineteenth century

One result of the Tokugawa peace was commercial growth, undermining samurai power. On several occasions after 1700 the shōgun's government tried to solve this problem by returning to the principles of Ieyasu. Austerity was encouraged and officials dismissed. More land was cultivated and townsmen were urged to return to their villages. Merchants made forced loans to warriors and censorship increased. None of these measures was successful for very long.

Parallel with this growing domestic crisis, Japan faced increasing threats from the Western powers. At first foreign ships were driven off and seclusion maintained, but this could provide no lasting solution. In 1853 Commodore Matthew Perry (1794–1858) led a US naval squadron to the Japanese coastline [6] and demanded stores and an opening up of diplomatic relations. After a year's delay his requests were granted and agreements were made with other powers.

From 1859 the Western powers traded at Yokohama, Hakodate and Nagasaki and diplomatic contacts increased. The Tokugawa made more agreements with foreign powers but many Japanese feared colonization. Foreigners were murdered and warships brought destructive retribution. A sense of national crisis covered Japan. The Tokugawa built warships and cannons and began modernization [8]. Yet their measures seemed insufficient for national survival. For centuries the emperors had reigned, mainly as figureheads, from their court at Kyōto; now many turned to the emperor for inspiration and in January 1868 warriors from the western provinces restored his power [Key]. The Tokugawa era was at an end; a modern imperial age had begun.

KEY

The Emperor Mutsuhito (1852–1912), aged only 16 at the time of the Meiji restoration (1868), played an important role in symbolizing its legality. In later years he was a focus for national unity and supported policies of modernization. He moved his court to Edo, which he renamed Tokyo.

5

5 After 1641 under the Tokugawa regime all Dutchmen trading with Japan were confined to the island of Deshima, Nagasaki (shown here), and their wives and children were forced to leave the country. Throughout the centuries, this small, secluded group of Dutch merchants provided Japan's only link with the West and supplied much vital information. Japanese scholars studied science from Dutch publications.

6

6 In July 1853 Perry's naval squadron arrived off Japan to demand the ending of isolation. These talks were unsuccessful but in 1854 Perry returned to sign Japan's first modern treaty.

7

7 The Tokugawa period (1600–1868) saw a growth in education and literacy. Village schools, such as this one at Okayama, gave a simple education to children from backgrounds other than the samurai.

8

8 Following the forced opening up of Japan in 1854, the Tokugawa government feared a Western invasion and ordered the construction of this furnace at Nirayama to produce European-style artillery, an example of military modernization.

Subsaharan Africa 1500–1800

The appearance of a Portuguese fleet off the west coast of Africa in the mid-fifteenth century marked a new and decisive stage in the history of the continent: the beginning of a long and tumultuous relationship between Africans and Europeans. The Portuguese were followed by the Dutch in the sixteenth century and then by British, French and other Europeans a century later.

Complementary systems of trade

The many trading posts which these Europeans established on the coast, and linked to the Europe-based worldwide system of trade, complemented an existing trans-Saharan commercial network. This had been operating for at least 500 years and was forged between Arabic-speaking people from North Africa and black Africans living south of the Sahara. However, the European and trans-Saharan commercial systems were separated by the vast distances of the West African hinterland and never clashed.

A surprising feature of this period in view of these twin commercial presences is the relatively little influence which these foreign cultures had on Africa as a whole. Many Africans in the Sudanic belt had become Muslims in the 500 years of Arab trading in North Africa – a process that set up stresses in African societies and eventually provoked the Holy Wars of the nineteenth century. These states continued to look north to the Muslim world for their external contacts. Equally, some of the savanna and forest peoples began to turn to the Christian newcomers in the south for their contacts with the outside world. However, many peoples in western Africa, and nearly all of those of central and eastern Africa, were entirely outside the direct influence of either Arabs or Europeans until 1800.

The purely passing effect of foreigners is well illustrated by the history of the East African coast. Here Arabs had established trading settlements at favourable harbours on the coast and nearby islands – for example, Mombasa and Zanzibar [3]. These settlements were part of a large and prosperous Muslim trading system in the Indian Ocean, which at the end of the fifteenth century was taken over by the Portuguese. By the end of the seventeenth century the East African settlements had reverted to Muslim control. Neither the Arabs nor Portuguese in all this time had any contact or influence with any but the peoples actually living on the coast, except in the Rhodesia/Zambesia area where gold spurred inland expeditions.

Human beings – "a most profitable trade"

When the Portuguese first arrived in West Africa, like European adventurers in the New World, they were interested in gold and luxury tropical products. But they and the Dutch, British, French and others who followed them quickly came to realize that there was only one profitable commodity in Africa – human beings [1]. The lands of South and Central America, the Caribbean islands and the southern parts of North America required large labour forces to exploit the silver mines and, more important, the tropical crops – sugar, coffee and cotton – growing there. Within a few decades the transatlantic slave trade came to dominate relations between blacks and whites in western and central Africa.

CONNECTIONS

See also
142 Africa in the 19th century
128 South Africa to 1910
60 African art

In other volumes
228 History and Culture 1

1

1 Although Africa was a major source of various goods, it was primarily the slave trade that established its link with the outside world as well as encouraging trade routes from within it. There was a steady flow of slaves to Muslim lands but most were shipped from west and west-central Africa for labour in the New World.

2 Before they were shipped to the New World slaves were assembled in barracks or "barracoon", where they were treated much as any other freight for shipping. The slave trade was so lucrative that European traders made little effort to develop large-scale trade in other commodities until the 19th century.

2

3

3 The great slave market at Zanzibar became the centre of Arab trading on the east coast after the Portuguese evacuation at the end of the 17th century. Here the Arabs traded with the native rulers who raided and enslaved their weaker neighbours to exchange them for guns and cloth.

4 The inhuman disposition of a human cargo is shown on this plan of the *Brookes*, a vessel of the late 18th century. The Atlantic crossing, the notorious Middle Passage, took a terrible toll of lives and was not as profitable as the sugar, cotton and tobacco carried back to Europe.

4

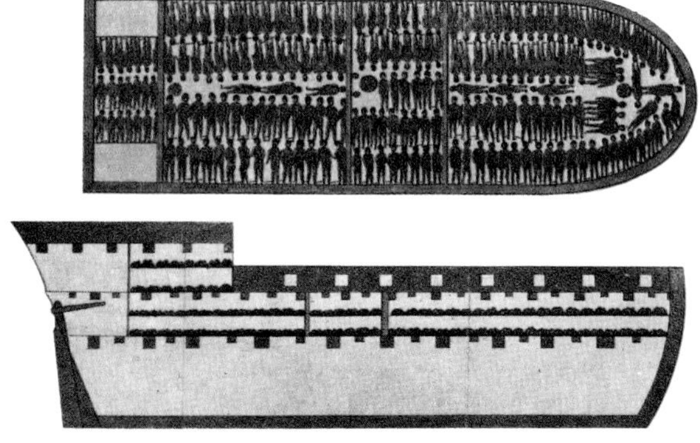

From 1451, when the first cargo in this barbaric trade was shipped across the Atlantic, until the early 1870s, when the slave trade finally came to an end, it has been estimated that almost ten million Africans arrived in the Americas – one of the largest migrations of peoples in history. The peak of the trade was the 50 years from 1760 to 1810, by which time most of the European nations had abolished the slave trade; during these years, over four million Africans were taken away from their homelands.

Most of the African slaves came from the inland regions – from Senegal right round the bulge of West Africa to the Angola region of west-central Africa. The African middlemen, who sold the slaves to the European "factories" on the coast [5], prospered from the trade, as did powerful raiding states such as Asante and Dahomey [6].

In the area of the Congo estuary and Angola, the Portuguese and their agents were more active in venturing inland in search of slaves. In the nineteenth century the East African slave-trading network, which up to then had been in Muslim hands, was tapped to supply slaves to Brazil and Cuba, where slavery persisted longest.

There is no doubt that the slave trade much increased the level of violence among many African peoples. This is especially true of the Niger delta region, neighbouring areas of southern Nigeria, in the interior of Angola and, later in the period, in east-central Africa where Arabs controlled the trade. In a number of instances the resulting breakdown of social order led, ironically, to increased interference by the Europeans who were responsible for the violence in the first place.

Purely African cultures

During the 300 years from 1500 to 1800, in the areas of Africa unaffected by the transatlantic slave trade, societies were developing under their own momentum. New and powerful states emerged, such as Ruanda and Buganda (in the fertile lands between the great lakes of East Africa) and the states farther south, on the plateaus of central Africa. The kingdom of Monomatapa faced the Portuguese in the Zambezi valley and the Rozvi state dominated the plateau.

KEY

Foreign influence – Muslim from the northern interior and European from over the sea – was a striking feature of African history between 1500 and 1800. African societies proved to be flexible but discriminating about these influences, a capacity reflected in their art. The kingdom of Benin, founded about 1400, was a rich source of art, mainly producing bronze sculptures similar to the figure shown here. Benin sculpture was based on a tradition that was more than a thousand years older than the kingdom itself, but the artists managed to assimilate influences from Western culture without departing from this venerable tradition.

5

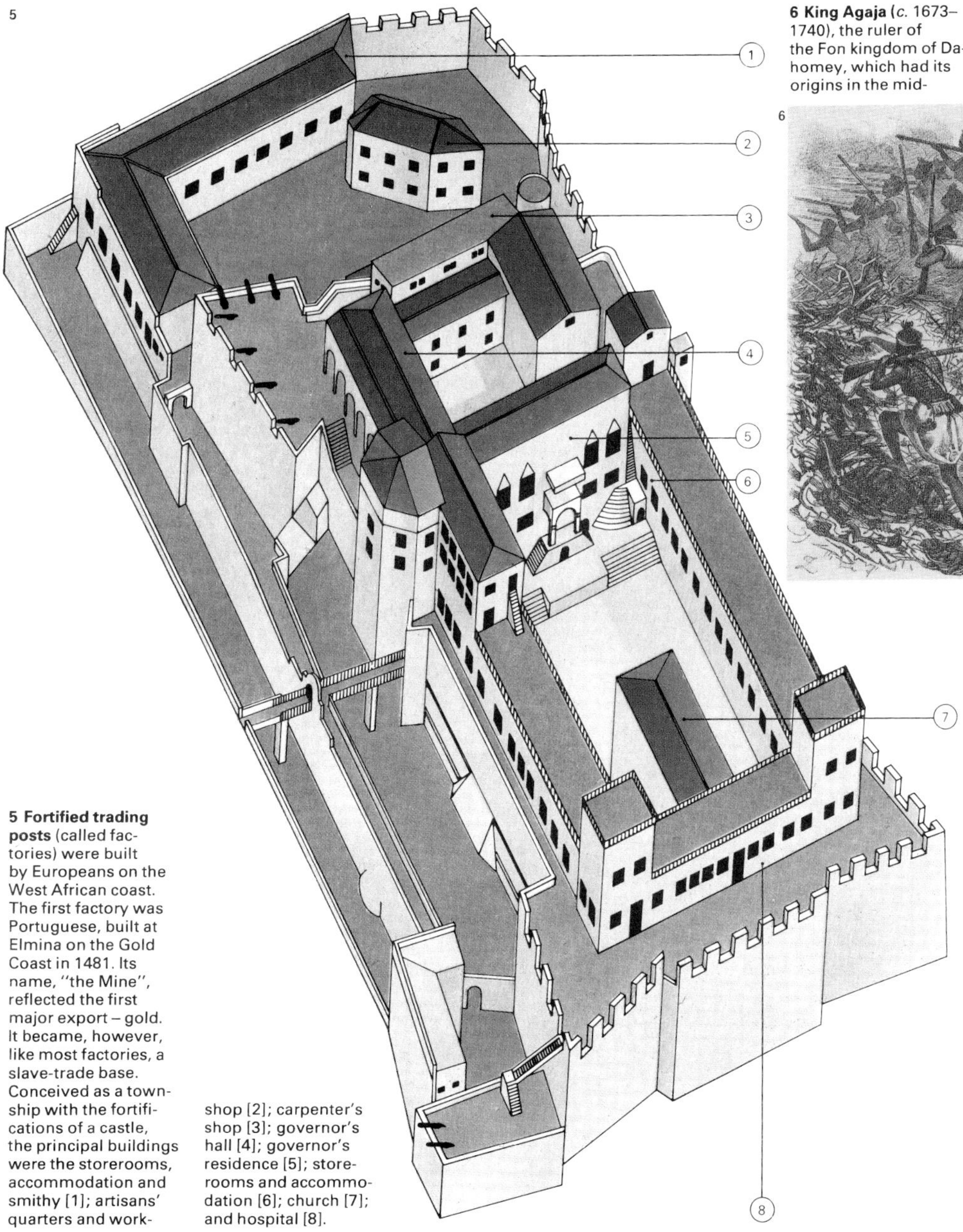

5 Fortified trading posts (called factories) were built by Europeans on the West African coast. The first factory was Portuguese, built at Elmina on the Gold Coast in 1481. Its name, "the Mine", reflected the first major export – gold. It became, however, like most factories, a slave-trade base. Conceived as a township with the fortifications of a castle, the principal buildings were the storerooms, accommodation and smithy [1]; artisans' quarters and workshop [2]; carpenter's shop [3]; governor's hall [4]; governor's residence [5]; storerooms and accommodation [6]; church [7]; and hospital [8].

6

6 King Agaja (*c.* 1673–1740), the ruler of the Fon kingdom of Dahomey, which had its origins in the mid-17th century, controlled a powerful state. He was able to press many people into its service, including these famous units of women soldiers, in the drive to establish the state against its rivals.

7

7 Kano in the north of Nigeria reached the height of its commercial power in the 18th century at the time of a Muslim revival. Usuman dan Fodio, leader of one of the principal Nigerian tribes, the Fulani, founded the Caliphate of Sokoto in 1807 and two years later took Kano.

African art

Present knowledge of the genesis and development of art in Africa is fragmentary and it is often extremely difficult to relate styles even within restricted geographical areas: for example, the bronze castings of Igbo Ukwu in Nigeria bear no stylistic resemblance to the castings of Ife or Benin, despite the proximity of these cultures.

The genesis of Benin art

Benin art of the Niger delta is unique for it is possible to correlate a vast body of works in brass, pottery, iron, ivory and wood with traditions recorded and handed down orally, the evidence being provided by present-day local ritual and political forms and European records since the late fifteenth century. By studying these sources, changes in style over the last five centuries can be dated relatively if not absolutely. The Portuguese, for example, who were the first Europeans to reach and trade with Benin, clearly had a profound effect on Benin art, although they certainly did not introduce the technique of brass casting. Firstly, they made available vast quantities of metal in the form of the bracelets with which they purchased pepper, slaves and ivory. This made possible the enormous expansion of the casting industry in Benin in the sixteenth century. Secondly, by their very presence, and perhaps by affording Benin artists the sight of exotic objects brought from Europe and the Orient in their baggage, the Portuguese traders stimulated them to invent and incorporate new forms and motifs in their art.

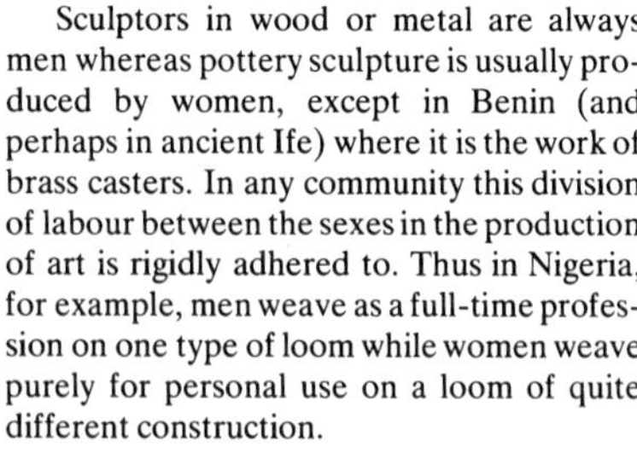

Today, as in the past, African art is the work of individual artists. Metalworking is always a full-time specialization whereas, among some peoples, woodcarving may be essentially a self-taught spare-time activity at which anyone may try his hand and which brings a man no special status in the community. Where, however, a woodcarver has been apprenticed to a master for several years he may belong to a professional guild and work more or less full-time at his art. A talented carver is likely to be recognized in his community. His work may be well known over a relatively wide area and it may be readily distinguishable from the less eminent work of others.

Sculptors in wood or metal are always men whereas pottery sculpture is usually produced by women, except in Benin (and perhaps in ancient Ife) where it is the work of brass casters. In any community this division of labour between the sexes in the production of art is rigidly adhered to. Thus in Nigeria, for example, men weave as a full-time profession on one type of loom while women weave purely for personal use on a loom of quite different construction.

Ephemera and the unfinished object

In addition to sculpture in wood, metal or clay, there are all kinds of less enduring forms in fabric, beads, basketry, feathers, leaves, wax and so on, which are often *ad hoc* creations of a completely ephemeral character. Even wooden sculptures are likely to be decorated and re-decorated by their owners using beads, seeds, bits of mirror, paint and so forth. Indeed the objects can hardly be described as finished when they leave the carvers, being liable to a variety of subsequent subtle transformations. This may perhaps reflect a greater preoccupation with process

CONNECTIONS

See also
58 Subsaharan Africa

In other volumes
228 History and Culture 1
234 History and Culture 1

1 Paintings and engravings on rocks all over the Sahara, like this example at Tassili, provide the earliest evidence of African art. They show a continuous development over the past 5,000 years in four main periods. In the earliest, wild animals are depicted, perhaps indicating a hunting way of life. In subsequent periods cattle, horses and camels are introduced. Rock painting is also found elsewhere, especially in southern Africa where it is probably the work of Bushmen. Murals are a common feature throughout Africa, and are to be found in the houses and shrines of most tribes.

2 The earliest known sculptures in subsaharan Africa are the pottery heads and figures of the Nok culture (*c.* 500 BC–AD 200) from the centre of northern Nigeria. Both human and animal figures have been found. Some of the heads are near life-size and are clearly broken from correspondingly large figures. The ability to fire large and complex pottery sculptures indicates a fairly high level of technology, which is shared only by the ancient Yoruba town of Ife (11th–14th centuries AD). The Nok culture also provides the earliest evidence of iron working in West Africa.

3 Iron working follows directly upon that of stone in subsaharan Africa. The earliest evidence of bronze casting comes from the Ibo village of Igbo Ukwu, Nigeria, where ninth century AD regalia and ceremonial objects such as this cylindrical stand have been excavated.

4 This brass figure was cast in ancient Ife although it was found far to the north in a village on the Niger. Several heads and figures of brass and pottery in this extraordinarily naturalistic style have been excavated at Ife together with other artefacts made of stone and glass.

5 This pottery head was found on Luzira Hill near the northern shore of Lake Victoria in Uganda. Nothing is yet known about its age or the people who made it. The unfired mud statuary of the Edo and Ibo peoples may be a related tradition. Pottery sculpture is found all over Africa, dating both from the recent past and from antiquity. Among the many ancient cultures in which pottery sculpting flourished was the Nok culture, Ife, and the Sao culture which grew up in the Lake Chad region.

rather than with purely static form in art.

It is often wrongly said that there is no art for art's sake in Africa. Of course, a sculpture may represent an ancestor, a god or one of its devotees, or even some impersonal magic form that is manipulated for curing the sick; but it can just as well be an ornament for a rich man's house. Masks serve to disguise someone impersonating a spirit or an idea, yet while some masked figures are charged with power or ritual significance [Key, 9], others are merely "costumed" entertainers devised for the delight of their owners.

African sculpture is, broadly speaking, concentrated around the two great river systems of tropical Africa, the Niger in West Africa and the Zaire or Congo in Central Africa. This curiously uneven distribution is partly explained by the distribution of the appropriate raw materials. Another factor is the more settled type of economy in these areas, whether agricultural or pastoral. People who are always on the move with their livestock, carrying all their belongings with them, are unlikely to see any need for sculpture. But even among some agricultural communities there are some for whom sculpture is unnecessary within the terms of their culture. In both West and Central Africa there are peoples who produce no sculpture either because they do not need it or because they are satisfied with the work acquired from neighbouring peoples. Equally, although a people may not have any sculpture of the kind that finds its way to a museum or art gallery, they may have a rich tradition of completely ephemeral forms that are dismantled immediately after use.

A multiplicity of art forms

It would be wrong to assume that because people have no sculpture they have no art. Sculpture is, after all, only one among many arts that flourish in Africa, including architecture, mural decoration, textile design, pottery, leatherwork, embroidery, basketry and, perhaps the most universal art of all, the decoration and adornment of the human body [10]. To the rich and varied visual arts of Africa must be added music and song, a unique gift for dancing, poetry and other forms of oral literature.

KEY

The art of Benin in Nigeria was created exclusively for the glorification of the king. The main forms of brass casting were memorial heads for the altars of dead kings and rectangular plaques mounted, in the 16th and 17th centuries, on wooden pillars in the palace. The early 16th-century ivory mask illustrated here was worn by the monarch as part of his ceremonial dress. Around the top of it is a tiara of Portuguese heads. The Portuguese first visited Benin in 1486. The clear derivation of Benin artistic tradition from that of ancient Ife disproves the theory that the Portuguese introduced brass casting to the Benin people. Ife long predates their arrival in Africa.

6 This pair of wooden doors was carved in 1915 for the palace of the king of Ikere-Ekiti in Yomtaland, Nigeria. Made by a famous sculptor, Olawe of Ise, they show the king receiving his first British administrator at the turn of the century. Wooden objects do not survive for long in Africa, with the result that much of the finest sculpture known was carved only within the past one hundred years.

6

7

7 Every king of the Bakuba people of Zaire had his portrait carved in wood during his installation rites, to house his spirit double. This custom was introduced by Shamba Bolongongo, 93rd king of the Bakuba, *c.* AD 1650: his portrait is shown here. Shamba Bolongongo was a great innovator, for he introduced weaving and textile manufacture to his people as well as initiating the custom of wood portraiture.

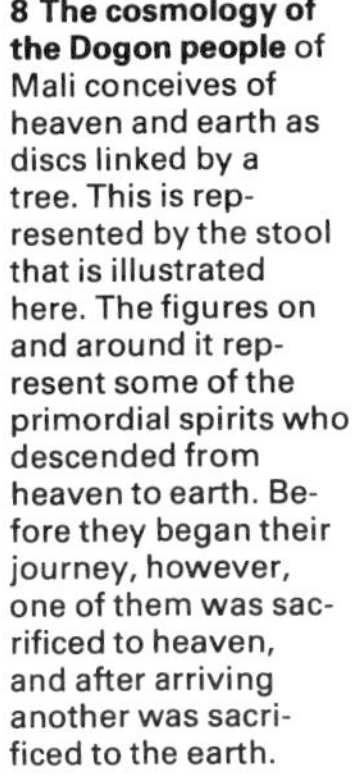

8 The cosmology of the Dogon people of Mali conceives of heaven and earth as discs linked by a tree. This is represented by the stool that is illustrated here. The figures on and around it represent some of the primordial spirits who descended from heaven to earth. Before they began their journey, however, one of them was sacrificed to heaven, and after arriving another was sacrificed to the earth.

8

9

9 These masks are worn during initiation rites among the Bwa or Bobo-Ule people of Upper Volta. They apparently represent particular animals, birds and spirits, the symbolic associations of which provide some indication of the meaning and value of the rites themselves. Conversely, masks also derive meaning from the wider context of the performance – which includes costume, movement, sacrifice, prayer and song – of which they are but a part.

10

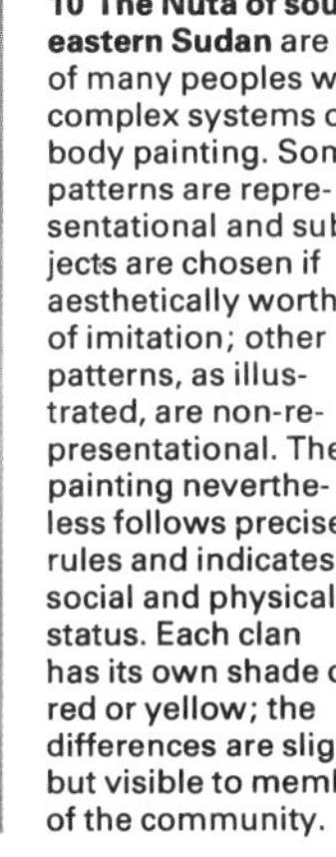

10 The Nuta of south-eastern Sudan are one of many peoples with complex systems of body painting. Some patterns are representational and subjects are chosen if aesthetically worthy of imitation; other patterns, as illustrated, are non-representational. The painting nevertheless follows precise rules and indicates social and physical status. Each clan has its own shade of red or yellow; the differences are slight but visible to members of the community.

Settlement of North America

The growth of the English economy in the late Middle Ages was achieved through increasing mastery of the seas. Between 1400 and 1600 English seamen ranged ever farther into the Atlantic, to Iceland, Greenland, Labrador and the northern seaboard of what is now the United States of America. Their search was primarily for fish. Discovery was a long and often discontinuous process; at times the English led the way, at others they trailed behind the Spaniards, Portuguese and French. Eventually the greater part of North America was to fall to the English, while Spain held the stronger empire in Central and South America, but the process of resolution was understandably slow.

The first emigrants from Europe

The settlement of the southern United States began in the sixteenth century: the first permanent city in North America – St Augustine, Florida – was founded by the Spaniards in 1565. They had explored and conquered the densely populated empires of Mexico and Peru (the population of Aztec Mexico when they arrived is said to have been as great as that of Western Europe). The English and French in the following century went to the West Indies and North America where they found vast, sparsely populated lands inhabited by semi-nomadic peoples living at subsistence levels. After 1700 free migration, as distinct from the importation of black slaves, was nearly all into the English colonies of the eastern seaboard, although most of these many new migrants were Scots, Irish, Germans or Swiss.

The first serious attempt to found a permanent English settlement on North American soil was made by Sir Walter Raleigh (1552–1618) at Roanoke Island off the coast of Virginia in 1584. Not all of the experience gained in voyages to and from this colony during the next six years was happy; some of it was indeed tragic, for the first settlers mysteriously vanished without trace. Raleigh's venture was partly a strategic move in the long sea war between England and Spain and, when his colony perished, the shoreline north of Spanish Florida was left open to other European powers.

The next attempt to establish an English colony in the area, the Jamestown settlement – established by the Virginia Company in 1607 [3] – was basically a commercial venture, although the aims of the company included helping to build a strong merchant fleet, training mariners for England's protection, spreading the gospel and planting a Protestant colony in a land still threatened by Catholic Spain.

Principal reasons for settlement

Trade and religion were the two principal motives for the founding of North American settlements [7]. Religious enthusiasts, hampered at home by the Inquisition in Spain and the Court of High Commission in England, were sometimes willing to venture into the unknown, but without the prospect of trade with Europe they could survive only in subsistence conditions. During the 50 years following the foundation of Jamestown, further colonies were established, mostly by the English. Plymouth was established in 1620 by the Pilgrim Fathers, who sought religious and civil autonomy from the English government, and Maryland by Lord Balti-

CONNECTIONS

See also
64 The American Revolution
18 England under the Hanoverians
46 British colonial policy in the 18th century

In other volumes
222 History and Culture 1

1

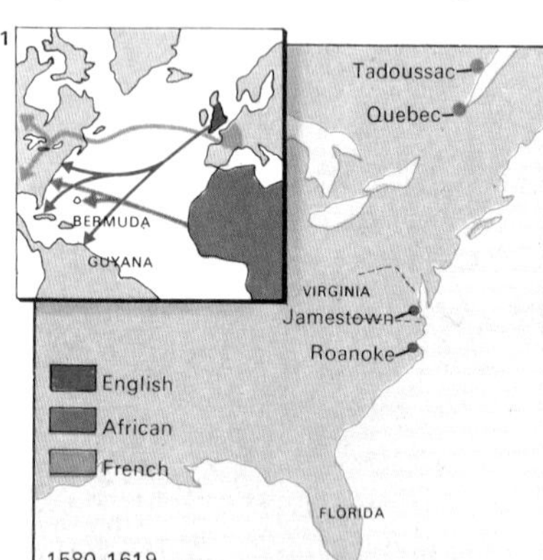

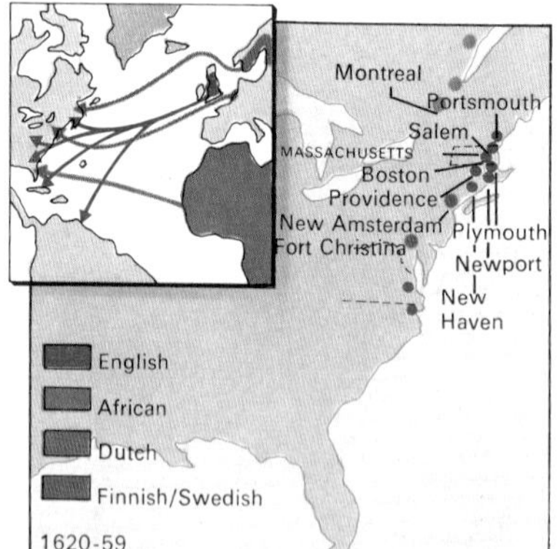

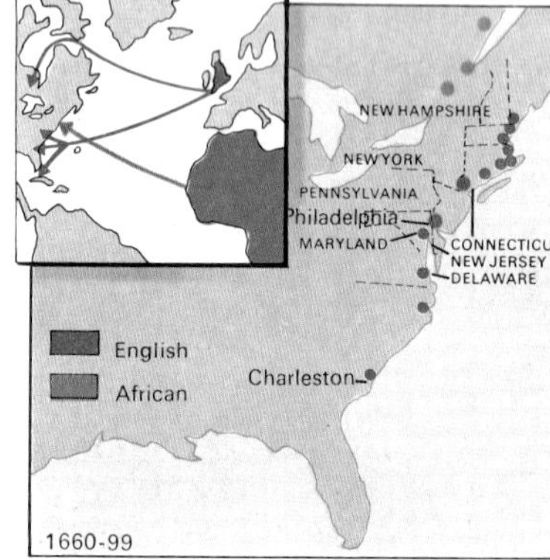

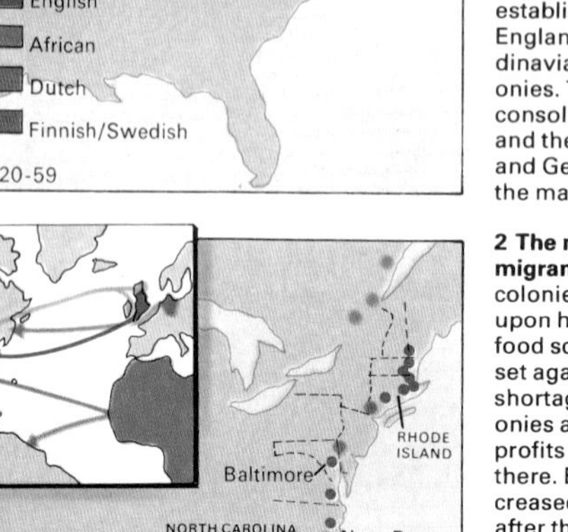

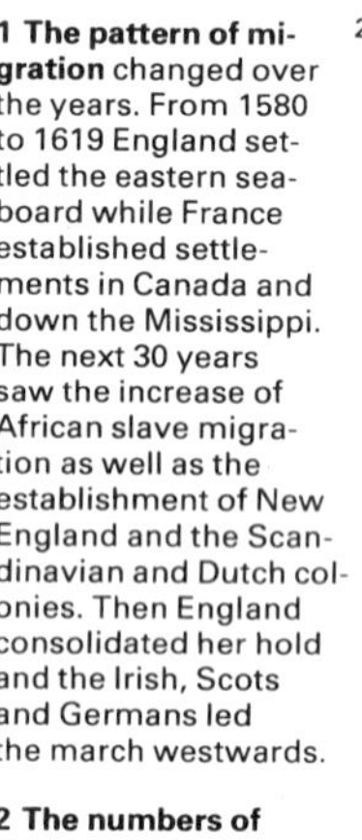

1 The pattern of migration changed over the years. From 1580 to 1619 England settled the eastern seaboard while France established settlements in Canada and down the Mississippi. The next 30 years saw the increase of African slave migration as well as the establishment of New England and the Scandinavian and Dutch colonies. Then England consolidated her hold and the Irish, Scots and Germans led the march westwards.

2

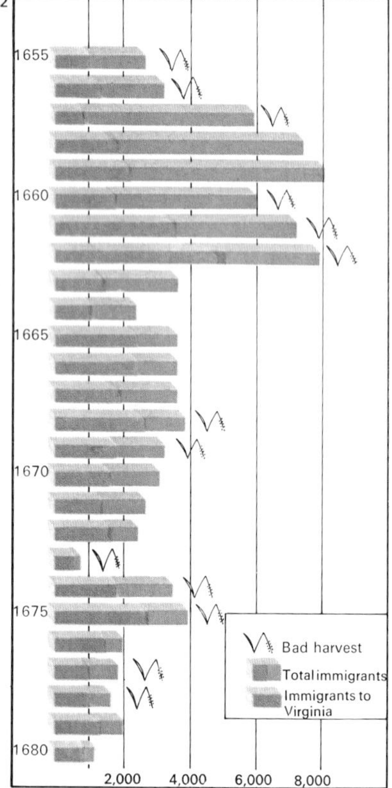

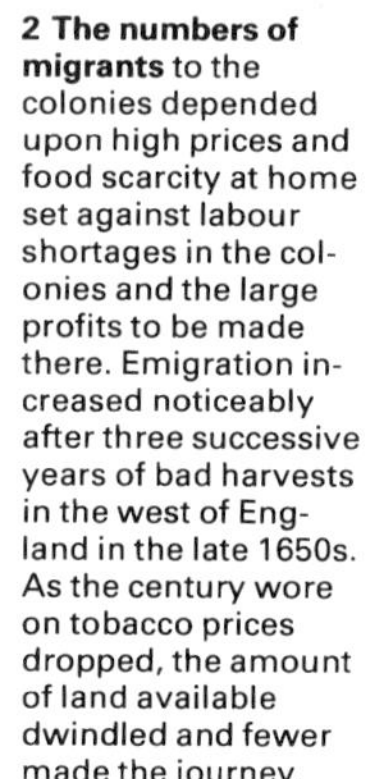

2 The numbers of migrants to the colonies depended upon high prices and food scarcity at home set against labour shortages in the colonies and the large profits to be made there. Emigration increased noticeably after three successive years of bad harvests in the west of England in the late 1650s. As the century wore on tobacco prices dropped, the amount of land available dwindled and fewer made the journey.

3

3 James I (1566–1625) granted charters to some merchants to colonize the eastern seaboard of North America. The London Virginia Company was allocated what is now Virginia and Maryland and the Plymouth Company the coast of New England as far as Maine. This company's charter was revoked and a royal colony, whose council's seal is shown, established in 1624.

4

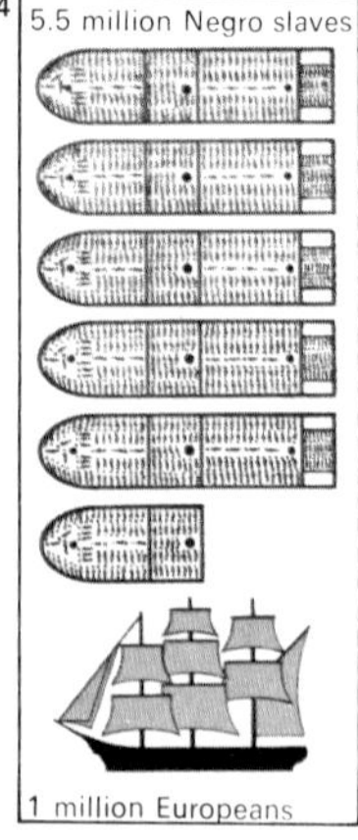

4 Six and a half million people had crossed the Atlantic to the New World by the 1770s. One million whites came from Europe – mostly from England, France, Germany and Spain; the other five and a half million were Negro slaves from West Africa, who were transported in appallingly cramped conditions in the slave ships. Chained flat to the decks they could cause little trouble and needed less food, thereby maximizing the profits of the traders.

5

5 Tobacco introduced to Virginia in 1612 became the main export by 1619 and with cotton was to remain the staple product of the Southern states, despite the repeated efforts of successive English governments during the colonial period to diversify their economies. The Northern states, at first a major source of furs and timber, developed their mineral resources, notably coal and iron, from the 18th century onwards, thus laying the basis for their early industrial development.

more, for Roman Catholics, in 1632. In 1625 the Dutch founded New Amsterdam, later renamed New York, as a trading post, to be followed by the Swedes and Finns.

French beginnings in North America stemmed from the trading activities of fishermen and fur trappers who established trading posts along the St Lawrence waterway. Then Samuel de Champlain (1567–1635) founded Quebec in 1608 and in less than 30 years the French had established posts as far west as Wisconsin. By 1660 a portage route from Lake Superior to Saskatchewan had been located and by 1720 New Orleans had been founded to guard the mouth of the Mississippi. Thus by the mid-eighteenth century the French had occupied, albeit sparsely, the whole of middle America, threatening the expansion of the English.

Influence of European events

Meanwhile, the Spanish North American empire, which included the whole coastline of the Gulf of Mexico as well as Florida, blocked English expansion to the south. But it was mainly events in Europe in the shape of the Seven Years War (1756–63) [11] that were to weaken France and Spain and to allow the English to fill the vacuum these two nations left in America. When George III (1738–1820) came to the throne of England in 1760, the French were confined to eastern Canada and the Great Lakes, while Spanish territory, vast in area although virtually unoccupied, stretched from Panama almost to the Canadian border west of the Mississippi. With the Treaty of Paris (1763) France lost all her North American possessions to Britain with the exception of the small island group of St Pierre et Miquelon, while Spain ceded Florida. Fearing a resurgence of French and Spanish power, however, the English set up a buffer zone west of the Alleghenies and east of the Mississippi.

Having consolidated their position, the British determined to exploit their possessions in North America. But it was the unwillingness of these colonies, now 13 in number, to submit to taxation without representation in Parliament that led to the American Revolution and Declaration of Independence in 1776.

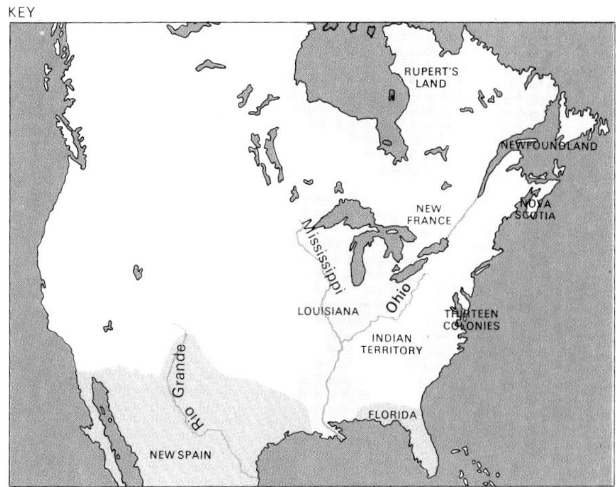

Conflicting claims to the North American continent were the subject of intense and bitter rivalries between France, Spain and England throughout the 18th century. By the Treaty of Paris England gained all France's important North American possessions; Spain was too much weakened to assert her claims.

6

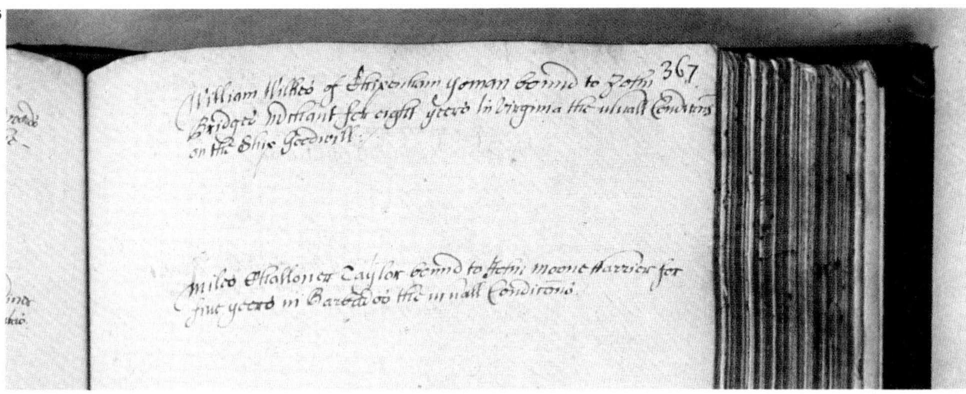

6 Indentured servants made up a large part of the total number of early emigrants. Orphans, petty offenders, political and religious prisoners, younger sons of impoverished landowners and young men and women who possessed a taste for adventure and a better life, bound themselves, or were bound for a term of years, to work for a planter in Virginia or the West Indies. In theory they were taught to become planters themselves and at the end of their term, usually four or five years, they were allowed to go free and were given 20 hectares (50 acres) of land and other essentials to start up on their own. The indentures shown were recorded at Bristol, July 1660. The first reads: "William Wilkes of Chipenham Yoman bound to John Bridges Merchant for eight years in Virginia the usual conditions on the Ship Goodwill".

7

7 Many Quakers left England in the late 17th century when they conflicted with laws passed at the restoration of Charles II on questions of worship, freedom from oaths and military service.

8

8 John Harvard (1607–38), an English clergyman and graduate of Cambridge University, founded Harvard College at Cambridge, Massachusetts, in 1636 within six years of the establishment of that colony.

9 Pocahontas (1595–1617), the daughter of Powhatan, an Indian chief in Virginia at the time the white man came, became a Christian and married John Rolfe, a prominent settler. This provided a period of peace.

9

10

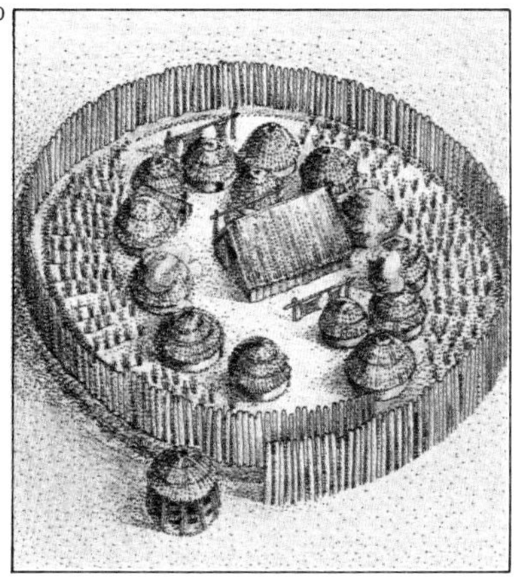

10 Indian villages, the homes of semi-nomadic hunters, bordered the rivers that flow into Chesapeake Bay and the creeks and inlets of New England. The early settlers bartered beads and trinkets for large tracts of land, much of it already cleared for cultivation, thus beginning the relentless process of Indian dispossession. Ports such as Baltimore and Fredericksburg were established around Chesapeake Bay by the 18th century.

11

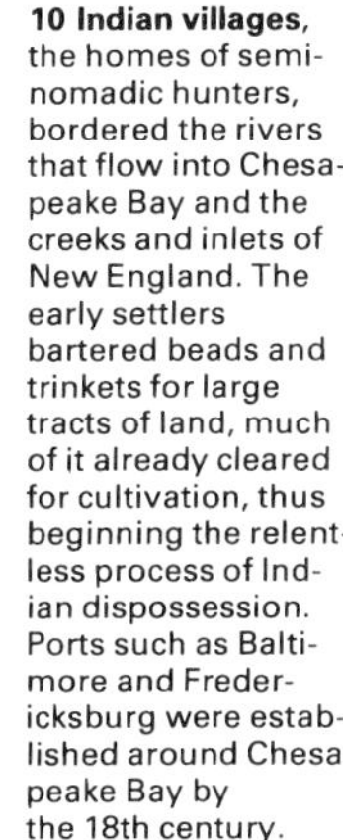

11 The British and French clashed on numerous occasions in the Seven Years War (1756–63). Regiments on both sides adopted uniforms designed more for splendour than efficiency or camouflage. Shown here are a trooper of the 10th British dragoons and an officer of the Regiment de Saint Germain. A significant part of the war was fought in North America, ending in defeat for the French. The 1763 Treaty of Paris that ended the war vastly increased Britain's territory in America.

The American Revolution

The American Revolution was both a rebellion and an act of nation-building. It was a political upheaval in which Britain's 13 Atlantic coast colonies in America gained their independence and formed the embryonic United States. The revolution was also the first national struggle in modern times for the rights of the individual and the establishment of democratic government.

The British colonies

The Treaty of Paris of 1763, which ended 70 years of colonial wars between Britain and France, gave the British complete victory in North America and control over vast new territories in Canada and as far west as the Mississippi. It caused fundamental changes in attitude both in Britain and in the 13 colonies. The colonists were now rid of the great external threat that had made them rely on Britain for defence. Since Britain was spending large sums to defend the new territories, it was felt that the terms of trade with the colonies should be revised so as to improve their profitability and to increase the local contribution to defence.

To achieve this a Sugar Act was passed in 1764 and a Stamp Act in the following year; and wider use was made of Admiralty courts in their enforcement [1]. The colonies reacted strongly, demonstrations and rioting broke out, and a congress was called in New York which defined the major objections: first that the acts had been imposed by the British Parliament in which the colonists had no representation, and secondly that the colonists, like all British subjects, should have the right to trial by jury, not by arbitrary courts. Such was the opposition that the Stamp Act was repealed in 1766. But in the same year, a Declaratory Act was passed which asserted that Britain had the right to legislate for the colonies if it so wished.

A year later, this right was put into force with a series of acts taxing glass, lead, paper and tea. Widespread unrest followed, climaxing with the "Boston Massacre" [4] in 1770. Most of the acts were repealed, but in 1773 another Tea Act was passed giving favourable trading terms to the East India Company. The colonists again objected and at the "Boston Tea Party" a cargo of tea was dumped into the harbour [3]. In Britain, acts were passed putting the government of Boston under direct British control.

First Continental Congress

When this became known, representatives of the colonies (except Georgia) met in 1774 at the First Continental Congress in Philadelphia [2], where a petition was drafted insisting that there should be no taxation without representation. The Congress also prepared an association between the colonies which would regulate their own trade. The British government, led by Lord North (1732–92), replied that a state of insurrection now existed in the colonies. Both sides prepared for war.

The first fighting took place on 19 April 1775 when Massachusetts militiamen fired on British troops at Lexington and Concord. An attempt by the militia to prevent the British improving their defences around Boston led to the Battle of Bunker Hill [7] on 17 June. A Second Continental Congress met and established an army with George Washington (1732–99) as its commander. As

CONNECTIONS

See also
62 Settlement of North America
46 British colonial policy in the 18th century
68 Pitt, Fox and the call for reform
88 The Industrial Revolution
148 USA: the opening up of the West
74 The French Revolution
94 The British labour movement to 1868
26 The English in Ireland

1

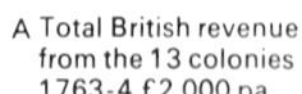

1 During the 17th century, the British colonies in North America had had the right to tax themselves embodied in their charters and had thwarted attempts by the British to obtain any more revenue from them. But in 1763, faced with heavy debts and the need to support a standing army in North America, Britain tried to relieve some of the burden by imposing a series of taxes on the colonies without consultation. The taxes fell far short of Britain's revenue expectations but they aroused the colonists in defence of their traditional rights and "Taxation without representation is tyranny!" became a rallying cry of the revolutionaries.

2

2 The Continental Congress, which met in Philadelphia on 5 September 1774, was a gathering of delegates from 12 colonies (Georgia did not attend until the following year) called to prepare a declaration condemning British actions. There was little talk of independence, but the government in Britain reacted strongly, treating the actions of the Congress as rebellion. When the Second Congress met a year later, fighting had broken out and it was rapidly accepted as the effective governing body of the rebels. Although it had no statutory powers, it managed to maintain its position of leadership. It was the Congress that took the vital steps to issue the Declaration of Indepedence and to move towards a federal constitution.

3

3 On 16 December 1773 about 50 colonists disguised as Indians boarded three British ships in Boston harbour and dumped their cargoes of tea overboard to discourage enforcement of a tea tax. British reprisals, including a commercial blockade of Boston, led to the calling of the Continental Congress.

4

4 The "Boston Massacre" was the first violent clash between colonists and British troops. Three men were killed and two seriously wounded when troops who had been jeered at and attacked by a Boston crowd, opened fire without orders.

royal government collapsed, the Congress took over as the governing body.

On 4 July 1776, the Congress institutionalized the break with Britain by passing the Declaration of Independence, which gave a valuable boost to American morale, but had little immediate effect on the precarious position of the ex-colonies with their coasts and trade blockaded by British sea power and with their small, ill-trained forces faced by professionals [5]. However, the British commanders made only fumbling attempts to seize the initiative and a force under General Burgoyne (1723–92) was surrounded and forced to surrender at Saratoga.

Victory for the colonists

This victory was crucial in persuading France to send a fleet to help the Americans in April 1778 and to declare war on Britain in July. With their naval communications now threatened, the British fell back from Philadelphia, and Washington was able to contain them around New York. The British then attempted to switch the centre of the war to the southern states of Georgia and South Carolina. Meanwhile, Washington was working steadily to build up the strength of his army, and when an expedition led by General Cornwallis (1738–1805) attempted to link up with British forces in the north, it was cut off and forced to surrender at Yorktown on 19 October 1781 [6].

This defeat convinced the British that the war must be ended. Negotiations were begun in Paris with an American delegation led by Benjamin Franklin (1706–90) and John Adams (1735–1826) and peace was formally ratified in September 1783.

Immediately after hostilities ended, steps were taken to forge a sense of American nationalism from the shaky wartime unity of the now independent states. A federal constitution, drawn up in 1787, became effective in 1789. A Bill of Rights was added in 1791 to protect the rights of individuals.

The success of the revolution encouraged and inspired democratic and libertarian movements elsewhere in the world during the following decades, particularly in Europe and notably in France, where revolution took place a few years later.

KEY

1 New Hampshire
2 New York
3 Massachusetts Bay
4 Connecticut
5 Rhode Island
6 Pennsylvania
7 New Jersey
8 Maryland
9 Delaware
10 Virginia
11 North Carolina
12 South Carolina
13 Georgia

Ticonderoga, Saratoga, Concord, Boston, New York, Philadelphia, Brandywine, Yorktown, Charleston, Savannah

0 200km

The 13 colonies in America were the seeds from which the United States grew. Resentful of British taxes and repressive measures, stirred by the attractions of liberty and independence, they joined together in 1776 to declare themselves an independent nation.

5 British "redcoats" were well-trained, professional soldiers who were generally superior in conventional battles to the imperfectly trained American volunteers. It was George Washington who kept the armies in existence despite repeated disappointments and who used the American skill in guerrilla tactics to wear down the British until they could be outmanoeuvred.

6

6 Lord Cornwallis surrendered to George Washington at Yorktown on 19 October 1781. His troops were trapped between superior American forces on land and a French fleet at sea and their defeat broke the will of the British.

7

7 At the Battle of Bunker Hill, outside Boston on 17 June 1775, the Americans twice drove back British assaults before retreating. The first major battle of the revolution, it was an expensive British victory, in which the Americans proved that they could fight.

8

8 Scottish-born John Paul Jones (1747–92) took the revolution to sea by raiding British shipping. Called upon to surrender when his vessel *Bonhomme Richard* was battered by HMS *Serapis*, Jones replied, "I have not yet begun to fight" and went on to capture *Serapis*.

9 A primary objective of the Constitution was to establish a balance of power between the executive (the president), the legislature (Congress) and the judiciary (Supreme Court) to prevent the emergence of tyranny. Much political power was reserved for the states, represented in Congress by senators.

9

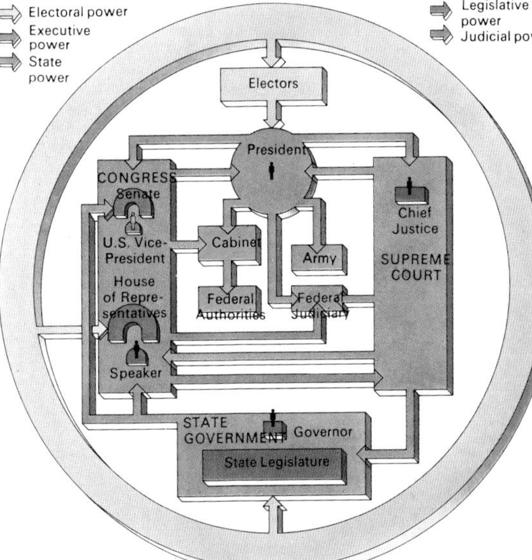

10

10 Thomas Paine (1737–1809) emigrated to Philadelphia from England in 1774 and soon became one of America's most influential revolutionaries. His pamphlet *Common Sense* and his *Crisis* papers profoundly stirred popular sentiment in the country with their impassioned pleas for liberty, condemnation of tyranny and powerful arguments favouring American independence. His tracts were often read to American soldiers to bolster morale during the war.

The early Industrial Revolution

Britain was the first industrial nation in the world. From the middle of the eighteenth century, a number of factors launched Britain into a period of self-sustaining economic growth by the first decade of the nineteenth century. However, the origins of the Industrial Revolution in Britain lay in the pre-industrial period; by the middle of the eighteenth century there was already a thriving commercial economy, with a growing population, developing agriculture, and expanding trade both at home and abroad.

Population growth

The growth of Britain's population from the mid-eighteenth century was not directly caused by industrialization although a large workforce was an essential factor in the development of industry. A run of good harvests in the first half of the century, low food prices, favourable climatic conditions, the decline of plague and a number of minor improvements in health all contributed to lower death rates and a consequent rise in population [2]. By the end of the eighteenth century, birth-rates began to rise, too, as people in the industrial towns were able to marry earlier and to have, and keep, more children. Unlike Ireland, where population growth led to impoverishment and, ultimately, to famine, Britain's commercial and agricultural prosperity meant that a growing population contributed to increasing demand for products of every kind. Increased consumption was a stimulus to industrial innovation and methods of production.

In the past, periods of agricultural expansion had been checked by harvest failure, population level and economic downturn. By the middle of the eighteenth century, the profits of thriving overseas trade enabled landowners to borrow capital to increase agricultural production [5]. With increasing demand and prices for foodstuffs, agricultural expansion followed. The enclosure movement grouped the old open fields and common lands into individual, more efficient units, on which more productive techniques could be applied, such as improved animal husbandry, new root crops and the first agricultural machines. Enclosure, secured through Parliamentary Acts, had affected about 20 per cent of the area of England by 1845. Capital was required to make the most of enclosure and it led to many smaller farms being amalgamated into larger holdings. Contrary to common myth, enclosure did not depopulate the countryside, but often increased the demand for agricultural labour.

Increased demand

The continued profitability of foreign trade [1], particularly as the colonies grew, provided the capital for increases in production to meet demand at home and abroad. One of the first industries to feel this increased demand was mining, with the need for more domestic and industrial fuel. Output was increased 400 per cent in the course of the eighteenth century through the use of steam pumping engines to keep mines from flooding. Coal was an important raw material for many industrial processes as well as the fuel for steam power. Coal and iron together laid the foundations for the development of industry [4]. The iron industry of the early eighteenth century depended on charcoal for smelting and had a relatively small output.

CONNECTIONS

See also
88 The Industrial Revolution
18 England under the Hanoverians
20 The agricultural revolution
90 The urban consequences of industrialization
92 The rural consequences of industrialization
94 The British labour movement to 1868
32 Exploration and science 1750–1850
158 Industrialization 1870–1914
16 Europe: economy and society 1700–1800
44 International economy 1700–1800

In other volumes
312 Man and Society
22 Man and Machines
30 Man and Machines

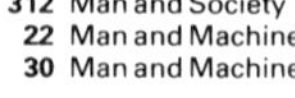

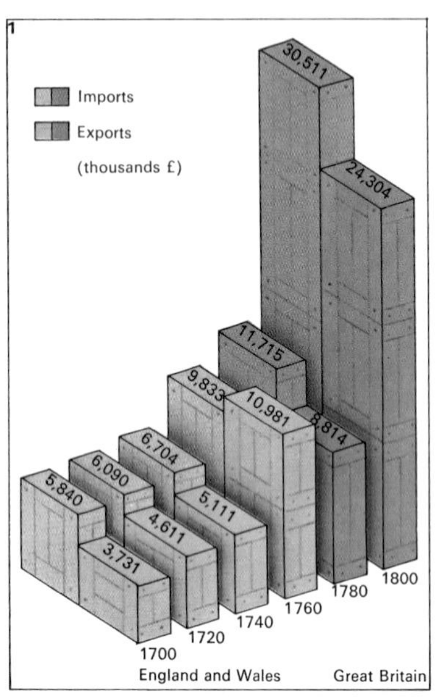

1 Industry was stimulated by growing demand, both at home and abroad. Britain's overseas trade experienced a rapid expansion from the 1680s, providing new market opportunities and the capital for investment in new techniques. New colonial markets acquired after the Seven Years War proved lucrative, as Britain engaged in the "Triangular Trade" carrying factory goods to Africa and the West Indies, transporting slaves across the Atlantic and bringing back colonial produce to Europe. Britain's largest export commodity in the first half of the 18th century was woollen textiles, but this was later overtaken by cotton.

2 Europe's population increased from the 1750s, and despite some appalling conditions in towns (here shown at one extreme in one of William Hogarth's Gin Lane pictures), mortality rates declined. The cause of this is not fully understood but may have been related to the end of plague epidemics after 1700 and improvements in hygiene after 1800, such as the availability of cheap soap, easily washable cotton clothing and improved water supply. Increased population because of earlier marriage and larger families provided a growing market for cheap industrial products and also the necessary ready supply of labour.

3 Mills driven by water provided the motive force for many processes before the Industrial Revolution, including grinding corn and spinning yarn. A flourishing woollen industry already existed in areas where water power was readily available, such as the Cotswolds, East Anglia and the West Riding of Yorkshire. Many early machines could be driven by water power and the first phase of industrialization was based almost entirely upon the use of water-driven machinery. Both the cotton and woollen industries developed on the slopes of the Pennines with abundant water power. It was only with the development of efficient steam power after 1776 that industry began to concentrate upon the coalfields and no longer had to depend on the hilly regions. The use of coal and invention of coke-smelting enabled industry to expand and escape the problems of a critical shortage of wood for fuel. However, the change to steam was gradual.

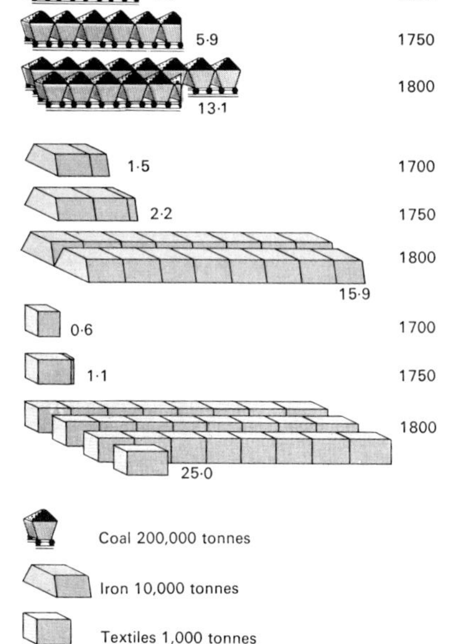

4 The most striking developments in 18th-century industry were shown in coal, textiles and iron production. Coal mining expanded with the rise of steam power, a growing population and improvements in communications. Wool output increased to meet domestic and foreign demand, but mainly using traditional processes. Cotton production grew dramatically with the use of machinery and steam power until it became Britain's principal export commodity. Iron production also increased rapidly with the introduction of coke-smelting. These developments were evidence of a broad expansion of techniques to meet opportunities presented by rapidly growing markets.

The discovery by Abraham Darby of coke-smelting at his Coalbrookdale works in the 1730s revolutionized the production of cheap iron and enabled it to be used in the first machines and iron structures.

Allied to these developments, there was a major advance in technological power following the patenting of the improved Boulton and Watt steam engines after 1774. They used much less fuel than earlier models. Beside pumping, Watt's steam engine of 1769 was harnessed to drive machinery.

Labour-saving machinery

After steam power, the most important innovations were associated with the growth of labour-saving machinery. They occurred most dramatically in the cotton industry, which witnessed technical breakthroughs in weaving (Kay's flying shuttle, 1733) then in spinning [Key] and gradually in other processes. The harnessing of steam power to machinery in the cotton industry led to the first factories in which the production processes were concentrated under one roof [7]. Although many factories still relied on water power [3], the development of the factory system in cotton foreshadowed the growth of the factory and the use of steam in other industries. Woollen production, for example, expanded mainly by using traditional methods such as water power. Gradually, however, the introduction of machinery and the use of steam power drew it towards the coalfields of the West Riding of Yorkshire.

Concentration of production needed both capital and cheap transport. Capital was provided out of the profits of agricultural improvement and overseas trade. Country banks, although subject to panics and bankruptcies, did provide a basic network of credit for industrial and agricultural development. By 1800 there were about 70 London banks and about 400 country banks, usually issuing their own notes. The Stock Exchange was founded in 1773.

Land transport remained slow and expensive for bulky products, in spite of the development of turnpikes. River transport was cheaper, but it was only with the development of the canal network that bulky products could be moved cheaply [6].

KEY

The use of machinery during this period greatly increased the production of goods. James Hargreaves's spinning jenny (1764) increased output of spun cotton.

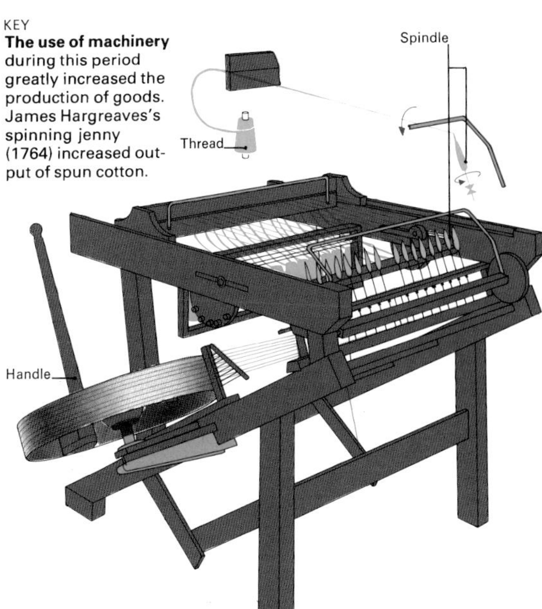

5

6

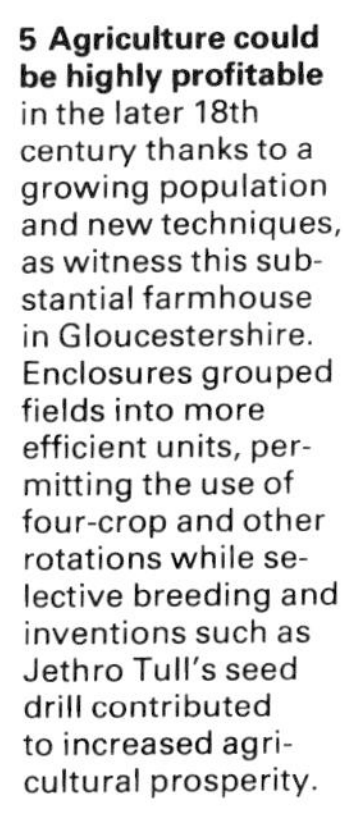

5 Agriculture could be highly profitable in the later 18th century thanks to a growing population and new techniques, as witness this substantial farmhouse in Gloucestershire. Enclosures grouped fields into more efficient units, permitting the use of four-crop and other rotations while selective breeding and inventions such as Jethro Tull's seed drill contributed to increased agricultural prosperity.

6 Transport developments played a vital part in the Industrial Revolution by widening markets and allowing production to be concentrated where goods could be brought by cheap bulk transport. Turnpikes and improved road surfaces increased passenger traffic by road, but the most important advance for industry came with the development of canals. The Bridgewater Canal between Worsley and Manchester was built for the Duke of Bridgewater by James Brindley (1716–72), an engineer who remained illiterate until his death. The canal, opened in 1761, halved the cost of coal in Manchester by reducing transport costs. In the "canal mania" that followed, an extensive canal network was built up and many early industries were based on it, giving them access to raw materials and markets.

7

7 A pioneer of the factory system, Sir Richard Arkwright (1732–92) built this cotton mill at Cromford, Derbyshire, which Joseph Wright of Derby painted in the 1780s. The first factories were built for the textile industry, where mechanization and the use first of water power, then of steam, made concentration of production essential. Factories increased in size as steam became the principal source of power. The words "factory" and "mill" were synonymous for a long while.

8

8 Josiah Wedgwood (1730–95) pioneered the large-scale production of pottery at his Etruria works near Stoke-on-Trent. He was a self-educated man and typical of those who made the Industrial Revolution.

9

9 Labour conditions were often poor in the early stages of the Industrial Revolution. Child labour was common, especially in the textile industry, with long hours of work, low pay and frequent accidents. Women also worked in the textile factories, where they made up half the workforce. Though women and children had worked on the land, these new industrial conditions provoked a series of Parliamentary enquiries in Britain and by the mid-19th century Factory Acts were passed, restricting hours of work and prohibiting women and children from certain areas of employment, such as work underground. By 1900 most other industrialized nations had also introduced some form of factory legislation.

Pitt, Fox and the call for reform

The age of the younger William Pitt (1759–1806) and Charles Fox (1749–1806) saw the beginning of the transformation that turned Britain from an agricultural society governed by a narrow oligarchy of the landed classes into an urban, industrial society with democratic rights for most of its inhabitants. During the 60-year reign of George III (reigned 1760–1820), economic and social change greatly enlivened political debate.

Party lines tended to harden during the latter part of the century, replacing the more fluid groupings of the time of Robert Walpole (1676–1745), and reflecting the rise of more divisive issues in politics, such as the American crisis, the power of the Crown, and the Wilkes affair. Out of these were born the demand for parliamentary reform and the emergence, for the first time since 1715, of something approaching a two-party division under the leadership of Pitt and Fox.

The power of the Crown

The accession of George III provoked a period of instability in British politics. The king's dismissal of the existing administration under the elder William Pitt (1708–78) and the Duke of Newcastle (1693–1768) was followed in 1762 by the elevation of the king's favourite, the Earl of Bute (1713–92), to lead the administration. These actions, as well as the pronouncements of the new king, reawakened fears that the Crown would attempt to dominate politics and that the mixed constitution of Crown, Lords and Commons, embodied in the Glorious Revolution of 1689 would be undermined.

In fact, George III was not aiming at the royalist reaction that his opponents feared. An inexperienced and obstinate young king, he wished to free the Crown from the domination of the group of politicians that had held power under George II (reigned 1727–60), especially the elder Pitt. The allegations that the king tyrannized his ministers and controlled a vast web of patronage were much exaggerated.

Nevertheless the resentments of the ousted Whig leaders were articulated in Edmund Burke's *Thoughts on the Cause of the Present Discontents*, published in 1770 [3]. Burke argued that the manipulation of patronage by the Crown permitted the monarch to dominate Parliament and rest his government upon a small group of "King's Friends", thus destroying the independence of the House of Commons.

At the beginning of George III's reign the attempts to exclude the MP John Wilkes (1727–97) [Key] seemed to suggest that the Commons was no longer an independent body or even representative of those who already had the vote.

Demands for reform

Hence the early years of George III's reign saw the rise of demands for reform. These were intended to reduce the influence of the Crown by removing the "rotten" boroughs and giving more seats to the large county electorates and some of the new manufacturing towns. The agitation for reform by the Yorkshire Association under Christopher Wyvill (1740–1822) and by John Wilkes's supporters in Middlesex and the City of London reflected feeling among small landowners and merchants. The war with America aroused still more dissatisfaction.

CONNECTIONS

See also
16 Europe: economy and society 1700–1800
18 England under the Hanoverians
20 The agricultural revolution
94 The British labour movement to 1868
160 The fight for the vote
28 Scotland in the 18th century
46 British colonial policy in the 18th century
48 India from the Moguls to 1800
64 The American Revolution
134 The story of the West Indies
136 The story of Canada
162 Ireland from Union to Partition

1 William Pitt, 1st Earl of Chatham, was secretary of state from 1756 to 1761 and the foremost politician of his age, known as the "Great Commoner". During his period in power he was absorbed in the Seven Years War (1756–63) and left the management of Parliament and elections to the Duke of Newcastle. Pitt kept free of party ties and showed no interest in parliamentary reform, despite his close friendship with Wilkes. His last political act was to plead for a policy of self-government under the Crown for the American colonies. He formed a second administration in 1766, but ill health forced him to retire from politics in 1768.

2 Charles James Fox was the effective leader of the Whigs during the last decades of the 18th century. Independent minded, a brilliant orator and a spendthrift who amassed huge gambling debts, Fox is remembered for his vigorous opposition to the Crown and his support for parliamentary reform and the anti-slavery movement. As a party leader he was not very successful, holding office only twice, in 1783 and 1806. His bitter opposition to George III deprived him of royal favour and kept him from power. In addition, his support for the French Revolution split the Whigs and lost him support, as did his opposition to Pitt's repressive acts in the 1790s.

3 Edmund Burke (1729–97) was one of the leading politicians and political philosophers of the 18th century. A Whig, he articulated the theory of "loyal" opposition to the government of the day, blaming the corruption and alleged oligarchic tendencies of George III's reign for political instability and the disorders of the Wilkes affair. He sympathized with the American colonists' struggle for independence from England, but was opposed to the French Revolution for destroying the historically established, traditional institutions of the country. He broke with Fox and the Whigs over this in 1791 and campaigned for war against France until his death in 1797.

4 The Gordon Riots in June 1780 were caused by opposition to the removal of legal penalties from Roman Catholics. In 1779, an extreme Protestant Association was formed by Lord Gordon (1751–93) to prevent what was believed to be growing Catholic power. Petitions and demonstrations were followed by a week of rioting and looting in central London after the Commons refused to debate their cause. Newgate prison was stormed and burned (shown here), property looted and the Bank of England attacked. More than 400 people were killed in the rioting and looting.

Its incompetent handling, leading to defeat, contributed in 1782 to the fall of Lord North's (1732–92) administration, which had held power since 1770.

The re-emergence of two parties

After a confused period with three ministries in under two years, William Pitt formed a government in 1784. Although he never used the word "Tory" himself, Pitt proved, over his long administration, to be the re-founder of the Tory Party. Fox then emerged as the leader of the Whigs and the effective opposition. The early struggles with George III had helped to sharpen party lines and legitimize opposition. Although the Whig and Tory parties were still more fluid than they were to become, Pitt and Fox provided leadership to a more coherent grouping of supporters than had been the case earlier in the century.

The passing of the "economical reform" acts in 1782 – reducing the number of officers in the pay of the Crown eligible to sit in Parliament – contributed to the waning of royal influence. The professionalization of the civil service under Pitt and his drive for greater economy further reduced offices and sinecures. George III's recurrent breakdowns into insanity contributed to the decline of monarchical power, culminating in his permanent incapacity in the last ten years of his life. Even so, the king retained sufficient personal influence to exclude Fox from office for much of the period and to support Pitt's administration. It was the king's obstinacy over Catholic emancipation that forced Pitt's resignation in 1801.

The last years of Pitt and Fox were dominated by the impact of the French Revolution and the wars with France. Pitt was forced to act against the threat of subversion in England with a series of repressive measures, culminating in the treason trials of 1794 – aimed at the radical Corresponding Societies – and the Two Acts of 1795. During those years, Fox alienated many of his parliamentary supporters by his support of the French Revolution at a time when its excesses shocked the majority of propertied opinion. Nonetheless, his opposition to the policies of Pitt and his brilliant oratory preserved the Whig's image as the party of reform.

KEY

John Wilkes achieved notoriety as one of the early champions of reform after he was arrested for criticizing George III in his *North Briton* newspaper, in 1763. Wilkes claimed immunity as an MP, but he was expelled from the House of Commons. In 1768 Wilkes was elected MP for Middlesex. He became a focus for popular discontent with the Government and was able to manipulate this to cause riots in London in 1768. Imprisoned, Wilkes was re-elected three times, each time being expelled by the Commons. In 1774 he was finally allowed to take his seat in the House, but his assertion of popular opinion and freedom in politics was not forgotten.

5

5 Poor harvests and high prices caused several waves of food riots in the late 18th and early 19th centuries. In particular, the wars with France from 1793 led to great hardship and many popular disturbances. In 1800, the price of corn was more than treble the price in 1790.

6

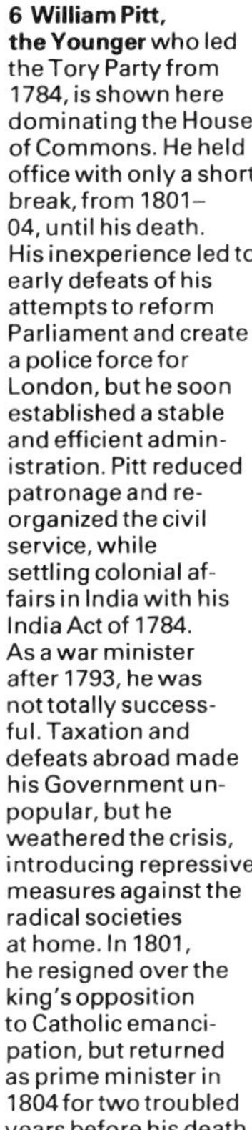

6 William Pitt, the Younger who led the Tory Party from 1784, is shown here dominating the House of Commons. He held office with only a short break, from 1801–04, until his death. His inexperience led to early defeats of his attempts to reform Parliament and create a police force for London, but he soon established a stable and efficient administration. Pitt reduced patronage and re-organized the civil service, while settling colonial affairs in India with his India Act of 1784. As a war minister after 1793, he was not totally successful. Taxation and defeats abroad made his Government unpopular, but he weathered the crisis, introducing repressive measures against the radical societies at home. In 1801, he resigned over the king's opposition to Catholic emancipation, but returned as prime minister in 1804 for two troubled years before his death.

7

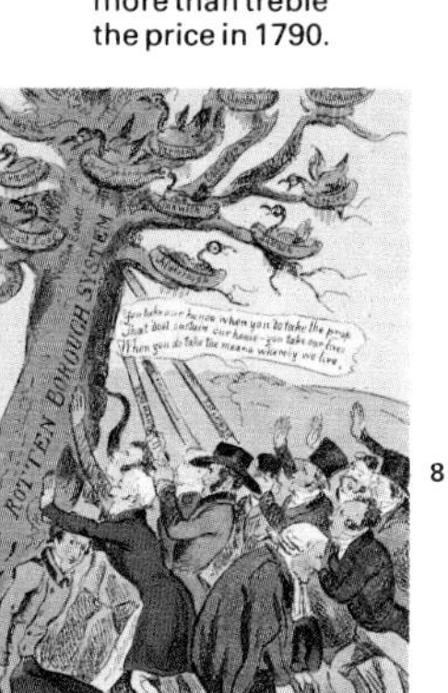

7 The movement for parliamentary reform gathered momentum in the last 25 years of the 18th century. This cartoon shows reformers attacking the "rotten" boroughs, the virtually uninhabited towns that still elected members to Parliament. Old Sarum was a notorious example of this – there a handful of voters returned two MPs. In addition, many seats were at the disposal of landed patrons, the so-called "pocket" boroughs. The larger manufacturing towns such as Manchester, Sheffield, Birmingham and Leeds were unrepresented, and the voting qualifications varied from town to town. The younger Pitt introduced a bill in 1785 to remove some of the rotten boroughs, but it was defeated in the Commons.

8

8 Agitation for reform culminated in the Reform Bill struggle of 1830–32. The Whig Government was returned in 1830 pledged to carry a reform bill. But rejection of the bill by the Lords in 1831 precipitated severe rioting in many parts of the country. At Bristol there were four days of riots and in Nottingham the castle was burnt (shown here) by supporters of the bill. The bill was finally passed in 1832.

Georgian art and architecture

In the eighteenth century British artists developed a new self-confidence and for the first time the taste of the general public was deliberately educated by art-theorists and connoisseurs. Nature and art were both looked at in new ways, leading to the enormous popularity of landscape gardening and to a much wider appreciation of the architectural styles of the past. Taste in painting developed beyond portraiture, and sculptors found new opportunities other than carving busts and funerary monuments.

The influence of Italy

The Georgian achievement was due to conditions that fostered a widespread interest in the arts and patronage of numerous competent, intelligent and ambitious artists. Continental travel, especially the Grand Tour, helped to educate rich young aristocrats. In Florence and Rome they could admire the remains of antiquity and the artistic achievement of the Renaissance. There they could start their own collections of works of art.

It was architecture that first felt the effect of these ideas and experiences. Palladianism was a deliberate system of rules and patterns for beautiful buildings, derived partly from antiquity but far more from the sixteenth-century Italian, Andrea Palladio. Inigo Jones (1573–1652) had been a follower of Palladio and gave a patriotic justification for imitating the Italian architect.

The leaders of Palladianism were Colen Campbell (1676–1729), whose *Vitrivius Britannicus* (1715, 1717,1725) first promoted the new style, which superseded the Baroque of Wren and his school; and Richard Boyle, 3rd Earl of Burlington (1694–1753), who was an architect, a patron of architects and theorists of architecture, and an arbiter of Palladianism. Through numerous books of pattern designs, English style spread far and wide, especially to America, where Palladianism flourished [7] long after it had become outmoded in the mother country.

Later Georgian architects

In the 1750s young architects began to study architecture outside the Renaissance tradition, exploring for the first time Greece and the Near East. They studied the fragments of antique interior stucco decoration hitherto largely ignored, and began to take more interest in long-neglected medieval buildings. Of this generation, the conservative William Chambers (1723–96) refashioned Palladianism with greater suavity; his major work was Somerset House, London, begun in 1776. Robert Adam (1728–92) was the prolific exponent of a new form of interior decoration [6]. Strawberry Hill [5], built from 1751 by Horace Walpole (1717–97), is the most famous piece of pioneering Gothic revival. At the end of the century the most original architect was John Soane (1753–1837), who concentrated on the problems of space and mass, and whose masterpiece was the Bank of England.

Georgian architecture achieved a high standard not only in the design and construction of country houses but in urban design. Of this period are the great squares of north London and the grandiloquent sequences of square, circus and crescent at Bath [3]. Most memorable are Regent Street and Regent's Park in London, the work of John Nash (1752–1835).

CONNECTIONS

See also
22 The Enlightenment
36 The Rococo
42 Neoclassicism
72 Origins of Romanticism
28 Scotland in the 18th century
30 Scots culture to 1850
80 Romantic art: figure painting
102 Romantic art: landscape painting
24 The English in Ireland
26 Irish culture to 1850
114 Victorian London

1 Thomas Gainsborough painted "Countess Howe" in the mid-1760s at Bath where he lived for 15 years. He studied the Old Masters in his clients' collections, in particular the portraits by Van Dyck, and developed an elegance of pose and delicate handling of paint that far surpassed his contemporaries.

2 "The Countess's Dressing Room" (1745) comes from "Marriage à la Mode" the second of four sets of satiric paintings by William Hogarth that deal with the absurdities of the everyday life of the contemporary middle-class household. They made Hogarth's name as an artist and a social commentator. Based on the methods of presentation of Restoration comedy, but with Hogarth's own dash of indignant venom, they widened the range of the visual arts. Much of his achievement lay in the caricaturing of faces. His successors in this vein were Thomas Rowlandson (1756–1827), the caricaturist, and James Gillray (1757–1815).

3 Bath was the most fashionable watering-place of Georgian England. Its coherent splendour was due to the vision of the architect and builder John Wood (1704–54), whose Queen Square brought to Bath the conception of the architecturally unified square, developed in London. The Circus was begun by him in 1754 and completed by his son John (1728–81), who was also responsible for the Royal Crescent (shown here), which was built 1767–75 and the Assembly Rooms at the spa itself, all of which were in a grand Palladian style. They also built important townscapes at Bristol and Liverpool. Georgian urban architecture was more concerned to build houses for the urban upper middle class than for the aristocracy.

Meanwhile, landowners created England's most important original contribution to the visual arts, the landscape garden [8]. William Kent (1686–1748), Lord Burlington's friend and protégé, was an important early exponent, but from the middle of the century the two busiest landscapists were Lancelot ("Capability") Brown (1716–83) and, from the mid-1790s, Humphry Repton (1752–1818), whose designs embody the taste for the "picturesque" – rougher, more irregular and "natural" than the shaven slopes and demure clumps of Brown.

Georgian painting

In painting, the first great figure of the century was William Hogarth (1697–1764). He is most famous for his paintings, and prints based on them, satirizing contemporary life [2]. He was a theorist of painting who emphasized the sinuous line as the basis of beauty, and was a delightful portraitist.

In 1768 the Royal Academy was founded with Joshua Reynolds (1723–92) as president. Its main aim was to train painters, sculptors and architects. Reynolds enshrined its philosophy that "history pictures" of noble themes from the Bible, classical mythology, and ancient or modern history, ranked highest in the types of painting.

In practice Reynolds made his career one of portraiture, and he developed a type of "historical" portrait in which the sitter, usually female, wore classical costume and took up a pose derived from the Old Masters. This was indeed the golden age of English portraiture, with painters of the calibre of Thomas Gainsborough (1727–88) producing fresh and sparkling portraits set in imaginary landscapes; the master of the English "conversation piece", or group portrait, was Johann Zoffany (1734–1810).

While the art of landscape was significantly developed by a series of minor artists, incipient Romanticism had already touched the work of the history-painter James Barry (1741–1806) and the Swiss-born Henry Fuseli (1741–1825). It is equally prominent in the engravings of the poet William Blake (1757–1827), which make a great contrast with the classical purity of the sculptor and illustrator John Flaxman (1755–1826).

KEY

"Croome Court", painted by Richard Wilson (1714–82) in 1758, immediately after his return from Italy, shows how much his study of Claude's landscapes, with their mellow colouring and well-controlled masses of foliage, had affected his approach to English landscapes. It also shows a typical Palladian house in a landscape garden setting. The central portico and corner towers with pyramidal caps had by this time become clichés of English country-house design. Its architect may have been Sanderson Miller, who was better known as a Gothic revivalist. The park with undulating greensward was laid by Capability Brown.

4

4 Pompeo Batoni (1708–87) painted many rich young English "milords" in Rome on their Grand Tours. They were often posed casually in front of a major classical monument. This portrait is of Cardinal York (1725–1807), a Stuart claimant to the throne.

5

5 Strawberry Hill, Twickenham, was built for the writer Horace Walpole who designed the house with his friends in a "Committee of Taste". The house grew slowly and irregularly, but it was generally based on engravings of medieval monuments.

6

6 The interior of Syon House, Middlesex, was remodelled for the Duke of Northumberland after 1761, and it was one of Robert Adam's first great opportunities to show how a modern noble could be housed with a splendour based on the antique palaces of the Roman emperors Hadrian and Diocletian. The anteroom incorporates marble column shafts physically brought from Rome by Robert's brother James (1730–94), who acted as his assistant. Most of Adam's interiors were more delicate than this one, but the brilliant colours are typical of his work, as was the free adaptation of the Mediterranean styles. He was perhaps the first English architect to try to create a vision that unified in style and proportion the exterior of the building, the interior and the furnishings.

7

7 Monticello, Virginia, was designed by Thomas Jefferson (1743–1826) the third president of the USA, for himself in 1769. Its central dome echoes Palladio's Villa Rotonda, a favourite model of the English Palladians. In this, it was typical of the intellectualism of much of the architecture of 18th-century America, where great stress was laid on clarity, simplicity of form, airiness and the human-based proportions.

8

8 The landscape garden at Stourhead, Wiltshire, was the personal creation of its owner, Henry Hoare (1705–85), and architect Henry Flitcroft. Begun in 1743, the garden was conceived as a pictorial union of water, trees and buildings in the spirit of a landscape by Claude. A walk round the lake passed many buildings: temple, rustic cottage, Pantheon (shown here), and a Gothic market cross at the end of the route.

Origins of romanticism

The reaction against the Enlightenment began early in the eighteenth century and was evident in many isolated ways that were to coalesce in the great pre-Romantic period of 1770–98. The Enlightenment had asserted the powers and worth of the individual, laying the philosophical foundations of an individualism that the Romantics elevated to subjectivism. Romanticism rejected the sterility of rationalism, exalting the emotions as the source of all truth.

Influence of Celtic mythology

One theme runs clearly through the mid-eighteenth century, that of Scandinavian and Celtic mythology and antiquities. Among many who explored these veins, Thomas Percy (1729–1811) translated runic poetry from the Icelandic in 1763 and Thomas Gray (1716–71) sought inspiration from Scandinavian sources for works such as *The Descent of Odin* (1761). But these were to be audaciously overtaken by the efforts of James Macpherson (1736–96). His *Fragments of Ancient Poetry* (1760) derived from Irish cycles that had found a way into Scotland and their success led to *Fingal* (1761) and the Ossian phenomenon [6], a timely invention of sublime Celtic lore that inspired the European Romantic movement.

Although romanticism quickly spread to painting and music, its origins and first expressions were predominantly literary – German and English. In 1765 Percy's *Reliques of Ancient English Poetry* brought the strength and freshness of the ballad into the literary domain; at about the same time a *cause célèbre* centred on Thomas Chatterton (1752–70), who devised medieval imitations alleged to be by Rowley, a fifteenth-century author, as well as writing his own poems. By his early suicide Chatterton became a symbol of the persecuted obsessive dreamer.

Genuine scholarship

While Chatterton indulged in his "pious fraud", genuine scholarship looked into the past to assist the revival of romance. *Letters on Chivalry and Romance* (1762) by Richard Hurd (1720–1808) and Thomas Tyrwhitt's edition of Chaucer's *Canterbury Tales* were influential in this.

Romantic curiosity was not merely academic. A new feeling for landscape took men on journeys to the wild Hebrides, to unknown mountains, in search of the physically "horrid and sublime" and, later, the picturesque and romantic. These pursuits would have been almost unthinkable at the beginning of the century under rational classicism. The latent awareness of nature was to bear fruit with the Lake Poets of the 1820's. Countries of the mind appeared in *Rasselas* (1759) by Samuel Johnson (1709–84) and *Vathek* (1786) by William Beckford (1760–1844).

In 1771 Henry Mackenzie (1745–1831) published *A Man of Feeling*, a novel of sentiment of no outstanding merit but attesting the influence of Jean Jacques Rousseau (1712–78), the man of nature and feeling *par excellence*, who, with Johann Herder [2] (1744–1803), was the great theoretical precursor of European romanticism. Rousseau's early involvement with the *encyclopédistes* turned into a conflict of head and heart and it is the supremacy of the heart that inspires *La Nouvelle Héloïse* (1761).

1 Horace Walpole (1717–97) converted a farm at Twickenham into a "little Gothic castle" and for 40 years added architectural detail, armour and stained glass, largely derived from chapels and cathedrals of Europe. A "Strawberry Hill Committee", consisting of Bentley the archaeologist, Walpole and others, virtually originated the revival of Gothic. Because of his influential social position, Walpole the antiquarian was an unconscious instrument of melancholy romanticism and the inspirer of many monastic country houses.

2 Johann Herder remains the most significant harbinger of the Romantic movement. His real achievement was to alter the course of Goethe's outlook. The young author's rococo ideas were replaced by concepts of spontaneity and originality and he was introduced to popular poetry, to Ossian and Shakespeare. Herder's own important statement of the *Sturm und Drang* movement lies in two essays written in 1773. He particularly sought to establish the *Volkslied* (the folk-song) as the only truly valid poetry.

3 Thomas Parnell's poem "A Night Piece on Death" (1721) initiated the morbid and baroque "Graveyard School". Robert Blair's "The Grave" (1743) in this pre-Romantic style was illustrated by the painter William Blake in 1808.

4 The new interest in Shakespeare owed much to Herder, whose essay *Shakespeare* (1773) celebrates him as an irrational genius, a philosopher of folk-poetry. This illustration of Lady Macbeth is the work of Henry Fuseli (1741–1825).

5 The poetic wonder of Goethe's old age, *Faust* (Part I 1808, Part II 1832), grew from a lifetime of reflection. The ultimate transformation of the medieval alchemist into the troubled Romantic scholar was Goethe's symbol of man in search of experience and salvation.

Its enormous success was followed by Rousseau's equally important *Emile* (1762), a novel that revolutionized the concept of education. The child, Rousseau maintained, should grow under the moral influence of nature's laws, protected from ready-made instruction, a theory that still echoes through contemporary thought. Rousseau, himself an anguished genius, exercised a profound influence on English literature and the French Romantic movement. His sensibility, sympathy with nature, lyricism and insistence on an immortal but complex soul fuse into the plausible dogma of the superiority of inspiration over rational thought.

The influence of Germany

Germany, however, can be regarded as the first home of romanticism and the one in which it took its most characteristic forms. Of its theorists, Gotthold Lessing (1729–81) was of prime importance. His contributions to literary periodicals dismissed the old classical forms, extolled Shakespeare as a model and drew attention to the resources of German folk-song. Shakespeare himself [4] was first translated into German by Christoph Martin Wieland (1733–1813), a move that further advanced the *Sturm und Drang* [2] ("storm and stress") movement, which embraced a number of young poets, including Goethe, and placed an overwhelming emphasis on intensity of passion.

Rousseau's counterpart, the critic Herder, was also paving the way for German romanticism. His advocacy of a return to nature – and to him Shakespeare was a natural phenomenon – and his precognition of Faust's "feeling of all", nurtured the young Wolfgang von Goethe (1749–1832) [8] at Strasbourg into inspirational rather than classical paths. Goethe's multifarious activities and literary achievements, and his fusion of both rational and Romantic elements mark him as the supreme Romantic figure in European literature.

Many strands were woven into later Romantic attitudes. None equalled the decisive impact of the French Revolution. Where madness and melancholia had been the escape route of the earlier Romantics, those who followed found a new freedom.

KEY

George, Lord Byron (1788–1824), was by temperament and tragic destiny an arch-Romantic. His voluminous poetry, reckless in its spontaneity and a ready vehicle for his disenchanted feelings, had a hypnotic effect throughout Europe. Its irony is however absent from his shorter love poems.

6

6 The misty Celtic world of Ossian was a rich and enormously influential vein in European romanticism. Ossian was a semi-legendary 3rd-century Irish poet-warrior (here dreaming a typical dream over his lyre); when Macpherson published his "translation" in 1762 he began a cult that spread throughout Europe. Despite some fierce academic criticism, few suspected that Macpherson had invented freely, with only passing nods to genuine Celtic lore.

7

7 Siegfried, a hero both of Germanic and Norse legend, is a principal figure in the *Nibelungenlied*. The imposing mythology of the Rhineland attracted much attention from early Romantic writers and painters seeking to establish a mystical German tradition.

8

8 Goethe's own disappointment in love was the foundation for his *The Sufferings of Young Werther* (1774). The hero's intensity of feeling and dramatic gestures (as in this illustration) inspired numerous imitators; when his suicide for love did the same Goethe's notoriety was assured.

9

9 The first substantial Romantic, who "lost his native country and conquered Europe", Lord Byron took Spain, Italy, the East and Greece as a background for his aristocratic individualism. A fervent ally of the Greeks, he died supporting their independence struggle.

The French Revolution

The prestige and apparent power of the absolute monarchy that Louis XIV (1638–1715) had built up disguised fundamental weaknesses that were to become serious under his successors. French society was increasingly divided into a small aristocracy jealously defending its privileges of wealth and partial exemption from taxation [2]; a growing middle class frustrated by its lack of political power and the incompetence of royal government; and the peasantry which did not own enough land for security from bad harvests and which hated the feudal dues it had to pay the aristocracy.

Calling of the Estates-General

During the reign of Louis XV (1710–74), royal prestige was damaged by a series of disastrous wars with Britain, and the government went deep into debt despite a general increase in trade and industry. Even success in helping the American colonists [1] at the beginning of the reign of Louis XVI in 1774 only highlighted the contrast between American ideals of liberty and democracy, and repression and privilege in France. An economic slump began in the 1780s and the state of royal finances became so bad that an attempt was made to tax the privileged classes. They refused to pay and the king was forced, for the first time since 1614, to call the Estates-General. When this met in 1789, the Third Estate – the bourgeoisie, or middle classes – swiftly tired of the actions of the aristocracy and clergy and on 17 June proclaimed itself a National Assembly [4A] with the intention of preparing a new constitution.

While this political crisis had been growing, a disastrous harvest in 1788 had brought many peasants and industrial workers close to starvation [3], and riots had broken out in many parts of France. When, on 11 July 1789, Louis (1754–93) dismissed his popular minister Jacques Necker (1732–1804), there was widespread protest.

Anti-royal feeling grows

The people of Paris stormed the Bastille [Key] on 14 July and there was a general breakdown of social order throughout France with aristocratic property being looted or seized. The National Assembly stripped away the privileges of aristocracy and clergy and the king had to leave Versailles for the Tuileries palace in Paris.

The political turmoil continued over the next two years with attempts to establish a new constitution and with anti-royal feeling growing. Confiscation of aristocratic and Church land and wealth gave the new government welcome financial help, but an issue of paper currency – the *assignats* – soon led to renewed inflation. In June 1791 the king attempted to flee abroad but was recaptured at Varennes. Popular hostility to him increased when the Emperor of Austria and the King of Prussia issued a declaration saying that the ancient rights of Louis would soon be restored. In September a new constitution [4B] was introduced setting up a legislative assembly and giving the king a strictly limited role. But tension rapidly grew between moderate constitutionalists and extreme anti-monarchists.

In April 1792, war was declared on Austria. As royalist armies backed by Austria and Prussia gathered on France's borders [6] the mob demanded that the

CONNECTIONS

See also
76 Napoleonic Europe
64 The American Revolution
22 The Enlightenment
16 Europe: economy and society 1700–1800
26 Irish culture to 1850
68 Pitt, Fox and the call for reform
108 the revolution of 1848
42 Neoclassicism

In other volumes
314 History and Culture 1
324 History and Culture 1

1

1 The Marquis de Lafayette became a popular hero when he led the French volunteers who helped the American colonists break free from Britain. With other aristocrats he joined the National Assembly in 1789, presenting a declaration of rights and organizing the National Guard. A moderate reformer, he became trapped between Jacobin extremists and the court and fled in 1792.

2

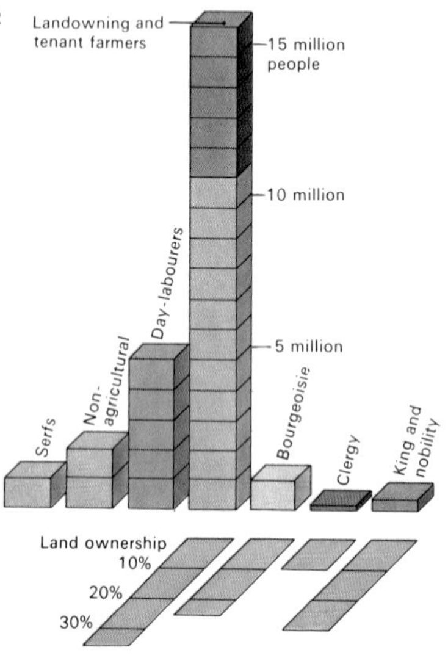

2 Unequal division of land with more than 40% owned by less than 3% of the population was a major grievance and fundamental problem of French society. As most of the nobility and clergy were largely exempt from taxation, the principal share of the burden fell on the bourgeoisie and the more prosperous of the peasantry.

3

●●●●●●●● (8 sous)
Price of 2kg loaf in "normal times"

●●●●●●●● ●●●●●●● (15 sous)
Labourer's wage — Amount of wages remaining

Year	Month	Amount of wages remaining	
1788	July	●●●●●●	
	August	●●●●●	
	September	●●●●	
	October	●●●●	
	November	●●	
	December	●	
1789	January	●	
	February	◓	
	March	◓	
	April	◓	Réveillon riots
	May	◓	
	June	◓	
	July	◓	Storming of Bastille
	August	●●●	

3 Prices rose steadily during the 18th century as a result of increases in population (more than 50%) and money supplies and relatively slow expansion of industrial and agricultural production. This had the effect of making the upper classes even more determined to hold on to their privileges, while the lives of the peasants and industrial workers became even more precarious. In "normal" times a loaf of bread cost a labourer about half a day's wage, but bad harvests in 1788 and 1789 lifted bread prices to the point where they almost matched wages. This precipitated political tension and rioting.

4 A

4 The meeting of the Third Estate as the National Assembly [A] on 17 June 1789, pledged to end feudal privileges, was the political start of the revolution. The constitution it produced [B] was a limited monarchy with power residing in a Legislative Assembly elected by citizens who paid direct taxation at least equivalent to three days' wages of a labourer per year. The 1791 Constitution also divided France into the local government *départements* that are still in use.

B

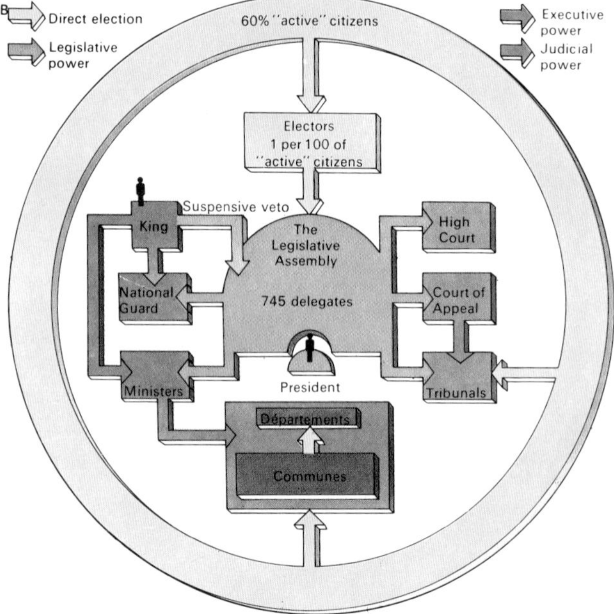

Assembly act against the king. In September Prussian armies invaded France, precipitating a massacre of captured aristocrats. An unexpected victory over the invaders at Valmy on 22 September relieved the pressure. On the same day France became a republic ruled by a Constituent Assembly which was elected by the extremist Jacobins, the most radical group to hold power during the revolution.

The king was put on trial and executed [5] in January 1793. In the following months, defeats by the *émigré* armies, pro-royalist risings in La Vendée and the south, and continuing economic problems prompted the Assembly to appoint a Committee of Public Safety to exercise emergency powers and to order total mobilization. A reign of terror began during which more than 40,000 "enemies of the revolution" were sent to the guillotine. All organized religion was officially abolished and replaced by worship of the Supreme Being.

By spring 1794 the republican armies had rallied; in June 1794 the counter-revolutionary armies were decisively defeated at Fleurus, and in July the Jacobin leader, Maximilien Robespierre (1758–94), who had been virtual dictator for a year, was overthrown and executed. A reaction set in with moderates seizing power. In 1795 a basically conservative constitution was set up headed by a five-man Executive Directorate.

Emergence of Napoleon

The Directorate made peace with Prussia and The Netherlands, but launched a major offensive against Austria by sending a young general, Napoleon Bonaparte (1769–1821), to campaign in Italy [7]. He was brilliantly successful during 1796, forcing Austria out of the war. He then led an expedition to Egypt to cut Britain's communications with her Indian Empire, but it was finally forced to abandon the campaign when Horatio Nelson (1758–1805) destroyed his fleet at the Battle of the Nile in 1798. Meanwhile the Directorate had become profoundly unpopular with all sections of the population and when Napoleon returned in October 1799 he was able to engineer a *coup* that gave power to three consuls [8], of which he was the senior.

KEY

The storming of the Bastille on 14 July 1789 was seen by contemporaries and later generations as the true beginning of the revolution. Although the political crisis began more than a year earlier, the rising of the Paris mob against this ancient prison and symbol of absolutism was of fundamental importance. It forced the basically middle-class National Assembly to ally with the people to prevent a royalist counter-attack and it led to uprisings in the provinces in which aristocrats' estates, were seized, land deeds destroyed and officials murdered. It paved the way for feudalism's downfall, transferring political power from the king to the legislature.

5

5 The execution of Louis XVI on 21 January 1793 followed the threat of an invasion of royalist *émigrés*. Popular opinion turned wholly against the king and the Jacobins were able to seize power and declare France a republic on 22 September 1792. Victory over royalist forces at Valmy gave them the self-confidence to try the king and his execution symbolized the break with the past system.

6

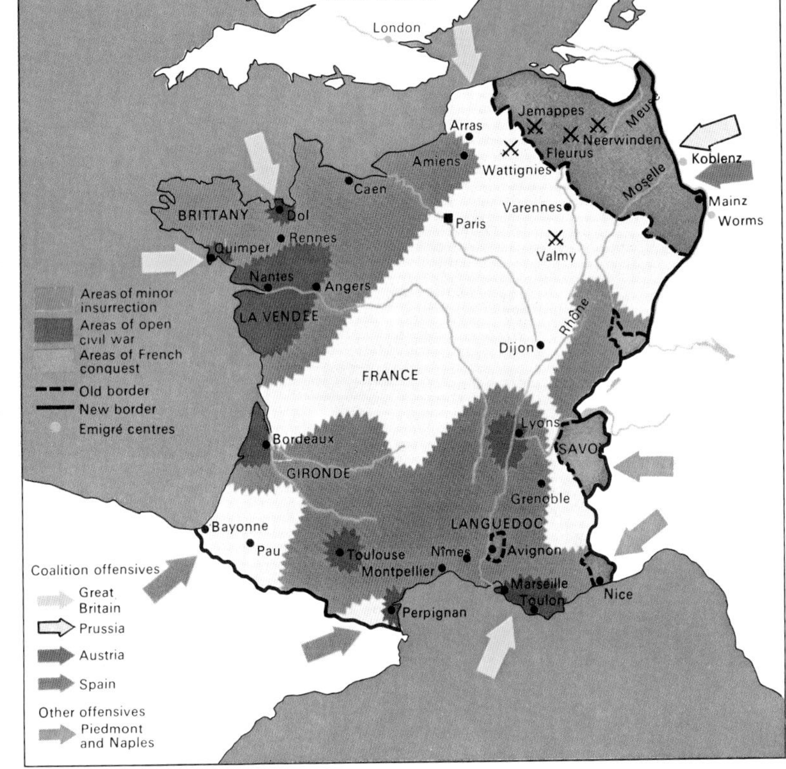

6 France's neighbours were antagonized by the gathering forces of the revolution. Aristocratic *émigrés* formed a nucleus of resistance and had support from Austria and Prussia. Their first invasion was halted at Valmy and the republic then counter-attacked, occupying Nice, Savoy and Belgium after a victory at Jemappes (November 1792), invading the Rhineland states and threatening Holland. After the king's execution, war was declared on Spain, Holland and Britain but military reverses followed, with a major revolt in La Vendée and enemy offensives in southern France, Belgium, Alsace and Britanny. Unprecedented emergency measures put down internal revolts, the invasions were repelled and Belgium and Holland were reconquered. By the end of 1795, France had made peace with all its enemies except Austria and Great Britain.

7

7 Napoleon Bonaparte led the French armies to attack Austrian territories in Italy early in 1796. The Directorate intended this to be a diversion while a major offensive took place in the Rhineland. But Bonaparte traversed northern Italy in an extraordinary series of victories, forcing the Austrians to make peace. The 27-year-old general changed the face of warfare by using shock tactics and mobility to harness the revolutionary zeal of armies raised by mass conscription.

8

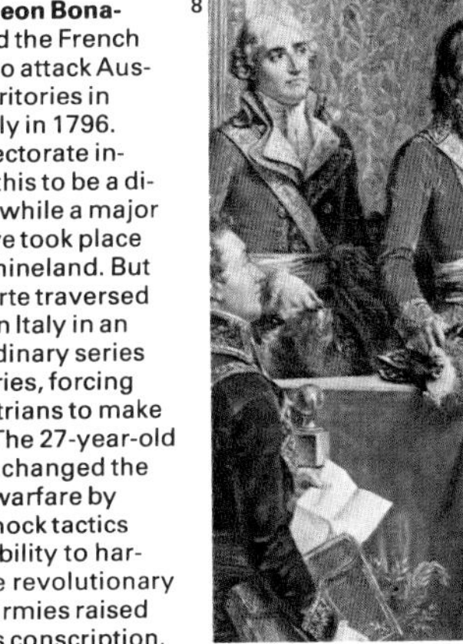

8 The installation of the Conseil d'Etat on 24 December 1799 made Bonaparte First Consul. With the prestige of his victory in Italy and the Egyptian campaign behind him, Napoleon was the most powerful man in the turbulent political scene at the turn of the century. The failure of the Directorate to solve internal problems had lost it all support and Napoleon hoped to use his widespread popularity to persuade the assemblies to vote him into power without any fuss. But they refused to do so and he had to use troops to drive them out and allow a small rump of supporters to vote through a constitution. This gave power to a first consul who was assisted by two colleagues and a senate nominated by the consuls. Napoleon then made use of a new device – the plebiscite – to obtain popular support. He announced that three million votes had been cast in favour of the new constitution and only 1,562 votes against it.

Napoleonic Europe

In 1800, Napoleon Bonaparte (1769–1821) [Key] became First Consul of France, then still menaced by hostile states. His new constitution confirmed the conservative policies of the Directorate and concentrated internal authority in his own hands. Once in power, he acted swiftly to achieve peace in Europe. After a surprise crossing of the Alps, he shattered Austrian power in Italy at the Battle of Marengo on 14 June 1800 and made peace with her at the Treaty of Lunéville. Russia, under the pro-French Tsar Paul (1754–1801), also ceased hostilities against Napoleon and in December joined Prussia, Denmark and Sweden in a French-inspired League of Armed Neutrality designed to weaken Napoleon's chief remaining foe, Britain, by blocking her trade with continental Europe.

War and peace

Although Paul was soon assassinated and succeeded by the pro-English Alexander (1777–1825), Britain made peace with France at the Treaty of Amiens in March 1802, agreeing to return all her overseas conquests except Ceylon and Trinidad, while Napoleon agreed to evacuate Holland and Naples. However, Napoleon soon aroused British suspicions by looking for new colonies, by refusing to evacuate Holland, and by extending French power in Germany. When the British realized that French markets would still be closed to their goods, and that Napoleon was building up Antwerp as a commercial rival, they refused to evacuate Malta and war broke out again in May 1803.

During the years of comparative peace between 1800 and 1803. Napoleon began the internal reconstruction of France which was to be his most lasting achievement. The Bank of France was established in 1800 and tax collecting centralized; the law was remodelled and codified, and a centralized secondary school system was set up. Napoleon's concordat with the Papacy extended his power – the Catholic Church gave up its claims to nationalized Church property in return for state support. In 1802 Bonaparte became First Consul for life with the power to nominate his successor.

The renewal of war identified Britain as Napoleon's most stubborn and dangerous enemy and at first he tried to defeat her by invasion. However the Royal Navy blockaded the coasts of France and Spain for more than two years and then under Vice-Admiral Horatio Nelson utterly destroyed the French and Spanish fleets at the Battle of Trafalgar on 21 October 1805 [2].

Military and economic strategy

Even before this interim defeat of his invasion plan, Napoleon, who had declared himself emperor [3] in 1804, had had to redeploy the Grande Armée to meet a renewed threat from Austria and Russia, who were now joined in a Third Coalition with Britain. In a swift campaign he smashed an Austrian army at Ulm on 20 October 1805, occupied Vienna and defeated the Russians at Austerlitz [5] on 26 December. At the Treaty of Pressburg with Austria, Napoleon gained complete control of Italy and unified much of Germany outside Prussia in the Confederation of the Rhine. Prussia felt obliged to intervene, but was defeated at Jena and Auerstadt in October

CONNECTIONS

See also
78 Nelson and Wellington
74 The French Revolution
82 The Congress of Vienna
84 European empires in the 19th century
108 The revolutions of 1848
170 Political thought in the 19th century

1

1800-1815

Total called up 1·3 million

Total army strength [including reserves] 2·5 million

= 10,000

FRENCH ARMY STRENGTH 1805 Total 400,000

Spain 300,000

Rhineland 200,000

France 100,000

Italy 60,000

FRENCH ARMY STRENGTH AND DISTRIBUTION 1808 Total 660,000

1 Conscription on an unprecedented scale laid the foundation for the armies that enabled Napoleon to dominate Europe. From 1800 to 1812, an average of 85,000 men were called up from France each year. The demand for military manpower grew increasingly onerous, especially in 1812 with the costly invasion of Russia. Total deaths during the Napoleonic wars were about one million, of which 400,000 were French.

2

2 Nelson's annihilation of the combined French and Spanish fleets at Trafalgar was the decisive event in the long naval war and convinced Napoleon that direct assault on Britain was impossible. Saved from invasion, Britain used her superb navy to blockade the coasts of Europe and her wealth to organize resistance to France. Napoleon was forced to extend his control of neighbouring states to stifle British trade and the hostility this caused finally brought down his empire.

3

3 As Emperor of the French, Napoleon used the trappings of imperial glory to consolidate his new dynasty. Most Frenchmen responded but some felt he had betrayed the egalitarian ideals of the Revolution.

4

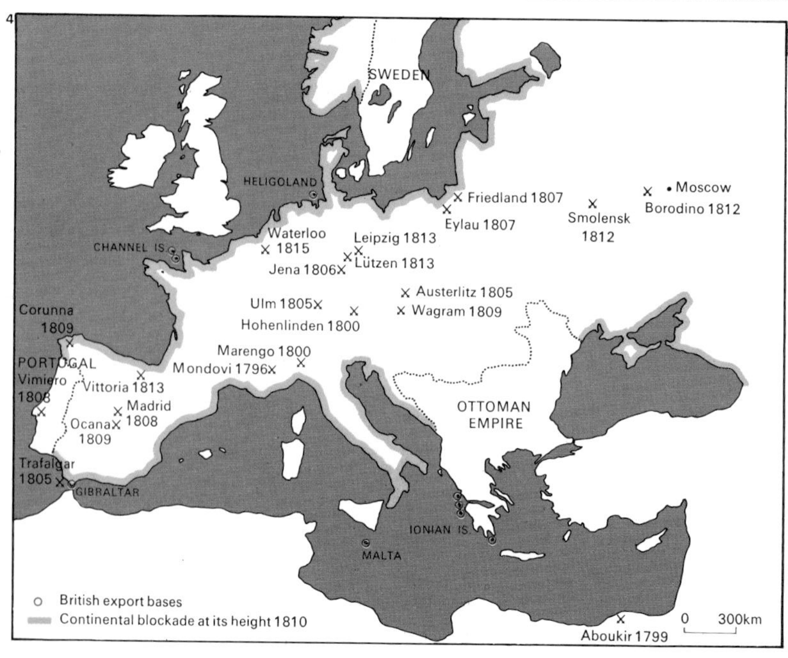

4 Brilliant victories in an almost continuous series of campaigns enabled Napoleon to establish France temporarily as the main power in Europe. In controlling the "traditional" powers of Austria, Prussia and Russia by a mixture of war and diplomacy, he enjoyed almost total success. But the need to extend and consolidate the Continental System led him to become trapped in a guerrilla war in Spain and then to launch the disastrous invasion of Russia.

1806. Napoleon occupied Berlin and defeated the Russians at Friedland in June 1807. Meeting the tzar at Tilsit, he persuaded him to enter an alliance with France against Britain, which once again remained Napoleon's sole effective opponent.

Napoleon now sought to defeat Britain economically by using force to prevent her trade with any part of Europe. Despite the power which his victories had given him, the British continually found ways of smuggling in their goods, and Napoleon had to try to extend his "Continental System" ever farther afield. The military presence [1] and resulting economic hardships made Napoleon's rule increasingly unpopular with his subject nations.

In 1808 Napoleon forced Charles IV (1784–1819) of Spain and his son Ferdinand to abdicate in favour of Napoleon's brother Joseph. The Spanish revolted and the British sent an army to support them. The Spanish campaign cost Napoleon more than 50,000 men and led to his first defeat on land. In 1810, Napoleon tightened up the Continental System by annexing Holland and the German coast. Europe was thrown into a commercial crisis that persuaded the tsar to end his alliance in December 1810.

Retreat from Moscow

In June 1812 Napoleon launched a massive invasion of Russia with 611,000 men. His troops reached Moscow, but lack of supplies and military reverses forced them into an undisciplined winter retreat, which left only some 10,000 men fit for combat [7]. In February 1813 Prussia declared war on France and Austria, and many subject states followed. Napoleon was defeated at Leipzig in October and the Allies pushed into northern France while the British invaded across the Pyrenees. Paris was occupied on 30 March 1814; Napoleon abdicated on 11 April and was exiled to the island of Elba.

On 1 March 1815, Napoleon took advantage of quarrelling among the Allies and the unpopularity of the restored Bourbons in France to re-establish his power. But defeat by the Duke of Wellington (1769–1852) at Waterloo [8] on 18 June 1815 led to his exile on St Helena where he died in 1821.

Napoleon Bonaparte, the Corsican-born general who made himself Emperor of the French, had the military genius to win France a short-lived supremacy over most of Europe. But it was his reforms of French society in codifying the law and rationalizing education and administration that were his greatest achievements. Some of them endure to this day.

5

5 Napoleon's victory at Austerlitz and the campaign that preceded it showed all the qualities of speed and decisiveness that made him one of the greatest generals the world has seen. Having force-marched the Grande Armée from the Channel to the Danube, he destroyed an Austrian army at Ulm and then pushed a Russian force back until it rejoined the main Russian army at Austerlitz. In the ensuing battle he used a combination of devastating artillery barrages and massive infantry assaults to sweep the Russians off the vital heights commanding the field of battle and into a precipitate retreat.

6

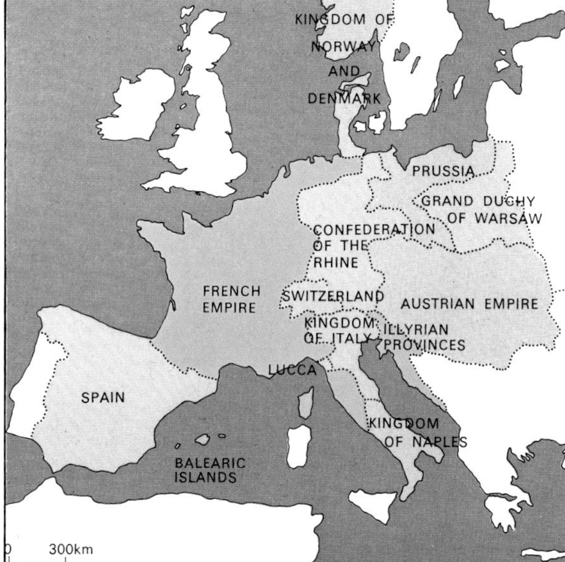

KEY

French Empire 1812

Dependent states 1812

French Allies 1812

6 Almost all Europe in 1812 was either ruled directly by Napoleon or members of his family, or allied with him. At the outset Napoleon had been able to draw on widespread support in Europe for the revolutionary ideals of overthrowing the old order. He furthered his own power by using the desire of neighbouring states for freedom, organizing many small states of Italy and Germany into dependent republics and setting up the Confederation of the Rhine that effectively ended the Austrian-dominated Holy Roman Empire.

7

7 The invasion of Russia was Napoleon's decisive error, celebrated by a Cruikshank cartoon. The Russians refused to make peace when Moscow was occupied and used scorched earth and guerrilla tactics to destroy the invasion armies and encourage subject states to rise.

8

8 Napoleon was finally defeated at Waterloo, near Brussels, by British and Prussian troops led by the Duke of Wellington and Marshal Blücher. An alliance of major European powers and conquered states had previously forced Napoleon to abdicate, but he had viewed exile on Elba only as an interlude. When the restored Bourbons had earned the dislike of most Frenchmen and the Allies were bickering among themselves at Vienna, he returned and marched to Paris with popular support. But the shock tactics of the Grande Armée met their match at Waterloo where the British infantry held firm against cavalry and infantry assaults until relieved by the Prussians.

Nelson and Wellington

For many centuries Britain opposed any European power that threatened to dominate continental Europe and from 1793 to 1814, with a short break in 1801–2, it fought to defeat the spreading power of revolutionary France. Lacking a large army, Britain had to rely on the traditional strategy of organizing alliances of other continental powers while using its naval supremacy to weaken France by blockade. Whenever possible, troops were sent to help anti-French forces, but Britain's major contributions to the ultimate defeat of France were a willingness to continue fighting, alone if necessary until new allies were found, and the use of a long-established prowess at sea.

Britain's weapons

The Royal Navy had long been recognized as the bulwark of British security but conditions of service were grim. The numbers of recruits needed to man the wartime fleet could only be maintained by forcible impressment [1] and the recruitment of convicts. Once enlisted, men were rarely allowed to leave.

In contrast to the conscript armies of Europe, the British army at that time was a small volunteer force numbered in tens, rather than hundreds, of thousands. Officers were able to buy their commissions, received no professional training and usually paid scant attention to the welfare of their men. By the end of the eighteenth century, however, efforts were being made to organize supply and medical services [2].

Nelson's great triumphs

Throughout the Napoleonic Wars Britain was fortunate to be served by a number of exceptional naval officers who proved to be both fine seamen and outstanding leaders. The greatest of these was Horatio Nelson (1758–1805).

At the outbreak of war Nelson commanded a ship-of-the-line in the Mediterranean and acquired a reputation as an active, able officer. During the Battle of St Vincent on 14 February 1797 his initiative in breaking the line of battle led to the capture of four enemy ships. For his part in the victory Nelson was knighted and promoted to rear-admiral. Wounded in several engagements, he lost an eye and an arm but his mental powers remained undiminished. In 1798, when Napoleon attempted to cut Britain off from India and its other eastern possessions by invading Egypt, Nelson annihilated the French fleet in the Battle of the Nile, fought in Aboukir Bay. Of the 17 French ships, 13 were captured or destroyed.

The victor of the Nile, now created Baron Nelson of the Nile, took command of the Mediterranean fleet in 1803. For the next two years, in a remarkable display of seamanship, Nelson off Toulon and Admiral William Cornwallis (1744–1819) off Brest kept the French fleet immobile. In 1805 the Toulon force managed to slip out and head for the West Indies meaning to return, link up with other forces and establish temporary command of the Channel so that Napoleon could invade Britain. But the French were forced into Cadiz while the British gathered outside under Nelson's command off the Cape of Trafalgar. When the combined French and Spanish fleet emerged it was utterly destroyed in battle [5] on 21 October 1805. Although Nelson was killed on the quarter-

CONNECTIONS

See also
76 Napoleonic Europe
82 The Congress of Vienna
68 Pitt, Fox and the call for reform
132 British foreign policy in the 19th century
74 The French Revolution

In other volumes
172 Man and Machines

1

1 The hated press-gangs, armed with cudgels, terrorized towns as they went ashore and roamed the streets in search of able-bodied men for the navy. Victims were forcibly seized and dragged aboard for medical examination. Volunteers were few, for life at sea meant separation from their wives and families for long periods, bad food, wretched conditions and brutal discipline; yet morale under Nelson was high.

2

2 Women were considered to be more a hindrance than a help in the army of Wellington's day, as implied in this drawing by Thomas Rowlandson. Some wives, but not many, were allowed to accompany their husbands on a campaign: the number was limited to between 2 and 6 per company of 100. Those women who did go received half-rations free. Some even took children as well. The women cooked meals, did soldiers' washing and acted as nurses. They had an eye for booty, too. Wellington once observed that "The women are at least as bad, if not worse, than the men as plunderers".

3

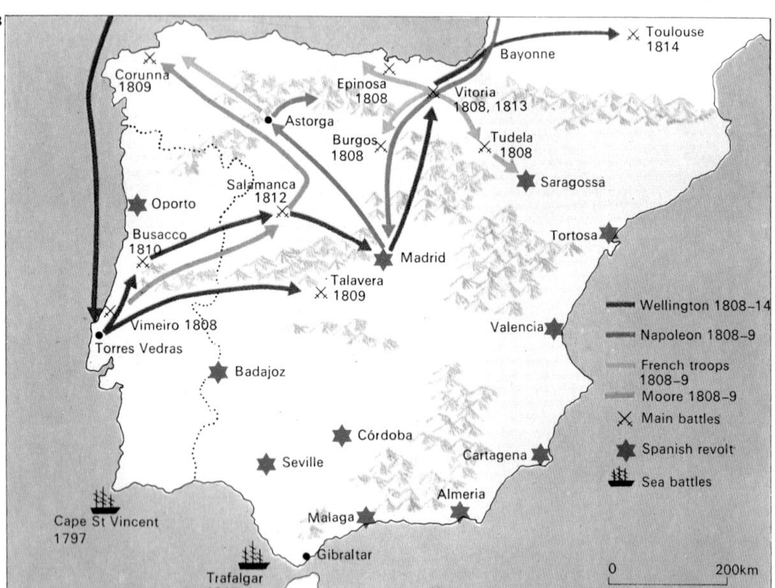

3 The French took Spain swiftly and compelled the British to leave. After Oporto fell (1807) Portugal appealed to Britain for aid and Wellington sailed with a force of 17,000. Napoleon ordered his commanders to drive the British into the sea, but the French themselves were expelled from the Peninsula and sent scurrying across their own border, with Wellington in pursuit. Napoleon later said that the "Spanish ulcer", with constant guerrilla activity and rioting, undermined his empire.

4

4 HMS *Victory*, Nelson's flagship at Trafalgar, was typical of the ships-of-the-line that formed the main battle fleet. Floating batteries with 60 to 120 guns firing in broadsides and a complement of 700, these slow, unwieldy vessels could remain at sea for years on end. Built at Chatham, and launched in 1765, *Victory* was 69.5m (227ft) long with a beam of 15.5m (52ft). She had more than 100 guns, the largest of which were two 68-pounders, 30 32-pounders and 28 24-pounders.

deck of HMS *Victory* [4] at the height of the engagement [6], he died knowing he had won a decisive victory.

The road to Waterloo

Nelson's success ended any hopes Napoleon had of invading Britain. The French emperor was therefore forced to try to destroy Britain by closing Europe to British trade. When Portugal and Spain refused to join the blockade, the French invaded. Britain was thus given the opportunity to intervene militarily. An expedition to Spain under John Moore (1761–1809) was compelled to retire but in August 1808 a second force under Sir Arthur Wellesley (1769–1852) [Key], later Duke of Wellington, landed in Portugal.

An Anglo-Irish aristocrat, Wellington learnt his soldiering skills in India from 1796 to 1805. After taking part in unsuccessful expeditions in north-western Europe in 1806 and 1807 he was given command in the Peninsula. There for the next three years he showed great skill in tying down vastly superior French forces [3]. He was always prepared to withdraw behind defences when necessary, but emerged to inflict a succession of defeats on the French. Finally in 1811 he launched a major offensive that cleared the Peninsula, winning major victories at Salamanca and Vittoria before invading south-west France in 1814 [7].

Napoleon abdicated and left for exile in Elba, but almost a year later he returned to France in an attempt to regain the throne. To meet this renewed threat Britain and the allies – Austria, Prussia and Russia – appointed Wellington to command a combined army gathered in Belgium. Despite being surprised by the speed of Napoleon's opening manoeuvres, Wellington held his ground against superior forces near the village of Waterloo [8] until the arrival of a Prussian army under Marshal Gebhard von Blücher completed a crushing victory.

For the second time Napoleon abdicated and went into exile – this time to St Helena, until his death in 1821. The victories of Nelson and Wellington, coupled with the nation's industrial and commercial supremacy, now made Britain the most powerful nation in the world.

KEY

"A Wellington Boot, or the Head of the Army": this 1827 cartoon shows the Iron Duke's distinctive profile and characteristic footwear. Taciturn and aloof, he affected to despise the troops he commanded as the "scum of the earth" but he based his tactics on their steadiness under fire. He chose defensive positions and relied on the discipline of his men to break the massive infantry and cavalry assaults of the French which had shattered most other adversaries. He hid an emotional nature under an icy manner and he cared for the welfare of his men. They repaid him with their respect and by beating the finest troops of Napoleon's *Grande Armée*.

5

5 At Trafalgar the British fleet went into action in two columns. Realizing that he was outnumbered 27 to 33, Nelson eschewed traditional tactics of the single line of battle, and succeeded brilliantly, capturing 19 enemy vessels.

6

6 Nelson's death overshadowed the triumph of Trafalgar. Hit on the shoulder by a musket-ball from a sniper, he was taken below decks where he died four hours later. A stern disciplinarian and a born leader, he displayed in battle great bravery and daring, tactical genius and shrewd judgment. His devotion to duty was absolute and the men he led revered him.

7

7 Wellington had a great welcome when he rode into Toulouse on 12 April 1814. The battle, he said, had been "very severe": combined deaths were 7,700. Victory, however, seemed complete when he learnt later that day that Napoleon had abdicated.

8

8 The Battle of Waterloo (1815) made Wellington a national hero. Napoleon had crossed into Belgium on 15 June and thrust back the Prussian army at Ligny but failed to rout them. Then on the morning of Sunday 18 June he attacked Wellington at Waterloo. Wellington had 67,000 men with 150 guns, Napoleon had 72,000 with 250 guns. The battle soon became a pounding match with few manoeuvres, but the arrival of the Prussians in the early evening brought swift and total victory.

Romantic art: figure painting

In the later part of the eighteenth century, the classical order was coming under attack in one area after another. Its most formidable antagonist, and one who is now recognized as the father of the Romantic movement, was the French philosopher Jean Jacques Rousseau (1712–78). He argued that feeling, not reason, should be the basis of belief and conduct and asserted, in opposition to classical theory, that art was not the servant of morality. In addition, British and German writers, such as William Blake (1757–1827) and Johann Wolfgang von Goethe (1749–1832), identified the unconscious as the source of art and poetry and made sincerity for the first time a test of artistic value.

Romanticism in French painting

The development of romanticism in the visual arts is easiest to see in French painting. This is because it was only in France that a tradition of state patronage of grand historical subjects was kept up and because one can trace in the treatment of these subjects a gradual progression, beginning with an almost pure Neoclassicism and leading through a steady undermining of classical principles to a more or less pure, but not undisputed, romanticism.

The subversion of classicism began in that temple of Neoclassical painting, the art of Jacques-Louis David (1748–1825), especially in his work done around the time of the French Revolution and under and for Napoleon. Not that he abandoned classical qualities of style – clarity, precision, economy and references to the Antique and to Nicolas Poussin – but in paintings such as "The Death of Marat" (1793) one can feel beneath the formal calm some of the emotions released by Rousseau; a sense of the precariousness of human life and institutions; an awareness of the power of fanaticism and of chance; a morbid fascination with violence. In part these qualities reside in the subject rather than its treatment, but the choice of the former is significant: one solitary man working, as he believed, for society and then being struck down by it. Throughout Romantic art, the image of a man alone defying and being ultimately defeated by some overwhelming force, whether of society, nature, the dark gods of unreason, or in extreme cases the universe itself, is one of the most potent of all symbols.

The poetry and horror of war

After David, the impulse towards emotionalism increased, stretching, although not yet breaking, the mould of classic form. Antoine Gros (1771–1835) [5] expressed the stirring poetry (as it was regarded at the time) of Napoleonic war, a poetry enhanced by its horror and destructiveness. Théodore Géricault (1791–1824) [6] represents a further stage still, in which the colour black becomes eloquent and dark shadows begin to bite into classic outline. With him, the solitary man may be not only the madman, the shipwrecked man or the man on a wild horse, but also the artist in despair.

Finally there was Eugène Delacroix (1798–1863) [7] over whose work there finally broke out a classic-romantic battle. While he never repudiated classical principles altogether (no Frenchman could ever quite do that), his supporters saw his work as directly opposed to that of his older contem-

CONNECTIONS

See also
102 Romantic art: landscape painting
72 Origins of romanticism
76 Napoleonic Europe

1 Henry Fuseli's "The Nightmare" (1782) is merely sensationalist by comparison with Goya's subtle comment on the horrors that may visit the mind asleep. It evokes a shudder in the spectator partly by drawing on legends of witchcraft (the incubus) and partly by its stirring of sexual fantasies. Yet it was a pioneering effort, not only in its theme – it seems to have been the first picture of a nightmare ever painted – but also in its reliance on (then) modern psychology.

2 In his painting of a vision – the damnation of Paolo and Francesca (*c.* 1824) from Dante's *Inferno* – William Blake treated a moral and poetic theme.

3 The execution of a group of Spanish insurgents in Madrid after their abortive rising against the French occupying forces was powerfully depicted by Goya in his "The 3rd of May, 1808", painted in 1814.

4 The Nazarenes, an early 19th-century group of German "pre-Raphaelites", were the first systematically to revive late medieval styles, as in this detail from Franz Pforr's "Rudolf of Hapsburg and the Priest" (*c.* 1810).

porary, the Neoclassicist Ingres (1780–1867) (who was not immune from romantic feeling himself). This "battle of the styles" was joined around 1830 and it was then that the word "Romantic" was applied to pictorial art for the first time. Romanticism in painting, as epitomized by Delacroix, was identified with colour, movement, breadth of handling and the uninhibited representation of violence.

It is much more difficult to discuss Romantic figure painting in other countries, as they had no continuous tradition either of patronage or outlook. In Britain, there was a growing interest in the irrational expressed by, for example, Fuseli (1741–1825) [1] and Blake [2] and this reached a climax before 1800, earlier than any comparable development in painting elsewhere; however, it was not followed up. In keeping with the date – the late eighteenth century – the style of Fuseli and Blake retains a strong link with Neoclassicism, in the use of forms derived from the Antique and Michelangelo and a dependence on outline. On the other hand, these forms were "pulled out" and given a sort of airy, boneless quality, which was employed by Fuseli to explore the psychological states of terror and nightmare and by Blake to express his "visions of Eternity". German Romantic painting took a different course in a turning back to the styles and subject matter of the Middle Ages and early Renaissance; this was carried out by a group founded by Johann Friedrich Overbeck and Franz Pforr who settled in Rome in 1810 and who called themselves the Nazarenes [4].

A portrayer of violence

There was no Italian painting to speak of in this period, owing partly to lack of patronage, but there was one very important artist in Spain: Francisco de Goya (1746–1828) [Key, 3] who is perhaps the hardest of all to classify. Formally, he was no Neoclassicist, yet his print "The Sleep of Reason produces Monsters" can be interpreted among other things as a warning of what happens when rational – that is, "classical" – order breaks down. No revolutionary celebrator of violence, he yet did not flinch from representing it, without traditional moral overtones, in its most terrible and bloody forms.

KEY

Reason attacked by the forces of irrationalism and the supernatural, and the solitariness of the individual in his journey through life, are two leading themes of romanticism. Both are reflected in Goya's original frontispiece to his series "Los Caprichos" (1799), a set of cryptic satires on contemporary *mores*. Both the frontispiece and its caption, "The Sleep of Reason produces Monsters", typically reveal only part of the artist's meaning. In a private note he explained that whereas fantasy abandoned by reason produces monsters, united with it she is the mother of the arts.

5

5 Violent action and the sense of "poetry" surrounding it was a major Romantic theme. The sense was that not only the results of conquest, battle and danger are desirable but that the experiences are exciting in themselves (seen in this light, defeat may seem as "poetic" as victory). The living embodiment of this ideal was Napoleon, whose military career was charted in pictures by Antoine Gros. In "Napoleon at Eylau" (1808) the emperor is shown displaying his humanity towards the defeated Russian troops (detail).

6

6 Romantic belief that artistic creation springs from pain and turmoil is depicted in realistic terms in Géricault's "The Artist" (*c.* 1818). Whether inspiration comes or goes, the artist is a lonely, tormented being.

7

7 "Sardanapalus" (1827) by Delacroix, a huge, sprawling exhibition of sex and violence, was the ultimate in French Romantic painting. The subject was taken from an oriental verse-play by – significantly – the English poet Byron.

The Congress of Vienna

Even before Napoleon Boneparte's first defeat, in 1814, the idea of an international diplomatic assembly to restore order in Europe was proposed by Prince Metternich of Austria (1773–1859). Intended to ratify decisions made at the first Treaty of Paris, the congress was announced and from September 1814 delegates from throughout Europe arrived in Vienna [Key]. From the start the congress was dominated by four great powers, Austria, Britain, Prussia and Russia, although Prince Talleyrand (1754–1838) soon skilfully gained an equal voice for France.

The distribution of rewards

It was hoped to prevent any one power from gaining more than its fair share of rewards, and to establish a balance of territorial interests. In fact Russia took the major share and established a dangerous foothold in Europe. From this time until the Crimean War (1854–6), fear of Russia was a dominant theme in European diplomacy.

At the Congress of Vienna, however, the immediate fear was that France might cause another European war. Three buffer states were created to hinder her expansion eastwards [1]. The Kingdom of Piedmont was strengthened; Belgium (previously the Austrian Netherlands) was joined with Holland in the Kingdom of the Netherlands; and the Holy Roman Empire (consolidated by Napoleon into the Confederation of the Rhine) became the German Confederation – 39 states joined in a weak *Bund* and dominated by an Austrian president.

Yet in the treaties of Paris of 1814 and 1815, France was generously treated. The frontiers of 1790 were restored and an army of occupation was installed only until France had paid an indemnity of 700 million francs to the Allies – a condition met by 1818. Although the monarchy was restored in the shape of Louis XVIII (1755–1824), he was obliged to reign under the Charter of 1814.

A new political settlement

In addition to the territorial changes, political settlement was considered essential for future peace. The French Revolution was largely blamed for the upheavals and wars of the previous generation. The best hope for stability seemed to lie in the restoration of the legitimate monarchs who had been overthrown. To try to prevent future disturbance in central Europe, the heads of state of the German Confederation were advised to offer constitutions to their subjects – advice which, for the most part, they subsequently ignored.

Finally, the Vienna settlement itself had to be maintained; to this end the four great military powers – Austria, Russia, Prussia and Britain – renewed their Quadruple Alliance and pledged to uphold the settlement, by force if necessary, for 20 years. Viscount Castlereagh [5], the British foreign secretary, in particular saw the alliance as fundamental to the maintenance of the balance of power in Europe, and the four powers agreed to hold periodic peacetime conferences to settle disputes and problems that might arise.

But the relative co-operation and harmony of views shown at Vienna did not continue in the four later congresses held between 1818 and 1822. Austria, Prussia and Russia had formed the Holy Alliance in September 1815. They rapidly adopted the view

CONNECTIONS

See also
84 European empires in the 19th century
76 Napoleonic Europe
78 Nelson and Wellington
108 The revolutions of 1848
182 British foreign policy 1815–1914
184 Balkanization and Slav nationalism
186 Causes of World War I
110 German and Italian unification

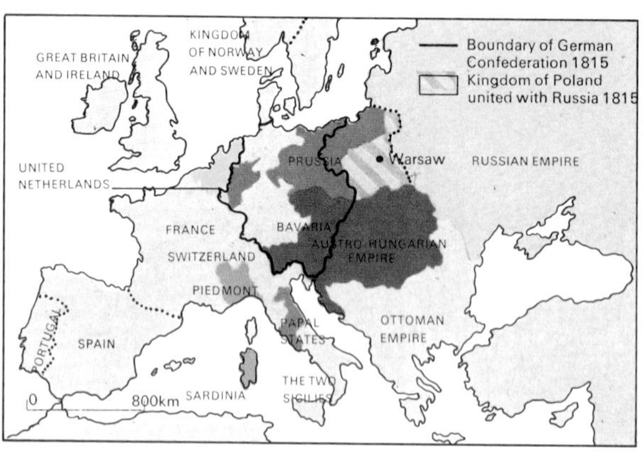

1 The map of Europe had to be redrawn after the 1815 Vienna settlement. The Hapsburg Empire received the Illyrian provinces and the two Italian provinces (Lombardy and Venetia) in return for the former Austrian Netherlands (Belgium). Sweden won Norway, which had been Danish; Russia kept her conquest, Finland, and dominated the new "puppet" kingdom of Poland. Prussia kept Polish Posen, received almost half the Kingdom of Saxony and an area of the Rhineland that included the iron and coal resources of the Ruhr. Britain consolidated her overseas empire and naval routes with the Cape of Good Hope, Malta, the Ionian Islands, Ceylon, Maritius, Tobago, St Lucia and Heligoland. Partly through these overseas acquisitions, Britain grew relatively remote from 19th-century European politics.

2 The diplomats at Vienna reached compromises over their territorial ambitions but there was to be no compromise with the new forces of liberalism and nationalism. Within 15 years unrest in Spain, Portugal, Italy, Germany and France showed the growing desire for constitutional restraints on the monarchies that had been restored. Nationalists were crushed in the Polish revolt of 1830, but they won independence for Belgium (1830) and Greece (where war with the Turks began in 1821). These threats to the Vienna settlement were the main topics discussed at the four subsequent congresses: Aix-la-Chapelle (1818), Troppau (1820), Laibach (1821) and Verona (1822). Greek independence was a blow, weakening Turkish resistance to the nationalist claims of her other Balkan states.

that the powers should intervene in the internal affairs of European countries where stability was threatened, a doctrine repudiated by Britain.

Britain therefore ceased to send official representatives to congresses after Aix-la-Chapelle. Finally Britain dealt the death blow to the congress system by forcing acceptance of Greek independence against the interests of Russia and the protests of both Austria and Prussia.

Consequences for Europe

The settlement reached by the Congress of Vienna shaped the following generation in Europe. The Continental powers were committed to upholding the status quo they had created, and they interpreted their obligations with a rigidity that turned the settlement into a straitjacket. Liberal revolts attempting to introduce constitutional limits to the powers of the restored monarchs were crushed almost without exception, although they were successful in France, Switzerland and Belgium in 1830 because it was neither convenient nor in the interests of all the powers to intervene [2]. The settlement had ignored nationalist feelings in the distribution of rewards and creation of buffer states, and there were revolts in Belgium and Poland and growing unrest in Italy and Germany. Furthermore, the old multi-national empires had been confirmed – the Hapsburg and the Ottoman (Turkish) in Europe.

The Greek revolt of 1821 proved disastrous for Turkey. Its success encouraged other Balkan states to push for independence and weakened the ability of Turkey, the "Sick Man of Europe", to resist. The Hapsburgs had added Croats and Italians to their multiplicity of nationalities. Nationalism anywhere was to be treated as an epidemic that could spread and destroy their empire. Metternich [3] used his skill at the Congress of Vienna, his influence in the congress system and his authority in the German Confederation and the whole of Italy to wipe out any symptom of nationalism. The Metternich system of repressive measures spread from the Baltic to Sicily. But the Congress of Vienna did succeed, in a formal sense in securing European peace.

KEY

In 1815 Napoleon was safely on St Helena and the waltz took fashionable society by storm. The monarchs of Europe danced to celebrate the restoration of their political power and the promise of armed backing by all powers. Five monarchs and the heads of 216 princely families arrived in Vienna for the peace negotiations and the festivities. Their fear of revolution and desire to restore the political situation of the 18th century meant that France was left intact.

3

3 Prince Clemens von Metternich was foreign minister of the Hapsburg Empire from 1809 until the revolution of 1848. To many he seemed the champion of autocracy, reaction and the police state.

4

4 A grand sleigh ride was included in one of the weekly programmes issued by the Festivals Committee responsible for entertaining the visitors. The expenses were paid by the emperor.

5

6 Viscount Castlereagh (1769–1822) was Britain's foreign secretary from 1812. Regarded as reactionary at home he proved too liberal for the congress system, which he had hoped would provide a diplomatic arena for peaceful change.

6

6 Frequent liberal and nationalist revolts threatened the settlement but were usually suppressed. Eugène Delacroix (1798–1863) won the *Légion d'honneur* for his painting "Liberty leading the People", after the successful French revolt of 1830.

European empires in the 19th century

The Austro-Hungarian, Russian and Ottoman empires were all deeply involved in the Balkan countries through most of the nineteenth century. The diplomatic and military conflicts between the three powers were a result partly of their own political ambitions and partly of aggressive national independence movements in the disputed areas.

The Serbian struggle for independence

It was in Serbia, one of the Ottoman provinces in the Balkans, that a subject nationality first challenged the political power of the Ottoman Empire. Turkish rule in Serbia, which had been conquered in 1389, had become particularly tyrannical at the end of the eighteenth century. The local military commanders (*dahis*) exercised a largely independent authority. In 1801 they executed the pasha of Belgrade, the sultan's own representative, and in 1804 they ordered the execution of 72 Serbian village elders. The Serbian uprising of 1804 under Karadjordje [3], a capable military leader, started off as a protest movement against the excesses of Turkish rule, but after striking military successes it developed into a movement aimed at winning full independence.

Russia offered some military and diplomatic support to the Serbs, to whom it was tied through the Orthodox religion and the Slav race, but it was chiefly a combination of Turkish weakness and Serbian resistance that enabled the rebels to remain independent for eight years. The Turks finally crushed the Serbian revolt in 1813 but within 18 months the Serbs revolted again, this time under the leadership of Milos Obrenović (1780–1860), a greater diplomat than Karadjordje.

Obrenović worked out an agreement with the Turks under which Serbia remained formally a Turkish province garrisoned by Turkish troops, but was allowed to share in the administration of justice, to maintain a militia and to summon a national assembly in the capital, Belgrade.

Serbia's struggle for independence was not fully consummated until 1878 when the Congress of Berlin [8] recognized it as an independent state. However, the example of the successful Serbian struggle had a powerful effect on the other Balkan nationalities, inspiring the growing nationalistic movements, especially among the other southern Slavs living under both the Ottoman and Hapsburg empires.

The unification of the Slavs

The effect was greatest in the Hapsburg Empire where many Serbs had fled from the Turks in the seventeenth century. The Orthodox Church was a powerful link between the Serbs in Serbia and the others outside. Fear of being crushed by the twin pressures of forcible germanization from Vienna and magyarization from Budapest brought the Croats and other Slavs, notably the Slovenes, closer together [7].

In the 1848–9 anti-Hapsburg revolution, the Croat general Josip Jelačić (1801–59) fought against the Hungarian revolutionaries with Serbian and Slovene support. But Vienna, after the successful crushing of the 1848–9 revolution, introduced a centralist, strongly germanizing rule. The existence of a semi-independent Serbia fired the imagination not only of the Serbs but of the Croats and Slovenes as well. Linguistic similarities

CONNECTIONS

See also
82 The Congress of Vienna
108 The revolutions of 1848
182 British foreign policy 1815–1914
110 German and Italian unification
168 Russia in the 19th century

In other volumes
220 History and Culture 1

1

1 Napoleon's victory over Austria at Marengo in June 1800 began the process of the Hapsburgs' expulsion from north-western and western Europe. Francis I was forced in 1806 to give up the title of Holy Roman Emperor which the Hapsburgs had held for many centuries. From then on Austria looked to the southeast.

2

2 Lord Byron, who raised an army in the cause of Greek independence, died of fever at Missolonghi in 1824. On 20 October 1827 the Turkish fleet was destroyed at the Battle of Navarino by Britain and France. In 1829 the Treaty of Adrianople recognized Greece's autonomy, and independence came in 1832.

3 Two of the most important figures in the Serbo-Croat independence movement were Ljudevit Gaj (1809–72) [A] and Karadjordje (Georgije Petrović) (1768–1817) [B]. Gaj founded the movement for the political and cultural emancipation of Croatia from Austria. Karadjordje led the uprising against the Turks in 1804. After the suppression of an uprising in 1813 he fled first to Austria and later to Russia.

B

3 A

4

4 Montenegro was conquered by the Turks in 1499, but a large area of its forbidding mountain territory remained outside their grip. From there Montenegrins like these raided the towns that the Turks held. Following the successful wars against the Turks in 1876–8, Montenegro was recognized as an independent state by the 1878 Congress of Berlin. As a result Montenegrin territory was increased by 70% and the population of the country almost doubled.

fostered the idea that all Serbs, Croats and Slovenes were one nation of Jugoslavs or southern Slavs. This idea was developed further in Pan Slavism, a nationalistic movement that agitated for the cultural and political unity of all the Slavonic peoples.

The effect of Russia's foreign policy

Russia saw these movements as instruments of its own drive towards Constantinople and access for its navy all year to ice-free waters. Meanwhile, with Prussia squeezing Austria-Hungary out of Germany since 1815, Austria developed a renewed commitment to its Balkan role. Because of its mistrust of the new nationalism of the Balkan Slavs, Austria in the first half of the nineteenth century also became a protector of Turkey. In response, Russia stepped up its support for Turkey's and Austria's enemies.

Turkey enjoyed the support of Britain, Russia's chief adversary; Britain was joined in the early 1850s by France. After a quarrel over the holy places of Palestine on 21 July 1853, Russia occupied the principalities of Wallachia and Moldavia, which were still under Turkish suzerainty, as a "material guarantee" for the concessions to her "just demands" in Palestine.

On 4 October 1853, Turkey declared war on Russia, as later did Britain and France, believing the integrity of the Turkish Empire to be at stake. Austria stayed neutral but in so doing harmed Russia and greatly increased the hostility between the two powers. The Russian forces were worn down in the Crimea [5] until Tsar Nicholas I died in February 1855. His successor Alexander II sued for peace.

The result of the Crimean War checked Russian ambitions in the Balkans, opened the Danube to international navigation and neutralized the Black Sea. The Turkish Empire's territorial integrity and independence were guaranteed and so were Serbia's liberties. In 1859 the election of Alexander John Cuza (1820–73) as Prince of Moldavia and Wallachia prepared the official union of the two principalities as Romania, which became formally independent in 1878. However, the Ottoman Empire continued to decline up to 1914.

KEY

Suleiman's Mosque still stands as a symbol of the once mighty empire of the Ottomans. In decline from the 17th century, the empire was still strong enough in the early 19th to resist Russian expansionism and maintain some power in Europe.

5

6

5 The Battle of the Alma on 20 September 1854 was the first big engagement of the Crimean War between Russia and Turkey, Britain and France. Following the Treaty of Paris in 1856, Russia's dominance in southeast Europe ended and Turkey gained a new lease of life under the joint protection of the European powers.

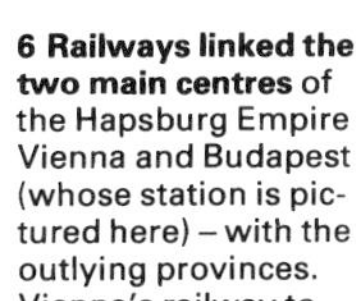

6 Railways linked the two main centres of the Hapsburg Empire – Vienna and Budapest (whose station is pictured here) – with the outlying provinces. Vienna's railway to the port of Trieste was built in 1854; her imports in 1869–73 increased by 83% compared with the preceding five years. Budapest was linked to Rijeka (Fiume) in 1873.

7

7 The coronation of Francis Joseph took place in Budapest on 8 June 1867. A dualist empire emerged as a result of a compromise (*Ausgleich*) between Vienna and Budapest in 1867: Francis Joseph was separately crowned in Vienna as emperor of the Austrian half of the dual monarchy and as king of its Hungarian half in Budapest. The Hungarians reached an agreement with Croatia in 1868, guaranteeing it special status and some autonomy within the Hungarian half of the monarchy. But the new Magyar nationalism was resisted by the Romanians, Croats, Serbs and Ukranians. In the Austrian half of the empire the Czechs led the autonomy struggle against pan-Germanism.

8

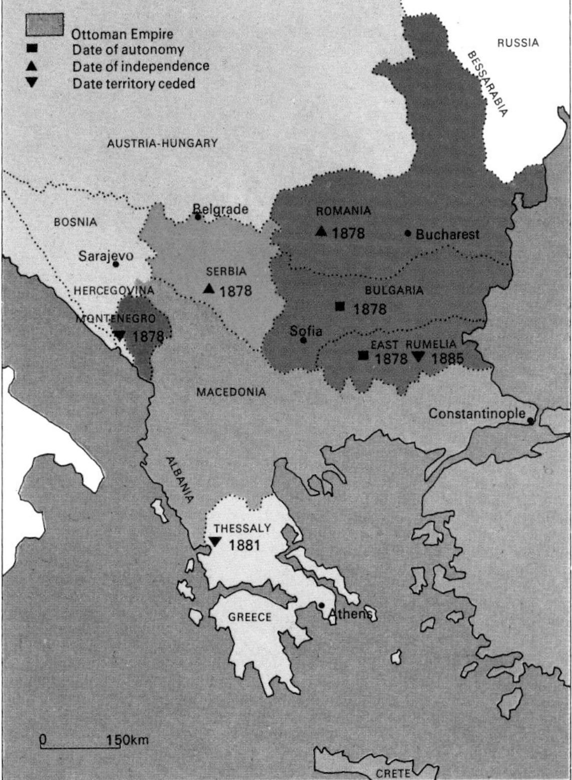

8 The Congress of Berlin produced an uneasy compromise that carried the seeds of future conflict. It gave Austria-Hungary control over the strategic province of Bosnia-Hercegovina but not the title to permanent occupancy. Serbia developed large-scale propaganda among its fellow Serbs and other southern Slavs in Bosnia-Hercegovina and other southern Slav-inhabited provinces of the Hapsburg Empire. In 1908 Austria-Hungary carried out the annexation of Serbia. Bulgaria, cheated of access to the Aegean and of Macedonia, nursed a grievance against Britain and other powers except Russia and Serbia. Romania gave up southern Bessarabia to Russia, which lost control of Constantinople.

Latin American independence

Most of the 20 republics that comprise present-day Latin America became independent between 1810 and 1824 – the period that began after juntas set up in major cities of the Spanish American Empire had refused to accept Napoleon's brother Joseph as their ruler and ended with the last significant battle for freedom, at Ayacucho in Peru.

Haiti had seized independence from France some years earlier, in 1804. The Haitians subsequently imposed their rule upon neighbouring Santo Domingo, which did not achieve freedom as the Dominican Republic until 1844. Brazil, the Portuguese Empire in America, became independent with very little bloodshed in 1822 and, the prince regent, Dom Pedro I, was crowned its emperor. Uruguay emerged as a separate state in 1828 after Argentina and Brazil had fought to claim it. Cuba remained a Spanish possession until the end of the nineteenth century, when the Spanish-American War (1898) led to its becoming independent although bound by close ties with the United States. Panama was a province of Colombia until 1903, when its inhabitants successfully revolted. Its new government leased in perpetuity to the United States (which had assisted the revolt) the strip of land 16km (10 miles) wide through which the Panama Canal, completed in 1914, was to be cut.

The consequences of independence

The independence of Latin America meant essentially that men of European stock who were born there replaced men from the Iberian Peninsula in positions of power and privilege. The social structure inherited from Spain and Portugal remained virtually intact typified by the *hacienda* or great landed estate. The Church, allied with the Crown in the colonial period, continued to exercise a strong conservative influence [5] and the military, greatly strengthened by the prolonged wars, was another privileged institution and one that prejudiced the establishment of effective civilian government.

The vast size of many of the new states, problems of communication, economic dislocation brought about by the wars, lack of experience in administration on the part of the new rulers and the illiteracy of the masses all contributed to make stable government extremely difficult. Few of the heroes of independence were able to govern successfully when peace came to their countries. Simón Bolívar (1783–1830) [Key], the greatest of them, died in self-imposed exile; José de San Martín (1778–1850) [6], the other outstanding liberator of Spanish America, decided to retire to Europe. The characteristic ruler of the new countries was the *caudillo*, or military dictator.

Relationships between countries

Relations between the Latin American countries following independence were generally neither close nor friendly. While Portuguese America remained intact (as Brazil), Spanish America had disintegrated along the lines of the old imperial administrative divisions. These divisions were the accepted basis for the new states, but there were often disputes over ill-defined boundaries.

Geography and history have combined to isolate the countries of Latin America from each other. Formidable physical barriers have been a major cause of this isolation, as

CONNECTIONS

See also
130 Imperialism in the 19th century
44 International economy 1700–1800
256 Latin America in the 20th century

1 On the eve of the wars of independence (*c.* 1800) Latin America was divided between Spain and Portugal. The newly independent states agreed among themselves to keep their national boundaries generally in line with the old colonial administrative divisions. But because these were often not clearly demarcated, territorial disputes inevitably arose. The Banda Oriental (the east bank of the Río de la Plata) had been a particular bone of contention between Spain and Portugal and continued to be such between Argentina and Brazil after independence. Following a war between these countries (1825–8) and diplomatic intervention by Britain, the disputed territory became a buffer state – the new Republic of Uruguay.

2 Britain's significant influence on the newly independent countries of Latin America was exerted primarily through commerce and finance. The massive inflow of British capital reached a peak from 1904–13, when it accounted for at least 20% of all British investment abroad.

2
Total investment
Total investment in government bonds
Total investment in economic enterprises
Share of government bonds and economic enterprises invested in Argentina, Brazil, Chile, Mexico, Peru, Uruguay
Figures in £ millions
25·3
179·4
425·7
540
999·2
682·8
251·3
316·4
123·0
194·4
56·4
1826 1880 1890 1900 1913

3 Joseph Bonaparte (1768–1844) was imposed on Spain by his elder brother Napoleon after the invasion of the Iberian Peninsula (1807–8). This forced the issue of Latin American independence. When the French deposed Ferdinand VII (1784–1833) of Spain and then threatened Portugal, the Spanish Americans at first pledged loyalty to Ferdinand but later declared for independence. The Portuguese royal family fled briefly to Brazil and the king's son stayed as regent of Brazil, declaring it independent in 1822.

4 Native Indians generally viewed Latin American independence as no more than a change of masters. Many who had been subject to the old forms of colonial bondage became *peones* (peasant labourers) on the great estates.

5 A church in Quito, capital of Ecuador, with an ornate and richly sculptured structure reflects the power and wealth of the Church in Latin America, both in colonial and modern times. But Church-state relations were generally uneasy following independence.

well as regionalism within individual countries. During the colonial era the viceroyalties, captaincies-general and presidencies into which the Spanish American Empire was divided were linked to the mother country rather than to each other. Since independence, relations with powers outside the region generally have been much more important than those among the Latin American countries themselves.

Colonial trading patterns continued after independence. Most countries had to rely on exporting one or two primary products and on importing manufactured goods.

Dependence on other countries

The new states of Latin America thus became economically and financially dependent upon powerful external countries. During the nineteenth century Great Britain was the major economic power in Latin America [2]. British capital played a key role in the economic development of Argentina in the latter part of the century. Her naval power forced Brazil to acquiesce in efforts to stamp out the slave trade. The eventual abolition of slavery itself was one of the main causes of the overthrow of the Brazilian emperor and the establishment of a republic in 1889.

By that time the United States had greatly increased its territory at the expense of Mexico, which it defeated in war (1846–8). Even earlier, in 1823, President Monroe (1758–1831) had enunciated his famous "Doctrine". This warned European powers against incursions or further colonization in Latin America and implied that the United States had a special relationship with Latin America. By the end of the nineteenth century the United States, with military strength, was able to compel respect for the Monroe Doctrine when its own interests were at stake. At the same time it promoted "pan-Americanism", embodying the idea that the countries of the Americas shared a community of interests and a special "system" of international relations: the inter-American system. A conference of the United States and Latin American countries in Washington (1889–90) set up the International Union of American Republics – renamed the Pan American Union in 1910.

KEY

Simón Bolívar, known throughout the continent as "The Liberator", was the greatest hero of Latin American independence. He played a leading part in winning freedom for his native land Venezuela, as well as Colombia, Ecuador, Peru and Bolivia, the country named after him. Bolívar brought together the first three of these countries in one state, the republic of Colombia, and he inspired the Congress of Panama (1826) with the principal aim of establishing a league of Spanish-American nations. But the league did not materialize: Greater Colombia split into its constituent states, and Bolívar died, deeply disillusioned, in 1830.

6

6 José de San Martín [right] was the outstanding liberator of southern South America. He assured independence for Argentina and gained it for Chile and part of Peru (including Lima, the capital). While the liberation of Peru, the last great stronghold of Spanish power, was incomplete, San Martín had a famous meeting with Bolívar at Guayaquil in Ecuador (July 1822) to discuss the future of Spanish America. San Martín then withdrew, leaving the field to Bolívar.

7

7 San Martín's "Army of the Andes" crossed the mountains through the Uspallata pass at a height of 3,799m (12,464ft) – an extraordinary military achievement. The army was on its way to liberate Chile, in co-operation with the Chilean patriot Bernardo O'Higgins (1778–1842). The Spanish forces in Chile were taken completely by surprise and routed at Chacabuco on 12 February 1817. In the following April a victory at Maipú, ensured the independence of Chile.

8

8 Bolívar [right] triumphantly accepts the surrender of the Spanish at the Battle of Boyacá (1819), assuring Colombia's independence.

9 Latin America in 1903 looked much as it does today. Mexico had long before lost more than half its national territory (the former Viceroyalty of New Spain) to the United States. Cuba and Panama had become nominally independent, although virtually protectorates of the United States, in 1902 and 1903 respectively. Paraguay had declared itself independent in 1842. Bolivia had lost its coastal territory to Chile in the War of the Pacific (1879–83) and was now landlocked. Central America had dissolved into its constituent states (Costa Rica, El Salvador, Guatemala, Honduras and Nicaragua) as early as 1838.

9

The Industrial Revolution

The first 70 years of the nineteenth century saw unprecedented economic development in Britain as forces unleashed at the end of the eighteenth century created the first urban industrial society. Population growth and urban development followed an acceleration of industrialization based on a great expansion of trade, the widespread application of the factory system to production and the harnessing of steam-driven machinery to an increasing range of processes. Steam power was also applied to transport with the development of railways and the first steamships. Urban life prompted Britain to develop many social and political institutions that were to become standard in other countries as the Industrial Revolution spread to Europe and the United States.

The British lead

Britain's economic development between 1800 and 1870 was startling, even compared with the progress of the late eighteenth century. There were giant increases in production. Output of pig iron grew 60 times, coal output ten times and total trade by the same amount. Britain maintained and increased her lead over other countries by advances in mechanization and factory production. In a real sense Britain had become the "workshop of the world" by the time of the Crystal Palace Exhibition [Key] in 1851 when great industrial expertise was on display.

Britain supplied a large percentage of the world's textiles, iron and machinery, and a massive increase in her export income was stimulated by the development of "free trade", especially during the 1841–6 ministry of Robert Peel (1788–1850). After 1850 trade expanded even more rapidly than it had in the first half of the century, encouraging further economic development. New industries such as steel (based partly upon the newly discovered Bessemer process) and shipbuilding began to balance Britain's dependence on exports of textiles [10] and iron products.

The development of railways after the opening of the Stockton and Darlington Railway in 1825 gave a major boost to the economy, making it possible to move bulky goods cheaply and stimulating the iron and steel industries. The railways served to concentrate production still further, as raw materials could be brought long distances and finished goods sent to ports many miles away. During the boom years of "railway mania" in 1845–7 a basic railway network covering the major towns, industrial areas and ports had been laid out by railway pioneers such as George Stephenson, Isambard Kingdom Brunel, George Hudson and Thomas Brassey. In addition, the development of railways played an important part in refining investment and banking procedures.

Financial organization

As the pace of industrial expansion quickened, the need arose for a more elaborate banking system. In Britain the less reliable "country" banks were more and more superseded by "joint-stock" banks after 1826. The Bank Charter Act of 1844 secured the role of the Bank of England as the central note-issuing authority and guarantor of the rest of the banking system. Company finance and formation were regulated by a series of limited liability and company acts in the

CONNECTIONS

See also
66 The early industrial Revolution
90 The urban consequences of industrialization
92 The rural consequences of industrialization
94 The British labour movement to 1868
96 Social reform 1800–1914
160 The fight for the vote
158 Industrialization 1870–1914
154 The impact of steam
130 Imperialism in the 19th century
164 Scotland in the 19th century
122 European architecture in the 19th century
148 USA: the opening up of the West

In other volumes
136 The Physical Earth
312 Man and Society
22 Man and Machines
30 Man and Machines

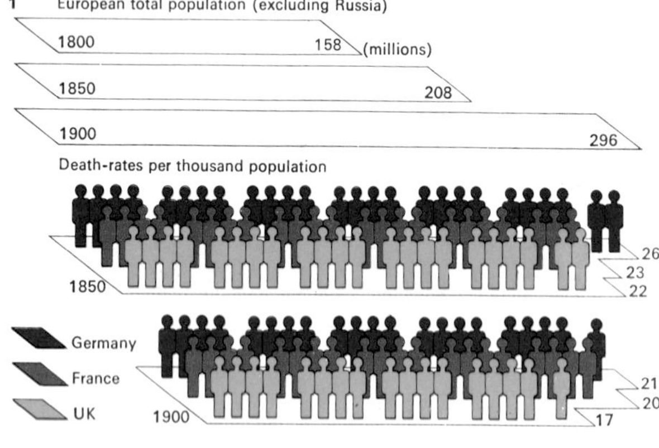

1 Europe's population rose steadily during the 19th century, mainly because of a falling death-rate through improvements in medicine, diet and living conditions. Birth-rates also tended to rise with industrialization and urbanization. As a result, the total population of Europe almost doubled in the course of the century, quickening migration from the countryside to the increasingly crowded urban centres.

2 Industrial output was rising in many parts of Europe by the middle of the 19th century. Germany and France began to take a significant share in producing iron, coal and textiles and smaller countries such as Belgium and Switzerland were also beginning to develop important industrial sectors. European industrialization still lagged behind that of Britain and was inhibited to some extent by Britain's marketing dominance.

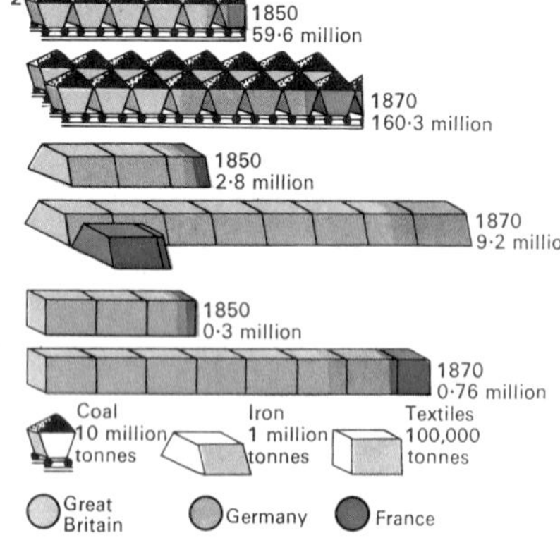

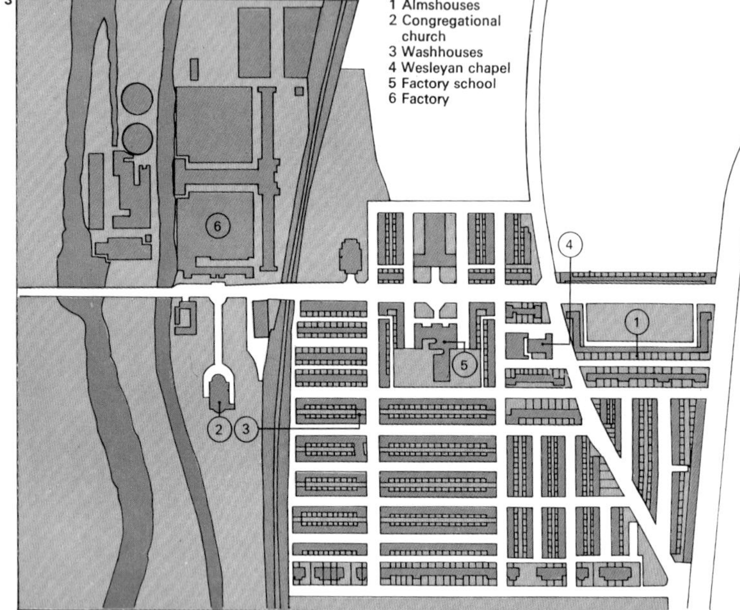

3 New industrial towns, such as Saltaire, in Yorkshire, England, provided shelter and adequate living conditions for large numbers of workers. By the middle of the 19th century factory owners and municipal authorities began to create some order out of the squalor of early factory towns. Regular grid-iron patterns of workers' housing were built, providing the basic amenities of sanitation and water.

4 Riots and strikes in England during the 1840s accompanied efforts by the Chartist movement to win urban workers the vote. Industrialization brought many such political movements and played a part in the European revolutions of 1848.

5

1835 ·25

1845 4·5

= 100km

1870 31·7

5 Railway expansion in Belgium between 1835 and 1870 was typical of the rapid developments that took place in Europe in the middle and late nineteenth century. British engineers, contractors and equipment were often employed in an effort to overtake the British lead. Although railways developed more slowly on the Continent, Britain had opened a major trunk route system for carrying goods and people by 1847. The diagram represents length of rail track laid.

middle of the nineteenth century. The growth of trade led to the expansion of the Stock Exchange and the rise of provincial exchanges [8] to deal in specific commodities. By 1870 Britain was not only the centre of the world's industry and trade but its financial capital. Personal wealth increased rapidly [7].

Population growth

Economic and industrial development was accompanied throughout Europe by population growth [1]. Britain's population increased most rapidly of all, doubling between 1801 and 1851. By the middle of the century Britain was no longer a predominantly rural nation, for more than half its people lived in towns [3]. In 1801 there were only 14 European towns with more than 100,000 inhabitants, but by 1870 there were more than 100.

Urban development brought with it a wide range of social and political problems. To deal with these Britain, as the first industrial nation, pioneered many social institutions fundamental to modern life. Measures to regulate public health, provide basic sanitary and housing amenities and preserve public order through the formation of professional police (the "Peelers") were copied by other countries. Similarly, the introduction of a reliable, cheap postal service [9], the rise of cheap newspapers and the development of cheap railway travel did something to offset the human misery that often accompanied urban development and industrial advance.

Factory Acts [6] regulated child and female labour, as well as hours of work, from the 1830s. Under early pioneers such as Robert Owen (1771–1858) and Robert Applegarth, industrial workers began to organize themselves into trade unions, political associations and the co-operative movement [4], in order to improve their status.

In Europe the gathering pace of industrial development was shown in the growth of railways [5], textile industries and iron and coal production [2] by 1870. Belgium, France and Germany made the largest strides, and although far behind Britain, both Germany and the United States were poised for rapid industrial development in the latter years of the nineteenth century.

KEY

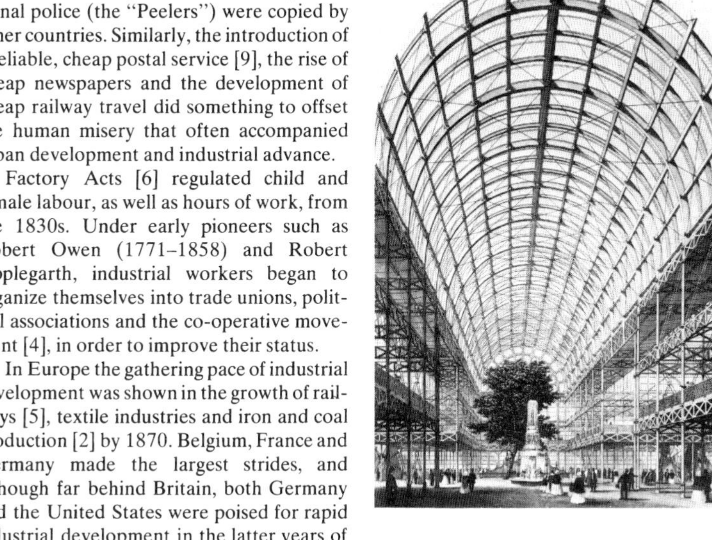

The Great Exhibition of 1851, in London, marked a high point in Victorian industrialization. Organized to show the progress in trade and manufactures achieved since the first days of the Industrial Revolution, it became a symbol of British manufacturing ingenuity and dominance of world trade, although it exhibited industrial goods from many other countries. It was intended to display the virtues of free trade (*laissez-faire*) as an agent of economic progress. To house it, a revolutionary building of glass and iron was designed by Joseph Paxton and built in only seven months. The Royal Society of Arts sponsored the exhibition with the backing of Albert, the Prince Consort.

6

Labour legislation | Labour prohibited | Length of working day

1833 Factory Act — Children under 9 20·30 – 05·30 — 9 hrs, 12 hrs

1842 Mines Act — Children under 10, Women

1844 Factory Act — Apprentices under 10 — 6½ hrs, 12 hrs

1847 Factory Act — 18·00 – 06·00 — 10 hrs, 10 hrs

1850 Factory Act — 10½ hrs

24-hour day · Labour permitted · Men · Women · Labour prohibited · Working day · Children 13-18 · Under 13

6 Exploitation of child and female labour, with long hours, low wages and poor conditions, was a major abuse of the Industrial Revolution. In the middle of the 19th century, humanitarian concern in Britain led to the passing of Factory Acts to protect women and children.

7

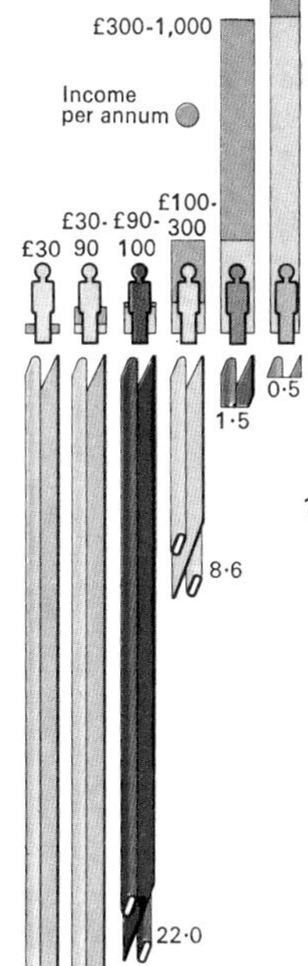

7 Incomes and social status in Britain changed with the rise of the middle and professional classes and the creation of a new class of manufacturers. But in the mid-19th century the largest group still earned less than £30 a year.

8

8 The Cotton Exchange in Manchester was one of a number of major commercial institutions set up throughout Britain to deal in particular commodities. The growth of large-scale industry and the demands of a more complex society forced rapid developments in finance and banking. The Stock Exchange, which had become the centre for financial dealings in the 18th century, continued to expand, doubling in size during the 1860s alone.

10

10 The cotton mill was the symbol of the 19th-century industrial town. Cotton was the most completely industrialized sector of the economy, being almost entirely mechanized, steam-powered and factory-based, and was one of the first industries to develop in Europe. Mills were gaunt, utilitarian structures, housing long banks of spinning and weaving machines, tended largely by women and children. Conditions were often dangerous with many accidents; hours were long, even for very young children, and discipline was strict. In Britain by 1851 over half of the population lived in urban rather than rural areas. Factory conditions improved only slowly.

9 A

B

POST OFFICE
LETTER BOX
Nº 1

9 A cheap postal system was one of the many new social amenities made possible by growing community wealth and a more ordered urban society. In Britain, the railway system permitted rapid movement of mail and a "penny post" was introduced by Rowland Hill in 1840 [A]. The British Post Office introduced the first of its distinctive red letter boxes in London in 1855 [B]. A telegraph system came into use in the middle of the century, with undersea cables providing the first international means of communication. By 1861, 18,000km (12,250 miles) of cable had been laid.

The urban consequences of industrialization

Pre-industrial Britain was a predominantly rural society in which there was only one large city, London, and few other large towns. In 1700 London had a population of more than half a million, but only six towns had populations of more than 10,000. Many parts of the country supported only villages and small market towns with populations of fewer than 5,000. Population growth from the mid-eighteenth century combined with the expansion of industry transformed Britain into a predominantly urban nation.

The growth of towns

By the middle of the nineteenth century, there were more than 70 towns in Britain with populations of more than 10,000, eight with more than 100,000 and Glasgow, Birmingham, Manchester and Liverpool had more than 250,000 inhabitants. By 1851 more than half the population lived in urban areas, compared with about a sixth in 1700 [1]. This growth continued until the eve of World War II, when more than four-fifths of the total population of Britain lived in urban areas. Only in the mid-twentieth century has the spread of urbanization in Britain been reversed. Continued suburban development and the growth of car ownership has permitted more people to live outside urban areas in the years since 1945 [5].

One major impact of population growth and industrialization was rapid urbanization. Population in Britain rose three-fold between the middle of the eighteenth century and the middle of the nineteenth, from more than 7.5 million to more than 21 million. Although population growth occurred in the countryside as well as in the towns, urban centres expanded both from internal increase and migration from rural areas. London received between eight and twelve thousand immigrants a year by the end of the eighteenth century. In addition, the redistribution of population was changed – new industrial regions such as Clydeside and Lancashire became principal centres of growth.

New industries often recruited substantial portions of their labour force from the surrounding countryside. Short-distance migration, of not much more than 30 or 40km (20 or 30 miles) in most cases, was the general rule within Britain. Some immigrants, however, did come from farther afield – from Scotland, Ireland, and rural Wales.

Local government created

The rapidity of growth is well illustrated by Manchester. A population of 75,000 in 1801 had grown to nearly 750,000 inhabitants by the eve of World War I. These tremendous increases in urban population almost completely swamped the provision of social amenities and local government. Until 1835, Manchester was still governed as though it were a rural parish, although it had 250,000 inhabitants. Slowly, the structure of local government was created to deal with these problems. The 1835 Municipal Corporations Act provided a basic framework for local government, and during the century most towns were given elected councils and the apparatus of local government [Key].

Conditions in the early industrial towns were often cramped, unhealthy, and insanitary [3]. Rapid expansion meant that families were crowded into cheap lodgings, cellars, and small courts. Piped water sup-

CONNECTIONS

See also
66 The early Industrial Revolution
88 The Industrial Revolution
96 Social reform 1800–1914
16 Europe: economy and society 1700–1800
68 Pitt, Fox and the call for reform
92 The rural consequences of industrialization
94 The British labour movement to 1868
114 Victorian London
122 European architecture in the 19th century
154 The impact of steam
158 Industrialization 1870–1914
164 Scotland in the 19th century
224 Britain 1930–45

In other volumes
192 History and Culture 1
294 History and Culture 1

1 In 1700 only an estimated 16 per cent of the population in Britain lived in towns of more than 5,000 people. The Industrial Revolution and its attendant dramatic population growth in the 18th century created a predominantly urban society by 1900, when 77 per cent of the population lived in towns. This growth of the new towns and cities within 200 years bore little relation to the pattern of towns in pre-industrial Britain. Instead the expansion was almost entirely dictated by economic necessity. Some of the most spectacular growth took place in parts of the country that had been least densely populated in the pre-industrial era, such as Lancashire, Yorkshire, north-east England, South Wales and the Lowlands of Scotland. These industrial regions dominated the UK economy until the economic slump of the Depression in the 1920s and 1930s.

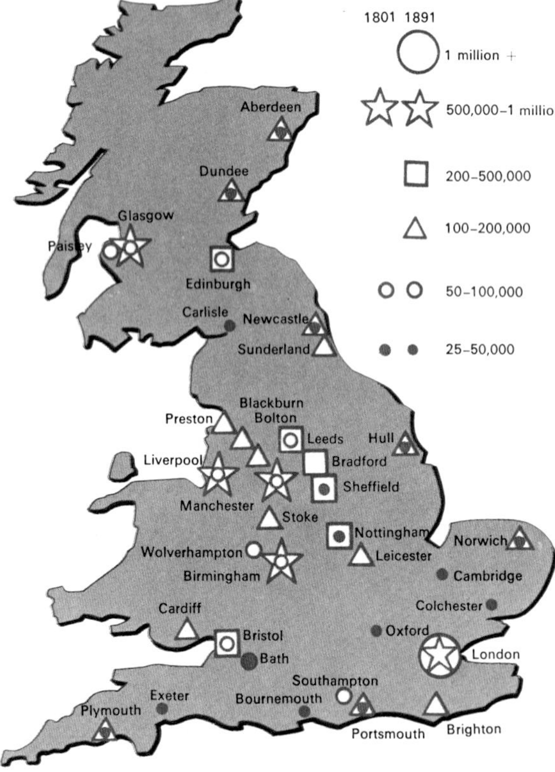

2 The human conditions behind the creation of the first industrial nation were tragic. The unprecedented changes wrought by the Industrial Revolution on Britain's demographic character brought an equally dramatic decline in the social conditions for the majority of the population. Glasgow, an expanding city of more than 100,000 people, had only 40 sewers in 1815. This horrific level of sanitation and hygiene caused an increase in the death rate, and the city's population level would actually have declined in the 1820s and 1830s had it not been supplemented by steady immigration.

4 Middlesbrough was literally a creation of the Industrial Revolution. In 1801 it was a tiny group of houses of only 25 inhabitants; but by 1901 the population was more than 90,000, with iron, and later steel, as the principal industry. Without the railways, in this case the Stockton and Darlington line, the town would probably never have existed. The carefully planned growth of the streets and houses, still evident today in this aerial view, was the product of its Quaker founders, who first recognized its great commercial potential at the terminus of the new railway line. In the space of 100 years, Middlesbrough had become one of the commercial prodigies of the 19th century.

3 Cramped "back-to-back" housing was constructed to accommodate the expanding populations of the early industrial towns. The growth of some old towns was actually restricted by local landowners who feared that their power would be undermined by the new industrial masses. This led to chronic overcrowding within the boundaries of the old towns. Only in the mid-19th century did the government begin to introduce legislation to clear and improve insanitary areas.

plies and sanitary services were often totally inadequate or non-existent, and resulted in disease and high mortality rates, especially among young children. In 1842 the average life expectancy for children of labouring families in Manchester was 17 years, compared with 38 in rural Rutland. Cholera epidemics in 1831–3, 1847–8, and 1865–6 helped to focus attention upon the need for improvement in sanitary conditions. The first Public Health Act was passed in 1848 and a Board of Health was set up to deal with some of the problems of the industrial towns. But industrialization was not responsible for all the squalor and overcrowding to be found in the towns. Pre-industrial London, for example, had had its unsavoury stretches.

Even when new housing was constructed it was often built cheaply by factory owners or speculative builders. Small, terraced houses, often without adequate light or ventilation, with poor foundations and of flimsy construction, soon infested by damp and vermin, created a legacy of slum housing that survived well into the twentieth century in many industrial towns. Indeed it was only after the destruction brought about by the blitz in World War II that extensive rebuilding of nineteenth-century slums in Britain's cities was carried out [7].

Social concern and planning

Towards the end of the nineteenth century philanthropists and social reformers, conscious of the destructive physical and social effects of industrialization put forward ideas for limited, planned towns and cities. Robert Owen (1771–1858) had attempted to create a "model" community at New Lanark and the first proper "garden cities" at Letchworth (1903) and Welwyn (1920) show a similar concern for careful regulation of the growth and structure of towns and cities. In 1895–6 the first industrial estate, at Trafford Park in Manchester, was built; railways, canals and other transport now enabled a separation to be made between work and home, and encouraged a concentration of industry that was socially and economically attractive. On a smaller scale, the houses built by knitting-machine pioneer Jedediah Strutt (1726–97) can be seen to this day.

KEY

Manchester Town Hall, designed by Alfred Waterhouse (1830–1905), symbolizes the civic pride of the urban civilization created by the Industrial Revolution. The nineteenth century saw the creation of local authorities to deal with the intense problems caused by uncontrolled growth.

5

5 Railways not only led to the spread of towns into the countryside, the creation of "suburbia", but they also resulted in the creation of holiday resorts for the industrial workforce. Blackpool and Scarborough are examples of seaside resorts that developed in the 19th century a short train ride away from industrial regions. Here holidaymakers are shown leaving London for Cornwall in August 1924.

6

6 Industrialization has created a more affluent society. Previously the predominantly agricultural population had been almost entirely dependent upon fluctuations in harvest levels. Until 1850 it is true to say that the overall standard of living did not decline, although it was subject to severe fluctuations and regional discrepancies. After that time, the standard of living of the population rose, with higher real wages, and kept to a more consistent level. This is shown in the provision of public amenities such as schools, roads and hospitals as well as in the level of personal consumption.

7

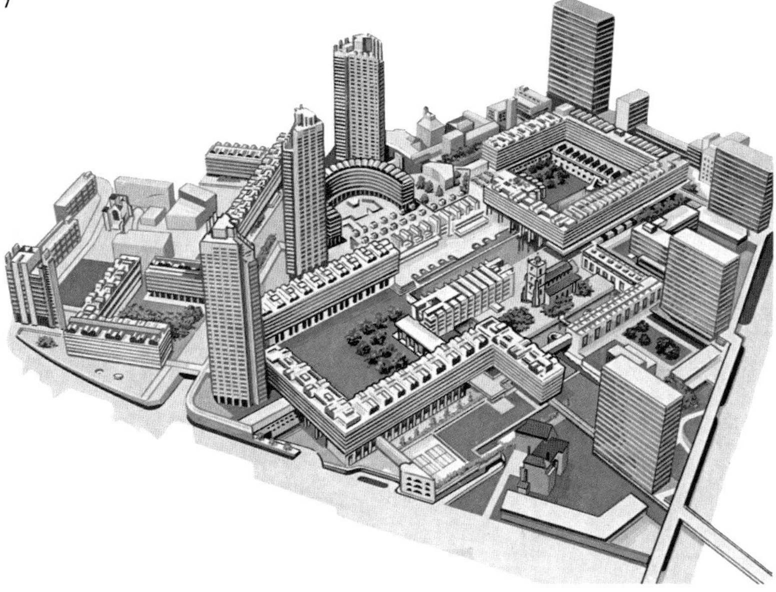

7 London's Barbican housing project is a fine example of the redevelopment that has taken place since the blitz destroyed large areas in many of Britain's cities. Historic features such as St Giles's Church have been sensitively incorporated into the scheme; and pedestrians and traffic have been separated. The complex also includes shops, a theatre, restaurants and a concert hall.

8

8 The Alton Estate at Roehampton in London illustrates one of the more successful attempts to rehouse the population of the overcrowded inner city areas in an attractive environment. Built between 1952 and 1961, the 11-storey blocks are carefully grouped among four-storey buildings and terraced houses with plenty of open spaces and trees situated on the estate.

The rural consequences of industrialization

The Industrial Revolution had profound consequences for agriculture and rural life. Population growth and increasing urbanization stimulated a demand for foodstuffs of every kind, which in turn made necessary a drastic expansion and development of agriculture. This involved the reclamation of marginal and waste land, the reorganization of landholding through enclosure, the introduction of new crops and techniques, the scientific breeding of healthier and bigger animals, and a more efficient, capitalistic type of farming. The result was a sufficient increase in domestic agricultural production to satisfy the demands of an expanding population until the last quarter of the nineteenth century, when cheap foreign foodstuffs became generally available.

Unemployment on the land

The expansion of agriculture was not sufficient to absorb all population growth on the land. Although the number of families engaged in agriculture rose from 697,000 in 1811 to 761,000 in 1831, many more were forced by sheer economic circumstances to swell the workforce of the industrial towns.

Those who remained were often faced with poor prospects. In the rural south, the system of subsidizing wages from parish rates, introduced by the magistrates of Speenhamland in Berkshire in 1795, discouraged farmers from paying economic wages. Moreover, population growth created conditions of chronic rural unemployment, which depressed farm wage levels to near subsistence level. The harsher New Poor Law of 1834 gave farm labourers the choice of low wages or even worse conditions in the workhouse. By the end of the century, the rural counties still had the highest levels of poverty in the country, often as bad as the worst urban slum areas. Cottage industry too, especially handloom weaving, was badly hit by competition from the factories. Although enclosure did not immediately reduce the agricultural labour force, often actually increasing the demand for labour, wages on the land remained persistently lower than those in industry.

By the turn of the century a drift from the land was accelerated by the depression in prices for farm produce. By 1901, less than ten per cent of the total labour force in the country was involved in agriculture [5].

"High farming" period

Mechanization had not played an important part in the agricultural improvements of the eighteenth and early nineteenth centuries. Seed drills and threshing machines had some success, but the latter aroused opposition in the "Captain Swing" disturbances of 1830–32. The mid-Victorian "high farming" period saw the introduction of more elaborate machinery, including the use of traction engines for steam ploughing. These machines were expensive and not suited to every type of soil, but many new types of apparatus were in use by 1870.

The introduction of the internal combustion engine in the twentieth century had a dramatic impact on farming. Tractors proved useful for a wide range of tasks and, by 1939, there were 55,000 in use. By 1945, there were more than 200,000 tractors working in Britain and more than 50,000 combine harvesters [8]. Electricity was also being

CONNECTIONS

See also
20 The agricultural revolution
66 The early Industrial Revolution
88 The Industrial Revolution
90 The urban consequences of industrialization
94 The British labour movement to 1868
96 Social reform 1800–1914
154 The impact of steam

In other volumes
196 History and Culture 1
158 The Physical Earth

1

1 The Nant-y-Glo ironworks in Wales in an early 19th-century picture presents a prospect soon to become too familiar — industrial pollution. While it was occasional and localized, pollution could be ignored or sometimes enjoyed as a "sublime" vision of hellishness. Despite the unhealthiness and squalor of the conditions in which they worked and were housed, to many in the most poverty-stricken areas industry brought a welcome opportunity to earn a living. Ironworks in South Wales and the coalfields in the valleys attracted labour from the surrounding regions and some men came on foot from North Wales. Factory life was even thought preferable to farming.

2

2 The map shows the routes of the earliest railways in Britain, initiated by the opening in 1825 of the famous Stockton-Darlington railway. The railways were in fact only the third wave of improvements in transport in Britain since the 17th century. The building of turnpike roads and of canals had already done much to transform communications and trade, and made travel itself more convenient and enjoyable – the 18th century was a golden age of British tourism. The success of the Stockton-Darlington railway – it more than halved the cost of coal in Stockton – initiated a railway boom, that bound the once distant provinces into an interdependent trading grid, establishing industries far from cities and ports.

3

3 John Kay, inventor of the flying shuttle in 1733, is wrapped in a sheet so as to make good his escape from the wrath of rioters at his window. The flying shuttle put out of a job those who previously had thrown the hand-shuttles, and enabled a loom to be worked by one weaver alone. By undermining the rural cottage industries, this and other inventions concentrated within the town the main sources of employment. These rioters were members of an urban workforce whose divorce from the land would soon be politically significant.

4

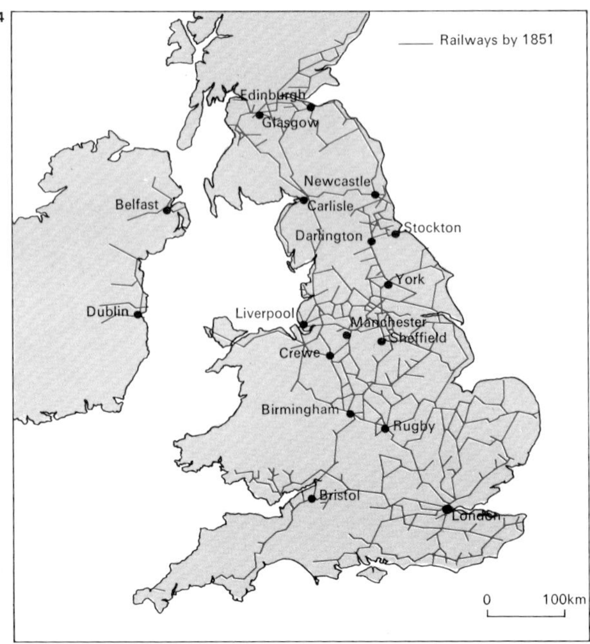

4 An expanding railway network was established by 1851. The bridges, tunnels and stations created by the railway engineers proved that the transformed landscape was nowhere inaccessible. But although the influx of trade brought whole new towns, such as Swindon, into being in the Midlands, the more backward parts — much of Wales, Scotland and Ireland — were unaffected. The new habits of leisure travel induced by railways could be seen in the success of the tours organized by Thomas Cook (1808–92).

used for milking and heating. Technology was applied to a wide range of farming techniques. Animal husbandry was now more scientific and embraced battery farming and complex fertilizers and feedstuffs. The dwindling workforce became much more highly skilled as manual labour was taken over by machinery.

Rural enfranchisement

Social relations on the land were much influenced by the changes in agriculture. Very gradually since the sixteenth century the rural "middle class" of tenant-farmers and yeomen was displaced by the larger landowners and farmers, who employed landless wage-labourers. The dominance of squire and parson was undermined by the enfranchisement of the rural worker and reorganization of local government. In 1884, most agricultural labourers received the vote. The Ballot Act of 1872 also removed them from the more obvious forms of landlord domination by introducing the secret ballot. The establishment of county councils in 1888, and parish councils in 1894, aided the decline in the influence of the landlord. The early successes of Joseph Arch's Agricultural Trade Union in the 1870s illustrated the permeation of union organization among the agricultural labourers; its progress was nevertheless much slower than among industrial workers.

Many rural areas were brought into the industrial age only with the coming of the railways in the late Victorian and Edwardian eras [2, 4]. The last phase of railway expansion brought branch lines to many hitherto untouched areas. This trend was reversed following the Beeching Report in 1963 which recommended cuts of 8,000km (5,000 miles) of railway. Subsequently, reduction in public transport further isolated many towns and villages. As a result, the motor car became a necessity for those living and working out of town.

The motor car also enabled city-dwellers to enjoy rural pleasures more easily. Large areas of land were set aside as national parks, to be preserved from urban encroachment, while other parts of the countryside were developed for tourists [7, 9].

KEY

A Yorkshire miner of 1814 retains a rural look against an early industrial background. Behind him the steam-driven pithead winding-gear brings coal and miners to the surface, and a Blenkinsop locomotive hauls tubs of coal as hills stretch behind. During those early days of industrialization, mining and textile communities were hardly different from farming villages, slightly larger but not yet obtrusive. With the expansion of industry accompanied by a rise in population, factories and housing began to encroach on the rural landscape so much that the countryside in many places became a mere interval between towns. Urban "sprawl" has continued to this day.

5 Figures in percentages

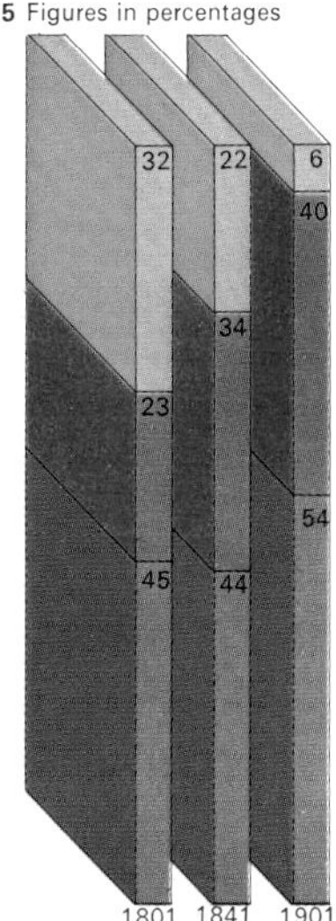

5 The proportion of workers employed in industry and services and in agriculture altered greatly between 1801 and 1901. This comparison reveals a clear drift away from the land into the cities; a trend that continues.

6

6 The destruction of the country house is a recent consequence of industrialization. In previous centuries the country house was the centre from which all agricultural wealth was created, and the unit of local power. Its architecture was an expression of its owner's local status and national role, even if his money came originally from trade or the colonies. After industrialization the country house lost its economic and political vitality and importance.

7

7 Traffic jams are a consequence of the countryside's role as the playground of the cities. One cause was acceptance by industry of holidays as periods of universal shutdown.

8

8 Industrialization of the farm itself is one of the inevitable consequences of the mechanization of the entire economy. The trend towards investment in machinery to do the work of many men has made machines such as combine harvesters a commonplace on the land. To obtain sufficient returns on capital outlay farms have had to expand greatly.

9

9 Giraffes quench their thirst at Longleat, Wiltshire, one of the most popular stately homes in Britain. If the opening of great houses and their parks to the public has made possible their upkeep, it has also spoilt them. The traditional English countryside has lost its essential rusticity, and even moors and mountains, once romantic dreamlands, are now "amenities".

The British labour movement to 1868

Craft organizations had existed for centuries in Britain, usually protected by a framework of paternalistic legislation that determined terms of apprenticeship and wages. With the growth of towns and industry during the Industrial Revolution in the eighteenth century, the old craft regulations came under pressure from employers who sought to free industry from rigid restrictions and to introduce labour-saving machinery.

Unrest and the Combination Laws

The wars with revolutionary France, which opened in 1793, were marked by high prices and labour unrest. Fearing the growth of radical ideas among the lower classes, the government passed the Combination Laws of 1799–1800 [Key]. These were the culmination of a series of laws against "combinations" in specific trades. The Combination Laws prohibited any association between two or more workmen to gain either an increase in wages or a decrease in hours. Unions were forced to operate in secret or under the guise of "non-political" Friendly Societies, which were recognized as legal in 1793.

The economic warfare between Britain and France in the latter part of the Napoleonic Wars brought trade depression and hardship to the growing industrial areas. In 1810–12 there occurred the most serious wave of Luddite disturbances [3], in which workmen under a mythical leader, "King Ludd", destroyed machinery which they saw as threatening their livelihood.

This violence was in large part the traditional reaction of workmen threatened with a decline in their living standards. The degree of union organization in the Luddite outbreaks is obscure, but some elements of union organization were undoubtedly present in Nottinghamshire. Further outbreaks of machine-breaking in 1816–17 and 1826 were also firmly repressed.

In the post-war years, continued distress and radical agitation for parliamentary reform made the government suspicious of trade union activities. Strikes in the factory districts took place in spite of the Combination Laws, most notably in Lancashire where the cotton spinners and weavers conducted an extensive strike in 1818. Elsewhere brickmakers and carpenters secured wage advances without being prosecuted.

Postwar agitation came to a climax in the St Peter's Fields meeting in Manchester of August 1819 [5]. The Peterloo Massacre, as it was dubbed by the radical press, helped to create a more sympathetic attitude towards working-class organizations. The writings of men such as William Cobbett (1763–1835) [4] were also creating a more self-conscious desire for improvement among workmen.

Growth in union membership

In the easier economic climate after 1820, the Combination Laws were attacked. A former tailor, Francis Place (1771–1854), devoted himself to the legalization of trade unions and, with the support of radical MPs, secured the repeal of the Combination Laws in 1824. Unions could now bargain about conditions although still surrounded by some restrictions. Attempts, in 1830 and 1833 to form a single national union failed. Many unions turned to "new model unionism", emphasizing their respectability and rejecting militant activity.

CONNECTIONS

See also

68 Pitt, Fox and the call for reform
160 The fight for the vote
210 The British labour movement 1868–1930
164 Scotland in the 19th century
20 The agricultural revolution
28 Scotland in the 18th century
64 The American revolution
90 The urban consequences of industrialization
96 Social reform 1800–1914
112 Queen Victoria and her statesmen
170 Political thought in the 19th century
212 Socialism in the West

1

1 Thomas Paine's (1737–1809) *The Rights of Man* (1791–2) was published in reply to Edmund Burke's (1729–97) criticisms of the French Revolution. It did much to stimulate popular radicalism. However, threat of prosecution forced Paine to flee the country for France in September 1792.

2

2 Disturbances broke out in England between 1830 and 1832 in which agricultural labourers protested against unemployment, low wages and the introduction of threshing machines. The unrest was not politically motivated, but was a reaction to growing poverty. Nine labourers were hanged and 457 transported.

3

3 Luddite rioters of 1810–12 and 1816–17 smashed factory machinery in protest against the introduction of new equipment in the hosiery and woollen cloth industries. The protesters claimed to be led by a "Ned" or "King Ludd" whose name was attached to public letters denouncing the introduction of the new machinery. The riots caused a series of harsh measures to be enacted by the government.

4

4 William Cobbett was the most influential of the radical critics and writers in the parliamentary reform movement. Of humble origin, he published a number of radical newspapers including, in 1816, the weekly *Political Register*, which soon had an estimated sale of 60,000 copies a week among working men. His hatred of the new industrialism is evident in his documentary *Rural Rides* (1830).

5

5 The Peterloo Massacre, so-called, was a tragic fracas that took place in August 1819. Manchester reformers called a meeting at which the radical demagogue "Orator" Henry Hunt (1773–1835) was to speak. But the local magistrates, fearing trouble, ordered the yeomanry to arrest Hunt at the meeting. When this failed, Hussars were sent in against the crowd of 60,000, and in the ensuing confusion 11 people were killed and more than 400 injured, including women and children. The incident was used by the government as a pretext for introducing a fresh wave of repressive legislation, the Six Acts, against "seditious assemblies" and politically "subversive" literature.

By the 1840s most unions consisted of skilled workmen and the bulk of semi- and unskilled workers still lay outside union organization. The conviction of the "Tolpuddle Martyrs" in Dorset in 1834 [6] for administering unlawful oaths showed the obstacles that could still face unskilled workers who tried to organize themselves.

Many unions took an ambivalent attitude towards the Chartist demands for the vote contained in the People's Charter [7]. Elite craft groups, such as the engineers or potters, were reluctant to align themselves with a movement tainted with violence and disorder. Some of the declining crafts, however, such as the handloom weavers, participated in Chartism as a desperate attempt to reverse their deteriorating situation.

With the decline of Chartism after 1848, the craft unions continued to consolidate their position. By 1852, the Amalgamated Society of Engineers had 12,000 members, centralized control, and high rates of subscription, which enabled it to wage successful strikes. Unskilled workers formed organizations, such as the Miners' Association of 1842, but still lacked the solidarity and strength of the skilled workmen.

From the period of model unionism, there was an improvement in the public image of the trade union movement. The Friendly Societies and Co-operative Movement, founded at Rochdale in 1844, were aided by middle-class sympathizers [9].

Co-ordination of union activities

In 1866–7, a short trade slump in the midst of improving conditions led to a number of strikes and some violence, notably at Sheffield. The "Sheffield Outrages" [10] led to a Royal Commission in 1867 on trade unions. The Commission recommended putting trade unions upon a firm legal basis and allowing them to secure their funds. These gains were established in the Trades Union Act of 1871. In 1868 the Trades Union Congress (TUC) was founded in Manchester with 34 delegates. In 1869 in Birmingham, quarter of a million trade unionists were represented at the TUC by 40 delegates and a "Parliamentary Committee" was established to represent trade union interests.

KEY

A FREE BORN ENGLISHMAN!
THE ADMIRATION of the WORLD!!!
AND THE ENVY of SURROUNDING NATIONS!!!!!

Repressive measures were adopted by the government against radical societies which arose following the French Revolution. Habeas corpus was suspended in May 1794, and some radical leaders were charged with high treason. In 1795, following an attack upon the king's coach in October, the Two Acts were passed. These restricted the right of free assembly and extended the law of treason to cover acts of speech and writing. The laws against combinations restricted the growth of trade unions. After 1815 the government again resorted to laws against meetings and radical propaganda, in the "gagging" Acts of 1817 and the Six Acts of 1819.

6

6 The precarious legal status of early trade unions was illustrated when six Dorset labourers were arrested in 1834 for swearing men into a union at the village of Tolpuddle. All were sentenced to seven years transportation. After demonstrations such as this, they were pardoned in 1836.

7

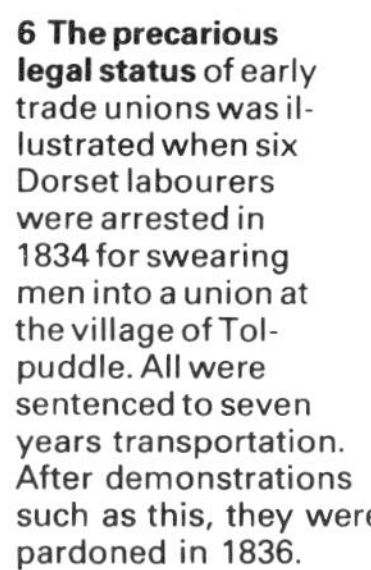

7 Chartism, expressed in the People's Charter, owed its origins to the failure of the 1832 bill fully to enfranchise the working man. The Charter demanded male suffrage, secret ballot, annual parliaments, equal electoral districts, an end to property qualifications for MPs and the introduction of official payment for them.

8

8 The Anti-Corn Law League, which was mainly composed of industrialists, was founded in 1839 to oppose the duties on imported corn that protected domestic producers. Although the League was campaigning for cheaper food in opposition to the power of the landed classes, the Chartists and the working classes did not fully support it. The Chartists argued that in reality the League wanted wages reduced by the amount that corn prices would fall if the Laws were repealed.

9

9 The first Co-operative shop, a non-profit making retail store, was one of a number of co-operative ventures in the 1830s and 1840s. By selling cheap and pure food it was the most successful.

10 The "Sheffield Outrages", a series of violent incidents directed at non-union members, led to the establishment of a Royal Commission to investigate the status of trade unions. In 1867 union status was further put into question by a ruling that they were defenceless against officials who absconded with union funds. Unions were represented on the Commission which recommended that they be given a legal basis.

10

Social reform 1800–1914

The rapid increase in population and new industrial towns during the Industrial Revolution created immense social problems in Britain. The new towns had grown uncontrolled, many lacked basic amenities such as sanitation and water supply, and the problems of poverty, ill-health, crime and bad housing were widespread. There was almost no schooling for most of the population. Child and female labour was regularly used in factories and mines [1], even for the most arduous and dangerous tasks. The prevailing ethic of laissez-faire that the state should not interfere with the workings of the economy or society held back any far-reaching legislation to improve working conditions.

Poverty and social concern

During the course of the nineteenth century some of these evils were diagnosed and brought to public notice by social commentators [8] and novelists such as Charles Dickens (1812–70), Mrs Gaskell (1810–65) and Charles Kingsley (1819–75). In addition, parliamentary enquiries were set up to examine social questions. The result was a considerable body of social legislation.

The Poor Law was a source of concern to nineteenth-century reformers. The existing system of "outdoor" relief, levied from parish rates, burdened the propertied classes, and Thomas Malthus (1766–1834) in his influential *Essay on the Principle of Population* (1798) had argued that it perpetuated poverty by encouraging population growth. Under the Speenhamland system, introduced in 1795, labourers' wages were subsidized out of parochial funds on a scale linked to the price of bread. But in the large industrial towns, the parochial organization of poor relief was totally inadequate to meet the strains of heavy unemployment.

In 1834 the New Poor Law was passed. It much reduced "outdoor relief". Instead of receiving charity, all able-bodied people requiring relief were forced to go into the workhouse, where a strict regime, including segregation of the sexes, even of married couples, was intended to deter all but the truly destitute [5]. In addition, poor law authorities were amalgamated to spread the burden of poor relief evenly.

The insanitary conditions of the great towns gave rise to considerable concern about public health. In the 1840s an inquiry showed that more than half the major towns in Britain had an insufficient or impure water supply. The cholera epidemics of the mid-nineteenth century acted as a spur to the public health movement. Edwin Chadwick's (1800–90) famous *Report on the Sanitary Conditions of the Labouring Population* in 1842 led to the creation of a central Board of Health under the Public Health Act of 1848. Individual towns were empowered to set up local Medical Officers of Health. In 1875 a Public Health Act laid the foundations for an overhaul of public sanitation.

Legislation on housing

Housing reform was left to piecemeal action. Lord Shaftesbury's (1801–85) [2] Lodging Houses Act of 1851 checked the worst abuses of "doss-houses". More important, however, was the Artisans' Dwelling Act of 1875 which gave local authorities the power to clear slums. A number of reforms of local government, especially the Municipal

CONNECTIONS

See also
88 The Industrial Revolution
90 The urban consequences of industrialization
94 The British labour movement to 1868
98 The novel and the press in the 19th century
114 Victorian London
158 Industrialization 1870–1914
160 The fight for the vote

In other volumes
274 History and Culture I

1

1 The use of child and female labour in factories and mines during the Industrial Revolution was widespread. In the early 1830s, nearly half the labour force in the cotton mills was under 21, and of the adults more than half were women. Hours and conditions were regulated only by the benevolence of employers, and a working week exceeding 90 hours was common until the 1833 Factory Act became effective.

2

2 Lord Shaftesbury was an evangelical churchman and a dedicated reformer. He is associated with the 1833 Factory Act and with legislation to prohibit the employment of children by chimney sweeps, in 1840, and of women and children in the mines, in 1842. But his overriding paternalism made him unsympathetic to franchise extension in 1867 and to too much state involvement in welfare.

3

3 The Corn Laws of 1815 protected British agriculture by prohibiting the importation of foreign wheat until the domestic price exceeded 80 shillings per quarter. These laws were widely opposed by the urban poor and also by the industrialists because it was generally thought that they forced up the price of food and wages. In the long term too, it was argued that protectionism would harm exports. In 1839, the Anti-Corn Law League was founded by Richard Cobden (1804–65) and John Bright (1811–89) to agitate for repeal. In attacking the privilege and sectional interests behind the laws, the league took on a reformist appearance. The Corn Laws were repealed in 1846.

4

4 No free public libraries existed before 1845. From the mid-century, however, many towns set up rate-assisted public libraries to provide access to books and newspapers for all classes.

5

5 Under the New Poor Law of 1834, workhouse conditions were to be made inferior to those of the poorest labourer outside, in order effectively to deter "laziness" and "vagrancy" among the poor.

Corporations Act of 1835 and the Local Government Act of 1888, provided the administrative machinery necessary to implement these measures on a local level.

Factory legislation began as early as 1802 when Robert Peel senior (1750–1830), introduced an act to limit the employment of children to under 12 hours a day. The 1819 Factory Act forbade the employment of children in cotton mills under the age of nine. Lord Shaftesbury's 1833 Factory Act further limited the working hours of all children under 18 years old and appointed factory inspectors to enforce this. Safety regulations and limitations on women's working hours were introduced by an act in 1844. This legislation was extended in the course of latter part of the nineteenth century to include all types of factories. In 1891, a consolidating act raised the minimum age for the employment of children to 11 years.

The rise of state education

Education remained a patchwork of private initiative and philanthropic effort for much of the nineteenth century. The Royal Lancasterian Association (1810) and the Anglican National Society (1811) founded hundreds of schools without any government involvement. State intervention began in the 1830s and the first government grant to education was made in 1833. In 1839 an education department was set up to inspect grant-receiving schools [6].

In 1870, Forster's Education Act provided virtually free elementary education for anyone who wanted it by setting up local boards empowered to establish schools financed, in part, from the rates. Education up to the age of ten years was made compulsory in 1880. In 1902, the Balfour Education Act created Local Education Authorities and thoroughly reformed the whole system of secondary education.

The growth of state responsibility for social welfare was embodied in the legislation of the Liberal governments after 1906, which went a considerable way towards creating a rudimentary "welfare state", with important, new measures such as the Old Age Pensions Act of 1908 and the National Health Insurance Act of 1911 [10].

KEY

Chronic overcrowding and grossly inadequate facilities characterized the new industrial towns that mushroomed during the Industrial Revolution. The sheer scale and complexity of the problems were quite unprecedented, and unnoticed until social reformers, philanthropists and the unavoidable pressure of events forced them upon public notice.

6

	Government grants (£ thousands)	Inspectors	Assistant inspectors
1839	40	2	
1851	150	23	
1858	663	30	16
1861	813	36	25
1865	637	62	18

6 The growth of education was a central feature of 19th-century reform. This diagram shows the rise in grants and school inspectors in elementary education between 1839 and 1865.

7

7 The Salvation Army, founded by "General" William Booth (1829–1912) in 1865, aimed at social as well as spiritual welfare. It provided soup kitchens, night shelters and many facilities for the destitute. Booth was particularly concerned at the adverse effects of urbanisation and the depopulation of the countryside. He hoped that through a system of rural re-education he could reverse this trend.

8

8 John Ruskin (1819–1900) art critic and reformer argued that art, ethics and social conditions were inextricably linked. Many of his proposals, such as pensions and state education, were later adopted.

10 The National Insurance Act of 1911 provided unemployment pay and free medical treatment in return for graduated weekly contributions to be paid by employers, employees and the state.

9

9 Private philanthropy in the 19th century very often preceded state action by many years. Port Sunlight, shown here, was built by the industrialist Lord Leverhulme (1851–1925) in 1888. It was the first village to be built on the garden city principles, then advocated as a means to eliminate the physical and moral effects of urban overcrowding by Ruskin and other social reformers. This is shown in the planned houses, open spaces, and the provision of public amenities.

10

The novel and the press in the 19th century

There were many technical innovations in printing between 1800 and 1900 that had important effects on newspapers and novels. The use of metal presses, steel engraving and, after 1848, of stereotypes and mechanical presses completely altered the production process. Marketing techniques changed too: circulating libraries [Key], railway station bookstalls and cheap reprints of successful titles helped to establish and satisfy a market that expanded with the rising population, increased literacy [7] and greater educational opportunities. In Britain, newspapers were hampered by taxation until 1855, but by the end of the century mass circulation newspapers had developed [4].

Changes in the novel

The novel never suffered taxation problems but was otherwise similarly affected by these changes. The huge problems of the new industrial cities [6] offered fresh subject-matter to be interpreted in the new intellectual climate of Europe after the French Revolution. Even two such early novelists as Jane Austen (1775–1817) [2] and Walter Scott (1771–1832) [1] reveal a distinct if conservative responsiveness to change. Jane Austen's domestic comedies, carefully structured in six novels, are at once amusing and deeply serious. Scott virtually invented the historical novel. His popular success brought him a considerable personal fortune.

Popular success was also enjoyed by his Victorian successors, William Makepeace Thackeray (1811–63), Anthony Trollope (1815–82) and above all by Charles Dickens (1812–70) [3] and George Eliot (1819–80). Dickens built up an astonishingly close relationship with his readers in his sentimental but brilliantly funny and sometimes despairing vision of city life. George Eliot, on the other hand, was provincial in her subjects and European in the range and discipline of her thought. The mid-century also saw the publication of the Brontë sisters' novels. Charlotte (1816–55) was the most successful, but Emily (1818–48), author of significant poems as well as of the novel *Wuthering Heights*, has since been more highly regarded. Important later novelists include George Meredith (1828–1909), George Gissing (1857–1903), Samuel Butler (1835–1902) and Thomas Hardy (1840–1928). Hardy's novels frequently express a passionate feeling for man's tragic involvement in nature and estrangement from it.

The novel in Western Europe

French fiction in this period was much more urbane and less prudish than English. The realists, Stendhal (1783–1842), Honoré de Balzac (1799–1850) [5] and Gustave Flaubert (1821-80), depicted French history and bourgeois life at great length and in minute detail. Romantic experience and attitudes, however, were given vivid expression in the works of Victor Hugo (1802–85) and George Sand (1804–76). Emile Zola (1840–1902) [8], leader of the naturalists, produced franker and more painfully pessimistic studies of the workings of heredity and environment in human affairs [6]. The enormous popular success of Alexandre Dumas the father (1802–70) and his historical romances was matched by that of his son of the same name (1824–95).

The giant figure of Johann Wolfgang

CONNECTIONS

See also
100 Poetry and the theatre in the 19th century
34 European literature in the 18th century
72 Origins of romanticism
30 Scottish culture to 1850
288 American writing: into the 20th century
290 Emergent literatures of the 20th century

In other volumes
210 Man and Machines

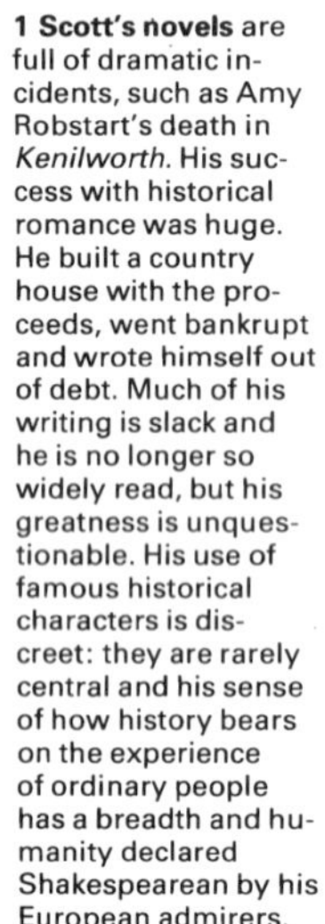

1 Scott's novels are full of dramatic incidents, such as Amy Robstart's death in *Kenilworth*. His success with historical romance was huge. He built a country house with the proceeds, went bankrupt and wrote himself out of debt. Much of his writing is slack and he is no longer so widely read, but his greatness is unquestionable. His use of famous historical characters is discreet: they are rarely central and his sense of how history bears on the experience of ordinary people has a breadth and humanity declared Shakespearean by his European admirers.

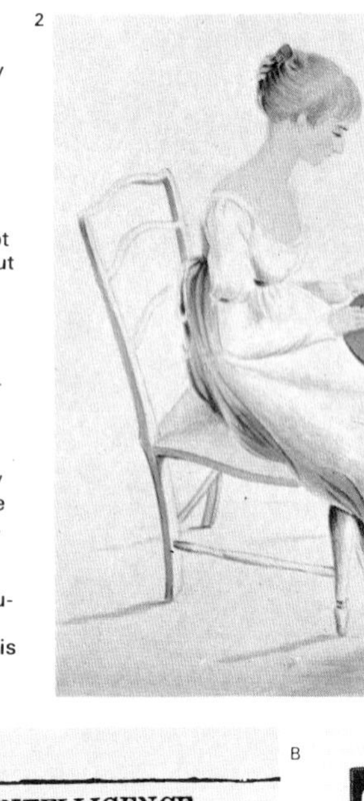

2 Jane Austen concentrated on witty, incisive descriptions of rural English society. Her sense of form had its roots in the classical English comedy of Congreve and Jonson. She was the first of a remarkable line of women novelists whose lives were otherwise provincial and obscure. During her lifetime she earned only £250 for six novels, but time has discriminated in her favour and she is now regarded as an immortal of English literature. Two of her best works are *Pride and Prejudice* (1813) and *Emma* (1816), both about ordinary people unaffected by world events.

3 Dickens's novels were published in frequent illustrated instalments, as with *Nicholas Nickleby*, the parts of which are shown here. Part-publication was common practice and allowed Dickens to keep in close contact with his public and alter plots if sales fell off; he kept the English-speaking world in agonized suspense over the death of Little Nell in *The Old Curiosity Shop*. Dickens was, however, a serious artist who influenced, among others, Dostoevsky.

4 A

LATEST INTELLIGENCE.

THE SIEGE OF SEBASTOPOL.

(BY SUBMARINE AND BRITISH TELEGRAPH.)

VIENNA, MONDAY MORNING.

The *Morgen Post*, which is a paper of no great authority, has the following :—

" CZERNOWITZ, Nov. 11.

"On the 6th the whole garrison of Sebastopol, amounting to 65,000 men, made a sortie.

" A furious battle ensued, which was not ended when the messenger left ; but the allies had the advantage."

We have received, at half-past 4 o'clock this morning, the following despatch, dated yesterday afternoon, from our correspondent at Vienna :—

" The news forwarded this morning relative to the sortie was but too true.

" Reliable information has been given me that the English suffered a very heavy loss, and had three Generals wounded.

" It is said that later intelligence has been received, according to which the Russians had at last been repelled with a loss of 3,000 men."

B

4 Mass circulation newspapers became possible after the development of new printing techniques and the ending of the newspaper tax in 1855. Serious, major journalistic innovations, like *The Times'* coverage of the Crimean War [A], probably had less influence on novels than "gutter press" sensationalism [B].

von Goethe (1749–1832) overshadows nineteenth-century German literature. In his wake the regionalist anti-romanticism of Theodor Storm (1817–88) and Fritz Reuter (1810–74) seems relatively less significant. Italian prose in this period was dominated not by one great man but by one great book, *The Betrothed* by Alessandro Manzoni (1785–1873), a patriotic Romantic who was greatly influenced by Scott. The task of modernizing the Italian novel fell to Giovanni Verga (1840–1922) and Antonio Fogazzaro (1842–1911).

The literary tradition in Russia

In some ways the most surprising national achievement in the evolution of the novel was that of Russia. The first major Russian novelists were Mikhail Lermontov (1814–41) and Nikolai Gogol (1809–52); their successors, Ivan Turgenev (1818–83), Fyodor Dostoevsky (1821–81) and Leo Tolstoy (1828–1910) [9] were to make a deep impression on Western European culture when their works were translated into French, German and English. Tolstoy's *War and Peace* and *Anna Karenina* are among the greatest of all literary works.

While Dostoevksy's intellectual perspectives are significantly modern it was Henry James (1843–1916) who introduced modern techniques into the novel. Although he spent most of his life in Europe, he remained in important ways American. The formal complexity and ironic indirectness of his work is also characteristic of Nathaniel Hawthorne (1806–64), Herman Melville (1819–91) and Mark Twain (1835–1910). In his own fiction James abandoned the convention that the author knows everything and selected one or two characters from whose point of view he told his story. Although most of his own novels are long, this technique led on to the writing of shorter, more economical works. As well as the artistic reasons for this development there were also strictly commercial ones: with the advent of cheap editions that readers could buy for themselves the great circulating libraries were in decline and publishers became less interested in length alone. The age of the Victorian novel was over.

KEY

The Temple of the Muses in Finsbury, London, was a rate-supported "public" lending library. Novel-reading was widespread in middle- and upper-class households by the middle of the 19th century. In Britain even the wealthy subscribed to circulating libraries. Consequently a novel's success or failure depended on the good will of these libraries, which had a vested interest in keeping books both expensive and "pure". "Society will not tolerate the natural in our art", complained Thackeray in his novel *Pendennis* (1848). But the libraries were not so easily resisted.

5

5 This remarkable sculpture of Balzac by Rodin suggests the bulk and force of Balzac's vision of human society. His 85 novels attempt to characterize almost every aspect of life in France between the 1789 revolution and the fall of Louis Philippe in 1848. By romantic standards an unrewarding subject for the novelist, but he made it lively and real.

6

6 *Germinal,* Emile Zola's outspoken novel describing the degrading conditions of life endured by miners in northern France, illustrates the freedom from prudery that French novelists enjoyed. Mrs Gaskell, Kingsley and Dickens had tried to depict the consequences of industrialization without dealing with human sexuality.

7

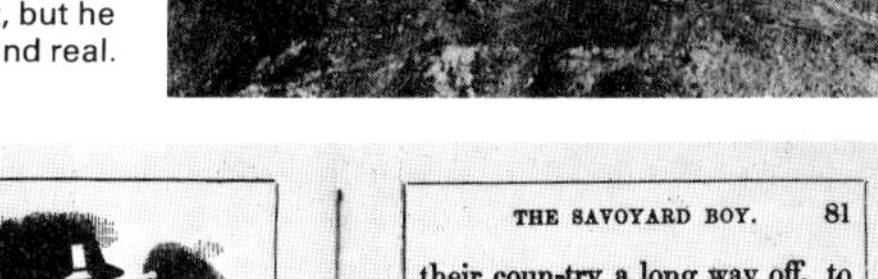

THE SAVOYARD BOY.

AH! here is a Sa-voy-ard boy, with his or-gan and his dog.
He has come to play us a tune to make us glad.
And what is that mount-ed on the dog's back?
Why, it is in-deed a lit-tle mon-key, the Sa-voy-ard's pet mon-key!
These poor boys come from

THE SAVOYARD BOY. 81

their coun-try a long way off, to gain a few pence.
Their own coun-try is not so rich as ours, and we ought not to de-ny them a few pence.
We may chance, some time or o-ther, to be left friend-less in a strange coun-try, and we shall then feel ve-ry glad if a-ny kind peo-ple take no-tice of us, and give us food, or mo-ney to buy food with.
We should al-ways give to those who are in need, and if we do so, we shall be sure to get help when we our-selves are in need.
You will some-times meet with per-sons who beg, and could do with-out beg-ging, but they like to live an i-dle life.

7 Green's *Universal Primer* injected heavy-handed moralizing into reading lessons. Urban, but not rural, literacy rates were fairly high in the 1840s. Total illiteracy ranged from 16% to 25% and about three-quarters of the working class was literate by the middle of the century. Most people read only "novels of the lowest character" and fears were expressed about whether "good" literature could survive.

8 Emile Zola in *Les Rougon-Macquart* attempted to follow the adventures of a family during the 1880s, calling it "a physiological history of the 2nd Empire". The series has 20 volumes with modern themes: in *The Dram Shop* (1877) the evils of drink; in *Nana* (1880) sex; in *Earth (1888)* the deadening brutality of peasant life. Naturalists believed writers should go beyond the surface detail of life.

8

9

9 Tolstoy's funeral was the first non-religious Russian funeral, yet he died with the reputation of a saint because of his religious and political devotion to the ideal "simple peasant" life. In *What is Art*? (1897) he had repudiated most of European literature including his own and Shakespeare's works, yet in his own ascetic and prophetic old age he demonstrated the same sort of passion and contradictory idealism with which he had invested his fictional heroes.

Poetry and theatre in the 19th century

The Romantic movement in poetry at the end of the eighteenth century stressed intensity of emotion rather than elegance and art, freedom of expression rather than stylistic rules. In England its most important forerunner was William Blake (1757–1827), who was less known in his own time than Walter Scott (1771–1832) or Thomas Moore (1779–1852). The rebellious spirit of the movement was epitomized in the life of Lord Byron (1788–1824) who, with Johann Wolfgang von Goethe (1749–1832), towered over European literature in the 1820s. Yet Byron's best work, *Don Juan*, is anti-Romantic in its sceptical wit.

William Wordsworth (1770–1850) and John Keats (1795–1821) better represent the actual changes in English poetry brought about by romanticism. Wordsworth's ideas about mind and nature [1] forced him to adopt an unorthodox style and subject-matter. His creed was to take "ordinary things" and show them in "unusual aspect", believing that intense joy could arise from deep harmony with nature. Percy Bysshe Shelley (1792–1822) wrote more directly of the power of joy as a reforming influence, as Keats stressed the power of beauty. The lyrical intensity of Keats's poetry deeply influenced later poets.

Lyricism, nature and the exotic continued to attract Victorian poets. Robert Browning (1812–89) used anti-lyrical effects, tough rhythms and difficult meanings but was always drawn to the exotic. Faith in joy and the senses waned however: both Alfred Lord Tennyson (1809–92) and Matthew Arnold (1822–88) wrote sombre, noble verse and Tennyson had earnest doubts about the relevance of his lyric gift. A reaction against undue moral earnestness came with Algernon Charles Swinburne (1837–1909) and the Decadents who stressed flagrantly amoral beauty. But later British poets, Gerard Manley Hopkins (1844–89), Thomas Hardy (1840–1928) and W. B. Yeats (1865–1939), remained deeply serious.

Poets in Europe

The writings of Samuel Taylor Coleridge reveal the Romantics' debt to Germany where Goethe, a champion of the *Sturm und Drang* movement, had established the concept of the suffering hero. But Goethe's work shows the difficulty of arbitrary distinctions between romanticism and classicism. He wrote the classical *Roman Elegies* as well as passionate lyrics to Charlotte von Stein. Similarly the Romantic 1827 *Songs* of Heinrich Heine (1797–1856) are balanced by his more sombre later poems. The Byronic mood was more influential in Russia where Alexander Pushkin (1799–1837) was a disciple [2], as was Mikhail Lermontov (1814–41).

In France, Victor Hugo (1802–85), poet, novelist and dramatist, led other Romantic anti-classicists including Alphonse Lamartine (1790–1869), Alfred de Musset (1810–57) and the young Théophile Gautier (1811–72). But Charles Baudelaire (1821–67), lyricist of moral decay [5], the boy-poet Arthur Rimbaud (1854–91) and Paul Verlaine (1844–96) are better seen as early Symbolists rather than late Romantics. Stéphane Mallarmé (1842–98) and the Symbolists tried to create a poetry of emblems to convey the meaning beneath the surface of things. In Italy, Giosuè Carducci

CONNECTIONS

See also
98 The novel and press in the 19th century
72 Origins of romanticism
282 Irish culture since 1850
288 American writing: into the 20th century
290 Emergent literatures of the 20th century

1 **The English Lake District** inspired some of Wordsworth's finest work. There, faced with sublimely rugged scenery, he experienced a sense of harmony with nature which he expressed as a moral force. He and Samuel Taylor Coleridge (1772–1834) published their *Lyrical Ballads*, including Coleridge's "Ancient Mariner", in 1798. Coleridge is remembered as much for his criticism as for his poetry.

2 **Pushkin, the first major Russian writer,** was exiled for writing epigrams against the Russian government. During his exile he read Byron and created a Byronic hero for his poem *Eugene Onegin* (1833), an immortal Romantic legend. He also wrote popular prose romances and experienced the conflict between patriotism and liberalism which became a common Russian problem later in the century.

3 **Riots followed the first performance** of Hugo's *Hernani* (1830), which broke with the rules of classicism. Hugo's romanticism was ardently supported by young French poets but a reaction against emotionalism and looseness of style soon followed as poets turned to subtler themes and more concise imagery.

4 **Gabriele D' Annunzio** worked on the script of the epic film *Cabiria* (1913) and was also a dramatist, novelist and flamboyant political leader both before and after World War I. His later support of Mussolini cannot detract from the sensuous poetry he wrote in the 1890s during his affair with the actress Eleanora Duse.

(1835–1907) led a reaction against undisciplined verse, but Gabriele D'Annunzio (1863–1938) sounded late in the century the authentic note of Romantic joy [4].

From melodrama to naturalism in drama

Romantic and post-Romantic drama generally fails as literature. The plays of Goethe show classical influences while his masterpiece, *Faust*, transcends categories. Hugo's triumph in France with the Romantic *Hernani* [3] is hard now to understand. Victorien Sardou (1831–1908) and Alexandre Dumas (1802–70) wrote successful comedies and romances. Later Alexandre Dumas the younger (1824–95) produced some solemn social problem plays and Edmond Rostand (1868–1918) poetic dramas. But the dominance of opera and of Shakespearean revivals [8] inhibited convincing representation of contemporary society.

Towards the end of the century, three major dramatists emerged as forerunners of modern theatre. The Norwegian Henrik Ibsen (1828–1906) moved from verse plays to a series of controversial and influential social dramas in prose such as *A Doll's House* [6]. August Strindberg (1849–1912), a Swede, drew on a tragic personal history to inject an element of psychosexual horror into his work. Like Ibsen, he finally moved towards symbolism. The Russian plays of Anton Chekhov (1860–1904) are notable for their formal grace, realism and insight into personal and social insecurities [7].

New directions in the theatre

The comedies of the Irish dramatist Oscar Wilde (1854–1900) [9] were old-fashioned in plot and characterization, but Wilde used his scintillating wit to parody cleverly the conventions of melodrama and Romantic comedy. Bernard Shaw (1856–1950), a champion of Ibsen, used similar techniques in constructing plays of social and moral ideas, at once amusing, humane and deeply thoughtful. In Paris, the anarchic farce *Ubu Roi* by Alfred Jarry (1873–1907) was already pointing towards expressionism, in which reality would be presented as a reflection of the mind, and towards the illogicality of the theatre of the absurd.

KEY

Shelley's death by drowning is immortalized in this memorial to him in University College, Oxford. A radical and passionate poet, Shelley connected the health of literature with the health of society and denied that poets had any obligation to express contemporary ideas of morality.

5

5 Charles Baudelaire in *Les Fleurs du Mal* (1857) foreshadowed symbolism by searching for significance in all things, not merely the respectable, and finding symbols of hollowness in the beauty that hid corruption. He was responsible for the European vogue of Edgar Allan Poe and was an important influence on English poetry, especially Swinburne and the poets of the 1890s.

6

6 Nora Helmer, here dancing the tarantella, was the central character of Ibsen's most controversial play, *A Doll's House* (1879), which was seen as a breakthrough in theatrical realism. His audience must have expected this drama of blackmail and wifely loyalty to end in a triumph of domestic virtue. But in a famous final scene Nora leaves her husband and children to seek her own identity.

7

7 Chekhov's *The Seagull*, as produced in 1898 by Konstantin Stanislavsky, was a landmark in drama. Stanislavsky taught actors to identify with the characters they played, a technique particularly adapted to Chekhov's plays which concentrates on the unfolding of character rather than on plot development or melodramatic situations.

8

8 Henry Irving (1838–1905), here playing Hamlet, led the idolatry of Shakespeare, who had become a national institution in England and an important influence in Europe by the 19th century. Shakespeare provided virtuoso actors with great parts. But heavy naturalistic sets led to tediously long intervals and to brutal cutting of the original text.

9

9 Oscar Wilde, who was imprisoned for homosexuality after a famous trial, was the wittiest dramatist of the 1890s and a leading poet of the English aesthetic movement. He was a master of paradox and an apostle of art for art's sake.

Romantic art: landscape painting

"Everything is becoming more airy and light than before, everything tends towards landscape", wrote the German painter Philipp Otto Runge (1777–1810) in 1802. His remark, although in one sense exaggerated, was truer than he perhaps realized, as landscape painting had become popular in Britain as well as in Germany at this time. In these two countries, especially the genre assumed a new role during the Romantic period. Previously it had been considered little more than a minor decorative form, despite its great seventeenth-century practitioners such as Poussin and Claude. Now, however, it was called upon to express feeling, not just for the outward beauty of woods, fields and skies but for nature's "inner life".

The German approach to landscape

In Germany where attitudes were more informed by philosophy than in Britain, nature was invested with an all-pervading spirit of an almost sacred character, not static but subject to growth and change, analogous to the spirit in man. To represent the changing states of nature as symbols of the varieties of human emotion was therefore the aim of Romantic landscape painting. As Carl Gustav Carus, a younger contemporary of Runge and follower of Friedrich, put it: "Just as the vibration of a string may cause another, similarly tuned, though of lower or higher pitch, to vibrate in unison with it, so congenial states in nature and in the human spirit may interact".

Runge's work was insufficiently developed (he died young), especially in landscape painting, to produce more than a fragmentary and eccentric, although highly interesting, reflection of these ideas [3]. With his visionary temperament and boldly original mind, he had something in common with his English contemporary, William Blake (1757–1827), although probably neither knew of the other's existence. The greatest German Romantic landscapist was Caspar David Friedrich (1774–1840) whose art is superficially more traditional, in that he represented natural views seen from fixed points in space and time. But for him, too, nature was only the physical manifestation of an inward life, a continuous process corresponding to the agitation of the artist's mind. He specialized in changing effects of atmosphere and light, depicting them with a refinement and air of gentle melancholy unlike almost anything else in art [4].

The British tradition

If the purest and most studied forms of Romantic landscape painting were produced in Germany, it was Britain that in this period had the longest and most varied landscape tradition. (In France, broadly speaking, what was new in landscape was not Romantic and what was Romantic was not new.)

Some of the qualities already discussed in connection with German landscape – the emphasis on mood, the concern with nature as a process rather than an order, the awareness of some spiritual entity concealed within nature's visible forms – are present in varying degrees in British painting too. They were intimated first in the calm and serene watercolour views of the Swiss Alps [1] by J.R. Cozens (1752–97) and in Thomas Girtin's (1775–1802) solemn watercolours of the Yorkshire dales [2]. They showed more fully

CONNECTIONS

See also
106 Romantic art: figure painting
72 Origins of romanticism
72 Realist painting in the 19th century
286 The arts in Wales

1

1 The emotional bond between man and nature is stressed by eliminating figures from the picture. J.R. Cozens's "Valley in the Tyrol, near Brixen" (*c.* 1783) offers the viewer an impression of stillness and of vast space. The innate beauty of alpine scenery was one of the early Romantic discoveries.

2

2 Stillness and quiet are features also of Thomas Girtin's "Kirkstall Abbey, Yorkshire" (*c.* 1800), but the setting is gentler. Besides exploring new types of scenery, the Romantics turned their attention to Gothic remains. Cozens and Girtin were two pre-eminent watercolourists.

3

3 Runge's "Morning" (1803), a baby lying in a radiantly lit paradisiacal landscape, symbolizes not only natural morning but also the dawn of the universe and the beginning of each individual life. More than the English, German Romantics dealt with the idea of nature in terms of symbols.

4

4 The symbolism of Friedrich's "The Cross in the Mountains" (1808), an altarpiece for the chapel of Schloss Tetschen, is at once more literal and more orthodox than that of Runge. Friedrich's theme is the impact of Christianity on world history and its gifts of faith and hope in God.

in the response of Joseph Turner (1775–1851) to the violence of storms at sea and his fascination with the brilliance of sunlight, in Constable's feeling for the moral and religious values inherent in ordinary nature, and in Palmer's assertion that "bits of nature are generally much improved by being received into the soul".

Movement and the sky

However, British landscape painting is, on the whole, less mystically inspired and more empirical than German. Its sense of the divine is diluted by being combined with more mundane preoccupations such as topography and the picturesque, the interaction of the ideal and the real, and the influence of the Old Masters. It is also more involved with the idea of the sketch – that is, both with "sketchiness", in the sense of breadth of handling (whereas German painting is very smooth and neat in handling), and with working direct from nature in watercolour and oils. The Romantic concern with transience is thus realized by British painters chiefly in terms of movement – through clouds being blown across the sky and wind whipping up the waves.

In both, indeed all, countries the sky is the focus of Romantic landscape painting; it was, as John Constable (1776–1837) called it, "the keynote, the standard of scale, and the chief organ of sentiment". Constable was not the only landscapist in this period, but merely the best-known to make sky studies, with notes on the back stating the date, the exact hour of the day and the direction of the wind. In his finished landscapes of the Suffolk countryside, in which he was born and brought up, he used the light of the sky to give vitality and poetry to his rendering of simple agricultural scenes [6]. Turner did the same with a much wider range of scenery and phenomena, finally almost dissolving form altogether in a haze of light and colour [8].

The next and final "Romantic generation" produced a type of landscape painting more overtly expressive of feeling, such as the apocalyptic and grandiose fantasies of John Martin (1789–1854) [7] or, at the opposite extreme, the intimate pastoral visions of Samuel Palmer (1805–81) [Key].

KEY

A sense of heightened mood is the chief common factor of Romantic landscape paintings. It is expressed by Samuel Palmer in terms of the pastoral genre, which he saw as part religious in "The White Cloud" (*c.* 1831–2).

5

5 The huge output of J.M.W. Turner (1775–1851) embraced the extremes of traditionalism and experiment and of agitation and calm. His early paintings, such as "The Wreck of a Transport Ship" (*c.* 1810) are predominantly dark in tone, acknowledging the Old Masters, but the vital role he always gave to light is evident. Shipwreck, an all too common instance at that time of the destructive powers of nature, is a frequent theme of Romantic art.

6

6 "Flatford Mill' (1817), one of Constable's best-known paintings, shows him at his most naturalistic. His sheer love of, and indentification with, the countryside he painted, the Suffolk of his childhood, make Constable a Romantic artist. His aim was to achieve truth to nature and its light and to combine this with the practical details of agricultural life at that time.

7

7 Mountain grandeur combined with the theme of the lone man defying his enemies are illustrated in Martin's "The Bard" (1817). The subject is from a poem by Gray lamenting the suppression of the Welsh bards – symbols of freedom and nationhood – by King Edward I.

8

8 The ultimate expression of the Romantic concern with light is in Turner's late work. In "Norham Castle, Sunrise", (*c.* 1835–40) his earlier preoccupations – old castles, hills, rivers, the head-on sunlight effects of Claude – remain, but only as traces suspended in colour.

Development of the orchestra

While by the sixteenth century there already existed a body of secular instrumental music, it was slight compared to the wealth of choral music sponsored by the Church. The growing patronage of secular works, especially opera and ballet, required accompaniment by instrumental groups, most of which were temporary, although several permanent ensembles had made their appearance by the early seventeenth century.

Various combinations of instruments had been popular from the sixteenth century. Some were of one family, such as the viol or recorder families, and were called consorts. A "broken" consort might include instruments of other families to lend a more lively character to a rather bland sound.

The development of opera in Italy led composers to more colourful use of instrumental groups. Claudio Monteverdi (1567–1643), in his opera *Orfeo* (1607), used an orchestra consisting of 15 keyboard instruments, brass, strings and woodwind. He left to the music director the choice of which instruments should play which parts of his music, with the exception of sections where he specified the use of trombones for music associated with Hades. His understanding of orchestral sound was not entirely new. About ten years earlier Giovanni Gabrieli (1557–1612) specified instruments to play parts in *Sacrae symphoniae* (1597), possibly the first work scored in such detail.

The string families

The improved quality and brilliant sound of the violin in the second quarter of the sixteenth century overwhelmed the viol family, although one descendant of the viol, the double bass, has survived. The violin family produced the sound that was to be fundamental to the symphony orchestra. Les Vingt-quatre Violons du Roi that played under the direction of Jean-Baptiste Lully (1632–87) was one of a few such ensembles to play in the courts of Europe.

Throughout the seventeenth century, composers used a thorough (through or continuo) bass in writing for orchestra. The continuo instrument, usually a harpsichord or an organ, "filled in" harmonies where there was no instrument free to play a certain part or in places where the part needed support.

By the late seventeenth century the four-part string orchestra – first and second violins, violas and cellos – was well established. The double bass at first played the cello line an octave lower. The instrument was regularly a part of the orchestra by the mid-eighteenth century, but was not of a standard form until the Italian model with four strings won general recognition in the late nineteenth century.

The woodwind families

To the strings various instruments were added until certain of them found lasting places in the orchestral establishment. First and second oboes and bassoons added woodwind tone in the seventeenth century. At first the bassoon took a bass role but later came to play tenor parts. From about 1650 oboes "doubled" the violin parts, but virtuoso players soon individualized.

The flute appeared in early orchestras. Its vertical forms, the recorders, were ousted by the oboes and the transverse flute, which owed much to French craftsmen and musi-

CONNECTIONS

See also
40 Baroque and classical music
106 Music: the Romantic period

In other volumes
248 History and Culture 1
82 Science and The Universe
50 Man and Society

1 The historical growth of the orchestra began in the late 17th century when the nucleus of the violin family became standardized. Other groups of instruments were added as they were developed and as influential composers [listed] wrote parts for them. Horns came into the orchestra from the hunting field, drums and trumpets from the army and trombones from the opera house in the 19th century.

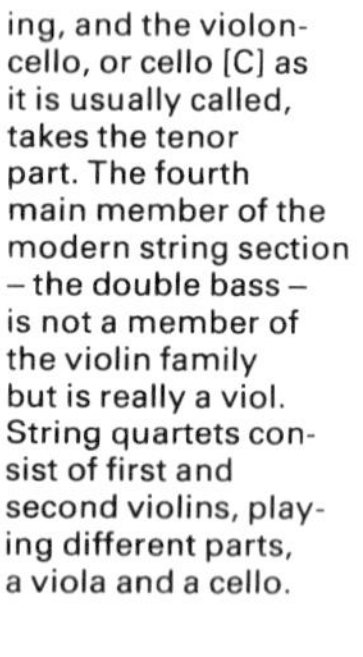

2 The string section of the modern orchestra is based on the violin family – violin, viola and cello. The violin role divides into first and second parts, although the instruments played [A] are identical for either part. The larger viola [B] plays a part that corresponds to the alto voice in singing, and the violoncello, or cello [C] as it is usually called, takes the tenor part. The fourth main member of the modern string section – the double bass – is not a member of the violin family but is really a viol. String quartets consist of first and second violins, playing different parts, a viola and a cello.

3 Tambourines combine the stretched membrane of drums with a jingle that has elements both of cymbals and of rattle instruments. Percussion instruments, including triangles and gongs, were late additions to the orchestra. For a long time, only timpani represented percussion and were usually combined with trumpets to give brilliance of effect. There are also some early instances of the orchestral use of bells by J. S. Bach and G. F. Handel.

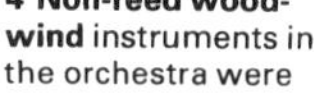

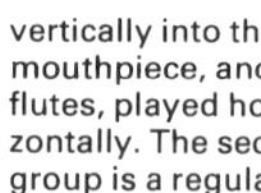

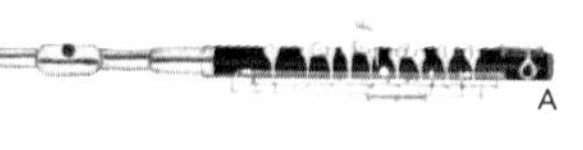

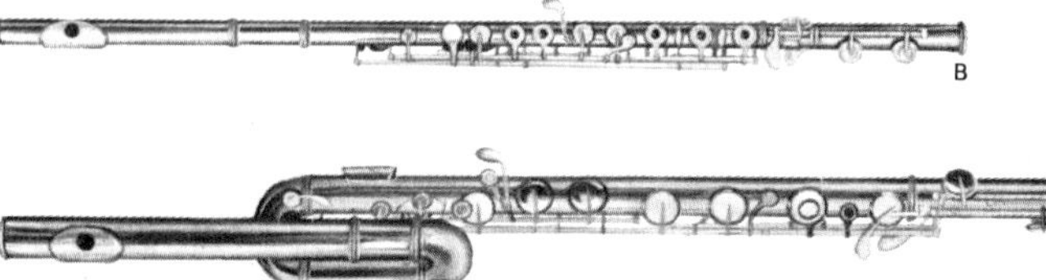

4 Non-reed woodwind instruments in the orchestra were originally of two kinds: recorders, played by blowing vertically into the mouthpiece, and flutes, played horizontally. The second group is a regular component of the modern orchestra. It includes the piccolo [A], concert flute [B] and bass flute [C]. Recorders are still used in orchestras where their characteristic tone colour is required in producing the authentic sound of the Baroque orchestra.

cians in improving their mechanics. The final addition to the orchestral woodwind families was the clarinet, "invented" by Johann Denner (1655–1707) in about 1700.

The principal woodwind instruments had some variants which gained regular orchestral places: the oboe's cousin, the cor anglais, with its deeper tone became a popular instrument with Romantic composers seeking fresh tone colours; the bass flute and bass clarinet were used occasionally and the piccolo added sparkle to wind sections.

The brass and percussion families

By the time Bach wrote his first *Brandenburg Concerto* (1721) the French horn had joined the orchestra. The trumpet had already won a place, usually playing at the top end of its range. French horns were often grouped in pairs with oboes in a woodwind section. Wolfgang Mozart wrote 19 of his first 40 symphonies for orchestras whose wind sections were of oboes and horns only. The trombones, used in late eighteenth-century opera orchestras, entered the symphony orchestra 50 years later.

The invention, in Germany, of valves for brass instruments in about 1815 meant that they could produce semitone scales throughout their ranges without fitting alternative lengths of tubing every time the music changed key. More percussion instruments were added as Romantic and post-Romantic composers explored the possibilities of orchestral colour.

The discipline of orchestras had not always been as high as the standards established by Johann Stamitz (1717–57) in Mannheim, whose orchestra was compared favourably by Mozart to his "rabble of players". By the early nineteenth century the leadership of orchestral performance was outgrowing the situation where the leader of the violin section controlled a performance. In 1820 Ludwig Spohr surprised orchestra and audience in London when he directed the orchestra with a baton, apparently for the first time. Hector Berlioz (1803–69) and Wagner pleaded for improved performance, and the middle classes, flocking to the new concert halls, made long rehearsal time and full-time orchestras financially possible.

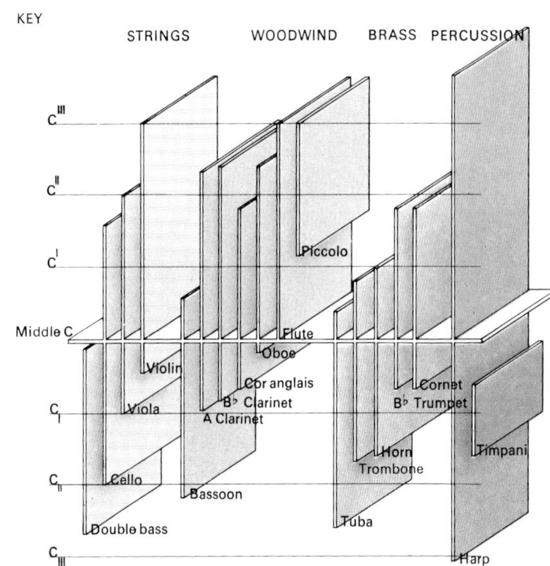

The pitch range of a symphony orchestra is fairly evenly represented across the various sections – strings [blue], woodwind [green], brass [beige] – except for percussion [brown].

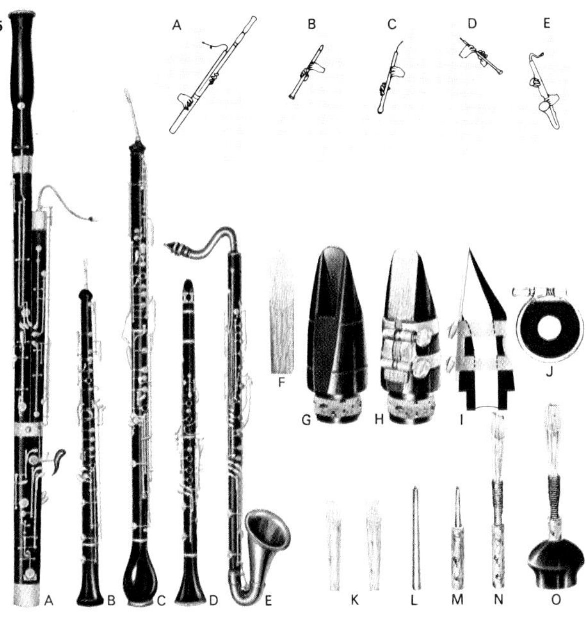

5 Reed woodwinds, shown in appropriate playing positions, are the bassoon [A], oboe [B], cor anglais [C], clarinet [D] and bass clarinet [E]. The clarinets have a cylindrical bore and are played by means of a single reed [F] fixed over a chamber in the mouthpiece [G] and secured by a ligature [H], shown in transverse [I] and cross-section [J]. The oboe family [A, B and C] is played by blowing through two pieces of reed [K] fitted round a brass tube [L]. The tube is placed in a cork cylinder [M] with the twin reeds whipped round it [N], (as shown in cross-section) with the complete double reed in position [O].

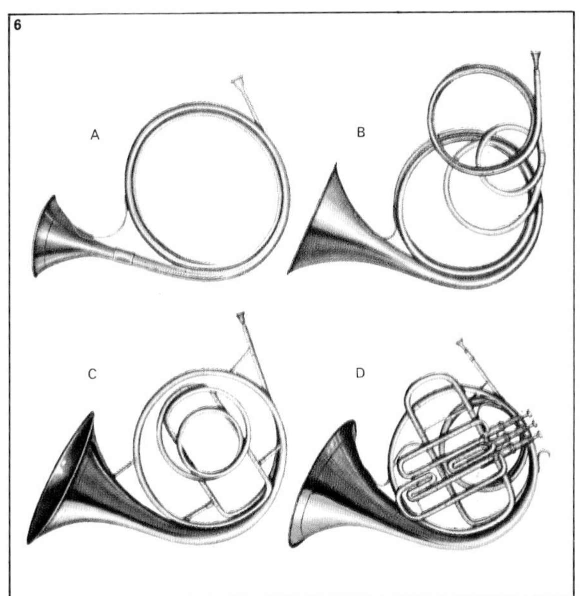

6 The French horn is a direct descendant of a coiled hunting horn [A] that originated in France about 1660. It changed its form because when it was first used in orchestras, players were obliged to change horns when the music changed key. Crooks, extra lengths of tube, were introduced [B] and later both crooks and a tuning slide were added [C]. The development of the piston valve [D], early in the 19th century, gave the horn a full range of semitones. Other members of the orchestral brass are the trumpet and trombone, with cylindrical bores, and at the lower end of the pitch range the tuba, which has a conical bore like the French horn.

7 Modern symphony orchestras are arranged generally on the pattern illustrated here. Conductors make minor adjustments to meet special demands of the music. This happens most frequently for Romantic and modern works where the sonorities are especially important.

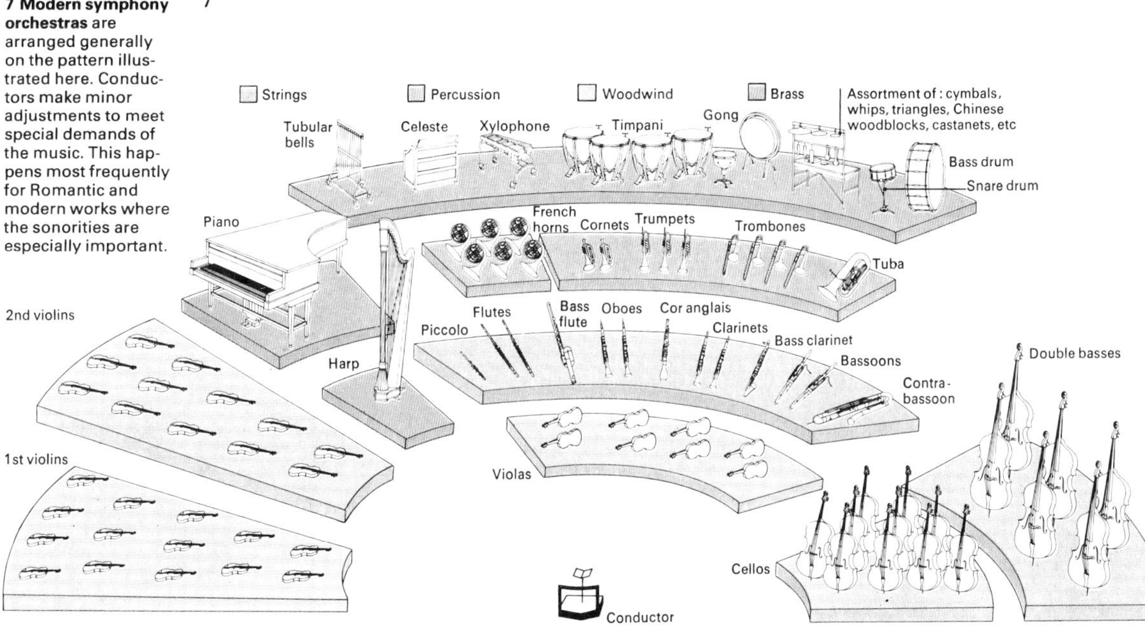

Music: the Romantic period

The nineteenth century saw the birth of the idea in Western music of the composer as an artist, instead of being merely a craftsman providing music for an employer – usually the Church or an aristocratic patron. Beethoven [Key], whose revolutionary stance was one of determined self-expression, was a central figure in the transition.

The influence of Beethoven

Having absorbed classical elements from Haydn and Mozart (whose last three symphonies to a degree prefigure romanticism), Beethoven embarked on a course that can be seen as a parallel in music to the emergence of the Romantic concept of the liberated individual. His third symphony, the *Eroica* (1803–4), is a pivotal work in this respect, and revealed an impetus that was to burst forth in the power of many of his symphonies, concertos and piano sonatas. The intense late string quartets are altogether a more intimate compression of similar emotional power.

Virtually all serious European music of the nineteenth century was to flourish under the far-reaching influence of this music, for Beethoven transformed the standard classical forms of sonata, symphony, concerto and quartet that he inherited by infusing in them a musically emotional intensity. Many subsequent composers took the liberation of individual emotional expression for their starting-point, rather than seeing it as the resolution of conflict through relentless and imaginative musical logic as in Beethoven; and at times Romantic music suffered the excesses of self-indulgent feeling.

The early Romantic giants

Carl Maria von Weber (1786–1826) is generally credited with being the first freely Romantic composer, and also the pioneer of German Romantic opera. His often superficial piano music was destined for the increasingly popular public concerts that encouraged virtuoso composers such as the violinist Niccolò Paganini (1782–1840) [7].

At the opposite pole were the private musical evenings given by Franz Schubert (1797–1828) [6] in Vienna. These "Schubertiads" united poets and musicians, and saw in particular the fashioning of the *Lied* (song), in which Schubert's accompaniment opened new worlds of melodic and harmonic enrichment of lyrics. Yet songs were just part of the general response music was making to literature in this period, offering both attractive distant realms of order, fantasy and heroism, and a framework for new ideas.

Felix Mendelssohn (1809–47) [3] made Shakespeare's *Midsummer Night's Dream* the subject of a concert overture when he was only 17, but his fresh-sounding music still drew much charm from a traditional eighteenth-century restraint and balance. Hector Berlioz (1803–69) [1], on the other hand, expressed a passion for the works of Shakespeare, Byron, Scott, Goethe and others, and they figure in many of his orchestral and dramatic works. His *Symphonie Fantastique* (1830) extends the idea of a literary "programme" to a love affair. Robert Schumann (1810–56) drew characters from Romantic writers.

More adventurously, Franz Liszt (1811–86) [7] wrote "masterpieces of music which absorb those of literature", to adapt his own words, and created the symphonic

CONNECTIONS

See also
40 Baroque and classical music
104 Development of the orchestra
120 Opera in the 19th and 20th centuries
270 Music from Stravinsky to Cage
72 Origins of romanticism

1 Romantic music, initiated by Beethoven, was still finding a powerful exponent in Rachmaninov in the 1930s. The portrait is of Berlioz, who represents the most intense expression of the movement.

2 The Royal College of Organists, established in London in 1864, was just one of the many conservatories and academies of music that proliferated in Europe in the 19th century. They owed their origin to the Italian *conservatorio*, where orphans were taught music. Notable conservatories were founded in Paris (1784), Vienna (1817), London (the Royal Academy of Music, 1822) and Leipzig (by Mendelssohn, 1843).

4 Tchaikovsky stood apart from the self-proclaimed nationalist composers of late 19th-century Russia in his constant use of established European forms such as the symphony, the concerto and the symphonic poem. Even so, his personal idiom was coloured by a characteristic Russian emphasis on style, minor keys and folk-like melody. His popularity today is high, based chiefly on his symphonies, concertos and ballet music.

3 Fingal's Cave in the Hebrides inspired Mendelssohn with a theme for a concert overture (1830) conveying his impression of the cave. Music describing scenery or literary subjects – "programme music" – was a commonplace of Romantic composition.

5 Johann Strauss the Younger (1825–99), conducted the orchestra and presided as musical director at such typical Viennese entertainments. Described as the "Waltz King", thanks to works such as "On the Beautiful Blue Danube" and "Tales from the Vienna Woods", he composed lighthearted music whose brilliance and gaiety captured the spirit of the Hapsburg capital during a 50-year period. He also composed a number of successful operettas.

poem (*Tasso, Mazeppa*) from the combination in music of both the narrative and psychological aspects of a story or poem. Many composers were to build on this format, most notably Richard Strauss (1864–1949). Liszt's brilliant piano music also often had an outside or literary impulse.

Nationalism and the Romantics

In 1848 revolutions throughout Europe were crushed but gave new directions to nationalist feelings that were finally to emerge in music. Frédéric Chopin (1810–49) [9] in exile had already used the mazurka and polonaise to express his nostalgia and hopes for Poland. In Bohemia, Antonín Dvořák (1841–1904) and Bedřich Smetana (1824–84) were to emerge as Czech nationalists, as would Edvard Grieg (1843–1907) in Norway and, following the early lead of Mikhail Glinka (1804–57), the "Mighty Five", headed by Modest Mussorgsky (1839–81), Alexander Borodin (1834–87), and Nikolai Rimsky-Korsakov (1844–1908). In Russia, Peter Ilyich Tchaikovsky (1840–93) [4] remained apart from this group in the United States, Edward MacDowell (1861–1908) and in Germany Richard Wagner (1813–83) were pre-eminent.

Wagner's use of native German myth to create a flowing music drama in place of traditional opera was eventually secondary to the pervading influence of his extremely lush, chromatic harmony and inspired use of the orchestra, almost the culmination of Romantic music. But composers such as Brahms [8], Anton Bruckner (1824–96), and possibly Gustav Mahler (1860–1911) harked back to classicism.

The final flowering of Romantic nationalism was seen in England with Edward Elgar (1857–1934) and Frederick Delius (1862–1934), in Finland with Jean Sibelius (1865–1957) and in France where, in 1871, a national society was founded under César Franck (1822–90) and Camille Saint-Saëns (1835–1921). The rising French school of Impressionist music culminating in the works of Claude Debussy (1862–1918) was to be a major signpost to the music of the twentieth century.

KEY

Ludwig van Beethoven (1770–1827) is portrayed as Janus, god of the new year. He looks back to the classical tradition which he transformed, and forward to 19th-century music and the Romantic composers who would be inspired by his achievement.

6

6 Franz Schubert, son of a Viennese schoolteacher, was a prolific composer during his brief life, writing nine symphonies, much chamber and piano music and an incomparable body of more than 600 songs. He gained little public recognation during his lifetime – his *C Major Symphony* was not performed until ten years after his death – and his last years were spent in Vienna, often in real poverty.

7

7 The close links between music and literature are underlined in this group portrait of several Romantic artists by Joseph Danhauser (1805–45). Liszt at the piano is playing to the authoress and mistress of Chopin, George Sand, who is sitting beside the novelist and dramatist Alexandre Dumas the younger. Standing [from the left] are the poet Victor Hugo, the violinist and composer Paganini and the opera composer Rossini. At Liszt's feet is the Comtesse Marie d'Agoult, with whom the pianist had a lengthy affair. Beethoven's bust is on the piano, Byron's portrait on the wall.

8

8 A silhouette of Johannes Brahms (1833–97) shows him going off to his favourite tavern, "The Sign of the Red Hedgehog". Brahms was a late Romantic composer who revitalized the tradition of classical forms that had culminated in Beethoven. Brahms's use of traditional devices such as the harmonic sequence and counterpoint, his emphasis on colourful harmony in structure and not only for effect, the stringent unity he sought within music, the independence of his pieces from poetic or literary interpretations, all show classical qualities. These, combined with his expansive rhythmic and lyrical romanticism – as in his songs – produce musical tension that is rich in feeling.

9

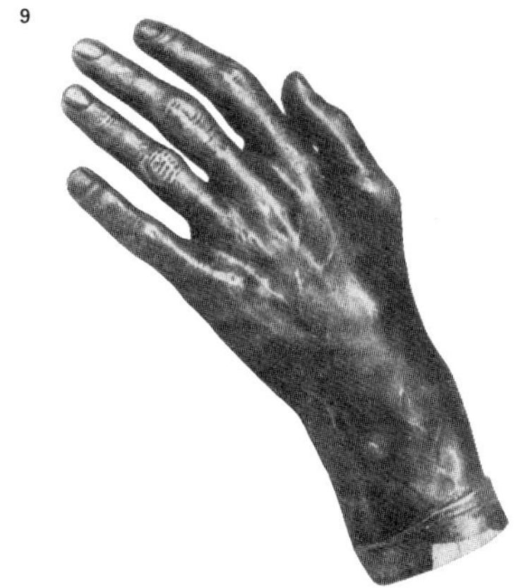

9 A cast made of Chopin's left hand testifies to the public enthusiasm and admiration evoked by his skill and sensitivity as a pianist. He was pre-eminent among 19th-century composers in his command of the modern piano's improved dynamic and expressive possibilities, and has been called "the poet of the keyboard". Early in his career he wrote music for piano and orchestra as showpieces with which to establish his reputation, but from the time he settled in Paris at the age of 21, having left his troubled native Poland, he concentrated on composing and playing short solo works, in the main for salon audiences. In all he wrote more than 150 such pieces before his death at the early age of 39.

The revolutions of 1848

In an age of revolution 1848 was the year of revolution. The governments of France, Italy and central Europe were all shaken by insurrection. Contrary to the belief of contemporaries, there was no overall plan, however, and lack of co-ordination was fatal for the revolutionaries.

Political reform through revolt

The roots of the risings throughout Western Europe were remarkably similar. The Industrial Revolution had unprooted traditional patterns of life and had created a new urban proletariat and a much enlarged bourgeoisie intent upon political power. Economic and social unrest was aggravated by the autocratic rule that was a legacy of the Vienna Settlement of 1815 and which provided a focus for the intellectuals who were agitating for political reform. People were hungry as a result of crop failures in 1845, 1846 and 1847 when bad corn harvests coincided with potato blight. Famine drove desperate mobs onto the streets prepared to demand any changes that offered hope.

Significantly, the centres of unrest were the great cities [2]. Many areas of Europe had recently experienced the Industrial Revolution and thousands had flocked to the cities only to live in squalor and work in conditions of frightening degradation. These people were hit by the second crisis of 1848 – an international credit collapse, which led to wholesale bankruptcies and unemployment. The unemployed joined the hungry on the streets. Finally, there was a psychological catalyst. The epidemic of revolution was accompanied by an epidemic of cholera, which spread panic and anger [7].

Wave of early successes

The first revolts erupted in Italy [5]. Once Louis-Philippe (1773–1850) had abdicated from the French throne in February [3] revolution took hold. In March the resignation of the apostle of European stability, Prince Metternich (1773–1859), Chancellor of the Hapsburg Empire, boosted the morale of the revolutionaries. Caught by surprise and overwhelmed by the extent of the outbreak, governments could not call on each other for help. Their only hope seemed to be to make concessions. Liberal constitutions [1] were granted everywhere and the Hapsburg emperor, the pope and the kings of France and Prussia fled from their capitals.

Simultaneously with the liberal revolts came an upsurge of nationalism. The Hapsburg Empire with its spheres of influence in Italy and Germany [4] seemed doomed. Hungary declared her independence, the Bohemians formed a nationalist movement and a Slav Congress met to consider a new deal for Slavs in the empire. In Italy, Giuseppe Mazzini (1805–72) called for a rising to form a new Italian state. At the same time King Charles Albert of Piedmont (1798–1849) sent an army to help the Lombards drive out the Austrians, hoping to form a north Italian kingdom. In the German Confederation an assembly met at Frankfurt to decide on a policy to unite Germany. These political moves showed the degree of hostility to the Vienna Settlement of 1815 and its legacy of repression.

In spite of all this, by the middle of 1848 the tide of revolution was stemmed. Early successes proved illusory. The Hapsburg

CONNECTIONS

See also
82 The Congress of Vienna
76 Napoleonic Europe
74 The French Revolution
110 German and Italian unification
170 Political thought in the 19th century
84 European empires in the 19th century
168 Russia in the 19th century
184 Balkanization and Slav nationalism

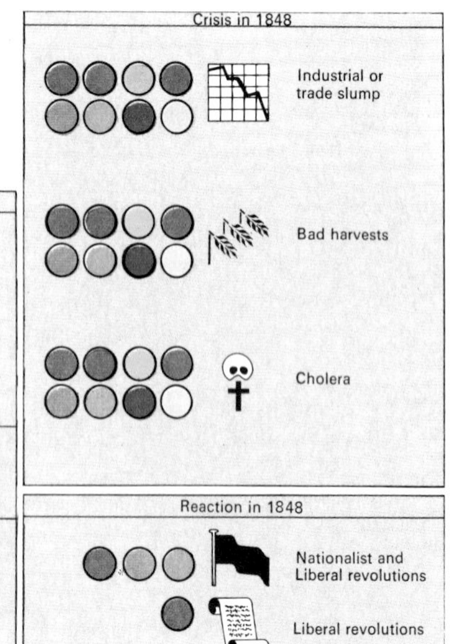
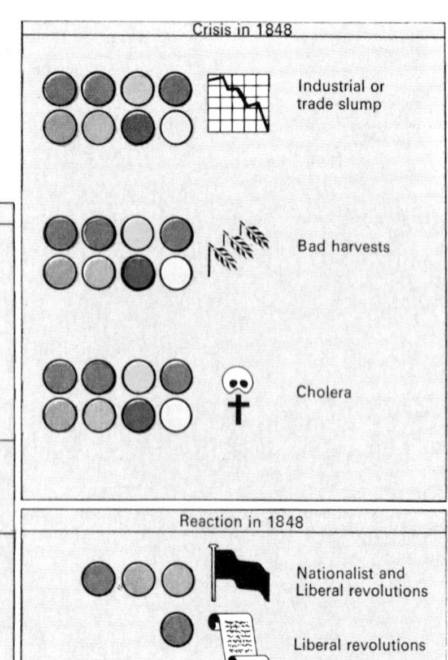

1 Uprisings occurred in most European countries in 1848, with similar causes but varying in intensity and effect. In Russia and Spain, political dissent lacked the concentrated support of the factory or city, while Belgium and Britain had already made political concessions in the face of heavy industrialization and urbanization, avoiding the violent confrontations of 1848.

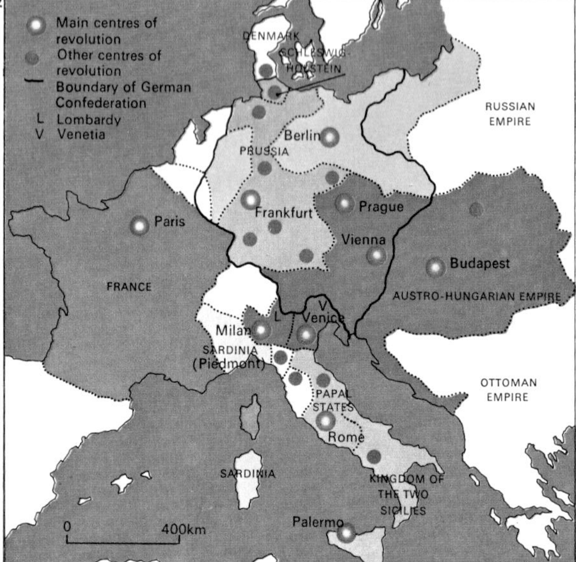

2 The revolutions of 1848 were urban; the peasants were apathetic or conservative. Political ideas spread quickly along the new railways, attracting city intellectuals, workers and businessmen.

3 Paris barricades in March 1848 were manned by middle-class liberals, working-class socialists and the unemployed. Shattered by his unpopularity, Louis-Philippe abdicated within a few days.

4 Liberal revolts in the 39 German states won constitutions that did not survive the repression of 1849. The impotence of nationalists in the Frankfurt Assembly was shown when they called Austrian and Prussian troops in to keep order.

Empire followed its historic policy – divide and rule – by exploiting deep divisions between the revolutionaries. Croats and Romanians who resented Magyar domination rose against Hungary's new leader, Louis Kossuth (1802–94). Their armies helped to do the Hapsburgs' work for them. In Italy, Charles Albert's forces were smashed by the Austrian army in two campaigns. Traditional loyalty to existing separate states deprived him of wide Italian support. Catholics hesitated to disobey the pope, who had forbidden violence against the Catholic Hapsburgs.

In Germany, at the Frankfurt Assembly, the intellectuals wrangled interminably and failed to decide on a form for the new Germany until it was too late. Everywhere the middle classes, who had provided the impetus and leadership for the revolution, were horrified by the forces they had unleashed [8]. Having seen revolution degenerate into anarchy, they welcomed the restoration of law and order. By 1849 all was quiet again. The forces of reaction seemed triumphant. Disorganized mobs [Key] stood no chance against the professional armies [6] of Austria, Prussia, Russia and France. The Hapsburg tradition of garrisoning each province with troops from other provinces had prevented any chance of soldiers siding with the revolution. In every area there was little hope of successful revolution since most of the population – the peasants – rejected it.

The legacy of 1848

There were a few significant gains, however. Serfdom was abolished in the Hapsburg Empire. Piedmont and Prussia kept their constitutions and eventually led Italy and Germany to unity in 1871. Governments learned to pay more attention to the material interests of their subjects and to pay lip service at least to more democratic processes.

But nationalists had learned that idealism and popular enthusiasm would not be enough. Their hopes would be fulfilled only if they could match their opponents' military strength. The revolutions of 1848 were followed by a period of cynicism and opportunism in politics and a use of armed force to settle grievances. Bismarck's age of "blood and iron" had begun.

KEY

Women on the barricades; the tricolour, symbolizing hopes of liberty, equality and fraternity; the red flag of socialist revolution; the flags of German, Italian, Hungarian or Bohemian nationalism – all these made a heroic display in 1848. But heroic slogans such as "Bread or Death" did not match an army.

5

5 Italian revolts for state constitutions, for republics in Rome and Venice and for a north Italian kingdom all collapsed by 1849.

6 Military saviours of the Hapsburgs (caricatured left to right) were Jellačić (1801–59), who led Croats against Magyars in independent Hungary, Radetzky (1766–1858), who successfully ended the Italian revolts, and Windischgrätz (1787–1862), who subdued Vienna and Bohemia.

6

7

7 The Paris sewers begun by Baron Haussmann (1809–91) during the 1850s were a response to criticism of governmental failure to stop cholera spreading in 1848, when fear of the disease acted as a catalyst in the revolutions. Haussmann also created wide boulevards to facilitate cavalry charges against future revolutionary barricades.

8

8 Karl Marx (1818–83), shown as Prometheus chained to his printing press, and Friedrich Engels (1820–95) published the *Communist Manifesto* early in 1848 as a doctrine and strategy for the Communist League. Although this made no contribution to the outbreaks of 1848, fear of socialism inhibited the revolution.

German and Italian unification

Italy and Germany were created in spite of limited popular support, strong communal loyalties to existing units and the proximity of two powers whose interests were endangered by their emergence as strong nations – the Austrian Hapsburg Empire and France. The new nations were the fruit of the ambitions of their strongest components, Piedmont and Prussia, and of the outstanding practitioners of the new *Realpolitik*, Camillo Cavour and Prince Otto von Bismarck.

The birth of modern Italy

As Prime Minister of Piedmont from 1852, Cavour (1810–61) [5] built up his state as a magnet to attract the rest of Italy. He made the new parliamentary democracy work, encouraged up-to-date agriculture and industry and linked the Piedmontese economy to that of Europe through a railway network and the modernized port of Genoa. He created a fair legal system and an efficient bureaucracy. With a competent small army and a king, Victor Emmanuel (1820–78), known to be a genuine Italian patriot, Piedmont became the focus of national hopes.

Outside help was vital to drive out the Austrians. France became a pawn in Cavour's game. In the Pact of Plombières (1858), Louis Napoleon – the French Emperor Napoleon III (1808–73) – promised him help in a future war. Cavour engineered an attack against Piedmont by Austria in 1859 and French troops were sent in. After triumphs at Magenta and Solferino, Louis Napoleon had second thoughts and withdrew his support from Cavour, but his help had been decisive. In the excitement of the victories, Parma, Modena, Tuscany and the Romagna demanded amalgamation with Piedmont. In return for the acquisition of Nice and Savoy, Napoleon backed plebiscites in Emilia and Tuscany and Cavour won an overwhelming majority in favour of the formation of a north Italian kingdom.

Matters might have rested there but for Giuseppe Garibaldi (1807–82). When the Sicilians rose in revolt against Naples in 1860, Garibaldi and 1,000 men went to their help. Within weeks, the Neapolitan army had been swept out of Sicily and Garibaldi marched in triumph through Naples itself. Rome was his next objective. But if Garibaldi attacked Rome then France and Austria might intervene to defend the pope, so Cavour sent a Piedmontese army to forestall any further advance. Garibaldi, in a dramatic gesture, gave up the south to Piedmont [6].

Only two areas of Italy now remained unintegrated. Venetia was held by Austria and Rome and its surrounding territories were held by the pope and a garrison of French troops [3]. In 1866, Victor Emmanuel joined Prussia in the Austro-Prussian war and was given Venetia. In 1870, France withdrew her troops from Rome to fight the Prussians and Victor Emmanuel became king of a united Italy [7].

Prussia and the "Iron Chancellor"

In northern Europe Bismarck (1815–98) had become Chancellor of Prussia in 1862, where he was faced with a Liberal majority hostile to his aims. But he managed to manipulate them and finally gain their support for unification and his policy of *Realpolitik*.

Together, the customs union, or Zollverein [2], and the growth of railways had

CONNECTIONS

See also
108 The revolutions of 1848
84 European empires in the 19th century
82 The Congress of Vienna
158 Industrialization 1870–1914
180 Europe 1870–1914
170 Political thought in the 19th century
130 Imperialism in the 19 century

1

1 Kaiser Wilhelm I of Prussia was acclaimed German Emperor at Versailles in 1871. He called it "The unhappiest day of my life"; he had wanted the less democratic title "Emperor of Germany" and left the room without glancing at the architect of the new Germany, Bismarck. Bismarck [centre] had crushed all opposition to German reunification by "blood and iron". It was his skill and vision that had created Germany; as Chancellor until 1890 he moulded its institutions and laboured to make it inviolable. Von Moltke (1800–91) [on Bismarck's left], Chief of the Prussian General Staff, was the strategist of the triumphs against Austria and France.

2

2 A potpourri of 39 states, the German Confederation, was united in the customs-free Zollverein in 1844. The Confederation was further extended by the Austro-Prussian War of 1866. Thuringia and Mecklenburg sided with Prussia, and joined; Hesse-Darmstadt and Saxony were annexed on Austria's defeat.

3 This caricature of Pope Pius IX expresses the disappointment felt at his failure to support liberalism consistently. As papal lands were lost, he increased the spiritual claims of the Holy See. In 1864 he condemned contemporary political doctrines and in 1870 went on to proclaim papal infallibility.

3

already removed most natural and artificial impediments to German integration and prosperity. Bismarck was determined to remove Austrian influence and unite Germany, and despite opposition in the Catholic south to the dominance of Protestant Prussia, in three wars he succeeded.

Bismarck's diplomatic skill ensured that each war was fought against an isolated opponent. In the Danish war of 1864, he fought ostensibly to free the two-German-speaking duchies of Schleswig and Holstein from Danish control. But by setting up a joint control of the duchies with Austria, the principal obstacle to German unification, he created an ideal situation for picking a quarrel with her.

The time was ripe in 1866. France's neutrality had been bought by vague promises of territorial concessions and Louis Napoleon had no time to realize his mistake [Key]. In a war lasting only seven weeks, the Austrian army was smashed at Sadowa. In 1867, Prussia dominated a north German confederation. A southern confederation was set up, but without Austria.

France realized too late the emerging danger on her eastern frontier. Vital reforms to her army had come too late and Louis Napoleon was outmanoeuvred by Bismarck in a diplomatic game over rival candidates for the throne of Spain. The hysterical reaction in both countries to the candidature of a nephew of the Hohenzollern king of Prussia provoked France to declare war on Prussia in July 1870. On 1 September, the French army capitulated at Sedan and all resistance collapsed by January 1871.

A wave of enthusiasm swept the south German states for unity with the north. On 18 January 1871, Wilhelm I, King of Prussia, was proclaimed German Emperor [1].

Death of a dream

Before 1848, Italian and German nationalists had dreamed of new states that would free their citizens, release their stifled talents and regenerate Europe. The new states of 1871 were created at a price. Liberalism was sacrificed to nationalism; cynicism opportunism and violence had triumphed – not idealism and liberty.

KEY

Napoleon III, Emperor of France, was exploited and outwitted first by Cavour, then Bismarck, in the unification of Italy and Germany. By offering help to Cavour, he hoped to gain Savoy and Nice and create a weak client state. In the event he almost missed his reward and saw the creation of a unified Italy. He was outmanoeuvred by Bismarck, realizing the threat too late, after Prussia had defeated Austria. By staking his authority on an attempt to force Bismarck to give up any future plans to put a Hohenzollern on the Spanish throne, he led France into the war with Prussia that brought his own empire to an end.

4

4 Mazzini's proclamation of a Roman Republic in 1849 left a legend of heroism to Italy. Giuseppe Mazzini (1805–72) had founded "Young Italy" to lead his countrymen towards democracy without outside help or compromise, and dreamed of a state that would "evoke the soul of Italy".

5

5 Camillo Cavour was never able to inspire the sense of moral crusade that was brought to the *Risorgimento* by Mazzini and Garibaldi. But he understood the politics, grasped the international context and had the skill to exploit the possibilities. Without the exertion of his skills Italy could not have been unified.

6

6 At an historic meeting in 1860 on the Naples road, Garibaldi gave to Victor Emmanuel the gift, in effect, of a unified nation – in exchange he took a sack of seedcorn.

7

7 Although Italy was united by 1870, political and economic development was uneven. Despite Garibaldi's dramatic exploits, southern Italy remained backward compared with Piedmont.

Victoria and her statesmen

Queen Victoria's reign, from 1837 to 1901, lasted longer than that of any other British monarch. During that time the party system and parliamentary democracy came to their maturity. The monarchy itself moved out of the arena of active politics, but achieved a new status as the neutral guardian of national stability. In 1830 even *The Times* had found it difficult to mourn the death of George IV; republicanism was a serious radical cry. By 1897, the year of the Diamond Jubilee [9], republicanism had been drowned in popular royalist enthusiasm.

The changing style of politics

Ten prime ministers served Queen Victoria [Key]. None of them was chosen by her in defiance of the wishes of the Commons. Each came to power by virtue of being the leader of his party, and cabinets were composed of members of the same party. That was a marked though gradual change from the eighteenth-century politics of connection. Party had replaced the Crown as the source of political power. After the 1832 Reform Act, both the Whigs and the Conservatives took steps to organize themselves into national parties. Elections lost much local colour and acquired national meaning.

The first half of the reign

It was not easy for the 18-year-old princess to step with confidence onto the crowded political stage. Victoria was fortunate to find a devoted tutor in her first prime minister, the debonair Lord Melbourne (1779–1848), then mellowed with age. To the man whom Caroline Ponsonby had flattered to deceive in marriage, Victoria brought a late spring in the autumn of his career. To her, he became as a father.

She was loath to part with Melbourne. But the weakness of her constitutional position was brought home to her by the Conservative victory at the 1841 elections. Loving Melbourne, she had learned to love the Whigs. Losing him, she learned to work as closely and fairly with his successor. Throughout her reign she kept herself fully informed on political developments; her opinions could never be treated lightly by her ministers.

Robert Peel, from 1841 to 1846 the first prime minister of a Conservative (as opposed to a purely Tory) Party, was a new breed of prime minister. His roots were commercial and his forte was economics. He had no sentimental attachment to the landed aristocracy. The squires on the backbenches, the heart of his party, found him uncommunicative and arrogant. He tried to turn the Tories into a party that worked to balance the claims of competing interests [1], instead of seeking to defend the exclusive interests of the land and the Church. He failed, split his party, and left it a minority for a generation.

Viscount Palmerston (1784–1865) was the beneficiary of this Conservative misfortune. He was prime minister for all but 14 months between 1855 and 1865. England was then enjoying the mid-Victorian boom; the standard of living was generally high and social problems unobtrusive. Palmerston believed that a government did best by doing as little as possible. His great interest was foreign affairs, the one sphere where the royal will still counted for something. He

CONNECTIONS

See also
132 The British Empire in the 19th century
96 Social reform 1800–1914
182 British foreign policy 1815–1914
162 Ireland from Union to Partition
114 Victorian London
68 Pitt, Fox and the call for reform

1 Robert Peel (1788–1850) sought an undoctrinaire approach to the problems of industrialization that brought violent Chartist unrest, and in 1846 after the Irish famine he alienated the traditionally Tory landowners by removing tariffs on imported corn, thus reducing the price of bread.

2 The Great Exhibition (1851) at the Crystal Palace, asserted Victoria's international standing early in her reign. Rulers from many parts of the world attended the festivities, which were originally conceived by Prince Albert to celebrate the wonders of industry and to promote peace.

3 Prince Albert (1819–61), married Victoria in 1840, and rebuilt much of the Kensington district of London for the Great Exhibition. Among the monuments erected to him was the Albert Bridge, shown here.

4 Disraeli became Conservative leader in the Commons in 1849. He passed the 1867 Reform Act in an attempt to outbid the Liberals for popular appeal, and founded the Conservative Central Office (1868) to organize the party in the country.

and Victoria clashed often and she sometimes won. In 1857 Palmerston made light of the Indian Mutiny; Victoria knew better, and it was on her initiative that troop reinforcements were sent to India which saved the British presence there.

Gladstone and Disraeli

Palmerston's death in 1865 allowed William Ewart Gladstone (1809–98) [6] to assume the leadership of the Liberal Party. He was the first prime minister to form four governments (1868–74, 1880–85, 1886, 1892–4). He was too single-minded, too earnest, and too radical to earn anything but Victoria's habitual distrust. But he was beyond question the giant of Victorian politicians. Under his premiership the Irish Church was disestablished (1869), secondary education made universal (1870), the secret ballot introduced (1872), and the agricultural labourer enfranchised (1884). His mission, he said, was to pacify Ireland, but his Home Rule Bill of 1886 was defeated by Liberal Unionists, led by Joseph Chamberlain (1836–1914), perhaps the greatest Victorian statesman never to be prime minister.

Gladstone's great rival was Benjamin Disraeli (1804–81) [4], already a tired man at the start of his main ministry (1874–80). His achievement was to help swell the tide of imperial sentiment by his rhetoric and, by making Victoria Empress of India in 1876, to exalt in the popular imagination the person and office of the monarch. And she for six years found in "Dizzy" an unfailingly courteous and amusing companion. He was her favourite prime minister. It was an extraordinary end to the career of one who, from his Jewish descent, his landless status, and suspect literary connections, had never become quite acceptable even to his own Conservative party.

His successor as Conservative leader, Lord Salisbury (1830–1903) [7], was in the purest Tory mould, the last great representative of the Cecil family that had risen to prominence under Elizabeth I. He was the ideal minister to preside over the Jubilee celebrations. He formed three administrations (1885–6, 1886–92, 1895–1902), and left politics just after the queen's death.

KEY

The queen, as head of state, appointed each new prime minister. Here she is shown giving the seals of office to Lord John Russell (1792–1878) in 1846 after the fall of Peel's ministry on the controversial issue of the repeal of the Corn Laws. In the 18th century, the monarchs chose their own ministers and exercised real discretion over dissolutions of Parliament. Victoria's power was less direct but her personal relations with her prime ministers were of great importance to the history of her reign.

5

5 The death of Albert in 1861 led Victoria to withdraw from social life and dress in mourning for many years. At this point her popularity reached its lowest ebb, but revived by the 1880s. She never recovered from the loss of her German-born husband.

6

6 Gladstone was the son of a Liverpool cotton merchant and retained a radical and evangelical outlook throughout his career. His stirring oratory made him the darling of the industrial masses, but he was a highly intellectual man who was MP for Oxford University. His political career, like that of Disraeli, began in The Conservative Party, in Peel's ministry; but he joined the Liberals after 1846.

7

7 Lord Salisbury was enigmatic and shy and an implacable foe of Irish Home Rule. He made the Conservative Party into the most powerful party in the state by 1902.

8

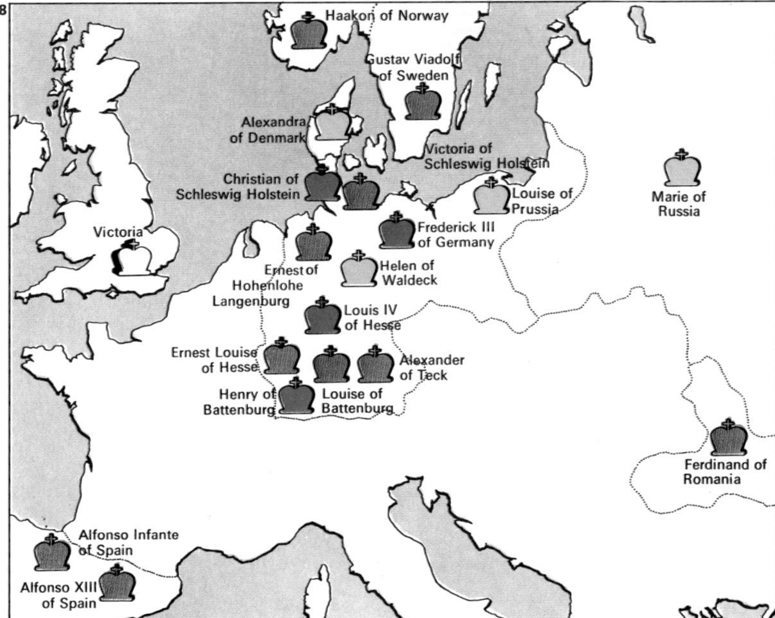

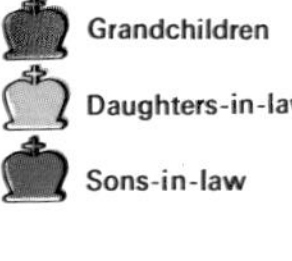

8 Queen Victoria was related to most of the royal houses of Europe by the marriages of her children and grandchildren. At her death there were 37 of her great-grandchildren living, and she became popularly known as the "grandmother of Europe". Her intimacy with foreign courts gave her a knowledge of diplomacy. But her constitutional position prevented her from exercising real influence over foreign affairs. She was restricted to the Crown's ancient right to be consulted, to encourage and to warn.

9

9 The Diamond Jubilee of 22 June 1897 was a grand imperial festival which was attended by representatives of Victoria's 387 million subjects. The queen was 78 and suffered from rheumatism and failing eyesight. The short service at St Paul's, which marked the halfway point of the royal procession from Buckingham Palace, was held outside the cathedral to avoid carrying the queen up the steps in a wheelchair. Like the jubilee of 1887 the Diamond Jubilee provided an occasion for a colonial conference.

Victorian London

In the nineteenth century, London became the biggest and richest city in the world, its population quadrupling to reach 6,586,269 by 1901 in Greater London (a term first used in the 1881 census). Its growth as the heart of a great commercial and military empire presented a spectacle both imposing and appalling. Between the plush and cut-glass elegance of the West End and the fever-ridden slums of Dickensian description lay a gulf the century could not bridge. Overwhelmed by the squalor in which many of the people lived, the critic John Ruskin in 1865 called London "rattling, growling, smoking, stinking – a ghastly heap of fermenting brickwork pouring out poison at every pore".

Commerical expansion

The port of London was central to the economic growth of the capital. The first large enclosed docks were completed in 1802. In 1885 the expansion of trade was marked by the completion of the Victoria Dock, 2 kilometres (1.2 miles) long. Although challenged by ports such as Hull and Liverpool, London remained the premier port, and 13 million tonnes of goods passed through it in 1880.

London was the centre of a host of industries associated with trade, refining and processing imported goods for distribution to the rest of the country or for re-export. In addition to brewing, distilling, tanning and food-processing the capital supported a shipbuilding industry that was overtaken by Newcastle and Glasgow only in the closing years of the century. By 1851 there were almost half a million workers engaged in manufacturing. Service industries employed nearly one million by 1861.

Even before the end of the eighteenth century, London had begun spreading out into rural areas of Surrey and Kent. The nineteenth century saw a rapid extension of this process as the City proper, the "square mile" formerly confined by the city wall, was given over to shops, offices and warehouses. In the West End, fashionable squares and town houses were completed. The growing middle classes built houses in suburbs such as Camberwell, Paddington and Clapham. Although the East End, [7] including Whitechapel, Bethnal Green and Stepney, continued to grow, the working classes too began moving to districts on the edge of the built-up area, such as Hammersmith.

Railways and transport

The growth of suburban London was greatly accelerated by the coming of the railways [1], which soon spread out into a dense network. The first underground line in the world, from Paddington to Fenchurch Street, was opened by the Metropolitan Railway Company in 1863 and in the first six weeks carried an average of 26,000 passengers a day. A first-class fare between Edgware Road and King's Cross was sixpence. In 1864 special workmen's trains were introduced with a maximum return fare of only threepence. Other lines soon connected the main line stations and all important parts of the metropolis. The first electrified line opened in 1890. By the end of the century horse-drawn buses and trams provided alternative transport [Key].

The central area of the capital was refurnished with a series of new public buildings,

CONNECTIONS

See also
88 The Industrial Revolution
90 The urban consequences of industrialization
96 Social reform 1800–1914
112 Queen Victoria and her statesmen
122 European architecture in the 19th century

In other volumes
294 History and Culture 1

1 Railways grew out of urban expansion but also created it. The London and Greenwich line, opened in 1836, was London's first steam-powered railway. By 1852, most of today's mainline stations were in being and there were several suburban lines. From 1863 the Underground system provided rapid transport in the metropolitan area. In 1880, between 150 and 170 million rail journeys were made in the city annually.

1A
Hampstead
Hackney
Stepney
Chelsea
Woolwich
Wandsworth
Croydon
0 5km
London 1769

B
Hampstead
Hackney
Stepney
Chelsea
Woolwich
Wandsworth
Croydon
0 5km
London 1888
Main railways

2 Cholera epidemics in the 1830s and 1840s gave an important stimulus to the public health system. Royal Commissions of inquiry led to the creation of a General Board of Health and a Medical Officer of Health in 1848, in spite of opposition from the City of London. Improvements in water supply and sanitation followed. Until the 1860s, London's sewers were discharged into the Thames.

2

3

3 A professional police force for London was created under the Metropolitan Police Act of 1829. Until then, London had only a few hundred professional police. The security of the capital largely depended upon an unco-ordinated band of watchmen and constables under several different authorities. The Act enabled a unified policing of the whole metropolitan area, with the exception of the City proper, using several thousand professional police officers

4 The Bank of England, established in 1694 to finance a war with France, was subsequently granted a monopoly of joint stock banking. While smaller banks were restricted to only six partners between 1708 and 1826, it became government banker and reserve bank for the whole country – a status ratified in the Bank Charter Act (1844). The fine buildings, completed in 1827, are imposing examples of the many public buildings erected in London during the 19th century. Others, in Victorian neo-Gothic style include St Pancras Station (1865–71) and the Royal Courts of Justice (1871–82).

4

5

5 London fashion set the pattern of taste and consumption for the country as a whole. The mass market in the capital stimulated the rise of large department stores in the 1880s along bustling streets such as Regent Street, shown here. During the 19th century small, family-run shops began to disappear. They either developed as chain stores, such as J. Sainsbury's, which first opened in Drury Lane in 1869, or they were replaced by large, independent department stores.

including the rebuilt Houses of Parliament (1836–67), the Royal Courts (1871–82), the Bank of England (1795–1827) [4] and the great museums in Kensington. Trafalgar Square was completed in the 1840s. Widened streets, notably the Embankment [8], imitated the boulevards of Paris and some of the old slums were cleared to make way for new streets such as Charing Cross Road. Sewerage [8], lighting, paving and water supply were gradually brought under control after the cholera epidemics of the 1830s and 1840s [2]. The establishment of a Metropolitan Board of Works in 1855 was important for unified planning.

Social reforms

The vitality and commercial prosperity of the capital were reflected only slowly in social reforms. As people crowded into the terrace houses thrown up by speculative builders around gasworks, breweries or warehouses, the parishes of the East End became spawning grounds for crime and disease. General William Booth (1829–1912), who founded the Salvation Army in 1878 to reconvert the slum dwellers called them the people of "darkest England". In that year, while many landowners were earning £100,000 a year and paying tax of only 2d in the pound, the average labourer's income was £70. Fashionable strollers [5] wore hand-sewn garments created by sweated labour paid at a rate of only 2d an hour.

In 1885, it was estimated that one in four Londoners still lived in abject poverty. Only after 1880, when primary education became compulsory, were the streets cleared of ragged children living on their wits.

Despite the ferocious penalties for even petty crime, an estimated 100,000 in London lived by thieving or swindling in the 1860s, and another 80,000 were prostitutes. Sensational stories of crime and capture by the Metropolitan Police Force [3] could be read daily in the "penny dreadfuls". This was the fog-shrouded city of Sweeny Todd the Barber and Jack the Ripper. Violent riots by the unemployed in 1886 and 1887 gave belated vent to the distress that went hand-in-hand with the music halls [6], gin palaces and imperial pomp of Victorian London.

KEY

St Paul's Cathedral, erected in more spacious days, looks down on the Fleet Street of 1900 when vehicles thronged London's streets and traffic jams had become common. In 1850 there were more than 1,000 horse-drawn buses at work in the capital as well as countless carts and wagons. Congestion was one indication of the need for a new form of urban planning. London was the first industrial metropolis to have to cope with the problems of public health, urban transport, housing and other services on a mass scale. The unprecedented difficulties created by its growth necessitated a dramatic increase in the powers of local government.

6

7

6 Music halls became immensely popular in the 19th century. After a licensing Act in 1843, music halls, unlike theatres, could serve alcohol. The first commercial halls were the Canterbury in Lambeth (1852) and the Oxford Music Hall in Oxford Street (1861). Forty halls were taking in custom in London in 1868, and as the century progressed, music halls came to be more widely accepted as an alternative to the theatre.

7 The East End of London remained notorious for its poverty and bad housing well into this century. Many of its inhabitants were immigrants who had come from Ireland and continental Europe.

8

9

8 Construction of the Embankment (started 1867) with railways, sewerage and other services was a rare example of unified planning for growth. An efficient London system of drainage and sewerage was delayed by a lack of centralized authority. In 1858, work began on a complete system of sewerage for the capital. This great engineering feat was completed in 1865 and cost £4 million, with 131km (82 miles) of pipe carrying 1,703 million litres (420 million gallons) of sewage each day.

9 London was the social centre of Britain. The London "season" attracted wealthy families up from the country to reside in the substantial houses they kept in town. Hyde Park (shown here) was a fashionable place of recreation.

Realist painting in the 19th century

Realism is the term used to describe the most characteristic style that arose in painting, particularly in France, between the end of Neoclassicism and romanticism and the beginnings of Impressionism. It belongs essentially to the years 1840 to 1870, although some paintings with realist tendencies were produced before this date, and the style continued to flourish until almost the end of the century. A significant event was the development, during the same period, of the new art of photography [Key].

The social context

Photography as the ultimate in pictorial realism was at once a challenge to painting and an echo of, and influence on, it. At first it was chiefly painting that influenced photography (many of the early photographers began their careers as painters) but from about 1860 onwards the influence began to flow the other way [8].

Realism grew as much from social as aesthetic motives, but the reasons for it were not the same in all countries. In Britain, where it began first soon after 1800, it succeeded chiefly because, in the nineteenth century, art for the first time became really popular with a mass public. The more traditional styles of painting, which depended for their appreciation on an educated few, fell out of favour. They were replaced by a new, more direct art [6] representing (within tasteful limits) things as they were, in a style based on the accepted models of seventeenth-century Dutch and Flemish painting and with a strong element of humorous or sentimental narrative which enabled pictures to be "read" like a novel.

The pioneer of popular narrative painting was David Wilkie (1785–1841), who was actually patronized by the aristocracy but whose art reached a wide public through exhibitions and prints. Wilkie was the most popular artist in Britain during the first 40 years of the century, and his approach [1] became the model, more or less, for all subsequent British Victorian artists.

The situation in France was different and Realism began there later. It was not a popular style as it was in Britain; rather, it was serious and committed, even subversive. Whereas in Britain Realism developed within the Academy, the home of official and aristocratic taste, in France it was conceived partly as an attack on the official historical art sponsored by the Ecole des Beaux-Arts, then the guardian of academic values.

Influence of Courbet

The leading French Realist was Gustave Courbet (1819–77), whose career ran from the mid-1840s to the early 1870s. He was aggressively bohemian and provincial, a democrat if not a revolutionary, and he founded the doctrine, later a Realist battle-cry, that the artist must be "of his own time". "Painting is an essentially concrete art", he wrote, "and can consist only in representation of real and existing things."

In contrast to British painters, Courbet played down the element of narrative and, for virtually the first time, represented ordinary provincial and working-class people in everyday terms. This was thought undignified. A picture such as "The Meeting" [2], which shows the rich bourgeois patron doffing his hat to the journeyman artist (Courbet himself), caused offence not only because of

CONNECTIONS

See also
158 Industrialization 1870–1914
80 Romantic art: figure painting
102 Romantic art: landscape painting
122 European architecture in the 19th century
118 Impressionism
42 Neoclassicism
30 Scottish culture to 1850

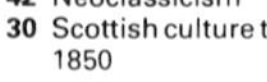

1 David Wilkie's "Chelsea Pensioners reading the Gazette announcing the Victory of Waterloo" (1822, detail), is an example of early "popular" Realism.

2 Gustave Courbet's "The Meeting" (1854), familiarly known as "Bonjour, Monsieur Courbet!", shows the artist being greeted on the road by his friend and patron, Alfred Bruyas.

3 The labours of the fields, previously depicted in pastoral scenes, were treated realistically by Jean François Millet. In "The Angelus" (*c.* 1858) he added an element of religious sentimentality which made the picture especially popular at that time.

4 The Pre-Raphaelite Brotherhood, founded in 1848, sought to combine fidelity to nature with the purity of spirit of the Italian painters before Raphael. These qualities are reflected in John Everett Millais' "Sir Isuoubras at the Ford" (1857).

5 William Holman Hunt (1827-1910) was a Pre-Raphaelite who in "The Awakening Conscience" (1853) turned his attention to personal morality, preaching a sermon to his middle-class audience on the evils and pathos of adultery. The girl starts up from her lover's lap on being reminded of her lost innocence by the tune he is playing and by the sunlit garden outside.

its reversal of the normal relationship between artist and patron but also on account of its apparent lack of any interesting subject.

With Courbet, French Realism began to take on a class-conscious, political tendency and to be identified with grim and sordid subject-matter. It is also noticeable that Realist paintings from this time onwards are normally dark in tone and drab in colour, resembling contemporary photographs. Although Courbet himself does not seem to have intended his work as social propaganda, the way towards this was now open and many painters took it. For instance Jean François Millet (1814–75) showed the hard life of a depressed peasantry redeemed only by the consolation of religion [3] and later, in England Hubert von Herkomer [9] specialized in painting the industrial working class.

The nineteenth century was the first in history to take work seriously as a subject for art and to treat it not in some symbolic guise, as had been done by artists in the pastoral tradition, but as a dedicated, often grinding and monotonous activity. Another interesting development was the realistic portrayal by the American Thomas Eakins (1844–1916) of the working lives and achievements of surgeons and inventors [7].

Morality, mythology and history

While Realism was usually identified in this period with modern life and dealt with questions of social rather than individual morality, there were exceptions to both these rules, especially in British painting. Pre-Raphaelitism was an English style of the late 1840s and 1850s that applied realistic pictorial aims to personal moral problems [5] and to religious themes [4]. In both cases it produced a sense of shock comparable to Courbet's paintings and for fundamentally the same reason: that art was being used to disturb its audience and not to please it.

Finally, Realism increasingly invaded the realm of historical and mythological painting, reducing that once noble and intellectual genre to the level of a make-believe voyage into past time, as in the languid reconstructions by Edward Poynter (1836–1919) and others of the daily lives of the ancient Greeks and Romans [10.]

KEY

The symbol of 19th-century Realism in art is, ironically, not a painting but a photograph. Photography, which began to become effective in the 1840s, fulfilled the Realist painter's wildest dreams, yet did so in a medium that was not his own and that dispensed with the arduous process of matching nature by means of brushmarks. In fact the two arts coexisted in an uneasy but mutually beneficial relationship for the rest of the century. The inventor of the first practical and successful photographic process was a Frenchman, Louis Daguerre (1789–1851), portrayed here in a "daguerrotype" by the English photographer, J. J. E. Mayall, in 1848.

6 "Derby Day" (1856–8, detail), by William Powell Frith (1819–1909) follows Hogarth and David Wilkie rather than either the Pre-Raphaelites or the French Realists. Yet it shows a characteristic side of nineteenth-century life – its energy and vulgarity – and has the contemporary feel of a magazine illustration. The picture also contains some witty character drawing and is composed with considerable skill.

6

7 The discovery of anaesthetics and antiseptics made surgical operations (as distinct from dissections) a possible subject for art. The opportunity was seized by the American Thomas Eakins in "The Clinic of Dr Samuel Gross" (1875).

7

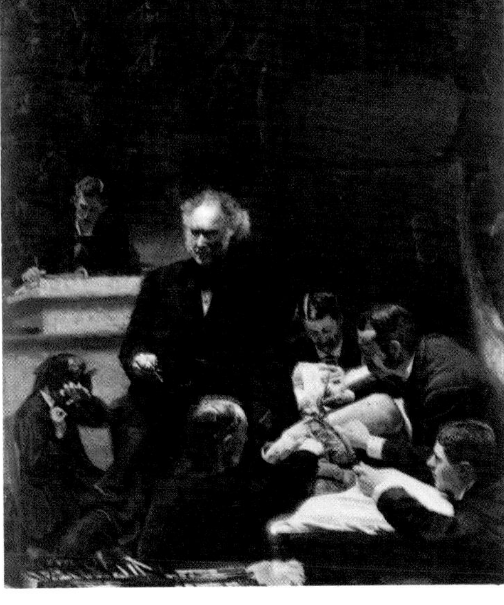

8 After 1860, cafés were a popular subject of the Realists, and cropping of the image to produce a casual effect, as in a photograph, became common. "Au Café" (1878) by Edouard Manet (1832-83) is an example of this. It also shows the reintroduction of vivacity and colour as the painter's technique moves towards Impressionism.

8

9 "On Strike" (1891), by Hubert von Herkomer (1849–1914), enters, with vivid realism, the world of the industrial working class. Its style reveals the direct influence of photography, although this is most evident in black-and-white reproduction. As with most life-size Victorian paintings, the original disappoints owing to its laboured execution.

9

10 "A Visit to Aesculapius" (1880), by Edward Poynter, shows what the grand style of history painting came to in the end. For all the correctness of the drawing, the intrusion of realism and the trivial subject – a parade of naked Victorian ladies with imaginary illnesses before a classical faith healer – make the picture embarrassing.

10

Impressionism

"Impressionism" was initially a derogatory term. A bewildered critic, Louis Leroy, first used the word in 1874 after being outraged by Monet's "Impression, Sunrise" [1] which was hanging in an independent exhibition of 251 paintings. The critics singled out eight painters as a distinct group that had apparently abandoned traditional form and content in favour of subjective impressions.

The Impressionist painters

The "father" of the Impressionist group was Camille Pissarro (1830–1903). Others were Paul Cézanne (1839–1906), son of an Aix-en-Provence banker and childhood friend of Zola; Edgar Degas (1834–1917), habitué of the race-course, theatre and ballet studio; Claude Monet (1840-1926) who was contemptuous of all Old Masters; Auguste Renoir (1841–1919), an ex-porcelain decorator at the Limoges potteries; Alfred Sisley (1839–99), son of a well-to-do English merchant living in Paris; Berthe Morisot (1841–95), pupil and sister-in-law of Manet; and Armand Guillaumin (1841–1927). Edouard Manet (1832–83), "leader" of the group and an undoubted influence on the younger painters in the 1860s, had refused to exhibit outside the official Salon.

Although not recognized as a group until 1874 the Impressionists had met and worked together during the preceding decade. Some of them trained together and some painted the same scenes side-by-side: Monet, Sisley and Renoir in the Forest of Fontainebleau in 1864; Monet and Renoir at "La Grenouillère" near Paris in 1869; and Monet and Manet at Argenteuil in 1874 [Key]. These artists also shared the experience of frequent rejection by the Salon, which made it very difficult to sell their paintings. Furthermore, with the exception of Degas none of them had received the Salon-approved training offered by the Ecole des Beaux-Arts. This freed them from current artistic conventions and allowed them to experiment.

Style and subject

The Impressionists refused to paint historical events in the tradition of J. L. David or idealized landscapes in the manner of Claude and Poussin. Instead they chose everyday subjects from the region of Paris and Normandy: Degas painted race-courses [2], Monet views along the Seine or the insides of railway stations, and Renoir painted figures in dappled shade [3, 4]. Cézanne [6] was the exception with his landscapes of Provence. These subjects could be directly experienced and immediately recorded on the spot by the artist; recourse to imagination was thus made superfluous and studied composition was impossible. Their paintings were impressions "in the sense that they portrayed not a landscape but a sensation produced by a landscape". This was not entirely innovatory. The determined use of everyday subject-matter by Courbet (1819–77) was an important precedent. Painting on location had been practised by French artists of the eighteenth century such as Claude Joseph Vernet and Valenciennes as well as by Corot (1796–1875), Constable (1776–1837) and the nineteenth-century masters of the Barbizon school. The Impressionists differed in that they produced not sketches but finished paintings out of doors.

Traditional artists working in studios

CONNECTIONS

See also
80 Romantic art: figure painting
102 Romantic art: landscape painting
116 Realist painting in the 19th century
174 Fauvism and Expressionism
176 Cubism and Futurism

1 Claude Monet's "Impression, Sunrise" provoked a painter companion of critic Louis Leroy to explode: "Impression – of that I was certain! I had to tell myself that since I am impressed it must contain some sort of impression."

2 A characteristic painting by Edgar Degas is "Provincial Racecourse" (1870–73) with its arbitrary composition, the "snapshot cropping" of horse and carriage and the uneasy spatial relationship between the foreground and the background.

3 Auguste Renoir's "Dance at the Moulin de la Galette" (1876) shows a Sunday afternoon scene at an outdoor dance at the foot of one of the surviving Montmartre windmills. It has been treated as an important landmark in the history of Impressionism ever since it was first exhibited in 1877. Its size, large by Impressionist standards, together with the many figures, made its execution on location technically demanding for the painter.

4 A detail from Renoir's "Dance at the Moulin de la Galette" shows rapidly placed dabs of bright green, blue, yellow and red jostling one another. Black is used not for shadow but only for a top hat; white is placed not for highlights but for a collar; and forms merge with their surroundings to express sunlight filtered through leaves.

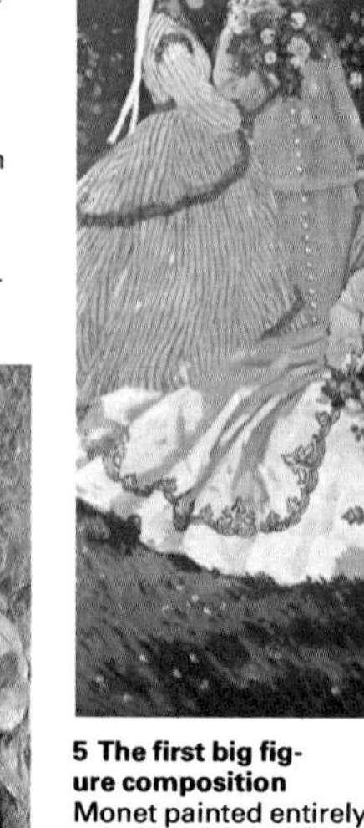

5 The first big figure composition Monet painted entirely out of doors and without any preliminary sketches was "Women in the Garden" (1866–7) painted in the artist's garden. His use of black and the harsh divisions between light and shade betray his early debt to the traditional Manet.

recorded light as tone. The Impressionists, by working out of doors, came to realize that light, whether ephemeral as for Renoir, or enhancing eternal forms as for Cézanne, was composed of colour. In developing this new-found relationship they quickly modified and increased the number of colours on their palettes; they eliminated black from shadow, substituting purple, and adopted the system of complementary colours proposed by the physicist Chevreul in the 1830s. Additionally, they lightened the grounds of their canvases, replacing the traditional brown and biscuit tones with the white and beige of the English watercolourists Bonington (1802–28) and Turner (1775–1851).

Latest developments and changes

The Impressionist style took time to mature. During the 1860s, for example, when Monet painted his "Women in the Garden" [5], extensive use of black and broad brushstrokes hindered expression of the movement of light. By the early 1870s, when Monet painted "Impression, Sunrise" and Pissarro "The Entrance to the Village of Voisins" [7], the Impressionist style can be said to have come of age and featured short comma-like brushstrokes, the banishment of black from the shadow and direct confrontation with the subject.

The opening years of the 1880s saw the group's stylistic unity crack. Their early champion, Zola, doubted whether "Nature seen through the temperament" could ever provide the recipe for a masterpiece. Sisley, Guillaumin and Morisot remained stylistically faithful, but Monet sought new subject-matter and more intense light on the Côte d'Azur and later embarked on his "serial" paintings [9]. Cézanne, in Provence, began an intense analysis of the relationship between colour and form. Renoir's lifelong concern with the figure brought him to rediscover the formal qualities of the classical nude. Degas reduced his compositions to exercises in two-dimensional patterning and Pissarro briefly adopted the divisionism of Georges Seurat (1859–91) who painted in small blobs of pure colour and whose scientific analysis of light lay at the centre of Neo-Impressionism [10].

KEY

The Impressionists' working method is amply demonstrated in Edouard Manet's "Monet at Argenteuil" (1874) which shows the spontaneity of his friend's method of work. Monet is sitting with his wife in his improvized studio, working up a finished picture without any preliminary sketches. Manet's picture was painted in a similar manner.

6

6 Cézanne, like his fellow Impressionists, sought to record his "powerful sensations in front of Nature" by working directly in front of his subject. In this photograph he is lifting a half-finished canvas of a favourite subject, the Mont Sainte-Victoire outside Aix-en-Provence, onto his easel ready for work.

7

7 Camille Pissarro, who painted "Entrance to the Village of Voisins" in 1872, had been in self-exile with Monet in London during the Franco-Prussian war of 1870–71. There he developed an admiration for the landscapists Turner and Constable, "who obviously shared our aims of *plein-air*, light and fugitive effects".

8

8 "Nymphs" (1918) is part of a series of monumental nudes that Renoir had begun in the early 1880s. Integrity of form has replaced a preoccupation with the dissolution of form by light. During a visit to Italy, Renoir wrote that he had become dissatisfied with the imprecise "blotting" technique of Impressionism and that he had discovered the grandeur and simplicity of Ingres and Raphael, both masters of the idealized nude.

9

9 "Rouen Cathedral – Morning Effect" is one of some 40 similar views painted by Claude Monet by 1895. In February 1892 and March 1893 the artist rented a room overlooking the west porch of the cathedral to observe the façade and make notes and sketches in different weather conditions and at varying times of the day. He worked up the finished paintings afterwards. As in his other series, Monet's choice of a static subject allowed him to turn his attention exclusively to the formal compositions that were created by the translation of light effects into colour.

10

10 Separate dots of primary colour fuse visually into the muted tones of a misty morning on the Seine in Camille Pissarro's "Ile Lacroix, Rouen" (1888). In 1885 Pissarro complained that his Impressionist paintings were "poor – tame, grey, monotonous – I am not at all satisfied". His reaction was to adopt the new "divisionist" technique of Georges Seurat whom he had met through Paul Signac. Although Pissarro's adherence to the style was short-lived, he did nonetheless produce a number of divisionist pictures.

Opera in the 19th and 20th centuries

About the turn of the nineteenth century the growth of a middle-class concert-going audience encouraged the practice of giving performances in public opera houses. The resulting demand for new repertory allowed many composers to specialize in opera for the first time and the public welcomed entertainment that was in many ways more attractive than the lavish court spectacles.

Serious and comic opera

The traditional streams of serious opera and comic opera remained distinct, but both were at that time seen regularly at the opera house. The naturally dramatic character of the form also began to reflect the political and social situations, however disguised, in which it was produced – *Risorgimento* or the hoped-for reunification of Italy in the works of Giuseppe Verdi (1813–1901), the national traditions in the Russian operas of Modest Mussorgsky (1839–81) and Mikhail Glinka (1804–57), or in the works of the Bohemian Bedřich Smetana (1824–84). Eventually a more realistic approach appeared in the *verismo* (realistic) operas of Pietro Mascagni (1863–1945), Ruggiero Leoncavallo (1858–1919) and Giacomo Puccini (1858–1924) later in the century.

When in 1791, the last year of his life, Wolfgang Amadeus Mozart (1756–91) composed a serious opera in Italian, *La Clemenza di Tito*, and a comic opera or *Singspiel* in German, *The Magic Flute*, it was the latter, using elements of mystery and folk-like humour as well as the vernacular, which foreshadowed later developments in German opera with its concern for unity of music and drama. Ludwig van Beethoven (1770–1827) in his single opera *Fidelio* (1805) made a rather earnest German contribution; but it remained for the deft touch of Weber (1786–1826) in *Der Freischütz* [1] and *Euryanthe* to shape the inheritance that Wagner would take up [4].

The influence of Rossini

This young German growth among the diverse branches of opera would have to struggle against the traditional dominance of the Italians and in particular against the prevailing fashion for the music of Gioacchino Rossini (1792–1868) [Key] which had swept through the opera houses of Europe in the early part of the century. *The Barber of Seville, The Italian Girl in Algiers* and *Cinderella* among his many comic operas demonstrate the exciting Rossini crescendo in the orchestral writing and brilliant vocal music that was such a crowd-pleaser.

The legendary virtuoso singers of the time, such as Maria Malibran, Luigi Lablanche, Giovanni Rubini and Giulia Grisi, encouraged the musicians who followed Rossini, such as Vincenzo Bellini (1801–35) and Gaetano Donizetti (1797–1848), to compose in the same vein. Bellini was more Romantic in *La Sonnambula* and *Norma*.

Grand opera and after

While the comic opera tradition in France was greatly weighed down under Rossini's influence, grand opera tradition found expression from 1830 onwards in the works of Giacomo Meyerbeer (1791–1864). In reaction to the eventual bombast of his *Les Huguenots* and *Le Prophète*, a lyric opera style emerged in the 1850s, represented by

CONNECTIONS

See also
106 Music: the Romantic period
270 Music from Stravinsky to Cage
274 Classical and modern Ballet

1 The casting of the magic bullets in the Wolf's Glen is the most famous scene from Carl Maria von Weber's *Der Freischütz*. The opera, literally "The Free-shooter", meaning a marksman who uses magic bullets, is regarded as a pioneer work of the Romantic era, and is notable for Weber's orchestral effects, particularly during the sinister action in the scene depicted. In *Der Freischütz* Weber reinforced the line of German opera leading from Mozart to Wagner.

2 Adelina Patti (1843–1919), the celebrated Madrid-born coloratura soprano, here in the role of Marguerite in Gounod's *Faust*, enjoyed an operatic career that spanned nearly 60 years. Patti was acclaimed as the last in a great line of prima donnas who were typical of 19th-century opera. She was noted as Rosina in *The Barber of Seville*, and Rossini himself arranged music for her.

3 Fyodor Chaliapin (1873–1938), the great Russian bass, gained world fame in the title role of Mussorgsky's *Boris Godunov*, which he was the first to perform outside Russia. His strong acting performances and resonant voice brought him world première roles in Massenet's *Don Quichotte* and Mussorgsky's *Khovantchina*. His New York success in the 1920s made him as admired as Caruso.

4 The knight Lohengrin arrives in a boat drawn by a swan in a scene from the first production of *Lohengrin* by Richard Wagner, given at Weimar in 1850 under Franz Liszt. *Lohengrin* represents a mid-point in the development of Wagner's music; it is the last in a series of operas with traditional elements from grand opera and set-piece numbers. His later operas – *Tristan and Isolde, The Mastersingers, The Ring of the Nibelung* cycle and *Parsifal* – would exemplify his ideas of opera as a continuous music-drama of endless melody, bound together by musical motifs representing characters, objects and ideas.

5 *Aida*, by Giuseppe Verdi (in a production at the Royal Opera House, Covent Garden), represents the full flowering of Italian opera in the 19th century, with its spectacle, colour, dramatic love triangle and tragic ending, all enriched by Verdi's dramatically apt music. The opera was commissioned as a festival work by the Khedive of Egypt to celebrate the opening of the Suez Canal, and first performed in Cairo in 1871. Set in ancient Egypt, the story tells of the Ethiopian slave girl Aida and her love for the Egyptian army officer Radames, who is tricked by her into betraying his country.

Charles Gounod (1818–93) and Jules Massenet (1842–1912). Hector Berlioz (1803–69) was simultaneously pursuing an independent course.

Meanwhile in Italy, Verdi [5] had *Nabucco* – his first major work – performed in 1842, and became identified with the cause of a united Italy. His operas began to show at the same time an emotional power and psychological insight that were to culminate in the magnificent dramas of *Rigoletto, Il Trovatore, La Traviata* and many others. His last two operas, *Otello* and *Falstaff*, emphasize continuity in the music rather than individual arias, a style that Richard Wagner (1813–83) [8] had developed independently in Germany in his music-dramas.

In Russia, the early emphasis on nationalist opera begun by Glinka now flowered in the works of Alexander Borodin (1834–87) Nikolai Rimsky-Korsakov (1844–1908), Mussorgsky and Peter Ilyich Tchaikovsky (1840–93) who wrote ten operas. *Carmen*, by the French composer Georges Bizet (1838–75), was another key work, beginning the trend of realism in operas. At the same time, another Frenchman, Claude Debussy (1862–1918), in his single operatic work *Pelléas and Mélisande* reacted against Wagnerian opera.

After Debussy the twentieth century was to see a wide range of operatic techniques and styles. With the dispersing of the Italian monopoly, operas were to appear in all languages: some, such as those of Richard Strauss (1864–1949), Leoš Janáček (1854–1928) and Sergei Prokofiev (1891–1953), followed national trends; others even delved back into the classics, as did Igor Stravinsky (1882–1971) for his opera-oratorio *Oedipus Rex* in Latin. Presentation ranged from the simple – two characters in *Bluebeard's Castle* by Béla Bartók (1881–1945) – to the representation of a city in the satirical *Mahagonny* by Kurt Weill (1900–50). Benjamin Britten (1913–76) made his mark as an opera composer of the highest international status with his *Peter Grimes* [9]. After some years of composing chamber operas, Britten returned to large-scale works with *Billy Budd* (1951), and *Gloriana* (1953) written for Elizabeth II's coronation.

KEY

The Barber of Seville is the best-known comic opera by Gioacchino Rossini, whose music dominated the world of opera in the early 19th century. In this characteristic scene from Act II the cheerful barber Figaro shaves the pompous old Dr Bartolo, who mistakenly believes he will marry the heroine Rosina, while Figaro is scheming another husband for her.

6

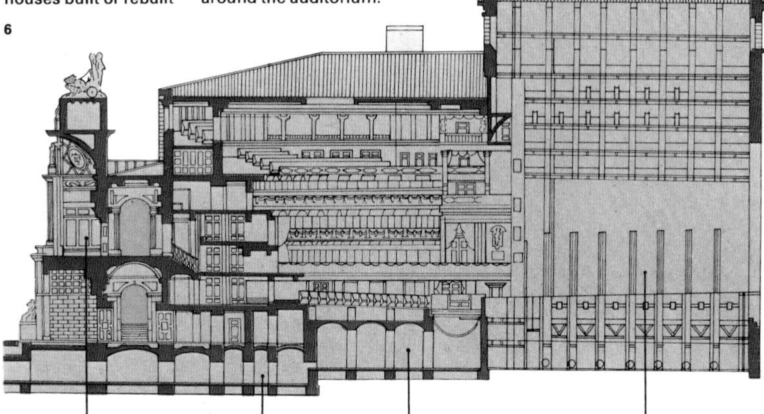

6 The Dresden Hoftheater (opera house) was designed and built between 1871 and 1878 to replace an earlier building that had been destroyed. Many European opera houses built or rebuilt in the 18th and 19th centuries (La Scala, Milan, 1778; Royal Opera House, Covent Garden, 1856) were modelled on the old Italian plan of tiers of boxes placed around the auditorium.

7

7 *Lulu* by Alban Berg (1885–1935) is one of the major operatic works of the 20th century. It tells the story of Lulu, a prostitute who ends up as one of Jack the Ripper's victims in a London street. Written in the 12-tone harmony developed by Arnold Schoenberg (1874–1951), Berg's teacher, *Lulu* was unfinished at Berg's death, with only two out of three acts published; it was performed like that two years later. The opera continues Berg's concern for human beings as victims of persecution – a theme that he first explored in *Wozzeck*, the study of an anti-hero – and is dramatically suited by the often harsh music.

8

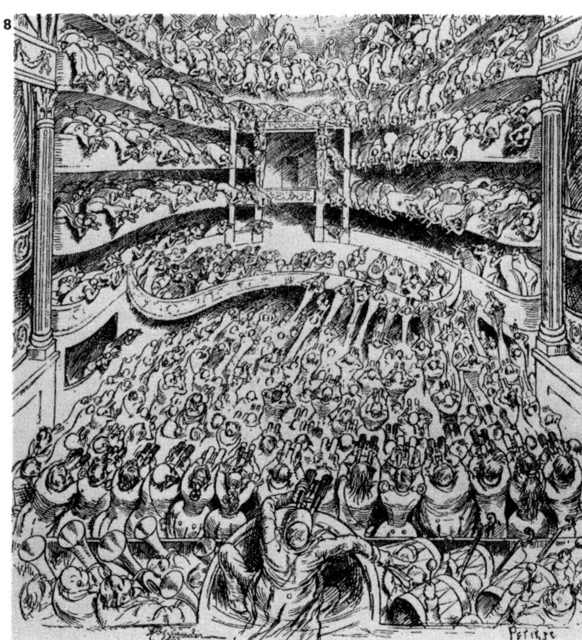

9

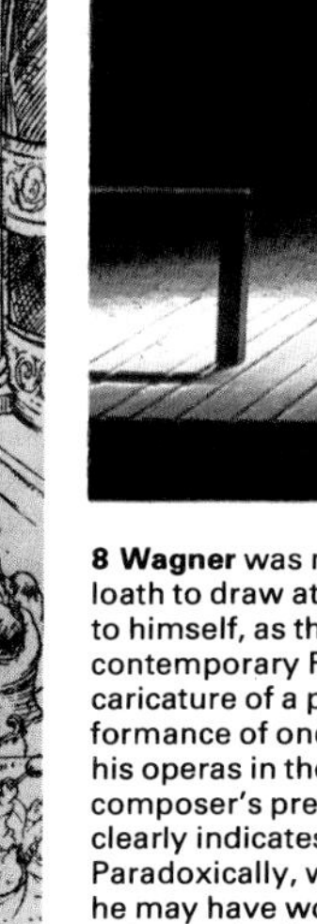

8 Wagner was never loath to draw attention to himself, as this contemporary French caricature of a performance of one of his operas in the composer's presence clearly indicates. Paradoxically, while he may have worn red velvet and called himself the apostle of a new religion (his music), he was the first to insist on dimming the lights and making latecomers wait so as not to disturb the audience. His opera house at Bayreuth remains a model of theatre planning.

9 *Peter Grimes*, by Benjamin Britten, scored a resounding success throughout Europe and America from its first performance in 1945, and started a new interest in British opera. Based on a poem by George Crabbe, its central character is the fisherman Peter Grimes, seen here with Balstrode, a retired skipper. Grimes is an alienated figure in his own community, and the situation is clearly reflected in Britten's spare yet attractively lyrical music.

European architecture in the 19th century

Industrial progress was the touchstone of the nineteenth century. With it came growth in population and prosperity, factors that created a boom in building. More buildings were constructed in this century than in any previous century, modern building types were born and new materials employed. Several styles and architectural theories jostled for supremacy. Likewise, a new professional man emerged: the architect.

In the eighteenth century the cataloguing of architectural styles began, with the careful recording of Greek and Roman remains. This was extended into the nineteenth century to cover Gothic, Italian Renaissance, Northern Renaissance and Byzantine buildings. Thus George Gilbert Scott could in 1857 change his Gothic design for the Foreign Office building, London [1], to an Italian Renaissance building more to the taste of the foreign minister, Lord Palmerston.

The selection of a style was by no means only a matter of individual taste: styles had associations. Commercial buildings were often of Italian Renaissance style to recall the wealth of such families as the Medici. The Houses of Parliament, London, designed by Charles Barry (1795–1860) and Augustus Welby Pugin (1812–52), were built in Perpendicular Gothic to reflect the period when the institution began to assume some importance.

Eugène-Emmanuel Viollet-le-Duc (1814–79), the French Architect, Gothic renovator and theorist, declared that "to believe that one can create Beauty by lying is a heresy". The moral connection made between beautiful architecture and truthful architecture was one that was enunciated by Pugin and by John Ruskin (1819–1900).

Truth and honesty

For Viollet-le-Duc, truthful architecture lay in the honest use of materials: "stone must really look like stone; iron like iron; wood like wood"; hence iron pillars must not be clad in stone, but left exposed and incorporated into the design, a point illustrated by his project for a market hall [3] and later by Hector Guimard (1867–1943) in his Sacré Coeur School, Paris (1895). Ruskin argued that "good" (that is, beautiful and moral) architecture could be produced only by a "good" architect who reflected in his work a "good" society. The material expressions of this abstract principle can be seen in the extensive building programme of High Gothic churches such as George Edmund Street's St Paul's, Rome [7], and in the University Museum, Oxford (1855–9).

Finally, it was good, honest design that played an important part in liberating the plan and elevation of the house – pioneered in Great Britain by Norman Shaw (1831–1912), Philip Webb (1831–1915) and Charles Voysey (1857–1941). The ideal of domesticity was allowed free expression in asymmetrical ground plans and unostentatious elevations, such as that seen at Broadleys, Lake Windermere, England [10].

New building materials

Industrialization not only provoked concern about the quality of society and its architecture. It also introduced new building materials such as cast and wrought iron, steel, plate glass and lightweight, fireproof, caustic bricks. These innovations not only permitted

CONNECTIONS

See also
178 Origins of modern architecture
88 The Industrial Revolution
158 Industrialization 1870–1914
154 The impact of steam
70 German art and architecture

In other volumes
216 History and Culture 1

1 Britain's growing prestige in the area of foreign affairs in the 19th century made the need for a new departmental building imperative. In the architectural competition held in 1857 George Gilbert Scott (1811–78) won with the Gothic design, later changed to an Italian Renaissance style.

2 New wealth from iron ore financed the building of Harlaxton Manor in Lincolnshire, England. Started by Anthony Salvin (1799–1881) and finished by William Burn (1789–1870), its size was dictated in part by the client's expanding art collection. The mainly Jacobean style was mixed with Elizabethan features.

3 Viollet-le-Duc's project for a market hall was published in his *Entretiens sur l' Architecture* (1863–72). There he advocated the use of exposed cast iron for pillars and roof supports.

4 Exposed cast iron was used by Karl Etzel in the Dianabad, Vienna (1841–3) to achieve the barrel-vaulted ceiling of the German "round arch". The design of the balcony supports echoes the main vault.

5 James Bogardus (1800–74) was neither an architect nor an engineer but rather a builder and inventor. In New York in 1848 he built a four-storey factory for his own use, made of cast iron which was screwed together on site. This was followed by a five-storey chemist shop and the Laing Stores, which is shown here. Although this building took but two months to erect, it displayed no trace of shabby prefabrication. Cast iron could adapt to different styles such as Gothic, Renaissance or Grecian with great ease and, when painted, the material could give the impression of stone. Bogardus pioneered the idea of bearing loads on cast iron columns rather than on walls.

the construction of "engineering" monuments such as Isambard Kingdom Brunel's (1806–59) Clifton Suspension Bridge, Bristol (1830–64), John A. Roebling's Brooklyn Bridge, New York (1869–83) and Victor Baltard's Les Halles, Paris (1853–8, now demolished), but also the construction of such "architectural" structures as Karl Etzel's Dianabad, Vienna (1841–3) [4] and H. P. F. Labrouste's Reading Room in the Bibliothèque Nationale, Paris (1862–8). Iron also lent itself to prefabrication. Buildings such as James Bogardus' Laing Stores, New York (1849) [5] were precast and screwed together on site, as was Joseph Paxton's Crystal Palace, London (1850–51).

Changes in society also brought aggrandizement, extension and specialization of traditional building types. With the increasing complexity and importance of central and local governments, government offices and town halls became monuments on a grand scale. Visconti and Lefuel extended the Louvre, Paris, in an ebullient neo-Baroque style in 1852–7, while across the Atlantic Alfred B. Mullet was adorning Washington, DC, with his neo-Roman State, War and Navy Department Building (1871–5). Growing public services required new buildings and many of them were gigantic, such as Giuseppe Calderini's Palazzo di Giustizia, Rome (1888–1910) [9]. Most were built after careful investigation of specialized requirements, as in P. J. H. Cuypers' Rijksmuseum, Amsterdam [6].

Railway stations

Most significant of all was the arrival of the railway. Railway stations, the symbols of the new industrial age, sprang up across the world. In some instances the station would be no more than a dominant engineer-designed shed with subordinate forecourt buildings, such as King's Cross, London (Lewis Cubitt, 1850–52), and F. A. Duquesney's Gare de l'Est, Paris (1847–52). In others the shed was masked by the forecourt structure, which often doubled as a grand hotel. Such was George Gilbert Scott's Midland Grand Hotel and St Pancras Station, London (1868–73) [8], which also set new standards in comfort, sanitation and mechanical innovation.

KEY

Charles Garnier (1825–98) built the Paris Opéra, at the end of the Avenue de l'Opéra, between 1861 and 1874. It is neo-Baroque in style and is one of many examples of wholesale urban improvement carried out in 19th-century Europe as the result of the growth of central government.

6

6 The Rijksmuseum, Amsterdam (1877–85), was erected to house the state collection of art. It was designed by P. J. H. Cuypers (1827–85), a leader of the Dutch Arts and Crafts reform movement in the 19th century.

7

7 St Paul's American Church, Rome (1872–6), was built by George Edmund Street (1824–81) to serve the religious needs of the American community. Its Italian Gothic style both shows a sensitivity to location and accords with the Ruskinian doctrine that Gothic was the most suitable style for church-building.

8

8 Architectural contests held for major 19th-century building projects reflected a faith in excellence that emerged from the workings of a free market economy, a desire for public accountability and a new professionalism in most careers. In May 1865 architects were invited by the Midland Railway Company to submit plans for a Grand Midland Hotel and station offices at St Pancras Station, London. A complex brief involved designing a building that would entirely mask a train shed erected two years earlier and also the planning of a type of building that had only recently been created – the grand hotel. George Gilbert Scott won the contest in January 1866 with a grandiose deisgn (shown here in the background) even though it added two storeys and involved the most expensive tender. Evidently the company wanted to advertise its services by making use of the prestige of Scott and the romance of his architectural conception.

9

9 A new national style emerged in Italy after Rome was established as the capital of a unified nation and new government buildings were needed. Their designs tended to be derived from the Renaissance or else the Baroque, a style that was the basis of Giuseppe Calderini's design for Rome's High Court buildings, prominently sited above the Tiber.

10

10 Broadleys, Windermere, England (1898), exemplifies a style of rural domestic architecture evolved by C. F. A. Voysey who sought to create an organic relationship between his houses and their natural surroundings. At Broadleys the scale is comfortable, with windows along the southwest façade designed for maximum sunlight and view. The service wing of the house is neatly tucked away.

Colonizing Oceania and Australasia

The voyage of Ferdinand Magellan across the "Peaceful Sea" in 1520 brought the Pacific Ocean to the attention of Europe. But it was 1565 before the Spaniard López de Legaspi (died 1572), sailing west from the New World, settled the Philippines, where Magellan had died [1]. Spanish rule, although challenged, was uninterrupted until the Spanish-American War of 1898, when the Philippines were ceded to the United States.

The Indies and Australia

Meanwhile to the southeast, as Portuguese power declined, the ships of the Dutch East India Company, founded in 1602, routed the pirates of the Malay Archipelago, seized control of the lucrative spice trade and paved the way for a Dutch colonial empire extending from Sumatra, Java and Borneo to Celebes, the Moluccas and western New Guinea [9]. The prosperity of the new colonies, largely derived from cloves, nutmeg, pepper and coffee, was set against a background of repression and bloodshed. In Borneo, where gold and diamond mining attracted Chinese immigration, Dutch rule was precarious; and not until 1701 did the British East India Company, formed in 1600, establish a factory or trading post in what later became a permanently divided island.

Commissioned by East Indies Governor Anthony van Diemen (1593–1645) to chart the western and southern shores of New Holland (Australia), Abel Tasman [2] in 1642–3 discovered Van Diemen's Land (later Tasmania), skirted New Zealand and later sailed along the southern coast of New Guinea into the Gulf of Carpentaria. More than a century passed before the British Admiralty dispatched James Cook (1728–79) [5] to take possession of any land in the south in the course of a scientific expedition to the South Seas. By sailing during 1768–71 from Cape Horn to New Zealand (which he charted as two islands) Cook finally exploded the theory that a great southern continent balanced the land mass of the Northern Hemisphere. He sailed up the east coast of Australia, claiming it for Britain, showed that New Guinea was a separate island and, in two later voyages, made other significant Pacific discoveries.

Britain was left to colonize the vast subcontinent of Australia in 1787, first as a penal settlement, later as rich sheep and cattle country. Population was concentrated in the east and south where Brisbane, Sydney, Melbourne and Adelaide were founded. Sparse settlement spread out as explorers trekked across the vast deserts of the interior [6]. The principal victims of white expansion were the nomadic Aborigines, their Stone Age culture based entirely on hunting, their clubs, spears and boomerangs ineffectual against firearms. Introduced diseases had an even more devastating impact. Guns and epidemics wiped out the native population of Tasmania and sharply reduced that of the mainland. The Aborigines were to have no share in new Australian prosperity, accelerated by later gold rushes [7].

New Guinea and New Zealand

Rumours of gold also drew prospectors to the great island of New Guinea in the mid-nineteenth century. Mineral resources proved negligible but traders and speculators stripped coastal forests of timber. In the mountainous interior, inhabited by plumed

CONNECTIONS

See also
126 Australia and New Zealand to 1918
32 Exploration and science 1750–1850
44 International economy 1700-1800
130 Imperialism in the 19th century

In other volumes
90 The Physical Earth
210 The Natural World
204 The Natural World
228 The Natural World

1

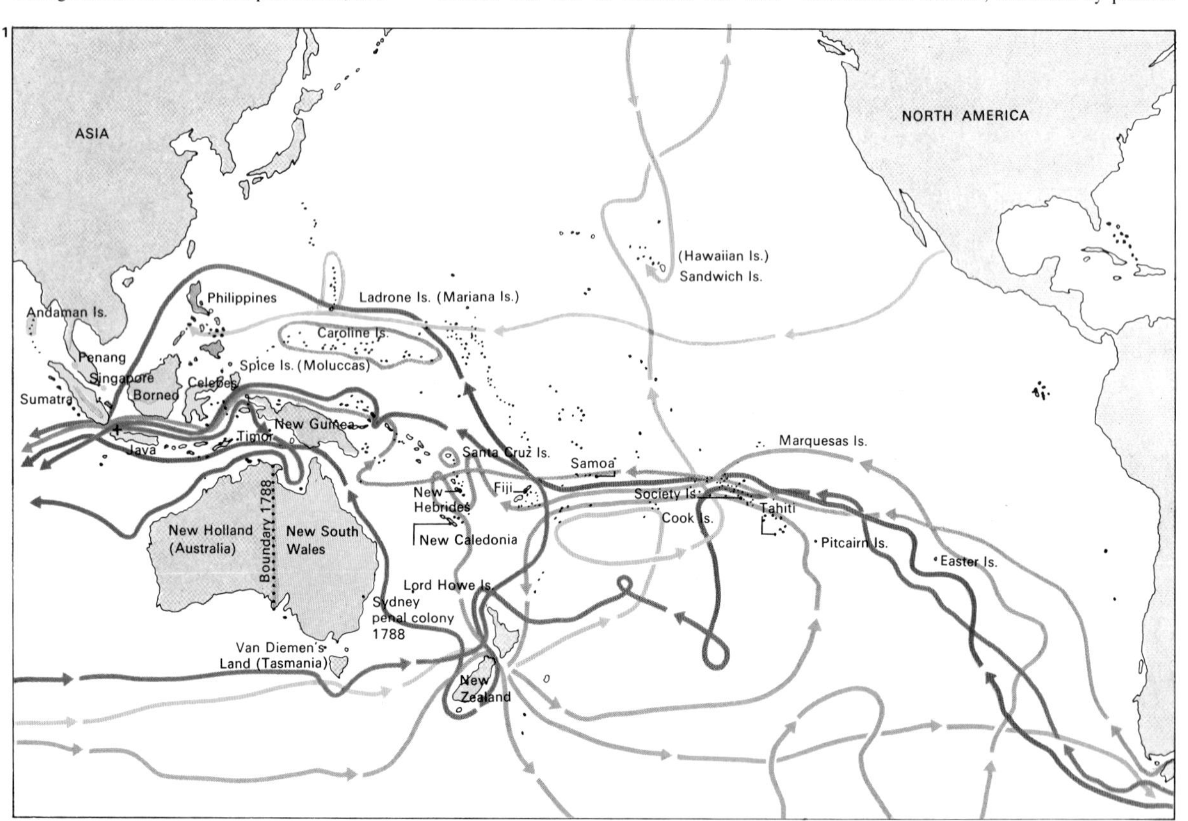

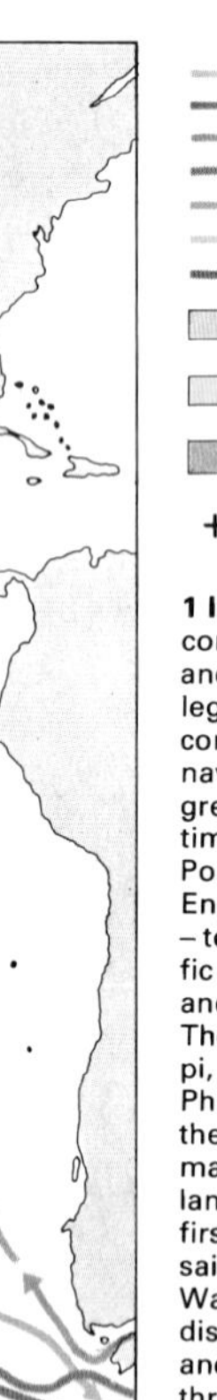

1 Imperial ambition, commercial rivalry and the search for a legendary southern continent motivated navigators of the great European maritime nations – Spain, Portugal, Holland, England and France – to explore the Pacific between the 16th and 18th centuries. They included Legaspi, who conquered the Philippines, Tasman, the discoverer of Tasmania and New Zealand, Bougainville, first Frenchman to sail round the world, Wallis, the English discoverer of Tahiti, and Cook, whose three voyages opened up most of the Pacific.

2

2 Abel Tasman (1603–*c.* 1659), an employee of the Dutch East India Company, touched on the southern shore of an island he named Van Diemen's Land (after the Indies' Governor-General). In 1865 the island was renamed Tasmania. He was deterred from landing in New Zealand by warlike Maoris. After discovering Tonga and the Fiji Islands he returned to Batavia where he was rebuked for "having been negligent in investigating the situation, conformation and nature of the lands and peoples discovered". An equally frosty reception greeted his second voyage along the south coast of New Guinea and north Australia.

3

3 William Dampier (1652–1715), formerly an English buccaneer, explored the coasts of Australia, New Guinea and New Britain, vividly describing lands and people.

4 Louis de Bougainville (1729-1811) set out on a round-the-world voyage of discovery in November 1766 in the frigate *La Boudeuse*. He sailed through the Straits of Magellan to the Tuamotus and Tahiti, which he claimed for France, unaware that Samuel Wallis (1728–95) had found it ten months earlier. He sighted and named islands in the Samoa and New Hebrides groups and would have reached the unknown east coast of Australia had he not been diverted by the Great Barrier Reef. Despite starvation and scurvy he had lost only seven men by the time he returned home in 1769. He also founded a settlement in the Falkland Islands.

4

5 The voyages of Captain James Cook were supplemented by careful and perceptive accounts of lands he visited and by scientific observations of great practical value. During his first voyage in *Endeavour* in 1768–71 he circumnavigated the two main islands of New Zealand, charted and claimed the east coast of Australia and returned home through the Torres Strait. In the second voyage he took *Resolution* to the Antarctic and discovered or rediscovered many Pacific islands. Finally, he visited Australia and New Zealand again in *Resolution* and discovered Hawaii, where he was killed in 1779.

5

and painted head-hunters, civilizations made little impact even when Holland, Germany and Britain annexed the island in 1884–5.

In New Zealand the Maoris, more advanced socially and culturally, were treated with more respect by European settlers. Whalers [10] and sealers were initially welcomed by the local population although disease took a terrible toll. The early nineteenth-century arrival of traders and missionaries in the North Island was followed by British annexation with Maori agreement in 1840 and rapid settlement of both islands. But misunderstandings over tribal rights to sell land to the colonists led to disputes as the Maoris realized the threat to their lands. They resisted in a series of fierce wars, particularly in the 1860s [8] but were defeated, lost most of their land despite nominal consultation and thus faced the future with a great deal of misgiving.

The Maoris had left their original homelands in Polynesia several centuries earlier. Other peoples – Micronesians, Melanesians and Polynesians – still inhabited the island groups of Oceania that were sighted (and often colonized) by Europeans between the sixteenth and nineteenth centuries [3, 4]. Dried coconut (copra), used for animal feeding and later for the extraction of edible oil, was the staple export crop. A few islands were commercially more rewarding – notably British Fiji with its forests of sandalwood; French New Caledonia, where nickel was found; and Hawaii, where a combination of American missionary work and enlightened local rule led to independence as early as 1843; a prosperous economy based on sugar and pineapples thereafter developed.

Cultural impact

Elsewhere, repression, missionary conversion, disease and "blackbirding" – the forced transport of native labour to work in the sugar and cotton plantations of Fiji and Queensland – all helped to destroy local cultures and tribal structures as white civilization spread. Colonialism also put a stop to more savage rituals – cannibalism, head-hunting and blood feuds – with a promise of improved education and a share in economic wealth and political power.

KEY

The Maoris of New Zealand, whose Polynesian ancestors paddled some 3,300 kilometres (2,000 miles) across the Pacific in about the 13th century, were unsurpassed craftsmen of dugout canoes, of which a model is shown here. Their war canoes, carrying up to 100 men, were elaborately carved by sculptors who also taught their pupils the magical and religious ritual associated with the craft. Paddled at full speed, they could overtake European sailing ships.

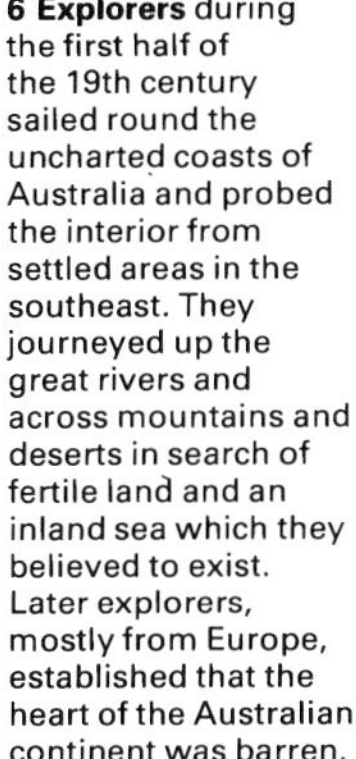

6 Explorers during the first half of the 19th century sailed round the uncharted coasts of Australia and probed the interior from settled areas in the southeast. They journeyed up the great rivers and across mountains and deserts in search of fertile land and an inland sea which they believed to exist. Later explorers, mostly from Europe, established that the heart of the Australian continent was barren.

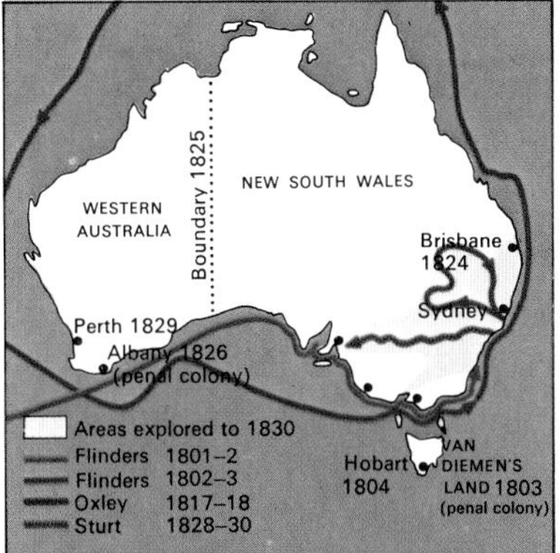

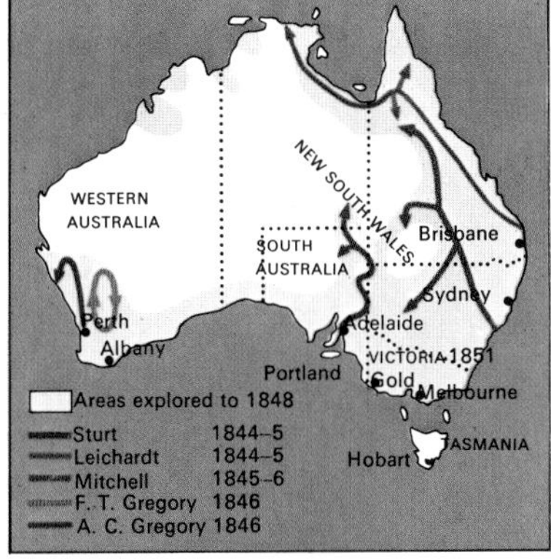

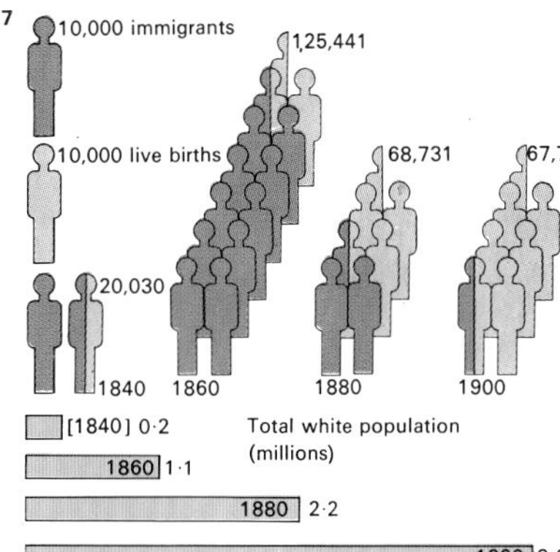

7 Australia relied initially on immigration to build up its population. An assisted immigration scheme was introduced in 1829 and up to 1860 immigrants accounted for over three-quarters of the population growth. The gold rushes of 1851–6 brought an even greater immigrant surge. Thereafter, the Australian birth-rate began to rise and overshadow a reduced flow of immigrants.

8 Maori gallantry against superior weaponry marked many battles during the 1860s when Crown attempts to satisfy the land hunger of New Zealand settlers without disrupting Maori tribal rights broke down in bitter disputes over land sales. Maoris defended redoubts such as this one above the Katikara Stream near Mt Egmont. In 1863 the fort was battered by naval guns and 350 troops routed 600 Maoris.

9 The fortress port of Batavia was the trade centre of the Indies in the 17th century when Aelbert Cuyp (1620–91) painted "The Return Fleet of the East India Company on the Roads of Batavia". Dutch naval supremacy and commercial enterprise, backed when necessary by guns, led to the establishment of a colonial empire that lasted 300 years. Batavia eventually reverted to its former name of Jakarta as the capital of the independent nation of Indonesia.

10 Whalers, along with traders and blackbirders, brought guns and disease to many Pacific islands in the 19th century. The profitability of whaling meant that fishing grounds were rapidly depleted, although the industry survived for many years. This somewhat fanciful print entitled "The North Cape New Zealand and Sperm Whale Fishery" may exaggerate the density of the whale population but typifies the old-style shore whaling practices which led to many coastal settlements.

Australia & New Zealand to 1918

Australia began as a penal colony for the overflow from British gaols, after the American War of Independence had closed off the main area for convict transportation. The First Fleet, under Captain Arthur Phillip (1738–1814) [1], arrived in Botany Bay on the eastern Australian coast on 18 January 1788, but the settlement soon shifted to a much better anchorage in nearby Port Jackson. There, at Sydney Cove, the colony of New South Wales was established on 26 January, which was subsequently commemorated as Australia Day.

The early colonies

Convicts provided the initial labour force for erecting a settlement and scratching a living from the poor soil in and around Sydney, but by 1815 a way had been found through the Blue Mountains to the fertile plains in the west. Free settlers, capital accumulated from shrewd trading or imported from England, illegal squatting on Crown lands, and merino sheep all contributed to a developing wool export industry. Wool and wheat exports paid for the necessary manufactured goods and, with land sales, helped to subsidize the passage of new settlers.

New settlements were established partly to pre-empt the French, partly by adventurers without authority, at points along the coast, on Norfolk Island, and in Van Diemen's Land (renamed Tasmania in 1853). The vast distances involved required colonial administrations separate from Sydney. These were set up in Van Diemen's Land in 1825, Western Australia in 1829, South Australia in 1836, Victoria in 1851 and Queensland in 1859.

New Zealand lies 1,920 km (1,200 miles) southeast of Australia. Its fertile, well-watered land had been occupied by Maori Polynesian tribes for more than 400 years by the time the colony was established at Sydney. Sealers, whalers, freebooters and missionaries soon made their way across the Tasman Sea, establishing coastal trading settlements among a warlike people numbering perhaps 200,000. Apart from the inroads of European diseases, the sale of muskets had a devastating impact on the Maoris, intensifying fierce inter-tribal wars in the 1820s. Britain annexed New Zealand in 1840 with the assent of most North Island chiefs at the Treaty of Waitangi [3] and made it a separate colony from New South Wales in 1841. Systematic settlement followed, inspired by the evangelical ideas of Edward Gibbon Wakefield [2]. Disputed land titles impeded initial development until Governor (later Sir) George Grey (1812–98) established order, although in the North Island the way was cleared for massive settler purchases only after Maori chiefs hostile to sales were crushed (1860–65). In the emptier South Island, pastoral settlement increased, boosted by a gold rush to Otago in 1861. Six provincial councils set up in 1852 gave way to centralized administration by a general assembly in Wellington, which replaced Auckland as the capital in 1865.

Expanding economies

In Australia, the discovery of gold [5], especially in Victoria in 1851, brought an influx of migrants, expanded domestic capital and investment, assisted social mobility and created problems of law and order [6]. It also

CONNECTIONS

See also
124 Colonizing Oceania and Australasia
248 Australia since 1918
250 New Zealand since 1918
132 The British empire in the 19th century
190 World War I: Britian's role
220 The Commonwealth
46 English colonial policy in the 18th century

1

1 Captain Arthur Phillip was the first governor of the colony of New South Wales (1788–92). He had to deal with the dregs of humanity sent to him from overcrowded British prisons, long, uncertain supply lines, and sandy soil unsuitable for crops. With such unpromising material he managed to set the fledgling colony on its feet. In spite of his repeated appeals for free settlers, Britain still sent more convicts.

2

2 Edward Gibbon Wakefield (1796–1862) developed in England a theory of colonization that was subsequently applied, with varying success, in New South Wales and Port Phillip district (1832–42), South Australia (after 1836) and New Zealand (after 1839). Crown lands were sold for agriculture to young people of good character, representing a cross-section of British society, from the nobility to labourers.

3

3 The Treaty of Waitangi, concluded on 6 February 1840 between Captain William Hobson RN (1793–1842) and Maori chiefs of New Zealand's North Island, gave Britain formal possession of both major and off-shore islands, while recognizing Maori land rights. Britain had been reluctant to declare sovereignty but by 1838 accepted the need for orderly relationships between Maoris and settlers.

4

4 A proclamation to the Australian Aborigines, dated 1816, asserted equal rights and punishments for black and white. But in general the aboriginal population of Australia suffered from the advent of the Europeans. Not only did they suffer hitherto unknown diseases, but encroachment of farming and mining on to their old hunting lands went unchecked. By 1900 Aborigine numbers had been dramatically reduced.

5

5 The discovery of gold in Australia in the 1850s brought a rush of immigrants, many of whom came from California after the end of its gold rush. And during the next 100 years Australia was one of the world's major gold producers. In the decade after the first important discoveries in New South Wales and Victoria in 1851, output, at nearly 25 million ounces, was 39% of the world total. Gold is found in all states, the largest producers being Victoria and Western Australia.

stimulated the founding of colonial constitutions for New South Wales, Victoria, Tasmania and South Australia and the achievement, between 1853 and 1860, of a large measure of democratic government by lower houses, restrained by upper houses with property qualifications. Frontiers were pushed inland; sheep flocks and wheatfields expanded rapidly and agricultural exports to Britain were increasingly supplemented by mineral products. In New Zealand, the Corriedale cross-bred sheep produced good meat as well as wool and refrigerated ships [7] carried meat and dairy produce on regular voyages to London.

Political developments

During the 1880s, falling export prices, the effects of over-borrowing for expansion, bank failures and a general depression produced strikes and class bitterness which encouraged the entry of labour into politics in both countries. With protective tariffs, industry was expanded rapidly in Sydney and Melbourne. New Zealand, under vigorous Liberal rule from 1891 to 1912 led the world in some aspects of social legislation, namely votes for women (1893), compulsory state industrial arbitration (1894) and old age pensions (1898).

The six Australian colonies became states in the federal Commonwealth of Australia on 1 January 1901, with responsible government based on universal suffrage. The second Prime Minister, Alfred Deakin (1856–1919) [9], like Richard John Seddon (1845–1906) [8] of New Zealand a staunch nationalist and imperialist, launched many of the policies of the new nation: restrictive immigration based on race; protectionist tariffs with British preferences; industrial arbitration by a court empowered to fix minimum wages; old age pensions, and naval and military defence.

Australia and New Zealand sent forces to South Africa to support Britain during the Anglo-Boer War (1899–1902). In World War I their troops again fought together under British command in the Dardanelles [10] and in France, both countries making a contribution out of all proportion to their size and forging for the first time, a mature sense of national identity.

KEY

"Shearing the Rams" by Tom Roberts (1856–1931) shows a sheep-shearing scene in Australia in the 1890s. Spanish merino sheep were first imported from South Africa in the late 1700s, and later from England. Since that time, the Australian economy has developed largely "on the sheep's back".

6

V. R.

NOTICE!!

Recent events at the Mines at Ballaarat render it necessary for all true subjects of the Queen, and all strangers who have received hospitality and protection under Her flag, to assist in preserving

Social Order

AND

Maintaining the Supromacy of the Law.

The question now agitated by the disaffected is not whether an enactment can be amended or ought to be repealed, but whether the Law is, or is not, to be administered in the name of HER MAJESTY. Anarchy and confusion must ensue unless those who cling to the Institutions and the soil of their adopted Country step prominently forward.

His Excellency relies upon the loyalty and sound feeling of the Colonists.

All faithful subjects, and all strangers who have had equal rights extended to them, are therefore called upon to

ENROL THEMSELVES

and be prepared to assemble at such places as may be appointed by the Civic Authorities in Melbourne and Geelong, and by the Magistrates in the several Towns of the Colony.

CHAS. HOTHAM.

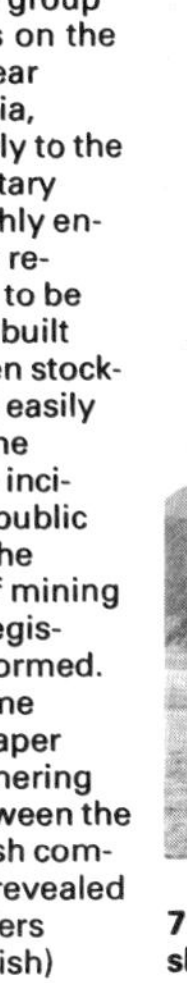

6 The "Eureka Stockade" of 1854 began when a group of gold miners on the Eureka field near Ballarat, Victoria, reacted violently to the police and military who were harshly enforcing the law requiring miners to be licensed. They built a rough wooden stockade which was easily overcome by the troops. But the incident captured public attention and the organization of mining and electoral legislation were reformed. Licences became easier and cheaper to obtain, simmering discontent between the Irish and English communities was revealed (the rebel leaders were mostly Irish) and every miner thereafter became eligible to vote.

7

7 The refrigerated ship *Dunedin* was commissioned in 1882 by the New Zealand and Australian Land Company to carry about 5,000 frozen lamb carcases from New Zealand to London. This followed an earlier successful trial shipment from Australia. As a result, trade from both New Zealand and Australia was opened up, and New Zealand lamb has been widely sold in Britain for nearly a century.

8

8 Alfred Deakin, journalist and statesman, initiated the irrigation movement in Australia and helped to form the Australian federation. He was a minister in the first federal government and prime minister intermittently between 1903 and 1910.

9

9 Richard John Seddon was Premier of New Zealand from 1893 until his death in 1906. A Liberal, he facilitated the granting of women's suffrage, old age pensions, free places in secondary schools and the passage of other social legislation.

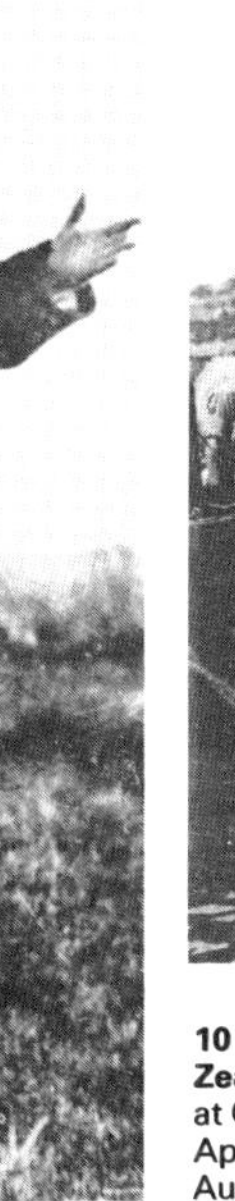

10

10 Australia and New Zealand came of age at Gallipoli on 25 April 1915 when the Australian and New Zealand Army Corps (ANZAC) went ashore at Anzac Cove. The cliff-face dugouts the troops occupied for eight months can be seen on the right. The campaign cost the Anzacs 44,822 casualties.

South Africa to 1910

Southern Africa is historically a conventional term for the countries lying south of Zaire and Tanzania, and it is not a separate entity from the rest of the African continent. Many of its peoples stem ultimately from Cameroon, filtering through Angola and Zaïre. There, in the fourteenth and fifteenth centuries, a number of kingdoms began to evolve. And, as populations increased and land grew scarcer, so a steady trickle of migrants set up new national groups from west to east right across Africa.

Early trade and commerce

In many areas the people mined copper and iron, and today Rhodesia and Transvaal are pitted with old workings. The Karanga Empire of the Monomotapa in present-day Rhodesia was especially favoured; by the twelfth century that region had begun to export gold – chiefly to Arabia and India – in return for cloth, beads, pottery and porcelain [4]. Slavery existed among Africans as it did among other peoples, but there was no extensive trade until slaves began to be exported in numbers by the Portuguese from Angola in the sixteenth century, despite frequent bans by the papacy. Although the Mani Kongo (King of Kongo) protested, by about 1530 some four or five thousand slaves were being shipped annually.

The Portuguese had established a fort at Sofala, Mozambique, in 1505 to control their gold trade. In 1507 they built a hospital, church, factory, warehouse and fort on Mozambique Island to serve primarily as a stop-over station for ships and their crews on the way to India.

Europeans did not occupy the Cape until the Dutch took it in 1652, to serve as a victualling station on the route to the East Indies. The British occupied the Cape on behalf of the exiled Prince of Orange from 1795 to prevent the Cape, like Holland, from falling into French hands. In 1802, the Cape was restored to The Netherlands under the Peace of Amiens. But the British returned in 1806, this time making their occupation effective.

It was at that time that Britons migrating to the Cape began an uneasy coexistence with the Afrikaners who had also absorbed Huguenot refugees in 1688 and 1689. Reaction to British rule took shape in the Great Trek of 1835, with Boer republics being set up beyond the frontiers. In 1843 Natal was annexed by the British, and in 1852 the independent Transvaal was set up; this was followed in 1854 by the establishment of the Orange Free State.

European exploitation

The real turning-point was reached with the gold and diamond rushes of 1869, out of which Cecil Rhodes (1853–1902) and his friends soon developed powerful controlling companies – De Beers and Consolidated Gold Fields. Rhodes was prime minister of Cape Colony 1890–96. Britain annexed the Transvaal in 1877, and also fought a series of small wars with Bantu peoples, of which the Zulu War (1879) was the hardest [6]. The annexation of the Transvaal led to war with the Boers in 1881, following which Britain recognized the republic. Shortly after, in 1883, Germany set up a post at Lüderitz Bay, and in 1884 annexed the whole of South West Africa.

CONNECTIONS

See also
58 Subsaharan Africa 1500–1800
132 The British Empire in the 19th Century
142 Africa in the 19th century
252 Southern Africa since 1910

1 The Iron Age building of Zimbabwe, the most famous of its kind in Rhodesia, was built by the Karanga, a Bantu-speaking people, in stages between the 4th and 15th centuries AD. They used patterned drystone walling and they modelled soapstone bowls and ceremonial bird figures. Zimbabwe was probably the religious centre for the *mwari* (spiritualist) cults, as well as the commercial centre — organizing mining and the gold, copper and ivory trades together with the Swahili middlemen. Relics recovered from the ruins include Chinese and Persian ceramics, 16th or 17th century Venetian beads and Arab glass and numerous African gold and iron ornaments.

2 Jan Anthonisz van Riebeeck (1619–77), first governor of the Cape, landed in Table Bay with about 90 men on 7 April 1652. The first winter months were testing ones for the little band. Illness laid low about half the work force, and there were many deaths. Also, food was extremely scarce for some time. In spite of setbacks, the first permanent fort was soon planned, and 100 men were engaged in its building. By 1662, when Van Riebeeck finally departed, the Cape had not only a fort but also a hospital, workshops, a mill, a granary, houses and fertile land under cultivation. Van Riebeeck and his companions are justly regarded as the founders of the Afrikaner nation.

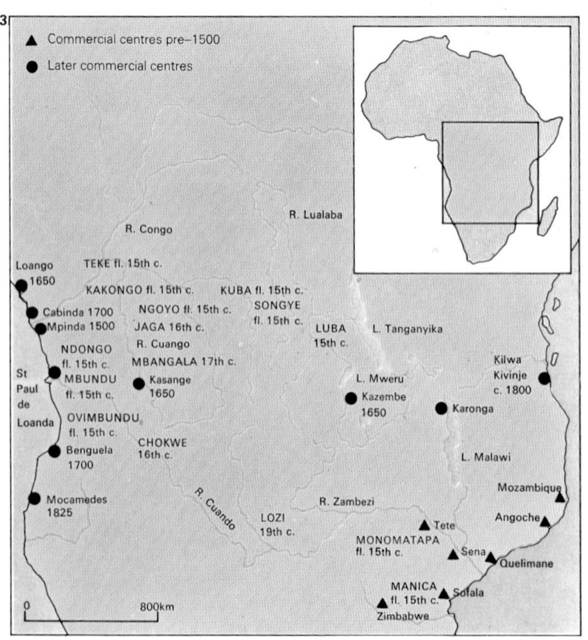

3 The river systems in southern Africa played an important part in the movements of early migrants. As early as the 10th century, Swahili settlements already existed on the east coast. In the 14th century, fresh waves of immigrants set up a series of kingdoms in the region of the present-day republics of the Congo and Zaïre and in northern Angola, which gradually extended from the west coast to the shores of the Indian Ocean.

4 Cape Town had a purely Dutch appearance until the mid-19th century. This picture dates from about 1888; the stucco-fronted brick houses were massively dominated by Table Mountain, 3.2km (2 miles) long and 1,070m (3,500ft) high.

The "scramble for Africa" was now at its height. There were minor British annexations in 1884. In 1885 Bechuanaland (now Botswana) was proclaimed a Crown colony, and part of Zululand was annexed in 1886. The Nyasaland Protectorate (now Malawi) was proclaimed in 1889, the flag was raised in what is now Rhodesia in 1890, Swaziland was annexed to Transvaal in 1893, Britain claimed Pondoland in 1894, and took over what was later Northern Rhodesia, now Zambia, in 1895.

The Boer Wars and after

None of this took place without African resistance nor friction between Boer and Briton and war broke out in 1899. Fewer in numbers and less well-equipped, the Boers showed themselves masters of guerrilla warfare, and only by resorting to a scorched-earth policy was Lord Kitchener (1850–1916) able to overcome them in 1901. The Peace of Vereeniging (31 May 1902) marked the end of the Boer Wars.

The Boers accepted British sovereignty and Britain promised them representative government and £3 million for restocking their farms. In the meantime, the British took over the valuable resources in gold and diamonds. An enlightened policy aimed at the conciliation of the Boers led to self-government in Transvaal and the Orange River Colony in 1906.

On 31 May 1910 the Union of South Africa came into begin as a federal state [9] and was given the status of a self-governing dominion on 1 July. The first parliamentary elections were won by the South African Party, and Louis Botha (1862–1919), a prominent Boer general, took office as the first prime minister.

Of the countries of southern Africa, the Union of South Africa stood alone with its wealth in gold and diamonds and, in the wetter regions near the Cape, a flourishing agriculture. By comparison, the Portuguese establishments in Angola and Mozambique were primitive and backward: and the British establishments in Southern and Northern Rhodesia and in Malawi, together with Bechuanaland, Basutoland and Swaziland, were at a pioneering stage.

KEY

The earliest inhabitants encountered by Europeans at the Cape were Khoisan (Hottentot) cattle herdsmen, who moved in search of grazing, together with groups of San (Bushmen) hunters under their protection, Both adapted to Afrikaner penetration.

5

5 Fort St Sebastian, on Mozambique Island, was begun by the Portuguese in 1558 and completed by them after 1595. Beyond the ramparts is the Church of Our Lady of the Bulwark, which was built about 1505.

6

6 British forces were crushed at Isandlwana, Natal, during the Zulu War. In January 1879, a mixed British and African contingent was overwhelmed by 24,000 Zulus. Almost all the 800 Europeans were killed. The next day, about 4,000 Zulus attacked some 110 men of the 24th Regiment at nearby Rorke's Drift. The Zulus were heroically beaten off, with 350 dead; the British lost only 17 men in the day's battle.

7

7 Johannesburg, by about 1900, was already a handsome city. First surveyed in 1886, the site was rocky, lacking in water, and uninhabited. But following the discovery of gold, it developed rapidly. By 1905, it had some 23,000 municipal voters.

8

8 Stephanus Johannes Paulus Kruger (1825–1904), was elected President of the Transvaal in 1883 and served until 1900. He was a consistently uncompromising fighter for Boer independence, and a lifelong and bitter opponent of the British.

9

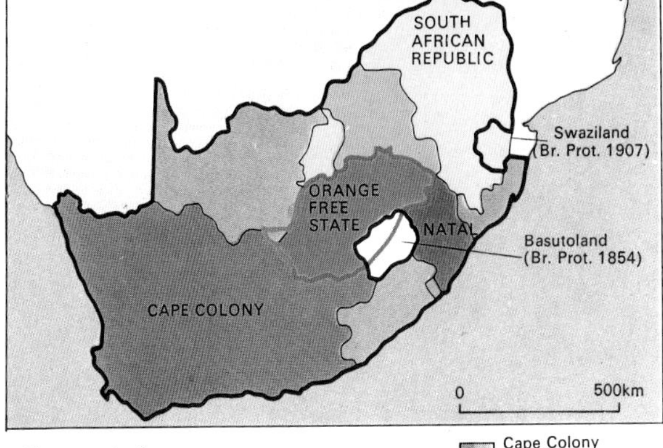

9 The growth of South Africa can be traced on this map, from its beginnings in Cape Colony under the Dutch East India Company to when it became the Union of South Africa on 31 May 1910 — a country with a parliament subject only to Westminster.

Cape Colony 1854, 1895
Natal 1854, 1895
Orange Free State 1854 boundary
Orange Free State 1895
S. African Rep. 1854, 1895
Boundary Union of South Africa 1910

Imperialism in the 19th century

The nineteenth century saw a major expansion in European control and influence over the rest of the world. Earlier, important empires had existed in the ancient world and the Spanish, Dutch and Portuguese had established extensive trading empires in the sixteenth and seventeenth centuries. But the nineteenth century was the period of Europe's greatest overseas expansion when European influence was introduced for the first time to a wide variety of races and peoples [Key]. By 1914, more than 500 million people lived under imperial rule [1].

The rise of Britain

In the course of the eighteenth and early nineteenth centuries, the older empires of Spain, Portugal and Holland entered a decline. A series of revolts freed the Latin American republics from Spanish domination and virtually ended the economic importance of the Spanish Empire. After a sequence of wars in the eighteenth century, culminating in 1815 with the defeat of Napoleonic France, Britain emerged as the strongest maritime nation with substantial colonies and many island possessions.

During the middle years of the nineteenth century, colonial expansion was relatively limited; Britain concentrated on consolidating her hold upon the colonies she already possessed, partly by conceding self-government to the most developed and responsible, such as Canada, and also by military force [7], as in the suppression of the Indian Mutiny of 1857–8. During this period Britain pursued a policy of "informal control", attempting to limit her commitments to those essential to the maintenance of trade, while avoiding large-scale involvements in governing new territories. Thus characteristic British acquisitions of the mid-nineteenth century were positions of strategic or commercial significance, such as trading rights in Singapore, purchased in 1819 from the sultan of Johore, and trade settlements on the African Gold Coast, bought from Denmark in 1850. The British attitude to India was somewhat anomalous. Although many Englishmen were prepared to contemplate the eventual secession of most of her white colonies, the prospect of India's becoming independent was never actively supported. After the suppression of the Indian Mutiny, the maintenance of India as a vital part of Britain's overseas interests became the lynch-pin of imperial policy.

The scramble for Africa

By 1870, there were stirrings in several parts of the world that had remained beyond European influence. Africa was being opened by the journeys of the great missionaries and explorers. Technological developments in weaponry and transport and advances in tropical medicine made it easier to penetrate the "dark continent". Once explorers had charted the routes it was inevitable that further European involvement in Africa would follow. The "scramble for Africa" began when, mainly for strategic reasons of safeguarding the main route to India, Great Britain occupied Egypt in 1882 [6]. Within 20 years almost the whole continent had been divided up between the major powers. Economic incentives, strategic concerns, and diplomatic rivalry all played a part in the expansion of European influence. However,

CONNECTIONS

See also
132 The British empire in the 19th century
52 China from 1368 to c. 1800
84 European empires in the 19th century
86 Latin American Independence
124 Colonizing Oceania and Australasia
138 The Expansion of Christianity
140 India in the 19th century
142 Africa in the 19th century
126 Australia and New Zealand to 1918
128 South Africa to 1910
182 British foreign policy 1815–1914
144 The opening up of China
146 Japan: the Meiji Restoration
154 The impact of steam
180 Europe 1870–1914
220 The Commonwealth

In other volumes
268 Man and Society

1 The colonial empires of the European powers were rapidly extended between 1800 and 1914. The British Empire, already with huge possessions, expanded in Africa and South-East Asia; France and Germany acquired big territories and Belgium, Italy, Portugal and The Netherlands also joined the scramble. Including the ex-colonies in America, European influence extended to 84 per cent of the world's land area by 1914.

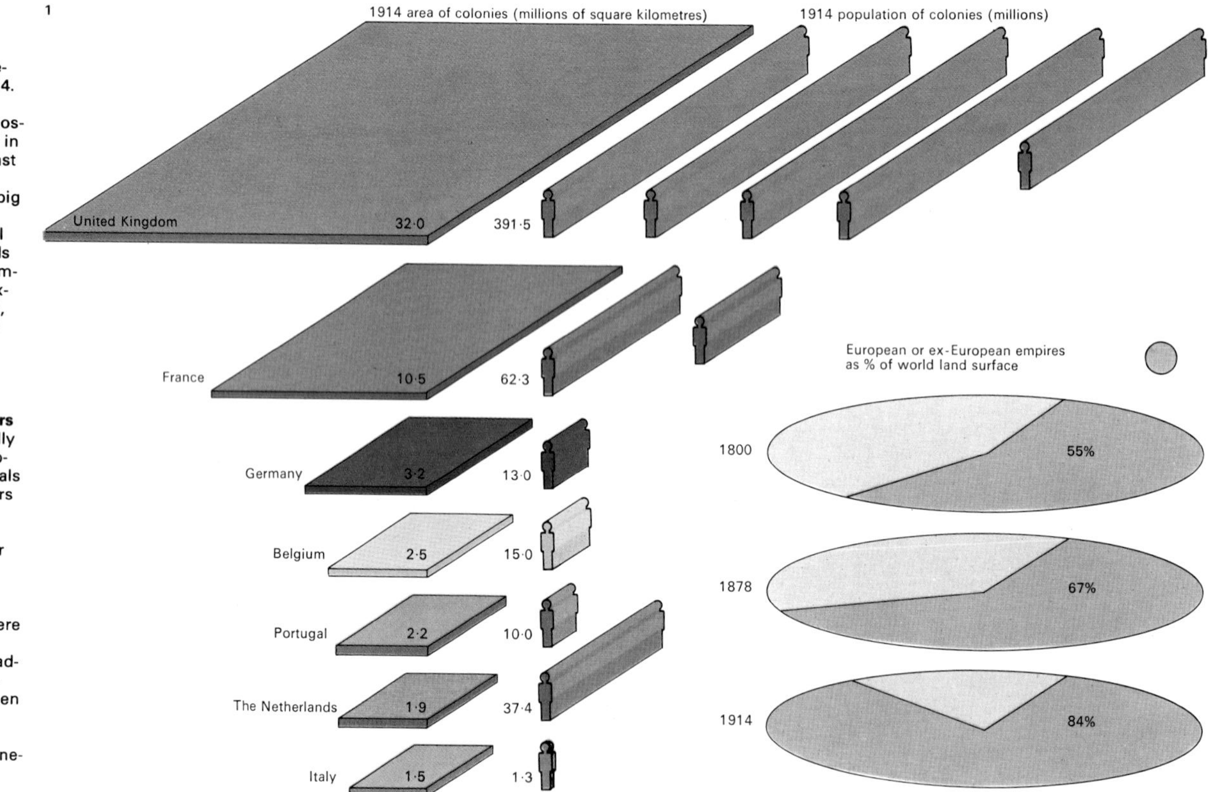

2 As trading partners colonies were usually more important suppliers of raw materials and food than buyers of imperial goods. Some of the territories acquired after 1870 hardly repaid the cost of running them. But Britain's "white" colonies were significant investment outlets and trading partners, particularly after 1900 when the volume of two-way imperial trade rose to more than one-third of Britain's total visible trade.

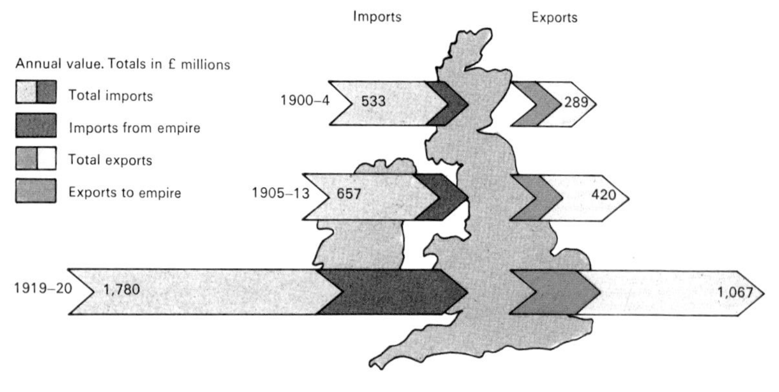

3 National rivalries for overseas territories, such as depicted in this cartoon of Britain, Germany, Russia, France and Japan dividing China, were often fanned by attitudes at home. In the 1870s the word "jingoism" was coined to describe a belligerent attitude fostered by the rise of mass-circulation papers. British disputes with Russia on the North-West Frontier of India and with France over Sudan in 1898 led to popular support for war, although ultimately it was averted.

the degree to which economic motivation accounts for the rapid expansion of the European empires between 1870 and 1914 has often been overstated. In contrast to the earlier phase of European colonialism, trade [2] now tended to follow the flag rather than act as a direct cause of territorial annexation.

Strategic and political considerations

In 1865, a British Parliamentary Committee was prepared to concede influence in the economically important area of West Africa in favour of strategic benefits in the economically poorer East Africa, with its ports on the Indian Ocean. In France, colonial development was largely a preoccupation of the government, a minority of businessmen, the military, and exploration groups, with little active support from the electorate. Similarly in Germany, Bismarck pursued a colonial policy for diplomatic and internal political reasons. As a result, the new territories acquired after 1870 tended to take only a limited part of the export of European capital [4] and population, and provided a relatively small volume of trade, supplying mainly tropical products such as rubber, cocoa and hardwoods.

Although the new imperialism was motivated primarily by political and strategic imperatives, it was fostered by a climate of approval for the "civilizing mission" of the European races. The benefits of trade, Christianity and European rule were considered obvious by many educated people in the imperial nations, providing powerful self-justification for the extension of colonial rule over "primitive" peoples. By the late nineteenth century, the glamour of imperial adventure [5] was taken up by the emerging mass-circulation press to foster "jingoism" and bring pressure to bear on politicians to support aggressive imperialism [3]. But until 1914, in spite of periods of acute tension and rivalry, the partition of Africa and expansion elsewhere was conducted without a major conflict between the European powers. A series of agreements and treaties defined areas of control and spheres of influence, leaving Great Britain with the largest overseas empire, followed in size by those of France and Germany.

KEY

European supremacy overseas was symbolized by Queen Victoria when she became Empress of India. The greatest imperial expansion of the 19th century, however, took place in Africa.

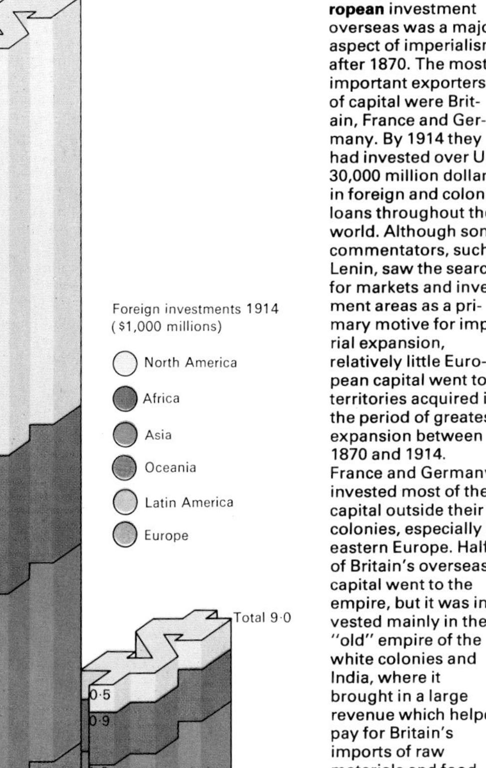

4 The growth of European investment overseas was a major aspect of imperialism after 1870. The most important exporters of capital were Britain, France and Germany. By 1914 they had invested over US 30,000 million dollars in foreign and colonial loans throughout the world. Although some commentators, such as Lenin, saw the search for markets and investment areas as a primary motive for imperial expansion, relatively little European capital went to territories acquired in the period of greatest expansion between 1870 and 1914. France and Germany invested most of their capital outside their colonies, especially in eastern Europe. Half of Britain's overseas capital went to the empire, but it was invested mainly in the "old" empire of the white colonies and India, where it brought in a large revenue which helped pay for Britain's imports of raw materials and food.

5

5 The death of Charles Gordon (1833–85), a British general, at the hands of Sudanese religious fanatics in Khartoum led to public outrage in England against governmental bungling. Gordon's bravery epitomized the romantic appeal of imperialism, which was seen as providing an outlet for heroism and adventure in exotic parts of the world, whether in seeking new colonies or in garrisoning and protecting existing ones.

6 The Suez Canal provided Britain with a reason to add Egypt to its empire in 1882. Constructed by a Frenchman, Ferdinand de Lesseps, the canal was opened in 1869, making a short route from Europe to India. Britain acquired the canal shares in 1875, following the bankruptcy of the Egyptian khedive. A nationalist revolt prompted Britain to intervene and take Egypt under effective control to safeguard the canal.

6

7

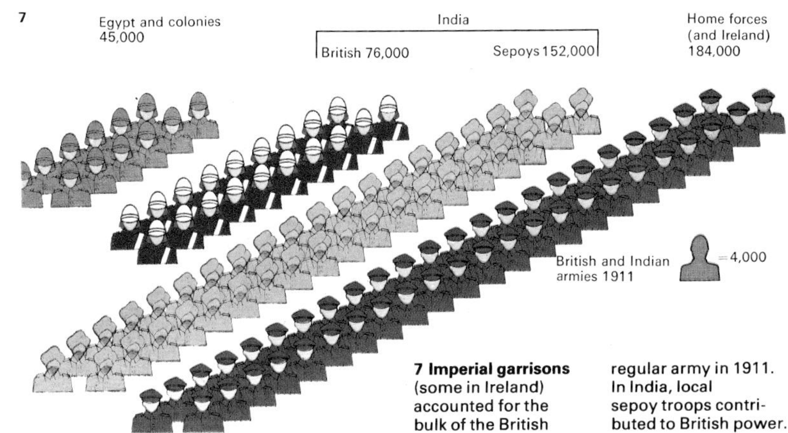

7 Imperial garrisons (some in Ireland) accounted for the bulk of the British regular army in 1911. In India, local sepoy troops contributed to British power.

The British Empire in the 19th century

In 1815 Britain was the world's greatest colonial power. Although it had lost the American colonies in the 1700s, it had asserted its claims elsewhere: in British North America, in India, in Southern Africa, and in the valuable sugar islands of the Caribbean. Outposts had also been established in Australasia; the New South Wales's penal colony began in 1788 and missionaries made contact with New Zealand Maoris in 1814.

The imperial debate

However interested missionaries and merchant traders were in the Empire, government circles and the population as a whole were doubtful of its value. The loss of the American colonies in 1776 and the successful rebellions of Spain's Latin American colonies in the 1820s suggested the notion that as colonies ripened to maturity they fell naturally from the mother tree.

The spectacular growth of trade between the United States and Britain after 1783 demonstrated that trade did not have to follow the flag. From 1815 until the 1870s, it was orthodox opinion that Whitehall should not impede the gradual devolution of the empire, and only in the last decades of the nineteenth century did Britain once more become a self-consciously imperialist power.

Lord Durham's epoch-making report of 1839, advocating a measure of self-government for Canada [2], set the tone of the early Victorian colonial debate. That debate was conducted not between imperialists and anti-imperialists, but between those who argued for an active policy of dismemberment and those who preferred to leave matters to the course of time. Even Benjamin Disraeli (1804–81), who in the 1870s was to sound the note of the new imperialism, in 1849 described the colonies as a millstone around the mother country's neck.

Between 1815 and 1870 only one-sixth of the £1,000 million in credit accumulated abroad was in the colonies. The cost of their defence lay heavily on the Exchequer, and the abolition of the slave trade in 1807 and of slavery in 1833 ended the triangular trade between Britain, Africa and the New World which had proved so lucrative in the eighteenth century. After the repeal of the Corn Laws in 1846, therefore, and the decline of protectionism, the old mercantilist system was dismantled.

In other ways, too, colonial ties were weakened. In 1852 New Zealand was granted a self-governing constitution and the Church establishment there was abolished. In Canada, a year later, the lands set aside for the support of the Church were given over to the disposal of the colonial assembly. In 1867 the four provinces of British North America became the united, self-governing confederation of Canada. In South Africa, responsible government was granted to Cape Colony in 1872 and to Natal in 1893.

India and the new colonies

The exception to this process of relaxation was India. After the Mutiny of 1857, the East India Company lost its share in the government of India, which was placed directly under the Crown department, the Board of Control. Indeed, there was everywhere a sharp distinction drawn between the white settlement colonies, extensions of British stock [3], and the coloured colonies, acquired

CONNECTIONS

See also
130 Imperialism in the 19th century
182 British foreign policy 1815–1914
46 British colonial policy in the 18th century
112 Queen Victoria and her statesmen
126 Australia and New Zealand to 1918
128 South Africa to 1910
134 The story of the West Indies
136 The story of Canada
140 India in the 19th century
142 Africa in the 19th century
216 Indian nationalism

1 The Colonial Office was established as a separate department in 1812, but began to exert great influence in government only with the appointment of "Mr Over-Secretary", James Stephen (1789–1859) in 1836. He remained there until 1847 as permanent under-secretary, and took the prevailing mid-century view that the colonies were a trust to be upheld, but also a liability bringing the mother country great difficulties and few rewards.

2 In 1849 English Canadians burned their parliament building in protest against the Rebellion Losses Bill, which compensated French rebels as well as English loyalists for the loss of property incurred in the Lower Canada rebellion of 1837. That rebellion, and its counterpart in Upper Canada, had been the occasion for Lord Durham's famous report (1839), which recommended granting responsible government to Canada.

3 Emigration, encouraged by reformers such as Edward Gibbon Wakefield (1796–1862) and Thomas Carlyle (1795–1881), eased the population problems in Britain. Most emigrants went to the Americas. The peak decades were the "hungry forties", when more than 1.5 million people left Britain, and the fifties, when the figure was more than 2 million. Many of the emigrants were people driven from Ireland by the great famine of 1845–7 and by the pressure of population in a rural subsistence economy.

4 Cecil Rhodes, the greatest imperialist-entrepreneur, built up a huge diamond and gold empire in southern Africa. He also established Southern and Northern Rhodesia as new colonies.

5 Rudyard Kipling was the bard of the Empire. His novels and poems were inspired by the glory of imperialism, yet "The White Man's Burden" stressed the awesome responsibility of empire.

by conquest and attracting few permanent emigrants [6].

There was also a distinction between the "formal" empire built up in the eighteenth century and the "informal" empire of the nineteenth century. Missionaries, traders and explorers went into Asia and Africa [Key], and governments were drawn, usually reluctantly, to follow them. The necessity of protecting British commercial interests lay behind the acquisition of Egypt, British New Guinea and North Borneo in the 1880s.

By the 1880s the private companies that carried British influence into the tropical zone had a semi-official sanction. Imperial administration then followed in the wake of commercial penetration. It was the financial empires of men such as Cecil Rhodes (1853–1902) [4] that drew Britain deeper into the whirlpool of southern Africa. The area west of the Transvaal became British in 1885, and the territories that became Southern and Northern Rhodesia were taken over. Kenya and Uganda became British protectorates in the 1890s.

Much of the impetus for this new era of expansion derived from the threat posed to Britain's former trading supremacy by the industrial competition from Germany, France, Belgium and America. Moreover, surplus industrial capital brought a quicker return in Africa than in Britain.

Consolidation and evolution

The imperial revival, sounded by Rudyard Kipling (1865–1936) [5] and cloaked in the language of civilizing mission, was not an issue that sharply divided the political parties. Voices were raised to argue that the trend of self-government in the white settlement colonies should be halted and the old empire consolidated as a bulwark against foreign competition. Proposals for a permanent imperial council and a revived scheme of colonial preferential tariffs came to nought. Six colonial conferences held between 1887 and 1911 marked the beginning of the general evolution of the Empire [8] into a commonwealth of self-governing states. The value of the conferences was shown by the speed with which the Dominions entered the war on behalf of the Empire in 1914.

KEY

The Empire-building of the 19th century was the product of a complex mixture of motives. It was often the work of private individuals – traders, business investors and missionaries, such as this one in Africa – who induced political control to follow in their wake. Quarrels between missionaries, who tended to defend native interests, the less selfless traders and the local populace, drew the government into official supervision of places such as Guinea and Bechuanaland which they would rather have left alone. Imperial ideas at home were very different from those of men overseas.

6

7

6 Malay House, photographed in the 1880s, gave governors a residence in the native style. British interest in Malaya began with the East India Company's acquisition of Penang Island in 1786 in search of goods to trade with China. In 1819 Thomas Stamford Raffles (1781–1826) set up a settlement in Singapore. This finally came under the official control of the Colonial Office in 1867.

7 The white colonies assisted Britain in the South African (Boer) War of 1899–1902. Australia became federated and self-governing in 1901. Its states are shown here as cubs supporting the British lion.

8 The British Empire at its greatest extent in 1914 was the largest empire in the history of the world. Although many of the smaller colonies were of little financial benefit to Britain, the larger colonies, especially the "white" ones, were important sources of cheap raw materials for the mother country. There were therefore many vital sea routes to be protected; many of the small islands and African coastal territories acquired in the course of the century had this strategic importance. The most important of these routes was the Suez route to India, which became central to British imperial strategies after Disraeli bought a controlling interest in the Canal for the country in 1875.

8

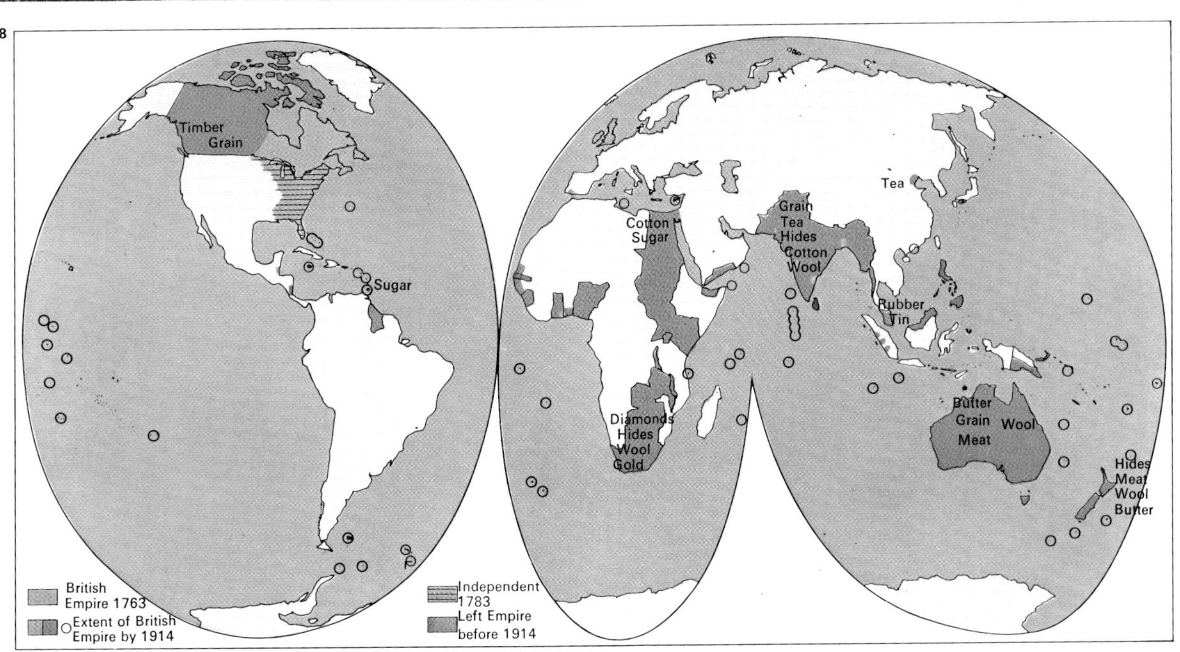

The story of the West Indies

During the four voyages of Christopher Columbus (*c.* 1451–1506) to the Caribbean (1492–1504) Spain asserted its sole right to colonize the region. At first this monopoly, sanctioned by the pope, went unchallenged; Spain settled Cuba, Puerto Rico, Jamaica and Hispaniola (now Haiti and the Dominican Republic). By the 1530s, French, Dutch and English seamen questioned the notion of the Caribbean as a Spanish domain. They began to trade illegally and attacked Spanish shipping and settlements.

Piracy and colonization

Prominent among the English interlopers were Sir John Hawkins (1532–95), who made three West Indian voyages (1562–8) with African slaves to begin England's involvement in the slave trade, and Sir Francis Drake (*c.* 1540–96), the most successful raider of all, who sacked Nombre de Dios in Panama (1572) and Santo Domingo in Hispaniola (1585). No attack, however annoying, seriously threatened to undermine Spanish hegemony, but they left a legacy of piracy and buccaneering [2].

Early seventeenth-century treaties with a war-weary Spain gave her confident rivals, so they believed, the right to colonize unoccupied islands without fear of molestation by Spain. These islands were in the eastern Caribbean and their fierce Carib inhabitants did not deter the new colonizers. The English settled in St Kitts in 1624 (sharing it with the French until 1713), Barbados (1627), Nevis (1628), Antigua and Montserrat (1632). The French took Martinique and Guadeloupe (1635); the Dutch, Danes and Brandenburgers settled elsewhere.

These islands had to provide tropical produce for the mother country, a duty they kept until independence. Originally they grew tobacco, cotton and indigo on small holdings worked by a farmer with a few white indentured servants, often Irishmen. But society changed drastically when Dutch entrepreneurs introduced the colonists to Brazilian techniques of large-scale sugar production using slave labour. By the 1660s lucrative plantation slavery [3] was ousting the small farmer and his indentured servants many of whom emigrated to North America; the West Indian population became predominantly African and the islands were bound to the fortunes of a single crop.

European exploitation

England's capture of Jamaica (1655) [1] attracted capital and planters to this largest British possession in the West Indies. At the same time the French were infiltrating western Hispaniola. Their new colony, St Dominique, became the world's largest sugar producer until its downfall in the Haitian revolution (1791–1803). These sugar colonies became the most prized imperial possessions in the eighteenth century. The remaining Windward Islands were settled in that period: Dominica, Grenada and St Vincent by Britain; St Lucia by France. By 1815 Britain had gained St Lucia, Spanish Trinidad and the mainland Dutch colonies of Berbice, Demerara and Essequibo – the last-named becoming British Guiana (now Guyana) in 1831.

Plantation life was remarkably similar in all sugar colonies. A planter or his deputy, supported by a few white or black overseers,

CONNECTIONS

See also
44 International economy 1700–1800
46 English colonial policy in the 18th century
220 The Commonwealth
246 Decolonization
254 Non-alignment and the Third World

In other volumes
222 History and Culture 1
234 History and Culture 1
278 History and Culture 1
276 History and Culture 1

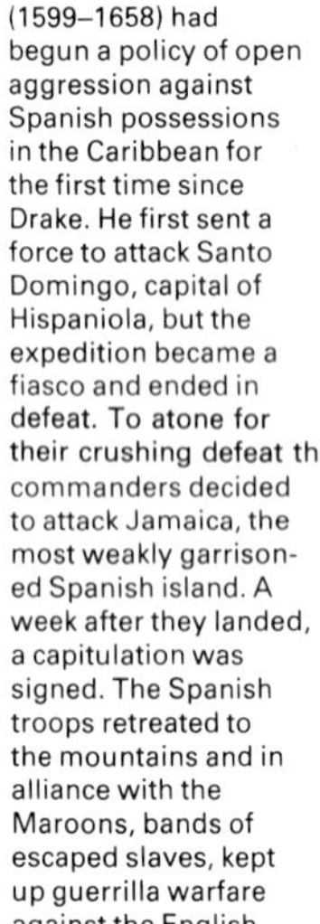

1 The British captured Jamaica with ease in 1655. Oliver Cromwell (1599–1658) had begun a policy of open aggression against Spanish possessions in the Caribbean for the first time since Drake. He first sent a force to attack Santo Domingo, capital of Hispaniola, but the expedition became a fiasco and ended in defeat. To atone for their crushing defeat the commanders decided to attack Jamaica, the most weakly garrisoned Spanish island. A week after they landed, a capitulation was signed. The Spanish troops retreated to the mountains and in alliance with the Maroons, bands of escaped slaves, kept up guerrilla warfare against the English until 1660.

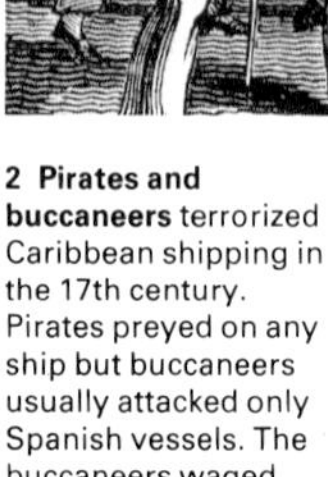

2 Pirates and buccaneers terrorized Caribbean shipping in the 17th century. Pirates preyed on any ship but buccaneers usually attacked only Spanish vessels. The buccaneers waged unofficial war even in peacetime, as well as enriching themselves, for England, France and The Netherlands, who supplied their manpower. Henry Morgan (*c.* 1635–88), one of the most notorious British buccaneers, looted and burnt Spanish-controlled Panama City in 1671. Three years later he was made Lieutenant-Governor of Jamaica. Around the year 1700 policies changed: the buccaneers were suppressed although piracy lingered on.

3 The plantation or great house where the proprietor or his deputy, the attorney, lived was usually the only substantial building on an estate, apart from the mill. The plantation houses that were built in the 18th century at the height of West Indian prosperity were often elegant mansions noted for their ostentatious and flamboyant hospitality. They were staffed by numerous house slaves. At a discreet distance from the planter's mansion and hidden from view were the crude huts and barracks that the field slaves retired to at the end of a gruelling day's toil.

4 The English Harbour, Antigua, was the largest of the two English naval bases in the West Indies in the 18th century. With Port Royal, Jamaica, it provided an important base for repairs and taking on fresh supplies that the French fleet lacked. This proved a serious handicap in the many wars that England and France fought over the sugar islands and trade in the 18th century. Many islands changed hands several times and nearly all of them were attacked by raiding forces in the bitter fighting.

ruled despotically over an enslaved workforce. There were field slaves, house slaves and craftsmen. Some arrived directly from Africa but an increasing proportion were "Creoles", born in the Caribbean. Uneasily in the middle were mulattos (the offspring of one white parent and one black).

Periods of adjustment

Humanitarian pressure [5] by reformers such as William Wilberforce (1759–1833) and, perhaps, more profitable opportunities elsewhere for British capital led to the ending of the British slave trade in 1807 and of slavery itself in 1833. There was stiff opposition from planters, who already faced competition from Brazil and Cuba and a new and efficient rival, European beet sugar. The planters sought a new source of labour [6] and finally found it in India. Between 1845 and 1917, 380,000 East Indians, as they are called, went as indentured labour to British Guiana and Trinidad. By 1970 they accounted for 51 per cent of the Guyanese population and 38 per cent of Trinidad's.

In the late nineteenth century, population pressure mounted as public health measures improved, West Indians began to emigrate. They went to Panama to build the canal and railways, to the plantations of Central America and Cuba, to the oilfields of Venezuela and to the USA.

Persistent poverty in the 1930s, made worse by world depression, led to rioting throughout the British West Indies and stimulated nationalist movements [9]. After World War II and the granting of universal suffrage the territories moved towards independence. A short-lived federation (1958–62) broke up through internal rivalries and countries became independent on their own: Jamaica and Trinidad (1962); Barbados and Guyana (1966); and Grenada (1974). The rest are self-governing with some powers still reserved to Britain.

Meanwhile the former British colonies, now members of the Commonwealth, have been re-defining their political positions. The sugar-producing nations were members of a cartel, formed to guarantee crop prices. Jamaica, Trinidad and Guyana are also members of the non-aligned nations.

KEY

Sugar production is both an agricultural and industrial process. Because the sucrose content declines rapidly, harvested sugar cane must be processed without delay. A sugar mill is therefore usually on the estate or near by. The molasses is separated and used in the manufacture of rum. Today, as for the last three centuries, sugar is the major crop in the Caribbean.

5

5 The Anti-Slavery Society, founded in 1823, brought order and direction to the efforts of religious sects and humanitarian reformers who led the early campaign against slavery. The society was one of the first pressure groups to be formed and more than 200 branches were set up. The society produced a lively magazine, and organized lecture tours by fiery campaigners and returned missionaries kept enthusiasm alight.

6

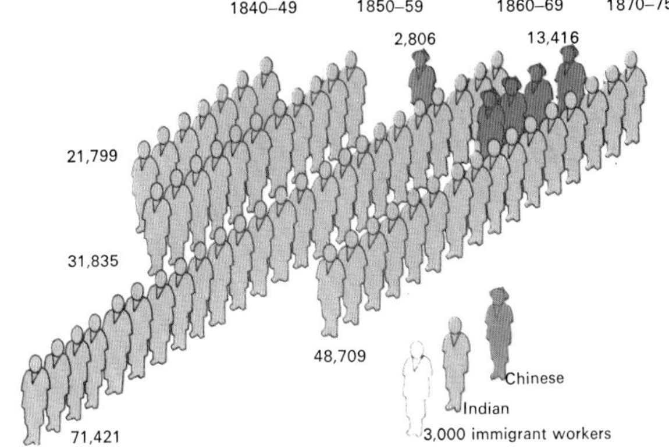

6 Immigrants came from a variety of places to take up jobs in the West Indies after the abolition of slavery led to a shortage of labour. East Indians were the most numerous, but there were also Africans liberated by the Royal Navy from ships smuggling slaves to Cuba and Brazil; Portuguese from Madeira; Chinese and freed blacks from the United States; and West Indians from overcrowded Barbados and the small islands. The Chinese and Madeirans soon moved from the estates mostly into commerce, while the Africans merged with the Creole population after about a generation.

7

7 About 115,000 West Indians arrived in Britain in the 1950s. After World War II the West Indies had too few jobs for its expanding population. It became increasingly difficult to migrate, as earlier generations had done, to foreign Caribbean islands or the United States because of new restrictions there. But from the mid-1960s West Indian immigration was reduced by restrictions introduced by the British government.

8

8 Tourism was seen by West Indian economists and politicians as a way of providing jobs and foreign exchange in the 1950s and 1960s. Their task seemed easier when the Cuban revolution of 1959 barred Havana, the traditional Caribbean playground, to North Americans. Since then beach hotels, usually owned by foreigners, have spread throughout the islands and facilities have been extended to cater for package holiday-makers as well as for the more usual, wealthy visitors.

9

9 Norman Manley, (1893–1969) typified a generation of nationalists who helped their countries to gain independence. Like Eric Williams (1911–) of Trinidad and Forbes Burnham (1923–) of Guyana, he was educated at an English university. He returned as a barrister to Jamaica where the unrest of the 1930s encouraged him to enter politics. In 1938 he formed the People's National Party, based on Fabian socialism, while a cousin, Alexander Bustamente, founded the rival Jamaican Labour Party. Since then these parties have been the main forces in Jamaican politics. Manley's son, Michael (1923–), followed him into politics and became Prime Minister of Jamaica in 1972.

The story of Canada

Indians and Eskimos inhabited Canada for thousands of years before the first Europeans set foot on its soil. The Indians, migrant from Asia perhaps 20,000 years ago, hunted and fished the vastness of the continent. In the Arctic region the life of the Eskimos, a branch of the same stock, revolved around seals, from which they obtain food, clothing, light and heat. These were the first human inhabitants, and the name of Canada itself comes from the Huron-Iroquois word *Kanata* which Jacques Cartier (1492–1557) noted during his explorations of 1534–5 [1].

The French influence

About AD 1000 the Vikings became the first Europeans to land in Canada. Little is known about these forays and some 500 years elapsed before details of frequent European contact began to be recorded. In 1497 John Cabot (*c.* 1450–*c.* 1500) under English patronage explored Newfoundland's coastline. He was followed by Cartier who explored the mouth of the St Lawrence River in 1534 and set up the first French settlements, but it was not until after 1600 that permanent bases were established. Samuel de Champlain (1567–1635) founded a base at Port Royal (present-day Annapolis, Nova Scotia) in 1605 and built a fort at Quebec four years later, thus laying the foundations for French settlement of what was known as New France. The new colony, however, was troubled by Indians and British settlers.

The rivalry between Britain and France in Europe and the Caribbean was also evident in North America, where both had colonies and trading posts [2]. The contest came to a head during the Seven Years War (1756–63). The British wrested Quebec [3] and then Montreal from the French whose position in the continent rapidly worsened. The Peace of Paris (1763) ratified the cession of all France's North American possessions east of the Mississippi, except for Louisiana.

Although the former French colonies now became British possessions, the French-speaking people of Quebec were allowed to keep their Roman Catholic religion and French civil law. British loyalists had flooded northwards during the American Revolution (1776–83) and in an attempt to avoid further friction between the two communities the British in 1791 created Upper Canada (present-day Ontario) and Lower Canada (present-day Quebec). Upper Canada was predominantly British, Lower Canada predominantly French.

The road to federation

Politically Canada resulted from a shotgun marriage of English-speaking and French-speaking settlers, but the English were more numerous and had the ear of the British governors. Furthermore, the nineteenth century saw a great influx of immigrants, most of whom came from Britain and the United States; few came from France.

Uprisings in Upper and Lower Canada in 1837 reflected social tension and growing frustration at the restrictions imposed by a system of government with officials appointed for life. Chosen to investigate, Lord Durham (1792–1840) recommended the granting of self-government in local matters and that Upper and Lower Canada be reunited, a union effected in 1840. Eventually, in 1867 the British North American Act

CONNECTIONS

See also
44 International economy 1700–1800
46 British colonial policy in the 18th century
62 Settlement of North America
64 The American Revolution
132 The British Empire in the 19th century
220 The Commonwealth

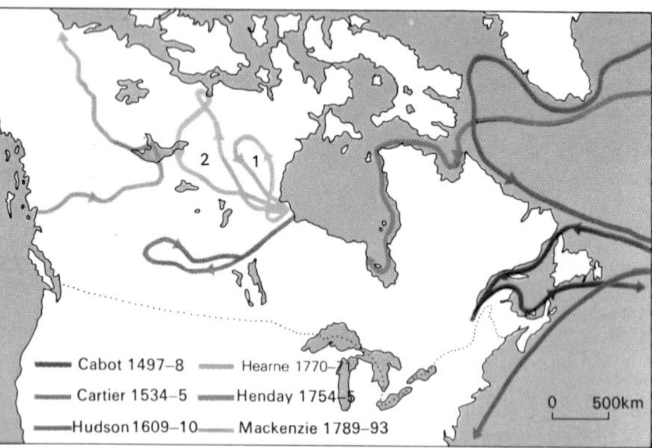

1 Recorded European exploration of Canada began with Cabot exploring the eastern coast and Cartier the St Lawrence River. Henry Hudson (*d.* 1611) in 1610 entered the great bay that bears his name. Samuel Hearne (1745–92) was the first white man to go overland from Hudson Bay to the Arctic Ocean. Anthony Henday reached the Rocky Mountains and Alexander Mackenzie (1755–1820) crossed to the Pacific coast in 1793.

2 Hudson Bay was the first site of the company that dominated the early economy of Canada. Founded in 1670, the Hudson's Bay Company had a charter from King Charles II (*r.* 1660–85) to seek a northwest passage to Asia, occupy land around the bay and trade with Indians. The company made huge profits in furs and also served as an outpost of the British Empire, extending its sway throughout the Canadian northwest.

3 The capture of Quebec in 1759 by Major-General James Wolfe (1727–59) led to the defeat of France and bolstered the British position in North America. Wolfe was only 32 when given command of an expedition to attack Quebec from the St Lawrence and gain it for Britain. He laid siege to the town in June but an assault in July failed. Yet on the night of September 12, British boats eluded the notice of the French sentries and more than 4,000 of Wolfe's men were able to scale the bluff overlooking the river to meet the French in battle next day on the Plains of Abraham. The contest was short and savage, and both Wolfe and the French commander, General Louis Montcalm (1712–59), were mortally wounded.

4 Canada came into being in 1867 with the union of Upper Canada (Ontario), Lower Canada (Quebec), Nova Scotia and New Brunswick. The western lands, bought from the Hudson's Bay Company, were organized as the North-West Territory and governed from Ottawa. Saskatchewan and Alberta were later created out of it. Newfoundland was more attached to Britain than to Canada and stayed separate until 1949.

5 John A. Macdonald (1815–91) first prime minister of federal Canada, is regarded as its architect. Born in Scotland, he combined a strong sense of nationhood with political opportunism. He survived a major scandal over receiving campaign funds from a railway contractor to win the 1878 election. He campaigned for a "national policy" of protection for industry by imposing high tariffs, a transcontinental railway and systematic development of the Canadian west.

set up a federal structure for the new nation [4] that was to enjoy a large measure of self-government and dominion status.

The growth of nationhood

In the twentieth century Canada has come to take an independent stance in international affairs. Recurrent fears of annexation by the United States had been an early spur to the evolution of a distinctive identity. The policies of such prime ministers as Wilfrid Laurier (1841–1919), Robert L. Borden (1854–1937), William Lyon Mackenzie King (1874–1950), Louis St Laurent (1882–1973) and Lester Pearson (1897–1972) were in sympathy with this development.

Canada was the first of the British colonies to assert its claim to full independence this century. In 1919 Canadians were forbidden to accept British titles; in the 1920s Canada opened its first diplomatic post abroad; when asked by Britain to send troops to Chanak in 1922, Mackenzie King refused to do so without first consulting parliament; Canada delayed its entry into World War II to stress the independent nature of the decision; and appeals to the Privy Council in Britain were brought to an end in the 1940s.

There has been a positive dimension to Canada's participation in world affairs. Canada was a founder member of the United Nations (1945) and the North Atlantic Treaty Organization (1949), and Prime Minister Pearson inspired the idea of a UN peacekeeping force in Suez in the mid1950s. Canada established diplomatic links with communist China in 1971, before most nations, and in July 1976 became the wealthiest non-member country to sign a trade agreement with the European Economic Community.

Domestically Indian agitation since 1945 for social and political recognition has led to a number of much-needed reforms including granting the vote to all Indians in 1960. In 1976 the victory of the Parti Québecois in Quebec's provincial elections brought to power a party with an avowedly separatist leadership. The possibility that Quebec may yet secede presents the most significant constitutional challenge to the federation since it was formed a century ago.

KEY

Pierre Trudeau (1919–) [right] addressed Congress during his visit to the United States in February 1977. It was an historic occasion: never before had a Canadian prime minister been invited to speak before both houses of Congress in assembly. Trudeau was able to assure his audience that he was implacably opposed to Quebec's secession, and that as long as the Canadian people desired it, the unity of the confederation would remain unimpaired. Trudeau's trip took place just one month after the inauguration of US President Jimmy Carter (1924–) [left] and the leaders discussed Canadian-American matters and broader world issues.

6

6 The Canadian Pacific Railway, completed in 1885, linked Montreal and Vancouver, quickened settlement and development in the west and helped to instil a spirit of national unity. A stupendous feat by any standards, the line was built by a private firm aided by bank loans and government land grants.

7

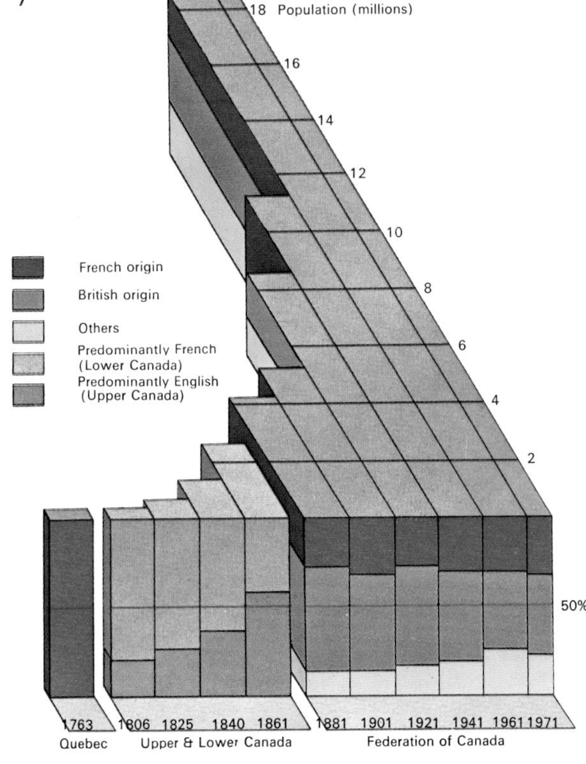

7 Fewer than 500,000 people lived in the British North American colonies in 1815, yet by 1850 the population boosted by immigrants topped 2 million. By 1901 it was 5 million. By 1971 it had reached 21.5 million, of which 18.2 million were Canadian-born, including (270,000 Indians, 17,500 Eskimos). Those born in Britain numbered 933,000.

8

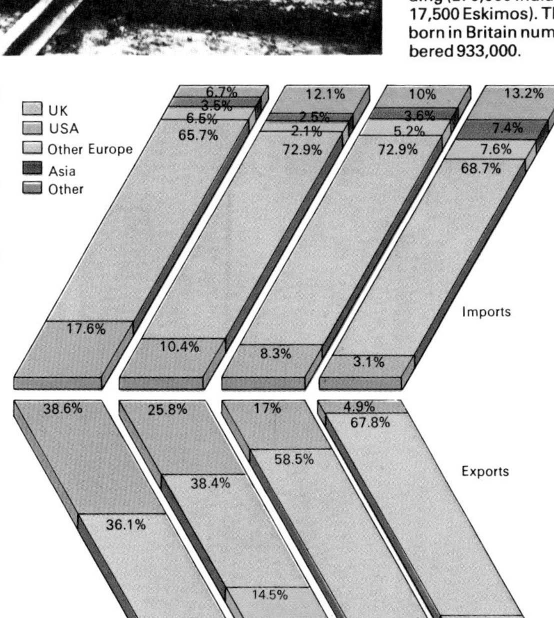

8 Canada was the sixth largest trading nation in the world in 1971 but her pattern of trade has changed dramatically during the century. In 1900 more than 50 per cent of exports went to Britain and only 38 per cent to the United States. By the mid-1970s 70 per cent went to the United States and 7 per cent to Britain. Japan and other EEC countries took as much as Britain. The United States supplies 70 per cent of Canada's imports; no other nation provides more than 5 per cent. The largest export earners for Canada are newsprint, wheat, lumber, wood pulp, nickel, aluminium, petroleum, iron ore and copper. Major import items include machinery, car parts, electrical goods, cars and tractors.

9

9 A major diplomatic crisis developed when President De Gaulle (1890–1970) of France raised the issue of separatism within Canada in 1967. On a visit to Montreal, he addressed a crowd with the slogan: "Vive le Quebec libre". Local separationists were delighted to have such an endorsement, but the government, here President Trudeau is satirized over the separation issue, and many more Canadians were simply affronted by what they regarded as meddling in their domestic affairs.

The expansion of Christianity

The spread of Christianity across the world has taken place in stages. The first saw the new religion spread from its birthplace in Palestine into the wider Roman world during the first few centuries of its existence. The second was the early medieval period when the faith survived the tumult of the Dark Ages and most of Europe became Christian. The third stage began in the fifteenth century when European civilization and Christianity turned to the oceans and the lands beyond Islam in the Near East.

The instrument of conquest

The founding of the Portuguese and Spanish empires in the Americas in the fifteenth century, and along the coastline of Africa, the Indies and the Pacific, gave an immense impetus to the advance of Roman Catholicism. The world was divided by a papal bull in 1493 into spheres of influence for the Catholic crowns of Portugal and Spain and the Church itself became an instrument of conquest and colonization.

In some instances whole populations in the newly discovered lands were forcibly converted and there were other abuses of colonial power. Often the Catholic missions were outspoken critics of these abuses, none more so than the Protector General of the Spanish Caribbean, Bartolomé de las Casas (1474–1566). Catholic advances were not, however, confined to territories formally ruled by Spanish or Portuguese governors.

The foremost Jesuit missionary, Francis Xavier (1506–52), was Papal Nuncio over the Portuguese Indian settlements. He went on to found a mission in Japan and died near Macao, in China. Another Jesuit, Matteo Ricci (1552–1610), was responsible for bringing Catholicism to China, where for a time it enjoyed the protection of the emperors and made many converts. Only in the eighteenth century did squabbles within the Church bring it into disrepute, so that Catholicism was repressed and by the end of the century the numbers of Catholics in China had become much reduced.

The success of the mission to Japan was less impressive. For 50 years from 1587 the Church was severely persecuted, and few Christians survived.

Protestant forms of Christianity were taken to those parts of the world where large numbers of Europeans settled – notably North America (except in French Canada, where the settlers were Catholics), and later South Africa, Australia and New Zealand; but the seventeenth and early eighteenth centuries were a time of dormant missionary activity. The main exception to the lack of interest was the work of the Moravian and other German Pietist groups; these were to inspire later Protestant missionaries.

Christianity and colonial activity

The second great spurt of missionary activity took place at the end of the eighteenth century and throughout the nineteenth, and was closely connected with the Protestant revival in northwestern Europe. The new Christian advance coincided with the increase of European colonial activity in the generally densely populated, tropical parts of the world, notably India and Africa, and with the ferment of the French and Industrial Revolutions. A spate of well-organized and often financially powerful societies were formed – the British

CONNECTIONS

See also
130 Imperialism in the 19th century
132 The British empire in the 19th century
304 Modern Christianity and the New Beliefs
144 The opening up of China
86 Latin American Independence
140 India in the 19th century
142 Africa in the 19th century

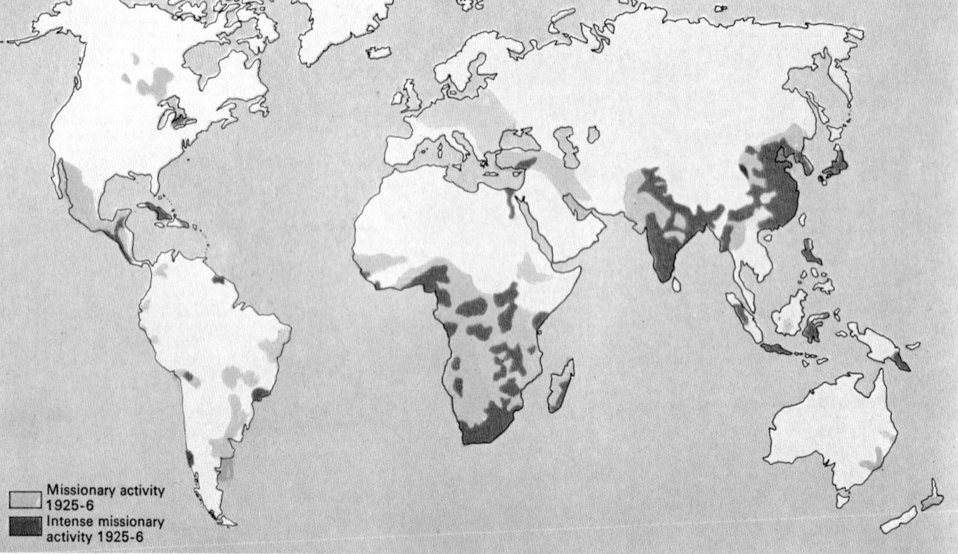

1 Christian missionary work greatly expanded from 1815 into the 20th century. As well as a revival of Roman Catholic missions there was an upsurge of Protestant activity, characterized by a notable degree of co-operation. This culminated in the International Missionary Council set up in 1921, which assisted and stimulated missionary activity throughout the world until it merged with the World Council of Churches in 1961. This map shows its activity in the mid-1920s. More than 1,000 million people were claimed to be Christians, in 1965 divided as follows: North America, 226 million; South America, 200 million; Europe, 515 million; Asia, 90 million; Africa, 90 million; Oceania, 7 million.

2 The Church in Brazil, as in the Spanish colonies, was closely linked to the state despite Rome's influence over the Jesuits. Portuguese churches, such as this one in Salvador, tended to be less oppulent than those built by the Spaniards.

3 This roadside shrine in Otovalo, Ecuador, symbolizes the assimilation of religion at grass roots level. The Church, although concerned with Indian welfare, aided their cultural decline by supporting their employment in the mines.

4 Christianity in Japan arrived with the Portuguese in the mid-16th century, but its presence became a source of suspicion within a few decades. In 1637, many thousands of Christian converts were massacred. The succeeding isolation of Japan was finally broken in 1858 and missionary work resumed, making notable contributions to education. Hugh Foss, one of the first missionaries, became Bishop of Osaka in 1899. He is seen here [left] with native clergy.

Baptist Missionary Society in 1792, which sent missionaries to India; the Nonconformist London Missionary Society in 1795; The Netherlands Missionary Society in 1797; the Church of England Church Missionary Society in 1799; the American Board of Commissioners for Foreign Missions in 1810; and the Wesleyan Methodist Missionary Society in 1813 (various Scottish Presbyterian societies came together about the same time). The interdenominational Basel Missionary Society, with support from Germany and Switzerland, was founded in 1815. Some of the great names associated with these Protestant missionary societies in Britain were the Baptist William Carey (1761–1834), William Wilberforce (1759–1833), who was also leader in the successful campaign for the abolition of slavery, and David Livingstone (1813–73).

Catholic revival in the nineteenth century

By the end of the nineteenth century, more than 300 such societies or boards existed. Catholic missionary activity, at first slow to revive, produced an effect as large as that of the Protestants – perhaps, in terms of numbers of converts, even larger. One of the foremost Catholic missionary societies was the mainly French White Fathers.

The result of all these remarkable missionary efforts, which continued from the nineteenth century into the twentieth, was the spread of Christianity over much of the tropical world. It did not make great advances in areas where other religions with claims to universality, particularly Islam, were strong. Indeed, at the same time as the spurt of Christian missionary activity at the end of the 1700s there was a revival of Islam, which made gains on the periphery of the older heartlands of the religion, especially in Indonesia and Africa [5].

Although most of the nineteenth-century missionary societies were rivals, the decline of European imperialism after World War I brought a wider, more international approach. These resulted in the formation of the World Council of Churches in 1948. This body did not include the Roman Catholic Church, but ties between Catholicism and the Council have grown increasingly close.

KEY

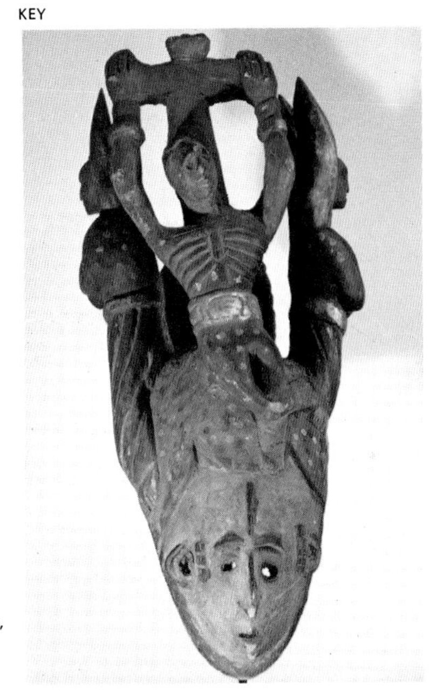

The Ibo of southeastern Nigeria, like a number of ethnic groups in Africa, were receptive to Christianity and education under British colonial and missionary influence in the 19th century. The Christian faith was often merged with existing faiths, which included belief in a creator god as well as numerous other deities and spirits. The mask illustrated reflects this assimilation. Carved in wood it depicts Christ on the cross flanked by angels. The Ibo mask is the basis of a long tradition still vigorously maintained. It is employed in various dramas such as the invocation of the gods, or initiation, as well as specifically Christian festivals.

5 A mosque in Malawi [A] stands in stark contrast to its Christian counterpart [B] and marks one point where Islam and Christianity competed for the souls of inhabitants in central Africa. The church built by the Church of Scotland mission in Blantyre, Malawi, in the late 19th century, symbolizes the permanence of its missionaries' work in the old British central Africa.

5 A

B

7 British dominions in India were the focus of increasing missionary endeavour in the late 18th and 19th centuries, initially centred on the work of medical doctors. After the pioneering work of Alexander Duff (1806–78) in Calcutta in the 1830s, Christianity became a central force in the education system established by the British. But it failed to make large inroads into the native religions, especially after the mutiny of 1857–8 led to a new realization of the importance of indigenous culture. Here St Thomas's Cathedral in Madras reflects the uncompromising application of Christianity in the Victorian mould.

6 Christian spires dominate the waterfront of Canton, in southern China. It was here that Jesuits arrived in 16th-century China after the successful pioneering work of Matteo Ricci. Canton became an important port of entry into China for later missionaries, who were able to establish colleges and hospitals there.

6

7

India in the 19th century

By the end of the nineteenth century most Englishmen regarded India as being as indissolubly linked to Britain as Yorkshire or Wales. The idea of an independent India was so remote as to be almost unimaginable. The creation of the great Indian Empire was largely accomplished between 1800 and 1860 and many Victorians saw it as Britain's supreme achievement, an essential part of Britain's rise to world power [Key].

British territorial conquests

After 1800 the British deliberately set about enlarging the territorial conquests that Clive had begun in the mid-eighteenth century [1]. By 1820 they had greatly expanded their holdings in south India and secured their position against the revival of native princes such as Tipu Sultan [2]. In the north of India the same process was carried on more slowly, but no less relentlessly, culminating in the conquest of the Punjab from the Sikhs in 1849 and the annexation of Oudh in 1856.

These great conquests were not inspired by simple avarice. They seemed to follow logically from the efforts of the East India Company (which was the instrument of British power in India until the British government's takeover in 1858) to protect itself against the threats to its trade. For with the decay of the Mogul Empire new states arose more unstable and less friendly to the company, forcing it to rely not on diplomacy but on its own armed strength. Once this process began it was difficult to stop. Raising armies in India required the company to control more land and more people, and to extract more revenue, the main source of which was the tribute traditionally paid by cultivators to their ruler. Thus each new war led to new annexations of land to pay for the company's armies and to ensure that the defeated rajahs and nawabs would not have another opportunity to attack.

Once India was fully under their control the British used its resources, and above all its army (paid for by the Indian taxpayer), for their own wider purposes in Asia, compelling the Chinese to open their ports to British trade [3]. Possession of India became indispensable to Britain's position as a great power east of Suez. But in India itself the British had to devise a system that would enable them to govern its vast area and huge population efficiently and cheaply. It was a novel problem: nowhere else had they attempted to rule people so different in language, culture and religion. And it had to be accomplished using only a very small number of British administrators [5].

The result was that for all the appearance of despotic power the British relied upon the co-operation of Indians: village administration was largely delegated to lesser Indian officials while the good will of rural notables – upon whom fell the main burden of keeping order in the countryside – was vital. This meant turning a blind eye to minor irregularities and preserving, where possible, the existing structure of local power.

Indian Mutiny: causes and effects

The extension of British control was not accomplished without violent reaction on the part of their Indian subjects, most notably in the mutiny of 1857–8 [4]. Although the mutiny arose initially from the refusal of Indian sepoys (soldiers) to bite open car-

CONNECTIONS

See also
48 India from the Moguls to 1800
216 Indian nationalism
50 Indian art: the Mogul age
132 The British empire in the 19th century
130 Imperialism in the 19th century
138 The expansion of Christianity
88 The Industrial Revolution

1

British possessions 1805
British acquisitions by 1858
British acquisitions by 1914
Dependent Indian states 1914
Area of mutiny 1857
KASHMIR
PUNJAB
BALUCHISTAN
RAJPUTANA
Delhi
OUDH
SIND
ASSAM
BIHAR
GUJARAT
BENGAL
BURMA
CENTRAL PROVINCES
HYDERABAD
CARNATIC
CEYLON
0 800km

1 British control of India developed from modest beginnings in small coastal trading stations into an empire that made Britain one of the greatest powers in Asia. Apart from direct administration of the great provinces Britain supervised nearly 600 princely states which were allowed wide autonomy but were carefully prevented from befriending imperial rivals or threatening the basic authority of the British.

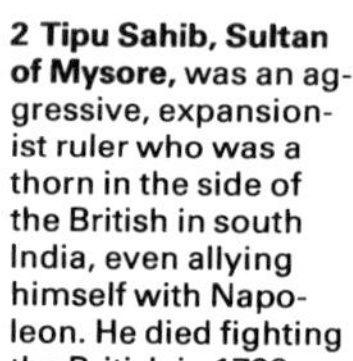

2 Tipu Sahib, Sultan of Mysore, was an aggressive, expansionist ruler who was a thorn in the side of the British in south India, even allying himself with Napoleon. He died fighting the British in 1799.

3 Indian opium was bought by the British in exchange for manufactures and sold in China for silks, spices and tea demanded by British consumers.

4 The Indian Mutiny of 1857–8 was marked by several fierce battles before British reinforcements arrived and suppressed the sepoys. Although the rising failed from a lack of concerted leadership it took Britain completely by surprise and left a legacy of distrust as well as denting the complacency of British attitudes towards the Indians.

tridges greased with animal fats forbidden to Muslims and Hindus, it swiftly became a much wider rebellion against the side-effects of company dominance: heavier taxation, displacement of Indian magnates from positions of authority and the introduction of laws that abruptly altered the old systems of landholding, rent-paying and tenancy.

For a time British authority all over north and central India swayed in the balance; Lucknow was overrun and Cawnpore besieged. The British restored their authority through the deployment of a large army, the systematic destruction of the hostile sepoy forces and savage punishment for those they considered rebels. But they learned their lesson. They realized that the mutiny had resulted from too rough a handling of the Indian gentry, from the anxieties that too much rapid change had aroused in the Indian population and from Indian fears that the British were planning to attack religious customs and practices.

After the mutiny the British were more careful and administration by the company was replaced by government rule. Headlong changes in law and in the economic character of rural life through new systems of taxation were slowed down or stopped altogether. The wholesale demolition of the remaining princely states was halted and the rajahs and princes were promised security in return for their swearing allegiance to Queen Victoria.

Stirrings of independence

By the later nineteenth century the whole spirit of British rule in India had changed. The British gave up the hope that social change and education would quickly and smoothly turn Indians into "brown Englishmen" and India into a modern society. Administrators [7] concentrated on keeping the status quo so as not to risk their power. This could not work for long. India had been opened up to the outside world and flooded with British goods and British ideas. In the big towns, economic change produced Westernized Indians who wanted a say in government. In 1885 men such as these founded the Indian National Congress and in doing so began, unwittingly, the long struggle for independence.

KEY

Stable British rule in India was underlined when Queen Victoria became Empress of India in 1876, at Benjamin Disraeli's suggestion. The event was depicted in a contemporary cartoon.

5

6

7

5 A British magistrate on tour represented a focal point of authority. Great value was attached to keeping in touch with local headmen and other important Indians in rural districts.

8

6 British and Indian troops on the North-West Frontier were deployed in large numbers in attempts to check the historical incursions of mountain tribesmen into the plains of northern India. When the British became rulers of India they were determined to subdue the unruly hillmen. They also feared that their great rivals in central Asia, the Russians, would try to undermine their power in India using Afghanistan as an ally. Desperate rear-guard actions, such as that depicted in W. B. Wollen's painting "Last Stand of the 44th Foot at Gandamuk", followed some Afghan campaigns.

9

7 Lord Curzon, Viceroy of India from 1898 to 1905, symbolized the pomp and circumstance of British rule. Although an untiring administrator he found the task of governing India frustrating and his autocratic ways were resented.

8 Indian economic life continued largely unchanged in villages during British rule. Better communications, however, did help to combat the scourge of famine and to stimulate the growth of large cities such as Calcutta and Bombay.

9 Simla became the summer capital of the British central administration in India after 1864. Lying in the Himalayan foothills, its bracing climate was a relief from the plains. It became a resort where British administrators, army officers and their families, isolated in their districts for most of the year, could enjoy a wider (and sometimes disreputable) social life. The hilly site became a status key: senior officials lived higher up.

Africa in the 19th century

The nineteenth century was a period of great and often rapid change for much of Africa, set in motion either by Africans themselves or by outsiders, especially Europeans. The partition of almost the whole continent among seven European states took place in the last 20 years of the century. The previous 80 years saw largely a continuation of trends already long established. Tiny trading "factories" (or castles) set up by European slave traders dotted the west coast of Africa, from Cape Verde to the Congo estuary [1]. On the southern tip of the continent, Britain had taken over the settlement of the Dutch East India Company at the Cape.

Foretaste of expansionism

The extension of European influence was gradual: in 1820–22 Egypt, technically an Ottoman dependency, conquered the Nilotic Sudan; in 1830 the French invaded the Ottoman dependency of Algiers and began the long, costly process of conquering it; and in the late 1830s Dutch farmers known as Boers trekked deep into the interior of southern Africa, away from British control.

With the abolition of the slave trade by most European countries late in the 1800s, trade in palm oil and other tropical products largely replaced it in West Africa. Only the French on the River Senegal expanded fairly deep into the interior; but missionaries were active, especially in areas settled by freed slaves, such as Sierra Leone [7].

Islam had so long penetrated what is known as the Sudanic belt of Africa that by the beginning of the nineteenth century it was thoroughly "Africanized". Much of this region was swept, from the eighteenth century on, by a wave of religious revival, spearheaded by holy wars, *jihads*, waged against black Muslims as much as against pagans. A *jihad* in 1804 rapidly conquered all the old Hausa city states (such as Kano) and beyond and led to the establishment of a huge new empire, the Sokoto caliphate, which survived until taken over by the British in northern Nigeria in 1903. Other Muslim empires were created on the middle Niger and in what is now Guinea and the Ivory Coast, where prolonged opposition was encountered by French invaders in the 1880s. South of the Sudanic belt, several great kingdoms, such as Ashanti and Dahomey, continued to expand and prosper (and offered vigorous resistance to the British and French respectively), while others began to disintegrate.

Rise of Ethiopia and the Zulus

In Ethiopia, the ancient Christian Amhara Empire, after a period of prolonged feebleness, slowly and painfully recovered during the reigns of three forceful emperors – Theodore (reigned 1855–68), Johannes IV (reigned 1868–89) and Menelik (reigned 1889–1911). These rulers asserted their power against that of the mighty landed aristocracy and the Coptic Church. Menelik [5] not only maintained his position against the powerful northern barons and greatly expanded the boundaries of Ethiopia in the south, but beat off an Italian attempt to conquer his state, at the Battle of Adowa (1896).

In 1818 in southern Africa, Shaka (*c.* 1787–1828) became king of a small group, known as the Zulu and, by revolutionizing the military and social structure of his people,

CONNECTIONS

See also
128 South Africa to 1910
58 Subsaharan Africa 1500–1800
132 The British empire in the 19th century
130 Imperialism in the 19th century
138 The expansion of Christianity
154 The impact of steam

1

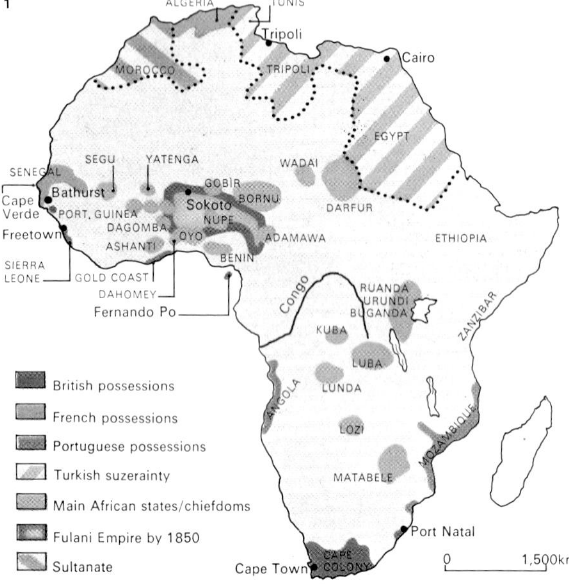

1 European possessions in Africa in 1830 were few. France had invaded Algeria (1830) and some Boer (Dutch) settlers were trekking out of the British Cape Colony into the hinterland of South Africa. Britain and France had a few tiny colonies in West Africa – Senegal, Sierra Leone and Gold Coast. Apart from Europeans on trading posts, only the Portuguese had old-established colonies – in coastal parts of Angola and up the Zambezi valley of Mozambique. Although the Egyptians had conquered the Nilotic Sudan in 1821, the rest of the continent consisted of African empires, kingdoms and peoples who still maintained their independence.

2

2 An idealized view of European influence appears in this picture of the British explorer John Speke (1827–64) with King Mutesa of Buganda. Men like Speke, who found the source of the Nile in 1858, played an essential role in opening up Africa to Europeans. Two Scottish explorers were James Bruce (1730–94) who went to Ethiopia and the Sudan, and Mungo Park (1771–1806) in West Africa early in the 19th century. Heinrich Barth in northern and western Africa, David Livingstone in central and eastern Africa and H. M. Stanley, who found Livingstone in 1872 and journeyed down the Congo, were dominant in the middle of the century.

3

4

3 The storming of Magdala, a mountain citadel in Ethiopia by British forces in 1868 was one of the most extravagant episodes in the history of relations between Europeans and Africans in the 19th century. An expedition under General Napier invaded Ethiopia to punish its emperor, Theodore (or Tewodoros) for briefly holding prisoner a British consul and some Europeans. After Magdala fell, the emperor dramatically committed suicide and the expedition then withdrew.

4 This Ethiopian village has hardly changed at all since the last century. Then, as a community of peasant cultivators producing little more than what was necessary for subsistence, it would have been typical of rural Africa.

fashioned a formidable and ruthless military state which rapidly conquered surrounding people. Offshoots of the Zulu, and other groups who copied their techniques, rampaged over much of southern and central Africa in mass population movements and tribal regroupings, known as the *mfecane*, the Time of Troubles.

Explorers and imperialists

During the middle years of the nineteenth century Africa was gradually becoming better known to Europeans through the efforts of many courageous travellers [2], such as the German scientist Heinrich Barth (1821–65), in the Sudanic regions, and the Scotsman David Livingstone (1813–73), whose travels were partly motivated by his concern over the ravages of the Arab slave trade in central and east Africa. The Welsh-American explorer Henry Morton Stanley (1841–1904) was more concerned with exploitation. In 1877 he completed an epic journey down the River Congo – and then sold his services to King Leopold II of the Belgians (1835–1909).

By this time bitter trading rivalries had grown up between Britain and France in West Africa, stimulated by British occupation of Egypt in 1882. Motivated largely by politics, a rush for African colonies began with Britain and France in the forefront, followed by Belgium and Germany, and with Portugal, Spain and Italy bringing up the rear. In many areas the conquest of Africa met with intense opposition and vicious wars of "pacification" were mounted. But resistance [3] was seldom more than local, and could be dealt with piecemeal.

In southern and central Africa the main impetus for British expansion was provided by Cecil Rhodes (1853–1902), who, from a base in the Cape Colony, appropriated a vast private empire for himself (as did King Leopold in the Congo/Zaïre). The two independent Boer republics of the Transvaal and Orange Free State were annexed in a war that fully extended the power of the British Army (the Anglo-Boer War, 1899–1902). By the turn of the century the whole of Africa, except for Ethiopia and Liberia, had been conquered by Europeans [9].

KEY

Moshweshwe (*c.* 1786–1870) was the founder of the Sotho nation (Lesotho) in southern Africa, and an example of how African rulers adopted practices and ideas introduced by Europeans. Moshweshwe emerged as leader of the Sotho, a small group of people who found refuge in the Drakensberg Mountains from the devastation produced in the interior of southern Africa by the Zulu and other warrior kingdoms in the 1820s. He was a man of peace, and ensured the protection of his people through wise diplomacy. Lesotho rapidly increased in prosperity, making use of European techniques. It was a British protectorate from 1868 until its independence in 1966.

5

5 Emperor Menelik (1844–1913) successfully maintained the independence of Ethiopia against European encroachment. In1896 his forces defeated an Italian invasion at the Battle of Adowa.

6

6 Mochudi in Botswana was one of several large towns to develop long before the coming of Europeans – notably in the Sudanic belt, Yorubaland in West Africa, Botswana and southern Africa.

7

7 Freetown, capital of Sierra Leone, was typical of European coastal towns in tropical West Africa. It was built in colonial style with churches, business centres and separate areas for whites and blacks.

8

8 Johannesburg in South Africa grew from a farm on the veld to a sprawling city by 1900. The discovery of gold in the Boer Republic of the Transvaal in the mid-1880s led to rapid development.

9

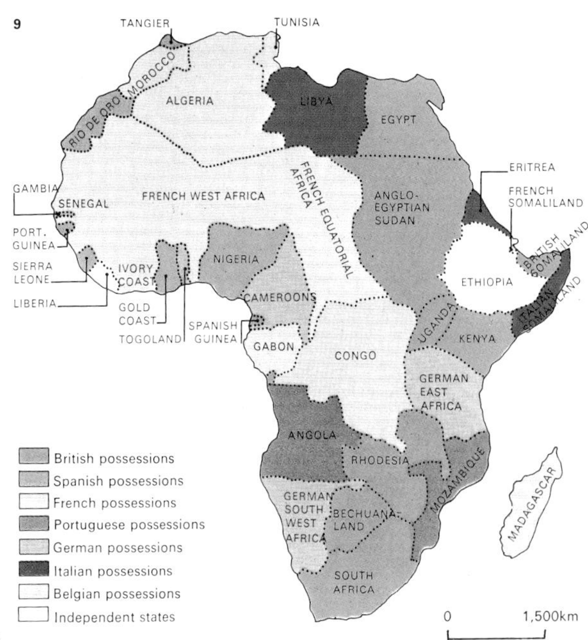

9 A map of Africa in 1914 shows how it had become partitioned among seven European countries. This partition was a rapid process taking place during the last 20 years of the 19th century. Only Ethiopia and Liberia remained independent of European rule. Although some territories were termed protectorates (like Uganda and Morocco) rather than colonies, Europeans were firmly in control. The four white-ruled colonies in South Africa had formed a Union in 1910 but remained a British dominion. Colonial boundaries drawn up entirely by Europeans were often merely straight lines on the map. This caused great problems when Africa regained independence.

The opening up of China

Two changes in China during the nineteenth century gave that country an impetus towards the revolution that flowered in the twentieth century. One that was not new in Chinese history was the decay of a dynasty – the Manchu (Ch'ing), founded in 1644. What was new, confusing and finally explosive was the challenge of Western power and technology.

The "unequal treaties"

The opening up by the West of the closed, Confucian, agrarian society of China began with the first "Opium War" of 1839–42, during which Britain crushed a Manchu attempt to stop illegal trade in opium through Canton, then the only point of Chinese contact with the money economy of the West. The resulting Treaty of Nanking (which also gave Britain a foothold in Hong Kong) was the first of the so-called unequal treaties. They eventually forced China to grant trade and territorial rights to Western powers, legalize the trade in opium and permit missionaries [4] to spread Christianity throughout the country. After pressure by France and Britain in 1856–60, China even had to grant Europeans a diplomatic quarter in Peking, implying equality with a country whose emperor had been a guardian of civilization for a thousand years and had always received tribute from inferior "barbarian" countries.

The disruptive impact of the West on the traditional pattern of Chinese life coincided with a chaotic situation in the countryside. Rural misery was accentuated by the massive population rise of the eighteenth century [1] combined with a weak and corrupt administration which neglected its duties to maintain grain reserves and irrigation. In reaction, China was swept by a series of risings against the Manchus, beginning with the Taiping Rebellion (1850–64). Virtually a civil war, this rising was suppressed only with the deaths of at least 25 million people in the lower Yangtze provinces [2]. Other rebellions soon followed, such as those of the Nien in north-central China and the Chinese Muslims in the southwest and northwest, which were suppressed by 1875.

Meanwhile, the Western-administered treaty ports, and the foreign missions spreading all over the country, steadily eroded Chinese sovereignty. In the 1860s a serious attempt was made to reinvigorate the dynasty. But this "restoration" failed to transform the conservative thinking of a court already influenced by the autocratic and dogmatic Tz'u Hsi (1835–1908) who became the Empress Dowager.

Slow technological progress

The "self-strengthening movement" that accompanied the restoration period began with the construction of arsenals, railways and dockyards in the 1860s [3] and went on with early moves for industrialization in the 1870s. But compared with Japan's speedy industrialization, China's was slow and unsure of its direction. Anti-Western feeling grew, often heightened by antagonism to Chinese "rice Christians" who took their own pickings from the privileges exacted by foreigners. Incidents in which Westerners were attacked embittered relations between the Chinese government and foreign powers. The need to learn from the West and to introduce fundamental changes was widely recog-

CONNECTIONS

See also
52 China from 1368 to c. 1800
130 Imperialism in the 19th century
146 Japan: the Meiji Restoration
168 Russia in the 19th century
214 East Asia 1919–1945

1 Pop. in 20 millions
1741
1851

1 China's population growth between 1750 and 1850 was immense, although the figures are unreliable. Growing land hunger in an overwhelmingly peasant economy coincided with worsening administration.

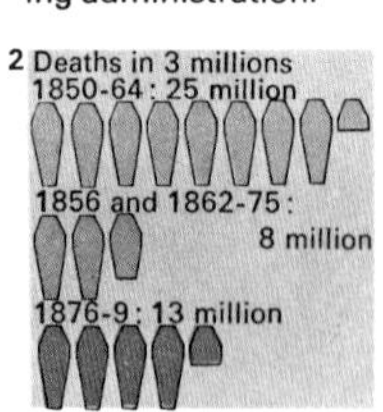

2 Unprecedented casualties were caused by the Taiping Rebellion (1850–64), two Muslim risings and a north China drought in 1876–9 which led to famine and millions of deaths by disease.

3 A "self-strengthening" movement aimed at increasing military strength to overcome Western power was launched in 1860. The build-up of armaments and improvement of railways were continued during the 1870s with moves to lay the foundation for a modern industry run by the mandarin class. Textile mills, a shipping company and an iron and steel works were established as well as smaller industries.

4 Missionaries to China
1870 350 missionaries
1910 4,000 missionaries

4 Some missionaries aroused Chinese hatred and stimulated nationalism. But some, such as the Welsh missionary Timothy Richard, also brought new ideas and won respect for their dedicated help.

5 Students to Japan
1900 500
1910 10,000

5 Chinese students began to go to America in 1872 but the real flow to Western universities began only after 1919. Until then, Japan was the source of modern thought for a generation of Chinese.

6 Li Hung-chang (1823–1901) became China's Foreign Minister after the former "Office for Barbarians" gave way to an office for "foreign matters" in 1861. He made his reputation commanding an army against the Taiping rebels and later revealed a talent for diplomacy which was acknowledged by Western powers with whom he negotiated from a position of weakness. Founder of the Chinese navy, he advanced China's interests by visiting Europe.

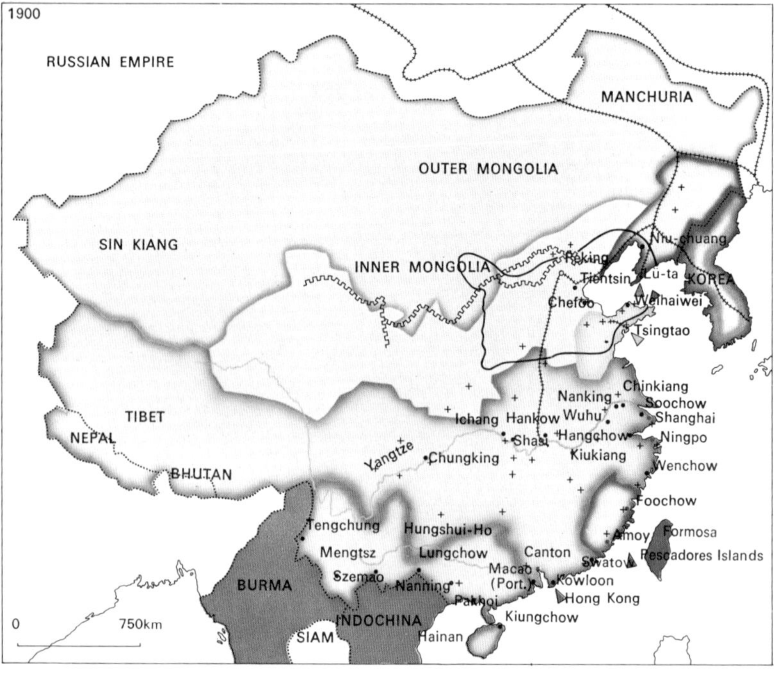

7 Foreign influence never extended to rule over China. But "treaty ports" such as Shanghai, Tientsin and others inland were administered, policed and taxed by foreigners. Chinese living in them were outside their government's jurisdiction. The diplomatic quarter of Peking itself was foreign administered until 1947. Towards the end of the 19th century key areas were divided up into "spheres of influence". Foreign missions and consulates abounded. While the coastal cities prospered, China's peasant economy suffered from foreign imports. Rural areas were drained of talent and the exactions of absentee landlords increased.

 15. 11. 2014

this day in 1908, the last of the Manchu emperors, Pu Yi, tured on the right, came to power, aged two

Oir na h-in

RAGHNALL MacilleDHUIBH
LEABHRAICHEAN ÙRA

THA Dòmhnall Iain MacÌomhair á Ùige Leódhais air a bhith a' sgrìobhadh agus a' foillseachadh sgeulachdan goirid fad a bheatha. Se *Caogad san Fhàsach* (Acair, £9.99) an treas cruinneachadh aige, a' leantainn *Camhanaich* (1982) agus *Grian is Uisge* (1991). Chan eil mi cinnteach ciamar a chaidh an lethcheud a thaghadh, ach is iongantach mur eil an t-ùghdar son tlachd a thoirt dhan leughadair agus cothrom a thoirt dhan chritig breithneachadh a dhèanamh air neart a chuid

Tha cuid nas toinnte. Tha ' Bogsa' mu dhrogaichean *hall natory*, 's tha ìomhaigheachd mar: "Nuair thàinig an t- dh'fhalbh an aois le William gu òige ùr." Tha 'Am Prògram' leth sgileil – tha murtair uabhasachadh nuair chì e drà air TBh a tha chearta choltach na rinn e fhéin. "Chunnaic e dealbh-chluich ged nach fha idir i." Chì na poilis cuideachd

'S tha cuid domhainn. Tha Aonar' 'na bhàrdachd de sg làn ìomhaigheachd den t-seò dh'ainmich mi. Eil e mu na h- eanan an-lar? Mu eachdrai nized only in 1895 when China was humiliatingly defeated by Japan [8].

In the treaty ports new middlemen in foreign trade were those patriots who knew what changes were needed. Sun Yat-sen (1866–1925), educated in Hawaii and Hong Kong, preached nationalism, and mandarins such as K'ang Yü-wei (1858–1927) backed the young Kuang Hsu Emperor (reigned 1875–1908) in reform edicts in 1898. But the Empress Dowager imprisoned her son and assumed power.

The end of the old China

Competing European imperialists now threatened to partition China – a scramble halted only by an American-inspired "Open Door" policy by which the Western powers agreed to restrain their territorial ambitions in return for open trade. Meanwhile fierce anti-foreign rioting broke out in 1900 when the court diverted a rising by the secret society of the "righteous and harmonious fist" against Westerners. Known as the Boxer Rebellion, this cost the lives of nearly 250 missionaries and thousands of Chinese Christians before it was suppressed by an international army.

The old China was finished however, outmoded and discredited. The archaic civil service examination system was abolished in 1905 and the Manchu dynasty hastily abdicated after a provincial revolt in 1911. The formula for a viable Chinese republic did not yet exist. A parliament headed by Sun Yat-sen immediately gave way to rule by a former Manchu commander, Yüan Shih-k'ai (1859–1916). A decade of rule by rival warlords followed.

The intellectual consensus needed for change was emerging, however. Sun Yat-sen [10] refounded his movement as the Kuomintang Party and thousands of students educated overseas [5] or at new universities were influenced by liberal teachers such as Ch'en Tu-Hsiu (1879–1942) and Hu Shih (1891–1962) [11]. When China's weak government accepted concessions to Japan imposed after World War I (in which China had taken little part), student protest on 4 May 1919 launched a revolutionary nationalism [11] that set China alight.

KEY

A common Western attitude to China in the nineteenth century was summed up in a cartoon of the Western powers shaking the corpulent body of China. After the 1840 Opium War contempt for the China of the Manchu emperors began to replace the admiration held by eighteenth-century Europe for the achievements of Chinese civilization. By the 1890s when nations such as Germany, Russia, Japan, Britain and France were scrambling for territorial rights, most Europeans thought "this rotten old hulk" would break up and be remade by Western enterprise. Few perceived the enduring strength of Chinese civilization beneath the decay.

8

8 After the Sino-Japanese War of 1894–5 China sued for peace (as seen in a Japanese drawing). Joining the Western powers in their demands on China, the Japanese had disrupted China's sphere of influence in Korea. In the war that followed, the Chinese were easily beaten and had to cede Formosa to Japan. This stimulated Chinese shame and nationalism more than earlier defeats inflicted by Britain and France because the Chinese had always regarded the Japanese as inferiors who had adopted Chinese culture a thousand years before. But Japan's modernization after 1868 sent its military and industrial power far ahead.

9 Chinese dislike of foreigners is shown in an 1891 cartoon of a pig as a Chinese Christian and goats as foreigners being slaughtered. Earlier in the nineteenth century, foreigners were almost unknown; most Chinese lived and died without seeing one. Christianity made little appeal to the Chinese.

9

10

10 Intellectual leaders played a vital role in changing Chinese attitudes to the structure of government and society after the old China was swept away in the turbulence that followed the death of the Empress Dowager in 1908. The next decade brought together strands of nationalism, cultural change and revolution. Sun Yat-sen [A] was an outsider to the Chinese classical tradition and the world of the mandarin. Affected by Victorian progress he wanted to modernize China. His magnetic personality built up a mixed following in the secret societies. Supported in Japan and welcomed in the West where his Christianity and good English helped, his tenacity finally won mass backing after 1919. Ch'en Tu-Hsiu [B] was a more revolutionary intellectual. When he founded the influential "New Youth" in 1915, he favoured "Mr Science and Mr Democracy" but by 1921 he had emerged as the first leader of the rising Chinese Communist Party.

11 A revolutionary consciousness was developed in China by the teaching of men such as Hu Shih [A]. He substituted the use of classical Chinese by writers, which had separated the educated classes from others, with the vernacular. A pragmatic thinker who studied in America, he remained the spokesman of Western liberalism but influenced the future communist leader of China [B], the young Mao Tse-tung (1893–1976), who was snubbed by professors when he went to Peking University as a library assistant. For Mao's generation of students 1919 was the year of revolutionary awakening.

11

Japan: the Meiji Restoration

Until the middle of the nineteenth century Japan had been closed to the outside world for more than 200 years. Only the Chinese and the Dutch were allowed limited trading access to one port, Nagasaki. It was Commodore Matthew Perry in command of a squadron of United States warships who, during visits in 1853 and 1854, cajoled a reluctant shōgunate – Japan's military government – into opening two ports to American shipping. Other powers soon followed the American lead; within a few years Japan's self-imposed seclusion was over.

Civil war and a new capital

The intrusion into Japan by the Western world mortally harmed the prestige of the Tokugawa Shōgunate which, under pressure, signed treaties granting extraterritorial rights and tariff privileges to the foreign powers [3]. The imperial court at Kyōto, universally revered but possessing no effective power of its own, became the focus of loyalty for those samurai (warriors) who called for the expulsion of the alien "barbarians". After some years of complicated domestic strife the shōgunate was overthrown in 1868 by an alliance of provincial lords and warriors from domains in southwest Japan. Their successful civil war was fought in the name of the youthful Emperor Meiji – "enlightened rule". He was installed in the shōgun's castle at Yedo which was renamed Tokyo and made the new capital.

By this political upheaval, known as the Meiji Restoration, governing powers were restored, although in name only, to the imperial house. It marked the beginning of Japan's transformation from a feudal society to a modern state. The new government, an oligarchy of relatively young samurai, resolved to bring Japan up to the technological level of Europe and the United States.

Japan's industrial revolution

Foreign teachers and specialists of every kind, skilled in the techniques of Western civilization, were invited to Japan; and Japanese in large numbers went abroad to study [2]. Remarkable progress in modernization was made within two decades [1]. The cotton-spinning industry provides a striking example. In the 1870s annual production, increasing yearly, barely exceeded 2,000 bales, but the figure for 1889 was 142,000 bales; ten years later it was 750,000 bales. Comparable growth occurred in many other sectors of manufacturing industry. Almost none of this early expansion was financed by borrowing abroad – instead the cost fell heavily on the rural areas.

Japan's industrial revolution was broadly completed by the eve of World War I and within the lifetime of some of the leading figures of the Meiji Restoration. Political change was symbolized by the Constitution of 1889 which established a diet (parliament) of two chambers. But the Meiji Constitution was authoritarian in letter and spirit. The upper house of the diet was non-elected and until after World War I members of the lower house were elected on a limited suffrage. Cabinet ministers were responsible only to the emperor, not the diet, and the war and navy ministers were always generals and admirals representing services strongly imbued with the samurai martial spirit.

The same spirit was also perceptible

CONNECTIONS

See also
214 East Asia 1919–1945
168 Russia in the 19th century
130 Imperialism in the 19th century
52 China from 1368 to *c.* 1800
56 Japan 1185–1868

1 The first Japanese railway line, completed in 1872, was built by British engineers and covered the 29km (18 miles) between the capital, Tokyo, and Yokohama. Railways played a particularly important role in the modernization of Japan for in pre-Meiji days there was very little wheeled traffic along the roads. Commerce between the main centres of the country was mainly seaborne. The growth of the railway system in 32 years was rapid: in 1886 there were 692km (430 miles) of track; in 1896, 4,007km (2,490 miles) and in 1906, 8,494km (5,278 miles). By 1918 the total was more than 14,480km (9,000 miles) of working track.

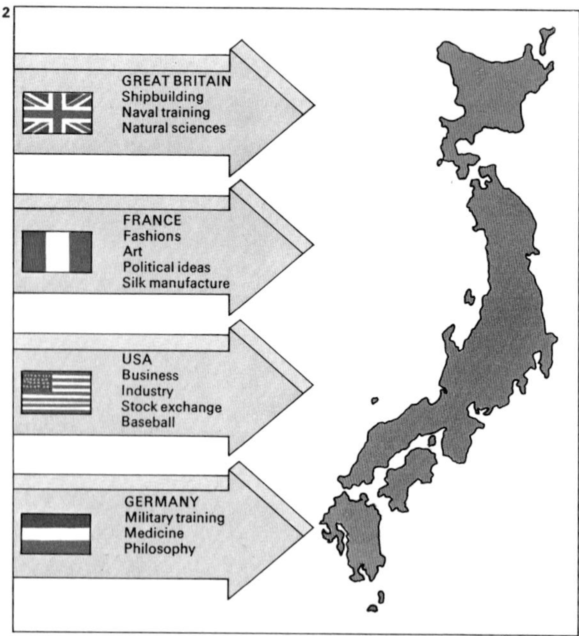

2 The greatest contributions to Japan's modernization were made by Great Britain, the USA, Germany, France, Russia and Italy. Britain trained the Japanese navy and influenced other maritime activities. The USA influenced such areas as business and education. France and Germany trained the army; Russia and Italy influenced the arts.

3 Japan's first important diplomatic mission abroad in 1871 was led by Prince Iwakura. In the United States and Europe the aim of the mission was to persuade the Western powers to revise the "unequal treaties" they had signed with Japan. But the Japanese had to wait nearly 30 years before they could secure treaty revision and thereby obtain tariff autonomy and the abolition of extraterritorial privileges. This picture of Iwakura's departure to Yokohama illustrates Japanese society obviously in a state of transition. Some of the men are wearing Western suits, while their companions still favour the traditional "top-knot" hairstyle and carry the samurai (warrior) sword.

among the people at large for the state education system gave great importance to loyalty and patriotism. The effectiveness of such indoctrination was illustrated by the events of the Sino-Japanese War of 1894–5 and the Russo-Japanese War of 1904–5.

Military and naval supremacy

In both struggles the Japanese surprised the world with their victories on land and sea, of which the most dramatic was the destruction of the Russian fleet off Tsushima in May 1905 by Admiral Togo (1847–1934) [5]. This masterly demonstration of naval supremacy won Japan acceptance as a great power.

The Sino-Japanese War had arisen from rivalries in Korea. Japan's victory gave her possession of Formosa and eliminated Chinese influence in Korea. In 1904 the reason for war was again largely Korea. Russia, occupying key points in Manchuria, seemed about to penetrate Korea, still nominally an independent state, although then dominated economically and politically by Japan. By the Treaty of Portsmouth (New Hampshire) in the United States, which ended the Russo-Japanese War, Japan acquired south Sakhalin and inherited Russia's lease of Port Arthur and her valuable rights and interests in south Manchuria. This setback for Russian power in the Far East sealed the fate of Korea, which was finally annexed by Japan in 1910 [7].

Emperor Meiji died in 1912 [6]. Due to ill health the new ruler (Taisho) was a mere figurehead. The Anglo-Japanese Alliance, first concluded in 1902 as a gesture of solidarity against Russian ambitions in Asia, brought Japan into World War I on the British side. Japanese forces captured Germany's leased port in China, Tsingtao, and occupied her island possessions in the Pacific. While the European powers fought each other, Japan partly extended its influence over a weak and divided China.

At the 1919 Paris Peace Conference, Japan was given a permanent seat on the Council of the newly created League of Nations, which amounted to full recognition of Japan's status as a world power. In the space of 50 years the aims of the early Meiji modernizers had been achieved.

KEY

The Japanese battleship *Kashima* (16,660 tonnes; four 12in guns) was built by Elswick shipyard and launched at Newcastle in 1905. Until the Russo-Japanese War all Japan's larger warships were built abroad, the majority of them in Britain. The last was the battle-cruiser *Kongo* (27,500 tonnes; eight 14in guns) launched by Vickers Armstrong, Barrow-in-Furness, in 1913.

4

4 Japanese aggression against Russia and Japan's reliance on foreign military aid was ridiculed in this Russian cartoon of 1904. Although Britain supported Japan, diplomatically and with arms, the United States was not in fact directly involved and in the following year President Roosevelt (1858–1919) acted as mediator between the two belligerents. Despite the confidence of the Russian defenders in this picture, their fleet was destroyed in May 1905.

5

5 Admiral Togo Heihachiro, a national hero after his annihilation of the Russian Baltic fleet at Tsushima in 1905, was revered almost as much in Britain as in Japan. Trained in England in HMS *Worcester*, he based his signal at Tsushima – "On this battle will depend the fate of our empire" – on Nelson's famous signal at Trafalgar exactly one century earlier. He was created count in 1907 and died in 1934, aged 87.

6

6 The death of the Emperor Meiji in 1912 marked the end of Japan's "Victorian age". The funeral procession in Tokyo took place at night, the coffin being carried on an ox-wagon from the palace to Tokyo Station. Enormous silent crowds kneeled respectfully as it passed. Interment was at Momoyama near Kyōto. A detachment of Royal Marines took part in the funeral procession.

7

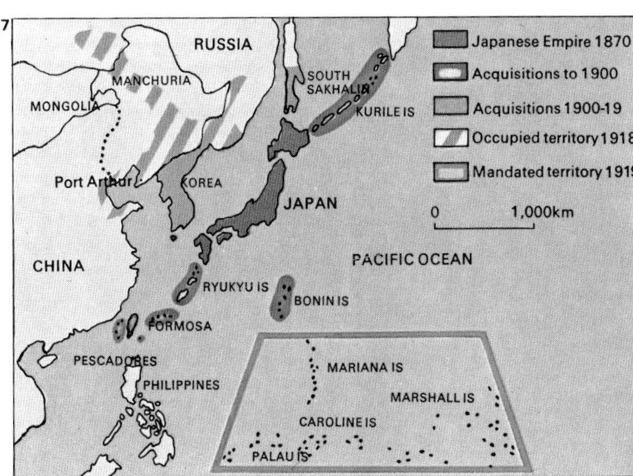

7 The expansion of Japanese territory at the end of the 19th and in the early part of the 20th century was a result of both war and treaties. In 1875 the Kurile Islands were acquired from Russia by treaty in exchange for abandonment of Japanese claims on Sakhalin. Formosa was won in the Sino-Japanese War; south Sakhalin, lease of Port Arthur and rights in south Manchuria in the Russo-Japanese War. Korea was annexed in 1910.

8

8 The Yawata Ironworks in northern Kyushu (completed in 1901) was for many years the main steel-producing plant in Japan. Production of iron and steel on a large scale began relatively late because Japan was deficient in natural resources such as iron ore and coking coal. The need for these materials, essential to Japan's industrialization, was one of the reasons for Japan's aggressive interest in both Manchuria and China.

USA: the opening up of the West

During the first half of the nineteenth century, the United States grew from a small cluster of 13 states huddled against the Atlantic coast into one of the largest nations on earth, extending from the shores of the Atlantic to the shores of the Pacific, and from Canada in the north to Mexico and the Gulf of Mexico in the south [Key].

Frontiersmen and settlers

The opening of the West began as a scattered penetration by hunters and explorers into the areas immediately adjacent to the coastal settlements. Even before the Revolution, men such as Daniel Boone (1734–1820), who crossed the Appalachians to scout out Kentucky, blazed trails through unknown regions. They and their successors drifted into the Shenandoah valley, the Alleghenies and the wooded wilds of Vermont. Probing ever deeper inland, frontiersmen reached the River Mississippi, the western limit of the territory won from Britain in the revolution.

Settlers followed, venturing westwards in search of land, livelihood and living space. Their numbers were swelled by migrants from Europe who, in addition, sought religious and political freedom. The settlers – their lives often imperilled by the Indians whose land they were appropriating – dotted the new areas with cabins, forts [4], communities, then towns. Gradually the western territories took shape.

To avoid a land scramble among the states Congress promulgated its precedent-setting North-West Ordinance of 1787. This was designed to promote an orderly development of self-government in the newly settled regions. Each "territory" was empowered to elect a legislature when its free male population reached a total of 5,000 and to claim statehood when its population had increased to the figure of 60,000.

From sea to shining sea

In 1803, the United States, barely two decades old, doubled in area. Napoleon, embroiled in a war with Britain, sold the vast Louisiana territory – extending from the Mississippi to the Rocky Mountains and from Canada to the Gulf of Mexico – to the American government for 15 million dollars. President Thomas Jefferson (1743–1826) immediately dispatched Meriwether Lewis (1774–1809) and William Clark (1770–1838) to explore this enormous acquisition [1], as well as the Oregon territory to the west. The prospect of a nation's extending "from sea to shining sea" began at last to materialize.

Pioneers penetrated beyond the Mississippi in ever-growing numbers. Among them were resourceful, independent, nomadic hunters who chose to make the western wilderness their home. Known as "mountain men", they ranged far and wide through the West, often acting as intermediaries between the Indians and white settlers and officials. They also served as scouts for the wagon trains of settlers who had to make long, hazardous journeys across Indian territory to lush, fertile valleys in the Far West [2].

To the south, thousands of Americans settled in the Mexican province of Texas. Refusing to accept Mexican authority, they rebelled in 1835 [7], setting up a provisional government. This paved the way for the American annexation of Texas a decade later

CONNECTIONS

See also
64 The American Revolution
150 The American Civil War
152 USA: reconstruction to World War I
158 Industrialization 1870–1914
154 The impact of steam

1 Lewis and Clark set out up the Missouri River, crossed the Rockies with the aid of Sacagawea, a young Shoshone, and reached the Pacific in their 1804–6 expedition to map the vast American heartland acquired from France in the Louisiana Purchase. The maps and drawings they made served both to establish American claims to the area and to encourage pioneers, although they failed to find the hoped-for portage route.

2 Covered wagons were the main vehicles used for long-distance travel by settlers penetrating the West. Wagon trains often consisted of more than 100 canvas-draped wood-framed "prairie schooners", which were usually drawn by from two to six yokes of oxen. A journey of migration up the Oregon Trail could take six months or more. Wagons crossed the central Rockies before turning north to reach Portland. Caravans would form in towns on the Missouri and the Mississippi. Seeking safety in numbers to cross dangerous territory, groups would elect leaders to consult with hired scouts about the route and to settle any disputes.

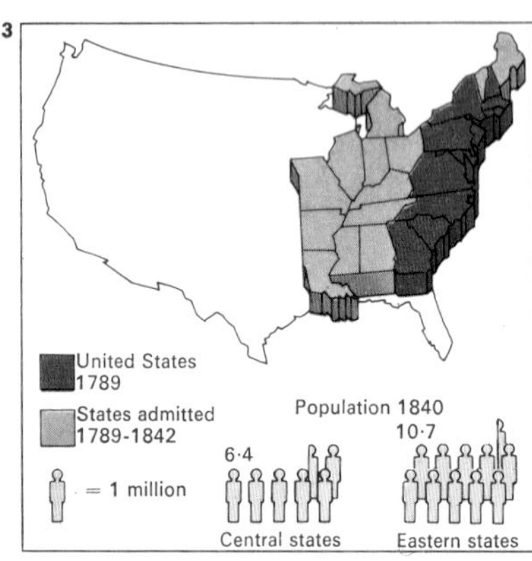

3 By 1842 the westward movement was well under way, opening up the fine farmlands of the new states. Meanwhile, a steady stream of European immigrants, particularly from Britain and Germany, converged on the northeast.

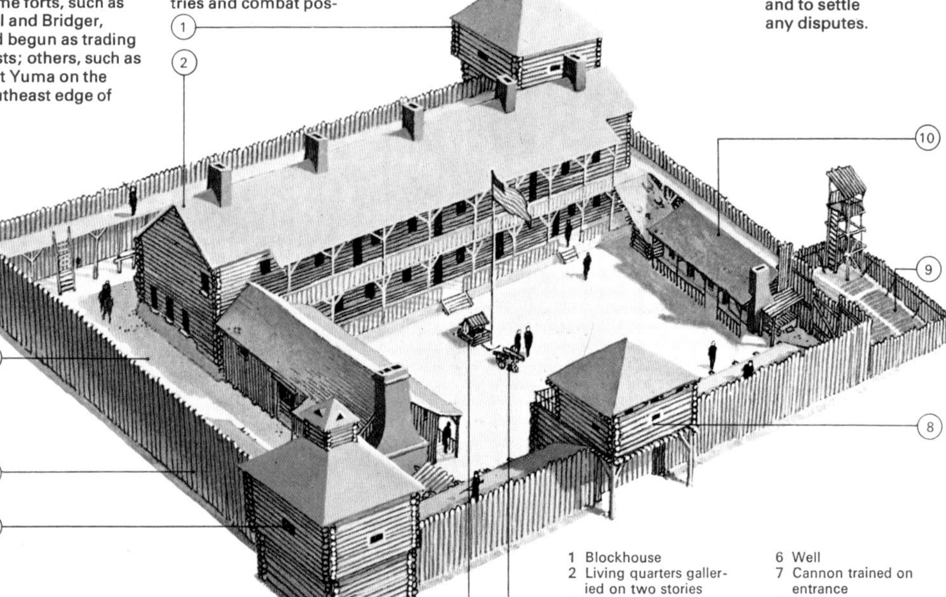

4 Forts were built along commonly traversed pioneer routes such as the Oregon Trail to protect travellers and scattered communities and provide refuge in the event of attack by Indians. Some forts, such as Hall and Bridger, had begun as trading posts; others, such as Fort Yuma on the southeast edge of the Rockies, became posting stations for mail routes. Manned by US Cavalry, the forts were rectangular enclosures up to 152m (500ft) long with timber walls up to 5m (18ft) high. Plank walks for sentries and combat positions were placed 1.2m (4ft) from the top with loopholes offering protected firing positions. Two blockhouses, at diagonally opposite corners, provided the main defence. Some forts became centres of thriving communities in the West as time passed.

1 Blockhouse
2 Living quarters galleried on two stories
3 Corral
4 Palisade
5 Loophole for small cannon
6 Well
7 Cannon trained on entrance
8 Gatehouse
9 Garden
10 Storehouse and kitchen

5 The Mormons, persecuted for their religious beliefs in the state of Illinois, set out in 1847, led by Brigham Young (1801–77), to found Salt Lake City.

and the Mexican War (1846–8), as a result of which the United States acquired vast areas of territory including New Mexico, Arizona and California.

Few events provided greater impetus for the opening up of the West than the discovery of gold in the Sacramento valley in 1848. Tens of thousands scurried to California to seek their fortunes [10], and communities sprang up overnight.

Impelled by different objectives, 148 Mormons had branched southwestwards from the Oregon Trail in 1847 to claim the inhospitable area around the Great Salt Lake [5]. There they sought a sanctuary to practise their newly founded faith without harassment. They transformed the stark Utah territory into flourishing communities by modern irrigation methods.

Dispossessed Indians

Sporadic settlement had left large areas thinly populated. In order to attract settlers to the Great Plains, Congress passed the Homestead Act of 1862, promising farmers free land for cultivation. Within five years of this significant event the settlement of the American heartland was well under way [6].

The relentless westward expansion was a disaster for the Indian peoples [9]. The 1830 Indian Removal Bill (authorizing removal of eastern Indians to locations west of the Mississippi) merely confirmed the right of settlers to dispossess Indians wherever they found them, including the regions beyond the Mississippi. Some tribes, notably the Creeks, Comanches, Apaches and Sioux, resisted the invasion, terrorizing isolated communities, attacking wagon trains and battling with the US Cavalry. Outnumbered and outgunned, they were swept aside, slaughtered or pressed back. Tribes were sometimes induced to cede their land for territory farther west – from which, later on, they were also expelled. They were relegated to reservations, and farmers, cattlemen [8] and miners moved in.

The coming of the railways sharply accelerated westward flow and settlement. In 1869 the first transcontinental rail link was completed [11] and the West's open spaces became significantly less remote. The frontier had passed into history and legend [12].

KEY

1 United States territory 1783
2 Louisiana purchased 1803
3 Ceded by Great Britain 1818
4 Florida purchased 1819
5 Texas annexed 1845
6 Oregon Country ceded 1846
7 California, Arizona and New Mexico ceded 1848
8 Gadsden Purchase 1853

By annexation, war, purchase or treaty, the United States increased its territory to include the whole subcontinental expanse in the space of 90 years between 1763 and 1853. In so doing, it prevented a resurgence of British or French influence and gave effect to the Monroe Doctrine of 1823 that America was no territory for colonization by any other power.

6

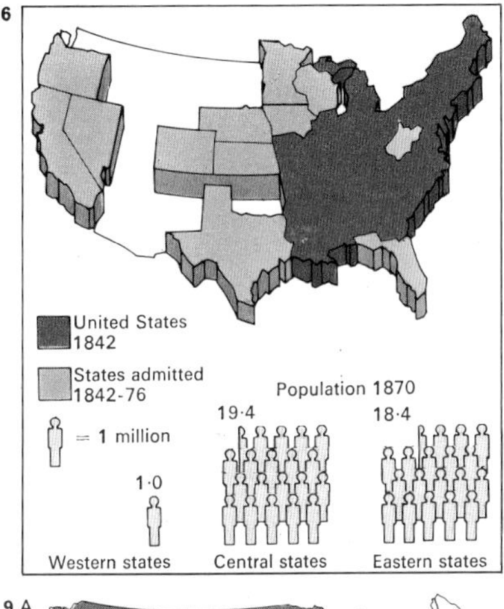

6 By 1876 Florida, the central and far western states had joined the Union as the spread of railroads allowed for more concentrated settlement. California, acquired in 1848, had achieved statehood two years later as the 1849 gold rush swelled its population to well over the 60,000 minimum required. At the same time rapid immigration continued from Europe (more than six million from 1840 to 1870), many of whom had fled from the Irish famine.

7

7 The Alamo, an old Spanish chapel in San Antonio, was the fortress in which about 150 Texans, rebelling against Mexican rule, held out for nearly two weeks in 1836 until all but two women and two children were killed. The Texans made good their independence later that year.

8

8 Cowboys, a hard-riding, hard-working breed, built the Texas cattle empires. Later they opened the range of the Wyoming, Montana and Colorado pastureland.

9 A

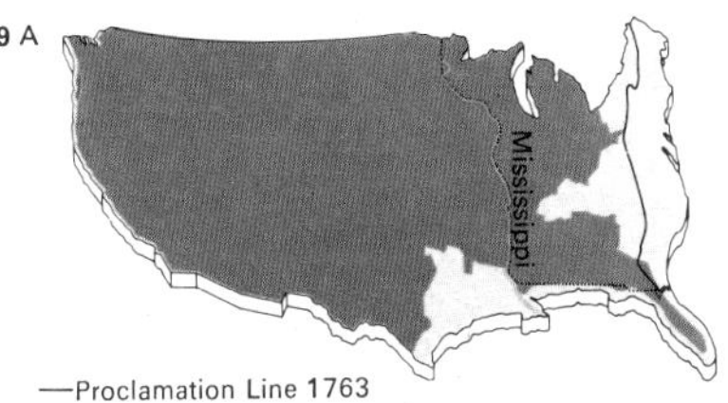

B

9 Indian land cessions were integral to westward expansion. The Proclamation Line of 1763 protecting Indian hunting between the Alleghenies and the Mississippi was soon passed by land speculators [A]. After independence, treaties with the Indians pushed them farther and farther west. By 1890 no Indian titles to land were left and the Indian population had been largely confined to reservations on poor land [B] or goaded to resistance and suppressed in the many Indian wars. A major campaign, in which General Custer was killed in 1876, followed a Sioux uprising led by Chief Sitting Bull [C] who attacked US Cavalry invading his hunting grounds.

C

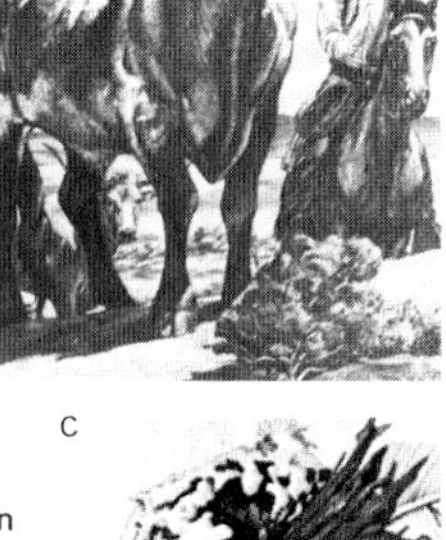

10

10 The California gold rush (1849) led to a frantic search for "pay dirt", which drew prospectors and then settlers to remote regions of the West. Miners alone numbered more than 5,000 by 1850.

11

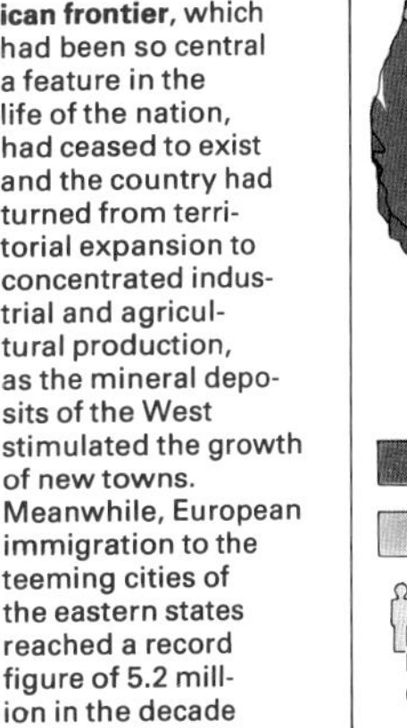

11 The continent was spanned by rail in 1869 when the Central Pacific and Union Pacific railways were linked by a golden spike at Promontory Point, Utah. By 1870 85,000km (52,800 miles) of rail existed.

12 By 1912 the American frontier, which had been so central a feature in the life of the nation, had ceased to exist and the country had turned from territorial expansion to concentrated industrial and agricultural production, as the mineral deposits of the West stimulated the growth of new towns. Meanwhile, European immigration to the teeming cities of the eastern states reached a record figure of 5.2 million in the decade 1880–90. By 1910 the total population was 91,972,266.

12

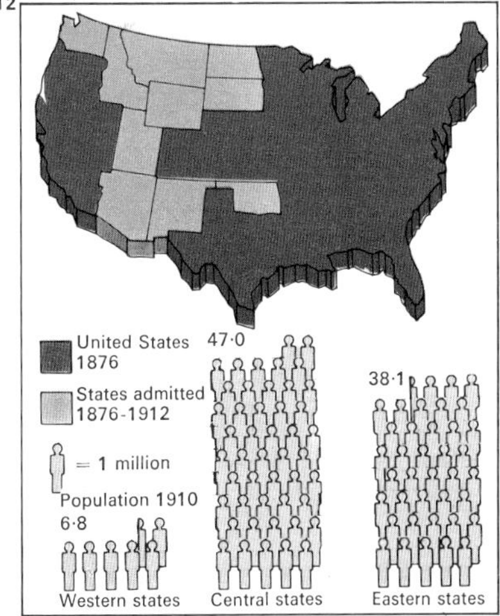

The American Civil War

The Civil War from 1861–65 was the bloodiest and bitterest conflict the United States has ever experienced. It was, President Abraham Lincoln said, a test of whether America could endure. Although the nation emerged from it intact, the "war between brothers" left a legacy of grief and hatred. It remains a vital formative influence on one of the strongest nations in the world.

Regional interests

The Civil War was kindled by a conflict of interests between the northeastern and southeastern sections of the country at a time when most of the West was still being settled. The North was a major manufacturing and commercial region while the South was overwhelmingly agricultural with "King Cotton" providing most of its wealth [2]. The North believed in strong central government to nourish its economic growth; the South insisted on "states' rights" to guard its regional interests. Tariffs, which the North demanded to protect its industries, were opposed by the South because they raised the prices of manufactured goods. Northern industrial expansion was able to accommodate growing numbers of free labourers, despite extremes of poverty and wealth. The South's plantation economy depended on a large workforce of black slaves [1] and it was on the slave question that North–South differences gradually came to focus.

By 1850 slavery had become the most important issue in American politics. The South considered the system proper as well as necessary; many in the North considered it abominable and held it responsible for the South's comparative economic backwardness. Congressional compromises patched up differences and delayed an open break, but the South continued to press for the extension of slavery into western territories. In the North abolitionists, of whom William Lloyd Garrison (1805–79) was the most eloquent, agitated against the "peculiar institution" of human bondage. The influential novel of Harriet Beecher Stowe (1811–96), *Uncle Tom's Cabin* (1852), dramatizing the brutalities of slavery, won support for the anti-slavery movement. The drift towards a violent resolution of sectional differences gathered momentum as hatred was whipped up by inflammatory speeches on both sides.

Both the Democratic and Whig parties, the two major national political organizations, were badly split over slavery and the Whigs proved unable to survive the internal divisions. From the ruins of their party there emerged in 1854 a new Republican Party whose presidential candidate six years later was a former Illinois congressman, Abraham Lincoln [Key]. Lincoln opposed the spread of slavery and foresaw its eventual disappearance as an economic and social system.

A month after Lincoln was elected president South Carolina, fearing an attack on the fabric of Southern society, seceded from the Union and was followed by Mississippi, Florida, Alabama, Georgia, Louisiana and Texas. On 8 February 1861 the secessionist states proclaimed the existence of a new nation, the Confederate States of America.

The war begins

Lincoln refused to recognize the dismemberment of the United States and appealed to the Confederate states to reconsider. Their

CONNECTIONS

See also
64 The American Revolution
148 USA: the opening up of the West
152 USA: reconstruction to World War I
44 International economy 1700–1800

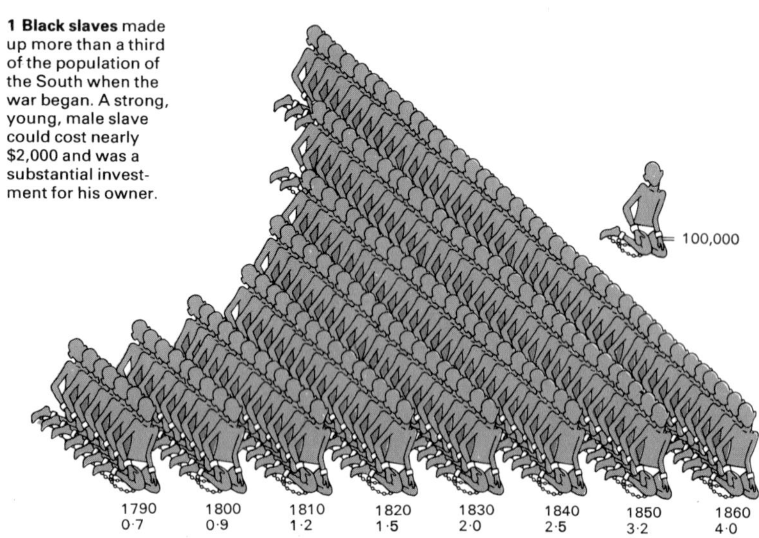

1 Black slaves made up more than a third of the population of the South when the war began. A strong, young, male slave could cost nearly $2,000 and was a substantial investment for his owner.

2

2 A stately mansion with stucco columns and verandas on the ground and first floors was the focal point of many Southern plantations. House slaves acted as servants while field slaves tilled the surrounding soil. Although there were fewer than 10,000 plantation owners who had 50 or more slaves in the 1850s, they wielded overwhelming political and social influence throughout the South.

4

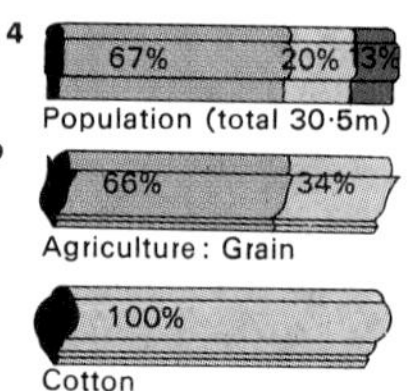

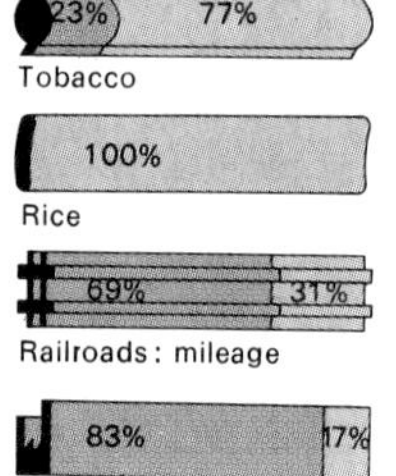

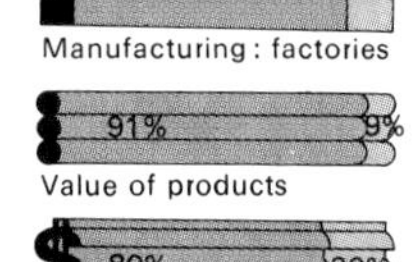

4 Outmatched in industrial capacity, and with a much smaller population, the South counted in vain on a collapse of Northern morale.

3

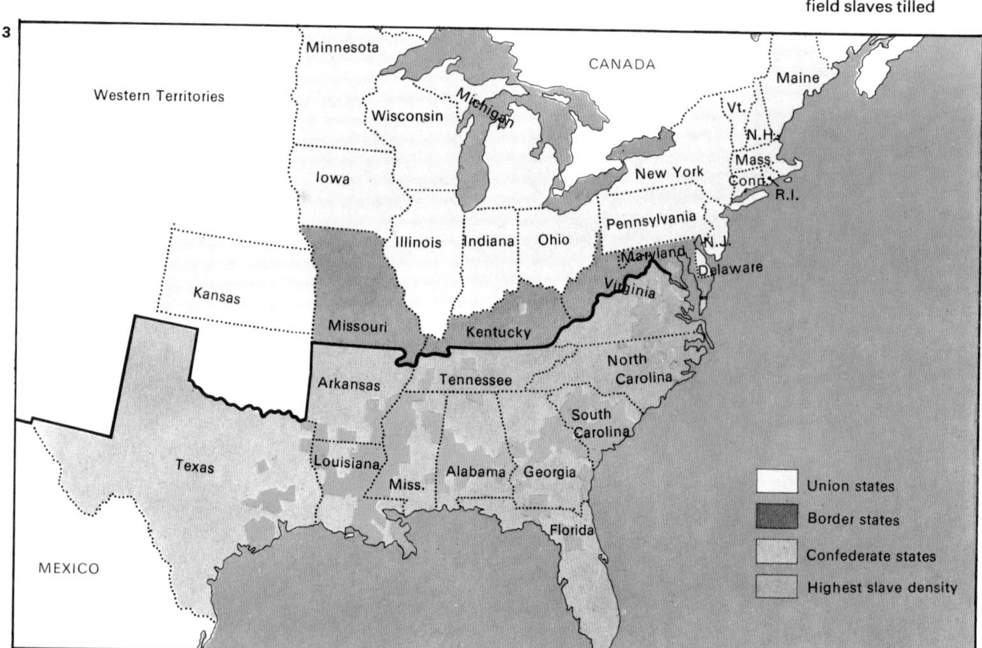

3 Loyalties to South or North crossed state lines and divided families during the Civil War. Three of Abraham Lincoln's brothers-in-law died fighting for the Confederacy. Of the 23 states, including California and Oregon, which were loyal to the Union, the most difficult decision fell to the "border states", the slave states of Kentucky, Maryland and Missouri. Their allegiance to the "Stars and Stripes" proved to be stronger than their purely regional interests. Although Virginia joined the Confederacy, the western part of the state chose the Union instead and gained statehood as West Virginia before the end of the war.

reply came at dawn on 12 April when Southern guns opened fire on Fort Sumter, a federal outpost in Charleston, South Carolina. Virginia, Arkansas, North Carolina and Tennessee soon joined the Confederacy [3]. Both sides mobilized. The Civil War had begun.

The North had distinct advantages because its industrial capacity was far greater [4]. The South's free population was less than a quarter of that of the North. The North controlled the navy and imposed an increasingly effective blockade of the South. The South's only, dubious, advantage, apart from the quality of its fighting men, was that it was defending its home ground, while the North had to launch an assault.

The first major battle quickly showed that there would be no easy Northern victory. Union troops tried to crash through Confederate lines at Bull Run, Virginia, and were driven back in panic to Washington. But Northern superior numbers and equipment soon began to tell. After a major Northern victory at Antietam Lincoln issued an Emancipation Proclamation, effective from 1 January 1863, declaring all slaves in the Confederate states to be free.

Southern attempts to rally to preserve slavery and the Confederacy met with increasingly confident and effective Northern onslaughts [6]. There was no recovery from a devastating Confederate setback at Gettysburg in July 1863 [7, 8]. General William Sherman's (1820–91) "March to the Sea" in Georgia the following year undermined the South's remaining capacity to fight.

Victory and its aftermath

With victory inconceivable and the bulk of his forces cut off, the Confederate commander General Robert E. Lee (1807-70) surrendered to the Union commander General Ulysses S. Grant (1822–85) at Appomattox, Virginia, on 9 April 1865.

The Civil War had cost the lives of 360,000 Union and 260,000 Confederate men as well as thousands of civilians. The South was in ruins. Despite Lincoln's plea for "malice towards none", the seeds of enduring bitterness had been sown.

KEY

Abraham Lincoln (1809–1865), United States president during the Civil War, believed the country could not survive "half slave and half free" but was determined to prevent the break-up of the Union. A self-taught lawyer of humble birth but great shrewdness, sincerity and common sense, he gained national recognition through public debates on slavery and was elected in 1860. Mild-mannered but strong-willed, he led the North with firmness and urged "charity for all" after the South was defeated. He was assassinated by an actor, John Wilkes Booth, in Washington on 14 April 1865, soon after starting his second term of office.

5

5 Jefferson Davis (1808–89), the champion of "states' rights" and the extension of slavery to western territory, was elected president of the Confederacy in 1861 and led the South until its surrender. He suffered from poor health and his relations with other Southern leaders were often strained. Although taken prisoner after the war and indicted for treason, Davis was never tried.

6

A

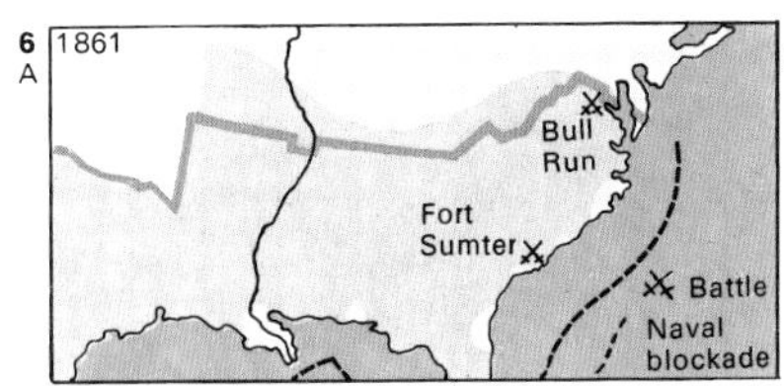

B

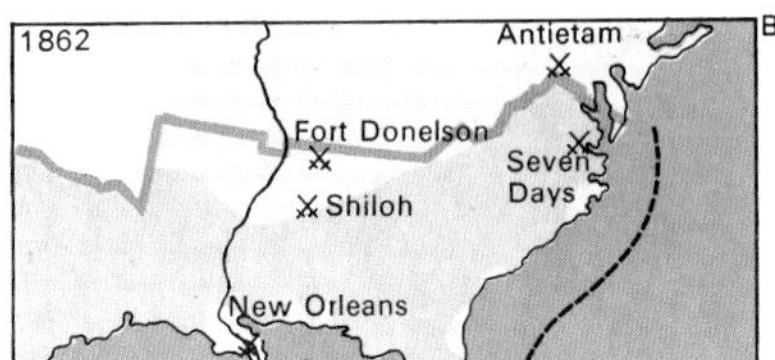

7

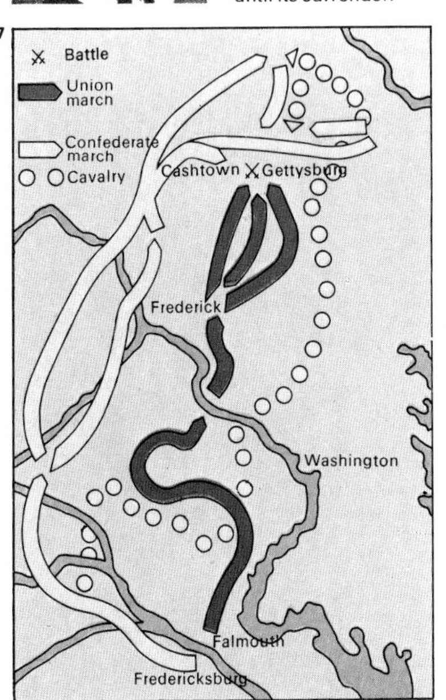

7 Gettysburg marked the turning point of the war in 1863 when a daring Confederate invasion of Pennsylvania was blocked in a ferocious three-day battle. General Lee, the Confederate commander, intended to await a Northern repulse near Cashtown. But a chance encounter between rival patrols precipitated the battle near the small town of Gettysburg on 1 July. Successful probing assaults on Union positions led Confederate officers to misread the situation and cavalry that was engaged elsewhere failed to scout the terrain. Finally three divisions of Confederate troops were sent into a withering barrage of artillery and rifle fire.

C

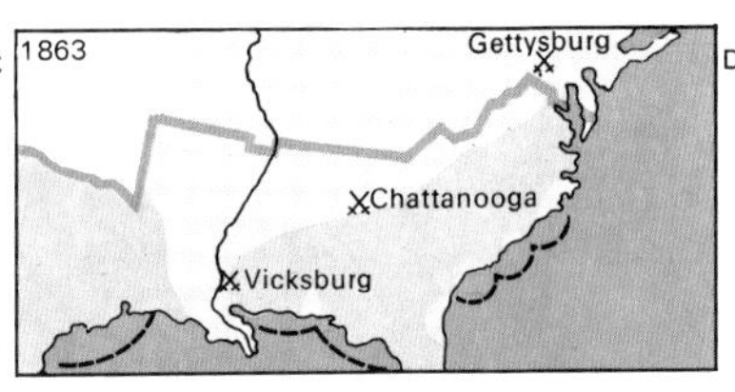

D

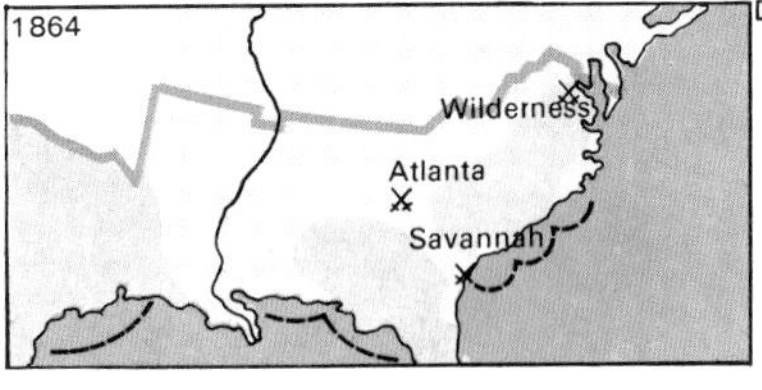

E

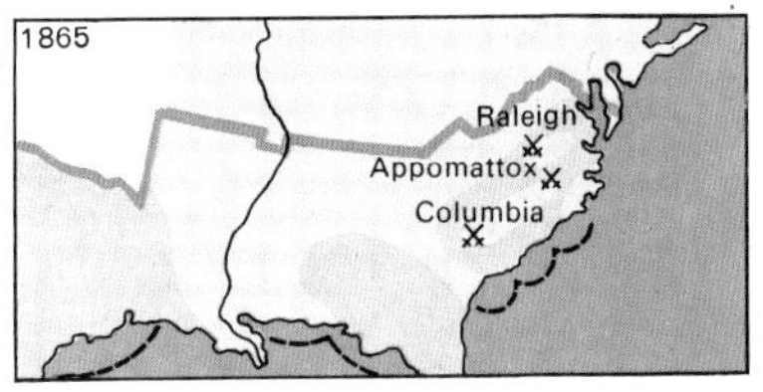

6 Erosion of Confederate territory was steady after an initial stalemate [A] in 1861 when the North realized it must blockade the South. In 1862, after victories westwards, the Union advanced from the north [B]. By May 1863 it controlled the Mississippi [C]. By the end of 1864 Sherman had split the South in two [D]. Surrender became inevitable in 1865 after further Union gains [E].

8

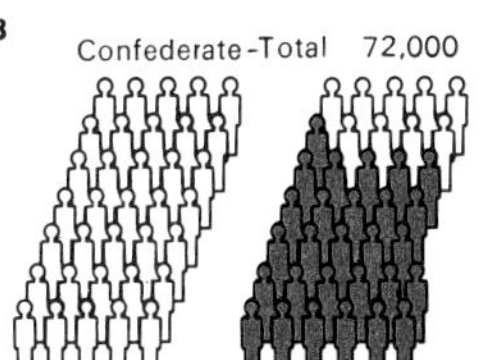

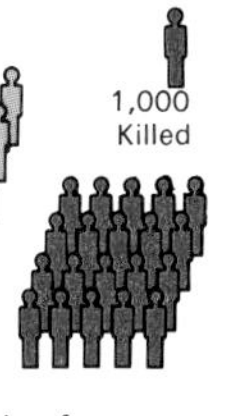

8 Battle statistics at Gettysburg are the subject of controversy, but it is likely that about 72,000 Confederate troops faced nearly 90,000 Union troops. This disparity need not have been decisive in view of earlier Confederate successes. While about 23,000 Union men were killed or died of wounds the South's losses were about 28,000 (higher than the official figure given at the time). The strength of the Union was not undermined significantly. But although its surviving forces escaped back to the South, the Confederacy had suffered a crippling and irrevocable loss. By 1865, with both sides conscripting men, the North had 960,000 under arms and the South only 450,000.

9

9 Union troops with a battery of 32-pounders near Fredericksburg were photographed by Matthew Brady (*c.* 1823–96), one of the first war photographers in history. The Civil War was also the first conflict in which telegraphy and railroad transport were used widely and the first in which ironclad naval vessels went into battle; in March 1862 the USS *Monitor* and the Confederate *Merrimac* fought each other at Hampton Roads, Virginia. An early submarine attack was launched by the *Hunley* which rammed its torpedo into the USS *Housatonic* off Charleston in 1864, with the loss of both ships.

USA: reconstruction to World War II

The United States developed from a predominantly rural nation at the end of its Civil War (1861–5) into the world's largest and wealthiest urban industrial power by the time of its entry into World War I (1917). Among the key factors responsible for this major transformation were a huge population increase; discovery and exploitation of enormous supplies of mineral resources; consolidation of the settlement of the Great Plains and most of America's vast western hinterland; and the sprouting of far-flung railway networks to service industrial, agricultural and population growth.

Problems of the South

America's development during this period was blighted by serious problems. Reconstruction of the defeated and devastated South after the Civil War [2] was retarded by residual North-South hostility. Northern military units policed Southern states to suppress lingering vestiges of rebellion. Carpetbaggers (northerners who migrated southwards for opportunistic or idealistic reasons) sought to govern and control sections of the ravaged South aggravating Southern animosity.

In the rest of the country, however, industrial development was rapid. Rich coal veins were worked along the Appalachian mountain spine and in the Monongahela, Ohio and Allegheny valleys. Vast deposits of iron ore were mined in the Great Lakes region. Copper, lead and other minerals were discovered and hungrily tapped [Key], as was oil.

Industrial growth was further intensified by a host of inventions [8] including commercially viable electric lighting, the telephone and rubber vulcanizing. The mechanization of agriculture through the invention of the reaper, thresher, mechanical harvester and other farm machinery enabled farmers to expand land cultivation. Between 1860 and 1910 farm acreage more than doubled and farm production more than trebled. Cattle kingdoms flourished on a wide stretch of open range from Texas to Montana.

A complex of railway networks reached out across the country linking industry, agriculture and their respective markets. By 1900, 310,000km (193,000 miles) of track criss-crossed the United States – more than in all of Europe at that time. By 1916 the figure was 425,000km (250,000 miles).

The rapid pace of development lent itself to the activities of aggressive entrepreneurs [5]. Men such as Scottish-born Andrew Carnegie (1835–1919), instrumental in consolidating the American steel industry, and John D. Rockefeller (1839–1937) who concentrated on oil, built personal fortunes through huge companies that could overwhelm competition, fix prices and benefit from large-scale marketing and speculation beyond the resources of smaller firms.

Population explosion

The giant companies [6] played a major role in the surge in America's gross national product, which rose from $7,000 million in 1870 to $91,000 million in 1920, despite economic fluctuations. The country's pool of labour, provided by a rapidly growing population, seemed bottomless. The number of Americans grew from 40 million in 1870 to 92 million in 1910. A flood of immigrants [3] from Europe throughout the period contri-

CONNECTIONS

See also
150 The American Civil War
158 Industrialization 1870–1914
148 USA: the opening up of the West
156 The foundations of 20th century science

1 **Members of the Ku-Klux-Klan,** hooded and robed, hold elaborate initiation ceremonies. The society was originally organized by former Confederate soldiers in 1866 at Pulaski, Tennessee, to maintain white supremacy in the Southern states after emancipation of black slaves had been confirmed by the defeat of the South in the Civil War. The Klan attracted many recruits to its ranks but its night-riding vigilante violence against blacks and northerners led to its dissolution in 1869. When it was revived in 1915 its anti-black policies were supplemented by anti-Catholic, anti-Jewish and anti-alien emphasis.

2 **In the unsettled years** after the Civil War, bands of outlaws roamed across the central states. One of the best-known figures was Jesse James (1847–82), here seated front left. He led a gang of bank and train robbers that included his brother Frank [front right] and four brothers of the Younger family – Coleman [rear left], James, Robert [rear right] and John. Jesse and Coleman had been members of Quantrill's Raiders – a band of Confederate mounted guerrillas – and they had no respect for Northern-controlled banks and railways. The James–Younger gang left a blood-soaked trail of robberies across the Mid-west. After John Younger was shot dead in a bank raid his brothers were captured and imprisoned. Badly shaken, the James brothers went into hiding. Three years later they went back to robbing trains. In 1882 Jesse was killed by Robert Ford, a new member of his gang who was tempted by the $10,000 reward.

3 **Health checks were largely superficial** for the more than 20 million immigrants who settled in the USA between the Civil War and World War I. At first they came mainly from Britain, Germany and Scandinavia, and later mostly from southern and eastern Europe, seeking religious or political freedom or escape from poverty. They formed German-American, Scandinavian-American and other intermixed ethnic islands in Great Plains agricultural regions; or they mined the natural resources, chopped down the forests and laid the rail tracks for the burgeoning American economy. By 1920 one out of every eight American citizens was of foreign birth.

railway through the heart of the troubled area. Following the suppression of an Ashanti rebellion by the British in 1900, the pacification of the Gold Coast was sealed by building a railway. Once colonial rule was established, railways reduced administrative expenses in the transport of personnel and stores. A train of the 1890s could do the work of 13,000 porters at five per cent of the cost.

Before the 1860s steamships had a voracious consumption of coal, which limited their range of economic operations to coastal and short sea routes. (The North Atlantic, with a large passenger traffic, was an exception.) Technical improvement came more slowly to the marine engine than the locomotive. In 1840 the 1,139-ton *Britannia* carried only 90 passengers and only 225 tonnes of goods because it needed 640 tonnes of bunker fuel for the Atlantic crossing. But the introduction of the compound marine engine in the 1860s made a 40 per cent saving in fuel consumption. In 1914 the 4,556-ton Cunarder *Bothnia* carried more than three times as much cargo as coal and had room for 340 passengers. With the opening of the Suez Canal in 1869 it was profitable to use compound-engined steamships in the Far East trade.

The use of steel in ship construction in the 1880s and the introduction of the steam turbine in the following decade drove sailing vessels off the sea lanes to Australia and New Zealand. These advances also led to the increase of freight carried in the world's steamships from 27 million tonnes in 1873 to 63 million tonnes in 1898.

Industry and agriculture

Before 1914 the use of steam power for driving textile machinery was still heavily concentrated in Western Europe and the USA, which together accounted for 80 per cent of factory textile production. But its dispersion had produced rapid advances in industrialization in India, Japan, Australia and Egypt. Steam power was used at all stages in the production of iron and steel, but dispersion of steam power outside the older industrial areas was slow. Steam's biggest impact agriculturally was on the processing of agricultural produce as in threshing [Key].

KEY

This steam-driven threshing machine displaced the primitive flail during the 1830s. This was steam's most dramatic contribution to agriculture – the one important incursion of steam power into an industry that remained largely unmechanized until the introduction of the petrol engine.

5

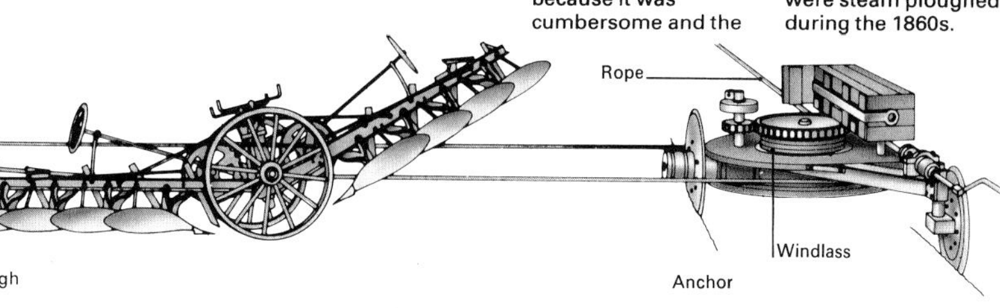

5 The agricultural application of steam power is indicative of nineteenth-century enthusiasm for its uses. Steam ploughing was never extensively used because it was cumbersome and the difficulties of fuelling with coal added to the cost of its operation. It was most popular in the cornlands of eastern England, where over 80,900 hectares were steam ploughed during the 1860s.

6

6 The American Civil War (1861–5) was the first major war in which railways played a decisive role. Here, a train bringing Union reinforcements to General Johnston has run off the track in the forests of Mississippi (1863). In Virginia, some railway tracks were blown up and relaid as many as six times during the fighting. The repair gangs worked in sight of the enemy's artillery.

7

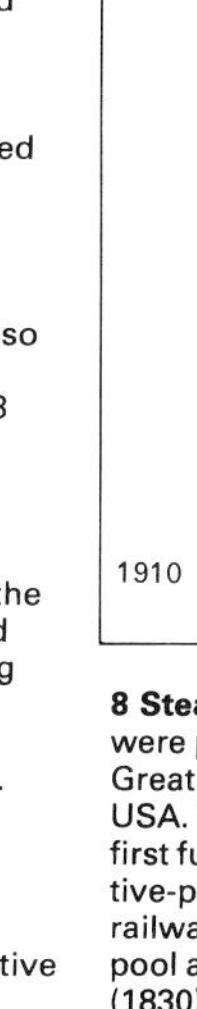

7 The Chilean railway from Valparaiso to Santiago was built between 1853 and 1864 and was the first important South American railway. Its construction through the Andes represented a great engineering feat, and was financed largely by British investment. Such railways brought development to remote areas, encouraged greater administrative centralization in previously disunited countries and focused nationalist aspirations.

8

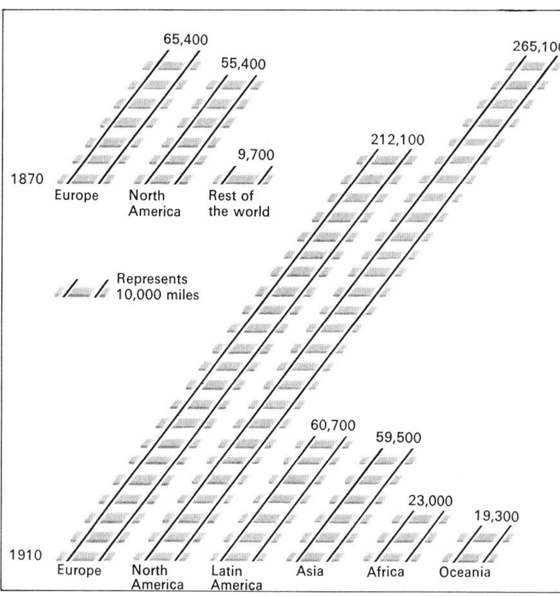

8 Steam railways were pioneered by Great Britain and the USA. The world's first fully locomotive-powered public railway, the Liverpool and Manchester (1830), was quickly followed by the Baltimore and Ohio and other lines. For the next 40 years railway building was mainly concentrated in Europe and North America, where capital and engineering skill were available, linking centres of industry and commerce. From 1832–8 railways were started in France, Belgium, Bavaria, Austria and Canada. After 1870, railways on the American continent were often built to open up new land and to develop its commercial potential. Railways in Japan and India dominated rail construction in Asia.

The foundations of 20th-century science

Many fields of science in the nineteenth century seemed to be marked by orthodox progress – that is, a series of discoveries that could be fitted into the existing view of nature. Yet, in retrospect some of these discoveries were to lead to fundamental changes in the scientific picture. The new fields of thermodynamics and electromagnetism suggested new concepts of energy, and the mathematical work of the German Karl Gauss (1777–1855) and Georg Riemann (1826–66), although purely academic in the 1850s was by the 1920s used as a means of describing the very nature of space.

Biology and medicine

The biological world also seemed straightforward until it was upset by Charles Darwin (1809–82) and Gregor Mendel (1822–84) [4]. When Darwin produced his *Origin of Species* in 1859, placing man among the animals in an evolutionary process that worked by natural selection, a storm of controversy broke that did not completely subside for more than a century. Mendel's work in the 1860s on inheritance factors went unnoticed at the time but in the 1900s was to help lay the foundations of genetics.

Medical science also progressed; Claude Bernard (1813–78) studied the chemical properties of the digestive system and the treatment of infection improved with the new bacteriological ideas of Louis Pasteur (1822–95) [3]. These studies, together with the introduction of antiseptics and anaesthetics, were to lead to advances in surgery and in the understanding of new ways to combat disease. Medical scientists also began to explore the realms of the mind, virtually uncharted until the late nineteenth century, and in the work of Sigmund Freud (1856–1939) [8] the foundations of psychoanalysis were laid and the important concept of the unconscious introduced.

The atomic theory

Chemistry finally broke its remaining ties with alchemy and its mysticism, becoming a true practical and scientific study according to the principles laid down by Antoine Lavoisier (1743–94). Its central advance was the atomic theory. Introduced in its modern form at the beginning of the nineteenth century by John Dalton (1766–1844), the theory propounded the view that all chemical changes were merely rearrangments of individual and indestructible atoms.

It took some time before the theory was accepted, since a considerable amount of independent evidence was deemed necessary, yet in the work of Amadeo Avogadro (1776–1856), Stanislao Cannizzaro (1826–1910) and Jöns Berzelius (1799–1848), the desired correlations and experimental proofs were found and the theory of the atomic nature of matter became established. Other discoveries followed not least in the field of organic chemistry. These advances were due especially to the work of Justus von Liebig (1803–73) on agricultural chemistry and fertilizers, and Friedrich Kekulé (1829–96) who discovered the ring-like structure of the atoms in a molecule of benzene and similar compounds. This has led to the development of plastics and many other petrochemical products, to the development of modern synthetic drugs and explosives and to a better understanding of

CONNECTIONS

See also
32 Exploration and science 1750–1850

In other volumes
302 History and Culture 1
32 The Natural World
30 The Natural World
62 Science and Universe
64 Science and Universe
66 Science and Universe
106 Science and Universe
186 Man and Society
64 Man and Machines

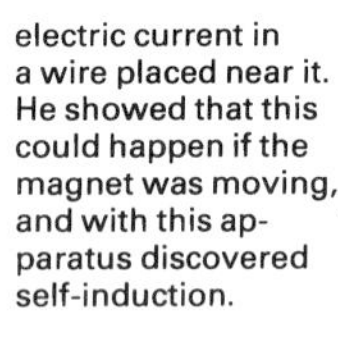

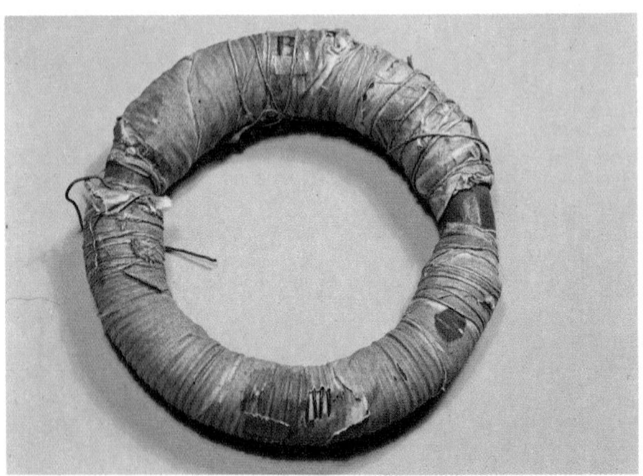

1 Faraday's "ring" was constructed for a central experiment in the work of Michael Faraday (1791–1867) on the nature of electromagnetism, in London in 1831. He knew that electricity flowing along a wire had an associated magnetic field and he argued that a magnet should create (by induction) an electric current in a wire placed near it. He showed that this could happen if the magnet was moving, and with this apparatus discovered self-induction.

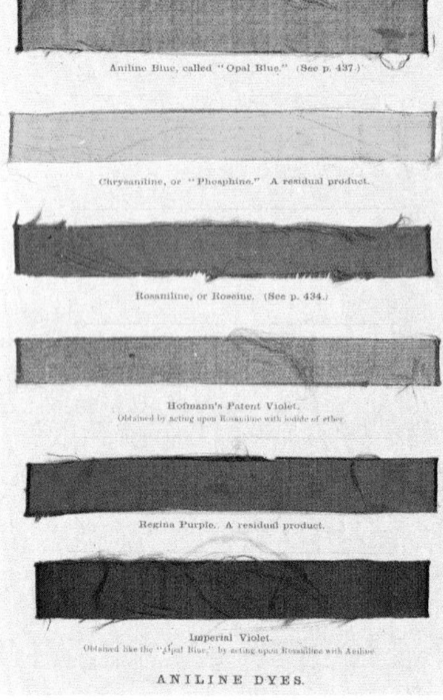

2 In the middle of the 19th century William Henry Perkin (1838–1907), in his laboratory, carried out research in organic chemistry and in particular into quinine and a substance derived from the coal-tar product aniline. By chance this led him to discover a mauve dye. Previously purple colours could be produced only from an expensive natural product. Other aniline dyes followed – some early samples are shown from the *Popular Science Review* (1864).

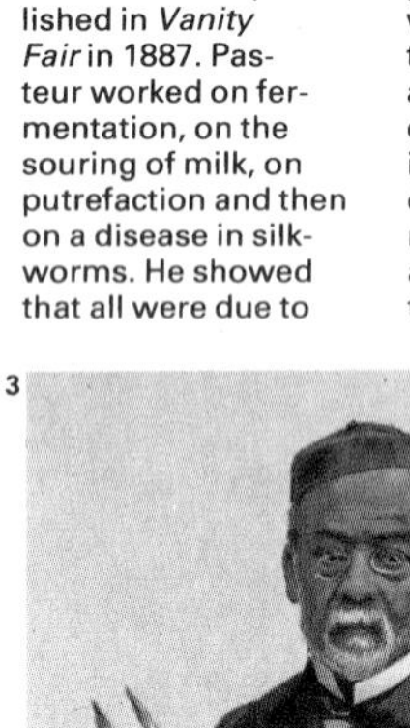

3 This illustration of Louis Pasteur is from a cartoon published in *Vanity Fair* in 1887. Pasteur worked on fermentation, on the souring of milk, on putrefaction and then on a disease in silkworms. He showed that all were due to the presence of micro-organisms and proved that these were airborne. He then studied other animal diseases and devised a method of immunization by inoculating a toxin to raise the host's resistance to more virulent types of the organism.

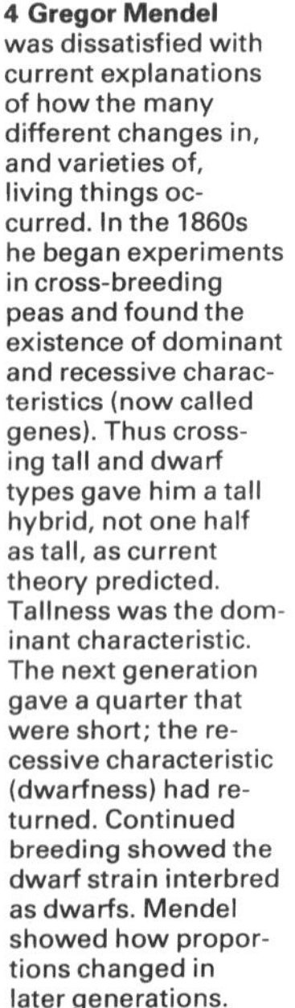

4 Gregor Mendel was dissatisfied with current explanations of how the many different changes in, and varieties of, living things occurred. In the 1860s he began experiments in cross-breeding peas and found the existence of dominant and recessive characteristics (now called genes). Thus crossing tall and dwarf types gave him a tall hybrid, not one half as tall, as current theory predicted. Tallness was the dominant characteristic. The next generation gave a quarter that were short; the recessive characteristic (dwarfness) had returned. Continued breeding showed the dwarf strain interbred as dwarfs. Mendel showed how proportions changed in later generations.

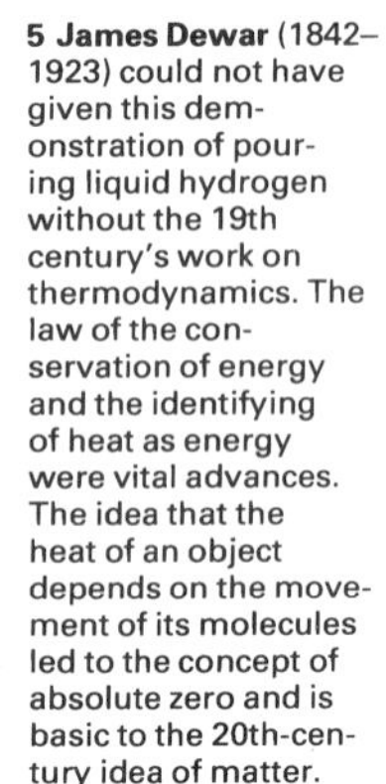

5 James Dewar (1842–1923) could not have given this demonstration of pouring liquid hydrogen without the 19th century's work on thermodynamics. The law of the conservation of energy and the identifying of heat as energy were vital advances. The idea that the heat of an object depends on the movement of its molecules led to the concept of absolute zero and is basic to the 20th-century idea of matter.

the chemistry of food and of all living things.

In physics important advances were made in the understanding of electricity. This began in 1800 with the construction of a cell or battery – the "voltaic pile" – by Alessandro Volta (1745–1827), by means of which a continuous flow of electricity was obtained for the first time. There was also research by others, especially Humphry Davy (1778–1829), on the chemical effects of electricity and Michael Faraday [1] on the connections between electricity and magnetism. His ideas were taken up and carried further by James Clerk Maxwell (1831–79) who studied the mathematical properties of the electromagnetic field, and predicted the existence of types of electromagnetic radiation other than light.

Experiments with light

Maxwell used the theory that light moves in waves, which had been developed by the experiments in interferometry by Thomas Young (1773–1829). Interest in the nature of life was intensified by the invention of spectroscopy, too, by Joseph von Fraunhofer (1787–1826) and later developed by Gustav Kirchhoff (1824–87). This proved to be a delicate means of chemical analysis and it soon became a revolutionary tool of astronomical research. In the nineteenth century it was generally assumed that light travelled through an invisible substance known as the ether, although no proof of its existence was available. An experiment conducted in 1887 by A. Michelson (1852–1931) and E. W. Morley (1838–1923) indicated that this substance did not exist. This experiment left physics in a state of confusion until the publication of the special theory of relativity by Albert Einstein (1879–1955) in 1905 [9].

In other fields, too, new studies at the end of the nineteenth century brought important breakthroughs. Work on the discharge of electricity through gases, in particular, led to astonishing results. William Crookes (1832–1919) discovered cathode rays and J. J. Thomson (1856–1940) [Key] the electron – together with the quantum theory of Max Planck (1858–1947), the cornerstone of atomic and nuclear physics.

KEY

J. J. Thomson was director of the Cavendish Laboratory at Cambridge. With a brilliant research student, Ernest Rutherford, he found that when a rarefied gas was bombarded by X-rays it became able to conduct electricity. From this Thomson was led to consider the nature of cathode rays, and this led him to discover that they were composed of electrons, the first sub-atomic particles, thus proving that atoms were not the smallest units of matter.

6 From 1800–09 Thomas Young revived the wave theory of light, which opposed Newton's particle theory. In the 17th century the Dutch physicist Huygens suggested light was due to waves pushing outwards (longitudinal waves) from a source; Young however used up-and-down (transverse) waves [1], illustrated in his *Course of Lectures on Natural Philosphy and the Mechanical Arts* (1807). Interference patterns [2] were explained by the wave theory of light.

6

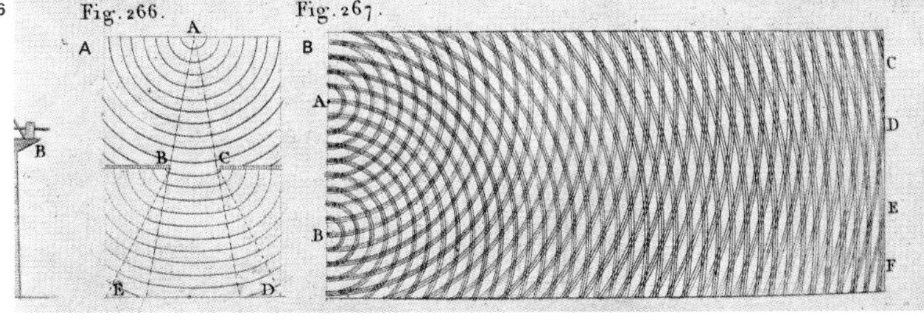

7

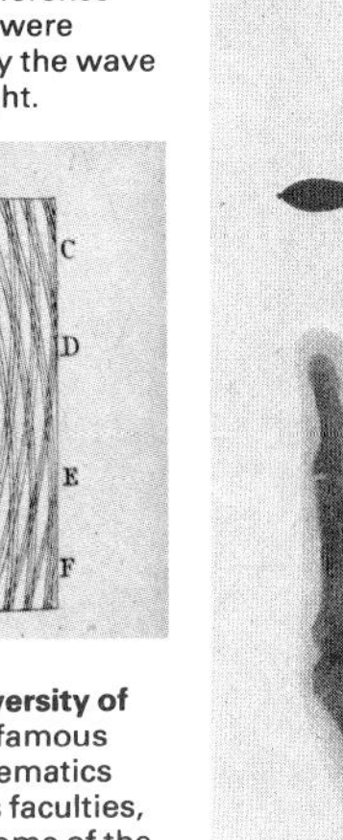

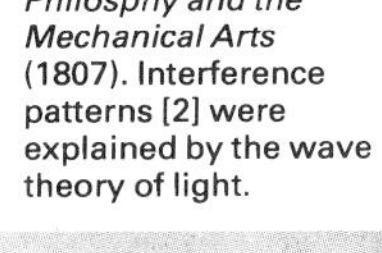

7 In 1896 it was discovered that uranium constantly emitted rays more penetrating than X-rays; such rays also made gases conduct electricity. This field was next studied by Marie and Pierre Curie. They found that other heavy substances emitted such rays (gamma rays). They analysed pitchblende, a uranium compound, and found it contained polonium and radium, the latter being a strong emitter. They realized the emission must be caused by some behaviour of the atoms in the materials; they called this behaviour "radioactivity". The illustration shows Marie Curie's hand photographed using a gamma-ray source.

8

8 Sigmund Freud, a Viennese, developed a theory of the "psyche", at first through the use of hypnosis in the treatment of hysteria and later through the technique of free association. Both probed the unconscious and its power to affect conduct. He also stressed the sexual motivation behind much human behaviour. Freud's ideas, which included the analysis of dreams, were taken up and modified by others, particularly Carl Jung.

10

10 The University of Göttingen, famous for its mathematics and physics faculties, produced some of the leaders in the massive expansion of physical science studies in late 19th-century Germany.

9

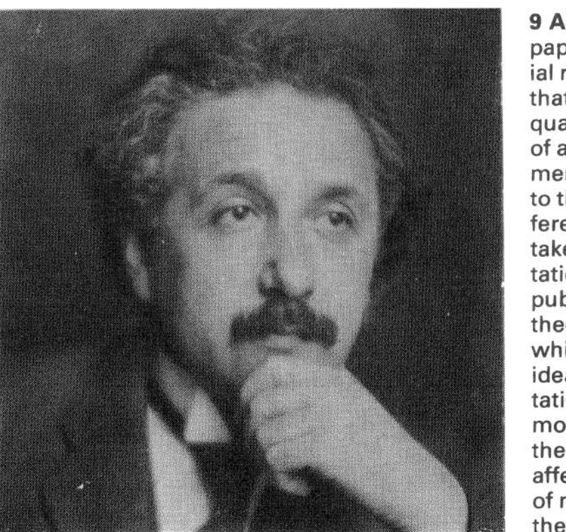

9 Albert Einstein's paper in 1905 on special relativity showed that not all physical quantities are capable of absolute measurement, but are relative to the same frame of reference. This did not take account of gravitation. But in 1915 he published his general theory of relativity which extended the idea to include gravitation and accelerated motion. Einstein's theory profoundly affected the outlook of modern science on the natural world.

Industrialization 1870–1914

The most striking feature of the latter half of the nineteenth century was the growth and spread of industrialization through Europe and into other parts of the world such as Japan and the United States. The rise of industrial economies in Western Europe had profound social and political consequences. With the rapid growth of cities and towns came the development of a more complex political society in which new groups of people – the middle and working classes in particular – began to group themselves and exert greater political influence than before.

Spread of industrialization

In 1850 the only country that could be described as having an industrial economy was Britain [3]. But industrial development spread to Belgium, France and Germany by 1870, and in the last decades of the century was becoming established in countries such as Sweden and Russia.

Belgium industrialized rapidly and by 1870 had one of the leading economies in Europe. French commerce, iron production and textile output were flourishing by the latter part of the century and between 1870 and 1890 French technical innovation played an important part in the development of many engineering products.

By 1900 the most important industrial economy to emerge on the continent of Europe was that of Germany. Her unification by 1871 was accompanied by an accumulation of capital and development of the transport network. From 1850 to 1880 Germany increased coal production tenfold and, with the acquisition of the iron ore fields of Alsace-Lorraine from France in 1871, output of iron and steel rapidly expanded [1]. Other countries, such as Sweden, Russia, Switzerland and Austria, began to share in these developments by 1900.

Technology and trade

European industrialization rested upon the application of a technology pioneered in Great Britain, but made use of more advanced techniques. The Bessemer process for making steel, invented in 1856, enabled the cheap production of a material that was stronger than iron. Steelmaking from the phosphoric ores common in Europe was made possible by the Thomas-Gilchrist process after 1878. Cheap steel could be used for machinery, shipbuilding and many other items of general use and provided the basis for the rapid expansion of engineering industries throughout Europe. By 1900, with scientific inquiry into chemical and electrical phenomena, other new industries appeared. The first electrical apparatus and industrial chemicals began to emerge, especially in Germany. Development of the internal combustion engine was well under way by the turn of the century and refinements in mechanical engineering provided the impetus for a flood of labour-saving products ranging from sewing-machines and vacuum cleaners to typewriters.

Trade expanded rapidly during the late nineteenth century, facilitated by the increasing use of iron and steel steamships. Imperialism stimulated the search for new markets and raw materials but the bulk of trade occurred between European and American markets. Cheap foodstuffs from North America after 1870 played an impor-

CONNECTIONS

See also
88 The Industrial Revolution
66 The early industrial Revolution
180 Europe 1870–1914
164 Scotland in the 19th century
152 USA: reconstruction to World War I
154 The impact of steam
186 Causes of World War I
130 Imperialsim in the 19th century
168 Russia in the 19th century
178 Origins of modern architecture

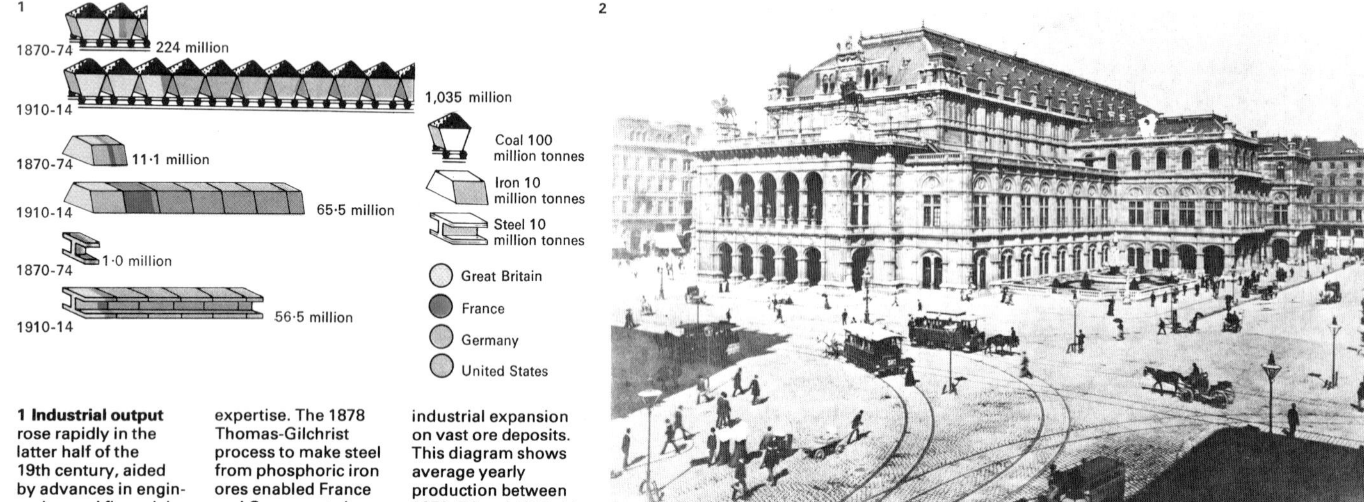

1 Industrial output rose rapidly in the latter half of the 19th century, aided by advances in engineering and financial expertise. The 1878 Thomas-Gilchrist process to make steel from phosphoric iron ores enabled France and Germany to base industrial expansion on vast ore deposits. This diagram shows average yearly production between 1870–74 and 1910–14.

2 Opera houses such as that of Vienna were part of an impressive urban culture created by the growing wealth of many European cities, which built concert halls, art galleries and museums together with municipal buildings and better systems of sanitation, lighting and street paving. Improved housing for better-off workers and the middle classes led to the first suburbs and mass transport by tram and railway.

3

	1850 Country	1850 Town	1870 Country	1870 Town	1890 Country	1890 Town
Britain	40%	60%	30%	70%	28%	72%
France	75%	25%	69%	31%	63%	37%
Germany	73%	27%	61%	39%	53%	47%

3 A population shift from the country to the cities proceeded rapidly as industrialization spread. In 1850, Britain apart, nearly three-quarters of Europe's people still lived on the land. But as Germany industrialized, its population ratio began to alter in the direction of Britain's – a trend followed by France towards the end of the century.

4 London celebrated the relief of Mafeking in 1900 during the Boer War with an outburst of national pride fuelled by widespread reporting of the war in the popular press. The rise of mass newspapers helped to create a powerful and excitable public opinion in the last decades of the century, when imperial adventures and colonial rivalry gave birth to "jingoism" expressed in bellicose literature, spirited demonstrations and songs.

4

5 The bicycle was the first "luxury" consumer product to gain a mass market. Heavily promoted by colourful advertisements such as this from the Michelin Building, London, it was sold in such numbers that manufacturers realized a huge new market had been suddenly created. Other mass-produced goods developed through advances in engineering and metallurgy, including sewing-machines, gramophones, typewriters and motor cars.

5

tant part in reducing European food prices while at the same time depressing local agriculture in what was called the "Great Depression". Established industries, too, were exposed to fluctuations with the rise of competing industrial economies. To protect their newly established industries France and Germany imposed tariffs. A more complex economic structure emerged with large trusts and cartels grouping related industries into large combines; joint stock companies supplanted many family firms and banking and investment institutions became more sophisticated. By 1914 London was the financial centre of the world, with large stakes in shipping, insurance and investment.

Far-reaching social changes

Industrial development and the continued rise of Europe's population associated with European urbanization [7] brought fundamental social changes. There was a great increase in middle-class wealth, often derived from investment in stocks and shares [8]. But even the poorest classes benefited from rising real wages. Living conditions in the growing towns and cities of Europe were often harsh and difficult, but were improving at record rates. Social welfare measures began to be adopted by some states, as in Bismarck's Germany, and philanthropy in countries such as Britain provided some relief for the most deprived. Emigration was widespread from the poorest countries, especially Ireland, Russia and the Austro-Hungarian Empire. Most migrants went to North America, although some went to British colonies, in particular to Australia.

An advance in living standards by 1900 was reflected in the emergence of the first aspects of mass consumer society [Key]. The rise of cheap newspapers, widespread advertising and selling of consumer goods such as bicycles [5], and the growth of mass entertainment in sport, music hall and holiday excursions, showed that the working classes were beginning to enjoy some of the fruits of industrialization. This was certainly the case with the large middle-class families [6] who gave the latter part of the nineteenth century a somewhat staid character that belied the changes at work in society.

KEY

Growing wealth for all sections of society in the late 19th century led to the first mass consumer market with the development of advertising and the growth of "chain" stores, among them Marks & Spencer, who opened a "penny bazaar" in Stretford Street, Manchester, in the 1890s.

6

6 Victorian families of all classes tended to be large because of a high birth-rate and declining mortality due to improved medical care, diet and general living standards. Among poorer sections of the community infant deaths from infectious diseases continued to be high and there were usually more pregnancies than surviving children. But upper-class family life was based on large units with many servants as well as children. Two or three servants were a bare minimum for a solid middle-class family. In the aristocratic households of the great country estates it was not unusual to find over 100 house-servants, kitchen staff and gardeners. The rambling Victorian house was often a viable living unit only when it could be maintained by numerous staff.

7

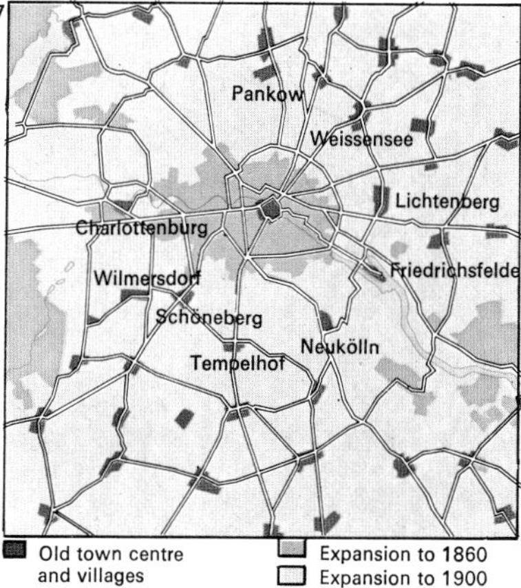

7 The growth of Berlin was typical of many nineteenth-century cities. Up to 1860 its expansion was mainly around the old city centre, but with the growth of the German state, and the development of Berlin as a capital city and industrial centre, it grew into a major European metropolis. As with many other cities, Berlin's rising population spread out to create surrounding suburbs, incorporating villages that had once been separate.

8

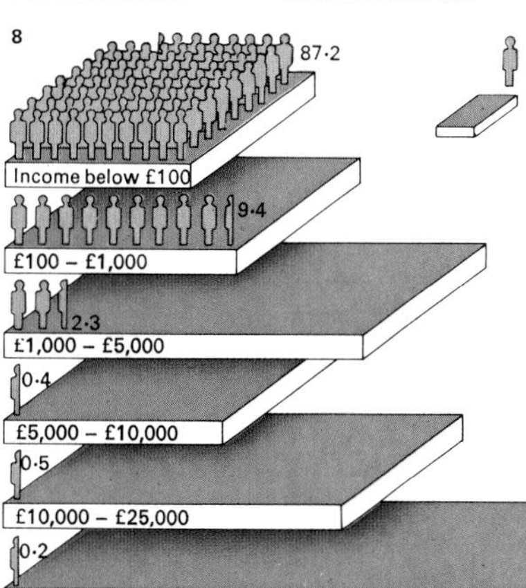

8 Much of the wealth created by the Industrial Revolution was concentrated in the hands of the upper classes. In Britain in 1911, as shown here, a tiny group of wealthy industrialists and aristocrats still disposed of a large share of the national income, although a growing proportion was held by the lesser industrialists and professional men.

The fight for the vote

The early nineteenth-century parliamentary system in Britain contained many anomalies. The right to vote was governed by a complex system of traditional rights and privileges that had hardly changed since the mid-seventeenth century. Many boroughs elected their MPs on a tiny franchise; some had become so reduced that they were known as "rotten" boroughs and election to the seat lay almost entirely within the power of the local landowner. Moreover, the dramatic growth and redistribution in population during the Industrial Revolution created an anomalous situation where large, thriving towns had no representation whatsoever in Parliament.

Twin aspects of reform

Parliamentary reform, therefore, had two major aspects; the progressive extension of the franchise, to encompass all men, and later women; and the redistribution of seats to rectify the anomalies of the "unreformed" House of Commons. In addition, the conduct of elections, the use of bribery, and the decisive power of individual patrons in the many "pocket" boroughs all formed part of the long-standing unreformed system.

Movements for reform began in the second half of the 1700s, when the radical demagogue John Wilkes (1727–97) whipped up much popular support in London in the 1760s and 1770s. Fear of disorder, following the French Revolution, and the vested interests of many existing MPs, held back reform for another generation. But reform and "radical" ideas were kept alive by men such as Henry Hunt (1773–1835), William Cobbett (1763–1835), John Cartwright (1740–1824), and Francis Place (1771–1854).

The growth of the manufacturing towns during the Napoleonic Wars created a demand for representation, seen in the formation of political unions in towns such as Birmingham and Manchester. Discontent with the Tory administrations brought the Whigs to power in 1830.

A bill was introduced in 1831 but was rejected by the House of Lords. This caused widespread unrest, including riots at Derby, Nottingham and Bristol. Under threat of the creation of new peers, the Reform Bill was passed in 1832. The First Reform Act replaced the existing confusion of voting qualifications with a more regular system. But the electorate rose to only 652,000 and power remained vested in the hands of the upper and middle classes. More significant was the redistribution of 143 seats from the worst of the insignificant rotten boroughs to the larger manufacturing towns, London, and the counties [1, 2].

The 1832 Reform Act was in many ways conservative. Even many Whigs regarded it only as a measure to cure the anomalies of the existing electoral system. Attempts by the Chartists to coerce Parliament into a further programme of radical reform was resisted by the propertied classes. Three mass petitions in 1839, 1842 and 1848 [3] in support of the Charter were ignored.

The vote for the working man

Growing prosperity brought more people within the 1832 franchise qualifications by the 1860s. With the increasing inevitability of a further measure of reform, the Conserva-

CONNECTIONS

See also
68 Pitt, Fox and the call for reform
94 The British labour movement to 1868
96 Social reform 1800–1914
112 Queen Victoria and her statesmen
170 Political thought in the 19th century
172 Masters of sociology
210 The British labour movement 1868–1930

In other volumes
270 Man and Society
278 Man and Society

1 The post-1832 "reformed" Parliament had members from the previously unrepresented manufacturing towns at the expense of the small "rotten" boroughs and some "pocket" boroughs.

Distribution of seats in the House of Commons Before 1832: 94, 100, 45, 419

After 1832: 105, 159, 53, 341

England and Wales counties
England and Wales boroughs
Scotland
Ireland

Industrial areas
1 member lost
1 member gained

2 The First Reform Bill was essentially a conservative measure. It rectified the anomalies created by the population changes in the previous hundred years and enfranchised the upper middle classes.

3 The Chartists, here shown at their last great meeting in 1848, demanded sweeping electoral reforms; but the movement died because of dissension and poor leadership.

4 Disraeli leads "the race for electoral reform" in this Punch cartoon. The Second Reform Bill was passed in 1867 by the Conservatives under Derby and Disraeli.

tive leaders, Lord Derby (1799–1869) and Benjamin Disraeli (1804–81), and the Liberal leader, William Gladstone (1809–98), juggled with the new proposals to win advantage for their parties. It was Disraeli who finally managed to keep his party together and who is credited with the Second Reform Act in 1867 [4]. This act extended the vote to about one million urban working men, a further redistribution of seats.

The Ballot Act of 1872 introduced secret ballot, and in 1883 the worst aspects of electoral corruption were made illegal. In 1884 the Third Reform Act was passed by the Liberals, which enfranchised agricultural labourers and increased the electorate from about three million to about five million. In the following year, another redistribution of seats removed the last proprietary boroughs. Finally, in 1918 all men over the age of 21 received the vote [8].

The suffragette movement

Women had been excluded from the vote in all the reform acts up to 1918. They still had very insecure property rights and were widely regarded as unfit to exercise the responsibilities of political power [6]. The Women's Social and Political Union was founded in Manchester in 1903 to fight for the vote, headed by Mrs Emmeline Pankhurst (1858–1928) [7]. Known as "suffragettes", they gradually gave up normal methods of demonstrations and propaganda and turned to violence, breaking windows, setting building on fire, chaining themselves to railings, and resisting arrest.

With the defeat of a moderate proposal for female suffrage in 1912, the campaign for women's rights was temporarily frustrated. World War I, however, advanced the status of women. They played an immense part in the war effort, working in munitions factories and previously male-dominated jobs. In 1918 women over 30 were given the vote; and this franchise was extended to all women over 21 in 1928.

Plural voting, through property or businesses in more than one constituency, was abolished in 1906; it finally disappeared in 1948 with the removal of university seats at Oxford and Cambridge.

KEY

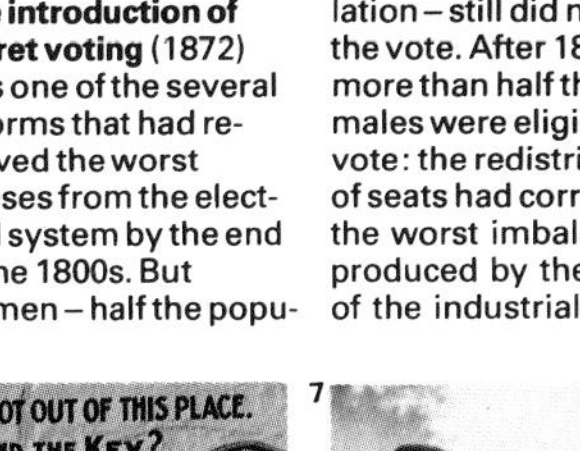

The introduction of secret voting (1872) was one of the several reforms that had removed the worst abuses from the electoral system by the end of the 1800s. But women – half the population – still did not have the vote. After 1884, more than half the adult males were eligible to vote: the redistribution of seats had corrected the worst imbalances produced by the growth of the industrial towns that occurred during the Industrial Revolution. One result of these and other reforms, such as the abolition of a property qualification for MPs (1858), was the rise of the Labour Party.

5 Ramsay Macdonald (1866–1937) [centre] formed the first Labour government in January 1924. The Labour Party achieved an electoral breakthrough in the 1906 general election, when they formed a pact with the Liberals. Labour's 30 seats at that election were a recognition of the growing power of a party that represented the interests of the newly enfranchised working classes.

5

6

7

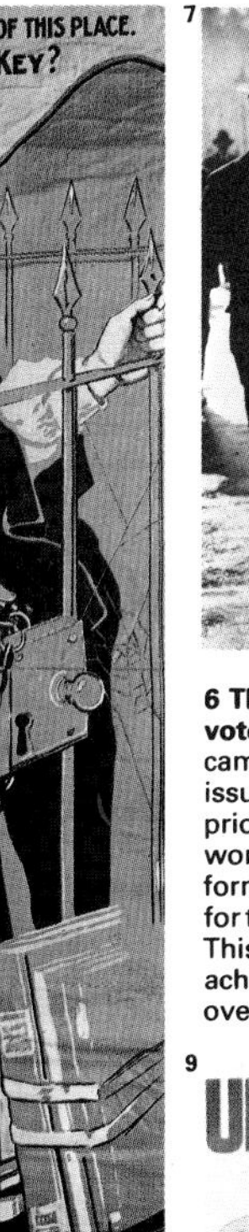

6 The question of votes for women became a prominent issue in the ten years prior to 1914 when women's groups were formed to campaign for the "suffrage". This was not fully achieved, for women over 21, until 1928.

7 Mrs Emmeline Pankhurst, leader of the suffragettes, is carried away by the police during a demonstration. After 1905, the suffragettes pursued a militant policy, which led to a number of arrests and imprisonments.

9

8

Voters as % population 20+ yrs
Voters as % population 17+ yrs

Pre-1832	1832	1867	1884	1918	1928	1948	1970
5	7	16	28	74	96.9	96.7	96.5

8 The electorate only gradually increased with the passing of the Reform Acts of 1832, 1867 and 1884. Growing economic prosperity brought many within the franchise qualifications without the need of legislation. In 1918, men over 21 and all women over 30 years old were granted the vote. In 1928 all women over 21 were given the vote. In 1948 the last remnant of plural voting was abolished and, as a result, the number of the electorate fell to some extent.

9 A contemporary election poster graphically portrays the conflict in 1909 between the Conservative-dominated House of Lords and the Liberal government; it reached a climax when the Lords rejected the government's budget. Two elections were forced, and on each occasion the Liberals were returned. In 1911 the primacy of the elected assembly was established when, under threat of the creation of more peers, a bill was passed restricting the powers of the Lords.

Ireland from Union to Partition

The legislatures of Dublin and London were combined on 1 January 1801 for reasons of state – British reasons, although the Union also suited those Protestants of the Irish Ascendancy who feared the rising forces of Catholicism and democracy. Other Irish Protestants opposed the measure, distrusting Westminster's will to preserve Protestant privileges, while Catholic leaders tended to favour Union, accompanied as it was to be by legislation to grant Catholics the right to sit in the Union Parliament.

Consequences of the Union

In the event, Protestant fears of the Union turned out to be as unfounded as Catholic hopes. Protestants continued to represent Irish constituencies in Parliament, the Anglican Church remained established and the separate Irish administration continued to favour Protestant interests.

To Catholics, the Union provided scant blessing. Their right to sit in Parliament was not conceded, the prime minister, the younger William Pitt (1759–1806) preferring to resign rather than jeopardize the war effort against France by provoking a constitutional crisis over King George III's (reigned 1760–1820) opposition to Catholic emancipation. Emancipation became, therefore, a principal issue of the Union Parliament: its denial completely disenchanted Catholics with the Union [1].

The land problem and Home Rule

The Great Famine of 1845–49 [2] stressed the enduring problem of nineteenth-century Ireland – the imbalance of its land and people. The Irish population had grown alarmingly from five-and-a-half million in 1800 to more than eight million by 1845. Crowded together in smallholdings subdivided into uneconomic units, increasingly dependent upon a potato diet, the Catholic labourers and tenant farmers presented a desperate spectacle. Without industrial alternatives, the peasantry had to remain on the land, exposed to periodic crop failures.

At Westminster, tenant and Catholic spokesmen tried to co-ordinate Irish MPs to deal with Irish issues, but, in practice, allegiance to the Liberal and Tory parties prevailed. But after the false start of Isaac Butt's (1813–79) Home Rule League (1873), Charles Stewart Parnell (1846–91) [5] welded together a disciplined Irish Party in pursuit of Home Rule.

As a result of the long-felt grievance over ownership, unsatisfactory tenancy arrangements, misguided legislation, a further series of bad harvests from 1877, and the organization of the Irish National Land League, rural discontent was brought to a new focus between 1879–82 [4]. Parnell yoked this to his parliamentary demands, while the shadowy Irish Republican Brotherhood (the Fenian movement) begun by James Stephens (1825–1901) in 1858 and now given direction by John Devoy (1842–1928) from America, lent clandestine support. Coercion proved an insufficient government response but the Liberal leader, William Gladstone (1809–98), accepting the logic of Parnell's position, attempted in vain to devolve a Home Rule parliament to Dublin [6].

Meanwhile, the Home Rule Party, split in 1890 and discredited by internal feuds, was being outflanked by other movements

CONNECTIONS

See also
24 The English in Ireland
68 Pitt, Fox and the call for reform
112 Queen Victoria and her statesmen
164 Scotland in the 19th century
282 Irish culture since 1850
294 Ireland since partition

1 **Daniel O'Connell,** (1775–1847), the first politician in the British Isles to mobilize mass support behind his cause, won Catholic emancipation in 1829. As MP for Clare, he alternately bargained for reforms amd attacked the Union itself. But support for his Repeal Association declined after 1843 when he refused to risk bloodshed in opposition to Westminster and the Union.

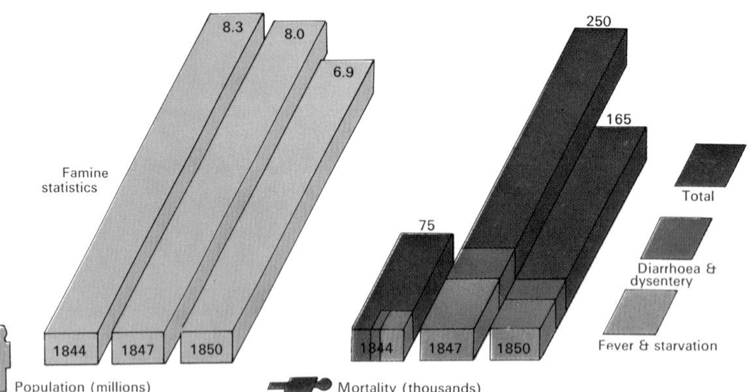

2 **The Great Famine of 1845–9** was a disaster on an unprecedented scale in Irish history. Total figures for deaths and disease disguise the famine's uneven impact, most severe in the West and least damaging in the North East. In 1845–6 government action relieved starvation, but renewed crop failure overwhelmed the shadowy administrative structure. The ensuing horror generated intense hatred against Britain.

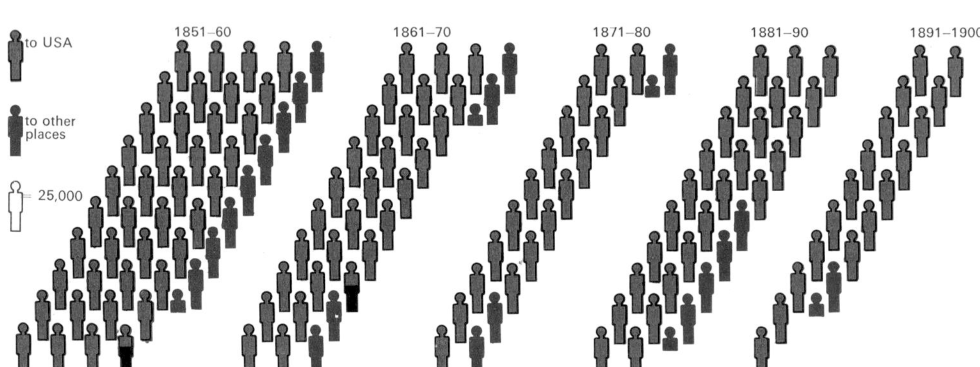

3 **Inhibitions against emigration** were broken by the Famine and a steady flow of emigrants began to leave Ireland. By 1911, when the population stabilized at nearly four-and-a-half million, more Irish lived in North America, Britain and the Empire than in Ireland. Their departure made possible a better standard of living in Ireland, and added an international dimension to Irish nationalism.

4 **Eviction of tenants was common** in the 1870s and 1880s when conflicts between tenant farmers and landlords were at their sharpest. Landlords did not consolidate sufficiently, being effectively restrained by popular opposition, but with tenants of tiny holdings unable to live, let alone pay rent, amalgamation into viable farms was the only economic solution. The Land League seized on the evictions to focus mass resentment against the landlord system.

5 **Charles Stewart Parnell,** MP for Meath from 1875, led 59 Irish MPs at Westminster by 1880, soon moulding them into a disciplined, salaried party (86 strong at its height), pledged to support Home Rule. Backed by constituency branches, mass Land League support and secret Fenian co-operation, he made Home Rule credible, and in 1886 won the Liberal Party over to this cause. Parnell lost Catholic support after he was cited in a divorce case in 1890.

working to "de-Anglicize" Ireland and to win complete independence. In 1906 Arthur Griffith (1872–1922) succeeded in mobilizing disparate political groupings into his own movement, Sinn Fein, dedicated to economic self-sufficiency and political withdrawal from the Union.

Yet, even while Sinn Fein gathered strength, the Home Rule Party, shamed into unity in 1900 under the leadership of John Redmond (1856–1918), received renewed authority from political circumstances in Britain. The return of the Liberals there in 1906 made Home Rule again a possibility.

Ireland divided

From 1912 onwards, tension grew first with the Protestant Unionists arming [7], then Home Rulers – the one to prevent, the other to enforce a bill expected to become law in 1914. Only the outbreak of World War I subsumed this minor quarrel within a mightier conflict. The operation of Home Rule was postponed until the end of the war.

Before that, however, republicans, socialists and other separatists had risen in 1916 [8] to proclaim an independent Irish Republic. They were quickly crushed and their leaders executed, but these groups re-formed in 1917 to merge under the Sinn Fein banner.

The Home Rule Party, compromised by its attachment to the British war effort and the indecisive leadership of the dying Redmond, could not be saved from humiliation in the post-war elections. But Sinn Fein, while winning 73 seats to the Party's 6, could not prevent the Unionists from winning 26 in the North East.

Prime minister Lloyd George (1863–1945) belatedly turned again to Ireland in 1919. In 1920 he created two Home Rule parliaments: one in Dublin for 26 of Ireland's 32 counties; the other in Belfast for the remainder in the North East. Reluctantly, Northern Unionists accepted this compromise, although they had been committed to preserving the 9-county Province of Ulster from Dublin rule. Contemptuously, Southern Nationalists, by now sworn to win a 32-county Irish Republic, refused either to accept the limited powers offered or the partition of the island involved.

KEY

POBLACHT NA H EIREANN.

THE PROVISIONAL GOVERNMENT OF THE IRISH REPUBLIC TO THE PEOPLE OF IRELAND.

IRISHMEN AND IRISHWOMEN: In the name of God and of the dead generations from which she receives her old tradition of nationhood, Ireland, through us, summons her children to her flag and strikes for her freedom.

Having organised and trained her manhood through her secret revolutionary organisation, the Irish Republican Brotherhood, and through her open military organisations, the Irish Volunteers and the Irish Citizen Army, having patiently perfected her discipline, having resolutely waited for the right moment to reveal itself, she now seizes that moment, and, supported by her exiled children in America and by gallant allies in Europe, but relying in the first on her own strength, she strikes in full confidence of victory.

We declare the right of the people of Ireland to the ownership of Ireland, and to the unfettered control of Irish destinies, to be sovereign and indefeasible. The long usurpation of that right by a foreign people and government has not extinguished the right, nor can it ever be extinguished except by the destruction of the Irish people. In every generation the Irish people have asserted their right to national freedom and sovereignty: six times during the past three hundred years they have asserted it in arms. Standing on that fundamental right and again asserting it in arms in the face of the world, we hereby proclaim the Irish Republic as a Sovereign Independent State, and we pledge our lives and the lives of our comrades-in-arms to the cause of its freedom, of its welfare, and of its exaltation among the nations.

The Irish Republic is entitled to, and hereby claims, the allegiance of every Irishman and Irishwoman. The Republic guarantees religious and civil liberty, equal rights and equal opportunities to all its citizens, and declares its resolve to pursue the happiness and prosperity of the whole nation and of all its parts, cherishing all the children of the nation equally, and oblivious of the differences carefully fostered by an alien government, which have divided a minority from the majority in the past.

Until our arms have brought the opportune moment for the establishment of a permanent National Government, representative of the whole people of Ireland and elected by the suffrages of all her men and women, the Provisional Government, hereby constituted, will administer the civil and military affairs of the Republic in trust for the people.

We place the cause of the Irish Republic under the protection of the Most High God, Whose blessing we invoke upon our arms, and we pray that no one who serves that cause will dishonour it by cowardice, inhumanity, or rapine. In this supreme hour the Irish nation must, by its valour and discipline and by the readiness of its children to sacrifice themselves for the common good, prove itself worthy of the august destiny to which it is called.

Signed on Behalf of the Provisional Government,
THOMAS J. CLARKE.
SEAN Mac DIARMADA. THOMAS MacDONAGH.
P. H. PEARSE. EAMONN CEANNT.
JAMES CONNOLLY. JOSEPH PLUNKETT.

On Easter Monday 24 April 1916, Irish republicans, socialists and other separatists rose in armed revolt against British rule in Ireland. The rebellion was quickly crushed, the last rebel strongholds surrendering to British troops six days after the republic had been proclaimed (the proclamation is shown here). In the fighting, 100 British troops and 450 Irish were killed. The rebel leaders were executed, notably Patrick Pearse (1879–1916) and James Connolly (1870–1916). Only Eamon de Valera (1882–1975) survived because he had been born in the USA. However these measures, in the aftermath of the rebellion, won Irish opinion to the republican cause.

6

6 William Gladstone, seen in the cartoon struggling with the Irish question, became absorbed with Irish affairs after 1886 and his unsuccessful first Home Rule Bill. Prior to that, in 1869, he had disestablished the Irish Church and passed Land Acts, in 1870 and 1881, which gave tenants greater security and legally fixed rents. In 1893 he introduced the second Home Rule Bill, which was rejected by the House of Lords.

7 Edward Carson (1854–1935) led the Ulster Unionists from 1910–20, pledging and arming them to resist Home Rule in any form. In 1914 his offer to accept Home Rule with the exclusion of Ulster was rejected. In 1916 he reduced this demand to only the six most Protestant Ulster counties. Although Carson preferred continued integration with the United Kingdom, he accepted the creation of a separate parliament for the six counties in Belfast in 1920.

7

8

8 The Easter Rising of 24–29 April 1916, led by the Irish Volunteers and the Irish Citizen Army, seized several public buildings in the centre of Dublin before surrendering to the British army. Although many Irish people were out of sympathy with the insurrection itself, support for the republican cause grew after the secret execution of the seven signatories of the Proclamation of the Irish Republic and eight other rebel leaders, and widespread arrests.

9

9 Michael Collins (1890–1922) (right) was a leader of the Irish struggle for independence. After 1916 he became involved in Sinn Fein politics and was elected to the Dail in 1919, becoming a leading member of the provisional government. Eamon de Valera (left) was the senior surviving officer of the 1916 rising and principal Irish leader. He became President of Sinn Fein in 1917, and President of the Republic and of Dáil Eireann (Irish lower house) in 1919.

Scotland in the 19th century

The political framework of nineteenth-century Scotland continued to be union with England but, by the end of the century, in place of the handful of privileged voters who had elected MPs under the old regime, something approaching a democracy based on adult male suffrage had been achieved. This had been the work of successive Reform Bills in 1832, 1867 and 1884. Eventually Scotland was electing 72 MPs from constituencies that gave weight to the Scottish urban population, and its share of the Westminster parliament – 670 MPs – fairly reflected the Scottish proportion of British population.

The beginnings of socialism

Following the triumph of the Whigs under Francis Jeffrey (1773–1850) and Henry Cockburn (1779–1854) at the time of the Great Reform Bill of 1832, Scotland settled down to become loyally Liberal – 53 MPs were elected under William Ewart Gladstone's colours compared to seven Conservatives in 1880, although the split in the party over Irish Home Rule in 1886 shook this allegiance seriously. At the same time there was a Scottish radical tradition to the left of this mainstream.

It surfaced at the time of the so-called Radical War in 1820, which was really a combination of a strike and a small abortive rising in the Glasgow area. It was seen again with the Chartists between 1838 and 1848, although the Scottish Chartists mainly disapproved of physical force and sought reformation through temperance and democracy. And in 1888 socialism struck root with the foundation of the Scottish Labour Party and the rise of Keir Hardie (1856–1915) [9] who later became the leader of the British Independent Labour Party (ILP) in 1893. Trade unionism grew rapidly, especially in skilled trades and among the cotton spinners and miners, although a Scottish Trades Union Congress (with 40,000 members) was not founded until 1897.

Generally speaking, there was little dissatisfaction over union with Britain, although because Westminster was increasingly obliged to legislate for Scottish affairs as the problems of industrial society became more complex (by reforming the Poor Law in 1845 and the school system in 1872, for example) there was more demand for specifically Scottish experts in the government. This was met in 1885 by the creation of a Secretary of State for Scotland and a Scottish Office based in London and Edinburgh. A small nationalist movement arose at the end of the century; few took it seriously, although Hardie and other early socialists also favoured Home Rule. For many Scots, Church politics were more significant than national ones; the Disruption of the Church of Scotland in 1843 [5] into the Established Church and the Free Church over the question of who should choose the ministers generated enormous excitement.

The prosperity of heavy industries

The Scottish economy in the nineteenth century was highly successful: to the original base of cotton textiles an important iron industry in the West Central Belt was added in the 1830s, and after 1870 the vitality of the shipyards and steelworks of Clydeside and of jute round Dundee prevented the country from slipping into recession [3]. A third of all

CONNECTIONS

See also
28 Scotland in the 18th century
260 Scotland in the 20th century
210 The British labour movement 1868–1930
160 The fight for the vote
88 The Industrial Revolution
90 The urban consequences of industrialization

In other volumes
298 History and Culture 1

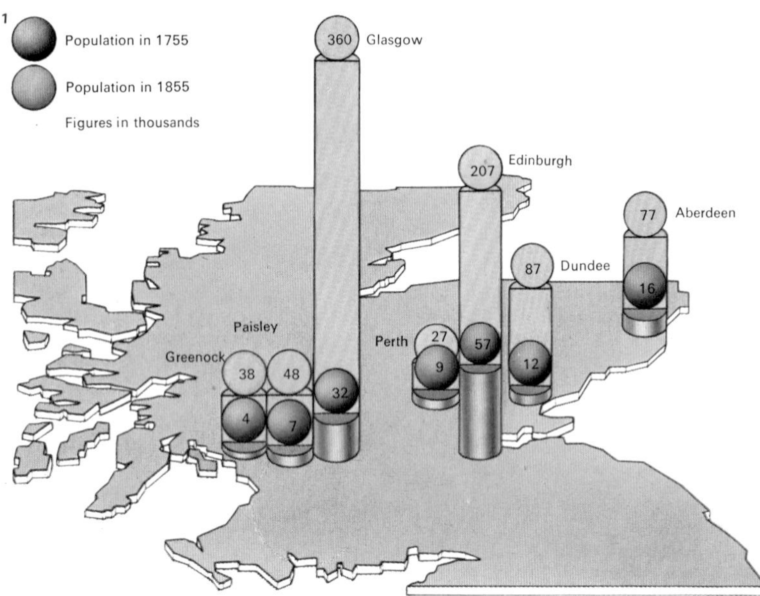

1 Industrialization in Scotland was accompanied by rapid urbanization, not so much in the foundation of new towns as in the very rapid growth of old ones: Glasgow, for example, had 23,000 inhabitants by about 1750, when it was already the second town of Scotland, but it had 329,000 by 1851. Nevertheless there were new towns, which often grew very fast. Airdrie, for instance, had a population of 1,200 in 1755, but with the development of iron and coal in Lanarkshire, it exceeded 13,000 by 1851. But primitive sanitation, unimproved from a previous era, menaced the growing towns and ominously increased mortality.

2 The Highlands' economy collapsed in the early 1800s and this resulted in a grim outlook for crofters such as these. Prices for three of their four main staples – cattle, kelp and fish – had fallen disastrously, and only wool was still viable. This meant that wealthy sheep farmers from the Lowlands began to introduce their animals into the crofters' fertile plots. As a result, the green summer pastures were quickly overcropped. The crofters themselves were usually evicted to the outskirts. Such evictions were sometimes executed considerately, but at other times the action was ruthless, causing great hardship.

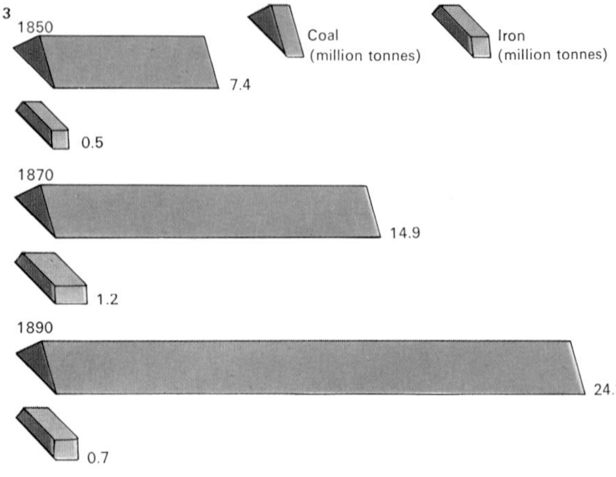

3 Economic growth in early Victorian Scotland was firmly based on Scottish natural resources, but with the invention of cheap steel after 1870 many of the ores had to be imported from various countries. The soaring indices of production point to an economy increasingly dependent on a narrow base of heavy industry. For example, by 1900 much of the metal went into the great ships being built in the yards along the Clyde.

4 The new castle at Balmoral was the apotheosis of Queen Victoria's love of the romantic. In 1856 she wrote of it in her diary: "Every year my heart becomes more fixed in this dear Paradise . . . and so much so now, that *all* has become my dear Albert's *own* creation . . .". The castle, standing about 80km (50 miles) west of Aberdeen on the banks of the River Dee, was bought by Prince Albert for the royal family in 1852.

ships built in Britain were being built on the Clyde by 1913. The Scots had earned much lower incomes than the English in 1800, but by 1900 the average working man on the Clyde was probably at least as well paid as the average English worker. This sense of prosperity made Glasgow an enormously self-confident business capital – few Scottish firms were controlled from elsewhere – but concealed the fact that Scottish wealth rested on a narrow base of heavy industries.

The improverished Highlands

The reverse side of the coin was the patchy nature of the wealth. Throughout the Highlands people were very poor: the population increased until 1841, far outstripping the growth of resources, and then collapsed after the potato famine of 1846. Thousands of small-scale tenants were evicted in the "clearances" to make way for sheep [2]; tens of thousands emigrated [6]. By the 1870s and 1880s over-intensive sheep farming had run down the fertility of the land and this, coupled with a dramatic slump in grain and wool prices, led to even further depopulation of the Highlands regions through migration.

Meanwhile those who left were partly balanced by those who arrived. These were Irishmen immigrating into the coalfields and factories of central Scotland where they generally had to take the lowest paid labouring jobs. The urban poor had a hard time; the slums of the great cities were probably the worst in Europe even when the economy was booming [Key, 1].

To the outside world, however, there were perhaps two main symbols of nineteenth century Scotland: Balmoral [4], where Queen Victoria gloried in a romantic view of the Highlands far removed from the unpleasant realities of the black houses of the Isle of Lewis; and Scottish science and medicine at the universities. Men such as Lord Kelvin (1824–1907) at Glasgow, or James Clerk Maxwell (1831–79), the physicist who was professor at Aberdeen, King's College, London and Cambridge, vied in their reputations with Joseph Lister (1827–1912) [7] and James Simpson (1811–70) who made surgery comparatively safe and painless.

KEY

The industries and slums of Clydeside were the central paradox of 19th-century Scotland. On the one hand they produced the greatest wealth Scotland had ever known – by 1906 wage rates were higher than in most of England. On the other hand, they had a population whose housing conditions were worse than any in Britain: even in 1911 two-thirds of the population lived in houses that had only one or two rooms.

5

5 Thomas Chalmers (1789–1847) was a widely influential theologian and preacher. He was for years head of the evangelical wing of the Church of Scotland, and then founded the Free Church which broke off at the Disruption of 1843. He abandoned the established church because of its traditional method of choosing ministers – he preferred democratic elections. Within Britain he was celebrated for his book, *Christian and Civic Economy of Large Towns,* which encouraged the middle classes to believe that the problems of poverty could be cured by generous philanthropic action with a rigorous inquiry into the personal and moral condition of the individual poor.

6

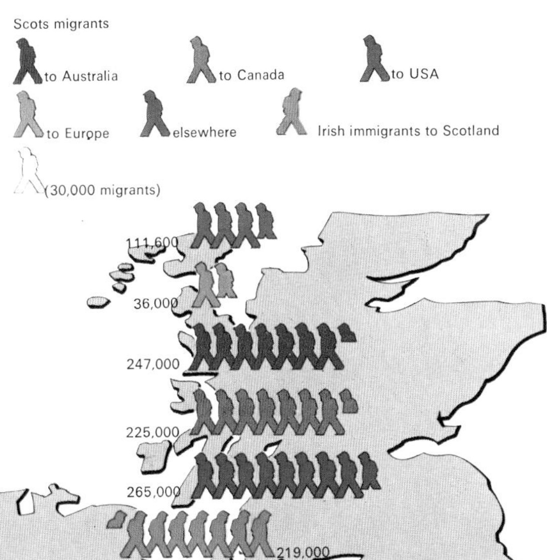

6 Nineteenth-century Scotland was like a bath with the taps full on and the plug out. There was a rapid natural increase accompanied by an inrush of Irish immigrants to the looms, mines and ironworks of the Central Belt. At the same time, many native Scots, especially Gaelic-speaking Highlanders unwilling to move to an unfamiliar urban life, chose to go to Canada, and also to Australia, the USA and New Zealand. This outflowing tide resulted not so much from lack of opportunity at home as from the enticement of kinfolk already abroad. Few European nations apart from Ireland and Norway lost so much of their natural increase.

7

7 Joseph Lister (1827–1912) founded modern antiseptic surgery. When he went to work at Glasgow Infirmary in 1861 he found that nearly half the amputation cases died of post-operative gangrene. Lister eventually began to realize that pus formed as a result of infection by germs. He ensured that hands, instruments and dressings were sterilized. This, together with his introduction of sterilized catgut and carbolic acid as an antiseptic, after 1865, dramatically reduced surgical mortalities.

8

8 The decision of a handful of crofters to resist eviction by force in 1882 alarmed the government, who sent a gunboat to Skye to put down the "rising". It was cheered by the peasants, who believed that Queen Victoria had come to hear their grievances.

9

9 Keir Hardie (shown here) and R. B. Cunninghame Graham (1852–1936) were the fathers of socialism in Scotland. Hardie became leader of the British ILP and was described as "the best-hated and the best-loved man in Great Britain". In 1928 Cunninghame Graham helped to found the Scottish National Party. Hardie was a confirmed pacifist and was fervently opposed to the Boer War. He also favoured women's suffrage, and founded *The Labour Leader,* a Scottish newspaper.

Wales 1536–1914

The Acts of Union (1536–43) decreed that Wales henceforth was to be governed "in like form" to England. Wales was given a definite administrative boundary and was also unified politically within itself [Key]. The most progressive of the Welsh gentry were happy to be subsumed in a common British citizenship and voiced their gratitude to the Tudors for bringing order, stability and prosperity to Wales.

The power of the gentry

The gentry were the most powerful element within society and the task of administering local government remained in their hands for some 350 years. Traditionally conservative, they supported the Crown through every event. During the English Civil Wars (1642–6, 1648) they fought for the king in order to protect their prosperity and security and, after the Restoration in 1660, they re-established a monopoly of influence on the society, economy and politics of Wales. Until the mid-nineteenth century political power lay in the hands of a narrow circle of landowning families, and the mass of society remained deferential to their will. Three developments – the growth of Nonconformism, the Industrial Revolution and the spread of political radicalism – undermined the foundation of this society.

From the sixteenth century onwards successive waves of Protestantism lapped over Wales. Much was achieved: Welsh became the language of religion, and the translation of the scriptures into the vernacular [1] fostered the growth of a Bible-reading public. With the coming of Methodism in the 1730s, Reformation ideas were propagated far more intensively [3]. In 1811, the Methodist movement was forced to sever its connection with the Anglican Church and, in the company of fellow Dissenters, spread widely into rural and industrial areas. Nonconformity became a popular movement so that by 1851 about 80 per cent of practising Christians in Wales were Nonconformists.

The Industrial Revolution in Wales

The second major factor that created modern Wales was the Industrial Revolution. Until the end of the eighteenth century Wales displayed the main features of a pastoral, pre-industrial economy: a primitive technology, a slow rate of technical development and a lack of capital. But the arrival of the Industrial Revolution after 1760 transformed the social and economic life of Wales. Financed largely by English entrepreneurs, industrial development focused on the chain of ironworks on the periphery of the South Wales coalfields and in north-east Wales, on the copper mines of Anglesey and the slate quarries of Caernarvonshire. The spread of canals and railways improved communications and hastened large-scale industrial expansion.

At the same time, population growth began to accelerate dramatically. It rose from 370,000 in 1670 to 586,000 in 1801. Small villages grew into booming towns: in 1801, Merthyr Tydfil, with a population of 7,705, was the largest town in Wales [4]. By 1861, 60 per cent of the Welsh people lived in industrial areas. The decline of the iron industry after 1850 was followed by the growth of new steelmaking processes and the massive expansion of the coal industry.

CONNECTIONS

See also
92 The rural consequences of industrialization
94 The British Labour movement to 1868
96 Social reform 1800–1914
160 The fight for the vote
262 Wales in the 20th century
286 The arts in wales

In other volumes
184 History and Culture 1

1

Y BEIBL CYSSEGR-LAN. SEF
YR HEN DESTAMENT, A'R NEWYDD.

2. Timoth. 3. 14, 15.

Eithr aros di yn y pethau a ddyſcaiſt, ac a ymddyriedwyd i ti, gan wybod gan bwy y dyſcaiſt.
Ac i ti er yn fachgen wybod yr ſcrythur lân, yr hon ſydd abl i'th wneuthur yn ddoeth i iechydwriaeth, trwy'r ffydd yr hon ſydd yng-Hriſt Ieſu.

Imprinted at London by the Deputies of
CHRISTOPHER BARKER,
Printer to the Queenes moſt excellent Maieſtie.

1588.

2

1 The first Welsh Bible (1588) resulted from a statute in 1563 which ordered that the translation of the Bible into Welsh should be undertaken forthwith. The work was duly completed by an erudite Denbighshire vicar, William Morgan (*c.* 1545-1604). The translation provided a literary standard for future generations and ensured that Protestantism would be propagated in the Welsh language.

2 The Sker House, a large, bleak edifice close to the Kenfig Burrows in Glamorgan, is a good example of the many new or remodelled buildings which were constructed by the Welsh gentry in the 16th century. The house was built on a former monastic grange by the Turberville family. The economic and political power of the gentry at that time was reflected in their imposing country homes.

3

3 Howel Harris (1714–73) was the moving spirit behind the growth of Welsh Methodism. A fiery evangelist, Harris provided the movement with inspired leadership and an efficient organization.

4

4 The massive Cyfarthfa ironworks, founded in the mid-18th century, became the focal point of the iron-smelting town of Merthyr. In keeping with much of the Industrial Revolution in Wales, the works were financed by English capital.

As the unparalleled resources of the Rhondda valleys were plundered, coal came to dominate the Welsh economy. By 1912, coal output in the mining valleys of South Wales was more than 50 million tonnes.

Nationalism and political radicalism

The third factor was the growth of political radicalism, inspired by the revolutionary ideals formulated in France. Many processes hastened these ambitions: the Welsh press created an articulate and informed body of public opinion; acute economic distress in rural and industrial communities encouraged class awareness and a growing interest in political reform; and a slanderous government report – *The Treason of the Blue Books* in 1847 – injected new life into radicalism and awakened a sense of nationhood. The extension of the franchise in the nineteenth century gave radical Nonconformists the opportunity to undermine the landowning monopoly, to remove religious disabilities, and to create cultural and educational institutions attuned to Welsh circumstances and aspirations. Between 1868 and 1918 Welsh Liberals voiced the ambitions of a new Nonconformist middle and working class, and the response which they evoked from the electorate enabled them to erode the power of the old Anglican squirearchy and to capture the overwhelming majority of parliamentary seats in Wales.

As political nationalism spread in from Europe and Ireland, a new effort was made to emphasize the distinctiveness of Wales and to press for national equality and justice [7]. In Parliament, a ginger-group of young Liberals, led by Thomas Ellis (1859–89) and David Lloyd George (1863–1945) [8], called for religious equality, educational opportunity and land reform. Eventually, many gains were achieved: the Church in Wales was disestablished in 1920; Welsh universities, a National Library at Aberystwyth and a National Museum at Cardiff were established; a Welsh department was created within the Board of Education; and the concept of Wales was firmly established. By 1914 it was no longer considered to be a mere geographical term with neither institutions nor pride in its own nationhood.

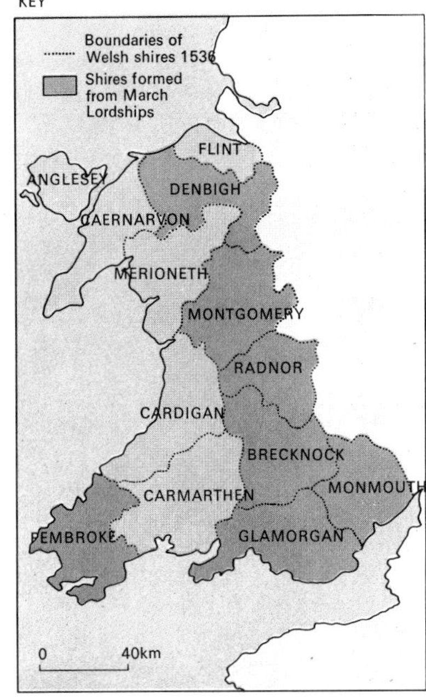

The Acts of Union incorporated Wales into England in order to achieve a more effective governance of Wales and the border area (Marches). Welshmen henceforth were to enjoy the rights and privileges of Englishmen; land was to be inherited according to the practice of primogeniture; and the whole of Wales was divided into shires — a framework that persisted until April 1974. English became the official language of law and government and English common law and methods of local administration were introduced. In return the new Welsh shires and boroughs could send 24 MPs to represent them in the English Parliament.

5

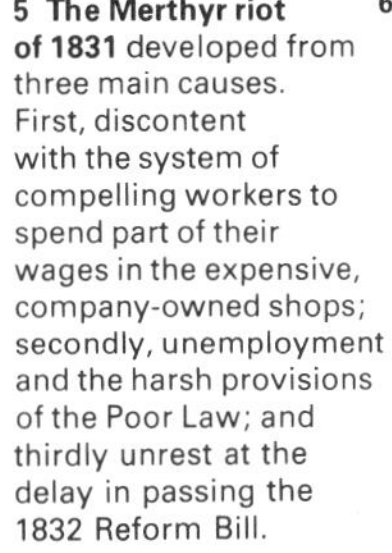

5 The Merthyr riot of 1831 developed from three main causes. First, discontent with the system of compelling workers to spend part of their wages in the expensive, company-owned shops; secondly, unemployment and the harsh provisions of the Poor Law; and thirdly unrest at the delay in passing the 1832 Reform Bill.

6

6 The Rebecca riots in the early 1840s occurred in separate places across south-west Wales. Disguised as women, small farmers protested against abuses of the turnpike system. They attacked the hated toll gates, burnt haystacks and threatened local magistrates. A government inquiry in 1844 resolved many of their grievances.

8

8 David Lloyd George, MP for Caernarvon boroughs from 1890, made his mark in politics as an enthusiastic champion of the rights of Welshmen, an enemy of privilege and as a "man of the people". As Chancellor of the Exchequer (1908-15) he introduced crucial social reforms, and his 1909 budget provoked an important constitutional crisis with the Lords. In 1916, he became the first Welshman to be appointed prime minister, which he remained until 1922. He earned a reputation as a courageous and decisive war leader, and a constructive peace-maker after World War I. His fertile mind and oratorial genius aroused widespread devotion and, equally, widespread dislike.

7

7 Michael D. Jones (1822–98) was one of the principal Welsh nationalists of the 19th century. He strove valiantly to persuade Welshmen to embrace a new, radical philosophy, to agitate for their political rights and to recover their self-respect and confidence. His determination to preserve national identity prompted him to establish a Welsh colony in Patagonia, South America, in 1865; this colony still exists as an isolated Welsh-speaking outpost today.

Russia in the 19th century

In Russia, since the time of Peter the Great (1672–1725), fundamental reforms have followed in the wake of war. For many years after the Crimean War (1854–6) [1], Russia was no longer regarded as a friendly power by Britain and France. Despite the fact that it had the largest land forces on the continent of Europe this war showed that Russia was no match for the Anglo-French alliance and that its effort to insulate itself from the political changes in the rest of Europe had proved to be a source of weakness rather than of strength. Finally, its economy and social order could not withstand the war. Russia if it wished to regain its position as a leading nation, had to imitate the Western powers and adopt their forms of government.

The emancipation of the serfs

Alexander II (1818–81), who came to the throne in 1855, was willing to introduce reforms. He warned the nobility that if reform did not come from above it would come from below. In February 1861 the Emancipation Act was ready.

The Act ensured personal freedom for millions of peasants and introduced the elective *zemstvo*, an organ of local government, which was to have an important say in the countryside. Other major reforms followed: in 1864 equality before the law, trial by jury and independence of courts and judges were introduced; legislation of 1863 and 1864 broadened the basis of education; the 1870 Government Act set up new municipal institutions; the 1874 army reforms established the principle of universal military service and reduced actual service from 25 years to six. But the peasants were still subject to customary law and had special courts; their freedom of movement was limited and they still paid poll taxes. Moreover, the Tsar did not grant a parliament

The emancipation disappointed most of the peasants and their supporters. Population increased from 70 million in 1863 to 155 million in 1913 (excluding Finland and Poland), aggravating rural poverty. Migration eased the situation slightly, but the problem of land hunger was exacerbated by the failure to introduce modern agricultural methods, obstructed by the communal system of land ownership. Much peasant dissatisfaction also stemmed from the poor quality of the land that they were allotted, and the high level of repayments they were forced to make to the government to compensate the former owners of the land.

Seeds of revolution

The inadequacy of Alexander's reforms aroused moral revulsion and anger among many sons and daughters of the gentry and others who had acquired some education. Disillusionment over the reforms at first encouraged nihilism. The nihilists believed that the existing order could not successfully reform itself and in Russia they contributed significantly to the tradition of revolutionary political movements. During the 1870s a more positive populism [5] or agrarian socialism developed which glorified the peasant as the repository of pure, untainted wisdom. Those who had received an education felt that they owed a debt of gratitude to the toilers who had made it possible.

Agrarian populism was difficult to convert into political action and the onset of

CONNECTIONS

See also
196 The Russian Revolution
198 Stalin's Russia
84 European empires in the 19th century
170 Political thought in the 19th century
180 Europe 1870–1914
184 Balkanization and Slav nationalism

1 Following the capture of Sevastopol and her defeat in the Crimean War, Russia became little more than a second-rate power. Britain and France had turned against her and exposed the backwardness of her economy and the brittleness of her army. The new tsar, Alexander II, was convinced that Russia had to imitate the Western powers if she was to beat them and so he favoured sweeping reforms.

2 Peasants received insufficient land as a result of the Emancipation Act (here being read out to Georgian peasants). They did not receive land freely, most of them having to pay a fixed annual amount to the state which in turn compensated the landlords with state bonds. Repayments were to extend over 49 years and were higher than the market value of the land warranted. The result was that the peasants had less land than before – in fact about 20% less in total – 23% of this in the black earth lands and 31% in the Ukraine. Former state and crown peasants received the best terms.

3 The execution of terrorists who planned the assassination of Tsar Alexander II by a bomb in March 1881, in the hope that the whole imperial edifice would collapse, sums up the impotence of revolutionary politics in 19th-century Russia. The acute disappointment felt by the peasants and intelligentsia after the Emancipation Act led to pessimism concerning the possibility of reform from above. Many radicals, known as populists or agrarian socialists, believed the peasantry would rise *en masse* and sweep away the hated autocracy. Some believed in the gradual awakening of peasant consciousness, moulded by radical idealists. Others were unwilling to wait for the uprising of the masses and adopted terrorist methods.

4 Georgy Plekhanov (1857–1918), the father of Russian Marxism, started his political life as a populist. He opposed terrorism, but had to flee the country for Geneva in 1880 during a wave of political repression and did not return to Russia until 1917. A brilliant writer and polemicist, his influence within Russia in the 1890s was immense. He initially supported Lenin, then opposed him.

5 Populism became the leading philosophical attitude in the 1870s. Its most significant leader was Peter Lavrov (1823–1900). Populism rejected the Industrial Revolution and favoured rural life.

industrialization in the late 1880s and the boom of the 1890s made it less relevant. Marxism, placing its faith not in the rural worker but in the urban, industrial worker, became a doctrine more in tune with contemporary Russian conditions. The Social Democratic Party, the forerunner of the Communist Party, emerged, although it still appealed for the most part to intellectuals rather than to the working classes.

The terrorist wing of the populist movement finally resulted in the assassination of Alexander II. But instead of collapsing, the autocracy struck back at its tormentors.

The end of the era

Alexander III (1845–94), who came to the throne in 1881, was ultra-reactionary. His policies reversed many of the liberal reforms of his predecessor and began a tradition of conflict between the *zemstvos* and central government that came to a head in 1905.

The succession in 1894 of Nicholas II (1868–1918) [8] occurred at a time of rapid economic advance [6]. The dynamic thrust of Sergei Witte (1849–1915), minister of finance from 1892–1903, kept the economy moving until the first years of this century. Then harvest failures and industrial crises produced civil unrest. The revolution of 1905-6 shook the autocracy to its foundations [10]. It could be suppressed only when the war against Japan had been lost and troops were released for internal duties.

The years 1903-13 were a golden era for industry and agriculture and this helped the government, led by Peter Stolypin (1862–1911), to resist the growing demands for political and social reforms, which were voiced in the Duma (a parliament forced on the Tsar by the crisis of 1905) by the Social Democratic and Kadet (liberal) parties. Thwarted in the Far East, Russia turned after 1906 towards the Balkans where, throughout the nineteenth century, it had supported Slav states against the decaying Ottoman Empire. But the Great Powers stepped in and blocked Russia's progress to the Mediterranean. The empire of Austria-Hungary was the main rival power in the Balkans and therefore Russia felt obliged to support Serbia against the empire in August 1914.

KEY

Servile labour was typical of the life of millions of Russians in the 19th century, but with industrial development and the population explosion, changes occurred. There were 412,000 barge hauliers on the Volga in 1830, but by 1851 this number had been reduced to 150,000. The steamship had gradually replaced them. There were approximately 40 million peasants (80% of the population) in Russia on the eve of emancipation and about half were in personal bondage to the gentry. Their plight dominated economic life in Russia.

6

6 In the 1890s the industrial development of Russia was improved by the opening up of new oil fields, including this one at Baku. Russia was the world's largest producer of oil until 1900, when the USA took the lead. Railway building was another dynamic force; by 1874 there were 18,220km (11,320 miles) of railway. A by-product of this was Russia's emergence as a major grain exporter. From the 1880s the state began to play an important role in the economy, guided by the policies of Sergei Witte. Development was concentrated in railway construction and in heavy industry.

7

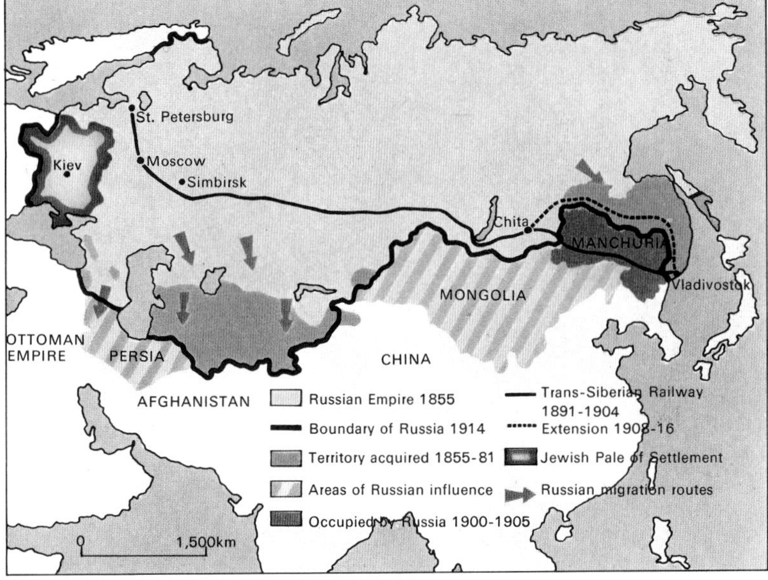

7 Russia's imperial advance was spectacular in the later 19th century. She colonized Central Asia and acquired territory which the Chinese still claim as their own. Russia's population explosion caused seven million peasants to move eastwards and cross the Urals. Meanwhile two million Jews emigrated to the USA and 200,000 more to Britain between 1880 and 1914. The Trans-Siberian Railway (built between 1891 and 1904) made a more active policy feasible in the Far East – that is, towards Japan – with the secondary aim of securing an ice-free port on the Pacific. Russia's eastward push and her influence in Manchuria alarmed the Japanese to the point of their going to war in 1904. Apart from her Far Eastern ambitions, Russia also greatly extended her influence in the regions on her southern borders.

8

9

8 The last of the Romanovs, Nicholas II, was a reluctant tsar. He came to the throne unusually young and made an inauspicious start in 1894. His mind lacked the cutting edge necessary to evolve a coherent policy and to see it through. Although Russia changed rapidly during his reign he did not move with the times and listened instead to reactionaries, including the monk Rasputin (1871–1916) who mystically influenced the empress. Here Nicholas [2nd from left] is with the Prince of Wales [far right].

9 An outstanding statesman, Peter Stolypin (1862–1911) introduced agrarian reforms. He swept away the commune and encouraged the peasants to consolidate their holdings and become farmers. But his autocratic methods lost him liberal support.

10

10 "Bloody Sunday" began as a peaceful demonstration on which troops opened fire in St Petersburg on 22 January 1905. Discontent had grown as the industrial boom of the 1890s gave way to a slump during the early years of the 20th century. Harvest failures aggravated the problem compounded by the defeat in war with Japan. Although unsuccessful, the subsequent revolution of 1905–6 did produce a constitution and a parliament (Duma).

Political thought in the 19th century

In the mid-nineteenth century most people with any political awareness would almost certainly have described themselves as either "liberals" or "conservatives". The conservatives would have had little difficulty in explaining what they were and what they stood for, namely the established order. They were firmly against radical change and followed the line laid down by Edmund Burke (1729–97) in his *Reflections on the Revolution in France*, published in 1790. This insisted that state and people alike were products of imperceptible, natural and organic growth and that artificial change based on general theories was self-defeating.

In the realm of practical politics, however, it was not quite so easy to preach and practise conservatism – particularly after the fall of the Austrian statesman Prince Metternich (1773–1859) in the revolution of 1848. Metternich refused to concede that any kind of change was permissible, if only as a tactical manoeuvre to prevent more radical developments, and was ultimately obliged to take refuge in England.

Metternich's downfall was one of the factors that encouraged the British prime minister, the astute Benjamin Disraeli (1804–81), to present the country with a Second Reform Bill. Meanwhile in Germany Prince Otto von Bismarck (1815–98) introduced universal suffrage and limited social welfare legislation. In France, Napoleon III (1808–73) had embarked on similar action.

The decline of liberalism

Liberals were distinguished by the belief that progress could be achieved by means of "free institutions". In Britain and France this usually referred to a freely elected parliament, with ministries responsible to it, an independent judiciary, freedom of speech and religion, freedom from arbitrary arrest and freedom to acquire and safeguard property.

In Russia a "liberal" might merely be someone who advocated a strong state council to advise the tsar. But even in France there were "liberals", including François Guizot (1787–1874), the statesman and historian, who believed that institutions were already as free as possible – a belief that made them seem highly conservative.

One of the most interesting themes of nineteenth-century European history is the decline of liberalism as a real political force. The main reason for the collapse was that, although the liberal ideal of making a framework of free institutions was born of the Enlightenment, once erected it became a bastion behind which the propertied classes defended their vested interests. The Continental turmoil of 1848 saw middle-class liberals deserting their ideals when faced with the prospect of sharing power with the lower-paid and less-educated sections of society.

The rise of socialism

The creed that began to appeal to many of those apparently abandoned by liberalism was socialism, and the greatest socialist thinker of the century was without doubt Karl Marx (1818–83) [Key]. The young Marx of the first half of the century drew his ideas from a wide variety of sources but the foundation of his beliefs was the conviction, derived from the German philosopher Georg Hegel (1770–1831), that history was progressive, had objective meaning and

CONNECTIONS

See also
172 Masters of sociology
22 The Enlightenment
72 Origins of romanticism
74 The French Revolution
88 The Industrial Revolution
94 The British labour movement to 1868
96 Social reform 1800–1914
108 The revolutions of 1848
110 German and Italian unification
160 The fight for the vote
210 The British labour movement 1868–1930
212 Socialism in the West
184 Balkanization and Slav nationalism
222 The rise of fascism

1 Appalling social conditions existed in 19th-century Europe as a result of the development and concentration of industry and a boom in population. By 1848 the "social question" was causing concern. Neither government nor individuals did much to tackle the problem. Chartism emerged as a force in Britain, while in Europe the old spirit of revolution was again showing signs of revival. But in the long term, a steady if slow increase in living standards was brought about not by political organization and agitation, as might be supposed, but by the unexpected growth of the economy.

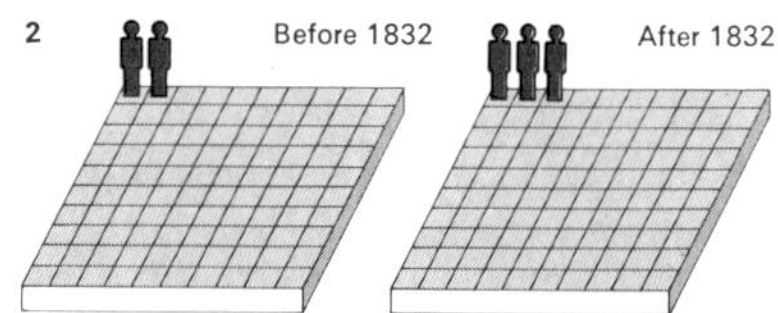

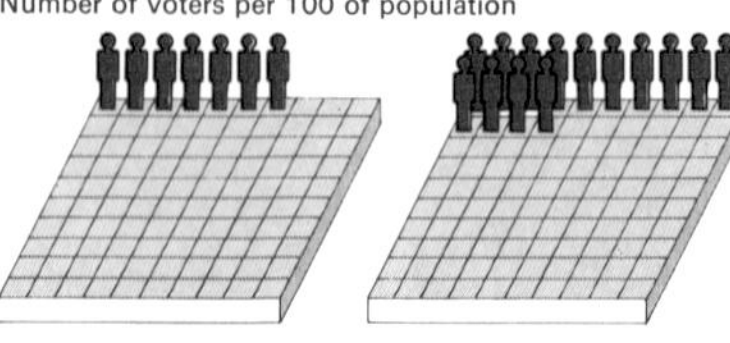

2 A new British electoral system was created between 1832 and 1885, based on a series of Acts of Parliament. The result was that by 1886 two-thirds of the adult male population of England and Wales, and three-fifths in Scotland, had the right to cast their vote in secret. The measures that brought this about were three Representation of the People Acts, a Ballot Act and two Acts to redistribute the seats and prevent corruption.

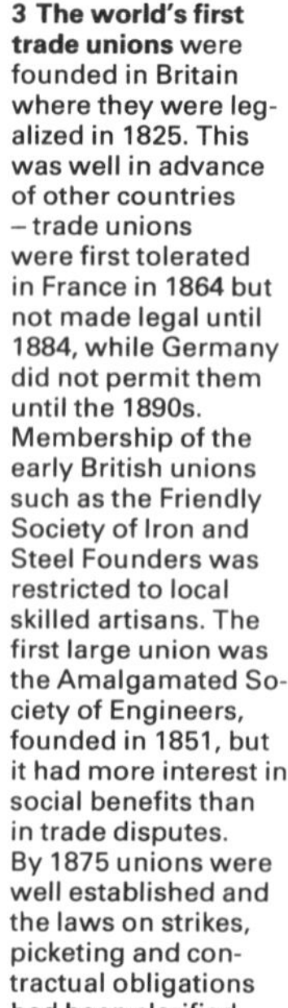

3 The world's first trade unions were founded in Britain where they were legalized in 1825. This was well in advance of other countries – trade unions were first tolerated in France in 1864 but not made legal until 1884, while Germany did not permit them until the 1890s. Membership of the early British unions such as the Friendly Society of Iron and Steel Founders was restricted to local skilled artisans. The first large union was the Amalgamated Society of Engineers, founded in 1851, but it had more interest in social benefits than in trade disputes. By 1875 unions were well established and the laws on strikes, picketing and contractual obligations had been clarified.

4 Mikhail Bakunin (1814–76), a Russian aristocrat, resigned his commission in the Imperial Guard to become Europe's leading anarchist. Not surprisingly his life was eventful: he was sentenced to death by the Austrians and the Prussians and was sent to Siberia by his own country. He escaped in 1861 and spent the rest of his life advancing anarchism in western Europe. Unlike the socialists, he believed that society could only be overthrown through individual revolt.

would reveal this meaning through a series of revolutionary jumps.

The *Communist Manifesto* of 1848 reflected Marx's faith in the success of the European revolutions of that year, but with their ultimate failure he laid more stress on the deterministic aspects of his thought. He predicted that bourgeois society would collapse as a result of its own internal contradictions. Capital, he said, would become concentrated in fewer hands until the oppressed workers would be forced to revolt against their exploiters. A "dictatorship of the proletariat" would then emerge, paving the way for such social harmony that the state could wither away. The Paris Commune [7] revived his faith in revolutionary activity and in the 1870s he even toyed with the possibility of a peaceful overthrow of the social system through the ballot box with the aid of a fully enfranchised proletariat.

The development of nationalism

It was not the thoughts of Marx, however, that dominated the nineteenth century. By far the greatest force was nationalism, which conquered both the liberals and the socialists.

In 1815 nationalism was still weak in Europe, but only 45 years later the philosopher and economist John Stuart Mill (1806–73) was to write that it was "in general a necessary condition of free institutions that the boundaries of government should coincide in the main with those of nationalities".

Meanwhile nationalism had developed in many ways. The German philosopher Johann Herder (1744–1803) had insisted before the end of the eighteenth century that men's minds were conditioned by their cultural environment and, especially, by their language. Other thinkers took up this theme at the beginning of the new century and subsequently gave rise to many linguistic revivals. European scholars compiled dictionaries and grammars; folk-songs and folk poetry were collected; national histories were written. This, in turn, stimulated political demands and national wars radically redrew the map of Europe. The rest of the world did not escape: frustrated nationalism led to adventures overseas and the great wave of imperialism.

KEY

Karl Marx was the father of modern socialism. His political views are outlined in the *Communist Manifesto*, his views on political economy in *Das Kapital* (*Capital*).

5 "The Republic", a symbolic painting by Daumier (1808–79), shows the idealism often attributed to such government. Before the French Revolution republics were considered as legitimate as any monarchy but after 1815 they went "out of fashion" and Europe grew more monarchical. As new states such as Belgium, Greece, Romania and Bulgaria were created, so too were new monarchies. Although monarchy was no longer divine it was the system of government most comprehensible to the ordinary man. It was argued that only monarchy could unite all groups and all classes. Even France was little different. It was ruled by kings or emperors for most of the century and the Third Republic was established by one vote in 1875 as the regime that "divided Frenchmen least".

6 The Geneva Convention of 1864 established the International Red Cross. This was a humane reaction to the suffering of soldiers in the wars of the 1850s but also reflected concern about the problems of war itself. Other aspects of this were the continuing attempts to regulate war by law and the strength of the international pacifist movements. Peace congresses were held frequently from the middle of the century onwards. By 1900 there was a belief current in Europe that some genuine progress had been made towards achieving permanent peace.

7 Napoleon's statue was overturned in 1871 to signal the founding of the Paris Commune, one of the significant events of 19th-century Europe. Socialists saw it as a vindication of their belief that only by resorting to force could workers hope to overthrow the rule of the bourgeoisie. Yet the truth, in retrospect, is more complex than legend and it must be conceded that national and sectional interests were involved in the tragedy. Paris had declared itself independent of the rest of France and had to be brought back into line before peace with Prussia was possible. The end of the Commune brought vengeance and bloodshed: 20,000 were killed and 50,000 arrested.

Masters of sociology

The development of sociology in nineteenth-century Europe was stimulated by the need to understand the birth of industrial society [1]. The traditional agrarian social order, apparently based on the squire and the Church, was in the process of dissolution. In its place a new order was emerging whose symbols were the factory and the vast, anonymous urban proletariat [2]. A previously integrated structure of culture and authority was giving way to a series of sharply differentiated economic cultures and to class warfare. In this atmosphere of uncertainty intellectuals began to search for explanations of what was happening to society.

The British tradition

In Britain the path of industrialization generally caused little concern. Until the end of the century most Englishmen felt that the factory represented an unequivocal force for good, which was taking their society towards perfection. This largely unquestioning acceptance of the notion of "progress" meant that Britain produced no original sociological theory. Indeed, the main British theoretical tradition was inherited uncritically from the optimistic Enlightenment of the previous century. Its tenets were that society consisted of autonomous individuals each of whom was naturally good; that an "invisible hand" lay behind human activity and pushed it towards conditions of freedom in which the individual could express his innate goodness; and that social science should proceed by reason to discover the objective laws by which the hand worked and so facilitate its operation.

The one man who added something new to these ideas was Herbert Spencer (1820–1903) [3C] who recognized that the orthodox interpretation of society assumed but did not explain change. Spencer, however, did not abandon the ideas of the Enlightenment but regarded them in relation to a model of social change owing much to Darwin's *Origin of Species.* He argued that societies were driven forward to more complex and higher forms by the struggle for survival between individuals, and that the struggle had produced in Britain a *laissez-faire* industrial society which was as yet the highest social form. Although Spencer's conclusions were controversial, his methodology was influential. For the next 50 years British sociologists sought to explain social institutions by their "history".

The French tradition

In France the aftermath of the Revolution produced a reaction against Enlightenment thinking. The Vicomte de Bonald (1754–1840) argued that society ought to be seen not as a collection of individuals but as an organic whole. Change in one part (as by one social group) was bound to upset the entire organism.

The organic tradition was continued by Auguste Comte (1798–1857) [3B], not only to order and control change but also to understand it. Comte held the Enlightenment view that there were objective, discoverable laws of social progress. But he insisted that these laws operated in the context of whole societies and not individuals. Men, through their conditioning in society, were made by laws they could not alter. They should recognize this fact and accept their assigned social position.

CONNECTIONS

See also
170 Political thought in the 19th century
212 Socialism in the West
266 20th century sociology and its influence
108 The revolutions of 1848
158 Industrialization 1870–1914
32 Exploration and science 1750–1850
22 The Enlightenment
98 The novel and press in the 19th century

In other volumes
224 Man and Society
262 Man and Society
264 Man and Society

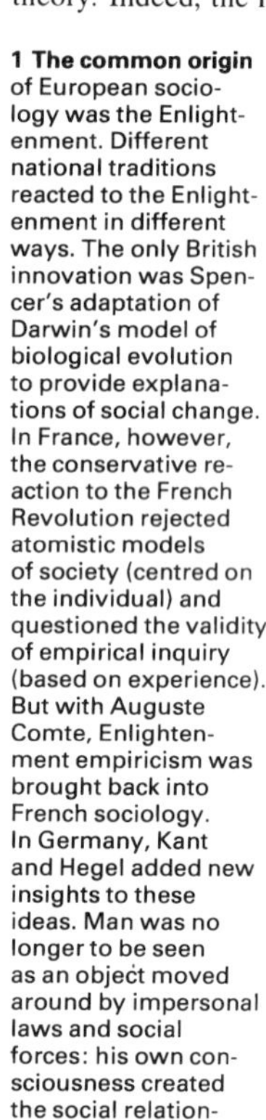

1 The common origin of European sociology was the Enlightenment. Different national traditions reacted to the Enlightenment in different ways. The only British innovation was Spencer's adaptation of Darwin's model of biological evolution to provide explanations of social change. In France, however, the conservative reaction to the French Revolution rejected atomistic models of society (centred on the individual) and questioned the validity of empirical inquiry (based on experience). But with Auguste Comte, Enlightenment empiricism was brought back into French sociology. In Germany, Kant and Hegel added new insights to these ideas. Man was no longer to be seen as an object moved around by impersonal laws and social forces: his own consciousness created the social relationships in which he participated.

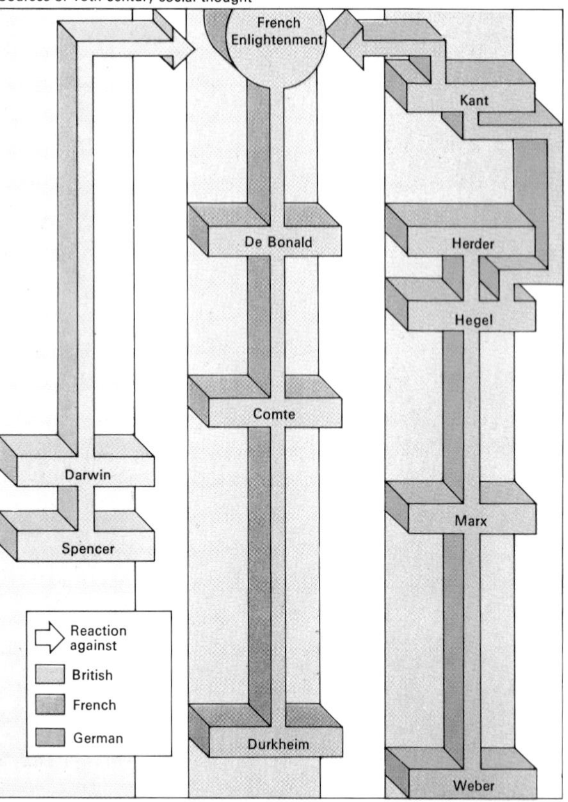

2 The Industrial Revolution dramatically changed the environment of European society. Millions of people were crowded into filthy, disease-ridden towns and were obliged to move to the new social and economic rhythms of factory labour. The obvious horror of mid-19th century urban life, illustrated by this Manchester slum interior, caught the attention of many early sociologists – Friedrich Engels (1820–95) for example – and produced some of the first exercises in applied sociology. Sociologists surveyed specific situations in the hope of finding remedies for major problems.

3 Major 19th-century sociologists included: Max Weber [A], who attempted to combine empiricism and neo-Kantianism in his *Protestant Ethic and the Spirit of Capitalism* (1905). Auguste Comte's [B] doctrine of positivism (to organize all knowledge into a consistent philosophy) is contained in *Système de Politique Positive* (1851–4). Herbert Spencer [C] amalgamated atomistic sociology and Darwinian evolution in *The Principles of Ethics* (1879–93).

Comte's "positivism" was most highly refined by one of the most influential individuals in all sociology, Emile Durkheim (1858–1917). The distinctive characteristics of French sociology included "methodological collectivism", which studied only phenomena that would reveal how men were conditioned by their society. There were also functional explanations whereby social institutions were described in terms of their functions within the entire social system rather than by their history. Lastly there was an emphasis on the need for order where change was regarded as the result of a malfunction in society.

The German tradition

In Germany the inheritance of Enlightenment rationality was joined by two other intellectual elements. The Kantian philosophical revolution (after Immanuel Kant [1724–1804]) held that the laws of nature existed only in men's minds; and the Romantic movement of Johann Herder (1744–1803) stressed the creative importance of language and culture.

The first great German theorist was G. W. F. Hegel (1770–1831), who saw social change as the product of human reason driven forward by its need to know and overcome the world around it. Hegel's theme was further developed by Karl Marx (1818–83) [4] who is perhaps best seen as a sociological Hegelian. Marx shared Hegel's view that the force behind social change was man's pursuit of rational understanding and control of his environment. But Marx's most important work resulted from his belief in the economic basis of social structure and in his suggestion of a sequence of social development.

The third major German theorist was Max Weber (1864–1920) [3A] who complemented Marx by adding an appreciation of the role of cultural values to Marx's work.

The principal achievements of the German tradition were "methodological individualism": an approach to society from the viewpoint of self-conscious human subjects; a combination of explanations from history and explanations from function; and the development of a theory of knowledge of the social sciences.

KEY

These men on strike in 1889 at the East and West India Docks in London symbolize the class and culture conflict produced by industrialization, which sociologists of the period tried to understand. It aggravated the division of culture along class lines and led to strife in every nation.

4

4 Karl Marx argued that human society developed in response to man's desire to satisfy his material needs. But needs themselves continued to develop. Eventually the prevailing form of social structure would no longer be able to accommodate these growing needs and so would break down, giving way to a new structure that permitted the continuation of need satisfaction. The final stage would be reached when bourgeois capitalism succeeded in concentrating wealth in a few hands and in impoverishing the masses. The starving proletariat, whose basic needs were not being met, would rise up and take over the means of production and create a society in which the forces of production and the social structure were no longer in conflict.

5
A

B

5 The interpretation of the European revolutions of 1848 and 1870 brought out the different perspectives of French, English and German sociology. For the French the revolutions (particularly the Commune of 1870 [A]) represented evidence of a deep-seated malfunction in society. For the British, they represented the just struggle of European society for individual, bourgeois freedoms against the tyranny of anachronistic, feudal governments. For the German Marxists the revolutions were a sign of the imminent destruction of the whole capitalist order: the cartoon [B] shows the French President Thiers (1797–1877) with a Prussian soldier looking down on the cauldron of Paris.

Fauvism and Expressionism

The technique of "divisionism" or "pointillism", meaning the building-up of a composition with a multitude of coloured points that merge in the eye of the spectator to produce the required colour, was pioneered by Georges Seurat (1859–91). Paul Signac (1863–1935) enlarged each point into a substantial block of paint so that there was no longer any question of such visual combination. In this way colour began to lose its representational function.

Liberation of colour

A different path to greater freedom of colour was taken by the Pont Aven school whose most important representatives were Paul Gauguin (1848–1903) [1] and Emile Bernard (1868–1941). They evolved what was known as "cloisonnism", based on the enclosure of forms within black outlines that bore the entire burden of expressing the shape of the object and did away with the need for shading with light or dark. This meant the painter could work in large flat areas of colour, producing the effect of a stained-glass window, the brilliance of which need not be diminished by the requirements of modelling.

The complete liberation of colour from form, so that it could act autonomously, was the hallmark of the group of diverse French painters known as Fauves ("wild beasts") who exhibited together in 1905 at the Salon D'Automne.

The most gifted of the Fauves, Henri Matisse (1869–1954), superficially adopted the divisionist technique in his "Luxe, Calme et Volupté" (1905), which Signac admired and bought. In fact Matisse was more concerned with the decorative possibilities of the style than with any analysis of colour.

At the end of the decade, he abandoned divisionism altogether and produced compositions dominated by areas of flat, unbroken colour of equal intensity [3], so emphasizing the picture surface in its own right rather than treating it as a kind of window through which the viewer looks.

The most important of the other members of the group were André Derain (1880–1954) and Maurice Vlaminck (1876–1958). The influence of Signac is apparent in the fragmentation of their brushstrokes [2], although the mood of their paintings owes far more to Vincent van Gogh (1853–90). At their best their pictures have an intense emotional force rooted in the immediacy with which the spectator feels he has shared in their creation.

Colour and emotion

This desire to transmit emotion links Derain and Vlaminck with Expressionism. However, Expressionism is not an historically precise term like Fauvism. It covers a whole range of art that, broadly speaking, is more concerned with expression than beauty and distorts the subject to that end.

The most significant influences on twentieth-century Expressionism were van Gogh and the Norwegian Edvard Munch (1863–1944), with his powerful and neurotic evocations of the tensions that underly daily life [Key].

The Brücke ("Bridge") group, founded in Dresden in 1905 and including Ludwig Kirchner (1880–1938), Karl Schmidt-Rottluff (1884–) [4] and Erich Heckel

CONNECTIONS

See also
118 Impressionism
176 Cubism and Futurism
204 Abstract art
280 Art and architecture in 20th century Britain

1 Paul Gauguin's "Taa Matete" (1893) simplifies reality into a colourful and decorative frieze. While the painter found stimulation in the life of Tahiti as a source of subject-matter, this work stems stylistically from Egyptian art in the figures' poses.

2 Fauve colour is intense in André Derain's "The Pool of London" (1906) and its lack of concern for realism is particularly noticeable in the portrayal of what must have been a very grey and dull scene. The high viewpoint contributes to the flattening of space and hence the reduction of the picture to a pattern.

3 In Matisse's "Harmony in Red" (1908–9), a large-scale decorative painting, the patterns on the wallpaper and the cloth are as important pictorially as the woman on the chair. The flatness of the composition is so extreme that the view from the window has been taken for a picture on the wall.

4 "Rose Shapiro" by Karl Schmidt-Rottluff uses certain Cubist conventions, such as geometrical forms and the stylization of facial features, to achieve a direct and pungent image unhampered by unnecessary detail. However, to suit his expressive purpose he also employs sharp perspective, as in the table or the window. Raw impact is gained by the rough canvas showing through.

(1883–1970), were influenced by the stark, violent colour of Fauvism, but the content of their art is fully Expressionist. (Even at their wildest, the French group essentially continued in the Impressionist tradition, transmitting joy in nature and light.) The work of Brücke painters was full of venom against nineteenth-century materialism and their paintings – particularly those of Kirchner – present a morbid and pessimistic view of contemporary society.

Abstraction and social criticism

Wassily Kandinsky (1866–1944) and Franz Marc (1880–1916) [5], who were working together in Munich in the years preceding World War I, were motivated by a similar rejection of materialism and pushed the distortion of the object towards total abstraction. Together they published an almanac, *Der Blaue Reiter* to which composers and critics as well as artists contributed.

Kandinsky arrived at a complete dissolution of the object in his work by a combination of Expressionist distortion and an emphasis on the picture surface by methods similar to those already used by Matisse.

After the war there emerged in German painting a new tendency which contemporaries called "New Objectivity". While reacting against the strident technique and colour of the earlier Expressionists, painters such as Otto Dix (1891–1969) and George Grosz (1893–1959) continued to employ distortions as a means of expressing their protest against the injustices of society.

Max Beckmann (1884–1950) is usually classed with these painters, but his work is more private in its imagery [7]. The Swiss Paul Klee (1879–1940), a close associate of Kandinsky, veered between abstraction and child-like fantasy [6].

The rise of Hitler effectively put an end to the development of modern art in Germany until after the war. But then the Expressionist tendency continued internationally. An outstanding practitioner today is the British painter Francis Bacon (1910–), whose painting owes little stylistically to the artists described above but, like their work, confronts the viewer with potent images of extreme situations [8].

KEY

Edvard Munch's "Evening on Karl Johann Street, Oslo" (1892) takes an everday subject that might have appealed to the Impressionists. However his concern is not the visual world, but the revelation of the anxieties of urban man. Notice how, in the background, the brightly lit windows have a sinister, almost monstrous presence; the obsessed eyes; and the way that the faces acquire a ghost-like quality in the gaslight. Munch's art was the expression of a neurotic personality, but he depicted more than a personal malaise. The oppression of bourgeois city life was a recurrent theme in Expressionism.

5

6

7

5 Franz Marc's "Fate of the Animals" is not just a forest scene but a comment on all the most threatening aspects of nature. To depict the particular was not enough for the Expressionists.

6 Paul Klee's picture "Senecio" (1922) is based on a kind of humanized geometry. Klee methodically investigated form.

7 In "The Night" (1918–19), Max Beckmann's art is shown to be Expressionist in its violence, but its power lies in ambiguity. This is far from the robust directness that characterized Brücke painting.

8

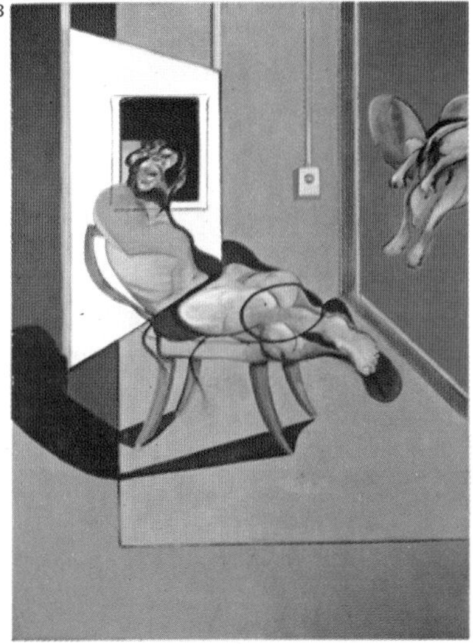

8 Francis Bacon's "Seated Figure" (1974) creates an uneasy atmosphere more by precarious postures and intense handling of paint than distortion or the use of violent colour, which have lost their real power as Expressionist devices.

Cubism and Futurism

Cubism was a term of abuse invented without understanding by a disgruntled critic, Louis Vauxcelles. It came to mean an international movement whose influence is still felt not just in painting but in sculpture and architecture. Pablo Picasso made it possible. He wanted to shock, a desire rooted in the philosophy of Friedrich Nietzsche (1844–1900) and in the demand for an individualistic assault on all conventions. Georges Braque (1882–1963) met him in 1907 and was indeed at first shocked by his work, but later responded positively with an ambitious painting "The Large Nude" (1908) [1].

The basic features of Cubism are present in this painting. First, the nude is distorted by fusing into a single image more than a single view of its parts. Second, it is treated as an arrangement of forms shallowly modelled in relief and not as a fully three-dimensional figure. Both these features followed from the shared conviction that painting should not imitate the appearance of things at any one moment (as in Impressionism) but should present the artist's accumulated idea of his subject, and that painting should be itself an art of flattened forms, not of three-dimensional illusions.

Between 1908 and 1911 a further feature was added to Cubist painting. Space was solidified, making the picture a single arrangement of flattened surfaces. Braque was the first to move in this direction, inspired by Paul Cézanne's attempt to treat the world as a mosaic of flat colour patches [2], and it was only in 1909, when Picasso also looked back to Cézanne that he followed Braque's lead with a series of landscapes whose skies appear as a crystalline structure almost attached to the buildings below them [3].

The invention of collage

The process of fragmentation followed by Picasso and Braque took them to the very edge of abstract art, but they always left recognizable details in their paintings because for them the real point was to create a flexible give-and-take between the spectator's appreciation of structure for its own sake and his remembered knowledge of the structure of figures and objects in nature. The invention of collage (material stuck on the canvas) in 1912 made possible both a flatter effect and a clearer reference to the objects of the subject. Although it was first developed by Picasso [5] and Braque, the painter who most clearly used collage to create a conflict between objects and pictorial structure was Juan Gris, the closest of their early allies [4].

Principal painters of Cubism

It was Picasso, Braque and Gris who developed the central line of Cubism, taking it further after 1918, but each in an increasingly personal way. For them Cubism was never a style with a single "look"; its basic principles lay behind innumerable variations.

These Cubists remained unconcerned with communication to a wide public even when their work began to sell during the 1920s, but from as early as 1909 Cubism was taken over by artists actively concerned with communication, who often took their themes from the most popular aspects of emerging industrial society. In Paris there were the painters Jean Metzinger (1883–1956) and Albert Gleizes (1881–1953), the Duchamp brothers, the husband and wife painters

CONNECTIONS

See also
202 Dada, Surrealism and their legacy
280 Art and architecture in 20th century Britain
174 Fauvism and Expressionism
204 Abstract art
118 Impressionism
178 Origins of modern architecture
158 Industrialization 1870–1914

1 The distortion in Georges Braque's "Large Nude" (1908) is almost fully Cubist. The buttock is presented both in profile and from behind, the left side of the back is pulled forward and the head is swung round to look out of the picture. Moreover, the figure is made even more fully Cubist by being treated in shallow relief, its surfaces defined with an angled brushstroke precisely like the surfaces of its equally flattened setting. One of the most important influences on Braque was Paul Cézanne, from whom this "hatched" brushstroke is borrowed.

2 Paul Cézanne (1839–1906) often painted the subject of this 1904–6 oil painting, "Mont Sainte-Victoire". He wanted to show how the volume and space was revealed by the fall of light on surfaces – warm colours where the sun struck, cool where it did not. This led him to break his paintings down into small dabs of colour, creating an effect that is both atmospherically spacious and flat. The Cubists, adopting this technique, also broke their surfaces down into small flat areas like the facets of a jewel.

3 Pablo Picasso (1881–1973) was on holiday in the Spanish Pyrenees when he painted this oil, "Landscape, Horta de San Juan" in 1909. Reversed perspectives flatten the roofs, while the sky is effectively solidified. Braque made similar although cooler landscapes in 1908–9. It was these landscapes and the still lifes influenced by Cézanne that prompted the critic Louis Vauxcelles to compare the work to "little cubes" which later became "cubism". However this work does not represent a fully developed version of Cubism in that it still uses traditional forms of pictorial space to denote three dimensions, albeit in distorted form.

4 Juan Gris (1887–1927) achieves a perfect balance between composition and subject in his 1913 collage "Violin and Engraving". He presents the objects aspect by aspect. These fragmentary aspects together form an idea of the objects which is complete. The still life is actually more fully represented than in traditional illusionist painting, yet each fragmentary aspect is firmly contained within a stable composition of vertical strips. Gris thus creates an artificial structure on a flat surface to make another structure in three dimensions – the still life. The tiny framed engraving is in fact a real engraving stuck on, and so important was Gris's subject to him that he even suggested that a future owner might change the engraving as if it were the decor of a real room.

Robert (1885–1941) and Sonia Delaunay (1885–), who tried to fuse an interest in colour with a Cubist sense of structure, and most impressive of all, the Norman painter Fernand Léger (1881–1955). It was Léger who successfully experimented with the flat cut-out colour planes of Cubist "collage" and the tangible modelling of early Cubist painting, adapting them to the task of communicating the sheer force of city life [6].

Outside Paris, Cubism initiated a spate of *avant-garde* movements: Vorticism in London, Russian and Czech Cubism, but most important of all, and an influence in modern art often the equal of Cubism itself, the Italian movement, Futurism.

The development of Futurism

Futurism was invented by the poet Filippo Tommaso Marinetti (1876–1944) who saw life as constant change and individuals as part of a dynamic system of forces caught up in progress. Modern experience heightened this vision – change was so dramatic, the machine capable of such speed. Umberto Boccioni, Carlo Carrà, Luigi Russolo, Giacomo Balla

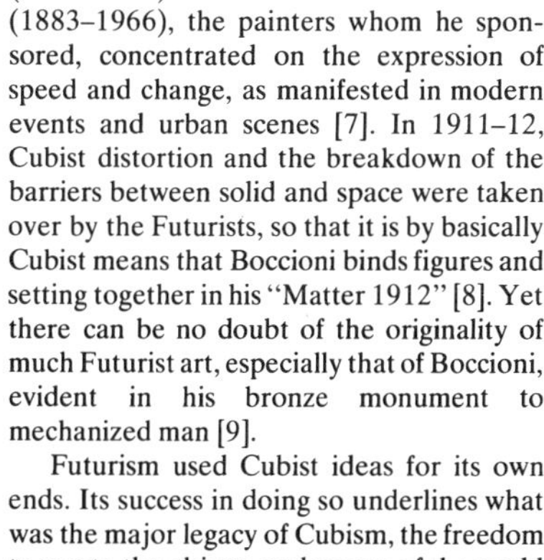

(1871–1958) and Gino Severini (1883–1966), the painters whom he sponsored, concentrated on the expression of speed and change, as manifested in modern events and urban scenes [7]. In 1911–12, Cubist distortion and the breakdown of the barriers between solid and space were taken over by the Futurists, so that it is by basically Cubist means that Boccioni binds figures and setting together in his "Matter 1912" [8]. Yet there can be no doubt of the originality of much Futurist art, especially that of Boccioni, evident in his bronze monument to mechanized man [9].

Futurism used Cubist ideas for its own ends. Its success in doing so underlines what was the major legacy of Cubism, the freedom to create the objects and scenes of the world in a fresh way. Cubism was not a style but a movement that made many styles possible because it allowed artists to paint ideas as well as what they saw. So infinitely adaptable has it been that it has led to developments as widely divergent as the geometric art of Mondrian and De Stijl and the so-called "pop art" of Robert Rauschenberg.

KEY

In Georges Braque's "Still Life with Fish" (1910) things are so fragmented and so absorbed into the overall linear scaffold that the painting is almost abstract art. The recognizable bottle and fish heads allow one to "read" the still life and also to see how much of the scaffold is the result of the distortion of observed things.

5 Picasso's collages, such as the 1913 "Violin", are less tidy than those of Gris. He collected junk and enjoyed the idea of making something out of otherwise worthless items. Here he pursues a series of paradoxes. The cut-out color prints of fruit (very realistic) sit on a piece of newspaper cut unrealistically to the rough shape of a fruit bowl. The solid violin head tops an utterly insubstantial body.

5

6 Léger, in his painting "Discs in the City" (1919–20), focuses on a combination of flat target discs, which appear mechanically geared together, suggesting the potential for movement. On either side there are scattered images from the city – robot men, crane derricks – which create the idea of an urban setting for the energy released by the colours and the whirl of the discs. The use of recognizable fragments to build a subject is Cubist; the subject and its dynamic interpretation have Futuristic overtones.

6

7

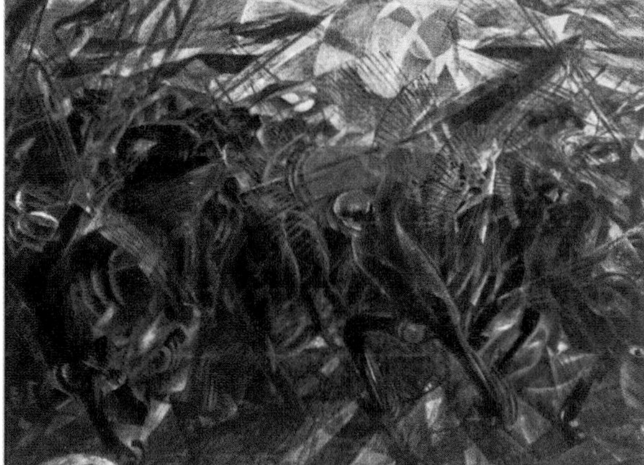

7 Carlo Carrà (1881–1966) in his oil "Funeral of the Anarchist Galli" (1911–12) combined riot and a funeral. The use of repeated images to represent the beating arms and legs is typical of Futurist painting at this time and generates a feeling of both psychological unity and violence.

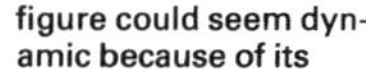

8 To the Futurists even an immobile figure could seem dynamic because of its unrelenting psychological mobility and potential for movement. Thus Boccioni painted his mother in a still pose in "Matter 1912", (detail, shown here).

8

9

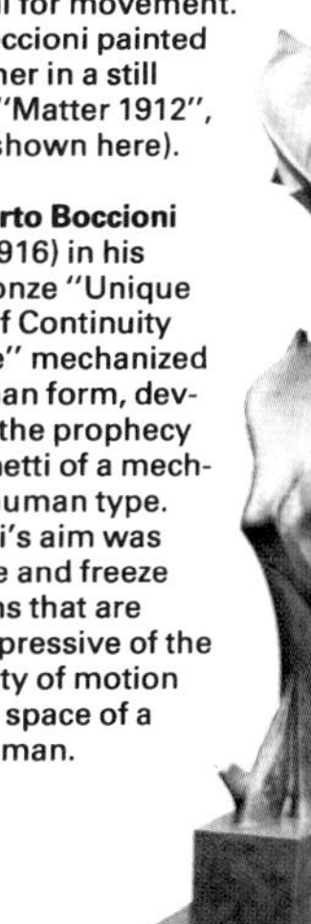

9 Umberto Boccioni (1882–1916) in his 1913 bronze "Unique Forms of Continuity in Space" mechanized the human form, developing the prophecy of Marinetti of a mechanized human type. Boccioni's aim was to define and freeze the forms that are most expressive of the continuity of motion through space of a striding man.

Origins of modern architecture

Modern architecture followed on the Industrial Revolution. Its styles were adapted to the discoveries of the engineers and to the mass production of materials such as iron and steel. The new movement began in France and Belgium and one of the most passionate advocates of a new style was Eugène-Emmanuel Viollet-le-Duc (1814–79), who claimed that iron construction must lead to new kinds of support and vaulting, and therefore a new architecture. The Paris Exhibition of 1889, for which the Eiffel Tower [1] was constructed, proved him right.

Art Nouveau in architecture

Towards the end of the century more and more buildings were being constructed in a freer, more naturalistic way. This Art Nouveau style was represented by Victor Horta (1861–1947) in Brussels, Hector Guimard (1867–1943) in Paris and Antoni Gaudi (1856–1926) in Barcelona, who all produced whimsical decorative styles which were free of backward-looking imitation. Horta in particular used decoration to underline slender elegance of iron construction [2].

However, there was another route towards modern architecture, often openly classical, and better attuned to the needs of mechanized production. This was the style developed by Louis Sullivan (1856–1924) and the Chicago architects after 1872, by Auguste Perret (1874–1954) and Tony Garnier (1867–1948) in France after 1900, by Adolf Loos (1870–1933) in Vienna and by Peter Behrens (1868–1940) in Germany.

Perret evolved the basic techniques of using reinforced concrete and his 1905 garage [3], now demolished, used concrete frames for posts and beams. Sullivan was capable of the most florid flights of decorative fancy, but Perret's dislike of surface decoration was shared by Garnier, Behrens and Loos, and stark simplicity became a feature of this anti-individualist route into modern architecture. The shape of the building became more important than decoration.

In 1907 German craftsmen and designers formed the Deutscher Werkbund, an organization that studied the problems and application of design. In 1914 the Werkbund held an exhibition at Cologne. Here the differences between the two main routes into modern architecture were thrown into high relief. On one side was the self-expressive architecture of Henri van der Velde (1863–1957), whose style was representative of Art Nouveau, and on the other was the architectural writer Hermann Muthesius (1861–1927), who stood for a functional modern style and a timeless "ideal" beauty. Somewhere in between was a former pupil of Behrens, Walter Gropius (1883–1969), his factory [Key], built for the exhibition, was geometrically simple and played on the effects of transparency and lightness produced by his steel-frame construction.

De Stijl and the Bauhaus

Yet the anti-individualist trend became dominant among those who searched for modern architecture during the 1920s. Headed by the painter Theo Van Doesburg a movement called De Stijl was founded in Amsterdam (1917). By 1923 its leading architect was Gerrit Rietveld (1888–1964), whose Schröder House, with its overlapping rectangles, its lack of complex curves, and its

CONNECTIONS

See also
122 European architecture in the 19th century
206 Modern architecture after 1930
208 Art and architecture in 20th century Britain

1 The Eiffel Tower was designed by Gustave Eiffel (1832–1923) for the 1889 Paris International Exhibition. He was already known as a daring engineer of some superb structures, including the Garabit Viaduct. When built, his iron tower was the highest structure ever known. Eiffel was also engineer of the Bon Marché department store in Paris (1876) and of the Statue of Liberty in New York, which was completed in 1886. Both of these use iron structures.

2 The Solvay House designed by Victor Horta, was built in 1895–6 for a rich Brussels manufacturer. It is a traditional double-fronted town house, but its combination of masonry and exposed iron construction was new, and so were many of its formal qualities. The projecting bays are glazed far more expansively than was usual, and their thin iron columns are shaped both to give the impression of growth and, where they carry weight, of gripping and lifting. Also the entire surface of stone gives the impression of a gentle swell, its flanks and the flanks of the bays curving outwards.

3 Auguste Perret's 1905 garage on the Rue Ponthieu in Paris uses a reinforced concrete frame. Perret, almost single-handed, developed the basic techniques for using this new material. He believed that reinforced concrete should be used like timber for frames and panels. The frame in this garage carries all the weight, as can be seen by the areas of glass. The building's structure is plainly expressed in the simple vertical and horizontal organization of the façade. The classical pillars, the small row of windows like a frieze above and the projecting cap are all features, however, of the Ecole des Beaux Arts tradition.

living-room window which turns a corner at first-floor level, pulled together many typically modern features [4]. Mies van der Rohe (1886–1969) was active in Berlin and between 1919 and 1928 Gropius headed the Bauhaus in Weimar and Dessau, which produced the first designs for "modern" furniture and fittings to be manufactured on any scale by industrialists. In France Charles-Edouard Jeanneret (1887–1965), known as Le Corbusier, took reinforced concrete architecture beyond the traditional classicism of his teacher, Perret. He got rid of the cornice, invented the horizontal window and avoided symmetry. His Villa Savoye [6] has all these features and an openness in its planning also found in the architecture of Gropius, Mies van der Rohe and Rietveld.

Towards an "International Style"

Planning as open as this was made possible by the fact that with steel or reinforced concrete construction internal walls were no longer needed to carry roofs, but it followed too from the desire for a more informal way of life coupled with a closer relationship between the house and nature outdoors. Frank Lloyd Wright (1869–1959), the most gifted of Sullivan's pupils, stressed the organic nature of architecture. He believed that a building, like a living organism, must "grow" out of its surroundings. Between 1893 and 1911 Wright built a number of small suburban houses (the "prairie houses") which were planned outwards from a central hearth. The open planning and flying horizontals of houses like the Willitts House (1902) [5] made a great impression on Gropius, the early architects of De Stijl and many others in Europe before 1914.

The individualistic modern alternative in architecture was never halted but during the 1920s the inventive styles of Gropius, Mies van der Rohe, Le Corbusier and De Stijl, characterized by an asymmetrical arrangement of simple geometric forms, by extensive glazing often turning corners, and by open planning, gave the illusion of a single modern style. It was this that led to the term "International Style", and the attempt, in the next decade, to create a modern international architecture based on shared convictions.

KEY

Walter Gropius's factory, built for the Cologne Werkbund exhibition in 1914, is one of the first wholly modern buildings. The office building with its sheets of glass wrapped around both ends and its glazing bars of steel establishes a clear rhythm. The horizontal slabs capping the short tower show the influence of Frank Lloyd Wright. Contrasts between transparency and opacity were to become common in this style.

4 A

B

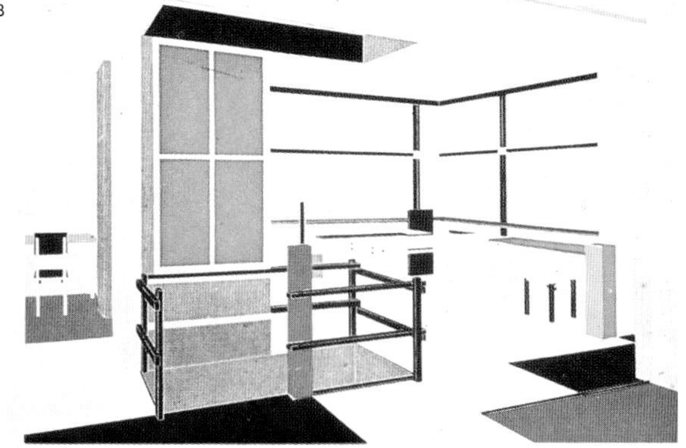

4 The Schröder House on the outskirts of Utrecht was designed by Gerrit Rietveld in 1924. The exterior [A] is brick built, with steel posts to support the projecting balconies only where necessary. Outside walls were white and grey, creating a neutral ground for the coloured horizontals and verticals. In the interior [B] large areas were in primary colours with white.

5 The Ward W. Willitts House in Highland Park, a suburb of Chicago, was built in 1902 by Frank Lloyd Wright. It was one of his "prairie houses". These were low, spreading houses, with open-plan interiors, terraces merging into the landscape and low, sloping roofs. The construction is traditional wooden post and beam. They were among the early Wright designs.

5

6 A

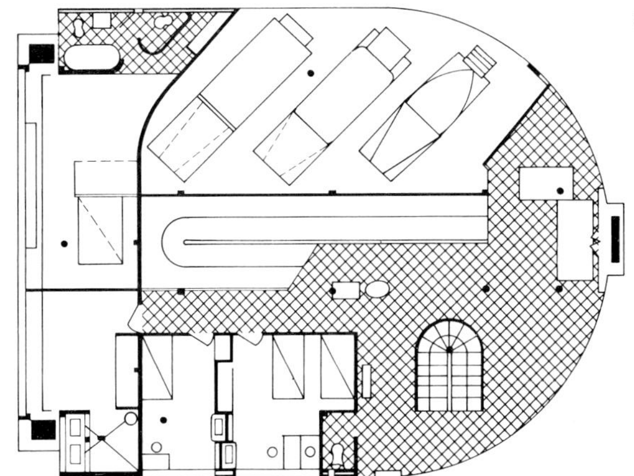

B

6 The Villa Savoye [B] at Poissy, east of Paris, was designed by Le Corbusier in partnership with his cousin between 1929 and 1931. The plan of the ground floor [A] shows the curved end of the structure and the way that car space is incorporated into the building to preserve the idea of a self-contained structure. Living quarters are mainly on the first floor, which is largely open plan, together with some services. There are four more rooms at ground level.

Europe 1870–1914

The period after the unification of Italy and Germany witnessed the consolidation and growth of the major nation states. Rising population, growing industrialization and stronger governments created a period of immense dynamism, but also intense national rivalry. The rise of democratic institutions in many parts of Europe and the development of trade unions encouraged more social legislation, such as welfare programmes. By the outbreak of World War I, socialist parties had appeared in many countries.

The rise of German power and influence

In terms of population, trade, industry and armed forces Imperial Germany was clearly the most powerful European state [4]. Its easy conquest of France in the Franco-Prussian War of 1870–71 testified to its military strength [1]. Following the war the German Chancellor, Count Otto von Bismarck (1815–98), sought to create a stable diplomatic environment in which a "satiated" Germany would be able to consolidate its gains and build up its international power and prestige. Germany's Dual Alliance with Austria-Hungary (1879) and the Reinsurance Treaty with Russia (1887) were designed to prevent those two countries clashing in the Balkans. Bismarck's diplomatic system survived recurrent crises over this issue [3] until his resignation in 1890.

The Dual Alliance became the Triple Alliance with the addition of Italy in 1882, and was faced by the Franco-Russian alliance of 1891. Great Britain joined France in the *Entente Cordiale* in 1904 and an Anglo-Russian treaty was signed in 1907, forming the Triple Entente. Bismarck's bequest became a dangerous system of alliances which was put under severe strain by imperial rivalry, Balkan crises and the instability of the Austro-Hungarian Empire.

Domestically many European states made considerable advances. In Britain extension of the franchise in 1867 and 1884 gave votes to many working men. France also operated a parliamentary democracy. Although still largely an autocratic state, Imperial Germany had the façade of constitutional government and political groups were developing rapidly, including a powerful socialist party. In northern Europe, the Scandinavian countries evolved along a largely peaceful path, often pursuing progressive social legislation.

In southern Europe parliamentary democracy existed only to a limited extent. Italy [8] was threatened by its own poverty and frequent periods of disorder and political instability. In the Iberian Peninsula a small middle class and the powerful hold of the Roman Catholic Church meant that politics remained oligarchic and backward. In eastern Europe, Austria-Hungary [5] remained an essentially monarchial state, troubled by severe national rivalries.

The conflict between Church and state

The growing power of the nation states and an increasing degree of state intervention in the areas of public education and welfare brought conflict with the Roman Catholic Church. The Church was attacked in many countries for political conservatism and opposition to liberal and national aspirations. In France the conflict was mainly about education, where the Church had great influ-

CONNECTIONS

See also
182 British foreign policy 1815–1914
158 Industrialization 1870–1914
130 Imperialism in the 19th century
110 German and Italian unification
186 Causes of World War I
184 Balkanization and Slav nationalism
212 Socialism in the West

1

1 The entry of Prussian troops into Paris at the end of the Franco-Prussian War of 1870–71 illustrated the power of the newly unified German state under the rule of the Hohenzollern dynasty and the direction of Bismarck. Domination of Europe by France as the greatest continental power was rudely supplanted by the growing industrial might of Imperial Germany, whose armies made efficient use of the German railways and artillery built by Krupps. In France defeat toppled Napoleon III's Second Empire and, after the Paris Commune, ushered in the Third Republic.

2

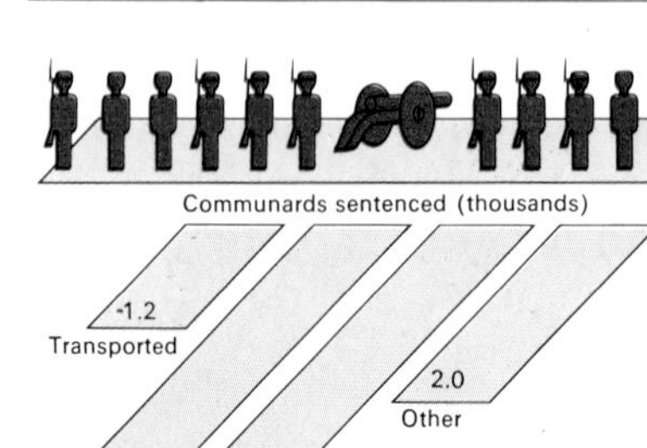

2 The Paris Commune followed privations endured in the siege of Paris during the Franco-Prussian War. When a new government at Bordeaux called in Paris rents, the lower middle classes and workers revolted and, although greatly outmanned and outgunned, they held the city from March to May 1871. They introduced a semi-socialist regime until savagely suppressed by government troops.

3

3 The great powers all attended the Congress of Berlin in 1878. A major source of conflict was the fate of the decaying Ottoman Empire and its Balkan dependencies, in which the interests of Austria-Hungary (represented by Karolyi, far left), and Russia (Shuvalov, right foreground, shaking hands with Germany's Bismarck), were deeply involved. The Congress recognized the independence of several Balkan states but denied them some of the territory they had just won from Turkey with Russia's help. Austria was allowed to occupy Bosnia-Hercegovina while France and Britain also made gains. The Congress however left all parties unsatisfied.

ence. Republican aims were advanced by the French statesman Jules Ferry (1833–93), who secularized education through legislation in 1882 and 1886. In spite of a period of relative amity between Church and state in the period that followed, known as the *Ralliement*, the Dreyfus affair [7] once again revealed the old tensions and led to bitter anti-clerical feeling. As a result, the concordat between the Papacy and the state was ended in 1905.

In Germany, too, between 1870 and 1880, Bismarck waged the *Kulturkampf* in which the Jesuits were expelled, religious orders dissolved, civil marriage made compulsory and other anti-Catholic legislation introduced [6]. In Italy, Belgium and other Catholic countries similar clashes occurred, although on a lesser scale.

Tariff reform became a pressing political issue in an era of growing rivalry in international trade and an influx of cheap foodstuffs from outside Europe. France protected its manufacturers by the Meline Tariff of 1892 and Germany built up its industry behind protective barriers. Even *laissez-faire* Britain witnessed a tariff reform campaign in 1902–5 by Joseph Chamberlain (1836–1914) which, however, failed to secure majority support among the electorate for protection of British and colonial goods.

Appeals to patriotism and nationalism

Several states sought to appease growing working-class demands by social legislation. In Britain, Benjamin Disraeli (1804–81) and later David Lloyd George (1863–1945) introduced social welfare. The latter copied the comprehensive social insurance schemes of Bismarck. In France, although anti-clericalism and other issues of the past could still create great passion, politics essentially constituted the safeguarding of vested interests and social legislation lagged. Governments everywhere tended to rally public opinion by stimulating patriotic feeling. Growing literacy, prosperity and communications also fostered intense nationalism. Conscript armies, equipped with the weapons of modern industrial economies, created war machines [Key] capable of unprecedented warfare.

KEY

A growing armaments industry towards the end of the 1800s produced weapons such as this German howitzer, which fired a 45kg (100lb) shell. Consolidation of nation states and the emergence of an intense patriotism was translated by conscription and industrialization into mass armies with which the nations of Europe faced each other in 1914.

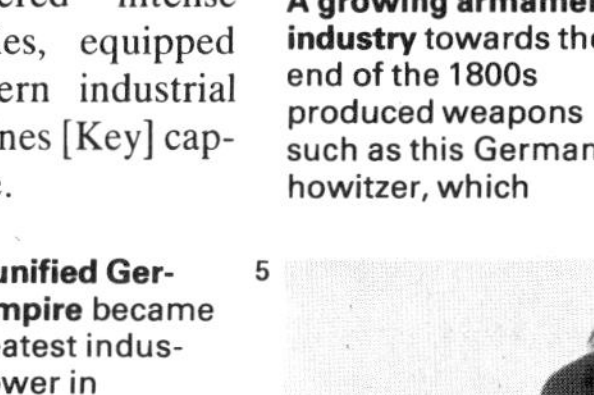

4

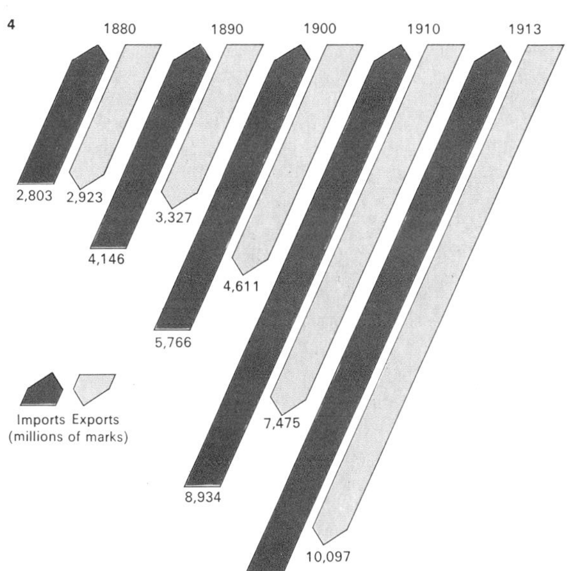

4 The unified German Empire became the greatest industrial power in Europe in the years before World War I, surpassing Great Britain in many branches of manufacture by 1900. From 1880 Germany's trade soared and both imports and exports increased more than threefold by 1913.

5

5 Elegant women, dashing officers – the outward glitter of "Gay Vienna" in the late 19th-century – masked a rich intellectual and artistic life that stemmed not only from the polyglot Austro-Hungarian Empire but also from much of eastern Europe. The culture it produced influenced the whole of Europe.

6

6 Count Bismarck was a master of diplomatic chess, countering the interdicts of Pope Pius IX (1792–1878) with anti-monastic legislation, as shown in this cartoon of the day. He presided over the unification of Germany, conducting both foreign and domestic policy with ruthless cunning until his resignation as Chancellor in 1890 after disagreement with the new kaiser, Wilhelm II. Groups such as the Catholics and socialists were subordinated to the interests of the state.

7

7 Caricatured as a traitor to France, Captain Alfred Dreyfus (1859-1935) was the centre of a bitter controversy after 1896, when it emerged that an army court had unjustly convicted him of spying for Germany. Dreyfus was a Jew and both anti-semitic and ultra-conservative groups tried to block a fair retrial. Anti-clerical and radical groups supported him with ultimate success and the issue showed the deep divisions underlying the apparent stability of France.

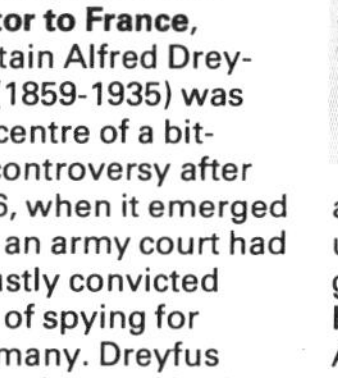

8 Giovanni Giolitti (1842–1928), five times prime minister of Italy between 1892 and 1921, managed to achieve periods of near stability and considerable industrial progress at a time when Italy was socially and economically backward. Parliamentary democracy was often difficult to introduce in recently unified states and in Italy political strikes and hunger riots were common before 1914.

8

British foreign policy 1815–1914

The years between the final British victory over the French at Waterloo in 1815 and the outbreak of European war in 1914 are known by the British as the *Pax Britannica.* They were not years in which Britain was entirely free from war, but, defended around the globe by the world's most powerful navy, it faced no direct threat to its security. During this period, Britain's foreign secretaries played upon a world stage, able to take an enlarged view of their duties and so to weave into their strategic considerations matters of very wide political import.

Protecting free trade and the empire

For most of the century Britain was able to conduct its foreign affairs with mere deference to the views of other powers. Britain's main strategic aims were to protect the empire, in particular the trade route to India, and to maintain the balance of power in Europe. Liberal statesmen tried to encourage the progress of liberal nationalist movements in various parts of the world. But in general, foreign secretaries did not interfere in foreign disputes.

Britain was, of course, favourably placed by the conquests of the eighteenth century and the strength of the navy [5] to look upon the world as its oyster. The idea of free trade came to dominate not only the Exchequer, but also the Foreign Office. British statesmen considered the world as a place in which all nations, freely trading with one another, would learn that commercial interdependence had made war obsolete as an instrument of national policy.

Only as a result of mounting fear of Russian influence in the Mediterranean did Britain intervene in the war of Greek independence [1] in the 1820s and the Turko-Russian quarrel that led to the Crimean War (1854–6) [2, 3]. The nascent power of Russia and the debility of Turkey, the "sick man of Europe", were eventually to turn the Balkans into a powder keg. For a century the "Eastern question" smouldered.

British liberalism abroad

The tendencies of the age were revealed in the 1820s, during the foreign secretaryships of Viscount Castlereagh (1769–1822) and George Canning (1770–1827). Their main achievement was to disengage Britain from the conservative Holy Alliance of the despotic northern powers – Prussia, Austria and Russia. At the Congress of Verona (1822) Britain refused to support intervention in Spain to put down the liberal constitutional government that had toppled the Spanish Bourbons. Nor would it aid the "reactionary" cause in Sicily and Portugal. In Latin America Canning gave his blessing and recognition to the revolts against Spanish and Portuguese rule that ended in the establishment of the independent nations throughout the continent. Canning also lent his support to the Greek patriots who fought to gain their independence from the Ottoman Turks. He died two months before the British navy destroyed the Turkish and Egyptian fleets off Navarino (Pilos, Greece) in October 1827, but in 1830 Greece became a fully independent nation.

In that year Palmerston [Key] began his first stint at the Foreign Office (1830–41), during which his most notable achievement was to assist Belgium to win independence

CONNECTIONS

1 **The revolt of the Greeks,** epitomized in this painting by Delacroix, was the first liberal cause of the century that took England away from the alliance that had defeated Napoleon. Whereas Austria and Russia opposed Greek freedom, Castlereagh and Canning supported the revolt, and English sympathizers went to fight for the Greeks against the Turks – among them the poet Byron (1788–1824), who died there.

2 **A Quaker deputation** led by Joseph Sturge on the eve of the Crimean War (1854) paid a special visit to Tsar Nicholas I to plead for peace. This was unofficial, and although the British cabinet was divided on the issue, public opinion clamoured for war. Radical MPs who denounced it, including John Bright (1811–89) and Richard Cobden (1804–65), lost their seats at the election of 1857, in which Palmerston was safely returned.

3 **The Crimean War** revealed the inefficiency of the army's organization and command. More soldiers died from disease than in battle. William Russell (1820–1907) reported the chaos in *The Times.*

4 **Giuseppe Garibaldi** (1807–82), the Italian nationalist leader, visited London in 1864 and received a great popular welcome, addressing crowds of 20,000 at the Crystal Palace. Several other Continental revolutionaries and nationalists had a similar reception, including the Hungarian Louis Kossuth (1802–94), who fled to England after the Russians had invaded Hungary following Kossuth's proclamation of Hungarian independence from the Hapsburgs early in 1849. Despite his dubious political ambitions, Kossuth was entertained by the foreign secretary, Palmerston. Support for Continental nationalist movements was a potent force in domestic politics in the 19th century; sympathy for the Italians' struggle against the Austrians took Gladstone, who had previously been a Conservative, into the Liberal Party in 1859. Garibaldi's visit to London in 1864 quickened the demand for parliamentary reform; this was met in 1867.

from the Netherlands. (British guarantees to Belgium had fateful consequences in 1914). In the East, Palmerston sought to uphold the territorial rights of Turkey. For a time peace was maintained, but in 1854 Russia and Turkey went to war and Britain and France entered on the side of Turkey.

The Crimean War and after

The Crimean War was ostensibly about the tsar's claim to protect Christians under Turkish rule in Europe; in fact it was about whether Turkey should maintain its empire in Europe as a bulwark against Russian aggrandizement in the Balkans. The British army suffered terrible losses, but, in the end, Constantinople and the Black Sea were preserved from Russian control.

Twenty years later, when Turkish misrule in Bulgaria threatened war once more, Benjamin Disraeli (1804–81) went to the Congress of Berlin (1878) and brought back "peace with honour". The *status quo* was upheld without war, but Turkey's failure to learn the lesson and put its house in order and the rising appeal of Slav nationalism throughout the Balkans was a bleak omen.

By the 1880s British security was being undermined. The scramble by European powers for colonies in Africa had begun and in 1882 Britain occupied Egypt. Germany was cutting into Britain's trading and manufacturing supremacy, and was politically worrying France. At the end of Victoria's reign, Germany started building up its naval strength.

As the German threat grew, fears of Russia receded. The Foreign Office was led to recast its priorities, and "splendid isolation" became a thing of the past. In 1904 Edward VII's diplomacy was instrumental in securing the Entente Cordiale with France. There were many people, among them Joseph Chamberlain (1836–1914), who hankered after a German partnership, but the current was flowing in the opposite direction. France was the ally of Russia and in 1907 Britain joined them in the Triple Entente. In 1908, when Bosnia-Hercegovina was annexed by Germany's ally, Austria-Hungary, against the wishes of Russia, the ground was prepared for World War I.

KEY

Viscount Palmerston (1784–1865) presided over British foreign policy longer than any other man in modern history. As foreign secretary (1830–41, 1846–51) and prime minister (1855–8, 1859–65), his policy rested on confidence in the global pre-eminence of mid-Victorian Britain. His forthright defence of British interests was expressed in the Don Pacifico debate (1850), when he used warships to protect a British citizen against the Greek government, and defended himself with the phrase *"Civis Brittanicus sum"* (I am a citizen of Britain), echoing the *"Civis Romanus sum"* of Imperial Rome. From that day until his death Palmerston was a national hero.

5

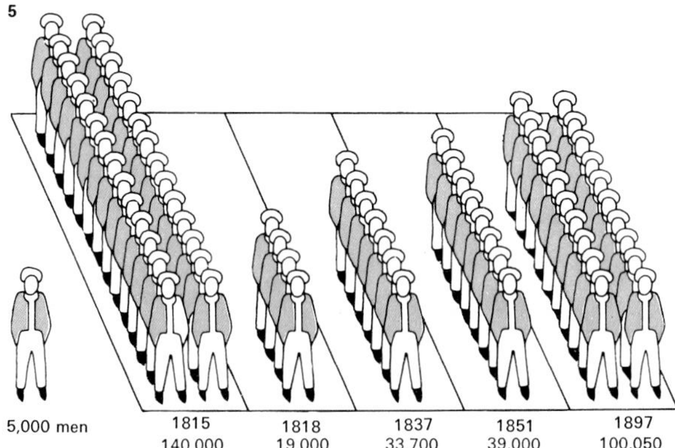

5 The British navy was the basis of the nation's power throughout the century. But the overwhelming victory at Trafalgar meant that the navy did not need to be large to maintain its ascendancy. It was only with the introduction of steam-driven battleships in the 1880s, and the start of the naval building race with Germany that the navy again employed as many men as in the time of Nelson.

6

6 The use of gunboats to quell local disturbances throughout the world, as in this expedition up the Nile to relieve Khartoum in 1884, was typical of the *Pax Britannica* as maintained by Palmerston.

7

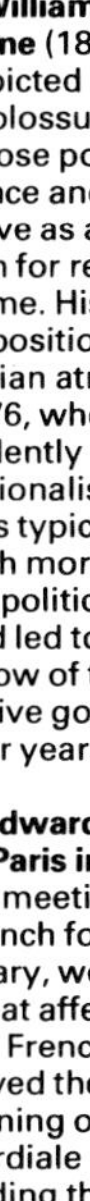

7 William Ewart Gladstone (1809–98) is depicted here as the "Colossus of Words", whose policies of peace and liberalism serve as an inspiration for reform at home. His stirring opposition to the Bulgarian atrocities of 1876, when the Turks violently put down a nationalist revolt, was typical of the high moral tone of his political feelings, and led to his overthrow of the Conservative government four years later.

8

8 Edward VII's visit to Paris in 1903, and his meeting with the French foreign secretary, won him great affection from the French. It also paved the way to the signing of the Entente Cordiale in 1904, so ending the enmity between the nations.

9

9 Naval strength was an important issue in the election of 1910, as this poster shows. HMS *Dreadnought* first of a powerful new class of battleship, was completed in 1906.

Balkanization and Slav nationalism

Austria-Hungary and Russia were the chief protagonists in the struggle to supplant the once powerful Turkish Empire, "the sick man of Europe" [3], as the dominant power in the Balkans in the second half of the nineteenth century. For Russia, the mastery of the Balkans would have served its historic aim: to gain control of the Straits of Bosporus and the Dardanelles with the city of Constantinople, and thus gain access to ice-free seas. Austria-Hungary's main concern was to prevent Russia from establishing itself in the Balkans as the protector of a cluster of small states, some claiming territory within the Hapsburg Empire. The Austro-Hungarian policy of blocking Russia's advance towards the Mediterranean was supported by both Germany and Britain.

Russian hopes dashed

In 1877–8, Russia fought Turkey on the side of Serbia and Montenegro in support of Slav Christians in the province of Hercegovina who had clashed with the Turkish authorities because they refused to pay taxes or to perform the customary labour services. A Turkish force sent against them in 1875 had been defeated with the aid of sympathizers from Serbia and Montenegro as well as from Austria-Hungary's Croat province of Dalmatia. The insurrection had then spread in 1876 to Bulgaria, where an estimated 12,000 to 30,000 Bulgarians were killed by Turkish irregulars in atrocities that aroused indignation throughout Europe.

Although Russian armies reached the outskirts of Constantinople in 1878, the diplomacy of Britain and Austria-Hungary frustrated Russia's main aim. At the Congress of Berlin [1], Russia secured territorial enlargement for Serbia and Montenegro and independence for Bulgaria. Austria-Hungary (which had stayed neutral) was allowed to occupy Bosnia-Hercegovina, Bulgaria was denied access to the Aegean, and the province of Macedonia, to which both Serbia and Bulgaria [2] aspired, was handed back to Turkey.

Serbian and Montenegrin successes in the war fired the imagination of all Slavs in the Austro-Hungarian Empire, but particularly those in the south: Croats, Slovenes and Serbs living outside Serbia proper in Bosnia, Croatia and Hungary. In Serbia itself the government covertly, and various non-official bodies overtly, gave money and encouragement to groups working for south Slav union. Serbian politicians and intelligentsia saw Serbia as the nucleus of a greater southern Slav nation [Key].

Revolutionary societies

Croats and other Slavs living in the Hungarian half of the Hapsburg Empire originally viewed the idea of a union with Serbia with suspicion, preferring a south Slav state under Hapsburg leadership. But alienated by Magyar dominance in Hungary, many of them became revolutionary towards the 1900s. Sensing the nationalist threat to their multi-national empire, the Hapsburgs redoubled efforts to control and subdue Serbia, in their view the originator of the monarchy's troubles. The annexation of Bosnia-Hercegovina in 1908 was the result. It was an attempt to pre-empt south Slav nationalism by simply incorporating a disputed area into the empire and thus, hopefully, neutralizing

CONNECTIONS

See also

84 European empires in the 19th century
108 The revolution of 1848
182 British foreign policy 1815–1914
110 German and Italian unification
186 Causes of World War I
180 Europe 1870–1914

1 **The Congress of Berlin** in 1878 drew up a Balkan settlement that was to last a generation. Dominant personalities were the British Prime Minister, Benjamin Disraeli (1804–81) and the German chancellor, Otto von Bismarck (1815–98). Under a treaty signed in July, Russia had to agree to the scrapping of the Treaty of San Stefano, made in March, giving her and her Balkan allies huge territorial gains. Under pressure from Britain, Austria-Hungary and Germany, victorious Russia agreed to limit itself to taking a strip of Bessarabia from Romania, Batum and Kars in the Caucasus and a part of Armenia. Romania's independence was formally recognized. Bosnia and Hercegovina were handed over to Austria-Hungary to administer. Britain was given Cyprus to keep as long as Russia kept Kars and Batum. Serbia and Montenegro received land that Bulgaria had gained earlier but remained cut off from the Aegean. Macedonia was returned to Turkey.

2 **San Stefano,** the name on the girl's flag in this Bulgarian poster, summed up Bulgaria's efforts to regain from her neighbours what she had won in the San Stefano treaty but lost at Berlin. To that end, Bulgaria fought and defeated Serbia in 1885, but was forced to withdraw after Austrian intervention. In October 1915 Bulgaria, allied to Austria and Germany, again fought Serbia.

3 **At Constantinople in 1876,** Sultan Abdul-Hamid II (1842–1918) proclaimed a constitution under pressure from Western-educated officials to reform the reactionary Turkish Empire. But he soon abrogated the constitution and it was only in 1908 that the Young Turk movement forced him to reissue it, summon parliament and abolish press censorship. When he prepared a counter-coup in 1909 he was overthrown and replaced.

4 **The German Kaiser, Wilhelm II** (1859–1941), here visiting Constantinople, played a major role in Germany's moves to acquire influence in Turkey as part of a larger extension of power in central Europe and the Mediterranean. Based on a concession granted in 1899 by Turkey to the German company of Anatolian Railways, a rail system was to be built all the way from Berlin to Constantinople and Baghdad as the key to a new German Empire.

it. Russia's weakness after her defeat in the disastrous war against Japan in 1904–5 enabled Austria-Hungary to escape without Russian retaliation.

The Balkan Wars

The Bosnian annexation initially turned the main thrust of Serbian nationalism south towards Albania and southeast Macedonia which Serbia, Bulgaria and Greece all claimed but which the Congress of Berlin had handed back to Turkey. Exploiting Turkey's preoccupation with its war against Italy in 1911–12, the four Balkan states – Greece, Bulgaria, Serbia and Montenegro – set up the so-called Balkan League and declared war on Turkey in October 1912 [6, 8] (the first Balkan War 1912–13). But the victorious anti-Turkish forces were again frustrated by great power diplomacy.

Germany saw Turkey as the strategic base of its own future thrust into the Middle East and beyond [4] to challenge its greatest rival, Britain. Under Austrian pressure the Serbs were denied access to the Adriatic by the establishment of Albania as a separate state. Serbia in turn quarrelled with Bulgaria over Macedonia and war broke out between them in June 1913 and lasted a month. Bulgaria was defeated by an alliance of all her neighbours including Romania.

But Hapsburg hopes of the situation becoming calmer in the wake of the Bosnian annexation were disappointed. Nationalist agitation [6] for a union of all south Slavs was boosted by Serbia's successes in the Balkan Wars. Assassinations by members of secret societies in Bosnia and elsewhere became commonplace. The apparent political impasse made Austria-Hungary's leaders think once again of a military solution. The idea was that if only Serbia, the hotbed of nationalistic agitation, could be subdued and neutralized, the rest of Europe would calm down. Germany's virtually unlimited backing of Austria-Hungary's policies strengthened the resolve of certain Austrian military and civilian leaders. The assassination of the heir to the Hapsburg throne, Archduke Franz Ferdinand, in Sarajevo in Bosnia in June 1914, by a revolutionary group based in Serbia, gave them the pretext for war.

KEY

The spirit of Slav nationalism is captured in "The Illyrian Revival" by a 19th-century Croatian painter, Vlaho Bukovac. Illyria was a hoped-for independent union of south Slavs under Croat leadership. But the Serbs, who managed to free themselves from Turkish rule in 1830 and became a kingdom in 1882, took the upper hand while Croatia and Slovenia remained part of Austria-Hungary.

5 The new Balkan states, formed as a result of Turkey's retreat from Europe, were in dispute with each other: Serbia and Bulgaria over Macedonia; Romania and Bulgaria over Dobruja; and Romania and Austria-Hungary over Transylvania. But the most explosive dispute was between Serbia and Austria-Hungary because of Serbia's support of terrorist activity among Slavs living under Austro-Hungarian rule. This was greatest in Bosnia-Hercegovina, a province of Croats, Serbs and Slav Muslims, which Austria-Hungary had taken over in 1878 and formally annexed in 1908 in order to prevent Serb agitation. Austria-Hungary also frustrated Serbia's attempt to gain direct access to the Adriatic by encouraging the formation of a separate Albanian state, which was proclaimed in 1912.

5

6

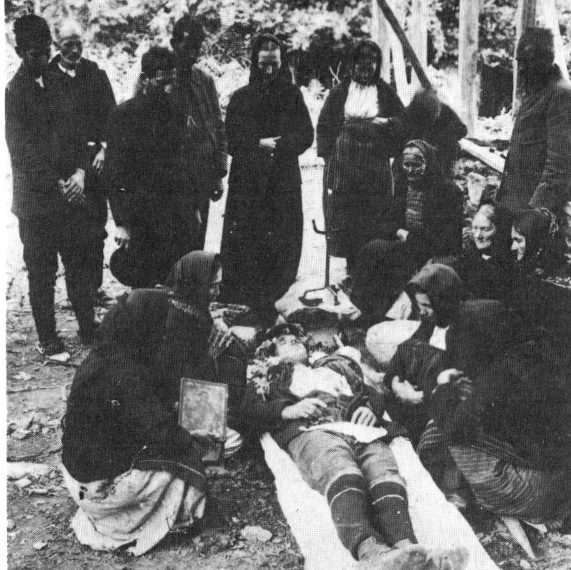

6 The first Bulgarian soldier to be killed in the first Balkan War is surrounded by mourners. Although fighting did not begin until October 1912, the seeds of the conflict were sown in a secret treaty concluded between Serbia and Bulgaria in March. They planned to attack Turkey and divide the spoils. According to this, Serbia was to have been given most of Albania. The war started when Montenegro attacked Turkey on 8 October. Bulgaria, Serbia and Greece then joined in, and soon the Turks were reeling under the combined onslaught. They asked for a truce in December.

7

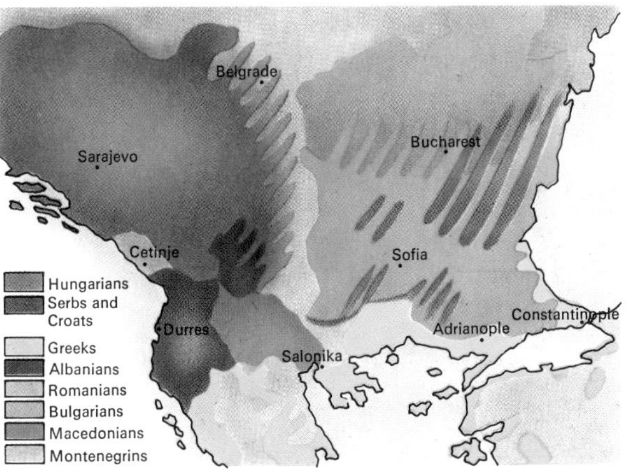

7 Peoples of many different races and religions inhabit the Balkans. The Croats have a Latin script and are Roman Catholics. The Serbs, Bulgarians, Montenegrins and Macedonians received their Cyrillic script, Orthodox religion and political tradition from Byzantium. Under Turkish rule the Orthodox Church retained its autonomy and was influential in the national revival of the Balkan peoples. Turkey left two enclaves of Islam in Europe: Bosnia and Albania.

8

8 Bulgarian troops (here shown near the Serbian border) moved against Serbia in June 1913, so starting the second Balkan war. The attack pre-empted Serbia's designs on parts of Macedonia, held by Bulgaria since the settlement to the first war. By August, however, Bulgaria had been defeated by an alliance of all her neighbours. In the settlement, Serbia gained most while Turkey, Greece and Romania also made teritorial gains.

Causes of World War I

CONNECTIONS
See also
158 Industrialization 1870–1914
184 Balkanization and Slav nationalism
182 British foreign policy 1815–1914
188 World War I
182 The Peace of Paris

During the 1890s Germany's ruling class, headed by the intelligent but vacillating German Kaiser, Wilhelm II (1859–1941), abandoned Bismarck's cautious foreign policy in favour of a more dynamic one designed to reflect Germany's industrial and military strength. Germany wanted a large colonial empire, not only for economic reasons but to enhance its prestige. To this end a law to expand the German navy, the first of many such laws, was enacted in 1898. The new navy was designed ultimately to challenge British naval supremacy [1] and to force Britain, faced with seemingly perpetual Franco-Russian hostility, to collaborate in a wholesale reallocation of colonial territory.

German diplomatic set-backs

The first set-back to Germany's "world policy" came in 1904 when Britain and France settled their colonial differences. Then, in 1907, Britain resolved its long-standing central Asian disputes with Russia, France's ally since 1894. In 1905 Germany, taking advantage of Russia's defeat by Japan, challenged France's increasing strength in Morocco [2] and coerced it into participating in an international conference in January 1906 at Algeciras to settle the Moroccan question on Germany's terms. However, Germany suffered a diplomatic defeat, for its plans for Morocco were supported only by Austria-Hungary. Moreover Germany's assumption that the Anglo-French *entente* would be wrecked by Britain's failure to support France proved to be similarly erroneous. Britain co-operated closely with France during the conference and, alarmed by Germany's aggressive policy, initiated unofficial Anglo-French military discussions.

Germany next proceeded to alienate Russia. In 1909 it insisted with a veiled threat of war that Russia recognize Austria-Hungary's 1908 annexation of Turkish Bosnia-Hercegovina and abandon support for Serbia's claim for compensation. International tension was further increased when, in a bid to secure colonial compensation from France, now almost in control of Morocco, Germany sent a gunboat to the Moroccan port of Agadir on 1 July 1911. Although during the following months Britain and France came close to war with Germany over the Moroccan issue, a Franco-German colonial compromise was signed in November. The crisis left a legacy of bitterness and hatred in both countries. As a result Germany, in 1912, further increased its naval strength and began to expand its army. It was followed inevitably in this action by every other Continental great power [3].

Instability in the Balkans

The causes of World War I were, however, more directly connected with events in the Balkans. In 1912 the Balkan League (Serbia, Greece, Montenegro and Bulgaria) drove Turkey out of most of its remaining possessions. The following year Bulgaria was defeated by its former allies, Greece and Serbia, and lost its Macedonian gains of 1912 to Serbia. Austria-Hungary was thus faced with a greatly enlarged and ambitious Serbia, determined that the Slavs within the Hapsburg Empire should come under this rule.

The cumulative effect of all these crises was to increase preparations for war: indeed, Germany had long since devised its blueprint

1

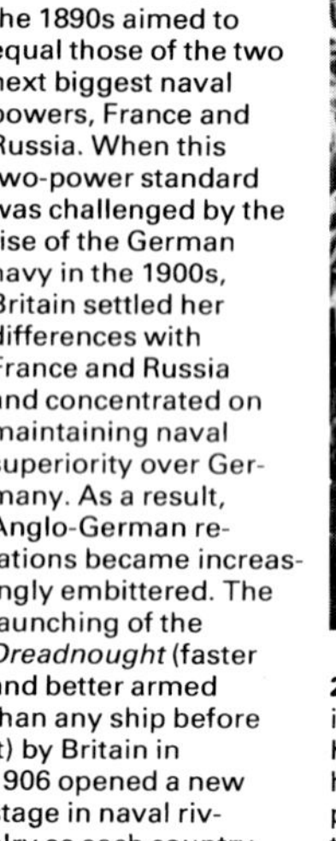

1 The British fleet in the 1890s aimed to equal those of the two next biggest naval powers, France and Russia. When this two-power standard was challenged by the rise of the German navy in the 1900s, Britain settled her differences with France and Russia and concentrated on maintaining naval superiority over Germany. As a result, Anglo-German relations became increasingly embittered. The launching of the *Dreadnought* (faster and better armed than any ship before it) by Britain in 1906 opened a new stage in naval rivalry as each country tried to build more such vessels than its neighbours. But Britain kept its lead.

2

2 Visiting Tangier in March 1905, the Kaiser pledged to uphold Morocco's independence. He hoped to protect German interests in Morocco (rapidly falling under French control) and to force France to recognize that its future lay in alliance with Germany. While the independence of Morocco was thus preserved until 1911, Germany's clumsy diplomacy drove France and Britain closer together.

3

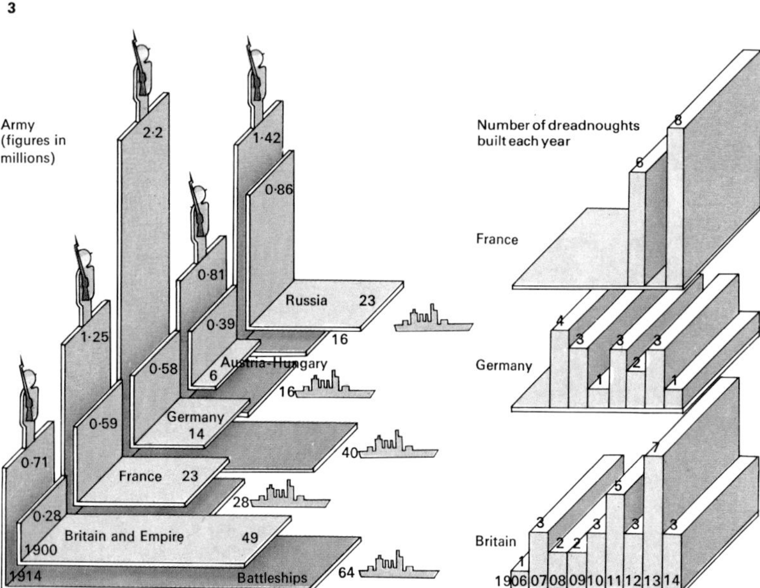

3 An armaments race between the Great Powers before 1914 both reflected and heightened European tension. In addition to building a large navy, Germany possessed the most formidable army in Europe. Although its size remained fairly stable from 1900–10, Germany's deteriorating diplomatic situation led in 1912–13 to increases in army strength which provoked the other Great Powers, except Britain (the only one with a volunteer army), to increase their own forces.

4

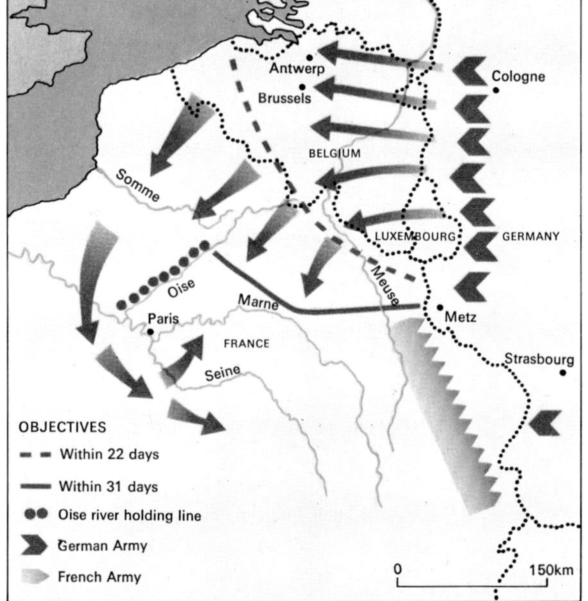

4 The Schlieffen Plan was based on a two-front war with Russia and France, which had been allies since 1894. It provided for a massive German assault through Holland (later excluded) and Belgium to outflank the French army. Meanwhile, Austro-German forces would defend the east until the main German army, having knocked out France, could be rapidly moved to meet the slowly advancing Russians. Violation of Belgian neutrality would risk British intervention.

for victory, perfected by Count Alfred von Schlieffen (1833–1913), Chief of Staff, in 1905, and amended by his successor, Helmuth von Moltke (1848–1916). The Schlieffen Plan [4] relied on the slowness of Russian mobilization and provided for a rapid thrust through Belgium to defeat France, leaving the German army free to move rapidly east to meet the Russians.

The assassination of the heir to the Hapsburg throne, Franz Ferdinand (1863–1914), at Sarajevo on 28 June 1914 [8] was the climax of a series of Serbian provocations towards Austria. Berlin feared that if Austria-Hungary failed to take the opportunity provided by the murder to bring Serbia within its orbit, its multi-national empire would collapse, leaving Germany isolated. Thus Austria was under German pressure to act against Serbia, with the promise of German military support should war ensue. Successive German diplomatic defeats, a sense of "encirclement" by Britain, France and an increasingly strong Russia, and deep divisions within German society all combined in 1914 to convince the German ruling élite of the desirability of war partly to preserve the idea of a German-dominated "*Mittel Europe*". Although apprehensive, German Chancellor Theobald von Bethmann Hollweg (1856–1921) gambled on both Russian and British neutrality [6] and hoped that the Austro-Serbian dispute could be localized, in spite of the rigid system of alliances that divided Europe.

The final steps to war

Austria finally [7] presented an ultimatum demanding the right to investigate Serbian terrorists and, when Serbia rejected this, declared war on 28 July 1914. Russia could hardly stand aside and, faced with growing pro-Slav feeling, Tsar Nicholas II (1868–1918) ordered mobilization. British mediation failed to persuade Austria-Germany to compromise. When France refused to leave Russia to fight alone, the Schlieffen Plan was activated and events proceeded rapidly [9] towards war between the Central Powers (Germany, Austria-Hungary and Turkey) and the Allies (Russia, Serbia, France, Belgium and Great Britain).

KEY

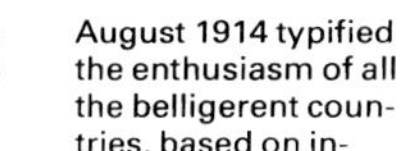

High-spirited French soldiers marching to the front after the outbreak of war in August 1914 typified the enthusiasm of all the belligerent countries, based on intense nationalism and a belief that the war would be short and glorious.

5 A wartime photograph of the Kaiser (centre) and his generals reflects his fondness for military life. Responsible for Germany's foreign policy, the ultimate decision to mobilize on 1 August was his alone.

6 Germany's Chancellor Bethmann Hollweg [left] and Foreign Minister Gottlieb von Jagow (1863–1935) misjudged the willingness of Britain to go to war over a "scrap of paper" guaranteeing the neutrality of Belgium. They gambled on diplomatic victory for the Central Powers when, with the promise of German military support, they encouraged Austrian action against Serbia in July 1914.

7 Count Leopold von Berchtold (1863–1942), the Austrian Foreign Minister, was convinced that the multi-national Hapsburg Empire would collapse unless Serbia was crushed. His opportunity was provided when Franz Ferdinand was murdered but although promised full German support, he encountered considerable opposition to his plans from the Hungarian government. This partly accounted for the delay in presenting the Austrian ultimatum to Serbia.

8 Gavrilo Princip (1893–1918) precipitated the chain of events leading to war when he shot the heir to the Austro-Hungarian thrones, Archduke Franz Ferdinand, and his wife while they were visiting Sarajevo, capital of Bosnia, on 28 June 1914. He was one of a group of Bosnian conspirators with Serbian support.

9 THE TIMETABLE OF CONFLICT JUNE 28—AUGUST 4 1914

1914	Austria-Hungary	Germany	Great Britain	Russia	France
June 28	Assassination of Archduke Ferdinand.				
July 5		Germany pledges support for action against Serbia.			
23	Austrian ultimatum presented to Serbia.				
25			Allied mediation rejected by Germany.		
27	Serbia accepts Austrian demands only in part.				
28	Austria rejects Serbia's reply and declares war.				
29		Germany fails to gain British neutrality.			
30				Nicholas II mobilizes Russian army.	
31	Austria mobilizes.	German ultimatum to France and Russia to stop war preparations.			
August 1		Germany declares war on Russia and mobilizes.			France mobilizes.
2		Germany invades Luxembourg, demands passage through Belgium.	Belgium appeals to Britain for support.		
3		Germany declares war on France.	British ultimatum to Germany to withdraw from Belgium.		
4			Germany fails to reply. Britain declares war.		

World War I

On 28th June 1914, the heir to the Austro-Hungarian throne, Archduke Franz Ferdinand (1863–1914), was assassinated in Sarajevo, Bosnia, by a pro-Serbian student, Gavrilo Princip (1893–1918), precipitating a chain of diplomatic manoeuvres that ultimately led to war. The Balkans had long been a centre of conflict. Serbian nationalism threatened the shaky Austro-Hungarian Empire, whose collapse would isolate her ally, Germany, in Europe. Russia, Serbia's ally, was also involved in the Balkans because whoever controlled them would be in control of Russia's main trade route.

The first battles on both fronts

Germany pressed her ally to take firm action and on 28 July Austria-Hungary declared war on Serbia. Two days later, Russia mobilized and Germany responded by declaring war on Russia on 1 August. Germany's Schlieffen Plan, drawn up to avoid a war on two fronts, necessitated an all-out attack through Belgium to knock out France, Russia's ally, quickly. Germany therefore declared war on France on 3 August and invaded Belgium the next day. As a result, Great Britain came to Belgium's defence.

By 9 September German forces had advanced to the Marne where the British and French were able to halt them. At the end of October each side faced the other in trenches running from the English Channel to the Swiss frontier. In the east, the vast, ill-equipped Russian army had lumbered into East Prussia where it was crushingly defeated on 20 August at the Battle of Tannenberg.

Throughout 1915 the Germans remained on the defensive in the west, allowing the Allies to exhaust themselves in a series of futile attacks, while launching a summer offensive in the east that hurled the Russians back more than 480km (300 miles).

Turkey had entered the war on the side of the Central Powers in October 1914. After a costly naval attack by the Allies, 75,000 Australian, New Zealand, British and French troops tried to open a new front at Gallipoli at the mouth of the Dardanelles. The expedition failed to achieve surprise, scarcely advanced from the beaches and suffered heavy casualties until withdrawn in December. Thus Russia was effectively cut off from Allied supplies.

By the end of 1915 both sides realized that the war was going to be a prolonged affair. On 21 February 1916, the Germans assaulted Verdun in an offensive calculated by General Erich von Falkenhayn (1861–1922) to exhaust the French, rather than to achieve a breakthrough. By the end of June nearly 600,000 men had died in this action, but the French managed to hold on. The Russians under General Alexei Brusilov (1856–1926) launched an offensive that gained some territory with terrible loss of life and the British under Field-Marshal Sir Douglas Haig (1861–1928) attacked on the Somme, suffering 20,000 dead on the first day and gaining less than 8km (5 miles) in five months' fighting.

The war at sea

At the beginning of the war, the Royal Navy had begun a blockade of German ports, turning back neutral shipping [6]. The Germans replied with submarine attacks [8], but had little success in 1915 and 1916 because

CONNECTIONS

See also
186 Causes of World War I
190 World War I: Britain's role
180 Europe 1870–1914
192 The Peace of Paris
184 Balkanization and Slav nationalism
190 What World War I meant to Britain
126 Australia and New Zealand to 1918
196 The Russian Revolution
212 Socialism in the West
208 The twenties and the Depression

In other volumes
164 Man and Machines
166 Man and Machines
170 Man and Machines
174 Man and Machines
176 Man and Machines
180 Man and Machines

1 Most of the fighting took place in Europe; the main battlefields were in northern France and Belgium, Poland, Russia and Italy. Overseas campaigns were fought in Mesopotamia and the Middle East and in the German colonies in Africa.

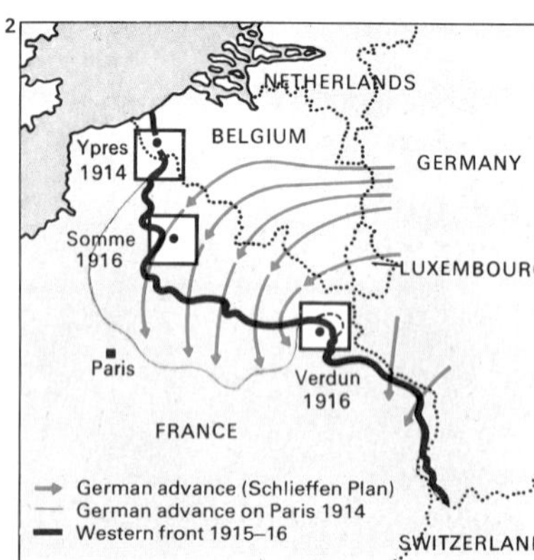

2 The Western front was the decisive battleground of the war. Once the Schlieffen Plan had failed to eliminate France, it was here that the bloodiest battles were fought, as both sides poured in men and materials to achieve the vital breakthrough. In 1918 the impetus of a new Allied offensive backed by the fresh American armies convinced the Germans that the war was lost even before the Allies reached them.

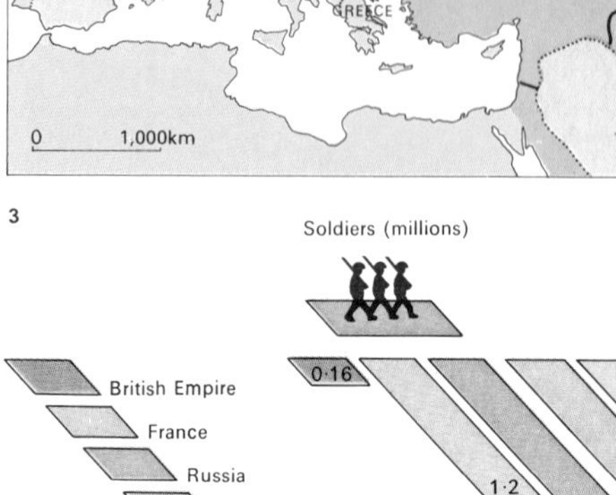

3 The strength of the two alliances was reasonably well-balanced, as what Britain lacked in troops she made up in naval strength. It was this balance that made World War I a war of attrition that was to result in horrific loss of life and massive destruction. Figures for troops quoted here are those of the standing armies. Mobilized forces were approximately: Britain 711,000; France 3.5 million, Russia 4.4 million, Germany 3.8 million (in emergency a maximum of 8.5 million could be raised), Austria-Hungary 3 million.

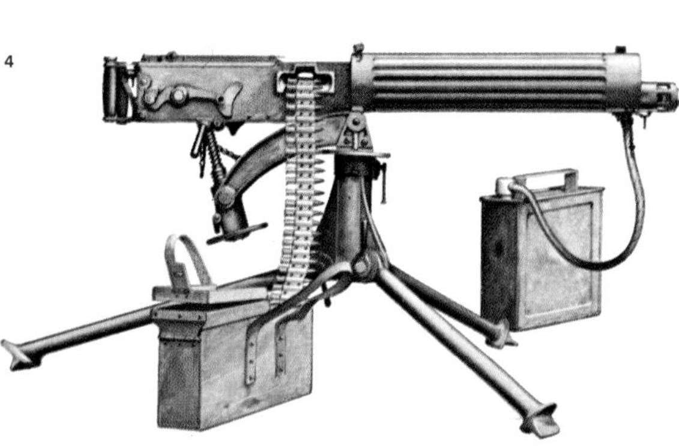

4 The generals of 1914 had been trained to think of mobile offensive warfare, but the relatively new British Vickers medium machine gun with its lethal effect on exposed infantry was among the armaments that upset their view. Once the exhausted armies had dug in, artillery and machine guns ensured that trench warfare would continue. Commanders tried for the rest of the war to break the stalemate, but massive infantry attacks proved hideously ineffective.

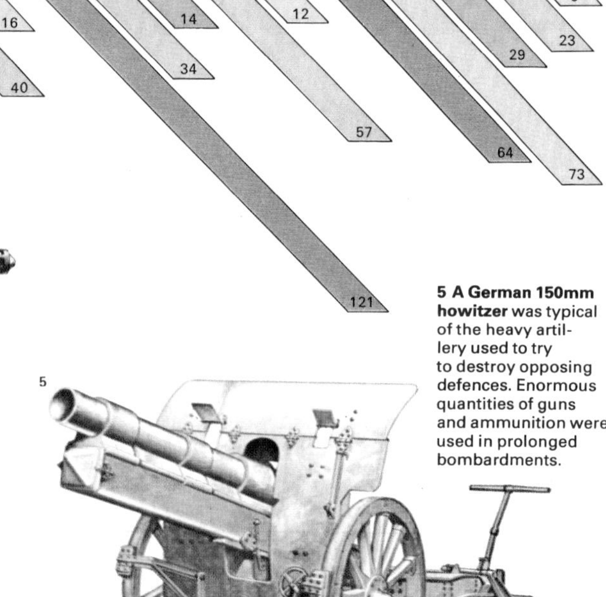

5 A German 150mm howitzer was typical of the heavy artillery used to try to destroy opposing defences. Enormous quantities of guns and ammunition were used in prolonged bombardments.

sinking neutrals was banned. The two great battle fleets fought only one major action, at Jutland on 31 May 1916. The outcome was inconclusive, but the German surface fleet remained in harbour for the rest of the war. During 1916 the blockade caused severe food shortages in Germany, which led to widespread unrest. On 31 January 1917, the Germans launched unrestricted submarine warfare, and by sinking US shipping pulled the United States into the war. Only the new convoy system prevented Britain from being economically strangled.

The final offensive and Allied victory

On the Western front the French began a series of unsuccessful offensives; elements of their army mutinied in May 1917, but were brought under control during June by Marshal Henri Pétain (1856–1951).

Tanks were used *en masse* at Cambrai on 20 November, but their initial successes were not followed up. Italy had entered the war on the Allied side on 26 April 1915 and fought inconclusively against Austria-Hungary until a massive defeat at Caporetto on 24 October 1917 almost knocked her out of the war.

In Russia the unpopularity of the war led to the overthrow of the tsar in March 1917. A provisional government launched another offensive but, after that had been thwarted, the Bolsheviks seized power in November and sued for peace. The Treaty of Brest-Litovsk in March 1918 gave Germany huge territorial gains in western Russia.

Aware that they must follow up success in the east with victory in the west before American help could arrive in force, the Germans opened a series of offensives under General Erich Ludendorff (1865–1937) from March to July 1918. They drove the Allies back to the Marne, but were again halted there. Then, strengthened by American troops, the Allies counter-attacked during August. A massive offensive launched on 26 September convinced the German High Command that the war was lost and they sued for peace. In early November anti-war and pro-Bolshevik risings took place, the kaiser abdicated on 9 November and an armistice was signed on 11 November. Austria-Hungary also collapsed in November after an Allied offensive.

KEY

For future generations World War I was to become a symbol of senseless slaughter and destruction. Not only did more than 10 million soldiers die, but the war affected every level of society in all combatant countries. Wholesale conscription was introduced and governments took dictatorial powers to control economies and to ration food and supplies. The war radically changed the map of Europe, sweeping away the German, Austro-Hungarian and Russian empires and setting up smaller states in Eastern Europe.

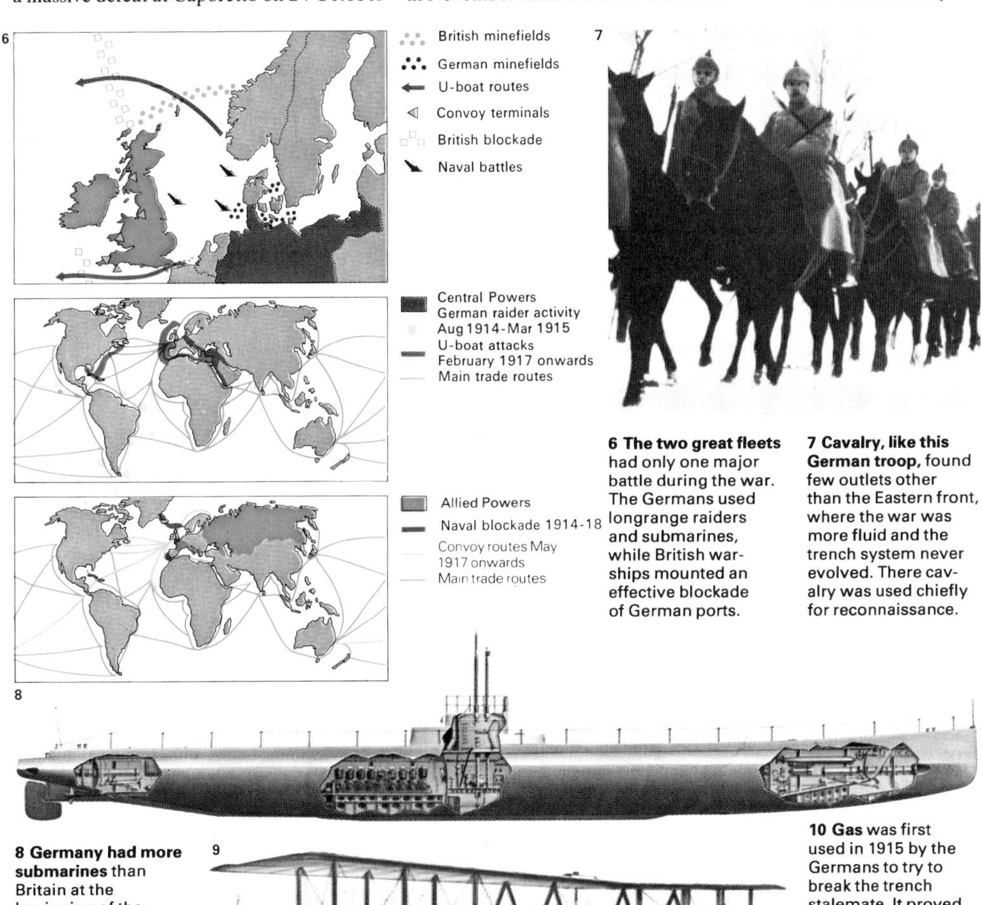

6 The two great fleets had only one major battle during the war. The Germans used longrange raiders and submarines, while British warships mounted an effective blockade of German ports.

7 Cavalry, like this German troop, found few outlets other than the Eastern front, where the war was more fluid and the trench system never evolved. There cavalry was used chiefly for reconnaissance.

8 Germany had more submarines than Britain at the beginning of the war. This is one of the Class 31-37 U-boats. It was 64.7m (212ft) long and fully submerged it weighed 880 tonnes. It was armed with 24 500mm (20in) torpedoes fired through four tubes. The attacks on British shipping were relatively ineffective during 1915–16. However, after 1917 the Germans came close to starving Britain into submission.

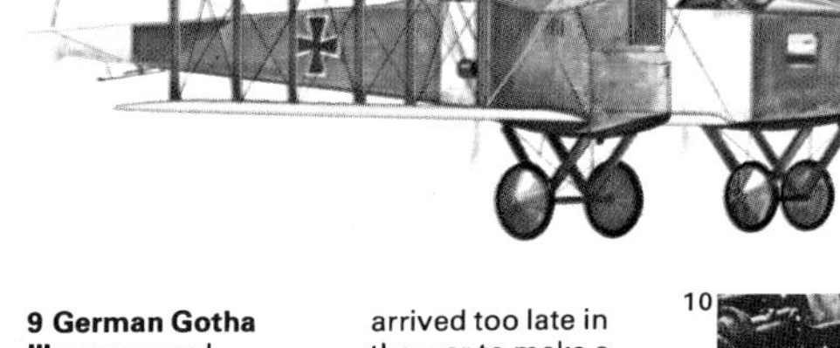

9 German Gotha IIIs were used for armed reconnaissance over the battlefields as well as for bombing. Developed to take over the Zeppelins' role in bombing English cities, they arrived too late in the war to make a significant difference. After their attacks on England (in which they claimed 857 lives) the Gothas were switched to the French theatre.

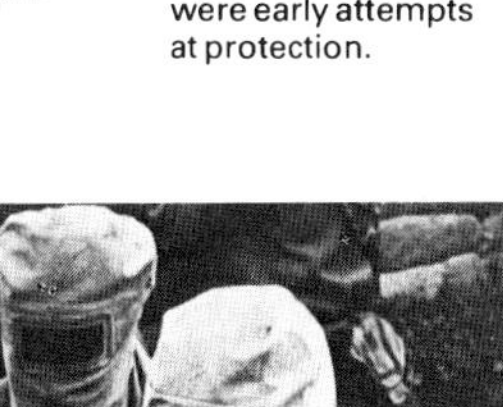

10 Gas was first used in 1915 by the Germans to try to break the trench stalemate. It proved inefficient, difficult to control and easy to detect. The masks these soldiers wear were early attempts at protection.

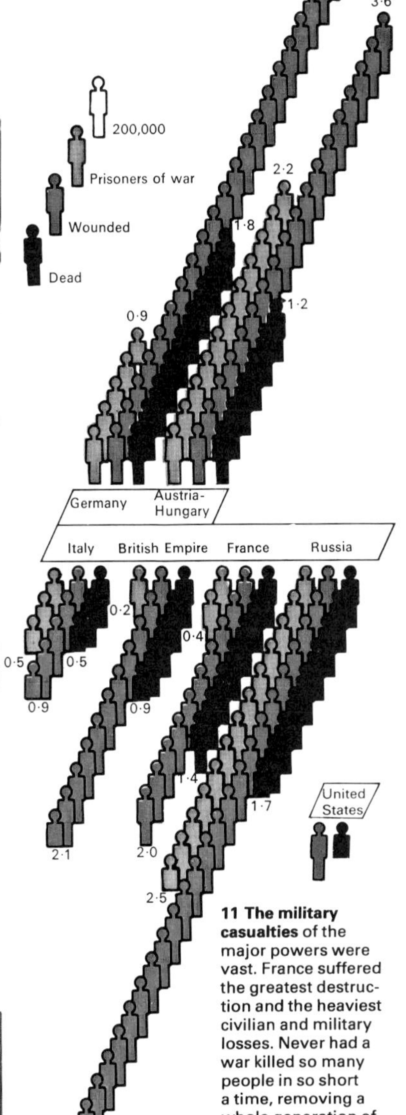

11 The military casualties of the major powers were vast. France suffered the greatest destruction and the heaviest civilian and military losses. Never had a war killed so many people in so short a time, removing a whole generation of young men and scarring Europe for the next 20 years.

World War I: Britain's role

Britain's small but professional expeditionary force of 100,000 men, commanded by Sir John French (1852–1925), landed in France on 14 August 1914, ten days after the declaration of war. With an insight that ran contrary to popular opinion, the War Minister, Lord Kitchener (1850–1916), was already telling the Cabinet that they would have to be prepared for a long struggle.

Initial reverses

After an initial clash at Mons, the BEF retreated. It stood fast at Le Cateau on 26 August, but suffered heavy casualties. On 5 September the Battle of the Marne began, with the Germans only 48km (30 miles) from Paris. The battle raged for seven days – by 14 September the Germans had withdrawn to the River Aisne and Paris had been saved. In October each side tried to outflank the other – the so-called "race to the sea" merely extended the line of trenches. By the end of 1914 the trenches ran from the North Sea to Switzerland; the British part of them from Ypres in Belgium to the River Somme [1]. That 80-mile strip was to account for almost 90 per cent of the 2,883,000 casualties the war cost Britain.

By 1918 the four original divisions had grown to more than 60 and from 1916 onwards Britain increasingly became the dominant partner.

Under pressure from both Germany and Turkey, Russia appealed to the British at the end of 1914 for some action to distract the Turks. The result was the Gallipoli campaign [4] which lasted eight months, cost 100,000 British casualties, and ended in evacuation of the peninsula. While the Allies were on Gallipoli, Bulgaria joined the Central Powers. On 5 October 1915, in anticipation of an invasion of Serbia [2, 3], one British and one French division landed at Salonika, in neutral Greece. They finally moved in September 1918, forcing the Bulgarians to sign an armistice.

The desert campaign and war in Africa

The Mesopotamian campaign [4, 5] at first made good progress. Sent out from India to protect oil interests in Kuwait, a force under Gen. Charles Townshend (1861–1924) got to within 28km (18 miles) of Baghdad, but then heat, disease and enemy harassment forced it into a defensive position at Kut-al-Imara. After holding out for five months, Townshend surrendered his force of 10,000 Indians and 2,000 British in April 1916.

From Egypt Gen. Archibald Murray moved into the Sinai and by the end of 1916 was close to Gaza, the nearest point of Turkish-held Palestine. He was twice beaten back and in June 1917 was replaced by Gen. Sir Edmund Allenby (1861–1936). A month later Capt. T. E. Lawrence (1888–1935), with a force of Arabs, captured Akaba.

Baghdad had fallen to an army under Gen. Sir Stanley Maude (1864–1917) on 11 March 1917, at a cost of 92,500 casualties. Instead of reinforcing Gaza, the Turks decided to counter-attack at Baghdad, and Allenby mounted a two-pronged attack against Beersheba and Gaza. By 9 December he was in Jerusalem. There was then a prolonged pause. In September 1918 Allenby advanced again, sweeping up through Damascus to Aleppo; Gen. William Marshall (1865–1939), who had taken over after

CONNECTIONS

Read first
188 World War I

See also
186 Causes of World War 1
192 The Peace of Paris
194 What World War 1 meant to Britain
252 Southern Africa since 1910
262 Wales in the 20th century

In other volumes
176 Man and Machines
170 Man and Machines
174 Man and Machines
164 Man and Machines

1 British infantry had to endure trench-feet, lice, flies and monotonous rations as well as regular shellfire when in the trenches. Out of the line they spent their time in working parties. Combat consisted of small-scale raids into enemy trenches and large set-piece battles. In the Battle of the Somme in 1916 (this is the front line at Ovillers), there were 420,000 British casualties in four-and-a-half months.

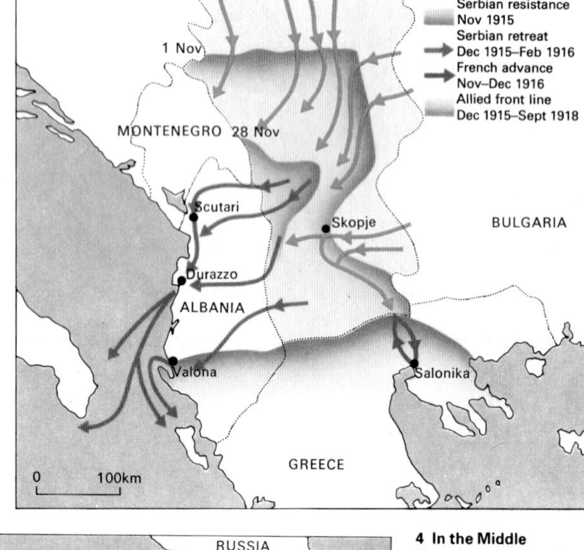

2 Serbia repulsed Austrian attacks three times in 1914. In October 1915 the Central Powers tried again, Austria and Germany attacking from the north and Bulgaria from the east. The Serbian army was forced to retreat across the Albanian mountains in appalling conditions. Of its 300,000 men, only 135,000 reached the Adriatic. Of 500,000 civilian refugees who accompanied the army, only 200,000 survived.

3 Belgrade was taken by the Austrians on 2 December 1914, but recaptured by Serbs under Gen. Radomir Putnik (1847–1917). Ten months later it finally fell. This painting by Oscar Laske shows the last day's resistance. One consequence of WWI was the creation, in 1918, of what became modern Yugoslavia.

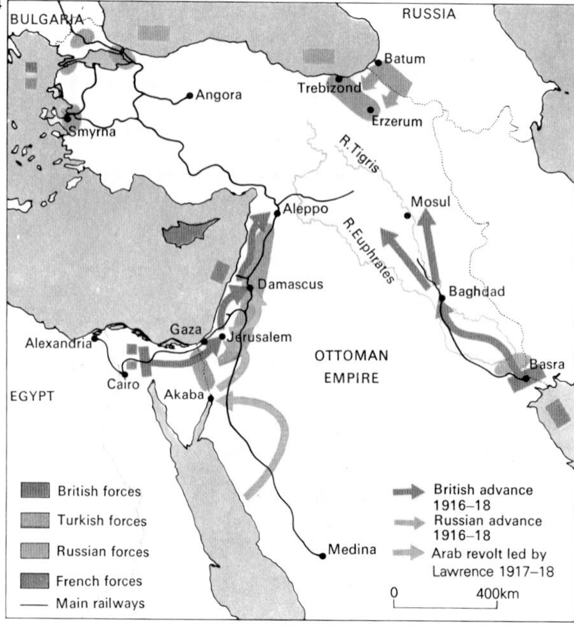

4 In the Middle East disease caused more casualties than enemy action. From January 1915 to the Armistice, 503,377 British troops went down with malaria, cholera, dysentery and other fevers, whereas only 51,500 were lost by enemy action. The eight-month campaign at Gallipoli in the Dardanelles, where the troops also suffered from disease, was an attempt to open a route to Russia via the Black Sea. An important consequence of its failure was that Russia was cut off from its foreign markets. One aspect of the desert war, later to be highly romanticized, was the exploits of T. E. Lawrence, who led an Arab revolt and led guerrilla raids against Turkish positions and the main railway.

Maude's death headed for Mosul. On 30 October Turkey surrendered.

Three weeks after the war began, a small British force accepted a German surrender in Togoland. In German Southwest Africa, Gen. Louis Botha (1862–1919), the Premier of the Union of South Africa, forced the Germans to surrender on 9 July 1915.

War in Europe

In August 1917 a decision by the German High Command to take the offensive on the Italian front [7] led to the Battle of Caporetto, fought between 24 October and 12 November. The Italians lost 305,000 men, 275,000 of whom surrendered, and five British divisions had to be pulled out of the Western Front and rushed to their support.

Cambrai, the first battle in which tanks were successfully used on a large scale, was yet another Allied attempt to break the deadlock that had existed since the beginning of 1915. In the three years since the Marne, the British had fought the First Battle of Ypres (October 1914, 58,000 casualties); Neuve-Chapelle (March 1915, 13,000); Second Ypres (April 1915, 59,000); Loos (September 1915, 60,000); the Somme (July-November 1916, 420,000); and Third Ypres (July-November 1917, 245,000).

The stalemate on land in those years had been offset to some extent by success at sea and in the air. The British blockade of Germany was extremely effective, whereas the German submarine campaign was restricted until late in 1916 by the fear of provoking the United States. When unrestricted submarine warfare was introduced, the British countered with the convoy system (the first sailed from Gibraltar on 10 May 1917) and improved anti-submarine technology. In the air, the Royal Flying Corps [6] received its first aircraft with synchronized guns in April 1916, and ended a ten-month period in which the German Air Services' 425 Fokker *Eindeckers* had created a reign of terror.

By 31 December 1917 there were 177,000 American troops in France, and less than a year later, at the Battle of Amiens (August 1918, 22,000 casualties) the end was in sight. At 11am on 11 November 1918 the shooting stopped.

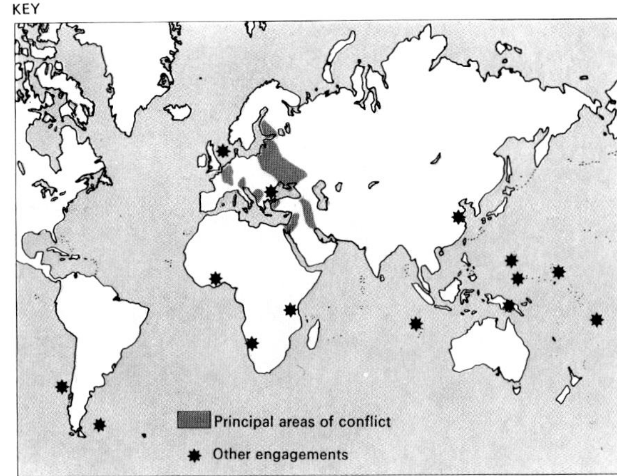

Britain's major concerns in WWI were France, Egypt, Gallipoli and Mesopotamia, but British and empire troops fought in the Pacific, Africa and even in China, where in November 1914 they joined the Japanese in the capture of Tsingtao. New Zealand took Samoa and Australia took New Guinea in the first two months; in November the raider *Emden* was sunk off the Cocos Islands. Other naval engagements included one at Dogger Bank in 1915, the historic Battle of Jutland in 1916, and the raid on the U-boats in Zeebrugge in 1918.

5

5 Australian and New Zealand cavalry, were part of Allenby's expedition to Gaza. The ANZACs (Australian and New Zealand Army Corps) also fought at Gallipoli, moving to the Western Front in 1916.

6

6 Captain Albert Ball, VC, was photographed in this SE5 at London Colney, Herts, in March 1917 and killed in it on 7 May. He was 20 years old. Ball shot down 44 German aircraft and, like Major Mick Mannock, VC, who with 73 victories was Britain's top World War I ace, was killed by machine-gun fire from the ground. The Royal Flying Corps sent 48 reconnaissance aircraft to France in 1914: by the end of the war the Royal Air Force (formed on 1 April 1918) had 22,171 serviceable aircraft. The war cost the air services 16,823 killed, of whom 12,782 were officers.

7

7 Italy joined the Allies in April 1915 and declared war on Austria-Hungary on 23 May 1915. Not until 1916 did it declare war on Germany. Many of the clashes between the Italians and the Austrians took place in the Alps: this Austrian gun is at a height of 3,860m (12,545ft). Of the five British divisions rushed to the Battle of Caporetto, two were withdrawn nine months later, but the troops who remained joined an Italian assault on the anniversary of Caporetto, in 1918, which led to the Austrians' seeking an armistice.

8

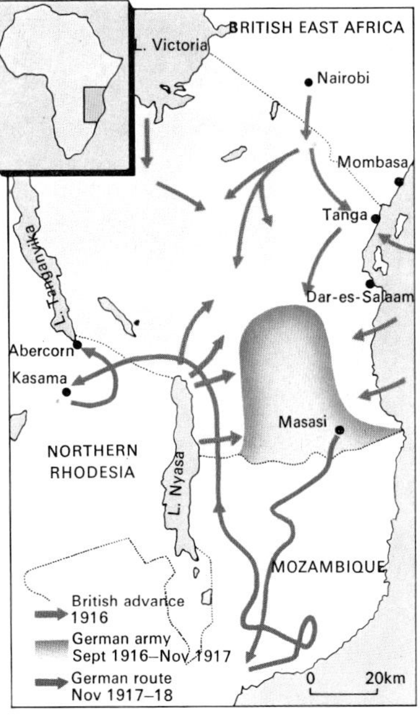

8 The East African campaign cost the British 19,000 casualties. That it was so protracted was due to the military genius of the German commander, Paul von Lettow-Vorbeck. who, with drastically outnumbered forces, fought on until November 1918.

9

9 Shorts and topees were standard uniform for troops in East Africa, and provided some relief from the intolerable heat. The torment of tsetse flies, fever and dysentery made conditions as bad, in their own way, as they were on the Western Front.

The Peace of Paris

The Paris Peace Conference, formally opened on 18 January 1919, was dominated by the five leading victorious powers of World War I – the United States, France, the British Empire, Italy and Japan. The defeated nations and Russia were excluded.

Conflicting demands

The French delegation, led by Prime Minister Georges Clemenceau (1841–1929), was obsessed with the long-term threat posed to France by Germany's larger population and superior industrial potential and demanded the imposition of a harsh treaty that would prevent any further German aggression against France. The French aims conflicted with those of the President of the United States, Woodrow Wilson (1856–1924) who, in his Fourteen Points (accepted with certain reservations by Britain and France on 4 November 1918), called for a peace settlement based on national self-determination and a League of Nations [2].

Britain's major demands had already been met with the surrender of the German fleet and the British occupation of most of Germany's colonies and the bulk of the Turkish Middle East. Despite pressure from Wilson that these areas should be administered directly by the League, they were retained by the British Empire under a complex League mandate system [4]. Thus the British Prime Minister, David Lloyd George (1863–1945), was in a position to mediate between the French and the Americans.

Italy demanded the satisfaction of its claims under the 1915 Treaty of London to the Tyrol, Trieste and a large part of the Slav-populated Dalmatian coast, including Fiume. The Italians were unable to persuade Wilson to agree to their claim to Fiume, which was assigned to Yugoslavia, leaving Italy with Trieste and the Tyrol [6].

Despite strong opposition from Wilson and the Chinese, Japan secured the former German concessions in Chinese Shantung promised to it by the entente in 1917.

Wilson's ideals compromised

The Republican victory in the November congressional elections in America undermined Wilson's prestige and he was forced to compromise on some of the Fourteen Points in order to secure the adherence of the other leaders to his League of Nations Covenant. However, neither he nor Lloyd George would accept France's demand for a Rhineland buffer state under French military control. This would have been a clear breach of the principle of national self-determination and, in Lloyd George's view, was likely to breed lasting German resentment. The French accepted a compromise on 14 April whereby the Allies were to occupy a demilitarized Rhineland, including the Rhine bridgeheads, for 15 years, with an Anglo-American guarantee to protect France against German aggression. The French were also given permission to exploit the valuable Saar coalfields.

Despite Wilson's strenuous opposition France also demanded massive reparations from Germany, not only to compensate for the immense destruction inflicted during the war but also as a means of weakening the German economy [7]. Lloyd George was, by the end of March, becoming concerned at the increasing severity of the Allied demands on

CONNECTIONS

See also
180 Europe 1870–1914
186 Causes of World War I
188 World War I
196 The Russian Revolution
208 The twenties and the Depression
212 Socialism in the West
146 Japan: the Meiji Restoration
214 East Asia 1919–1945
222 The rise of fascism
184 Balkanization and Slav nationalism

1 The new East European states emerged from the wreckage of the German, Austro-Hungarian and Russian empires. Although founded on the basis of national self-determination, they also included alien minorities like the Germans in Czechoslovakia. They were a source of constant unrest after 1919. Britain and France divided the former Ottoman Middle East between them, but both faced rising Arab nationalism and, in Britain's case, increasing Arab-Zionist conflict in Palestine. In the Ottoman empire nationalists formally established the Republic of Turkey in 1923.

2 The Allied leaders (from left to right) Lloyd George, Orlando, Clemenceau and Wilson, were bitterly divided by conflicting policies and temperamental differences. The peace settlement they eventually imposed on Germany was soon condemned by their countrymen and they did not remain in office for long after it. The treaty was signed in 1919 by all the great powers except the United States.

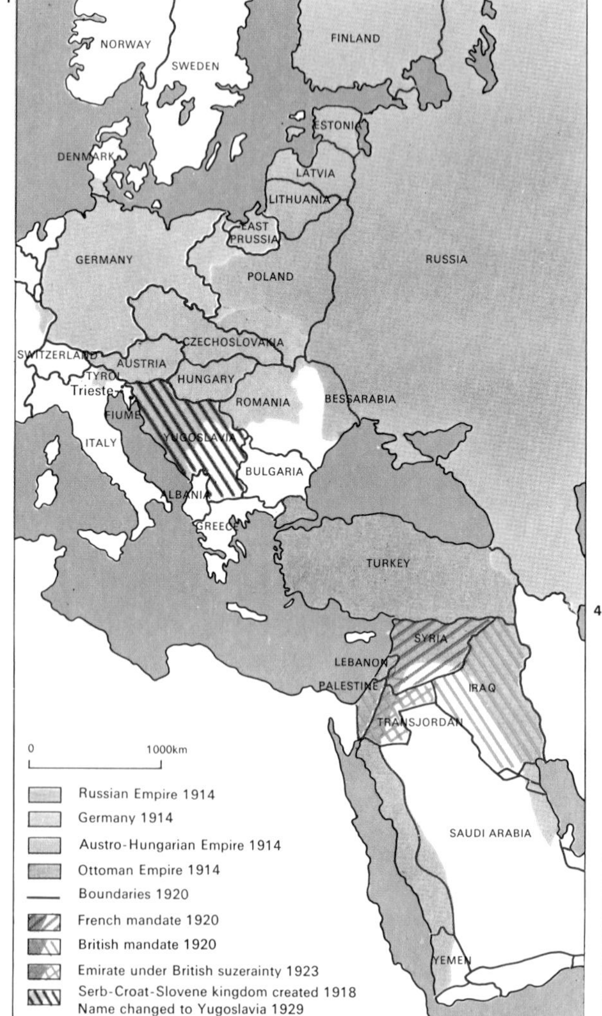

3 Germany's losses and gains from 1919 to 1938 are shown on this map. The Supreme Council had endeavoured to settle Germany's frontiers on the basis of nationality. Its territorial losses, in the east particularly, were a cause of Germany's detestation of the treaty. Allied disunity and weakness in 1938 enabled Hitler to incorporate Austria into the Reich and to annex the German Sudetenland.

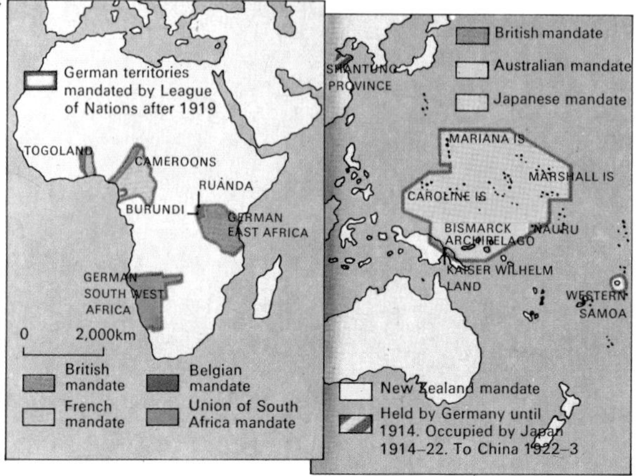

4 Germany lost all its colonies at the end of World War I. Woodrow Wilson hoped that the captured German colonies would be administered directly by the League of Nations. This idea was opposed by the British Dominions and Japan, which had conquered them. A compromise was reached by a system of "A", "B" and "C" mandates: "C" was virtually indistinguishable from annexation. Thus Wilsonian idealism was again frustrated by the other powers.

Germany which he feared would hinder Germany's economic recovery and lead to the creation of a Bolshevik Germany.

Eventually a compromise was reached that left the total sum owed by Germany to be determined by an inter-Allied reparation commission by 1921. Meanwhile Germany was forced to accept responsibility for causing the war. The Allies also imposed a substantial measure of disarmament on the German army, navy and air force.

The eventual compromise

The three leaders could not agree about the settlement of Germany's eastern borders. France supported large territorial gains at Germany's expense by the newly established East European states, especially Poland. After a long struggle Lloyd George managed to reduce Poland's acquisitions by insisting on a League-controlled free port of Danzig, the reduction of the Polish corridor and a plebiscite in Upper Silesia. Czechoslovakia retained the German Sudetenland. Austria, stripped of its former empire, was forbidden to unite with Germany. In 1920 Hungary lost all its non-Magyar lands to its neighbours and a severe peace was imposed on Turkey [1].

The Allies finally presented the draft treaty to Germany on 7 May, giving her 15 days to draw up counter-proposals, the bulk of which were rejected. After further delays Germany signed the treaty at Versailles on 28 June 1919. It was widely regarded in Germany as a dictated peace and a betrayal of Wilsonian principles. Failure to apply the principle of self-determination to the distribution of the German countries of the former Austrian Empire, in particular, was a major German grievance and one that gave the German nationalists and Hitler's Nazi Party valuable propaganda against the Weimar Republic in the 1920s [3].

The United States Senate rejected the treaty and the League covenant, and the United States retreated into isolationism. France thus lost the Anglo-American guarantee and became even more determined to insist on German compliance with the treaty, especially the reparations clauses. This intransigent attitude led to considerable friction with Britain.

KEY

The armistice was signed on 11 Nov 1918 at the French headquarters in the Forest of Compiègne. Marshal Foch signed for the Allies. It secured Germany's evacuation of occupied territories and a complete cessation of all hostilities.

5

5 British troops marched along Whitehall, London, in July 1919 in a "Peace Procession" that marked the signing of the Treaty of Versailles. But disillusionment soon replaced enthusiasm.

6

11 Nov 1918	German armistice
18 Jan 1919	Paris Peace Conference opens
22 Jan	Council of Ten sets up League of Nations Commission
3 Feb	League of Nations Commission meets
14 Mar	Formation of Council of Four
10 Apr	Council of Four appoints Reparation Commission
22 Apr	Italians withdraw
26 Apr	Conference accepts League of Nations Covenant
6 May	Italians return
7 May	Draft treaty is presented to Germans
29 May	Germans present counter-proposals
16 June	Allies reject most of the counter-proposals
28 June	Treaty of Versailles

6 A quick conclusion to the peace conference was essential, to pemit European reconstruction. Although attended by most nations and governments, it was soon dominated by Britain, France and the USA.

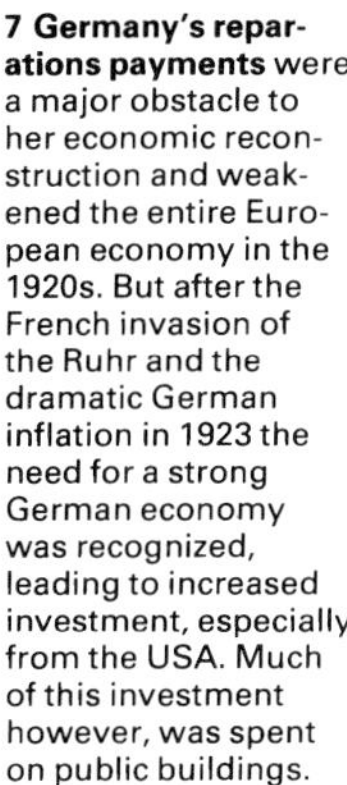

7 Germany's reparations payments were a major obstacle to her economic reconstruction and weakened the entire European economy in the 1920s. But after the French invasion of the Ruhr and the dramatic German inflation in 1923 the need for a strong German economy was recognized, leading to increased investment, especially from the USA. Much of this investment however, was spent on public buildings.

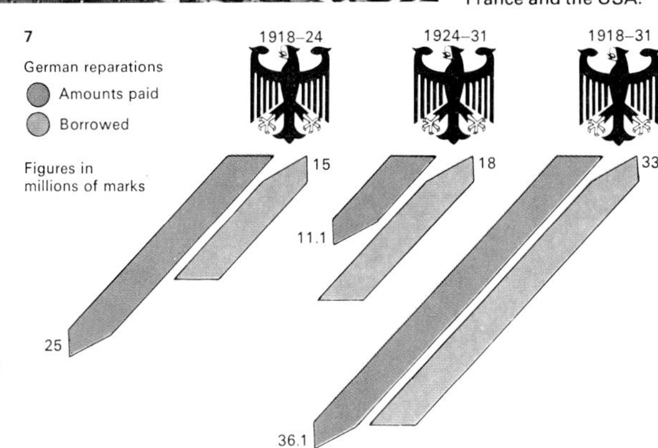

8

8 The League of Nations was intended by Woodrow Wilson to be the foundation of a new and peaceful world order. However, the USA's refusal to join in 1920 and the exclusion of both Germany and Soviet Russia (until 1934) reduced its prestige. After the admission of Germany in 1926 the League was fairly successful, until its failure to prevent the Japanese conquest of Manchuria in 1931–2 and the Italian conquest of Ethiopia, 1935–6.

9

9 Hitler's rise to power in 1933 and the Treaty of Versailles are not directly connected, but the treaty was used as an important element in Nazi propoganda against the Weimar Republic and the Social Democrats in the 1920s. Hitler's appointment as chancellor was the product both of luck and calculation. His opportunity was provided by general discontent and the economic depression, and the inability of Weimar politicians to cope with either.

What World War I meant to Britain

World War I is seen as one of the turning-points in British history — but it would be wrong to suggest that before the war all was tranquillity and security, a last "golden age", and that after it all was uncertainty and depression. Major political, economic and social changes were already taking place in Britain and the empire before 1914. They would have overturned the old way of life anyway; the war merely speeded them up, and made their effects far more shattering than they would otherwise have been.

Optimism and disillusionment

In 1914, Britain had effectively been at peace for almost a century; the Crimean (1854–6) and Boer wars (1899–1902) had had little effect on the population, and had seemed only minor interruptions in the growth of Britain's power. Several generations had grown up who knew little of war and were convinced of the superiority of their country and race. But all levels of British society were becoming more aware of the German threat to British naval and commercial supremacy in the years before 1914, and the hostility that this caused goes some way towards explaining the enthusiasm with which war was greeted. More than 500,000 men volunteered in the first few weeks, and during the following year 125,000 men a month went gladly to the front [1].

Early hopes that the war would be over by Christmas 1914 faded as both sides dug in. A static war of attrition ensued. By mid-1916 the fighting men were disillusioned by the squalor of the trenches and the mass slaughter. Because new battalions were formed on a geographical basis, whole towns and villages in Britain were almost depopulated by the fighting. On 1 July 1916, the first day of the Battle of the Somme, nearly 20,000 British soldiers were killed: individual battalions suffered heavily, the 10th West Yorkshires, for example, losing almost 60 per cent of its strength. At home there were some shortages and a few air raids [4], but the civilian population never really understood what it was like at the front. At the start of the war the government established a Press Bureau with the task of censoring newspaper reports and the true progress of the war was concealed from the public. Instead, the mass of public opinion was coloured by propaganda stories of atrocities.

The economy and government control

The unforeseen demands that the war placed on the British economy forced the state to intervene more actively than ever before. Although attempts were made after the war to retreat from this, active state involvement was never lost. The need for vast supplies of munitions, and the inability of private industry to produce them, led to the creation of a Ministry of Munitions in May 1915 with considerable directive powers. In 1916 British Summer Time was introduced to prolong daylight working hours. The need to ensure adequate food supplies led, in December 1916, to the establishment of county committees to direct agriculture and the creation of a Ministry of Food. In 1918 rationing was introduced.

The war brought an end to the free trade policy that Britain had struggled to maintain since the 1840s. The McKenna duties of 1915, putting a tariff on luxury imports,

CONNECTIONS

See also
188 World War I
190 World War I: Britain's role
160 The fight for the vote
192 The Peace of Paris
208 The twenties and the Depression
210 The British labour movement 1868-1930
218 British foreign policy since 1914
224 Britain 1930-45

1

1 Voluntary recruiting at first resulted in more men than could be adequately trained or equipped. The outbreak of war was greeted with overwhelming enthusiasm by all classes. Hatred of the Germans was whipped up by an almost hysterical press, and the chance of adventure and glory after long years of peace brought men flocking to join the forces. With no conscription, Britain had to rely on volunteers and, in spite of massive losses, the supply of new recruits was adequate for more than a year. But the authorities felt obliged to introduce conscription by May 1916 in order to reinforce the depleted ranks.

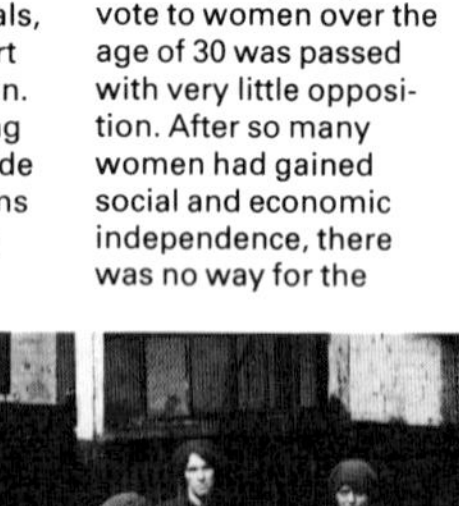

2 These women working in a factory in 1917 testify to the sexual revolution that took place on the home front during the war. As more and more men volunteered or were drafted into the forces, their places in the munitions factories, shops, offices, voluntary services, hospitals, schools and transport were taken by women. By thus ably replacing men or working beside them, women's claims for equality of status and rights were so widely accepted that in 1918 an Act giving the vote to women over the age of 30 was passed with very little opposition. After so many women had gained social and economic independence, there was no way for the conventional barriers to be re-erected once the war was over. This radical change in attitudes was reflected later, in the 1920s, in extremes of fashion and a degree of permissiveness in social behaviour

2

3

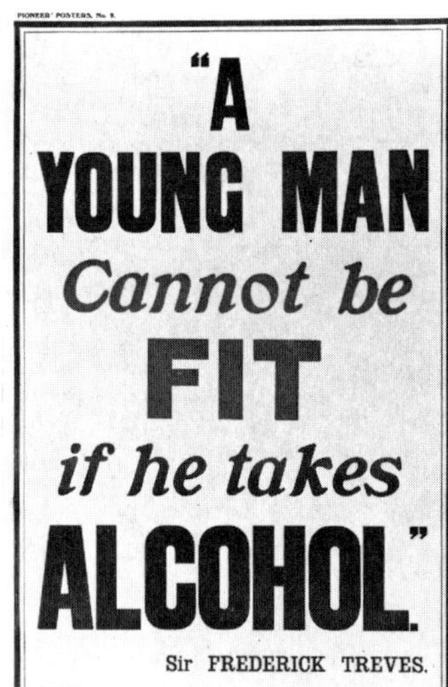

3 Watered beer and afternoon closing of the pubs were introduced by the government because it was felt that the national consumption of alcohol was impairing the war effort.

4

5

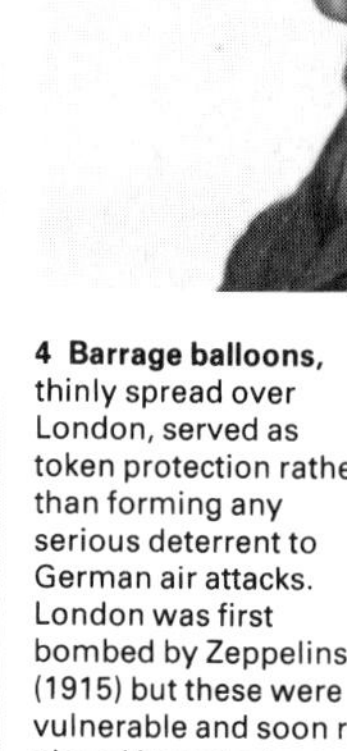

4 Barrage balloons, thinly spread over London, served as token protection rather than forming any serious deterrent to German air attacks. London was first bombed by Zeppelins (1915) but these were vulnerable and soon replaced by aeroplanes.

5 Wilfred Owen (1893–1918) and other young poets such as Siegfried Sassoon (1886–1967) and Robert Graves (1895–   ), who had fought in the trenches, wrote about the horror and despair of the experiences through which they had passed.

were retained after 1918, and were followed in 1921 by a Safeguarding of Industry Act to protect certain industries against foreign competition. On the outbreak of war, the Bank of England was authorized to issue banknotes not backed by gold, and there was a rapid and lasting rise in rates of income tax, which themselves had a much more progressive structure. The national debt rose from £650 million in 1914 to more than £7,000 million in 1918.

Shortages of labour caused by the demand for troops made workers realize their strength. Trade union membership rose from 4.1 million in 1913 to 6.5 million in 1918 and 8.3 million in 1920. Similarly, the widespread recruitment of women into industry broke down prejudices and strengthened the cause of the suffragettes [2]

The peacetime boom and slump

In November 1918 there was little evidence of any widespread demoralization caused by wartime losses – rather a pride in having come through an unprecedented trial. David Lloyd George (1863–1945), who had become prime minister of a Liberal-Tory coalition in 1916, took the opportunity to hold a general election which swept the coalition back into office. There was a brief restocking and rebuilding boom, but by spring 1920 it had degenerated into speculation and collapsed [6]. The economy slumped and the numbers of those unemployed rose to more than two million in June 1921 [8].

The government attempted to correct the economy by cutting public spending, wages and prices, all of which only made the problem worse. The war had accelerated the decline of Britain from the industrial and commercial supremacy it had once enjoyed. Traditional export markets had developed their own industries and major exporting sectors of the British economy, such as cotton, coal and shipping, were permanently reduced [7]. The war had given impetus to some new industries, such as chemicals and motor car manufacturing. But these tended to be developed in new regions, far from the traditional centres of industry where the misery and hopelessness of long-term unemployment were at their worst.

KEY

This document is a translation and facsimile of signatures from the original treaty of 1831 guaranteeing the independence and neutrality of Belgium, which was confirmed by the six Powers in the famous treaty of 1839, the breaking of which by Germany is responsible for the present war with the British Empire.

The "Scrap of Paper"

ARTICLE II.

Her Majesty the Queen of the United Kingdom of Great Britain and Ireland His Majesty the Emperor of Austria, King of Hungary and Bohemia, His Majesty the King of the French, His Majesty the King of Prussia, and His Majesty the Emperor of all the Russias, declare, that the Articles mentioned in the preceding Article, are considered as having the same force and validity as if they were textually inserted in the present Act, and that they are thus placed under the guarantee of their said Majesties.

ARTICLE VII.

Belgium, within the limits specified in Articles I., II., and IV., shall form an independent and perpetually neutral State. It shall be bound to observe such neutrality towards all other States.

PALMERSTON
British Plenipotentiary
SYLVAN VAN DE WEYER
Belgian Plenipotentiary
SENFFT
Austrian Plenipotentiary
H. SEBASTIANI
French Plenipotentiary
BULOW
Prussian Plenipotentiary
POZZO DI BORGO
Russian Plenipotentiary

The little "scrap of paper" was a contemptuous phrase used by the German Chancellor in 1914 to describe the 1839 treaty that guaranteed Belgium's neutrality. As a gesture, the Germans asked permission to go through Belgian territory on their way to Paris. King Albert I (*r.* 1909–34) of the Belgians replied: "Belgium is a nation, not a road", but German troops had already crossed the frontier. During the critical days before the Germans invaded, some sections of British opinion were opposed to Britain's participation in a continental war. But this act of aggression against "brave little Belgium" united the country in its determination to forcibly intervene against Germany.

6

100
90
80
70
60
50
1914 16 18 20 22 24

6 Britain's gross national product enjoyed a brief boom immediately after the war as industry restocked and changed over to peacetime products. But drastic cuts in government expenditure, the loss of export markets and the erosion of favourable economic conditions, such as free trade, led to a severe slump.

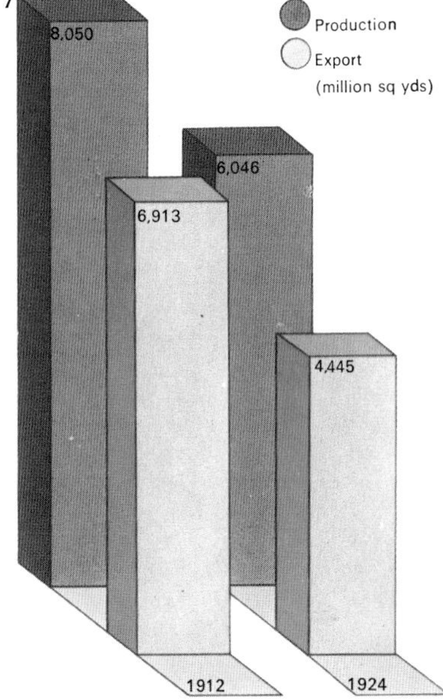

7 Cotton production and exports in the decade from 1912 to 1922 show a postwar slump that was typical of several major British industries. After the war they discovered that many of their markets had disappeared forever. It was a failure to replace the jobs in these industries that was the basic cause of lasting unemployment.

8

BOARD OF TRADE
LABOUR EXCHANGE

8 Unemployment was non-existent during the war, but after 1920 an intricate system of reliefs had to be built up in response to a fundamental change in the attitude of the public. Before the war the unemployed had been resigned to their fate as an inevitable fact of life. But after the war men expected the government to find them jobs, or to support them adequately until the necessary employment was available.

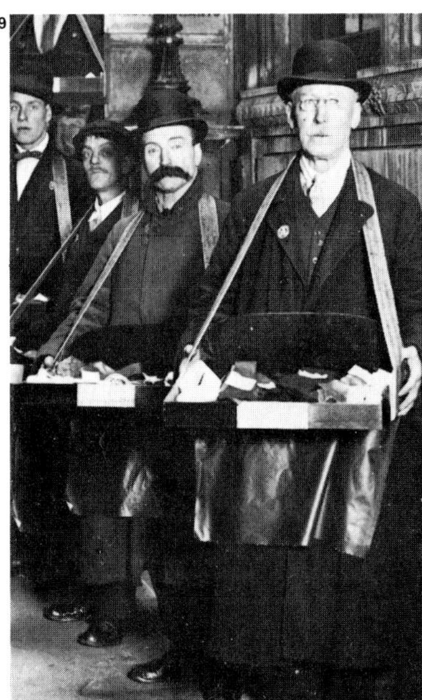

9 Ex-servicemen hawking their wares in the streets in 1920 symbolized the disillusionment and despair that broke down all the old certainties of British society. There was a dawning and bitter realization that the prodigious feats of government organization and direction that had helped to win the war did not seem to be winning the peace. The poor no longer accepted their fate as inevitable or unalterable, while the middle classes saw their income and status being steadily eroded by higher taxes. The frivolities of the "Gay Twenties" stemmed from a widespread desire to ignore doubts and difficulties that seemed insoluble. It amounted to enjoying life for the moment and letting tomorrow look after itself.

10 Striking coal miners in Wigan formed part of a "triple alliance" of miners, engineering and transport unions who were prepared to call a national strike. There was little industrial strife early in the war, but various government Acts, such as the Munitions of War Act of 1915 (which set wage levels and enforced arbitration), led to widespread strikes in 1917. The government modified its approach, but when in 1921 rising unemployment coincided with a withdrawal of government subsidies, support for a minimum wage, and removal of state control over the mines, the triple alliance was born. But the government compromised, the transport and engineering workers withdrew their support, and the threat of a general strike was ended. The miners came out alone, but within three months they were defeated, and returned under worse conditions than could have been reached by negotiation. This was followed by the political excitement over the collapse of Lloyd George's coalition government in October 1922.

The Russian Revolution

Russia went reluctantly to war in 1914. Her army was in no condition to face imperial Germany and early enthusiasm for the war waned with a shattering defeat by the Germans at Tannenberg within a month of hostilities commencing. But only the Bolsheviks vehemently opposed the war, the five Bolshevik deputies in the Duma (Parliament) being banished to Siberia. Their leader, Vladimir Ilyrich Lenin (1870–1924), nevertheless saw the defeat of imperial Russia as the surest way of furthering revolutionary goals.

Impact of the February Revolution

The longer hostilities lasted, the more incompetent the imperial administration appeared. It was astonished by the revolution in March 1917 (dated as February by the old-style calendar), but then so were its opponents. Power was transferred, by hungry peasants, disenchanted aristocracy and mutinous troops, from Tsar Nicholas II (1868–1918) [3] to a provisional government that was intended to be a temporary, caretaker administration until a Constituent Assembly adopted a constitution and appointed a legal government. The first provisional government (there were four in all) fell because of its failure to end the war.

Peace and the redistribution of land were closely connected. If Russia left the war, the soldiers (who were mostly peasants in uniform) would descend on the countryside and demand more land; if the peasants were granted land while war continued, the soldiers would desert to seize their portion. The government had also to contend with the emergence of genuinely democratic institutions, the soviets (councils). The most famous of these were in Petrograd and in Moscow, but they sprang up spontaneously everywhere after the revolution. Despite support from the moderate socialists – the Mensheviks and the Socialist Revolutionaries (SRs) – the provisional governments were violently opposed by Lenin and the Bolsheviks. During July, armed workers and soldiers tried to seize power in Petrograd [4, 5]. Denounced for accepting German money, Lenin was forced to flee to Finland when the demonstrations were unsuccessful. On 22 July, Alexander Kerensky (1881–1970) became premier and tried to restore order in the capital [2, 6]. But Leon Trotsky (1879–1940), a leading figure in the Petrograd soviet, organized armed insurrection under the cover of soviet legitimacy. Lenin slipped back into Russia and on 7 October (25 November, old style) he and his Bolsheviks [7] swept away Kerensky.

The October Revolution and after

Some workers hoped that the new Russia would be ruled by the soviets but events soon dictated otherwise. Given their narrow political base (there were fewer than 300,000 Bolsheviks in November 1917), Lenin and his supporters faced widespread opposition on every front [8]. There were those who advocated a revolutionary war to advance socialism in the rest of Europe; there were Bolsheviks who wanted money abolished and a socialist economy overnight; there were the peasants who wanted to be left alone with the land now redistributed; and there were the dispossessed of the former regime.

The treaty of Brest-Litovsk in March 1918 ended the war with Germany; in the

CONNECTIONS

See also
168 Russia in the 19th century
198 Stalin's Russia
212 Socialism in the West
170 Political thought in the 19th century
172 Masters of sociology
188 World War I

1 Russia paid a fearful price in human life for her incompetence in waging a long modern war. More than 15 million men had been mobilized by mid-1917. About 1.7 million men perished on the battlefield, 4.9 million were wounded and 2.4 million were taken prisoner. Russia was superior in strength to Turkey, Bulgaria and Austria-Hungary, but was outmatched by their ally, Germany.

2 Alexander Kerensky played a major role in shaping policies of the provisional governments in 1917. He was a minister in the first two provisional governments, prime minister from July onwards, and after he had suppressed an army revolt in September he also took over as commander-in-chief. His failure to solve the twin problems of land and peace paved the way for Lenin's victory in October.

3 On 15 March 1917 Tsar Nicholas II, shown here with his family, was persuaded to abdicate and the first provisional government was formed in Petrograd.

4 Demonstrations during the "April Days", 1917, against the war led to the fall of the first provisional government and the resignation of Foreign Minister Milyukov (1859–1943). But Russia's war effort continued, and in the soviets support for the Bolsheviks grew at the expense of the moderates. Calling for peace and a complete transfer of power to the soviets, further demonstrations in June showed the growing influence of the Bolsheviks and the declining support for the provisional government.

5 Clashes broke out in Petrograd on 16–18 July 1917 when armed workers demonstrated for "All power to the soviets" but were suppressed by the government.

6 General L. G. Kornilov (1870–1918), Kerensky's commander-in-chief, marched his troops on Petrograd in August 1917. This was seen by Kerensky as a right-wing attempt to take power and he turned to the Bolsheviks for help. The plot dissolved but it emphasized the growing political divisions that Kerensky could no longer bridge.

7 The Winter Palace was taken by the Bolsheviks on 7 November 1917. Lenin had secretly returned to Petrograd to forward Bolshevik plans for the overthrow of the provisional government, the collapse of which seemed imminent as unrest mounted. With the almost bloodless seizure of the palace, Kerensky fled and other members of the provisional government were arrested.

summer of the same year civil war broke out between the "Reds" (the Bolsheviks) and the "Whites" (anti-communists). In the autumn, the Allies intervened in an attempt to re-establish the eastern front [9], and soon began assisting the Whites. Hostilities lasted until the end of 1920 and revealed two victors, the Red Army and the Communist Party. During this time the Bolsheviks murdered the imprisoned tsar and his family. The Reds had the advantage of a claim that they were defending Russia from invasion. The desperate measures needed to secure military victory alienated many workers and peasants. Although desertions from the Red Army were frequent, Trotsky was successful in forging Soviet military might, but democracy fell victim to the needs of the hour. Lenin fashioned a new force to rule the country, the Communist Party of the Soviet Union. Aided by the feared Cheka (a secret police force), the Party and the military were willing to obey Lenin and his colleagues, while the soviets would not.

The bloodshed and exhaustion of seven years of war left Soviet Russia racked by revolt. Lenin gave way to the peasants and in 1921 introduced the New Economic Policy (NEP) which temporarily relaxed socialism in favour of some private ownership. The "commanding heights of the economy" stayed in state hands but agriculture, employing 80 per cent of the population, was on a market basis. The economy thus gradually recovered under the NEP.

The emergence of the new Russia

Soviet Russia had to be satisfied with less territory than the old empire held. The borderlands – Finland, Estonia, Latvia, Lithuania, Poland, part of the Ukraine and Bessarabia – were lost. But in the three independent Transcaucasian republics, following the British evacuation of Transcaucasia in December 1919, the way was clear for the Bolsheviks to take over. By April 1921 Transcaucasia was back in the fold thanks to the activities of the Red Army.

There remained the problem of succession. Lenin expected Trotsky to succeed him but ultimately Joseph Stalin (1879–1953) proved the more ruthless politician.

KEY

Vladimir Ilyich Ulyanov (known as Lenin) was born in Simbirsk (now Ulyanov) on the Volga. He was mainly in exile from 1900, but returned for the Revolution of 1917.

8

8 "Comrade Lenin sweeps the world of its rubbish" in this early Soviet cartoon. Peace and land were the two major demands Lenin promised to meet and immediately on seizing power the Bolsheviks put into effect the land policy they had adopted from the Socialist Revolutionaries. Land was later nationalized, but in 1917 most peasants still regarded it as their own. During the civil war grain was forcibly requisitioned to feed the Red Army and the cities. The peasants planted less and there was famine and disease. Finally Lenin capitulated and introduced the New Economic Policy in 1921.

9

9 Civil war divided Russia from 1918–20, threatening Bolshevik rule. In March 1918 Germany had forced Russia to a disadvantageous peace settlement at Brest-Litovsk. However, Allied troops then came to Russia to prevent German forces occupying key centres. After Germany's defeat they stayed and aided the Whites during the civil war. The Bolsheviks, who had demobilized the imperial army by granting land to the peasants and seeking a separate peace, had to create a new force, the Red Army. Trotsky, the father of the Red Army, was a brilliant military leader. The Reds had to contend with Greens (anarchists), Poles and dissident nationalities and the British, Americans, Japanese and French scattered around the country. The cartoon shows Uncle Sam about to release the dogs of war: the White leaders Denikin, Kolchak and Yudenich.

10

10 The famine that devastated the Volga region in the winter of 1921–2 claimed about 5 million lives and came on top of a virtual collapse of the Russian economy in 1921. By the end of 1920, the defeat of the Whites and the withdrawal of the Allies was complete. But seven years of war had left Russia in chaos, and popular unrest was fermented by inflation, shortages of food and fuel and the increasingly autocratic measures introduced to deal with internal and external threats to the infant Soviet state. Lenin introduced the New Economic Policy in 1921 to stimulate economic reconstruction and to placate the peasants by allowing a limited market economy with greater freedom of production. The period of the NEP was also one of considerable freedom expressed in the arts.

11

11 Lenin's death in 1924 followed a stroke in 1922, at which time a troika of Zinoviev (1883–1936), Kamenev (1883–1936) and Stalin had been established to continue the leadership. Lenin distrusted Stalin, whose main rival for the succession was Trotsky. But by skilfully playing off various factions and by control of the party mechanism, Stalin isolated Trotsky by 1925 and moved towards personal domination and ultimate dictatorship.

Stalin's Russia

The Soviet Union's evolution between 1917 and 1953 was dominated by two men, Vladimir Ilyich Lenin (1870–1924) and Joseph Stalin (1879–1953) [Key]. While Lenin was alive he was the main driving force behind events. Nevertheless there were other important personalities such as Leon Trotsky (1879–1940) [1], Nikolai Bukharin (1888–1938) [2], Mikhail Tomsky (1880–1936), Grigorij Zinoviev (1838–1936) and Anatoli Lunacharsky (1875–1933), to name only a few. All made an original contribution to Soviet development. Lenin, a man of outstanding intellectual ability, would listen to an opposing point of view if it came from one of his supporters, but had noticeably less respect for the views of his outspoken political opponents.

The policies of Stalin

Lenin realized the importance of consolidating the revolution, Stalin developed and extended the means. He sanctioned the revolutionary violence of the Cheka and extended the primacy of the party in state affairs. His doctrine of "Socialism in One Country" meant that all foreign communist parties became subservient to Soviet interests through the Comintern (Communist International). Furthermore, he continued to hold the show trials of a number of so-called counter revolutionaries. The first took place in 1922 and were directed at the Socialist Revolutionaries.

Nevertheless, there were major differences between the two men. Stalin was an intuitive anti-intellectual. His intellectual insecurity did not permit him to envisage a policy and then take on his opponents in open debate. Instead he sought to outmanoeuvre them in labyrinthine intrigue. Lenin was good at placing labels, often misleading ones, on his opponents; Stalin was a past master at the art. Lenin used the Cheka and the show trial against non-Bolsheviks; Stalin used them against the Communist Party as well.

The achievement of power

Stalin built up his power by his administrative skills and filled the leading party bodies with workers, but he did take the precaution of first briefing them on how to vote.

Stalin's journey on the way to supreme power can be divided into three stages, the completion of each marking a significant step forward. The first, terminating in 1928–9, saw him with almost total control over the apparatus of the Russian Communist Party which, because of the events of the immediate post-October period, had inherited the dominant role in the state. Victory over the party was not sufficient to permit Stalin to reach out to every corner of the Soviet Union. This he did during the 1930s when collectivization and industrialization transformed the scene. The peasants lost their land and their livestock and were brought under complete state control [3]. The foundations of great industrial advance were laid with heavy industry, vital for defence, receiving top priority [4]. A terrible massacre of real, putative, imaginary and potential opponents of Stalin's dictatorship took place. No one was secure, whether top party official (a major target were the Old Bolsheviks, those who had seized power with Lenin in October 1917), military leader, writer, peasant, worker, engineer or foreign

CONNECTIONS

See also
196 The Russian Revolution
306 Europe and confrontation to détente
254 Non-alignment and the Third World
234 The division of Europe
240 The Soviet Union since 1945
242 Eastern Europe since 1949
226 Causes of World War II
212 Socialism in the West
168 Russia in the 19th century
192 The Peace of Paris

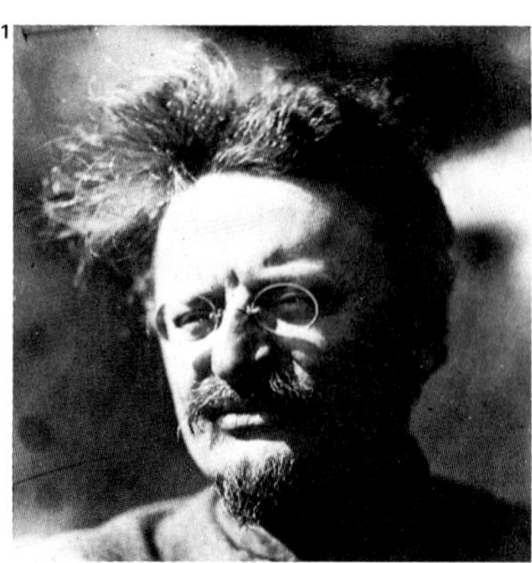

1 An outstanding theorist, Trotsky was, however, a poor politician, ill at ease with the minutiae of government. Although expected to succeed Lenin, he was inept at intrigue and was defeated. It was his failure to perceive the machinations of his fellows that soon led to his exile and death. He was an unequalled speaker, but his independent, critical attitude was not tolerated by Stalin.

2 Lenin called Bukharin "the darling of the whole party" and its "most valuable and most powerful theorist". Bukharin was the leading party writer on economic subjects. He sided with Stalin against Trotsky, Kamenev and Zinoviev and was a leading defender of the New Economic Policy. He was swept aside at the end of the 1920s when collectivization became the new official policy.

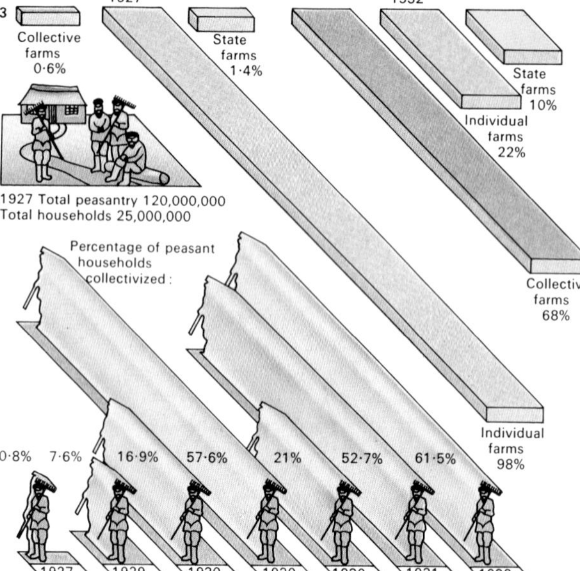

3 The New Economic Policy was a compromise on the way to socialism. It permitted the blossoming of private farming and since four out of five Soviet citizens lived in the countryside there was a risk of the capitalist ethic proving attractive. Lenin had preached co-operation and Bukharin ably elucidated his views after 1924. When agricultural production climbed back after 1924 to the level of 1913, the Soviets were faced with a choice – allow private agriculture to develop and provide the basis for overall economic growth, or socialize agriculture and base economic growth on industrial development. They chose the latter out of fear that private agriculture could overturn the socialist state and Stalin wanted food supplies for the urban worker.

4 Soviet power was insecure without a strong industrial and military base. Ambition ran riot as the first Five Year Plan got under way in 1928. Production goals were pushed up in the belief that a revolutionary spirit could perform miracles. Heavy industry was favoured at the expense of light industry and agriculture. Wonders were performed, but at appalling cost. Enthusiasm waned after the first plan and labour discipline became severe with saboteurs and counter-revolutionaries unmasked everywhere. Living standards dropped as millions flooded to the cities where accommodation was primitive. Both food and clothing were also in short supply.

5 The tractor was the symbol of Soviet power in the Russian countryside. The collective farm or *kolkhoz* became the dominant enterprise in socialist agriculture after 1928. Much virgin land was brought into cultivation in the 1930s and *sovhozes* or state farms were usually set up in new areas. Collective farm peasants were permitted a small, private plot and some animals. They received a share of the produce in proportion to the net income of the *kolkhoz*.

communist leader living in exile in the USSR. More than ten million people perished, including the great majority of the class of *kulaks* (well-to-do peasants).

When this period ended Stalin was master of all he surveyed in Soviet Russia; he controlled the party, the government and the police. Through the agency of foreign communist parties he could influence the internal politics of other countries. The third phase, which began with the outbreak of World War II and ended with Stalin's death in March 1953, saw Stalinist Russia reach the peak of world influence.

Stalin exhibited great tactical skill in the 1920s in overcoming his competitors one by one. In 1923–4 he allied himself with L. V. Kamenev (1883–1936) and Zinoviev against Trotsky; in 1925 he sided with Bukharin against Kamenev and Zinoviev; in 1926–7 still with Bukharin (who realized too late that Stalin's allegiance was merely tactical) against Trotsky, Kamenev and Zinoviev and finally in 1928–9 he was strong enough to oppose Bukharin, Tomsky and Rykov (1881–1938) by himself. By 1929 Trotsky was in exile and the others living on borrowed time. Most were to perish in the purges of 1936–8 [6, 7]. Trotsky, exiled in Mexico, was murdered by Stalin's executioner in 1940.

Russia's development

Russia's industrial effort in the 1930s made great progress. The bases of a thriving heavy industry were established and were to prove of vital importance when war came. Stalin took a long time to learn foreign affairs [11]. He indirectly helped Hitler gain power in Germany; then saw the danger and launched the Popular Front, inviting the collaboration of all democratic forces. He again put his faith in National Socialist Germany in 1939 and almost paid with the annihilation of the USSR after the German attack of June 1941.

Stalin's war record, except for the opening days of the war when he lost his nerve, is admirable. He led by example and his ruthlessness steadied his armies. Stalin played a vital role in the victory of the Allies. But had he allied Soviet Russia with Britain and France in 1939, it is possible that Germany would not have attacked Poland.

KEY

"Lenin is the Marx of our time" was the slogan when Vladimir Ilyich Lenin [left] was alive. Soon a new form appeared: "Stalin is our Lenin". Stalin became the main interpreter of Marx's chief Russian disciple. Those who threatened his supremacy were soon removed from positions of power.

6

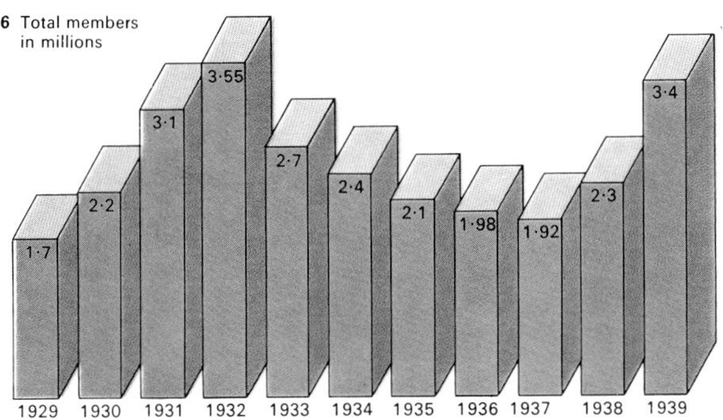

7

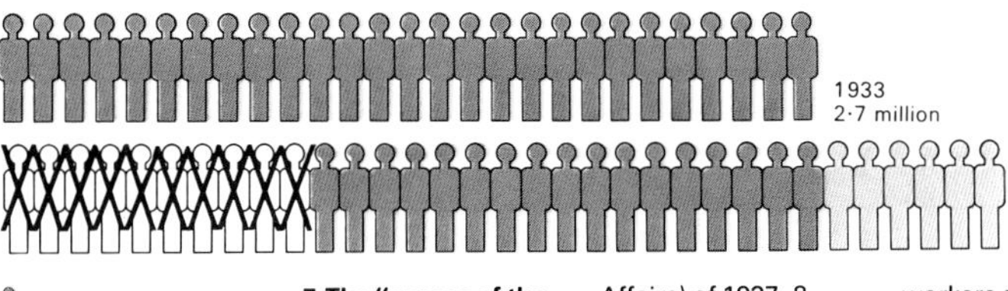

7 The "purges of the 1930s" were in fact composed of many different operations; these gathered momentum and reached a crescendo in the "Yezhovschina" (named after Yezhov, the head of Internal Affairs) of 1937–8. The first purge was launched by the party in January 1933. In 1935 a "verification of party documents" was ordered. About one member in five was expelled, including recent workers and peasant recruits. About 9% had been purged before the "show trials" ushered in the devastating Great Purge of 1936–8, when millions perished in the party and populace alike.

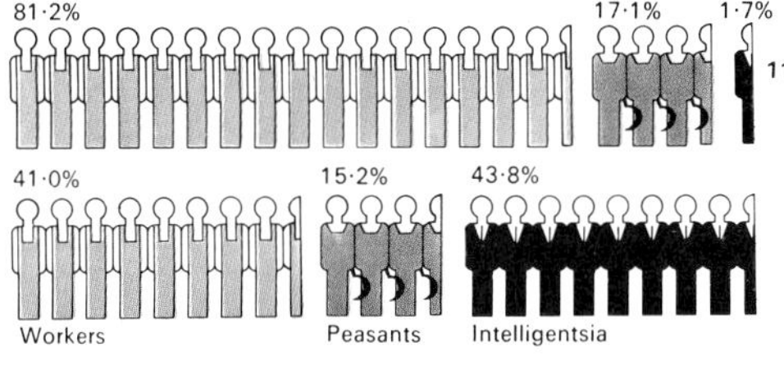

6 The composition of the Communist Party's membership changed markedly between 1929 and 1939. In 1929 four-fifths of party members were workers; ten years later that proportion had dropped to two-fifths. The difference was made up by a massive recruitment of the intelligentsia, partly because of a campaign to recruit the "best" people, partly because a party card was needed to qualify for a number of important posts in industry and the administration. The proportion of peasant members remained fairly constant. The dramatic change in membership reflects the policy behind the 1933–8 purges as well as their effects.

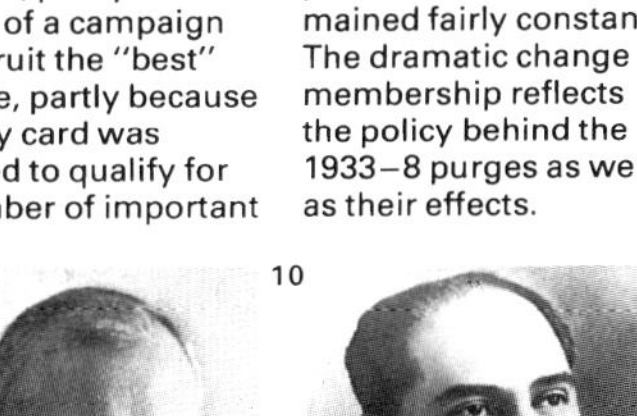

8

8 Vyacheslav Molotov (1890–) became a full member of the Politburo in 1926. He was instrumental in shaping the non-aggression pact with the Nazis and he remained a loyal servant to Stalin. He was also involved in party construction.

9

9 Nikita Khrushchev (1894–1971) was on the Moscow party committee between 1932 and 1938, when he took the key post of First Secretary of the party in the Ukraine. He became a member of the Politburo in 1939 where he backed Stalin.

10

10 Lazar Kaganovich (1893–) became a full member of the Politburo in 1930. He headed the Moscow party committee 1930–35 and was minister of transport 1935–44. A loyal supporter of Stalin, he retained favour during the years 1930–53.

11

11 Stalin misjudged fascism in the early 1930s but when he realized the danger he launched the "Popular Front" policy in 1935. All progressive forces were to unite against the common enemy and posters declared "Let's mercilessly rout and destroy the enemy". This policy did not deter Germany, and Stalin, thinking he understood Hitler, signed the pact of August 1939. Stalin intended to intervene opportunely in the impending war when Hitler had become over-committed on the Western Front. Stalin was so thunderstruck by the invasion of June 1941 that he lost his nerve and failed to provide resolute leadership during the first days of the war. The failure of the German *Blitzkrieg* in 1941–2 to overrun the USSR meant that the war of attrition, which Germany could not win because of inadequate resources, became inevitable. Major battles were Moscow and Stalingrad.

Origins of film

Moving pictures began as a technical novelty based on the brain's inability to detect a fractional gap between a rapid series of still photographs. Nobody at first suspected that a toy would become the most significant medium of communication, entertainment and art of the twentieth century.

The illusion of movement

Asian shadow plays, European magic lantern shows of the eighteenth and nineteenth centuries and devices such as Emile Reynaud's Praxinoscope, which projected images from a spinning drum, were early methods of producing the illusion of movement on a screen by back-lighting and magnification. After the patenting of the Eastman roll film, the way was open in 1891 for W. K. L. Dickson of the Edison laboratory to photograph vaudeville acts at 46 frames a second on a perforated film and run them back in a peepshow machine the Kinetoscope. Louis Lumière (1864–1948) and his brother Auguste (1862–1954) combined this with magic lantern techniques to project the first public cinema show in Paris on 28 December 1895.

Within the next five years the techniques of double exposure, fast and slow motion, reverse projection, fades, dissolves and close-ups were all discovered, many of them by Georges Méliès (1861–1938) who became fascinated with film magic after he observed a bus turn into a hearse when his camera jammed while he was recording a street scene. Filming from a fixed position, Méliès pioneered a cinematic fantasy linked to the artifice of the theatre [1].

One of the first film-makers to recognize that the camera could move freely and build up stories by a kind of visual shorthand was Edwin Porter (1869–1941). His two 1903 films, *The Life of an American Fireman* and *The Great Train Robbery* [2], had a sensational impact on the public, largely as a result of editing innovations.

The early film industries

Beginning in cheap nickelodeons patronized by the poor migrant populations of America's cities, the American film industry quickly became a medium of mass entertainment, turning out simple one- or two-reel morality sketches. The fledgling European industries sought greater artistic respectability by using well-known stage actors in film versions of the classics. Comedians such as Max Linder (1883–1925) at the big Pathé studios in France began to break away from this literary form of cinema, establishing a basis for the style of visual slapstick that would soon make Charles Chaplin [Key] and Buster Keaton (1895–1966) famous. In Italy, the nine-reel *Quo Vadis?* (1912) and the still more spectacular *Cabiria* (1914) introduced the idea of long feature films [3]. In America, D. W. Griffith (1875–1948) [4] was encouraged to make *The Birth of a Nation* (1915), a massive interpretation of the Civil War which incorporated techniques he and others had been developing since 1908. Working almost like a novelist, Griffith composed his story by inter-cutting close and long shots, flashbacks and parallel action to create powerful emotional effects.

The creative influence of Griffith and such directors as Thomas Ince, the huge popularity of Mack Sennett's slapstick Keystone comedies [7] and the damaging

CONNECTIONS

See also
268 Hollywood
276 Cinema as art
270 Music from Stravinsky to Cage

In other volumes
272 Science and The Universe

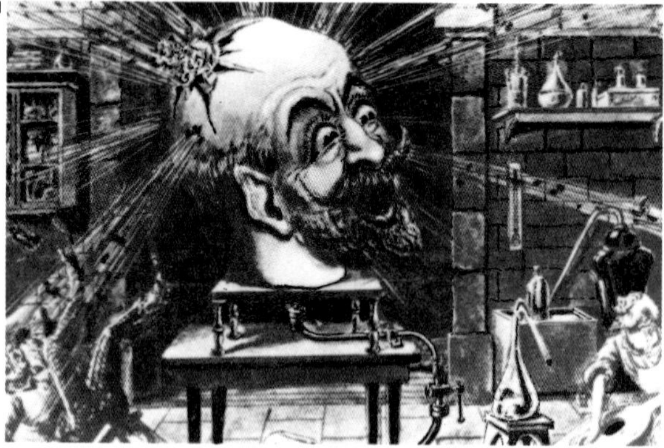

1 Georges Méliès's poster for his 1901 film *The Man with the India-rubber Head* captures the delight in novelty and magic that led Méliès to pioneer many cinematic tricks. A theatre proprietor, he saw film mainly as a marvellous new medium for illusions such as the pumping up of a head to the point of explosion. He also had considerable gifts of fantasy and comedy, best seen in *Journey to the Moon*.

2 Edwin Porter's juxtaposition of images in *The Great Train Robbery* created a sense of movement and speed that transformed film technique and led to endless imitations. He switched from outdoor shots of bandits making off with their loot [top] to interior shots of a telegraph operator being released and bursting into a dance hall to alert the local citizenry and lead the pursuit [bottom]. The freedom with which Porter moved his camera and edited separate shots into a thematic relationship astonished audiences who were used to watching only the progression of events within a fixed scene. His "Western" ran for 10 minutes in 15 shots.

3 The first genuine film epic was *Cabiria* (1914), running four hours and deploying massive scenic resources to capture the splendour of ancient Rome. It was the culmination of the Italian cinema's prewar interest in historic themes, which had already produced the nine-reel *Quo Vadis?* – a film that convinced American producers that the public wanted long feature films instead of one- or two-reelers. *Cabiria*'s director, Giovanni Pastrone, achieved an impressive silent spectacle.

4 David Wark Griffith [centre] was the acknowledged master of silent film in 1919 when he was photographed with two of the highest-paid stars of the day, Mary Pickford and Douglas Fairbanks. In *The Birth of a Nation* and *Intolerance*, Griffith used the resources of cinema on a scale and with an emotional intensity that profoundly influenced most other directors. Although he experimented boldly, his greatest achievement was the expressive quality he brought to the telling of a screen story.

effect of World War I on the European film industries, all combined to make the fiercely competitive American industry dominant by 1920. Hollywood [5] became the cinematic capital of the world and profited from the escapist hedonism of the jazz age. By the mid-1920s it was financially powerful enough to raid Europe for the talented directors who had emerged in a postwar flowering of film art.

Under the influence of its Expressionist painters, Germany was pre-eminent with directors such as G. W. Pabst, Ernst Lubitsch, Fritz Lang and F. W. Murnau, whose *The Last Laugh* (1924) astounded Hollywood by its technical versatility and psychological penetration. Surrealism, Dadaism and Impressionism all had an impact on the more subtle cinematic tradition of France where Abel Gance had led developments and where René Clair and Luis Buñuel were beginning work.

In Sweden, Mauritz Stiller and Victor Sjöstrom had explored the possibilities of filming in natural settings. And in Russia, Lenin's declaration that the cinema was "the most important art" led to the first nationalized film industry. To express the message of the Soviet revolution, montage editing was brilliantly refined by Vselvod Pudovkin and by Sergei Eisenstein (1898–1948), who moved his film forwards in a series of shock "cuts" [6].

The arrival of talkies

With the spread of radio, Hollywood recognized by the mid-1920s that to hold its audiences the cinema would need greater depth than could be provided by mime backed with titles and orchestral or piano music. Workable sound systems had existed from the early days of film, but recording and amplifying systems now became efficient enough to make synchronized sound-on-film fully practicable. Although businessmen feared re-equipment costs and problems in marketing English-language films abroad, the success of *The Jazz Singer* (1927) precipitated events in Hollywood [8]. Within three years the American film industry switched to talkies and was amply rewarded with a 50 per cent rise in audience numbers.

KEY

Charles Chaplin's little tramp (poignant in *City Lights*) was the first immortal screen character. Beginning in 1913, Chaplin (1889–) touched millions with his superb artistry.

5

5 Hollywood began, according to screen legend, in a barn in which Cecil B. De Mille [seated on a humble box] directed *The Squaw Man* in 1913. The barn (later enshrined in the Paramount lot) had in fact been used for earlier short films, but *The Squaw Man* was the first big commercial success filmed there and Hollywood soon began to mushroom.

6

6 The Odessa steps sequence from *Battleship Potemkin* (1925) is a landmark in the history of films as art. Sergei Eisenstein intercut shots of advancing Cossacks with close-ups of the impact of their bullets – a blinded woman, a runaway pram – to convey the drama and horror of revolution.

7 Film stunting and slapstick were born in the Keystone studio of Mack Sennett (1880–1960) who in 1912 got together former music hall comedians, acrobats, cowboys, daredevils and other uninsurables, put them into policemen's uniforms and sent them on an endless series of surrealistic escapades as the Keystone Kops. Sennett's one-reelers were fast, furious, funny – and dangerous. They invariably ended in a chase sequence with cars, trams or other vehicles careering into screen infinity.

7

8

8 Al Jolson, blacked up for his part in *The Jazz Singer* (1927), told audiences "You ain't heard nothin' yet" and the line became immortal as the first sound dialogue to be heard in a feature film. Sound had been tried as early as 1900 and Fox studios had produced a dialogue "short" before *The Jazz Singer*, but it was the huge popularity of the songs and the few lines of dialogue in the Jolson film that persuaded Hollywood businessmen that the daunting cost of re-equipping for sound would be recouped at the box-office. The first all-dialogue film was Warner's *The Lights of New York*, in 1928. Sound revolutionized the cinema, but it was some years before technical improvements restored the artistry achieved in the silent era.

Dada, Surrealism and their legacy

World War I had a twofold impact on the development of twentieth-century art. The centres of activity moved from France and Germany to New York and neutral Switzerland. Meanwhile the rejection of established artistic values (postulated by the Cubists and the Expressionists) acquired a new political relevance in the light of the war, which many intellectuals saw as the logical culmination of the whole ethos of the nineteenth century.

Shock the bourgeoisie!

Dada, a complex international movement, was essentially an attack on both artistic and political traditions. There remains some controversy as to the origin of the name, but it was certainly in use by the middle of 1916 to describe the activities of the Cabaret Voltaire in Zurich, which included performances and recitations intended to outrage the conventional. One of the early associates was the Franco-German artist Jean (Hans) Arp (1887–1966). A refugee from the war, he was making wood reliefs based on organic forms so simplified as to appear ridiculous; in his own words they were "designed to show the bourgeois the absurdity of his world". Meanwhile in New York, Marcel Duchamp (1887–1968) [1] was questioning established artistic procedures – and, by implication, the context in which they operated – by exhibiting "ready-mades" such as a bottle-rack or urinal.

After the end of the war Dada spread to other centres. Its varied guises had in common nihilism and a desire to shock by whatever means possible.

The collage technique developed by Picasso and Braque was employed by many Dadaists for their own subversive ends. Kurt Schwitters (1887–1948) made art from rubbish [2] and Max Ernst (1891–1976) assembled fragments of photographs and engravings to create irrational compositions. This latter method was to lead Ernst back to the art of painting when, in 1921, he embarked on a series of paintings in an illusionistic academic manner that presented suggestive and disturbing juxtapositions of images [3].

As the Dada manifestations died down, a group of writers and painters including Ernst and Arp assembled in Paris around the poet André Breton (1896–1966). While sharing Dada's disgust for bourgeois values, they rejected its nihilism and adopted a strongly positive philosophy inspired by the psychological theories of Sigmund Freud (1856–1939).

They believed that society repressed man's true nature and that in both life and art it was necessary to give full rein to the imagination. In the *First Surrealist Manifesto* of 1924, Breton defined Surrealism as "pure psychic automatism" and decreed that Surrealist literature was to be achieved by writing without conscious control.

Chance and imagination

When applied to painting, this procedure led Surrealism away from illusionism. Joan Miró (1893–) in his paintings of 1925 laid down highly diluted paint in a fairly arbitrary fashion that would simply suggest the lines of a more controlled composition [5]. Ernst sought inspiration in the textures of wood grain which he transferred to the surface of the paintings by rubbing. Both painters were exploiting chance in order to provoke the

CONNECTIONS

See also
174 Fauvism and Expressionism
176 Cubism and Futurism
204 Abstract art
270 Music from Stravinsky to Cage

1 Marcel Duchamp's "Bicycle Wheel" (1913) was the first of his "ready-mades", attacking the almost religious reverence given by society to original art works. An artist's choice of object, in this instance reflecting an interest in movement shared with the Italian Futurists, was enough reason to give it artistic status, to place it on a pedestal, here represented by a stool. By 1964, this rebel had been accepted sufficiently for an edition of replicas of this work to sell well.

2 Kurt Schwitters, a Hanover Dadaist, made his "Merz" pictures, such as "Das Sternbild" (1920), from rubbish. They contrast strongly with the neat collages of the Cubists. Their texture is rich, their design strong.

3 Max Ernst's "Two Children Threatened by a Nightingale" is an early (1924) Surrealist attempt to render in paint experience of dreams. Ernst uses both paint and wood relief, the latter breaking out of the illusionist space of the picture surface to spill across the frame, perhaps an analogy for the transition from the nightmare to the waking world. Birds are an obsessive motif in the art of Ernst. The threat of the nightingale represents here both an external menace and fear at discovery of sexuality.

4 "Song of Love" (1914) by Giorgio de Chirico (1888–) is one of his mysterious scenes that anticipate the dream pictures of the Surrealists. However, he rejected their psychoanalytic interpretation of his art and was concerned with creating a heightened "metaphysical" awareness of reality without any desire to shock in the Surrealist manner. Indeed, our surprise at the juxtaposition of the plaster head and the rubber glove stems from the confounding of expectation.

5 Miró's "Birth of the World" with its exquisite contrasts of hard-edged shapes and abstract washes shows how effective the Surrealist notion could be that the painter should free his imagination by painting without any preconceived idea of the end result.

imagination to broaden into new directions.

By the end of the decade Surrealism had returned to the illusionism of the earlier Ernst, largely as a consequence of the impact made by the work of Salvador Dali [Key] (1904–) who painted sensational subject-matter deriving from psychoanalysis in a highly accomplished academic manner. The Belgian René Magritte (1898–1967) rejected automatism in favour of the presentation of startling visual paradoxes [6].

Generally the Surrealists of the 1930s tried to express the unconscious by highly conscious artistic means. At the same time they turned increasingly to the making of "object-sculptures" in order to create disturbing images in a more tangible form than was possible in the most illusionistic painting.

Examples include Miró's combination of a stuffed green parrot, an artificial leg and a bowler hat, and Maeret Oppenheim's fur-lined teacup. This tendency reached its climax in the Paris Surrealist exhibition of 1938. A total Surrealist environment was created here with a row of fantastically garbed mannequins; for the setting Duchamp covered the ceiling with coal sacks and the floor with moss.

World War II forced Breton and many other Surrealists to flee to the USA. Young painters there, most notably Arshile Gorky (1905–48), took great interest in the automatic aspect of Surrealism and the way in which in the works of Ernst and Miró semi-abstract forms could bear a potent sexual charge [8]. This led Gorky to a highly personal manner of loosely painted, contorted forms which influenced Jackson Pollock (1912–56), whose "action painting" was in itself a form of automatism.

Surrealism today

After the war Breton returned to Paris and continued to organize Surrealist exhibitions, but the real heritage of the movement lay elsewhere. The "combine-paintings" of Robert Rauschenberg (1925–) [9] have the irrational rightness of the best Surrealist objects, while the Dada assault on art and society has been continued by, for example, the self-destructive machines of Jean Tinguely (1925–).

KEY

Salvador Dali's "Rainy Taxi" (1938) had live snails climbing over the face and chest of the mannequin. The disturbing eroticism is characteristic of Surrealism and Dali's particular adeptness here, plus his genius for publicity, marked him in the public mind as the Surrealist leader long after he had left the movement.

6 René Magritte's "On the Threshold of Liberty" raises paradox to a point where our notions about the way we understand a picture are much undermined (just as Surrealist philosophy threatened established morals by making desire, although violent or perverse, the final criterion of all value judgments). The openings in the surrounding space are contradictory. The woman's torso and the cloudy sky cannot both be real. Which, if any, represents liberty? Perhaps they are all just painted panels on an imprisoning wall? The only way to find out for sure would be by firing the cannon.

6

7

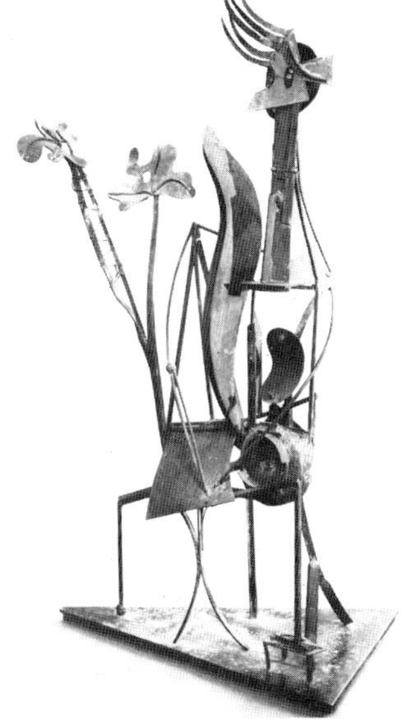

7 Pablo Picasso was never a member of the Surrealist group, but his "Woman in a Garden", unifying as it does images of both flowers and the female body within one structure, was the kind of metamorphosis that appealed strongly to the Surrealists. They were eager to claim him as an antecedent and ally because of his great prestige, his love of visual metaphor and the strong erotic content of much of his art. These factors became particularly apparent in the late 1920s, when his work turned towards great violence of expression after a period of serene classicism. He had found an ideal vehicle for fantasy in the new sculpture.

8

8 "Agony" (1947) was painted by Arshile Gorky. Armenian-born Gorky was strongly influenced by Miró, whom he met as a refugee in New York during World War II. By 1944 Gorky had arrived at a more original style based on automatic procedures (highly diluted paint dribbled down the canvas) while retaining suggestions of organic form. His final works, such as "Agony", were more tightly handled, and brought to abstraction unprecedented emotional force.

9

9 Robert Rauschenberg's "Monogram" (1961) is almost a posthumous compendium of Surrealist preoccupations. The unusual stuffed animal was sometimes presented as a "found object" in Surrealist exhibitions; here, as usual, it is given an unexpected context. The accumulation of letters and images on the base suggests the "Merz" pictures of Schwitters, and the Surrealist automatism survives in the vigorous smears of paint. The implications of the juxtaposed goat and tyre and the red paint on the face make it clear that the subject is a traumatic birth. Veiled presentation of taboo themes was a constant element in the shock tactics adopted by the Surrealist group

Abstract art

Abstract art is the most dramatic manifestation of the attempt by twentieth-century painters to overturn the assumption that art must represent appearances. By 1900 photography had already begun to replace realistic painting. The developing use of photography coupled with new ideas about the expressive potential of painting and sculpture resulted in the genesis of abstraction.

The beginnings of abstract art

Between 1910 and 1918 abstract art evolved in several places. In Munich, Wassily Kandinsky (1866–1944) achieved almost total non-representational painting in 1912. He possessed a first-hand knowledge of the work of Gauguin, Van Gogh and the Neo-Impressionists as well as a profound admiration of "primative" Bavarian glass-painting and Russian icons. He worked spontaneously, abstracting from images inspired by landscape, legend and biblical themes [1].

It was in Amsterdam and Moscow that artists first made works that were composed of "pure" forms without being consciously abstracted from nature. In Moscow Kasimir Malevich's (1878–1935) "Suprematist" compositions of 1915–19 [2] were the product of an attempt to define an "alphabet" of simple geometric shapes which, set on a white background, seemed to be imbued with movement in infinite space.

At the same time Vladimir Tatlin (1885–1953) launched Constructivism with dynamic constructions of glass, metal and wire, sometimes suspended across corners. These works were free of any mystical content. They led Tatlin to an art based on the tangible qualities of materials assembled in space. By 1921–2 Tatlin, joined by the Russian painter and typographer Alexander Rodchenko (1891–) and others, was making structures directly related to engineering; celebrations of an emerging socialist industrial society. Many of the structures were inspired by Tatlin's own wooden model for a metal structure (which was never built) taller than the Eiffel Tower, his "Monument to the Third International" [3].

Piet Mondrian (1872–1944) in the Netherlands was the other artist to arrive at an abstract art that was not abstracted from natural objects. His friends in the Amsterdam-centred de Stijl movement geometrized observed forms, but Mondrian began to compose works with straight black lines and colour patches during the years 1917 and 1918. He dabbled in theosophy, but behind his rigorously ordered paintings lay a more rational idea of life. His philosophy of life became so fixed that the style he arrived at in 1920–21 remained unchanged for nearly 20 years [4].

Biomorphic and geometric styles

Alongside these developments the Alsatian artist Jean (Hans) Arp (1887–1966) introduced organic forms in an abstract style called "biomorphic". He made a series of painted wooden reliefs when he was a Dadaist in Zürich (1916–18). His early Dada truculence led him to give his reliefs and sculptures comical titles like the 1926 "Navel Shirt and Head" [6], and his links with the Surrealists in Paris after 1924 ensured a strong biomorphic line in Surrealism, with Joan Miró (1893–) and Yves Tanguy (1900–55) as its best-known exponents.

CONNECTIONS

See also
174 Fauvism and Expressionism
176 Cubism and Futurism
202 Dada, Surrealism and their legacy
206 Modern architecture after 1930
280 Art and architecture in 20th century Britain

1 Kandinsky's 1911 "Composition IV" is abstracted from a fairy tale scene. In the centre is a blue mountain crowned by the jagged outline of a castle. To the left riders fight, their mounts leaping at each other over a rainbow. Although the forms can be interpreted thus, in 1913 Kandinsky wrote that he meant us to read no narrative into them. The story is a conflict of abstract elements, shrill yellow against deep blue, swelling curve against angular, linear action.

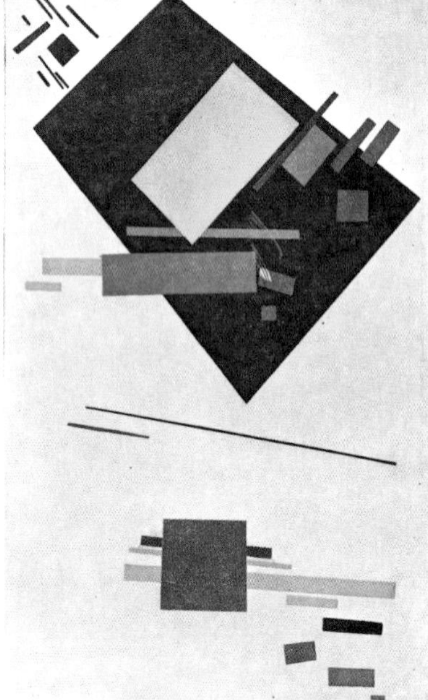

2 In 1915 Kasimir Malevich exhibited a simple black square on a white ground. The painting shown here, "Suprematist Painting" (1915), combines geometric shapes which by their over-lappings, their different sizes and their colour, create the illusion of movement in space.

3 This is a reconstruction of Vladimir Tatlin's "Monument to the 3rd International" (1919) which influenced sculpture as much as architecture.

4 "Composition I with red, yellow and blue" is one of the paintings with which Piet Mondrian established his complete abstract style in 1921. Mondrian held that life was change, and that change was created by the reconciliation of opposing forces. He therefore deliberately reduced painting to a conflict of the most basic visual oppositions.

Henry Moore in moving from his strongly figurative work of the 1920s to a highly abstracted style (1931 onwards) took Arp's direction a step further [7].

After 1922 Russian Constructivism moved in a utilitarian direction, and the mystical art of Malevich was left to die. At the Bauhaus in Germany, Laszlo Moholy-Nagy (1895–1946) backed Constructivist developments and Kandinsky too moved towards geometry. In France artists such as César Domela (1900–) and Jean Gorin accepted the more static line followed by Mondrian, while in London, from 1933, Ben Nicholson (1894–) developed a geometric style [5], as did Burgoyne Diller and Fritz Glarner (1899–) in New York.

Abstract Expressionism

Neither geometric nor biomorphic abstraction died during the 1940s, but in New York there was a further major development in abstract art – Abstract Expressionism. This was not a style but rather a group of individual styles, the most influential artists being Jackson Pollock (1912–56) after 1943 and Willem de Kooning (1904–) after 1947. Behind this development lay the Surrealist emphasis on the creative process itself coupled with a desire to break with the confining strictures of geometric abstract art and Cubist structure.

After 1947 Pollock's "drip paintings" [8] focused attention on the movement of the painter's hand and decisively challenged the tight shapes of twenties and thirties abstraction, both geometric and biomorphic. Thus an entirely new kind of abstract painting was created. Among the artists to follow Pollock's direction without sacrificing individuality was Franz Kline (1911–62), who in 1950 began to produce black-and-white paintings, such as "Chief" [Key], which were in effect hugely magnified brush drawings. They evolved out of calligraphic figurative drawings done over the previous few years. Less explosively exciting, but equally free of the shaping and the spatial structures of Cubism and geometrical abstraction, were the huge expanses of colour produced by Clyfford Still (1904–), Barnett Newman (1905–70) and Mark Rothko (1903–70).

KEY

"Chief", by Franz Kline (1950), is one of the artist's earlier black-and-white large abstracts. Up to the late 1940s Kline was painting city scenes, but he made a rapid change. The resemblance to Chinese calligraphy is misleading: Kline stated clearly "I paint the white as well as the black, and the white is just as important". His later experiments with colour did not lead him back to a full use of it again.

5

5 Ben Nicholson carved and painted this "White Relief" in 1935. In 1933 he visited Mondrian in Paris and the painting's tight geometric forms, its simplicity and its exact balance are in sympathy with Mondrian. There is, however, a personal feeling for the wood from which the shapes have been cut and a pleasure taken in the depths of surface that is peculiar to the artist.

6

6 "Navel Shirt and Head" (1926), a painted plywood relief by Jean Arp, anticipates his later free-standing sculpture. The first true three-dimensional work, "Head with Three Annoying Objects" appeared in 1930. Arp wanted to make things that seemed alive and were the product of hand, eye and intuition, yet that repeated the form of no known living things. He never abstracted from observed forms and disliked the term "abstract art". His connections with the Surrealist movement were made possible by his dislike of reason and calculation and by his spontaneous way of working.

7

7 In 1931 Henry Moore (1898–) began to use bones, flints and pieces of wood as the inspiration for his sculptures, which were evocative of the human figure. He made sheets of drawings to explore the figurative possibilities in these natural forms, arriving at images which he then carved in stone or, more rarely, as in this small "Figure" of 1931, in wood (beech). Several are more abstract than this.

8

8 Jackson Pollock's first attempts to create "automatically" without the intervention of conscious control, used archetypal symbols from Jung as their starting-point. But here in "Autumn Rhythm" (1950) he did not require the impetus of symbolic imagery, producing by the swift action of hand and arm sweeping trails of paint which cross over one another to form a whirling mesh of movement. For some critics, paintings such as these are expressive through the action of the painter's hand they recorded, hence the description "action painting"; for others, they were significant for the new type of abstract composition – having no sense of object or background – that they introduced.

Modern architecture after 1930

In France, Germany and The Netherlands leading architects developed, during the 1920s, a new architectural vocabulary that was to become known as the "International Style". Its main features are the asymmetrical arrangement of simple geometrical forms, extensive glazing that often turns corners and an open plan – all features that were possible because of developments in the use of steel, reinforced concrete and glass. Within a few years the style had spread throughout Europe and across continents, not only in homes, but for other types of building such as Owen Williams's [1] 1935 Pioneer Health Centre in Peckham, London, Alvar Aalto's Paimio Sanatorium in Finland (1929–33) and Howe and Lescaze's Philadelphia Savings Fund Society Building, USA (1932).

Mass housing and its architecture

World War II brought architecture effectively to a standstill in most of Europe; in contrast, South America – especially Brazil – was able to assimilate and develop the International Style. Le Corbusier (1887–1965) had visited Brazil in 1936 and vitally influenced Lucio Costa, whose team's building for the Ministry of Health and Education (1937–43) in the city of Rio de Janeiro marks the beginning of modernism in South America.

In 1945 some 40 million new homes were needed in Europe, largely in London, Berlin, Warsaw, Rotterdam and other cities devastated by the war; the rapid building of new homes was the priority and stylistic innovation had to take a back seat. The general formula was the repetition of large-scale units, a system that found less favour in the 1950s. Examples of this change are the Hansa District development in West Berlin and the Alton Estate in Roehampton, London. In building complete new towns the architect has increasingly taken on the role of planner.

The Unité d'Habitation, Marseille (1946–52) [Key], by Le Corbusier, and the Lake Shore Drive Apartments, Chicago (1948–51) [2], by Mies van der Rohe (1886–1969), are the two major monuments of the postwar period. In contrast to the smooth, white concrete finish of the 1920s, Le Corbusier, still working in concrete, chose to exploit the nature of the material by leaving it as it was found after the removal of the wood in which it was cast. The acceptance of this treatment – roughcast concrete – has radically changed the appearance of world architecture. Nevertheless, the importance of the Marseille structure was as an ideal for mass housing, that was adopted and adapted in countries such as England, Scandinavia, South America, Japan and India.

Structure and function

The Lake Shore Drive Apartments demonstrate primarily a search for precision in glass and steel in which the importance of the structure overrides its function. This search has given rise to the anonymous tower block that can accommodate home or office with no exterior differentiation.

By the mid-1950s skyscraper building was not confined to the United States, but was also gaining wide acceptance in the New World, with notable developments in Venezuela and Mexico. With Lever House, New York (1951–2) [5], Skidmore, Owings and Merrill had established the characteristic

CONNECTIONS

See also
178 Origins of modern architecture
176 Cubism and Futurism
204 Abstract art
280 Art and architecture in 20th century Britain

In other volumes
48 Man and Machines

1

1 Health preserved rather than ills cured was the theme of the Pioneer Health Centre (1935), London, designed by Owen Williams. Amenities included a swimming pool, sewing and reading rooms and facilities for medical research. The building's wavy form and its use of new materials symbolized the optimism of this new ideal. Contemporary British developments tended to be much more tentative.

2

2 Two identical towers, at right-angles to each other, make up the Lake Shore Drive Apartments, Chicago, by Mies van der Rohe. The building has 26 floors with apartments grouped round a central core for stairs and lift. The structure is a load-bearing steel frame, fireproofed with concrete and filled in with glass. The black frame and pale blinds behind the glazing add interest to an impersonal façade. Unlike the Unité of Le Corbusier the major characteristics are a highly efficient technology and simple forms – an important development in building using factory-produced components and setting a new standard to which architects aspired for the next 20 years. Of immense significance was the concept of a home in an anonymous box with flexible planning and mechanical services.

3
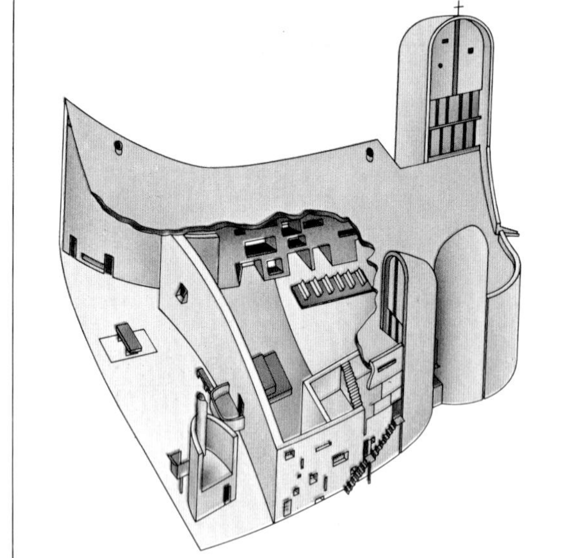

3 Chapel of Notre-Dame-du-Haut, Ronchamp, is the climax of Le Corbusier's tendency to Formalism seen previously in the Villa Savoye and the roof of the Marseille Unité. While other forms had been constrained within a geometric framework, here the amoeboid curves dominate. Built of reinforced concrete and set among hills, the design of the chapel echoes the forms of the surrounding landscape. With the Unité, the chapel provided an alternative to Miesian austerity. The "irrationalism" it seemed to presage was seen by architects and critics as a deliberate gesture against the right-angle and straight line, which had come to represent both honesty and rationality.

4

4 The Jaoul Houses, Paris, designed by Le Corbusier, are two homes on one rectangular site. Their use of crude materials, rough brickwork and ribbon slabs of concrete inside and out, presaged International Brutalism. Stirling and Gowan's 1956 flats at Ham Common, London, were the first significant reinterpretation of the idiom. The shallow arches on the exterior reflect tiled vaults inside, a feature often imitated for decorative effect.

office block format of a tower block on a podium, often with a plaza below to create an urban environment at street level. European skyscraper building has never been as convinced or refined, but important examples include Ponti's 36-storey Pirelli Building in Milan (1957–61).

In 1953 Brutalism – possibly derived from Corbusier's *béton brut* (roughcast concrete) – was first aired as a concept by Alison and Peter Smithson and demonstrated in their Secondary School, Hunstanton, England (1949–54). At that stage Brutalism's main aim was a search for visual honesty. It required the form of the building to reflect its purpose and refused to conceal functional items such as plumbing and electrical ducts. International Brutalism, from about 1958, shifted the emphasis from the image of the building as a whole to a greater concern with details, based not on Mies but on Le Corbusier and his Jaoul Houses, Paris (1954–6) [4]. International Brutalism was distinguished first by the rugged use of materials, for example in Tange's Town Hall, Kurashiki, Japan (1958–60) [7], and secondly by the separation of different functional elements as in the Engineering Building at Leicester University, England (1963) [9], by Stirling and Gowan. But a shift towards Formalism, in which the emphasis is more on aesthetics than on making plain the function of the building, was to mark a distinct trend in the architecture of the 1960s.

Aesthetics and function

Formalism and Brutalism are not entirely opposed; their differences are complementary and it is only with the free forms in the work of the masters that the Expressionism of the 1920s is evoked – in, for example, the chapel at Ronchamp (1950–55) [3] by Le Corbusier and the Guggenheim Museum (1957–9) [6] of Frank Lloyd Wright (1869–1959) in New York.

The Formalism of the 1950s and 1960s tended to be within a strict geometric framework and often referred back to earlier designs, as in Oscar Niemeyer's public edifices in Brasília (1958–60) [8], which exploit the curves of traditional Brazilian Baroque architecture.

KEY

Design for living by Le Corbusier – in Unité d'Habitation, Marseille, 337 two-storey flats, are slotted into a massive concrete frame with amenities added to complete the community. Bright coloured balcony side walls accentuate the raw concrete and huge supports (*pilotis*) leave the ground free for recreation and movement.

5

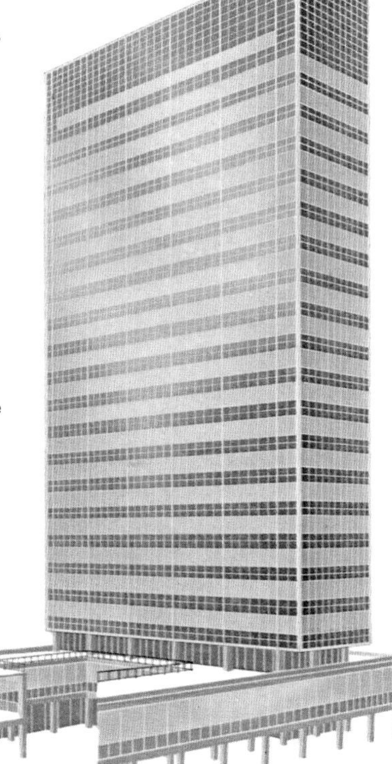

5 Lever House in New York, designed by Gordon Bunshaft of Skidmore, Owings and Merrill, is an important example of the glazed curtain wall. The main tower, at right-angles to the street, rises from only a portion of the structure, thus creating its own space in the city; the podium stands on pillars, leaving the court free for pedestrians. Although skyscraper buildings depend on a highly developed technology, the economic necessity of such structures – using a minimum amount of ground space, at a premium in cities such as New York – was an equally powerful motivation.

6

6 The Solomon R. Guggenheim Museum, New York by Frank Lloyd Wright, is designed as a continuous spiral ramp surrounding an open well. The spiral widens in diameter as it rises towards a glass dome 28m (91ft) above the ground. The dome is a main source of natural light on the exhibits. Wright believed that the curving walls of the museum, which are quite plain inside and out, were the best surface for showing pictures and, although there has been some criticism of the building, it is widely felt that it is apt for its purpose. It is cast in concrete with a smooth finish that recalls buildings of the International style. Although, as with Le Corbusier's chapel at Ronchamp, the fluidity of the shape appears to be a comment on industrial values, the building could not possibly have been realized without the technology of the 20th century.

7

7 Kurashiki Town Hall, designed by Kenzo Tange, is constructed from the roughcast concrete of International Brutalism and incorporates Corbusian features. Japanese architecture, surprisingly, has favoured rugged Le Corbusier rather than the clean lines of the Miesian idiom. The Town Hall façade is nevertheless an essentially Japanese interpretation of the Brutalist manner.

8

8 Brasília: the Senate, Secretariat and Assembly building by Oscar Niemeyer was designed for the new capital. The design combines technological sophistication, the curved lines of Brazilian Baroque and simple grandeur – a successful attempt to transcribe modern architecture into something peculiarly Brazilian. Although impressive this monumental Formalism illustrates the dangers of architectural design lacking scale or texture.

9

9 The Engineering Building, Leicester University, shows how different functions of a building can be stressed by its forms – an idea inherent in International Brutalism. The picture shows, from left to right, laboratories, lift shaft, administration tower and one of two projecting lecture theatres. The lack of symmetry reflects strong contemporary interest in a variety of geometric shapes – an affinity with the early 20th century.

The twenties and the Depression

The years from 1919–38 were dominated by an economic depression that troubled Europe for most of the time and affected the rest of the world most heavily in the 1930s. The aftermath of World War I was notable for an attempt to return to "normalcy", a term coined by the American President Warren Harding (1865–1923), and in Britain the immediate postwar years witnessed a boom in industrial production and living standards. After 1922, however, trade and industrial activity fell off, creating unemployment in the major heavy industries of the British economy [1]. Germany, the other great industrial economy of Europe, was unable to recover from the effects of the war and the impositions of the peace settlement [2]. The result was to depress the economy of Europe, which needed the prosperity of German industry. With the problems of inflation, political instability and the heavy reparations to contend with, the German economy did not begin to make a major recovery until the mid-1920s.

The war had left the United States as the major creditor nation, supplanting the position Britain had once held. A large proportion of the world's gold reserves had accumulated in Fort Knox, providing the basis for a large-scale expansion in American output. The growth in credit and consumption which these gold reserves allowed enabled a boom in manufacturing output to take place [3].

The twenties saw a wave of prosperity in the United States. It combined with a sense of release after war years to create the hectic atmosphere of the "roaring twenties". To a lesser extent this was felt in Europe towards the end of the decade, when an economic revival helped to popularize American music, dances and films.

Aspects of social life

Socially, the twenties had a paradoxical air. On the one hand, the end of the war heralded new freedoms, particularly for women. They had worked in many new occupations during the war and began to reap the benefits in terms of political and social emancipation. Fashions became more practical, there was a greater knowledge about birth control and there was a wider range of job opportunities. The twenties in America also saw Prohibition, which restricted the sale of alcohol, and created a boom in illicit alcohol.

Crisis and deflation

The more optimistic economic climate of the late twenties was, however, brought to an end by the Wall Street Crash of October 1929. The American boom had already begun to falter by the summer of 1929 with a downturn in the economic indices. The slide in share prices that followed became a panic [5]. In America, unemployment soared as credit dried up, consumption declined, and bankruptcies and redundancies multiplied. Compounding the Depression, agricultural prices fell disastrously for farmers in many other countries. World unemployment doubled within a year; in the United States it reached six million by the end of 1930.

For two years the Depression deepened throughout the industrialized world. By 1932, more than 12 million people were out of work in the United States and whole communities were at a standstill. The impact of

CONNECTIONS

See also
212 Socialism in the West
222 The rise of fascism
194 What World War I meant to Britain
224 Britain 1930–45
248 Australia since 1918
250 New Zealand since 1918
158 Industrialization 1870–1914
258 Evolution of the Western democracies
300 The World's monetary system
268 Hollywood

In other volumes
312 Man and Society

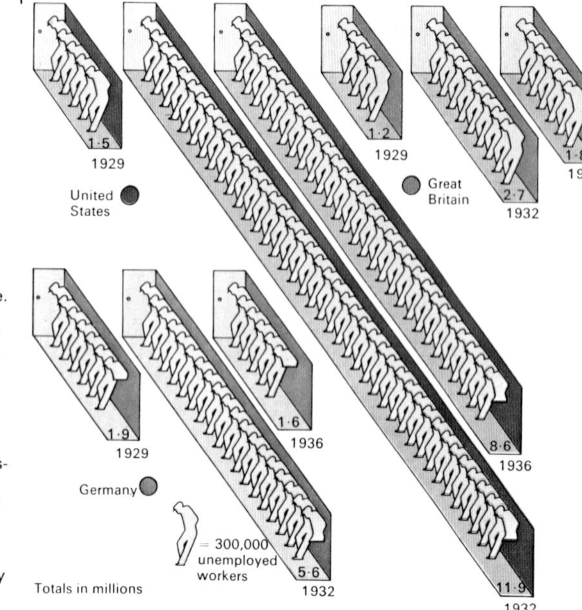

1 After World War I Britain suffered from the decline of her basic industries and the rise of competition, while Germany needed several years to recover from the war and reparations. The USA enjoyed a boom period in the twenties, which was brought to a halt by the Great Crash of 1929 and the decline in financial confidence and world trade. It brought a dramatic rise in unemployment in the industrial West, as shown here.

2 German paper marks equivalent to one gold mark

16 · 46 · 75 · 1757

Jun 1921 · Dec 1921 · Jun 1922 · Dec 1922

2 The German economy was thrown into severe difficulties by the effects of the war and the peace settlement. The loss of major industrial areas and reparations depressed the economy and created preconditions for inflation. With French occupation of the Ruhr because of Germany's default of reparation payments, massive inflation was triggered off, wiping out all savings, until a loaf of bread cost millions of marks.

3

1921	1923	1925	1927	1929	1931	1933

58 · 88 · 90 · 95 · 110 · 75 · 69 · 100

American industrial production

3 The American economy boomed in the twenties with a rapid growth of heavy industries. The industrial production index here shown is based on an average index of 100 for 1935–9. Rising consumption and easy credit fuelled the boom until 1929.

4 The motor car industry grew to major importance in the interwar period. Although invented and produced before 1914, cars remained expensive luxuries. By 1932, the assembly lines and conveyor belts, which had created the cheap, popular cars for a wider market, had come to a halt, leaving thousands jobless.

5 Thousands rushed to sell their shares on Wall Street in the panic selling of 1929. In two months share values had declined by a third and a paper loss of $26 million was registered. The growth in the American economy had been accompanied by a major speculative boom in share prices, involving small investors and large trusts. By 1929 industrial production began to peak and share prices slumped, causing the panic.

the Great Crash was equally disastrous on European economies, many of which depended on United States credit.

Current economic thinking decreed that a crisis of this kind could be cured only by a harsh dose of deflation, to balance budgets, reduce surplus capacity, and ride out the storm. In Germany the government of Franz von Papen (1879–1969) applied ever tougher doses of deflation and this pattern was followed in Britain, under the National Government of Ramsay MacDonald (1866–1937), and in the United States under President Herbert Hoover (1874–1964). Although the British economist J. M. Keynes (1883–1946) was in the process of formulating alternative policies, in which emphasis would be placed upon increased government spending and rising consumption to revive economic activity, his radical views were not generally available.

Political repercussions

The Depression had important political repercussions. In the United States dissatisfaction with the performance of President Hoover and his management of the economic crisis was reflected in the victory of Franklin D. Roosevelt (1882–1945) with his promise of a "New Deal" [7]. In Britain, the effects of the deepening depression in 1930 brought about a financial and political crisis for the Labour Government of Ramsay MacDonald. A National Government was formed after the 1931 general election, with a massive Conservative majority, but under the leadership of MacDonald and a small group of Labour followers. In Germany, the mounting unemployment and fear of social breakdown engendered support for the Nazi Party and undermined the basis of the Weimar Republic [8]. France was affected later than the rest of Europe because her large agricultural sector disguised unemployment and her industrial base was smaller than that of other countries.

Although the Depression dominated the thirties in Europe and the United States, recovery began in 1933, so that by the outbreak of World War II some considerable advances had been made in living standards in the period as a whole for those in work.

KEY

Drought and low prices for farm produce forced many farmers and their families to migrate from the American Midwest to California. Their hardships, immortalized in John Steinbeck's *The Grapes of Wrath*, symbolized the Depression.

6

6 In Britain, the Depression led to "hunger marches", such as that of 1936 when 200 men from Jarrow marched to London seeking work. In America, unemployed ex-servicemen marched to Washington in 1932. The action of police in dispersing them and leaving some dead caused much resentment.

7

7 Under Roosevelt's "New Deal" a number of ambitious projects were started to bring work to the unemployed and to stimulate the economy. The Tennessee Valley Authority sought to revitalize the economy and living conditions of a whole region by prestige projects such as the Hoover Dam, shown here.

8

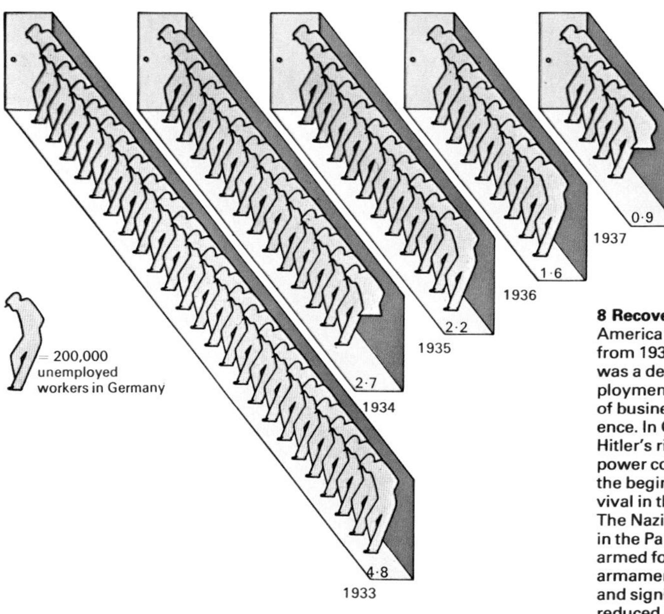

8 Recovery began in America and Europe from 1933. There was a decline in unemployment and a return of business confidence. In Germany, Hitler's rise to power coincided with the beginning of a revival in the economy. The Nazis created jobs in the Party, the armed forces and the armaments factories, and significantly reduced unemployment, as shown here.

9

9 Franklin D. Roosevelt brought a new period of prosperity to the United States after the worst years of the Depression when he became President in 1933. He won a landslide victory over Herbert Hoover on a programme for a "New Deal" for America, consisting of welfare legislation, public works, agricultural aid and planning, and an end to Prohibition. Roosevelt's confident style was almost as important as his legislation, bringing a measure of optimism and stability to the business and commercial world. His "fireside chats" on the radio helped to reassure the public that the government was acting to help the ordinary people. He went on to be elected for a second and third term. He died in office in 1945.

The British labour movement 1868–1930

The driving force behind the British labour movement in the latter half of the nineteenth century was the trade unions, which had been given restricted legality in 1825. Until the advent of the so-called "new unionism" in the 1880s, most trade unions were associations of skilled workers of varying political allegiance. Nonetheless, by the 1880s they had established a relatively secure position for themselves. In 1871 trade unions had been given legal recognition and in 1875 peaceful picketing was legalized.

New unionism

The period from 1875 to 1900 saw rapid growth in trade unions. This resulted partly from the rising prestige of the Trades Union Congress (TUC) which was founded in 1868, and partly from the efforts of a generation of "new unionists" who preached a much more militant form of trade unionism and organized semi-skilled and unskilled workers, such as dockers and gas workers, into new, industrial unions [Key]. These unions were prepared to take strike action with much less hesitation than before [2]. The result was the growth of working-class solidarity, an increasing dissatisfaction with the Liberal Party and the spread of genuinely socialist ideas among working men.

The growth of socialism had been demonstrated in 1888 when James Keir Hardie (1856–1915) and R. B. Cunninghame Graham (1852–1936) founded the Scottish Labour Party. It was given national expression in 1893 when Hardie [3] founded the Independent Labour Party (ILP) with the aim of encouraging trade unionists and socialists to join forces for the creation of an independent political party with working-class representation in Parliament. A non-revolutionary path to socialism was also sought by the Fabian Society which was founded in 1884. Among its best known exponents were Sidney (1859–1947) and Beatrice (1858–1943) Webb and the writer George Bernard Shaw (1856–1950). In 1900 the Fabians, with the ILP, the Marxist Social Democratic Federation and trade unionists, set up the Labour Representation Committee (LRC). Its aim, to quote Hardie, was to form a distinct Labour group in Parliament. Its first secretary was James Ramsay Macdonald (1866–1937).

The LRC's programme was a moderate one – it avoided commitment either to socialism or to the class war. As a result, in 1901, it lost the support of the Marxist Federation, but it did gain considerable trade union support, largely in reaction to the Taff Vale decision by the House of Lords in 1901 which found trade unions liable for losses incurred through strikes. In 1906, therefore, the LRC saw 29 out of 50 of its candidates elected to Parliament; later that year, the LRC was renamed the Labour Party.

The growth of the Labour Party

From 1906 to 1914 the Labour Party supported the social reforms of the Liberal governments, which in turn passed legislation benefiting the trade unions. The Trade Dispute Act of 1906 reversed the Taff Vale decision of 1901 and the Trade Union Act in 1913 allowed trade unions to support the Labour Party financially. Nonetheless from 1910 to 1914 trade union militancy increased [4] as a result of rising prices and the spread,

CONNECTIONS

See also
94 The British labour movement to 1868
96 Social reform 1800–1914
160 The fight for the vote
164 Scotland in the 19th century
208 The twenties and the Depression
212 Socialism in the West
224 Britain 1930–1945

1 The London match girls came out on strike in 1888. Their appalling working conditions had previously been exposed by the Fabian lecturer Mrs Annie Besant (1847–1933) in her paper *The Link*. With her help and that of other socialists, the match girls were eventually victorious and won recognition for their union. This was one of the first examples of the wave of "new unionist" activity and organization that spread among the semi-skilled and unskilled workers from 1889. It clearly indicated the bad conditions that had to be endured by these people who made up by far, the bulk of the British working class.

2 The London dock strike (1889), the first major action of its kind by unskilled workers, lasted five weeks. It ended in victory for the dockers who won their claim for a basic 6d (2½p) an hour (the "dockers' tanner"). The most significant aspect of the strike, however, was the widespread support won by the dockers from skilled workers and other sectors of the community. The dockers advertised their case skilfully and thus notably advanced the cause of working-class solidarity. Their militancy also highlighted the spread of socialism among British workers.

3 James Keir Hardie was one of the leading and best-loved figures in the British labour movement. Born in Lanarkshire, Scotland, he worked as a coal miner from the age of ten, and in 1886 formed the Scottish Miners Federation. He was the first chairman of the Scottish Labour Party (1888), and in 1892 became the first workers' representative in Parliament when he was elected as an independent Labour MP. Through his tireless efforts he was involved in the foundation of the Independent Labour Party in 1893, and the Labour Representation Committee, in 1900. He lost his seat in 1895, but was re-elected in 1900 as Labour MP for Merthyr Tydfil, south Wales, which he held until his death.

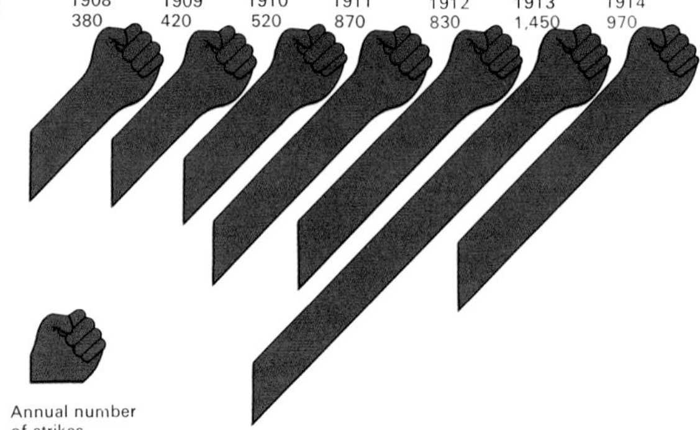

4 Industrial unrest characterized the years 1911–14. In 1908 there were 380 strikes; in 1913 there were 1,450. Dockers, seamen, railwaymen, and miners all struck between 1911 and 1914. There were militant and bitter conflicts and the men often held out for long periods in support of their demands. The strikes were prompted by various factors – the restoration of trade unions' legal immunity in 1906, falling standards of living, the apparent failure of the Labour Party to protect the interests of the working class, and the growth of Marxist and syndicalist ideas among working men. With the onset of the war in 1914, unrest declined because most union leaders and members chose to back the war effort.

from France and the United States, of syndicalist ideas that advocated a general strike to destroy capitalism.

The Labour Party continued to co-operate with the Liberal Party in Parliament and during World War I Arthur Henderson (1863–1935), who succeeded MacDonald as leader of the Labour Party in 1914, sat in the war cabinet of the coalition government. Various other Labour members also held administrative posts. By 1918, however, the Labour Party stood for a more independent policy, and influenced by events in Russia, adopted a more socialist constitution.

After the war the Labour Party soon became the second party in the country. Disillusionment, unemployment, and political strife within the Liberal Party meant that the Labour Party became the official opposition in Parliament in 1922. In 1924 Ramsay MacDonald became prime minister at the head of a minority government. His administration lasted only ten months. Publication of the so-called "Zinoviev letter" – instructions for a communist uprising in Britain apparently sent by Gregori Zinoviev (1883–1936), chairman of the Communist International – severely damaged the Labour Party. Although the letter was later proved to be forged, Labour fell before the Conservatives in November 1924.

The second Labour Government

In 1926 the trade unions challenged Conservative rule when the TUC supported the General Strike on behalf of the miners [7] but the government successfully resisted the challenge and in 1927 outlawed general strikes and attempted to reduce trade union subscriptions to the Labour Party.

In 1929, with the onset of the Depression, Labour returned to office with Ramsay MacDonald once again at the head of a minority government. His cabinet was divided over economic policy. Because socialist legislation was impossible in the midst of the economic slump, in 1931 MacDonald formed a coalition national government. In doing so he forfeited the support of the Labour Party, whose parliamentary representation dropped sharply in the 1931 general election.

KEY

THE RIGHT TO WORK MANIFESTO.

FELLOW CITIZENS—

UNEMPLOYMENT IS STILL ON THE INCREASE. No section of the workers are exempt. The official figures—published in September—show that very nearly 9 per cent. of the skilled workers of the country are workless. These figures—serious as they are—do not reveal the whole facts, since they do not include that greater mass of suffering men and women—the unorganised and unskilled workers.

NEITHER THE CENTRAL GOVERNMENT NOR THE LOCAL AUTHORITIES ARE TAKING EFFECTIVE STEPS to meet this desperate condition of affairs. Theirs is a policy of "wait."

UNDER THESE CIRCUMSTANCES, we once more call upon the people everywhere—those in work as well as the workless—to make their voices heard and their presence felt. Statistics and facts do not move the official mind. It is time, therefore, that the workers of the nation roused the Government with that historic utterance of the late Sir Henry Campbell-Bannerman: "Stop your fooling and get to business."

PARLIAMENT WILL MEET IN OCTOBER. Before that time, every member must be made to understand by his constituents that his first duty at the opening of the Session is to insist upon the immediate amendment of the Unemployed Workmen Act, 1905, which is a mockery and a delusion for those who are suffering, and an excuse for inaction by politicians.

THE ACT MUST BE MADE COMPULSORY, so as to place upon the State the responsibility of providing work with adequate financial support.

IF IT IS A QUESTION OF COERCION FOR IRELAND, money for a war, or for battleships Parliament can find it, and pass Acts in less than 24 hours. The demands of the starving unemployed deserve as prompt action.

SUFFER NO LONGER IN SILENCE. You have rights, and the most sacred right of all is your right to earn a honest living for yourselves and families.

TO COMPEL THE GOVERNMENT to recognise this right you must—

1. ORGANISE.—Let those who are in work come to the rescue of those who are workless. It may be your turn next. Call upon the Local Distress Committee to open at once the Unemployed Registers. Form representative and determined deputations and wait upon the local authorities. Insist upon something being done and done quickly.

2. DEMONSTRATE.—Bring the unemployed out into the open, in order to convince the public of their desperate need.

3. CALL UPON THE RELIGIOUS BODIES of the nation to support the agitation for the right to work and the right to live, since only by so doing can they justify their professions.

REMEMBER!

Governments help more quickly those who show that they are ready to help themselves. The silence of the Unemployed means the neglect of their needs. Therefore, we call upon every right-thinking man and woman to join with us in our attempt to put an end to those terrible conditions of life—hunger, cold and homelessness—to which Unemployment condemns so many of our fellow men with their wives and little children.

SIGNED ON BEHALF OF THE RIGHT TO WORK NATIONAL COUNCIL.

GEO. N. BARNES (Chairman)
J. KEIR HARDIE
J. RAMSAY MACDONALD
GEORGE LANSBURY, Treasurer

MARY R. MACARTHUR
ANNIE COBDEN SANDERSON
EDWARD R. PEASE
HARRY QUELCH
} Executive.
FRANK SMITH, Hon. Sec., 10, Clifford's Inn, London, E.C.

All who are in earnest in this locality should communicate with—

By the 1870s trade unions had achieved legal recognition. Until that time unions had followed no specific political viewpoint, but from the 1880s the movement took a new turn. Disillusioned with the Liberal Party and influenced by socialist ideas, the "new unions" increasingly stressed the political role. They demanded a legal minimum wage, an 8-hour day, and the right to work. Although union militancy continued until well after World War I – until its defeat in the General Strike of 1926 – with the establishment of the Labour Party by 1906 union activity increasingly followed more conventional, parliamentary channels.

5

5 Tom Mann (1856–1941) was one of the leading "new unionists" of the late 19th century. In 1881 he joined the Amalgamated Society of Engineers and by 1886 had become involved in the socialist movement. In that year he published a pamphlet arguing that a more militant attitude should be taken by trade unionists. In 1889 Mann helped to organize the London dock strike and from 1894–7 was secretary of the Independent Labour Party. He emigrated, and in 1902 was active in the Australian labour movement. In the 1920s, after his return to England, he became a founder of the British Communist Party, feeling that the existing unions could not be militant enough.

6

6 Labour exchanges were introduced into Britain in 1910 by Winston Churchill (1874–1965), then Liberal President of the Board of Trade. Advocated by the Poor Law commission of 1909 and by the economist William Beveridge (1879–1963), labour exchanges were intended to provide a service for workers seeking employment and for employers seeking labour. They also prepared the way for a national system of social insurance. Initially, they were not as effective as had been hoped. Registration of unemployment was not compulsory so that only one-third of vacancies were filled through the nationwide exchanges.

7

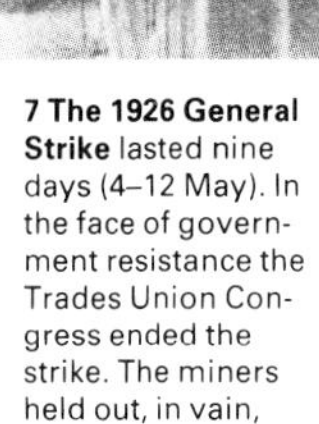

7 The 1926 General Strike lasted nine days (4–12 May). In the face of government resistance the Trades Union Congress ended the strike. The miners held out, in vain, until August.

8

8 Ramsay MacDonald was the Labour Party's first prime minister. In 1894 he joined the Independent Labour Party and was its chairman from 1906 to 1909. He helped to found the Labour Representation Committee and in 1924 became the first Labour Party premier. In 1929 he again became prime minister but was rejected by the Labour Party when he formed a coalition national government in 1931, the only way he saw of keeping Labour in power.

Socialism in the West

Socialism developed from a group of thinkers, especially Robert Owen (1771–1858), Henri de Saint-Simon (1760–1825) and Charles Fourier (1772–1837), who criticized industrialism because of the suffering and hardship it caused the working class. But it was not until the mid-nineteenth century that socialism developed a mass following as a direct result of the growth of industry in different parts of Europe and the related rise of an urban working class.

Early developments

As the first industrial nation, Great Britain took the lead in the development of workingmen's organizations [7]. Despite legal restrictions and occasional persecution, such as the transportation to Australia of the Tolpuddle Martyrs in 1834 for trade union activity, unions flourished by the middle of the nineteenth century, especially among skilled workers. The political ideas of this "labour aristocracy" were largely Owenite, emphasizing co-operation and reformist political activity. Attempts to establish a Grand National Consolidated Trades Union had failed by 1834, and following this the Chartist movement attempted to enlist the mass of factory operatives in the cause of political rights, which were enshrined in the "People's Charter", presented to Parliament and rejected three times. Under reformist leaders British trade unions concentrated upon securing gradual concessions in the political and social sphere during the period of prosperity after 1851.

In Europe the slower progress of industrialization hampered the growth of organized socialist movements. Trade unions remained illegal in France until the middle of the nineteenth century and socialist support was divided between the followers of revolutionary leaders, reformists and anarchists. Although workers participated in the overthrow of Louis-Philippe (1773–1850) in 1848, there was no organization to unite them. In Germany, too, the workers who supported the revolution of 1848 remained divided and dominated by middle-class liberals. The German risings of 1848 did, however, see the emergence of Marxism in the *Communist Manifesto*. Written by Karl Marx (1818–83) and Friedrich Engels (1820–95), the manifesto provided a coherent intellectual basis for many later socialists.

The First International

Although socialist ideas played little part in the revolutions of 1848, and Chartism was defeated in Britain in the same year, they did mark the emergence of the first important mass movements of workers in Europe. In 1864 socialist groups came together in the First International. Although racked by dissension, the International provided a vehicle for Marxist ideas and encouragement to socialist groups throughout Europe. In France in 1871 the rising of Parisian workers and the lower middle classes in the Commune was proof of the growing strength of socialist ideas. The International was liquidated in 1876, following quarrels between the anarchists and Marx. In the less developed parts of Europe, especially Spain, Italy and Russia, anarchist ideas propagated by Mikhail Bakunin (1814–76) had a strong appeal and led to risings in Spain and terrorist acts in Russia [2].

CONNECTIONS

See also
170 Political thought in the 19th century
208 The twenties and the Depression
222 The rise of fascism
210 The British labour movement 1868–1930
218 British foreign policy since 1914
196 The Russian Revolution
180 Europe 1870–1914
188 World War I
172 Masters of sociology
158 Industrialization 1870–1914
258 Evolution of the Western democracies

In other volumes
278 Man and Society

1

1 Two reformers, Sidney (1859–1947) and Beatrice Webb (1858–1943), adapted socialism to the cause of social reform which they sought to achieve gradually through democratic procedures. They formed the Fabian Society in 1884. It attracted many middle-class and intellectual figures such as George Bernard Shaw. The British Labour Party adopted the ideals of "Fabianism" for its philosophical basis.

2

2 In Russia, anarchism inspired the opponents of the tsarist regime in a campaign of terrorism, including the assassination of Alexander II in 1881. Anarchism grew out of the ideas of Pierre Proudhon (1809–65) among others. It rejected all authority in its search for a self-governing ideal in which men could totally fulfil themselves. The most famous 19th-century exponents were Russians, especially Mikhail Bakunin and Prince Peter Kropotkin (1842–1921). In France, anarchism became blended with trade unionism, and in Spain anarchist groups played an important part in the political upheavals of the early 20th century, including the Spanish Civil War.

3

3 The years before World War I were marked by labour militancy and violent strikes throughout Europe and the USA. In Britain there was a wave of bitter disputes and troops had to be called out in South Wales during the coal strike of 1912. The trouble was caused by the rise of organized labour, the spread of militant ideas and a slight downturn in living standards after a period of improvement.

4

5

6

4 Jean Jaurès (1859–1914) was a most eminent French socialist. A successful politician and moderate Marxist, he brought unity to the fragmented socialist groups in France before being assassinated by a fanatic for opposing the war with Germany in 1914.

5 Polish-born Rosa Luxemburg, with Karl Liebknecht, led the Marxist "Spartacist" movement which sought to end the 1914–18 war through revolution. They were both assassinated by reactionary troops in Berlin during the revolt of 1918–19.

6 Like these Londoners, liberals everywhere protested in 1927 at death penalties imposed on two US anarchists, Nicola Sacco (1891–1927) and Bartolomeo Vanzetti (1888–1927). Many believed their conviction, for murder, was politically motivated.

After 1870 the German socialist movement became the most powerful in Europe. In 1890, in spite of laws restricting its operation, the Social-Democratic Party was the largest in the Reich. Although divided between Marxist and "revisionist" groups, the socialists continued their rise up to 1914. In the aftermath of Germany's defeat an alliance between the social-democrats and the army was formed to set up the Weimar government and to frustrate the challenge from the Marxist "Spartacists" led by Karl Liebknecht (1871–1919) and Rosa Luxemburg (1871–1919) [5]. In France the socialist movement remained fragmented. French workers turning aside from party politics were attracted to syndicalist ideas of control being achieved by workers through strikes.

The Second International

The Second International, formed in 1889, was severely divided between reformist and revolutionary groups, and was not strong enough concertedly to oppose World War I. Nonetheless, by 1914 socialism was a powerful political force in Europe and had also spread to Latin America and the United States. Although it was never as strong in the USA as in Europe, a socialist candidate for the presidency, Eugene Debs (1855–1926), polled 900,000 votes in 1912, while the militant Industrial Workers of the World (IWW) mounted a series of bitter strikes. The war caused a breakup of the international socialist movement because its members had to choose between patriotism and allegiance to the socialist ideals.

The Russian Revolution led to a revival of left-wing militancy in the aftermath of World War I, but the inter-war period saw the socialist parties of Britain, France and Germany playing a prominent part in parliamentary politics, and the triumph of socialist parties in Scandinavia. Although the Depression and the rise of fascism led to suppression, as in Germany, Italy and Spain, they also led to a revival of socialism in middle-class and intellectual circles. The Spanish Civil War [9, 10] provided a rallying point for the left and the triumph of the Allies in World War II left socialist parties in a prominent position in nearly all the countries of Europe.

KEY

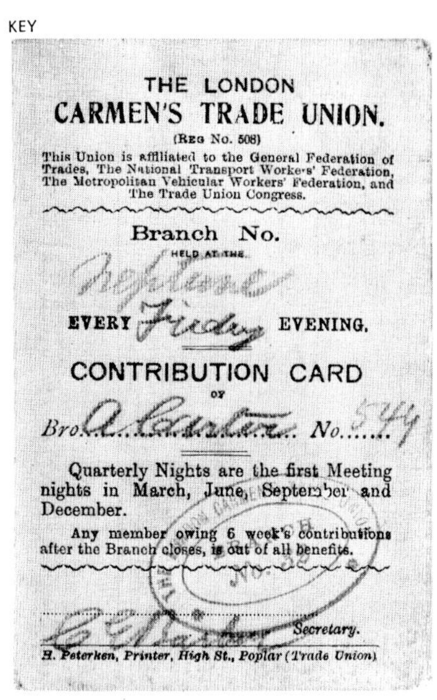
THE LONDON
CARMEN'S TRADE UNION.
(Reg No. 508)
This Union is affiliated to the General Federation of Trades, The National Transport Workers' Federation, The Metropolitan Vehicular Workers' Federation, and The Trade Union Congress.
Branch No.
HELD AT THE
EVERY Friday EVENING.
CONTRIBUTION CARD
OF
Bro. No.
Quarterly Nights are the first Meeting nights in March, June, September and December.
Any member owing 6 week's contributions after the Branch closes, is out of all benefits.
Secretary.
H. Peterken, Printer, High St., Poplar (Trade Union)

By 1914 the trade union movement, representing millions of working people, was a growing force in the major industrial countries. The years between 1900 and 1914 saw an increase in the number and intensity of union strikes. Generally, employers still disputed the right to strike and often still challenged the unions' right to exist. Bitterness and hostility underlying strikes often led to open violence. However, trade unions were often narrowly sectional in their interests while generally supporting a socialist political stance. Contribution cards, such as this for the Carmen's Trade Union, were proof of full union membership.

7

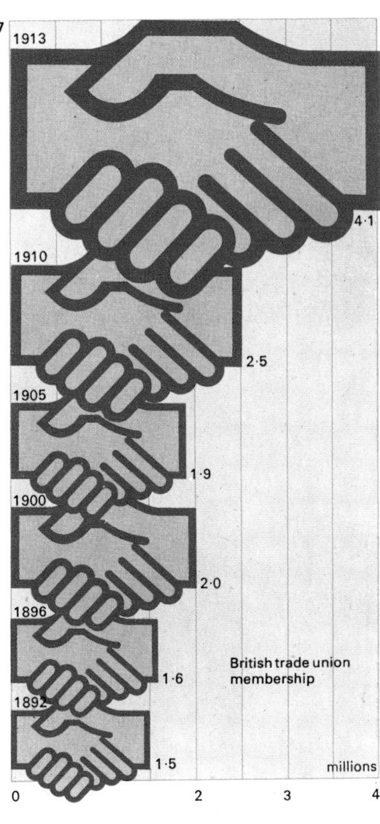

7 Before 1914 there was a surge in trade union membership because of industrial development. In Britain the number of unionists more than doubled between 1905 and 1914, mainly as a result of the organization of unskilled and semi-skilled workers such as the dockers and railway workers, as opposed to the "labour aristocracy" who had created the unions. In 1893 the Independent Labour Party (ILP) was formed, later to become the Labour Party (1906).

8

8 The concept of the general strike became widespread in the early years of the 20th century under the influence of syndicalist ideas. In Britain, the reformist character of the Trades Union Congress, formed in 1868, made it reluctant to use the general strike as a weapon, but in 1926 it called the General Strike in support of a bitter dispute in the coal industry. After a tense confrontation with the Conservative Government of Stanley Baldwin (1867–1947) the strike was defeated. Because of the Government's fear that food supplies would be looted, imports were collected from the London docks by armed convoys.

9

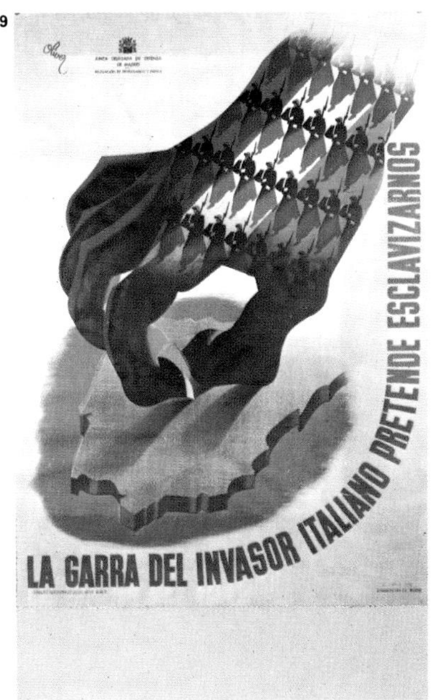

9 The Spanish Civil War (1936–9) was a rallying point for left-wing forces in Europe. The attempt by Franco's Nationalist forces to topple the Spanish Republic with aid from Italy and Germany resulted in cooperation between many divided communist and socialist parties. Although the war was a complex battle between various Spanish groups, it seemed to many socialists to symbolize the threat of fascism and the need for a united front.

10 As a result of widespread concern for the Spanish Republic among left-wing groups, the International Brigade was formed to fight in Spain. It was drawn from many different nationalities and consisted mainly of Communist Party members, trade unionists and sympathetic intellectuals. The Brigade was recruited through the Communist Party, which organized training, equipment and transport to Spain. The volunteers played an important part in preventing an early victory by Franco's forces and his German and Italian allies, but they suffered heavy casualties. Their role symbolized the wider significance of the civil war and its emotive appeal for a whole generation.

10

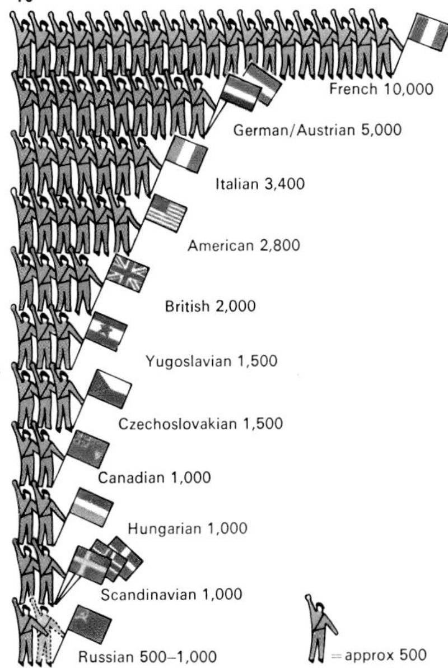

East Asia 1919–45

The history of East Asia from 1919 to 1945 is dominated by two related themes: the rise of Chinese nationalism in the 1920s and the spread of Japanese imperialism after 1931. Both developments were influenced by Western imperialist presence in the region. Chinese nationalism was complicated by the diverging interests of the two major political parties, the Nationalist Kuomintang (KMT) and the Chinese Communist Party (CCP).

Rise of Chinese nationalism

The year 1919 is a watershed in Chinese history. Demonstrations against the Paris Peace Conference's granting of former German concessions in China to Japan – which the Chinese government accepted – developed into an unprecedented national movement [1]. Sensing the revolutionary mood, Sun Yat-sen (1866–1925) reorganized his Nationalist Party into the disciplined KMT. With both a socialist ideology and a party-dominated army under Chiang Kai-shek [Key], the KMT received help from the Comintern and collaborated with the fledgling CCP formed in 1921. Both parties sought to end the division of China and its exploitation by foreign powers.

These privileges were little diminished by the Washington Conference (1921–2) which achieved only partial withdrawal by Japan. Chinese dissatisfaction coalesced with labour unrest, particularly in the treaty ports, culminating in a 15-month strike and boycott of foreign trade in Hong Kong in 1925–6. Against this background, Chiang Kai-shek led a northern expedition to unite China under the National Government set up in Canton. In 1927 Chiang clashed with party leftists, especially the communist bloc within the KMT. Purging the areas under his control [3], he succeeded in reunifying the KMT at the expense of the left and the CCP, setting up his own government in Nanking and bringing Peking and much of China under his control in 1928.

By 1930 extension of Nationalist authority put Chinese nationalism and Japanese imperialism on a collision course. Japanese privileges secured in Manchuria since 1905 were threatened by China's reassertion of its sovereignty there. Not only was Manchuria a buffer against Soviet ideology and military power, it also represented a considerable economic investment and had a million Japanese subjects.

Japanese imperialist expansion

Japan of the 1920s was characterized by paternalistic capitalism with limited democracy at home and co-operation with the great powers abroad. But in the 1930s ultra-nationalism and militarism fostered ideas of an autonomous economic empire as an answer to the Depression, which had exacerbated tensions in Japanese society. As confidence in politicians waned, popular support grew for the militarists who were close to Emperor Hirohito [5]. Japanese officers in Manchuria used the Mukden Incident of 1931 [4] to create a situation that led to the establishment of a Japanese puppet state, Manchukuo, in 1932. Expansion southward in 1935–6 was designed partly to create a subservient North China to protect Japan's rear in the event of war with the USSR.

Japan's encroachment brought a temporary truce between the KMT and CCP in

CONNECTIONS

See also
244 China: the People's Republic
228 World War II
146 Japan: the Meiji Restoration
144 The opening up of China

1 The May 4th Incident in 1919 was a demonstration by 3,000 students in Peking, protesting at the Paris Peace Conference that left Japan in control of German possessions it had seized in China. Spreading protest forced government changes and foreshadowed a new Chinese nationalism.

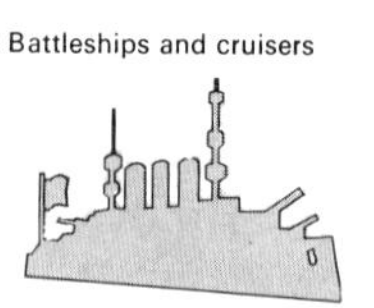

2	Battleships and cruisers	Destroyers	Submarines
Built by Japan 1919–20	4	17	9
Built 1921–24	12	34	32
Built 1925–28	8	21	19
Built 1929–32	7	17	12

2 Japanese naval power grew rapidly in east Asia after 1919 despite the 1922 Naval Treaty limiting replacement of capital ships by the US, Britain and Japan to a 5:5:3 ratio. Ratios for auxiliary ships set in 1930 were: heavy cruisers 10:10:6; light cruisers and destroyers 10:10:6; submarines, parity.

3 Communists were massacred in Shanghai on 12 April 1927 when Nationalist troops, police and secret agents disarmed workers and pickets and dissolved labour unions. The culmination of a power struggle between the left and right wings of the KMT, the purge spread elsewhere with more massacres of the Chinese left-wing and communists.

4 Japanese troops marched into Manchuria after the Mukden Incident of 18 September 1931. Acting without the authority of their government, Japanese forces occupied Mukden using the pretext of a bomb on the Japanese-run South Manchurian railway and a skirmish with Chinese patrols. The speedy occupation of Manchuria (shown here) followed.

1936. Chiang had dislodged the communists from their southern rural bases and forced them to undertake the Long March [6]. But the CCP leader, Mao Tse-tung (1893–1976), urged on by Russia, now sought a united front against Japan and Chiang was forced to agree. When full-scale fighting broke out in 1937, the powerful Japanese army forced the KMT to retreat to Changking in the southwest. The fall of Nanking in December [7] was followed in 1938 by the announcement of Japan's "New Order" with Japanese army rule in occupied parts of China and a puppet government in Nanking (1940).

Japan's empire in World War II

To secure access to South-East Asian raw materials and to block Western aid for Chiang, Japanese troops entered Indochina in 1940 and moved southward in 1941. America, Britain and Holland responded with a near total embargo on exports to Japan in July 1941, reducing oil supplies by 90 per cent. Japan soon put into operation its contingency plan to achieve economic self-sufficiency by force. Allied to Germany and Italy, and envisaging the imminent collapse of Britain and China, it tried to eliminate American interference by sinking the Pacific Fleet at Pearl Harbor on 7 December 1941.

By August 1942 Japan had seized a vast oceanic and continental empire [8]. It was not until early in 1944 that Allied sea power reversed these successes. While the Chinese Nationalists and communists tied down large numbers of Japanese troops in a war of attrition and Allied supply lines were restored in Burma, American amphibious offensives in the Philippines and Gilberts established bases from which air power could be brought to bear on Japan itself. In 1945, after atomic bombs had destroyed Hiroshima (6 August) and Nagasaki (9 August), Japan agreed to unconditional surrender on 2 September [9].

Japan's defeat left China divided between a Nationalistic administration gravely weakened by the war and the communists who had gained in strength. Japan was transformed under American occupation into a democratic state. In east and southeast Asia, the old empires were never to recover their shattered prestige and power.

KEY

Chiang Kai-shek (1887–1975) was the leading military aide of Sun Yat-sen by 1919. After Sun's death in 1925 he dominated the Kuomintang and became president of a largely reunited Republic of China in 1928. But his authority was contested by the Communist Party and threatened by the Japanese. Recognized by the Allies as China's wartime leader, he secured the abolition of extra-territorial rights in China in 1943 and in 1945 a seat for China in the UN Security Council. Renewed postwar conflict with the communists led to his defeat and the withdrawal of his government to Formosa (now Taiwan) in 1949.

5

5 Emperor Hirohito (1901–) came to the Japanese throne in 1926, having been named regent in 1921. Under the Meiji constitution his position was both sacred and sovereign, although there is little evidence to show the part actually played by the Emperor in Japanese policies.

6

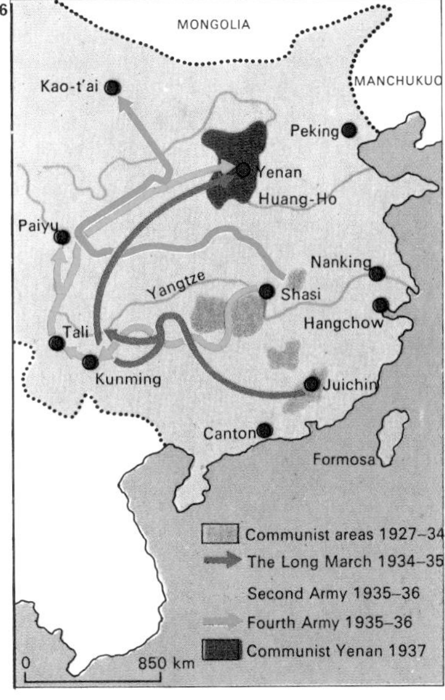

6 In the Long March, about 85,000 communist soldiers and 15,000 officials left Kiangsi province under pressure from Chiang Kai-shek, in 1934. A year later 30,000 survivors regrouped near Yenan after a march of 8,000km (5,000 miles). The communist 2nd and 4th armies also had to regroup in the north.

7

7 The fall of Nanking, Chiang Kai-shek's capital, on 12 December 1937, was followed by the massacre of some 100,000 people by Japanese troops. Known as the "rape of Nanking", this atrocity was revealed at the International War Crimes Tribunal in Tokyo. The city's fall came after three months of stubborn opposition by Chiang's army to the advance of the Japanese.

8

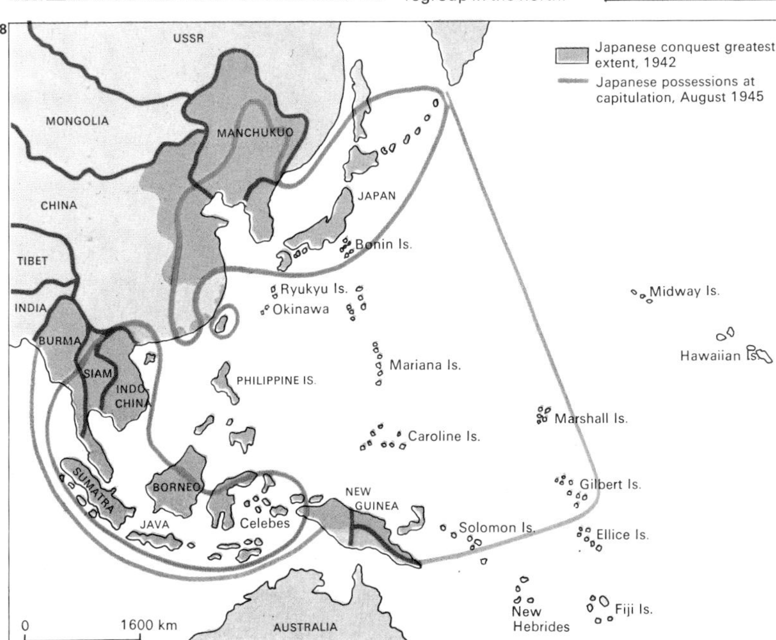

9

8 Japan's territorial acquisitions in World War II reflect its initial aims: to conquer China before dealing with the USSR and to control the south-west Pacific. Later the military priority shifted to include invading India in preference to defending Pacific islands. Before the Allies entered the war against Japan, China traded space for time. Once deep in Chinese territory, Japanese troops, although controlling most industrial areas, were surrounded by a hostile countryside.

9 Japan's surrender was signed aboard the USS *Missouri* in Tokyo Bay on 2 September 1945 with General Douglas MacArthur representing the Allies. The Japanese decision to surrender on 14 August 1945 came from the Emperor.

Indian nationalism

In 1900 British rule in India appeared more secure and more permanent than ever. Lord Curzon's years as viceroy (1898–1905) emphasized the determination of the British to remain the governors of India. The greater efficiency of the administration, the maintenance of peace and order, and the spread of railways [3] and the telegraph, all seemed to confirm Britain's grip on India, while in the wider world British foreign policy was geared to the retention of the Indian empire as the second great base (after Britain itself) of British world power. Yet within fifty years that same Indian empire had been split up and the British rulers dismissed.

Growth of nationalism

Part of the reason for this reversal lay in the growth of a nationalism which drew support from Indians all over the subcontinent. This nationalism had risen from modest beginnings in the late nineteenth century with the foundation of the Indian National Congress Party and was at first approved by the British for its attempt to break through the divisions of caste, religion and region that stifled efforts to modernize India. But before long they came to see it as a potent threat to British power and a stimulant to disorder and anarchy. Anti-British terrorism before 1914 made many officials deeply hostile to the call of nationalists for more Indian participation in government. The British believed that the Congress was the tool of ambitious and unscrupulous westernized Indians, seeking not independence and unity but self-advancement, regardless of the poor.

The first great triumph of Indian nationalism came in the years immediately following World War I when Mahatma Gandhi (1869–1948) [Key] emerged as a charismatic leader pioneering the technique of non-co-operation and non-violent resistance to the government through peaceful demonstrations and refusal to pay taxes. Gandhi was helped in showing the British that many Indians rejected their authority by the effects of India's involvement in World War I. Higher taxation, the recruitment of thousands of Indians for the army, and the use of that army to defend Britain in northern France united Indians of diverse interests in the belief that the British were placing new and unfair burdens upon them and breaking the terms on which British rule was accepted. They turned for protection to the Congress Party. To the British, Gandhi's campaign was deeply worrying. Some of them believed a second mutiny was imminent (the first mutiny in 1857–8 had resulted from unrest amongst the sepoys [soldiers], but was suppressed by the British): and it was in a climate of panic that the notorious shooting of unarmed Indian demonstrators – the Amritsar Massacre [1] – occured in 1919.

Divisions among the Indians

For all its successes between 1918 and 1922, Indian nationalism faced enormous problems in trying to destroy British power. Once India had settled down after the war and its aftermath, non-co-operation fizzled out. Many Indians were profoundly suspicious of the politicians who ran the Congress Party. The rural landowners who wished to keep the social status quo disliked the urban and westernized Indians who dominated the nationalist movement. They feared that if

CONNECTIONS

See also
140 India in the 19th century
220 The Commonwealth
246 Decolonization
254 Non-alignment and the Third World

1 At Amritsar on 13 April 1919, British troops shot dead over 300 unarmed Indians during an illegal demonstration. The Indians were forced to apologize publicly after the riot because the British thought this would encourage them to be orderly and respectful. Here a Sikh is arrested.

2 By origin, Gurkha soldiers, still a distinctive element in the British army, were mountain tribesmen from Nepal who were defeated by the British in the Gurkha wars of 1814–16. They became famous for their endurance, loyalty and courage, and for their *kukri*, deadly broad-bladed curved knives that they carry.

3 By 1948, India had the fourth largest railway system in the world. The railways had originally been constructed to serve British purposes, to help control India's vast expanses cheaply and efficiently, and to open up the hinterland to trade. But they also helped to unify India economically and politically and thus lay the foundations for an Indian nationalism on the subcontinent.

4 Jawaharlal Nehru (1889–1964) was the first prime minister of independent India. Educated in England, he emerged as a leading figure in the Congress Party in the 1930s and as Gandhi's heir.

5 Mohammed Jinnah (1876–1948) was the architect of Pakistan, which resulted from the partitioning of the old unified India into two states. He believed this was the only way to safeguard Muslim interests.

6 As well as the provinces which they ruled directly, the British retained ultimate power over nearly 600 autonomous princely states. The Victorians adopted the durbar (traditionally in India a gathering of vassals to do homage to their ruler) to symbolize the allegiance of the Indian princes to the British monarch. The 1911 durbar was attended by King George V in person.

such men became strong enough to throw out the British their next target would be the conservative gentry, still so powerful in the countryside. And not all Indians wanted democracy and one man, one vote. Hindus living in areas where the majority was Muslim, and vice versa, were fearful that popular government would threaten their interests and maybe their lives.

This meant that the British still had an advantage. They were willing to give Indians a greater say in running their internal affairs so as to avoid trouble; and they found that delegating some power to some Indians was a convenient method of preventing all Indians from combining against them. They hoped by this to keep India united in a federation which in international matters would still be tightly bound to Britain; and they wanted to go on using the Indian army.

These clever calculations were swept away by two "accidents". The first was the outbreak of World War II – once again involving India – which aroused more resentment among Indians than World War I. Meanwhile British prestige was undermined by humiliating defeats by the Japanese. The second "accident" was the resolve of the leaders of the large Muslim communities of north India to insist upon the creation of Pakistan as a separate Muslim state.

Independence

Thus India gained independence in a way quite unintended by the British. The division of the subcontinent wrecked the delicate mechanisms of federalism through which they had planned to influence India in her international role as a pillar of the Empire-Commonwealth. Deprived of Indian help, the British Empire east of Suez withered away in less than 20 years. In India itself, independence left vast problems unsolved: the overpopulation of the countryside; the failure to increase food production sufficiently; the desperate poverty of village and city alike. The British had lacked the means and the nerve to modernize Indian society properly. The victory won by Indian nationalism in 1947 was, therefore, only a beginning. The building of a modern nation state lay ahead.

KEY

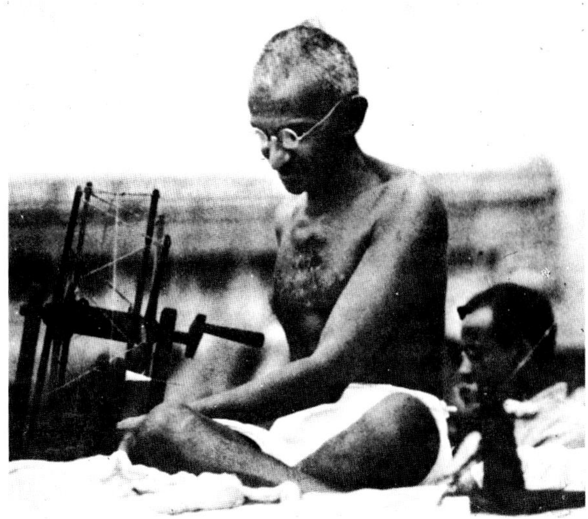

Gandhi adopted the symbol of the spinning wheel in 1920. He believed that India should make her own cloth, thus threatening British textile exports to India and giving Indians the self-confidence necessary for independence.

7

8

7 The Indian Army was an enormous asset for the British in the defence of their vast empire. In World War I, Indian troops fought on the Western Front, while in the period 1939–45 they were used in Burma, the Middle East, the Mediterranean and North and East Africa. The loss of their services after independence in 1947 placed a great strain on Britain's own military resources.

8 Lord Louis Mountbatten (1900-) was the last Indian viceroy. His political gifts, much needed in the transition of India and Pakistan to independence, had been shown in SE Asia in World War II.

9 Political democracy in India was extended in the reforms of 1919 and 1935 when the British gave the Indians the right to participate in government. Universal suffrage was achieved in 1947.

10 As independence drew nearer, the tensions between different communities in India became acute. The most serious were those between Hindus and Muslims, especially in areas where their numbers were almost equal and the proposed partition caused great bitterness. The aftermath of bitter riots in Calcutta in 1946, when at least 4,000 people died in the communal fighting, is illustrated here.

9

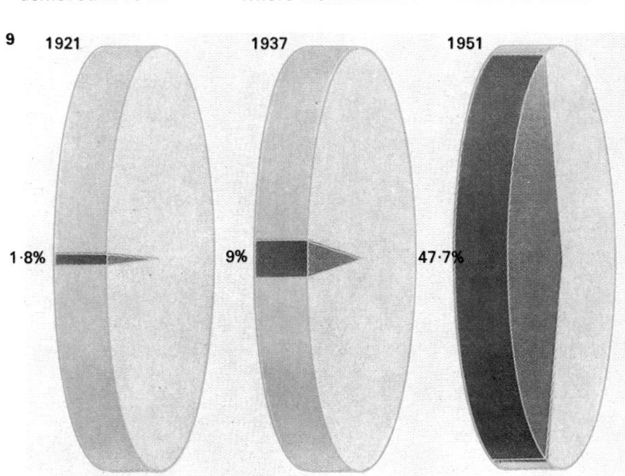

10

British foreign policy since 1914

Britain's aim in World War I (1914–18) – to prevent the domination of Europe by any single power – was achieved by the peace treaties imposed in 1919 on Germany and its allies. But then the Allied coalition that won the war dissolved: the United States withdrew into isolation, Russia, under communist control, campaigned against the West, and France disagreed with Britain on the treatment of Germany [1]. Between 1925 and 1930 British governments welcomed a short-lived reconciliation between France and Germany, helped by the flow of American money into Europe. But by 1931 the world was hit by a grave economic crisis.

The age of appeasement

The economic slump propelled Adolf Hitler (1889–1945), leader of the National Socialists, into power in Germany in 1933 and ended the liberal regime in Japan. This provided Britain with two major foreign-policy problems in the 1930s: the satisfaction of German pressure for revision of the Treaty of Versailles 1919, especially its reparation and disarmament clauses, and the expansion of Japanese militarists into Manchuria, which they annexed in all but name in 1931–2, and into China proper in 1933–7.

Britain was handicapped in dealing with these problems by three factors: First its economic weakness, expressed in long-term unemployment; second, a public opinion stunned by the losses in World War I and nervous about rearmament; finally the sheer inability of Britain and France, with no help from isolationist America, to control Germany and Japan, as well as a restless Italy under Benito Mussolini (1883–1945).

A confrontation became unavoidable after the Munich Agreement in September 1938 [5]. Britain and France thereby agreed to German occupation of the Sudetenland, which was Czech territory accepting that this was Hitler's "final" demand; but on 15 March 1939 German forces cynically abrogated the agreement by taking over the rest of Czechoslovakia. Almost the only benefit derived by Britain from this short-sighted policy of appeasement was time to build up its pitiably weak defences. When Hitler confidently invaded Poland on 1 September 1939, the British government finally honoured their treaty obligations and declared war on Germany two days later.

Policy during World War II

After the collapse of France in June 1940, Britain faced Hitler's Europe alone. The military situation was transformed when Germany invaded the Soviet Union on 22 June 1941 and when the United States entered the war after the Japanese attack at Pearl Harbor on 7 December 1941. But the diplomatic situation was complicated. The United States' President, Franklin Roosevelt (1882–1945) agreed with the British Prime Minister, Winston Churchill (1874–1965) on general war aims in the Atlantic Charter, signed in August 1941, but had no interest in preserving the British Empire. At the Yalta conference (4–11 February 1945) the two met Soviet leader Joseph Stalin (1879–1953) and Roosevelt seemed to side with Stalin on imperial questions against the British.

Churchill, an old opponent of communism, willingly accepted Stalin's territorial claims in eastern Europe, but he was worried

CONNECTIONS

See also
182 British foreign policy 1815–1914
192 The Peace of Paris
194 What World War I meant to Britain
214 East Asia 1919–1945
220 The Commonwealth
224 Britain 1930–1945
226 Causes of World War II
228 World War II
234 The division of Europe
236 Britain since 1945: 1
238 Britain since 1945: 2
246 Decolonization
258 Evolution of the Western democracies
306 Europe from confrontation to détente

1

1 On 11 January 1923, French and Belgian forces, despite British protests, occupied the German industrial Ruhr (until August 1925) as a penalty for alleged non-payment by Germany of coal reparations. In the aftermath of World War I, Britain and France disagreed over the treatment of Germany. Britain, intent upon economic recovery, wanted Germany leniently treated; France demanded strict enforcement of the Treaty of Versailles.

2

2 At the League of Nations, Geneva, in March 1925, Britain rejected a major attempt to enforce the peaceful settlement of international disputes in refusing to sign the Geneva Protocol. France sought to strengthen the League's powers of collective action against aggression by providing for compulsory arbitration of disputes. But Britain was wary of the protocol's absolute commitment to armed intervention.

3 Winston Churchill was a backbench MP during the 1930s, and an outspoken critic of the National Government's policy of appeasement towards Germany. Public opinion favoured a vague pacifism in the face of German rearmament and growing Italian and German aggression. Such was the desire for, if not faith in, peace, that British rearmament did not seriously begin until after 1938.

3

4

4 The Spanish Civil War was fought between the Nationalist rebels on the one hand, helped by Hitler and Mussolini, and the Spanish Republican Government on the other, from 1936 to 1939. The war seems to have presaged the later conflict waged between Fascism and democracy in World War II. But in 1936, the British Government was chief sponsor of an international agreement for non-intervention which was signed by all the major powers. This was adhered to by all countries except Italy, Germany and the Soviet Union – the last-named sent some aid to the Republican forces. Public opinion in Britain was divided: some, illegally, went to Spain to fight, mostly for the Republicans, whose British International Brigade numbered about 2,000 men.

5 The Munich Agreement of 29 September 1938 typified Britain's policy of appeasement. Britain, France and Italy agreed that the Sudeten region of Czechoslovakia should be ceded to Germany. British Prime Minister Neville Chamberlain (1869–1940), shown here on his return from Munich, made an agreement with Hitler to consult on any future Anglo-German questions. A year later Britain and Germany were at war.

5

how far into western Europe Soviet influence would penetrate and whether the United States would help to resist it.

Loss of world power

Britain's Labour government of 1945–51 hoped fervently for co-operation between the Soviet Union and the West after the war. But when the East-West cold war developed with disagreements about the revival of Germany and about Soviet communization of Eastern Europe, Labour ministers took Britain first into the Brussels collective defence pact of March 1948 with France and the Benelux states – Belgium, Holland and Luxembourg – and then into the North Atlantic Treaty Organization (NATO), signed on 4 April 1949, with Canada, the United States and nine other states of Western Europe. At the same time, Britain received economic assistance from the United States through a £1000 million loan in December 1945 and then through the Marshall Aid programme of 1948–52.

Britain strove continuously to moderate East-West tensions, by urging restraint on the United States during the cold war, but its credibility was undermined by recurrent balance-of-payments difficulties, and later by severe unemployment and inflation. These called into question the basic assumption of British policy after 1945: that Britain remained a world power, not with the strength of the United States or the Soviet Union, but still with an assured presence at conference "top tables".

In January 1968 the Labour Prime Minister, Harold Wilson (1916–), decided to terminate the East-of-Suez role by December 1971, ie, the maintenance of British forces in the Persian Gulf and at Singapore. The effect of this was to reduce Britain to the level of an essentially European and Mediterranean power.

The change in Britain's international position was symbolized in January 1973 by its entry into the European Economic Community. By that time Britain's empire, which once embraced a quarter of the world's population, had been gradually transformed into a loosely knit Commonwealth of politically independent states.

KEY

Anti-British feeling in Cyprus (1955–60) typified the strains of decolonization in areas where Britain's handover of power after World War II was complicated by divisions in the local community. In other colonies, the transition to independence was often peaceful. But areas of violence included mandated Palestine, India, Egypt, Kenya, Malaya and Aden.

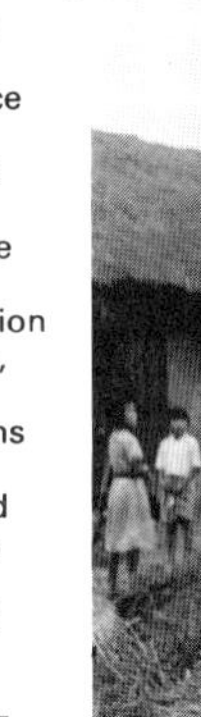

6

6 British troops were sent to Korea in 1950 as part of a United Nations force to repel a North Korean communist invasion of South Korea. The UN force had been formed despite the opposition of the Soviet Union, which supported North Korea's claims to South Korea. At that time the cold war had reached its height and the UN was deeply divided by the East-West tensions that had emerged since 1945.

7

7 The nationalization of the Suez Canal in July 1956 by the Egyptian Government was part of a policy that aimed to unite the Arab world and end foreign control. Britain and France tried to internationalize the Canal and when this failed they attempted to seize the Canal by armed force. The troops seen here were landed in November, but as a result of US pressure a ceasefire took place within two days. The incident was a major blow to the international prestige of Britain and France. Anthony Eden (1897–1976), the British prime minister, who had pressed for the use of force, resigned in the following January.

8

8 The Anglo-American "special relationship" was a principal feature of British foreign policy after 1945. Two of its chief exponents were Harold Macmillan (1894–) [left], British Prime Minister (1957–63) and John F. Kennedy (1917–63) [right], US President (1961–63). Here they are shown after talks in Washington in 1961 that were aimed at controlling the spread of the H-bomb and increasing unity among the countries of the Western alliance.

9

9 Ian Smith (1919–), Prime Minister of Rhodesia, unilaterally declared independence from Britain on 11 November 1965 after rejecting British terms for granting independence. Smith, shown here following discussions with British Prime Minister Harold Wilson in October 1965, wanted to maintain white supremacy in Rhodesia, although the white population was outnumbered 22 to 1 by black Africans.

10

10 The leaders of France and Britain, Georges Pompidou (1911-74), [right] and Edward Heath (1916–) [left] cleared the way for Britain to enter the EEC, in January 1973. When the Community was first formed in 1957 Britain refused to join, fearing the EEC's supranational powers. Two subsequent applications for membership, in 1961 and 1967, were both blocked by Pompidou's predecessor, President De Gaulle. Final talks had begun after De Gaulle's resignation, in 1969.

The Commonwealth

The Commonwealth [1] is a free association of nations comprising Great Britain and (in 1976) 35 other sovereign states that were once colonies or dependent territories within the British Empire.

The origins of the association go back to Britain's relations with her colonies of European settlement. For these she established during the nineteenth century a system in which they would acquire independence by stages [4]. Crown government concentrated power in the governor and his council, which might be advisory or executive. Later there might also be a legislative council or assembly that lent itself to constitutional progress as the proportion of elected members grew in relation to the nominated members. Further advances would occur when the executive council, or government ministers, became responsible to the representative assembly, and when the indigenous council and assembly acquired powers of internal self-government over all but special and external matters. Finally these, too, would be won when full independence was granted.

Perhaps the first significant step in the transition from empire to commonwealth was a report by Lord Durham (1792–1840) [2], in 1839, stating that one of the causes of contemporary unrest in the Canadian provinces was the lack of harmony between the executive and the legislature. The remedy, according to the report, was to choose executive ministers from the majority group in the representative assembly.

From colonies to dominions

There were limitations on such local government: the management of foreign trade and relations, the disposal of unoccupied public lands and the amendment of constitutions remained with the British government.

These limitations were gradually removed, but that on the conduct of foreign relations remained until the twentieth century (by which time the self-governing colonies – Canada, Australia, New Zealand, South Africa and Newfoundland – were known as "dominions"). The precise status of the dominions was undefined until a committee under Lord Balfour (1848–1930) produced a report (1926) which stated, "They are autonomous Communities within the British Empire, equal in status . . . although united by a common allegiance to the Crown, and freely associated as members of the British Commonwealth of Nations."

In those colonies that were not primarily of European settlement, constitutional advance was very much slower. In the 1930s colonial rule in Africa and Asia was assumed to have a long future ahead of it. World War II, however, helped to stimulate nationalist pressures, especially in India, which had already acquired some international recognition as a state. The struggle for independence in India had produced a number of powerful leaders, such as Mahatma Gandhi [3], who acquired worldwide fame for their defiance of imperial authority, usually by non-violent resistance. In 1947 and 1948 independence was granted to India, Pakistan and Ceylon.

Nkrumah's "self-government now"

The concept of gradual progress was destroyed after Malaya and the Gold Coast became independent in 1957. In the Gold Coast, the nationalist leader Kwame

CONNECTIONS

See also
132 The British empire in the 19th century
218 British foreign policy since 1914
216 Indian nationalism
142 Africa in the 19th century
130 Imperialism in the 19th century
246 Decolonization
252 Southern Africa since 1910
248 Australia since 1918
250 New Zealand since 1918
134 The story of the West Indies
136 The story of Canada
254 Non-alignment and the Third World
258 Evolution of the Western democracies
298 The United Nations and its agencies
302 Underdevelopment and the world economy

In other volumes
316 Man and Society

1 A quarter of the world's land surface is covered by the Commonwealth, which also embraces a huge number of languages and dialects as well as numerous religions. The feature common to all of the members is the historical accident of settlement, annexation or conquest by Britain; and British institutions and the English language also remain important elements in the modern association. Since 1965, there has been a permanent central secretariat, based in London. It organizes conferences and spreads information. But the Commonwealth has very little unity in international affairs. The dates of independence of each country are shown on the map.

2 Lord Durham's report in 1839 on the unrest in the Canadian provinces led to the introduction of responsible government in Canada and later in other colonies of settlement. This report opened the way for the development of independent parliamentary governments linked by a common allegiance to a single crown.

3 The strong, moral leadership of Mahatma Gandhi (1869–1948) lay behind much of the nationalistic agitation against the British in India after 1918. His asceticism, aura of holiness and his use of fasting and passive resistance often embarrassed the British Government and actively involved the masses in India in the campaign for independence.

Nkrumah [5], who became prime minister in 1952, refused to recognize any impediment to the early transfer of power, advocated positive action to cripple the forces of imperialism and popularized the slogan "self-government now". When the Gold Coast gained its independence, British Togoland joined it to form the new nation of Ghana. In 1960 Ghana became a republic, with Nkrumah as president. The "wind of change" speech by the British prime minister Harold Macmillan (1894–) in 1960 reflected the new attitude of the British Government towards Africa.

The newly independent countries chose to remain in the Commonwealth, despite the fact that they lacked the ethnic and common historical origins of the older members. They believed that their participation in the Commonwealth would bring them economic and diplomatic benefits and enhance their international influence. But they did not feel the special attachment to the monarchy that the older members had felt, and in 1949 India was the first state allowed to retain its membership as a republic while accepting the British monarch merely as the symbol of the free association of the members.

All members participate in the Commonwealth system of consultation and co-operation that covers a multitude of activities at governmental level. Periodically the heads of government meet [8] in conferences.

Expansion and the loosening of old bonds

Nevertheless, the old bonds of Commonwealth are not as strong as they used to be and much of the informal intimacy of earlier years has been lost as the association has expanded. Disillusionment with the Commonwealth was apparent among many members in the 1960s and 1970s – notably in Britain. The modern Commonwealth, however, continues to function as a flexible system of co-operation between states, enabling its member countries to confer with one another in an unusually frank, friendly and relaxed manner. As an association it represents the fulfilment both of the nationalist aspirations of colonial peoples and of a policy of constitutional evolution pursued by the imperial power.

KEY

George V was called "King of Great Britain, Ireland and the British Dominions beyond the Seas, and Emperor of India". His title illustrated the extent of his authority and also emphasized India's special position in the British Empire.

4 The character of colonial government varied in detail from one territory to another, but each was expected to follow much the same series of stages on the way to independence. And the advance, through responsible government, to dominion status by the old European colonies of settlement became the model. But in the final period of decolonization, the stages were not always as clearly defined as earlier.

4

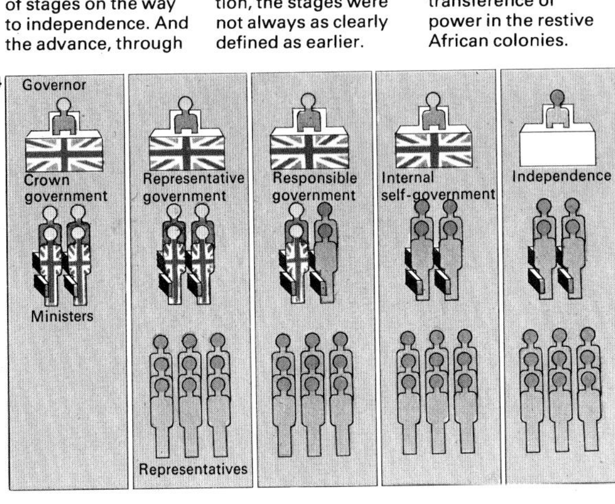

5 By campaigning strongly for "self-government now" in the Gold Coast, Nkrumah (1909–72) helped destroy the concept of gradual transference of power in the restive African colonies.

5

6

6 Conflict – like this in Bangladesh – sometimes followed independence. National unity could be endangered and often parliamentary democracy was replaced by one-party rule and an autocratic president.

7 Until colonial territories in Africa and elsewhere became independent after World War II, the Commonwealth "family" – here assembled in 1926 – was a small, intimate group, all subscribing to British traditions and acknowledging one Crown.

7

8

8 As membership of the Commonwealth grew, the informality of heads of government meetings became more difficult to maintain. But the members still felt that the meetings were valuable for the discussion of problems and improvement of mutual understanding. When other capitals (here Singapore in 1971) began to offer to be host, this further emphasized that the modern Commonwealth was no longer "British" but a unique, worldwide association representing many races, creeds and cultures.

The rise of fascism

Fascism developed in the years between World Wars I and II to become a major ideological and political force in many European countries, most notably in Italy and Germany. Expressed as an intense nationalism, often with strong social and collectivist overtones, it had the support of many different groups of people in countries that were suffering from, or seemed threatened by, a total breakdown of both their economy and their society.

Fascist ideology

Although fascism shared many characteristics with reactionary nationalism and more conservative, authoritarian regimes, it had distinctive characteristics of its own. These were derived from its rejection of nineteenth century individualistic liberalism.

Fascist ideology embraced many thinkers, often distorting and misapplying their ideas. Indeed, fascism was never to formulate a clear ideology in the same way as Marxism, but remained open to a number of different interpretations, in which the component elements received varying emphasis. Among the most important contributors to fascist ideas were Friedrich Nietzsche (1844–1900), who stressed the need for dynamic "supermen"; Henri Bergson (1859–1941), who stressed instinct above reason; and Georges Sorel (1847–1922), who emphasized the moral value of action.

Italian fascism and Mussolini

Italy emerged from World War I disappointed and frustrated by her war losses and the failure of the Versailles settlement to fulfil the treaty promises that had induced her to enter the war. Unemployment, strikes and violence [1] provided the background to the breakdown of parliamentary government. Right-wing groups, such as that led by Gabriele d'Annunzio (1863–1938), seized the port of Fiume on the Adriatic coast in 1919 in defiance of the Versailles settlement. In city and countryside riots, estate seizures by the peasants and countless sit-in strikes created a menacing and unpredictable revolutionary atmosphere.

In this situation Benito Mussolini (1883–1945) [Key], an ex-socialist schoolteacher, organized anti-socialist *fascios* to combat left-wing groups by strong-arm methods. He received support from diverse conservative elements and by 1921 there were more than 800 branches of his "blackshirts", the *Fasci di combattimento*. Taking advantage of the disorganization of left-wing forces, he organized a "March on Rome" which ended with his installation as premier in October 1922.

Mussolini concentrated on liquidating and terrorizing opponents, establishing the Fascist Party in power and building up his personal position. Press, courts and unions were brought under his control and he established a concordat with the Roman Catholic Church. He inaugurated public works, such as the draining of the Pontine marshes, and mounted a drive for self-sufficiency for Italy. Increasing state intervention marked Mussolini's economic policy after 1925 as he tried to create a "corporate state" in which industrialists and workers co-operated for the good of the nation. Combined with his expansionist foreign policy, demonstrated both in the Abyssinian War and also his

CONNECTIONS

See also
226 Causes of World War II
224 Britain 1930–45
218 British foreign policy since 1914
212 Socialism in the West
208 The twenties and the Depression
192 The Peace of Paris

1 A

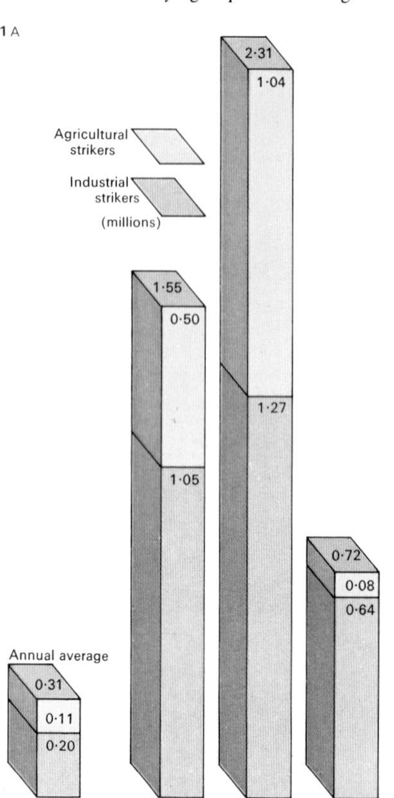

B

1919 0·9
1920 20·6
1921 249·0
(totals in thousands)

1 In 1919 Italy was crippled by war losses, inflation and unemployment. Fascism grew in response to conservative fears of a left-wing revolution, fuelled by the mounting toll of strikes [A]. Fascist membership [B] rose from under 1,000 in 1919 to 249,000 in 1921, taken mainly from the middle classes. Aided by industrialists, landowners and the army, Mussolini took power in 1922 after a threatened coup.

2

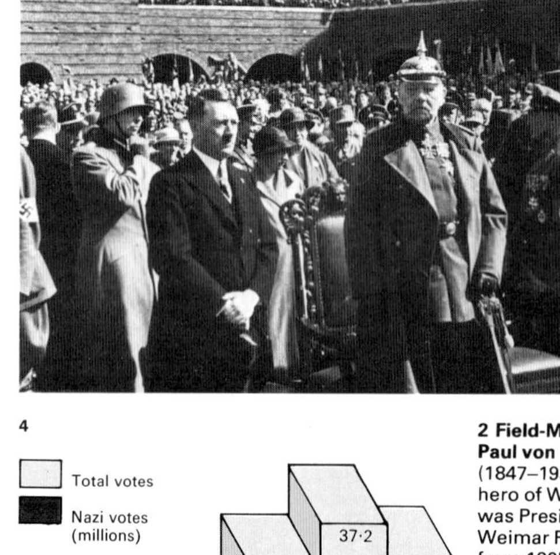

2 Field-Marshal Paul von Hindenburg (1847–1934), a national hero of World War I, was President of the Weimar Republic from 1925. Under nationalist pressure, he made Hitler Chancellor in 1933.

3

3 The Nazis based much of their propaganda upon virulent anti-Semitism, in which the Jews were used as scapegoats for Germany's economic difficulties. The Nazis conducted boycotts of Jewish shops, attacked synagogues and assaulted individuals but were unable to adopt formal measures until Hitler's accession to power in 1933. Anti-Jewish laws promoted emigration and denied civil rights. Many Jews had left Germany or were in camps by 1939.

4

Total votes
Nazi votes (millions)

	May 1924	Dec 1924	May 1928	Sept 1930	July 1932	Nov 1932
Total votes	29·7	30·7	31·2	35·2	37·2	35·7
Nazi votes	1·9	0·9	0·8	6·4	13·8	11·7

4 The fluctuation in votes for the Nazis reflected the economic fortunes of the Weimar Republic. In May 1924 the Nazis gained 1.9 million votes and 32 seats in the Reichstag. With the recovery of the Weimar Republic from its postwar difficulties and the inflation of 1923, the Nazi vote declined to its lowest point in 1928 when they held only 12 seats in the Reichstag. Under the impact of a renewed depression after 1929, and with the rise in unemployment and the polarization of the middle classes, the Nazi vote rose rapidly. By 1932 the Nazis were the largest party with 13.8 million votes. Although they lost votes, Hitler became Chancellor in January 1933.

involvement in aiding Francisco Franco (1892–1975) in Spain, Mussolini's policies not only antagonized other European nations but also exhausted Italian resources.

Hitler and German fascism

In Germany the Nazis (National Socialist Party) were founded in the disillusionment and economic chaos in the years following World War I. Joined by Adolf Hitler (1889–1945) [Key] in 1919, who expanded and transformed it, the party gained some seats in the Reichstag [4]. In 1923 Hitler tried, unsuccessfully, to overthrow the Bavarian government in a *putsch* in Munich, for which he was imprisoned.

Votes for the Nazi Party declined as the Weimar Republic recovered in the middle and late 1920s but the onset of the worst phase of the Depression after 1929 swelled party ranks with the young, the unemployed, and frightened middle-class and conservative elements. For Hitler and some of his followers, anti-Semitism [3] formed an important part of the programme, the Jews being cast as scapegoats for Germany's misfortunes and as intruders in a purely Aryan Germany.

Support for the Nazis, however, seemed to have reached its peak towards the end of 1932 and the party was running into financial difficulties as funds from major industrialists dried up. In January 1933 Hitler was put into office through a coalition with the right-wing Nationalist Party, who hoped to control him. After the Reichstag fire [6], Hitler was able to assume dictatorial power. The rule of terror through the Gestapo gave the regime a more vicious character than Mussolini's in Italy. Like Mussolini's fascism, however, Nazism also offered an aggressive foreign policy and a solution to unemployment through public works and rearmament [5].

Fascist parties grew up in many other countries. In Spain [7], the Falange provided support for Franco, while in Eastern Europe the Romanian "Iron Guard" and the regime of Admiral Horthy in Hungary had strong fascist elements. In Western Europe the blackshirts of Oswald Mosley (1896–) in Britain and the *Croix de Feu* in France [8] appeared, temporarily, to threaten the overthrow of democratic government.

KEY

By 1934 Italy and Germany were ruled by fascist dictators. Mussolini (right) assumed power much earlier than Hitler (left), but the latter dominated international politics in the 1930s. The Rome-Berlin Axis was formed in 1936.

5

5 Hitler aimed to satisfy public opinion by cutting unemployment and creating a prosperous Germany. Public works, such as the building of the autobahn network, then the most extensive in the world, provided an advertisement for the regime in reply to the criticism of its domestic and foreign critics, and also served the purposes of the military. To increase vehicle building capacity, while also providing a cheap automobile for the population, the "people's car" or Volkswagen was launched in 1938. By the late 1930s, however, living standards had begun to stagnate as arms expenditure rose.

6

6 Hitler's rise to power was only half completed with his accession to the chancellorship. He awaited the opportunity to introduce emergency laws to strengthen his position and this was offered when a young Dutchman, Marinus van der Lubbe, set fire to the Reichstag on 27 February 1933. The Nazis were suspected of starting the fire, but it appears they merely took advantage of it to promulgate emergency decrees, banning rival political organizations, imprisoning opponents and vesting power in Hitler and the Nazi Party. Although the Nazis failed to achieve a majority, they were supported by the Nationalists.

7

Aid to Spanish fascists 1936-9

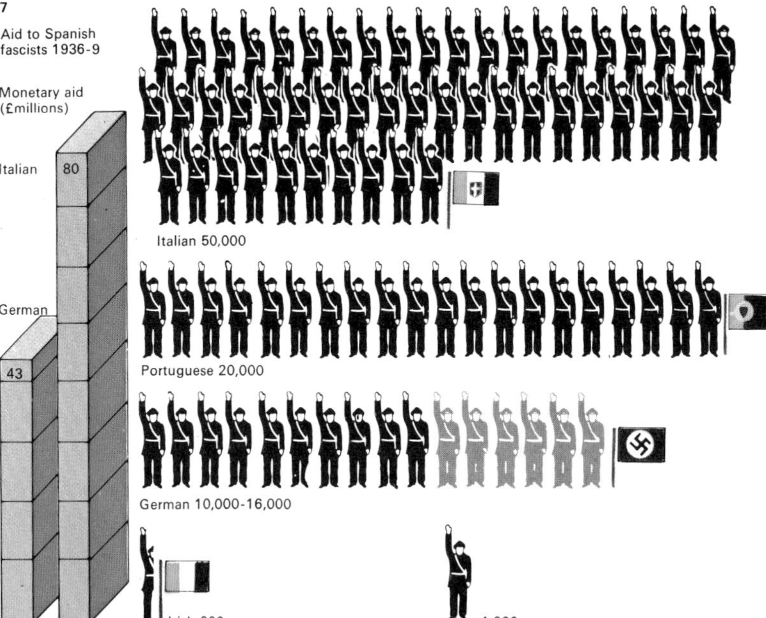

7 By 1936 both Italy and Germany were expanding their influence in international politics. The outbreak of civil war in Spain provided diplomatic and military advantages for both countries. Mussolini hoped to gain military bases in the western Mediterranean. By 1937 Italian war production was beginning to show signs of strain. Hitler hoped to sow dissension between Britain and France, while binding Italy closer to him. He used Spain as a training ground for his air force, including the "Condor Legion", a force of 6,500 men consisting mainly of air force units but with a few supporting ground units. From 1937 Spain became a mere side-show.

8

8 Political instability in France promoted anti-Semitism, particularly in magazines such as *Le Cahier Jaune*.

Britain 1930–45

Between 1930 and 1945 Britain experienced the deepest economic depression in its history and the massive mobilization of resources required for total war. In 1929, when a Labour government was elected under Ramsay MacDonald (1866–1937), Britain was already suffering from depression in its staple heavy industries: coal mining, iron and steel, textiles and shipbuilding.

Consequences of the Depression

The Labour government was pledged to tackle the problem of unemployment, which stood at more than one million insured workers [4]. No sooner was the government formed, however, than the Wall Street crash plunged the major western industrial economies into deeper depression. By 1931 the government was faced with more than two-and-a-half million unemployed and a heavy drain on its resources to meet the cost of unemployment benefits. The Labour government had little to offer as a solution to the economic depression. Radical voices, such as that of Oswald Mosley (1896–), a junior member of the Labour government, and Lloyd George (1863–1945), leader of the Liberals, offered solutions along the lines later advocated by John Maynard Keynes (1883–1946), but were ignored in the pursuit of orthodox economic policy. This dictated that the government should curtail its expenditure and raise business confidence in the hope that normal trading conditions would begin to reduce unemployment. The recommended cuts in expenditure included a reduction in unemployment benefit.

In 1931, the Labour cabinet was deeply divided over implementing the cuts. The government was forced to resign over the issue, but MacDonald and a group of Labour MPs joined with the Conservatives and Liberals to form a coalition, the National Government. A general election was then called, which led to a resounding victory for the new administration [1].

The National Government introduced cuts in government expenditure, especially in unemployment benefit and the pay of state employees such as teachers and civil servants. Gradually the coalition was converted into a Conservative administration which triumphed at the general election of 1935.

In spite of the absence of major economic initiatives from governments in office after 1931, the economic situation began to improve from 1933 onwards. Unemployment reached a peak of almost three million in the winter of 1932–3 and remained at more than a million until the outbreak of war in 1939, but it was falling from 1933–4. Revival was concentrated in a range of new industries such as electricity supply, motor vehicles [2], consumer durables and chemicals. These industries brought increased employment to the southeast and the Midlands, while the older industries of the "distressed" areas remained depressed and only slowly began to recover.

Political unrest and social change

The rise in prosperity in some areas helps to explain the failure of the extremist parties to obtain greater support before the war. Oswald Mosley [8] formed the British Union of Fascists in 1932, after leaving the Labour Party and adopted the style of continental fascist parties. The party espoused radical

CONNECTIONS

See also
194 What World War I meant to Britain
208 The twenties and the Depression
210 The British labour movement 1868–1930
212 Socialism in the West
218 British foreign policy since 1914
226 Causes of World War II
232 The home front in World War II
236 Britain since 1945: 1
260 Scotland in the 20th century
262 Wales in the 20th century

1

1 Unemployment was the major issue of the early 1930s. In October 1931 the National government, formed the previous August, sought a mandate from the electorate for its economic policies – designed to deal with the Depression. Under Ramsay MacDonald, the ex-Labour premier, the National government campaigned for a restoration of business confidence and reduced unemployment. In a mood of deep national crisis, the electorate swung heavily towards the National candidates. Only 46 Labour MPs were returned, compared with 554 National government MPs. Every Labour ex-cabinet minister lost his seat, except George Lansbury (1859–1940).

2

2 Mass production methods, pioneered in the United States, were adopted in Britain during the interwar years. They brought the first cheap motor vehicles within the reach of the middle classes. By 1939 there were nearly two million motor vehicles in Britain and the "motoring revolution" had begun. Car production for the home market increased each year up to the war, with the exception of 1932.

4 The thirties witnessed a rapid growth in commercial air transport and routes were set up across the world. Imperial Airways, a government-subsidized amalgamation of several privately-owned companies, was established in 1924. One of its main aims was to routes throughout the empire. Airmail was as important as passenger services; by 1938 Imperial Airways carried all first-class mail to the empire.

4

3

3 The communist-led National Unemployed Workers' Movement organized several "hunger" marches on London in the thirties to protest about the plight of the unemployed. The marches however, had little effect on government policy.

5

5 Private house building expanded greatly in the 1930s, and was a principal factor in the economic recovery during the last half of the decade. Despite government cuts in building programmes, private investment in housing boomed, especially in the thriving regions of the Midlands and the south-east. Nearly three million houses were built between 1930 and 1939. This expansion in building led to a boom in other industries such as electrical and household goods.

economic ideas but earned a reputation for violence and anti-semitism that cut it off from mass support.

The thirties witnessed the rise of new social patterns, with an enormous growth of suburban living, a housing boom [5], slum clearance, and ameliorative social legislation. Opportunities for leisure activities, such as the cinema [6] and dance halls, expanded and provided cheap entertainment. Another influencial and inexpensive source of entertainment was the radio. The BBC broadcast hours of popular music daily and did much to enhance the reputations of some of the great dance bands of the 1930s. The rise of the football pools, with their lure of instant wealth, was another social phenomenon of the times.

There was a profound distrust and loathing of war in the thirties. Peace movements flourished and the governments of Baldwin and Neville Chamberlain (1869–1940) pursued a policy of appeasing the dictators. But rising international tension led to gradual rearmament from the mid-1930s, helping to revive the economy.

The experiences of the Depression and thirties followed by total war helped to create a new mood in Britain. The Beveridge Report of 1942 advocated a high level of employment and the creation of a welfare state. Even before the end of the war, the Butler Education Act of 1944 made free secondary education available to all.

Postwar optimism

World War II witnessed an acceleration of many of the trends evident in British politics and society before 1939. The war further stimulated new industries as well as reviving the old ones, and led to widespread recognition of social problems such as poverty and unemployment. Widespread and vigorous debate about the nature of postwar British society paved the way for a Labour victory at the 1945 general election. The Labour government inherited considerable good will from the electorate. Demobilization caused far less resentment than it had in 1918 [10] and Labour's programme seemed to meet the demand for new policies and an avoidance of mass unemployment.

KEY

Edward VIII (1894–1972), came to the throne on 20 January 1936 with considerable popular support, accumulated during his years as Prince of Wales. Public interest in his life showed the widespread devotion to the monarchy even during the worst years of the Depression. But the king's continuing relationship with an American divorcee, Mrs Wallis Simpson (1896–), precipitated a constitutional crisis following her second divorce, in October 1936. The king wanted to marry her, but the prime minister, Stanley Baldwin (1867–1947), advised that she was unacceptable as a queen. In spite of considerable popular support for Edward, he abdicated in December.

6

6 The cinema was one of the most important forms of cheap mass entertainment in the 1930s. By 1939 there were 5,000 cinemas in Britain and more than 20 million cinema tickets were sold each week.

7

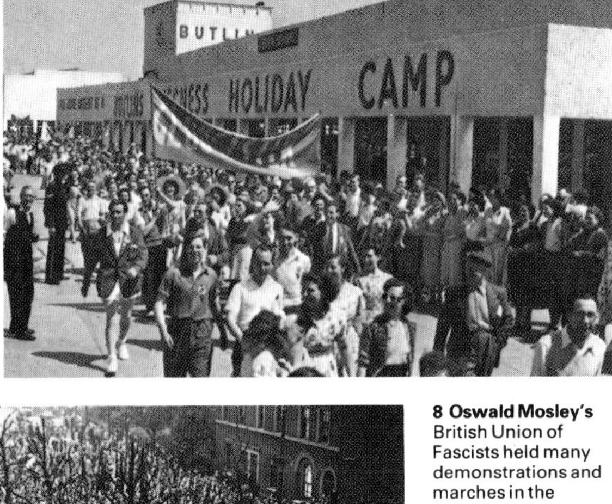

7 Rising living standards for those in work as well as a more widespread introduction of paid holidays contributed to a growth in holiday-making. The first holiday camps were opened in 1937.

8

8 Oswald Mosley's British Union of Fascists held many demonstrations and marches in the years before the war. Their use of uniforms and violent methods aroused widespread hostility, as did, more particularly, their anti-semitism. In 1936 the Public Order Act was passed forbidding the use of uniforms and strengthening police powers against political demonstrations and mass meetings.

9

9 Britain was slow to rearm in the thirties. Limited rearmament was undertaken from 1934, mainly in the airforce and navy, although German expenditure on arms was sometimes exaggerated. The government delayed thoroughgoing rearmament until after 1938 on the assumption that public opinion, as manifested in the Peace Ballot (a house-to-house poll) and by-election results, would not stand for sterner measures.

10

10 The immense war-effort in Britain put eight million people into uniform in World War II. Over 300,000 members of the armed forces and, on the home front, about 60,000 civilians lost their lives in the conflict. In contrast with 1919, demobilization went relatively smoothly, although in the Far and Middle East, British troops often became involved in local police-keeping and occupation duties, such as in Cyprus, that continued for some time after 1945.

Causes of World War II

The inter-war years in Europe saw the rise of fascist dictators [Key] in Italy and Germany. Their nationalistic and expansionist policies increasingly undermined the credibility of diplomatic negotiation.

The rise of the dictators

World War I had left a bitter legacy in the crippling reparations and arbitrary divisions of territory that were features of the Treaty of Versailles (1919). Its effects were influential in the rise to power of Benito Mussolini (1883–1945) and Adolf Hitler (1889–1945). Italy had suffered losses in World War I and disappointments in the peace settlement at Versailles, and Mussolini owed a large part of his support to a policy of militant nationalism which was bound to create tensions in the postwar world [6]. Hitler also gained support from a policy of extreme nationalism that was determined to reverse the penal aspects of the Versailles Treaty and unify the German-speaking peoples in eastern Europe territorially [4].

The isolationism of the United States meant that the major initiative for peace lay with France and Britain as the two strongest European powers. Both nations were fearful of renewed war. They felt that war in 1914 had arisen out of the diplomatic system's inability to cope with international crises, so they believed that they must negotiate with the dictators.

During the 1920s faith was placed in the League of Nations and the pursuit of policies of disarmament – policies that foundered on mutual distrust among the great European powers. By the early 1930's it was increasingly clear that the League of Nations was unlikely to act as a guarantor of peace. Japan's invasion of Manchuria and then, more seriously, the Abyssinian crisis (1935–6) and the Spanish Civil War (1936–9) were patent indications that the League was incapable of restricting international aggression by powerful states.

A policy of appeasement

For much of the 1930s, statesmen in both France and Britain believed that Hitler's policies were designed solely to satisfy Germany's legitimate demands for revision of the Versailles settlement. In spite of Germany's reoccupation of the Rhineland in 1936 [2] and virtual control of Austria [1], Britain in particular maintained the hope that war could be averted by concession. The efforts of both Stanley Baldwin (1867–1947) and Neville Chamberlain (1869–1940) to negotiate with Hitler were supported in large part by a populace afraid of another war and resentful of expenditure on armaments in a period of economic depression. Left-wing forces in Britain were convinced that policies of disarmament must be pursued to lessen the risk of war. Chamberlain was operating from a position of weakness when Hitler was busy rearming [3]. France was also beset by weakness; internal political divisions prevented a firm foreign policy and the country's losses in World War I inclined it to follow a defensive policy, enshrined in the construction of the Maginot Line.

Although Hitler's long-term aims cannot be determined with certainty, he exploited the confusion and weakness of the Western European powers to reverse the Versailles Treaty and further his plans for conquest in

CONNECTIONS

See also
228 World War II
192 The Peace of Paris
222 The rise of fascism
212 Socialism in the West
218 British foreign policy since 1914
224 Britain 1930–45
208 The twenties and the Depression
198 Stalin's Russia

1 

1 Chancellor Dollfuss of Austria was murdered in 1934, on Hitler's orders, as the first stage of Germany's Austrian annexation. Virtual control was achieved in 1936; the take-over came in 1938.

2 

2 The Versailles Treaty had excluded German forces from the Rhineland. In March 1936 German troops reoccupied it in defiance of France and Britain; neither was prepared to risk war to prevent this.

3 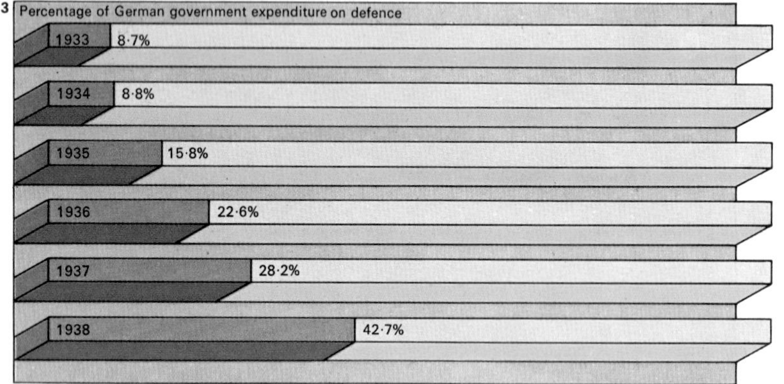

3 Expenditure on defence increased fivefold in Hitler's Germany between 1933 and 1938. Spending reached a peak in the latter stages of World War II. Germany started rearming immediately after Hitler came to power, but this drive became dominant only after 1936. Then the adoption of a four-year plan for rearmament directed more of the German economy to war than was the case in any other European country.

5 

5 Germany's aim of absorbing German-speaking parts of Czechoslovakia – the Sudetenland – almost plunged Europe into war. At Munich in 1938 Britain and France virtually sacrificed Czech industry and defence in an attempt to appease the Germans.

4 Hitler's *Mein Kampf* was written while he was in prison following his abortive "Beer Hall" revolt in 1923. It contained a demand for *lebensraum* (living space) for the German peoples in the east. Expansion into eastern Europe and the USSR had long been a part of right-wing German thinking and Hitler adopted it as a major feature of his political policy. Its true place in his plans is much debated, but his conquests in eastern Europe by diplomacy and ultimately by war, backed by propaganda like this poster, fulfilled his professed policy.

4 

the east at a later date. The reoccupation of the Rhineland was followed by the Austrian *Anschluss* and demands for the cession of the German-speaking Sudetenland from Czechoslovakia [5]. After threatening war, Hitler was placated by an agreement in 1938 that virtually dismembered Czechoslovakia in return for promises not to occupy the non-German-speaking areas of the country. Chamberlain's surrender was hailed as a triumph that had avoided war. But the occupation of Prague in March 1939 broke the illusion upon which appeasement had been based – that Hitler's demands were limited and had been satisfied.

The influence of peripheral powers

Resistance to Hitler had been confused by suspicion of the Soviet Union's intentions. Coming out of isolation in the mid-1930s, the Soviet Union was concerned to prevent an alliance of Western European states against her, but became increasingly fearful of the rise of fascism in Germany with its implied threat to herself. The USSR sought to bring the Western powers into an anti-fascist alliance, but was frustrated by the faith in appeasement and widespread mistrust of the USSR in conservative circles. The actions of Britain and France over Czechoslovakia encouraged the USSR to form a non-aggression pact with Germany in 1939.

In the Far East, the rise of a militantly aggressive Japan provided an additional strain upon the fragile peace [7]. Japan's occupation of Manchuria (1931–2) and its war with China from the mid-thirties illustrated the weakness of the League and increased Japanese self-confidence and territorial ambitions.

Britain's guarantees in 1939 to Poland and Romania were a last attempt to restrain Hitler's actions. But he had agreed with the USSR to dismember Poland on the pretext of annexing the "Polish Corridor" [9]. Hitler probably expected Britain and France to back down once again as they had over Munich. Instead they presented Hitler with demands to withdraw. When the British ultimatum expired on 3 September 1939, Britain declared war on Germany, and France followed suit a few hours later.

KEY

Hitler riding into Vienna at the head of German troops symbolizes the domination of Europe by the dictators. While Mussolini was backing Hitler in the west, Japanese economic expansion threatened the stability of the Far East.

6

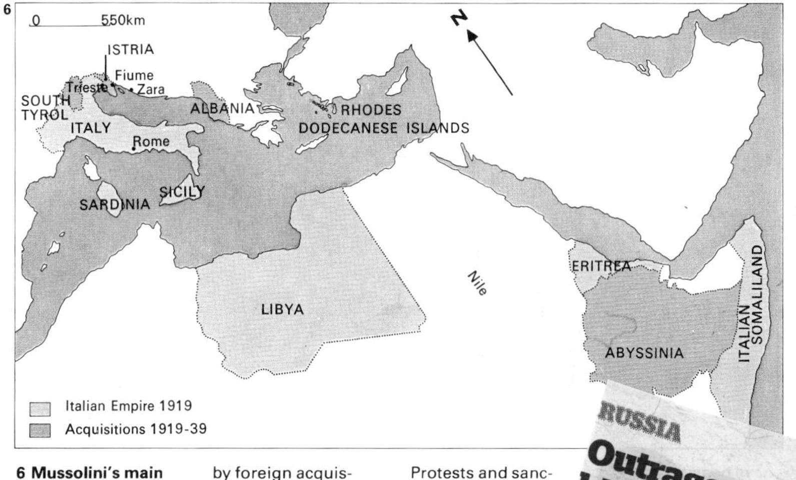

6 Mussolini's main aim from the time of his appointment in 1922 was to increase Italy's prestige and to consolidate her "Great Power" status by foreign acquisition and aggressive diplomacy. After 1922 Italy tightened her grip on Fiume, the South Tyrol and the Dodecanese Islands. Protests and sanctions from the Leagu of Nations did not pr vent war with Abyssinia in 1935 a the country's rapic annexation. Italy

RUSSIA

Outrage as diplomat blames Poles for war

Poland has called Sergey Andreev, the Russian ambassador, for a meeting today after he put some of the blame for the Second World War on Poland. Mr Andreev said "Polish policy led to the disaster... because during the 1930s Poland repeatedly blocked the formation of a coalition against Hitler's Germany."

28.9.2015

7 Matsuoka [left] the Japanese foreign minister from 1940 to 1941, was largely responsible for Japan's Tripartite Pact with Germany and Italy. Japan had already joined Germany in an Anti-Comintern Pact in 1936. Throughout the 1930s Japan pursued an aggressive foreign policy: in 1931 she had taken Manchuria, increasing tension in the Far East where the League of Nations was virtually powerless. In 1937 Japan went to war with China, seizing a large area of the Chinese mainland. Europe and America failed to resolve or control the conflict, encouraging Japan to further aggression.

7

8

8 The appeasement policy of Britain and France arose out of fear of renewed war and belief that the dictators' demands could be met by negotiation and concession. But concern grew that such "weakness" just provoked more demands.

9 German troops symbolically destroyed the Polish frontier when they invaded Poland in August 1939. Polish access to the Baltic had been guaranteed by Britain and France, who therefore declared war on Germany on 3 September.

World War II

On 1 September 1939, German troops invaded Poland. Britain and France were pledged to support Poland and declared war on Germany two days later. Using revolutionary *Blitzkrieg* ("lightning attack") tactics the Germans defeated the outdated Polish army in 18 days and the country was partitioned between Germany and the Soviet Union, with whom Germany had just signed a non-aggression pact. A British army crossed to France but did not attack, and a "phoney war" lasted until the spring.

German and Japanese victories

Germany overran Norway and Denmark in April 1940 and then on 10 May invaded Holland, Belgium and Luxembourg, which had been neutral. As the Allied armies swung forwards to meet them German tanks burst through the "impassable" Ardennes and reached the English Channel. The Allied army to their north was forced back into the Dunkirk region, and 338,226 British and French troops escaped to England by sea between 29 May and 3 June. Most of France, except for the southeast under the puppet Vichy regime of Henri Pétain (1856–1951), was occupied by the Germans.

Germany's leader, Adolf Hitler (1889–1945), expected Britain to make peace, but she fought on defiantly under the leadership of Winston Churchill (1874–1965). The *Luftwaffe* (air force) of Hermann Goering (1893–1946) then attempted to destroy the Royal Air Force (RAF) so that an invasion of England could be launched. But the Germans were defeated in the Battle of Britain fought between August and October 1940.

Taking advantage of the French Atlantic ports, German submarines intensified their attacks on British sea routes and in the next two years came near to strangling Britain [7].

Italy entered the war in June 1940 but suffered serious defeats in Greece and Libya. Germany sent forces under General Erwin Rommel (1891–1944) to help the Italians in North Africa and swiftly overran Yugoslavia, Greece and Crete in April and May 1941.

On 22 June 1941, in breach of the earlier pact, German troops swept into the Soviet Union [4], achieving total surprise. After five months they were just 30km (19 miles) from Moscow but were halted by bitter winter weather and stubborn Russian resistance. On 7 December 1941, in the second major onslaught of the war, Japan launched a surprise attack on the US fleet at Pearl Harbor.

The first half of 1942 saw the Axis forces (Germany, Italy, Japan and minor allies) at the height of their powers. In the Pacific the Japanese captured the Dutch East Indies, Malaya, Burma, the Philippines and many Pacific islands [5]. In the Soviet Union a German offensive advanced on Stalingrad and the Caucasus. In North Africa the British had been driven back to the borders of Egypt.

The turn of the tide

A series of crucial battles later in 1942 and in 1943 gave the initiative to the Allies. In the Pacific, Japanese naval power was shattered at the Battle of Midway on 4–7 June 1942; and on 7 August US marines landed in Guadalcanal in the first of the amphibious assaults by which US naval power under Admiral Chester Nimitz (1885–1966) pushed back the Japanese. In bitter weather

CONNECTIONS

See also
230 World War II: Britain's role
226 Causes of World War II
232 The home front in World War II
222 The rise of fascism
192 The Peace of Paris
208 The twenties and the Depression
198 Stalin's Russia
234 The division of Europe
214 East Asia 1919–1945

In other volumes
236 Man and Machines
164 Man and Machines
166 Man and Machines
170 Man and Machines

1

A

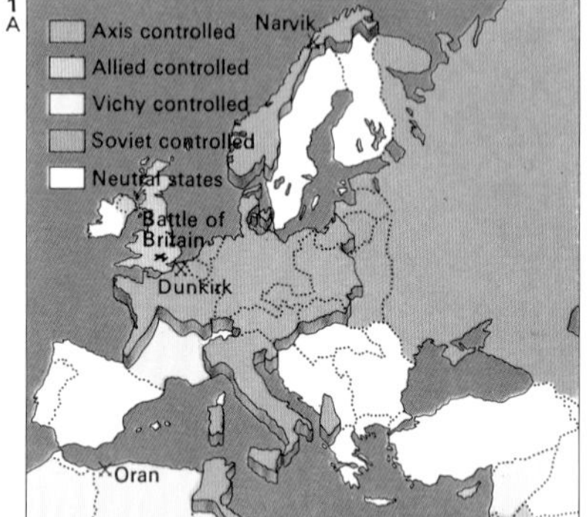

B

C

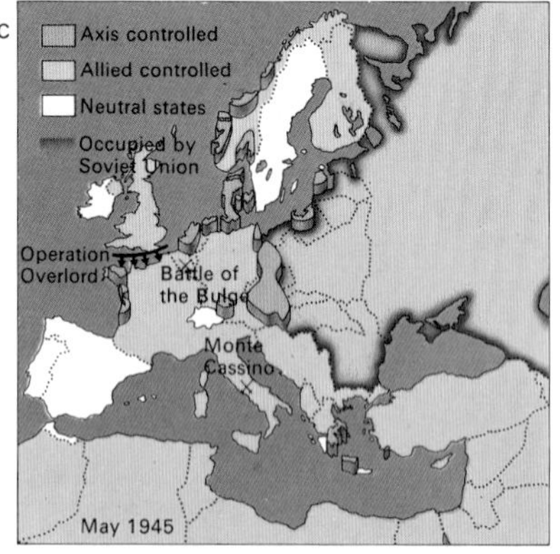

1 The main theatre of war was in Europe, as it was in World War I. [A] By June 1940 the Axis powers controlled almost the whole of Western Europe and Germany then broadened the conflict by attacking the Soviet Union a year later. [B] Axis conquests reached their peak in November 1942. [C] By May 1945 Russian counter-offensives and Allied landings in France and Italy had defeated Germany.

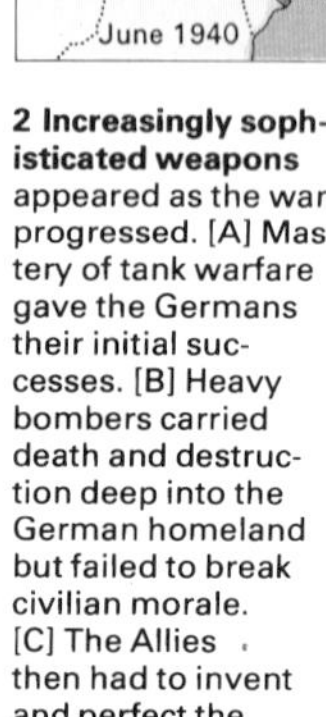

2 Increasingly sophisticated weapons appeared as the war progressed. [A] Mastery of tank warfare gave the Germans their initial successes. [B] Heavy bombers carried death and destruction deep into the German homeland but failed to break civilian morale. [C] The Allies then had to invent and perfect the techniques of amphibious warfare in order to invade "Fortress Europe".

2

A

B

C

3

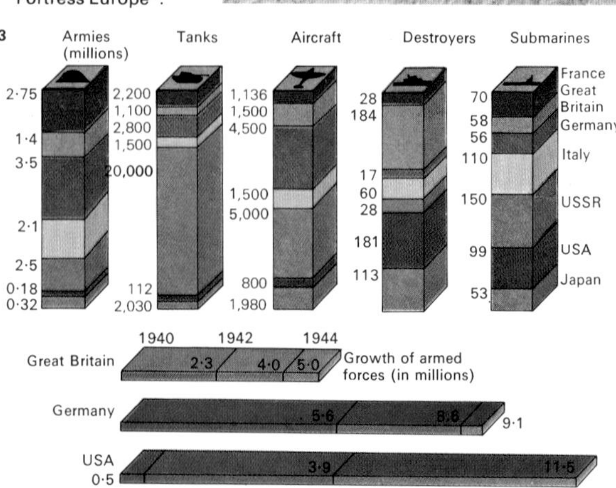

3 This comparison of military power at the outbreak of war shows that, although Germany had more aircraft in 1939, France and Britain together were in fact stronger in men and equipment. The vast size of the Soviet forces shows how advantageous the Stalin-Hitler pact was to Germany. Manpower was needed on a a massive scale and the lower half of the diagram shows the growth in armed forces, which was particularly appreciable in the United States.

4

4 The turning-point of the war in Europe came when Hitler attacked the Soviet Union in 1941 and failed to deliver a swift knock-out blow. The key battle took place at Stalingrad where, after weeks of frozen siege, the German 6th Army was forced to surrender. Germany was committed to a war on two fronts, with a possible counter-attack from Britain in the west and a war of attrition against the vast Russian reserves available in the east.

in the Soviet Union 110,000 men of the original German army of 270,000, fighting at Stalingrad, surrendered on 31 January 1943. The remaining 160,000 men had been killed. In North Africa the victory of General Bernard Montgomery (1887–1976) at El Alamein in October 1942, and an Allied landing in Algeria, forced the Axis troops back into Tunisia where 250,000 surrendered on 12 May 1943. In the Atlantic, Allied sonar and radar, more escorts and long-range aircraft led to increased U-boat losses.

The beginning of the end

The last major German offensive in the Soviet Union was halted at Kursk in July 1943 and the Red Army pushed forward during the autumn and winter. The Allies under Field-Marshal Harold Alexander (1891–1969) invaded Sicily on 10 July 1943 and landed in Italy on 3 September. The RAF had made its first "1,000-bomber" raid on Germany in May 1942 and, with the arrival of the United States Army Air Force in mid-1943, massive day and night raids were mounted for the rest of the war.

On D-Day, 6 June 1944, Allied forces under General Dwight Eisenhower (1890–1969) landed in Normandy and crossed France and the Low Countries to reach the Rhine by November. In Italy, Rome had been captured on 4 June, while a Soviet offensive begun in the same month drove the Germans out of the Soviet Union and swept into Poland and the Baltic states. In the Pacific, American forces destroyed the remnants of the Japanese fleet at the battles of the Philippine Sea and Leyte Gulf, and invaded the Philippines in October 1944. In Burma, the British defeated a Japanese attempt to invade India and counter-attacked successfully.

The Allies crossed the Rhine in March 1944 and drove deep into Germany. A Soviet assault under Marshal Georgi Zhukov (1896–1974) began in January 1945 and reached Berlin in April. Hitler committed suicide and on 4 May Germany surrendered.

On 6 August US forces dropped the first atomic bomb on Hiroshima, Japan [8]. A second bomb on Nagasaki forced Japan to surrender on 14 August 1945.

KEY

1914-18	17	total dead (millions)
1939-45	37	45

1939-45 casualties: major powers

Germany 7 / 35; Japan 2·6 / 12; Italy 0·8 / 3·3; USSR 100 / 75; Yugoslavia 12·8 / 4; France 3·5 / 2·5; UK 0·62 / 3·26; USA 3

Axis; Allies; Civilian = 100,000; Military = 100,000

World War II was the most destructive and wide-ranging war in history: the dead may have totalled 45 million. Military casualties were only slightly higher than in World War I, but massive bombing and German policies against civilians in the occupied territories meant that civilian deaths were far higher.

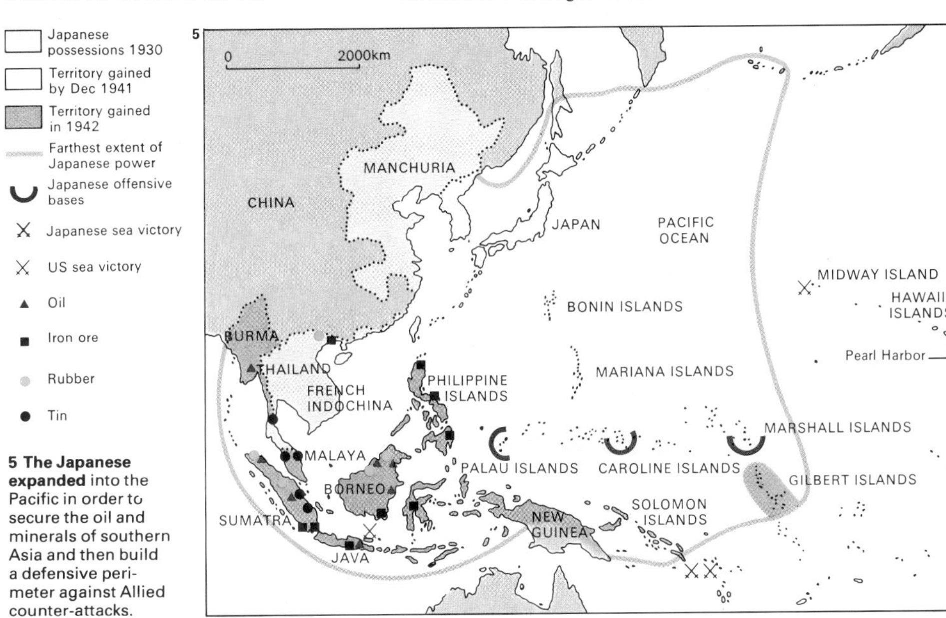

5 The Japanese expanded into the Pacific in order to secure the oil and minerals of southern Asia and then build a defensive perimeter against Allied counter-attacks.

6 The Allied counter-offensive in the Pacific depended largely on a unique naval campaign in which carrier-borne aircraft played a decisive role. Quickly mastering this new type of warfare, the US Navy was able to destroy the Japanese fleet, bypass enemy-held islands and cut off Japan from its vital supplies. Major land campaigns took place only in Burma and the Philippines.

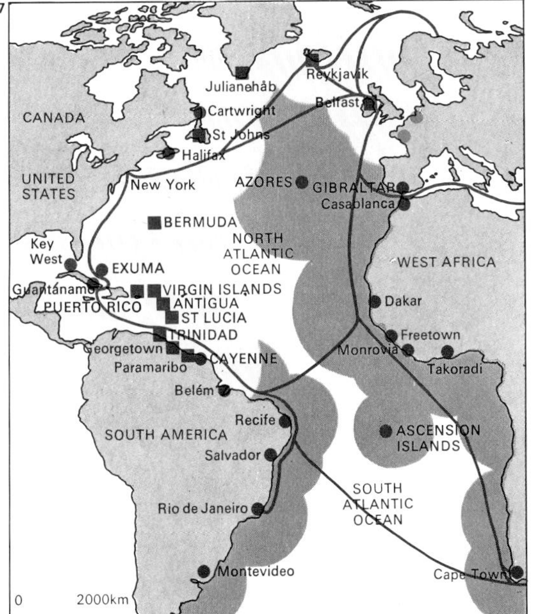

Allied air cover
Sept 1939—May 1945
Added 1942—May 1945
US air bases
Sept 1939—May 1945
Added 1942—May 1945
U-boat bases
Allied convoy routes

7 The Battle of the Atlantic was a crucial one for Britain once the threat of a German invasion had been removed. The German U-boats hoped to starve Britain into submission, thus eliminating the possibility of a counter-attack in the west. In 1941–2 the U-boats almost succeeded in their aim, and it was not until anti-submarine measures had been intensified and improved that the U-boats were eventually mastered.

8 Hiroshima was devastated by the first atomic bomb. By later standards this was a very small bomb of less than one kilotonne, but it was enough to obliterate an entire city and kill more than 78,500 people in the space of one minute. A new era of warfare threatening total annihilation had been unleashed on mankind.

828,352 (citizens of) Russian, Polish and other Soviet-annexed countries
83,874 not registered
82,090 Yugoslavian
22,467 German
21,967 other
total 1,038,750
(main destinations)
86,346 United Kingdom
123,479 Canada
185,056 other places
328,851 United States
182,159 Australia
132,109 Israel

9 By the end of the war more than a million displaced persons were living in refugee camps throughout Europe. The majority were Soviet citizens or citizens of countries annexed by the USSR. The diagram shows where the east European refugees came from and where the International Refugee Organization succeeded in settling them.

World War II: Britain's role

Britain's involvement in World War II was global. Though its principal areas of concern were Europe, North Africa and the Far East, the Royal Air Force flew missions on the Russian front and Royal Navy ships fought engagements off South America. And where Britain itself was not heavily involved – notably in the Pacific theatre – Australians and New Zealanders fought alongside Britain's American allies.

Early campaigns

True to the British tradition of losing every battle except the last, the war opened disastrously. Hitler's *Blitzkrieg* through Poland, the Low Countries and into France [1] wrecked the British Expeditionary Force of ten divisions: by what seemed a miracle at the time, all but about 25,000 to 30,000 men got back to Britain but the BEF left behind all its heavy equipment.

With the Battle of France lost, the Battle of Britain opened on 10 July 1940 with Goering's Luftwaffe directing its efforts against convoys in the Straits of Dover. The convoys had to be stopped. Phase two, which began on "Eagle Day", 13 August, was aimed at RAF fighter bases in Kent. On 7 September, having lost 225 aircraft to the RAFs 185 in just eight days, the Luftwaffe turned aside to attack London. On 15 September – "Battle of Britain Day" – it lost between 56 and 60 aircraft to the RAF's 26. Chastened, Goering switched to night attacks against English cities, and Operation Sea Lion, the invasion of Britain, was first postponed and finally abandoned altogether after the invasion of Russia. The blitz, during which some 30,000 people were killed, continued until mid-April 1941.

The dark days

A consequence of the fall of France was that the German navy was able to operate U-boats from France's west coast. In the first year of the war the U-boats never numbered more than 60, but they sank nearly a million tonnes of merchant shipping. The Battle of the Atlantic reached its peak in early 1941, but by the summer of 1943, the convoy system and American involvement prevailed.

In the Balkans Hitler's seven-week campaign through Yugoslavia and Greece ended with the British being ejected from Crete [2] in May 1940.

Germany's assault on Russia on 22 June 1941 [4] offered Britain a breathing-space. Britain could do little to help Russia beyond offering supplies: and the route to Murmansk and Archangel was, in the winter, the worst sea-route in the world.

While the Battle of the Atlantic and of the Russian convoys was under way, Britain was losing in the Far East. The Japanese attack on the US base at Pearl Harbour on 7 December 1941 was followed by the ignominious fall of Singapore [5], the loss of the battlecruiser *Repulse* and the battleship *Prince of Wales*, and a threat to India. The Japanese were within 200 miles of Australia: Australia and New Zealand, with most of their troops in North Africa, had to turn to the United States for protection.

Towards final victory

The tide began to turn in 1942. In North Africa, after a see-saw series of battles in which the Italians and Erwin Rommel's

CONNECTIONS

Read first
226 Causes of World War II
228 World War II
232 The home front in World War II

See also
192 The Peace of Paris
222 The rise of fascism
224 Britain 1930–1945
198 Stalin's Russia
214 East Asia 1919–1945
234 The division of Europe
248 Australia since 1918
250 New Zealand since 1918

In other volumes
164 Man and Machines
168 Man and Machines
170 Man and Machines
174 Man and Machines
176 Man and Machines
180 Man and Machines

1 The Allies and Germany faced each other in the West at the outbreak of war more or less evenly matched in numbers. Britain and France had 122 divisions against Germany's 136, and 3,254 armoured vehicles against 2,574. But the Allies still pursued outmoded ideas of positional warfare, and made poor use of their armoured divisions. German armour was used to optimum advantage and coupled with air power to form the spearhead of the *Blitzkrieg*. This was designed to burst through and surround the enemy rather than fight head-on battles. As a result Paris fell in only four weeks.

2 German paratroops, here entering a JU52 transport aircraft, proved decisive in the capture of Crete, the final phase of Hitler's Balkan campaign. Bernard Freyberg (1889–1963), commander of the New Zealand Division, was in charge of all the forces on the island. Using 1,390 aircraft, the Germans forced the British to withdraw on 27 May, after three weeks' stubborn resistance.

3 The arrival of Erwin Rommel (1891–1944) and his Afrika Korps in February 1941 rescued his Italian partners from being completely overrun by British and Commonwealth forces in North Africa. Twice Rommel reached the frontiers of Egypt, engendering Allied nightmares of an Axis victory that, together with an advance in the Caucasus, could have completed a successful pincer movement.

4 Hitler's invasion of the Soviet Union began brilliantly with the German armies using tactics that had been perfected in Poland and France. Deep-thrusting columns destroyed more than a million enemy troops, but stiffening Soviet resistance and the onset of merciless winter conditions prevented the Germans from achieving the swift victory they needed. Despite some further successes in 1942, the Germans were catastrophically defeated at Stalingrad where they lost 300,000 men. Thereafter they could not hope to match the Soviet Union's apparently inexhaustible manpower and were steadily pushed back.

5 The Japanese captured 85,000 men at Singapore in February 1942: it was the largest surrender in the history of the British Army. Complacency about the Japanese threat had led Britain to neglect already inadequate defences, but even so British and Commonwealth troops outnumbered the Japanese who swept through Malaya and Burma. The Japanese relied on their mastery of jungle warfare to outflank British troops, who had virtually no jungle training. Singapore was approached and attacked from its lightly defended landward side and fell in a matter of days.

Afrika Korps [3] got to within 60 miles of Alexandria, and Australian and British units distinguished themselves by stubborn resistance in the isolated pocket of Tobruk, Montgomery won the Battle of El Alamein [6]. The battle opened on 23 October; on 4 November Rommel's Afrika Korps began to retreat; four days later Anglo-American forces landed in French North Africa. With victory in North Africa in the spring of 1943, Italy became the next objective. Sicily was invaded in July 1943 and mainland Italy – by the British 5th Army at Salerno – in September. American insistence that the Pacific and Burma campaigns be given priority in late 1943 meant, however, that the Italian campaign was drawn out. Of the Commonwealth troops who had fought in Africa, the New Zealanders went on to fight in Italy, notably at Monte Cassino [7] and the Australians returned home to help push the Japanese out of the Pacific.

The Mediterranean campaign ended effectively with the capture of Rome on 4 June 1944, although the German resistance in Italy did not end until May 1945.

The invasion of Normandy began on 6 June 1944, British troops landing on the coast near Caen and Bayeux and Americans farther west. After an initial period of close fighting in France, the Allies broke out and swept towards the Rhine. An attempt to speed matters by an airborne landing at Arnhem in The Netherlands [8] failed, but in the spring of 1945 renewed offensives resulted in Germany's surrender on 7 May.

In the Far East Slim's [9] "forgotten" 14th Army had been confronting the Japanese in Burma while Americans, Australians and New Zealanders island-hopped towards Japan following the Battle of Midway, an American carrier-fleet victory that ranked with Stalingrad in strategic importance. Two atomic bombs, on Hiroshima and Nagasaki, ended the war against Japan.

Britain lost far fewer men in World War II than in World War I – 300,000 dead against 750,000. Civilian casualties were higher: 60,000 against 1,500. The legacy of the war was an enormous economic debt – £4,198 million – the loss of an empire and, in compensation, an industrial leap forward.

KEY

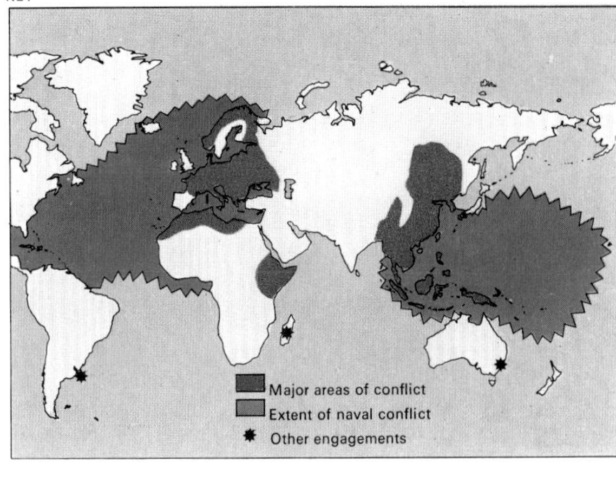

World War II began in Europe, but developed into a global conflict with campaigns in Africa, Asia and throughout the Pacific and Atlantic Oceans. Italy proved a weak member of the Axis in 1940, but Germany and Japan enjoyed a series of victories in the first three years. Thereafter, Allied superiority in potential manpower and industrial capacity steadily grew. More than any previous conflict, this was a war of technology, with developments in tanks, aircraft, submarines, radar – and eventually the atomic bomb – helping to influence strategic and tactical thinking.

6

6 The turning-point in North Africa came in July 1942 when the overstretched Afrika Korps failed to break through British 8th Army positions around El Alamein. Three months later, substantial Allied reinforcements enabled the new commander, Montgomery, to begin an offensive that secured North Africa.

7

7 The ruins of Monte Cassino monastery in Italy saw some of the most savage fighting of the war. The Allies believed that the Germans had turned the monastery into a strongpoint: their decision to bomb it not only provoked a controversy, it defeated its own end — the rubble was easier to defend than the intact monastery.

8

8 British paratroops experienced nine days of bitter street fighting – and final failure – at Arnhem in September 1944 when, with American and Polish forces, they attempted to capture 17 canal crossings and major bridges in Holland. Of the 35,000 troops involved, more than 17,000 became casualties. Four of the five major bridges were taken; the bridge at Arnhem was not. The plan devised by Montgomery – dashing, and contrary to his usual style – might have shortened the war had it worked.

9

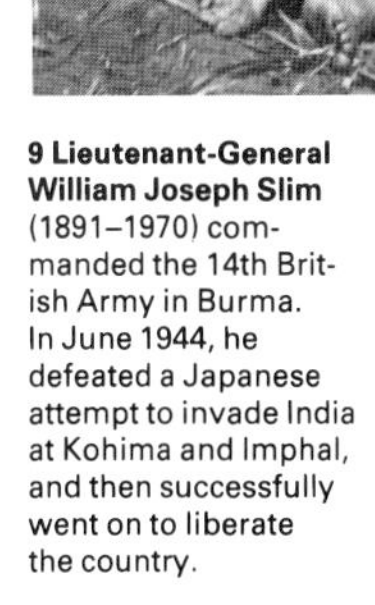

9 Lieutenant-General William Joseph Slim (1891–1970) commanded the 14th British Army in Burma. In June 1944, he defeated a Japanese attempt to invade India at Kohima and Imphal, and then successfully went on to liberate the country.

10

10 The Japanese in New Guinea suffered their first major setback on land in September 1942 when Australian forces defeated an attempt to capture Port Moresby. After savage fighting in atrocious conditions, the Australians successfully counter-attacked. Throughout 1943 and early 1944, a series of small-scale but brilliant combined operations were mounted as part of a wider Allied offensive in the southwestern Pacific. These isolated and neutralized a whole Japanese army.

The home front in World War II

World War II has often, and accurately, been described as "The People's War". No previous conflict in history had so directly involved the civilian population of the combatant countries or caused them so much privation and death.

Civilian involvement in war

Even before war had been declared civilians had become involved through conscription, introduced in Germany in 1934, in Great Britain in June 1939, and in the United States on a selective "unlucky dip" basis in 1940. Once the war began even those civilians who escaped being called up into the armed forces found themselves in varying degrees directed into home defence (Local Defence Volunteers, later the Home Guard) [2] or civil defence or into essential work in factories [8] and vital services such as transport. In every combatant country (except the United States, which could meet almost all the demands made upon it) the share of the national resources allocated to civilians was by the end of the war sharply reduced to give priority to the fighting men.

Although they vastly outnumbered the soldiers, the civilians were in far less danger. Even in Germany casualties among civilians, including those caught up in military operations, were estimated at no more than 700,000 compared with 3,500,000 servicemen who died. The figures for Britain were 62,000 against 326,000 and for Japan 260,000 civilians compared with 1,200,000 servicemen; the United States had virtually no civilian casualties. But the civilian's life was far more at risk than in any previous war. Although each country claimed at first to be directing its bombers against only military objectives, such restraints were soon abandoned [11]. But bombs were not the main cause of civilian deaths. Under German occupation, far more deaths were caused by disease, famine and mass murder.

Civilian daily life

Even within occupied Europe daily life varied enormously between different countries. In Denmark, Hitler's "model protectorate", the standard of living was far higher than in Britain. In France, if one had access to the black market, it was also possible to live reasonably well. But in Holland by the winter of 1944–5 people were living on tulip bulbs and in the Channel Islands only the arrival of Red Cross parcels prevented starvation. All the occupied countries shared some shortages and discomforts. Fuel, both for heating and transport, was scarce [5]. Everyone's life was encompassed by curfews, permits and the fear of being rounded up as a suspect or forced-labour "volunteer".

In the countries still under arms the civilian population was encouraged to believe that a vast gulf separated them from their counterparts in enemy lands. Civilian experience in Germany probably had more in common with life in wartime Britain than in any other country. Both suffered the upheaval of evacuation [Key] and of long nights in shelters.

Almost all necessities were either rationed or hard to find and although the German system of control was more complicated and less efficient than the British there were many similarities between them. Household textiles and clothes, for example,

CONNECTIONS

See also
230 World War II: Britain's role
228 World War II
236 Britain since 1945: I
258 Evolution of the Western democracies
214 East Asia 1919–1945
208 The twenties and the Depression
240 The Soviet Union since 1945

1 **Saucepans were collected** for aluminium for making aircraft after a British Government appeal in 1940.

2 **Britain's Local Defence Volunteers** were formed in May 1940. Here two railway men are briefed.

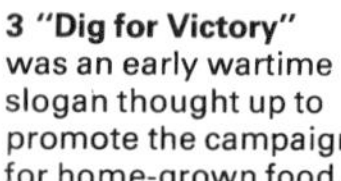

3 **"Dig for Victory"** was an early wartime slogan thought up to promote the campaign for home-grown food.

4 **Air raids on London** – the Blitz – began on 7 September, 1940, and lasted until mid-1941. In the opening phase the capital was bombed on 57 consecutive nights. In the first four months, 13,339 people were killed and 17,937 injured.

5 **Refugees flooded on to the roads** of Europe as the German armies advanced. This Frenchman's horsedrawn vehicle laden with goods was one way of overcoming the petrol shortage; bicycle taxis were also common.

6 **Nazi military bands** like this one, photographed in the Place de l'Opéra in Paris in June 1941, often played in public in the occupied countries. Ostensibly a goodwill gesture, they were also a symbol of German strength.

were rationed on a points system in Germany in 1940 and in 1941 the same system was used in Britain. To find consumer goods of any kind, from babies' prams to furniture, necessitated a long search and in both countries as coal was diverted to the war factories people struggled in winter to keep warm.

Food rationing made the deepest impact on most people and here, as in other spheres, the Germans probably suffered most. The same basic items were rationed in both countries: meat, butter, fats, bacon, cheese, sugar, jam, milk and eggs. But in Germany one also had to part with coupons for bread and potatoes, both plentiful in Britain [3], and there were no "lend lease" supplies from America to help fill empty stomachs.

American soldiers arriving in Europe readily admitted that "back home they don't know there's a war on". Even in the United States there was, in theory, some rationing – of many canned goods, sugar and coffee – but in practice there was no real shortage of any type of goods. The Japanese fuel shortage prevented their indulging in the constant ritual bathing demanded by tradition. Japan also suffered a near-breakdown in the railway and island-ferry transport systems, and by 1945 food supplies had shrunk to no more than 1,300 calories a day, less than half the normal minimum.

A new prosperity

If the war years brought unprecedented hardship to civilians they also brought many benefits. In both Germany and Britain, due to the fairer sharing out of food supplies and full employment, poorer families lived better than they had ever done before. Everywhere, rigid price controls kept the rise in prices within limits; by 1945 the cost of living was only a third higher than before the war.

For factory worker and farmer alike, in every combatant country, these were boom years. The new prosperity masked deeper long-term changes. The drift from country to city was accelerated; there was increased pressure for urban amenities to be extended to the countryside; it was demonstrated that full employment was not an impossible dream and everywhere people demanded a fairer social order after the war.

KEY

Evacuation was carried out by the British Government at the beginning of the war because of the fear of massive air-raids. Nearly 1,300,000 people, mainly children, left the cities.

7

7 War savings were encouraged by all countries to stop inflation. This US Victory Bond was designed by Walt Disney.

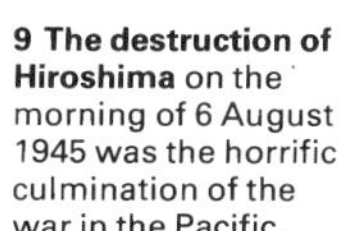

8

8 The mobilization of women was greatest in the USSR. Here ammunition is stacked to repel the Leningrad siege.

9

9 The destruction of Hiroshima on the morning of 6 August 1945 was the horrific culmination of the war in the Pacific.

10

11

12

10 Propaganda was used by both sides, both offensively and defensively. This German poster warns against careless talk.

11 Allied bombing devastated the non-military city of Dresden. These are the ruins of the church of St Sophia.

12 "Traitor" warns this German poster. German propaganda techniques were generally more sophisticated than the Allies'.

The division of Europe

The cold war is usually thought of as a global struggle between the two Great Powers that had emerged by the end of World War II. These two powers, the Soviet Union and the United States, were initially by no means equal; the United States was far superior in terms of economic capacity, air power, and in the fact that she possessed nuclear weapons before the Soviet Union. However, the Soviet Union had an important advantage – the ability to threaten Western Europe with the might of her army. It was because of this Soviet threat that the United States was obliged to come to the rescue and defence of the Western European countries.

East-West misunderstandings

This is the traditional view of the origins of the cold war and it derives from an interpretation whereby Stalin's Russia overran eastern Europe between 1945 and 1947 and seemed to threaten Western Europe too. Against this a different view has been suggested by some historians. They say that the USSR which had in the past been invaded many times from the West, was still afraid of her titular allies at the end of World War II. In this view, the Stalinist takeover of Eastern Europe was a defensive reaction to possible attack.

These views are contentious, but it is fairly clear that mutual misunderstanding between the Soviet Union and America played a large part in bringing about the division of Europe [5]. When Churchill (1874–1965), Roosevelt (1882–1945) and Stalin (1879–1953) met at Yalta in 1945 [Key], Soviet suspicion of the Western timing of a Second Front gave way to Western suspicions over Soviet intentions in the East – particularly towards Poland. Thereafter the powers failed, through a series of increasingly acrimonious conferences, to reach agreement on Germany. The process of division inevitably took over.

The division of Germany

At first the American forces had not intended to stay long in Germany. They did not expect the Soviet troops to remain either. The victorious powers were supposed to supervise German reconstruction only until they could all agree on its future as a united country. All four – through their foreign ministers, Ernest Bevin (1881–1951), Georges Bidault (1899–), Vyacheslav Molotov (1890–) and Secretary of State George Marshall (1880–1959) [4] – administered Berlin equally. But the picture changed, partly because of Soviet dominance in Eastern Europe and in the Soviet zone of Germany, which was rapidly organized as part of the Soviet system. Also important was the Soviet reparations policy, which seemed to threaten the economic ruin of the West by leading to the total collapse of any German economy [2]. Between 1946 and 1948 it became clear that a German economic revival was necessary for Western Europe's recovery.

At first the United States had hoped to include Eastern as well as Western European countries, and certainly the whole of Germany, in a vast programme for European recovery based on American aid. This plan, the European Recovery Programme, or "Marshall Plan" of 1947, was rejected by the Soviet Union but was still applied to the western zones of Germany. Applying it there

CONNECTIONS

See also
228 World War II
240 The Soviet Union since 1945
242 Eastern Europe since 1949
306 Europe from confrontation to détente
218 British foreign policy since 1914
298 The United Nations and its agencies
198 Stalin's Russia

1

1 US and Soviet troops met at Torgau, Germany, on 25 April 1945. But already Russian resentment over delay in the Second Front, and US distrust of Soviet motives, heralded the cold war.

2

2 The Russians dismantled German industry so thoroughly that it caused hardship in the Western zones and was halted, despite the fact that the USSR's reparations claims had at first been accepted.

3

3 James Byrnes (1879–1972), the US Secretary of State, attended the 1946 Paris Conference which was to draft peace treaties with Italy, Romania, Finland, Bulgaria and Hungary. Achieving only part of its aim, the conference also showed up disagreements over Germany.

4

4 Marshall, Bevin, Bidault and Molotov made a futile attempt to agree on the German question, in 1947.

5 A
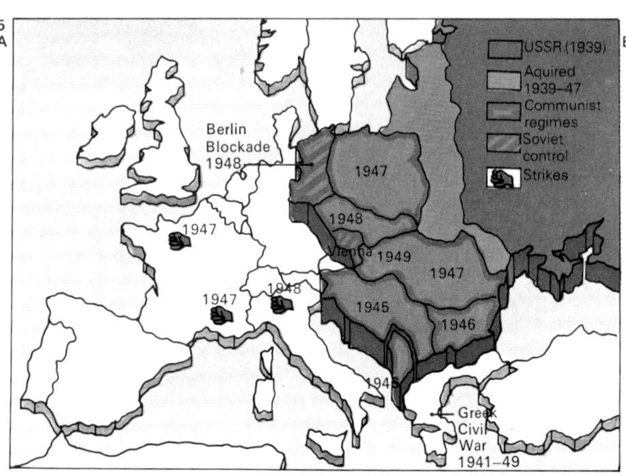

B
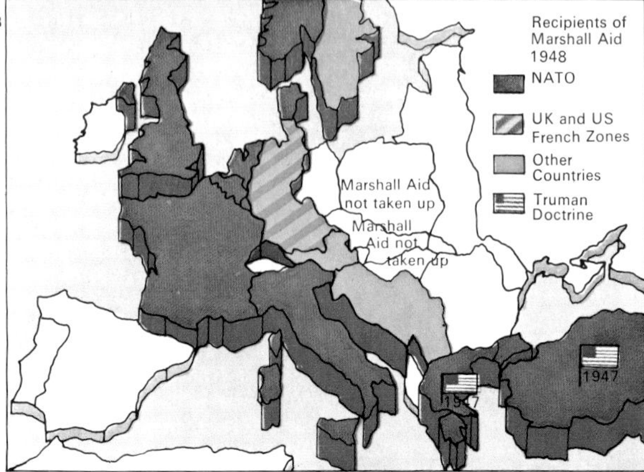

5 From the Western point of view [A] it appeared that a vast Soviet army had taken over Eastern Europe, reduced it to Stalinist rule and was poised ready for a westward advance. From the East [B], the superior economic power of the Western world backed by American nuclear weapons seemed ready to disrupt the defensive system that the USSR was trying to create. Each seemed to be threatening the other and so the cold war escalated.

meant the introduction of a separate and reformed West German currency.

After the currency reform in West Germany the USSR began the blockade of Berlin. The blockade [6] lasted for nearly a year, from 1948 to 1949, and was a turning-point in the history of Europe. It came when the division of Europe was complete – for in February 1948 the Soviet Union had completed its take-over of the eastern countries by a coup against Czechoslovakia.

The birth of NATO

It was against this background that the decision was taken to form NATO (North Atlantic Treaty Organization) [8, 9] – a long-term alliance by which the United States was pledged to the defence of Western Europe. The original (1949) members of NATO were the USA and Canada and the principal nations of Western Europe [10]. Greece and Turkey joined in 1951 and West Germany in 1955. Meanwhile West European countries began to recover and to co-operate. They had already sketched some form of co-operation in defence (in the West European Union before NATO was founded) but equally important was the Organization for European Economic Co-operation (OEEC), formed in 1948, in which the United States supported the West European countries in creating a system of mutual prosperity. And from 1949 onwards the Europeans began to pool their resources in a system of co-operation that was eventually to form the European Economic Community.

In the east the Stalinist system of almost total control exercised through the Cominform was challenged only by Yugoslavia, although later a more co-operative pattern was established after 1949 through the Council for Mutual Economic Assistance (or COMECON). But the early contrast between Western co-operation and the Eastern dictatorship reinforced the division of Europe and the rigidity of the cold war.

Before this, in 1950, the Korean War had broken out and seemed to confirm the necessity of NATO. As a result, by 1955 West Germany was invited to join. When she did so, it meant that Germany could not be united, and the division of Europe was complete.

KEY

The three leaders of the Grand Alliance, Churchill [left], Roosevelt [centre] and Stalin, met at Yalta in February 1945. France was not invited. It has often been argued that Europe was divided into two blocs at this meeting, but the "Big Three" agreed on little beyond the final arrangements necessary for temporarily dividing Germany.

6

6 The Berlin blockade was the first great confrontation of the cold war. It arose from restrictions imposed by the Russians on Western access to Berlin. For months the city was maintained by an airlift. However, the outcome depended as much on the refusal of West Berliners to accept Soviet economic help in return for political surrender. The blockade divided Berlin and completed the division of Germany.

7

7 The Allied Control Council (shown here in 1948) governed Germany from 1945 to 1948. It did not establish a central government for the whole country, but served to resolve disagreements arising through the separate governments of the different zones. When the three Western powers decided to introduce a new currency in West Germany, the Russians walked out and the Council came to an end.

8

8 The foreign ministers of NATO countries gathered in Washington to sign the NATO Treaty before the Berlin blockade was over. Events in Europe had seemed to confirm the aggressive intentions of the USSR and the need for a firm Western response. Also, the economic recovery of Western Europe depended on a security guarantee. By committing the US to a long-term defence arrangement, NATO superseded European attempts to ensure security.

9 NATO was formed to offset the Soviet military presence in Europe. The forces committed to NATO were too weak to be anything but a stop-gap in the case of an emergency until, that was, the full resources of all the signatories could be mustered. In 1955 West Germany became a fully independent state and a member of NATO. In May of that year, partly in response to that event, the Soviet Union set up the equivalent defence organization of the Warsaw Pact.

NATO membership 1955

9

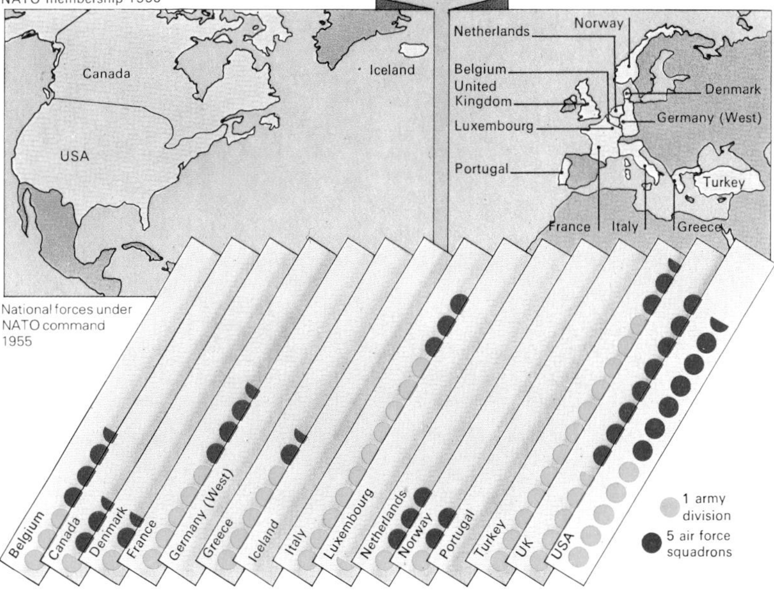

10

10 By 1955 West Germany had made an amazing recovery. At Paris in 1954 the powers met to determine the extent of her entry into the European community. Konrad Adenauer (1876–1967), seen here with other leaders, had worked for this since becoming Federal Chancellor of West Germany in 1949.

After six years of war Britain's return to peacetime conditions needed a prolonged period of adjustment. Despite the remarkably united and disciplined war effort, the country's economy had been overstrained, and Britain was not in a position to shoulder properly the burdens of occupying its zone of Western Germany while also playing its part in achieving some kind of peace settlement in the East as well as in the West.

Labour victory and the Welfare State

Although Britain still ranked as one of the "Big Three" powers when the war ended, along with the United States and the Soviet Union, it soon became clear that it was no longer in the super-power league. At the 1945 general election, the bulk of the electorate showed that it was more interested in the approach to peacetime reconstruction offered by the Labour Party than in the continuation of Britain's role in big-power politics which it associated with Winston Churchill (1874–1965), linked as that would have been with a period of Conservative rule. A landslide victory for Labour deprived the country of the world figure who had been – not just for the British but for millions elsewhere – the personification of resistance to Nazism and Fascism. Clement Attlee (1883–1967) became prime minister.

Ernest Bevin (1881–1951) as Foreign Secretary supplied something of Churchill's bulldog quality in the negotiations that began to shape the peacetime settlement. At the same time he and others undertook the vast work of decolonization, starting with the granting of independence to India and Pakistan in 1947.

Domestic changes were almost as dramatic as those taking place outside Britain. The government's brand of socialism stressed nationalization of various sectors of the economy as the way forward, while greatly extending the state health and medical services and education, creating a "Welfare State" [4]. The Bank of England was nationalized in 1946 and in 1947 the railways and the coal mines were also taken under state control. The steel industry was also nationalized, in 1947, after a constitutional crisis brought on by Conservative opposition in the House of Lords, whose power to delay bills was subsequently reduced. What affected people most directly was the massive reorganization of the Health Services [2], accomplished by Aneurin Bevan (1897–1960), in order to provide medical and hospital treatment and prescriptions and also dental and other services "free", or at minimal rates.

The government had inherited a wartime economy. It continued rationing (not completely ended until 1954) and also policies of heavy taxation and wage restraint. Despite a large increase in exports, the country (or rather the sterling area as a whole) had an almost chronic deficit with the United States, which forced a devaluation of the pound from $4.03 to $2.80 in September 1949.

Conservative rule

Long-drawn-out opposition by the British Medical Association to the Health Service reforms, and bitter wrangling in Parliament over steel, indicated that Labour's popularity was waning. At the 1950 election Labour was returned to power with a reduced majority,

CONNECTIONS

See also
218 British foreign policy since 1914
220 The Commonwealth
224 Britain 1930–45
234 The division of Europe
238 Britain since 1945: 2
246 Decolonization
258 Evolution of Western democracies
260 Scotland in the 20th century
262 Wales in the 20th century
280 Art and architecture in 20th century Britain
294 Ireland since partition
306 Europe from confrontation to détente

1

1 A landslide victory brought Labour to power in 1945 with 393 seats against the 213 won by the Conservatives and their allies. The Conservatives and most foreign observers had assumed that Churchill, with his great wartime prestige, would carry them to victory. But the electorate was moved by Labour's promises of employment, housing and welfare and the proposals for nationalization of basic industries and state planning of the massive reconstruction that lay ahead. Years of wartime organization had left the people with a collectivist legacy that gave a strong appeal to Labour's socialist programme.

2

2 The centrepiece of the new Welfare State was the National Health Service, whose creation was the work of Aneurin Bevan. For the first time medical attention, prescriptions and many other services, generally became free or available on low charges. Some 3,000 hospitals were taken over under the scheme. While the hospital consultants welcomed the proposal, most of the doctors, organized by the British Medical Association, were bitterly opposed to it, as depicted in this contemporary cartoon. Bevan fought a long battle with the doctors, who saw in the scheme threats to their independence; but when the service began over 90 per cent of the doctors enrolled.

3

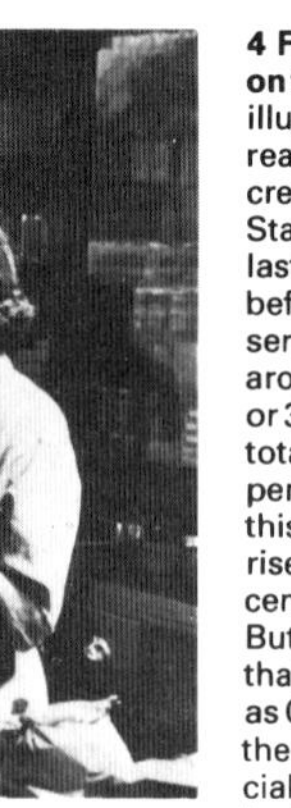

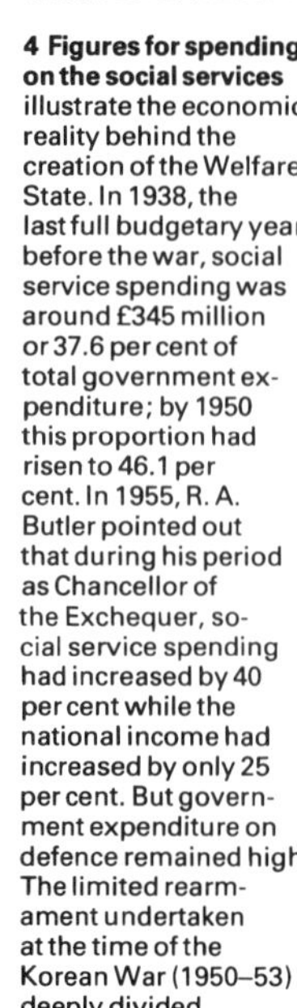

3 Rationing in the postwar period was more severe than in wartime. Until its defeat in 1951, the Labour government pursued an unpopular programme of austerity to rebuild the economy and finance government expenditure. Abroad things were serious; in The Netherlands and the British zone in Germany there was near famine and there was a lack of raw materials all over the world. But ironically a higher percentage of each age group in the London area in 1946 was classed as of "excellent nutrition" than in 1938, and this was true of the country as a whole. Rationing began to be reduced after 1948; in 1949 clothing and furniture were freed. Meat was the last item to disappear from the ration books, and that took place in 1954.

4 Figures for spending on the social services illustrate the economic reality behind the creation of the Welfare State. In 1938, the last full budgetary year before the war, social service spending was around £345 million or 37.6 per cent of total government expenditure; by 1950 this proportion had risen to 46.1 per cent. In 1955, R. A. Butler pointed out that during his period as Chancellor of the Exchequer, social service spending had increased by 40 per cent while the national income had increased by only 25 per cent. But government expenditure on defence remained high. The limited rearmament undertaken at the time of the Korean War (1950–53) deeply divided the Labour Party.

4

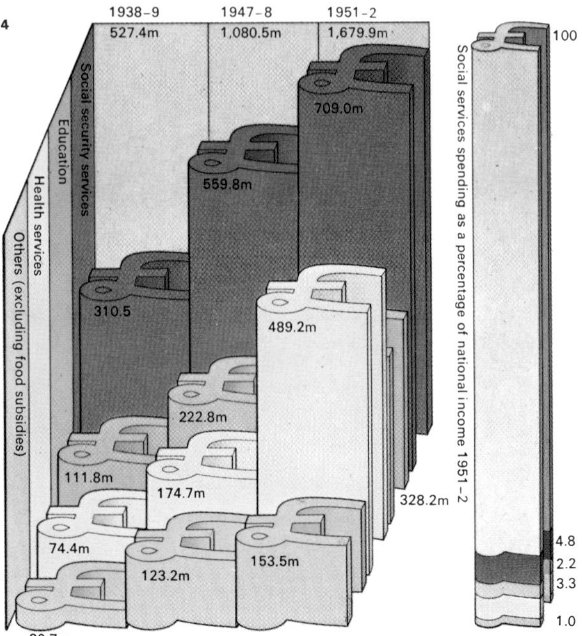

and at the following election in October 1951 the Conservatives under Churchill won a majority of 26. With this they de-nationalized the steel industry in 1953 (it was later re-nationalized by Labour in 1967).

The Chancellor of the Exchequer, Richard Butler (1902–), introduced a series of measures designed to improve the balance of payments and to increase domestic consumption. In the 1955 election the Conservatives were returned with an increased majority. The party was now led by Anthony Eden (1897–1976) who had taken over the leadership after Churchill had resigned through ill health.

In addition to maintaining an independent nuclear deterrent [8] and continuing national service (until 1958), the government favoured British influence and defence commitments overseas on a scale that the economy could no longer support. The failure in 1956 of the Suez operation against Egypt, when the collusion of Britain and France with Israel was opposed by America, made it clear that Britain could no longer continue the stance of a world power.

Meanwhile, thanks in part to vast infusions of dollars through US loans and Marshall Aid, the economy had a run of good years. The standard of living rose, and the working classes, like most of the population, had "never had it so good". These words of Harold Macmillan (1894–) [9], who became prime minister following Anthony Eden's resignation in 1957, can serve as a motto for this final phase of the 13 years of Conservative rule (1951–64).

Loss of confidence

In the early years of the 1960s the economy took a downward turn, however, and successive pay policies introduced by the Conservatives failed. Britain's application for membership of the European Economic Community was vetoed by France in 1963, and the Beeching report proposing a one-third reduction in railway services also undermined Macmillan's popularity. Sir Alec Douglas-Home (1903–), who took over the premiership after Macmillan had resigned because of illness in 1963, could not restore confidence.

KEY

The Festival of Britain in 1951 was conceived as marking a new era of reconstruction following the destruction of World War II. Opened by George VI (*r.* 1936–52) on 3 May, it attracted 8.5 million visitors to the Festival Hall and other sights on the south bank of the River Thames.

5

5 The 1950 election had returned Labour to power with a majority of only five. In 1951 under increasing pressure, the government resigned and an election gave the Conservative Party a majority of 26. The Conservatives presented an attractive alternative after the prolonged austerity of the preceding years.

6

6 The coronation of Queen Elizabeth II (1952–) in June 1953 was taken by many to symbolize a new "Elizabethan Age" with the promise of great prospects for Britain in the postwar world. The event was televised worldwide and thousands of cheering spectators lined the streets to watch the colourful procession.

7 A new youth culture emerged in the 1950s, alongside the beginnings of rock 'n' roll music, which presaged predominant youth cultures of the 1960s. Like the music, the new style was aggressive and uncompromisingly youthful and reflected the new affluence of the postwar period. Styles included those worn by "Teddy Boys", who affected Edwardian-style suits, string ties, and duck's-tail haircuts.

7

8

8 Ban-the-bomb demonstrations were frequent after the CND (Campaign for Nuclear Disarmament) was founded in 1958. Many public figures shared this widespread concern.

9

9 Harold Macmillan (centre) was prime minister for six years from 1957–63 until he retired from the Conservative leadership because of ill health. During that time the country had a period of prosperous and efficient government, although the economic problems that dominated British politics in the 1960s became evident during the final years of his term of office.

Britain since 1945:2

The British general election of 1964 initiated a period of Labour rule broken only by nearly four years of Conservative government under Edward Heath (1970–4). The period as a whole was one of increasing economic difficulty for Britain. It failed to maintain its competitive position against trade rivals despite its entry in 1973 into the European Economic Community (Common Market), an action that was reaffirmed after a referendum in 1975 [8]. Only on the "invisible" side of its trading account (banking, brokerage, insurance and other services) did Britain maintain its position, thereby alleviating the effects of the frequent deficits in its balance of payments.

Trade union militancy

Trade imbalances were offset by loans that became ever more massive, despite a few better years when repayments were made, notably during the period 1967–70. Among reasons for the weakness of trade were the increasing productivity of competitors, and their greater ability to adopt new methods and machinery both for older industries and for the new high-technology enterprises. In contrast, British management found it difficult to secure the co-operation of trade unions in introducing modern plant and reducing labour costs. This failure was coupled with successful union pressure for increased wages and reduced hours of work, backed by go-slows and strikes.

In Parliament the Labour Party was increasingly polarized between left-wing socialists of the Tribune Group and some Marxist-oriented MPs on the one hand, and those who pursued a moderate social-democratic line on the other.

Among the latter were Harold Wilson [1], prime minister 1964–70 and 1974–6, and James Callaghan, who followed him as prime minister. Wilson coped skilfully with the divisions in his party but at the cost of compromising over some important issues to the point where governmental authority was eroded. The continuing high cost of defence, together with growing education, health and pensions services, imposed burdens which the weakening position of the country in productivity and trade made it difficult to meet. This weakness was reflected in the tendency of inflation, which had been chronic but manageable (three to five per cent), to increase to, at times, more than 20 per cent. As a result, sterling weakened against other currencies [4].

Devaluation of sterling

Labour's fine ideals in 1964 of modernizing Britain and moving it steadily towards socialism, were soon obscured by the fight to "save the pound". Desperate efforts were made to maintain the exchange rate of the pound at $US2.80 by large-scale borrowing from abroad – but to no avail. A seamen's strike in 1966 hastened the loss of confidence in sterling and the pound was devalued to $2.40 in November 1967. Attempts were made to bolster sterling by an incomes policy that restricted wage increases to certain ceilings or percentages. But the Labour programme for pursuing this objective, formulated in the White Paper, "In Place of Strife", failed in 1969 in the face of union militancy and left-wing opposition.

The Conservative government from

CONNECTIONS
See also
134 The story of the West Indies
218 British foreign policy since 1914
224 Britain 1930–45
236 Britain since 1945:1
246 Decolonization
252 Southern Africa since 1910
258 Evolution of the Western democracies
260 Scotland in the 20th century
262 Wales in the 20th century
280 Art and architecture in 20th-century Britain
294 Ireland since Partition
306 Europe from confrontation to detente

1 Harold Wilson (1916–) became prime minister of a Labour government with a majority of only four in 1964. He consolidated his party's position at a further election in 1966. The youngest MP to attain cabinet rank in the first post-war government (at the Board of Trade), he took over the party leadership after the death of Hugh Gaitskell in 1963. Although his flexibility enabled him to hold together the left and right wings of his party, his hopes of modernizing the British economy and the trade union system were dashed first by the weakness of sterling and secondly by union opposition to sweeping changes.

2 Holland Park Comprehensive made news as a large purpose-built (1958) school in a fashionable part of London to which some public figures sent their children. With its sixth form block, ten science laboratories, ten art studios, seven workshops and three gymnasiums, it summed up the aspirations of a new-style education system based on the principle of giving all children, no matter what their background or means, equal opportunity. As some academic standards slipped, however, Labour's policy of replacing grammar, secondary modern and grant-aided schools by comprehensives provoked increasingly fierce controversy.

3 Mick Jagger and the Rolling Stones, seen here at a 1969 concert in Hyde Park, London, were the most aggressive, irreverent, anti-establishment rock group to appear in the entertainment world of the 1960s. Their appeal was less broadly based than that of the Beatles, whose popularity with virtually all age groups helped to break down some of the traditional barriers of class and accent in Britain. The driving music of the Rolling Stones was directed more frankly at youthful rebellion. It chimed in with trends of the times, reflected in the increasingly open treatment of sex and violence in films and on television, and the use of drugs as stimulants on a scale previously unknown.

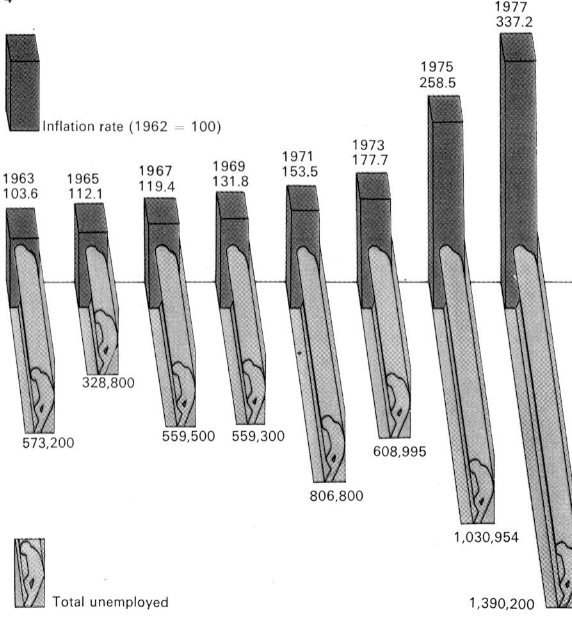

4 Inflation had been at the rate of 3% to 4% from 1945 until 1964, fuelled by the steady increase of government spending and the outpacing of production increases by wage rises. The rate of inflation jumped when the 14.3% devaluation of the pound in 1967 put up the prices of imported goods. From 1973 onwards, price rises imposed by the Organization of Petroleum Exporting Countries (OPEC) doubled the price of oil. Soaring costs and wages in 1975 brought inflation to 20%. Government efforts to hold down spending, together with some company failures, led to a rise in the number of unemployed to more than 1 million, the worst level of unemployment since the 1930s.

1970–4 fared no better in attempting to control inflation, and in fact worsened the position by dismantling some of Labour's controls, only to return to an incomes policy. Obdurate union resistance to pay restraints was exacerbated by the Industrial Relations Act, which established three-phase statutory wage and price controls. A coal-miners' strike [6] early in 1972, involving power-cuts when the miners obstructed coal deliveries to the generating stations, led to the treatment of the miners as a "special case". The 25 per cent pay rise they received breached the incomes policy. The Government attempted to counteract an overtime ban by the miners the following winter by introducing an emergency three-day working week to save fuel. But when in 1974 it sought a mandate for a firm line against union pressure for higher wages, a general election resulted in the return of Labour to power by a small majority.

Despite hopes of future prosperity through the development of North Sea gas and oil resources, the pound continued to sink, impelling a return to an incomes policy under the chancellorship of Denis Healey. To restrain a rising rate of unemployment [4], the government was forced to back some ailing firms with public money [7].

Political and social strains

Economic weakness aggravated political problems. Proposals were made for devolution of some powers to local assemblies in response to demands for greater autonomy and even independence by parties in Scotland and Wales. In Northern Ireland, terrorist activity by the Irish Republican Army (IRA) and counter-terrorism by Protestant extremists led in 1972 to direct rule from Westminster, supported by large-scale and continuing army operations.

Socially, the strains of a further influx of Commonwealth migrants [5], major changes in patterns of education [2] and a shift in economic power from the older to the younger generation met with mixed success during the period after 1964. Pop music groups such as the Beatles and the Rolling Stones [3] were associated with a new image of London as the "swinging" capital [Key].

KEY

Carnaby Street, with its boutiques and shops specializing in fashionable clothes, colourful posters and the latest pop records, became a symbol for the new "swinging" London of the 1960s. Together with King's Road, Chelsea, it provided, a visual idea of a city that had cast off the imperial trappings of the past. London was now the capital of the youth orientated societies of the affluent Western countries, a youth whose tastes and demands needed up-to-the-minute satisfaction. One of the most popular of Carnaby Street's emblems was the Union Jack itself, converted from the national flag to almost anything, from a lively T-shirt to a plastic shopping bag.

5

5 Immigrants from the West Indies and Asia provided staff for medical, transport and postal services and for certain industries, notably textiles, after World War II. But their rising numbers and limited prospects brought social strains while problems of housing and education led to government measures to regulate their entry during the 1960s. The entry restrictions were partially waived to accommodate Asians holding British passports expelled from Uganda in 1972. Community services and immigration liaison offices were set up in several cities to help with their integration and improve race relations.

6

6 Striking miners in 1972 supported wage claims of up to 47% made by the National Union of Miners at a time when the Conservative Government was hoping to bring inflation down from 6% to 5%. Rejecting increases of between 7% and 8%, miners picketed generating stations until power shortages forced the government to set up an inquiry. The strike, from 9 January to 28 February, ended with acceptance of increases averaging 25% recommended by a court of inquiry. Further miners' claims in 1973, were resisted by the Government, but its handling of the economy led to defeat in the 1974 election.

7

7 Rolls-Royce engine manufacture was threatened in 1971 when the company's financial problems forced it to seek assistance from a Conservative government pledged to leave "lame duck" industries to their fate. The government had to take over those parts of the company essential to defence. A similar crisis in 1975 in the American-based Chrysler company obliged the Labour government to inject £162.5 million to save the jobs of car plant workers.

8

8 A last-ditch fight against Britain's entry into the EEC was defeated when a referendum in 1975 produced a 67% vote in favour of continued membership of the Community. A large section of the Labour Party, particularly the left-wing Tribune Group, had opposed Britain's joining in 1973.

9

9 Arabs shopping in London became a new feature of life in the capital during the 1970s, reflecting rising incomes in the oil states of the Middle East, particularly Saudi Arabia and Kuwait. At the same time sterling balances held by the oil states became a key factor in Britain's management of her currency reserves. Arab investors in the UK tended to favour buying real estate, such as the Dorchester Hotel, rather than shares in British industry.

The Soviet Union since 1945

The USSR at the end of World War II had lost more than 20 million of her citizens and four and a half million homes. Some 65,000km (40,000 miles) of railway track were laid waste, thousands of industrial and agricultural machines crippled and livestock vastly depleted. Reconstruction was a formidable task. Joseph Stalin (1879–1953) reintroduced five-year planning, and soon he declared many ambitious targets over-fulfilled.

Costly progress

By the time of Stalin's death the Soviet Union had acquired nuclear weapons and had far surpassed prewar production in iron, steel, coal, oil and electricity. It achieved these targets at the cost of great sacrifices by its own people and those of Eastern Europe, whose resources were in effect put at Moscow's disposal after 1945.

Life was hardest in the countryside where under-investment, low prices for compulsory deliveries, high taxes on private plots and doctrinaire administrative measures hampered production. By 1953 agricultural output per capita was below that of 1928.

The onset of the cold war together with Stalin's attempts to contain the effect of Tito's independent line in Yugoslavia increased tension within the USSR. Stalin's "personality cult" reached its peak in the postwar era when purges were revived. Stalin's paranoia towards the end of his life and the sense of fear and suspicion he created around him were publicly expressed in January 1953 when he unjustly accused nine eminent doctors, most of them Jewish, of having murdered the deputy premier, Andrei Zhdanov (1896–1948). In his last days not even Stalin's closest intimates and advisers were safe from his secret police.

Collective leadership

After Stalin's death on 5 March 1953 [1], "collective leadership" was proclaimed and accordingly the new Premier – Georgi M. Malenkov (1902–) – relinquished the position of senior Party Secretary ten days after assuming it. Soon the Kremlin doctors were released, their "plot" having been exposed as a fabrication. Curbs on secret-police power were dramatized by the secret trial and execution of Lavrenti Beria (1899–1953), the reorganization of his ministry and the progressive release from labour camps of an estimated 10–12 million people.

There were serious rivalries between Stalin's successors. Premier Malenkov and First Secretary Nikita S. Khrushchev (1894–1971) [2] disagreed over economic priorities and the implications of nuclear warfare, but while Khrushchev exploited their differences to engineer the removal of Malenkov from the premiership in February 1955, he subsequently endorsed many of Malenkov's proposals.

The trend towards relaxing domestic and foreign policies alarmed Foreign Minister Vyacheslav Molotov (1890–), especially after the 20th Communist Party Congress in February 1956 at which Khrushchev denounced Stalin and envisaged different roads to socialism. The subsequent turmoil in Poland and Hungary confirmed Molotov's fears. He spearheaded a revolt, culminating in the Party Presidium's vote for Khrushchev's dismissal in June 1957. However, in the Central Committee meeting that

CONNECTIONS

See also
306 Europe from confrontation to détente
242 Eastern Europe since 1949
234 The division of Europe
198 Stalin's Russia
254 Non-alignment and the Third World

In other volumes
104 Man and Society

1 Stalin's funeral, on 9 March 1953, drew crowds of Russians to Red Square in Moscow. Not everyone mourned. Some grieved for the man who had transformed their country into a powerful state, but others counted the cost. Without abandoning police control or strict censorship, Stalin's successors eradicated the "personality cult" and the rule of terror. Stalin's body was removed from the Lenin Mausoleum in 1961.

2 Nikita Khrushchev joined the Communist Party in 1918 and became a loyal executor of Stalin's policies. As First Secretary in the Ukraine in 1938, he administered the purges with fervour. Later, as the Party chief, he dismantled the cruder forms of terror and moved fitfully towards *détente.* His caprices and bombast infuriated all, yet his consignment to oblivion in 1964 was widely regretted.

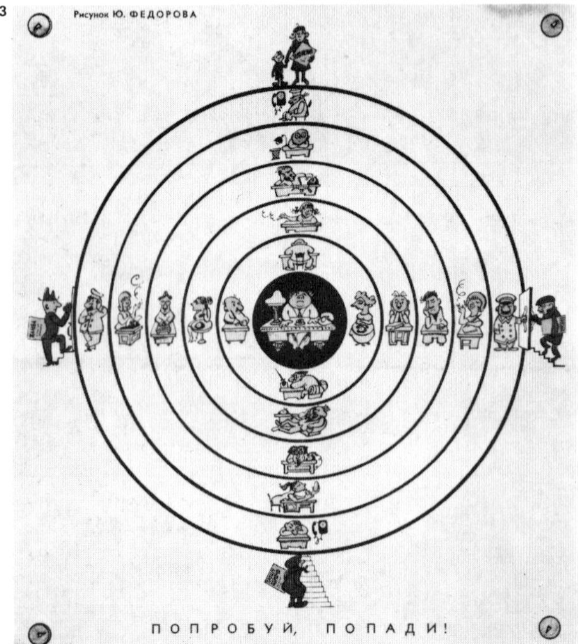

3 This Soviet cartoon satirizes the inadequacy of curbs on bureaucracy. Attacks on officialdom are tolerated, but criticism of higher officials and Party policy is banned.

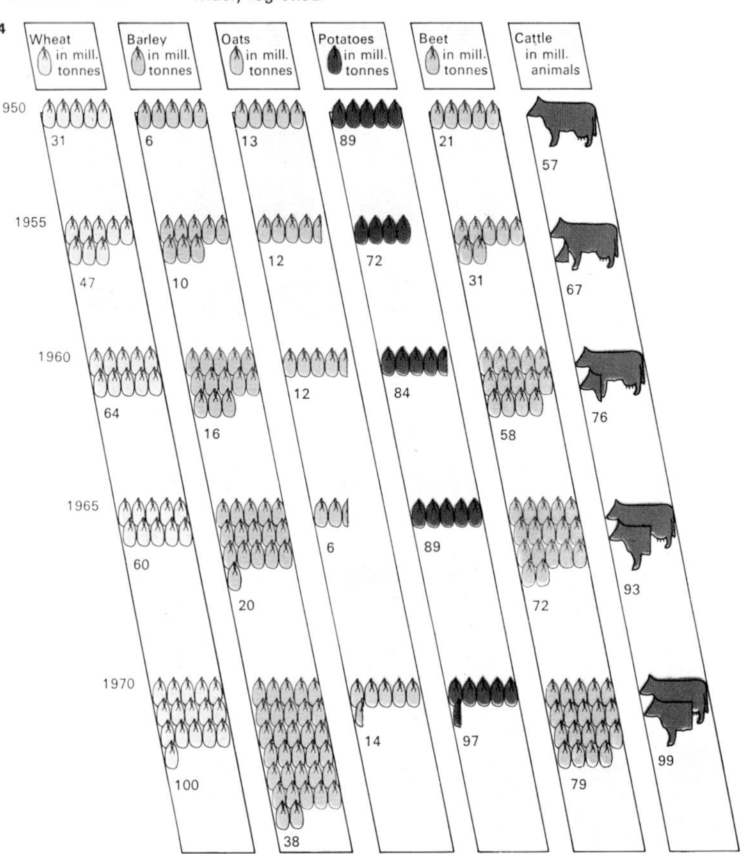

4 Output of Soviet agriculture has been disappointing since collectivization. Price incentives and doubling of investment in the 1960s raised productivity. Today, with over 500 million acres under cultivation and a quarter of the workforce engaged in farming, Soviet agriculture is still unable to meet the population's demands for a better diet.

followed, Khrushchev's opponents were themselves defeated. Nikolai Bulganin (1895–) remained, but lost the premiership within a year to Khrushchev himself.

Khrushchev's elevation meant improved material conditions. He cut the working week, reduced wage differentials, diminished the stringency of Stalin's Draconian labour laws and gave greater priority to consumer needs. But over-centralized, often incompetent planning plagued economic development. Notwithstanding industrial performance [7] and the Sputniks and other space triumphs after 1957 [Key], agricultural production [4] remained disappointing despite increased investment. Khrushchev's failure aroused resentment and, in October 1964, he was dismissed.

A decade of stable government

Despite policy disagreements, the post-Khrushchev leadership has been remarkably stable. Leonid Ilyich Brezhnev (1906–), First Secretary of the Central Committee, Alexei Kosygin (1904–), Chairman of the Council of Ministers, and Nikolai Viktorovitch Podgorny (1903–), President, having held office for well over a decade. The USSR has advanced militarily to achieve virtual strategic parity with the USA, while the rift with China, begun under Khrushchev, has widened. Economic progress has been less spectacular. Central planners have resisted complete decentralization, but they have permitted some degree of autonomy. In agriculture massive investment and concessions such as the relaxation of restrictions on private plots have helped to boost production, although major problems remain, despite military and space successes.

Yet there are signs of strain and nonconformity in the monolithic society of the USSR. Alcoholism is one problem, dissidence another. Outspoken intellectuals and writers, such as Sinyavsky and Daniel, along with protesters at political events such as the Soviet intervention in Czechoslovakia in 1968, suffer a harsh official response. But administrative measures have failed to silence the nonconformists or stem the clandestine circulation of *samizdat* (illegal typescripts).

KEY

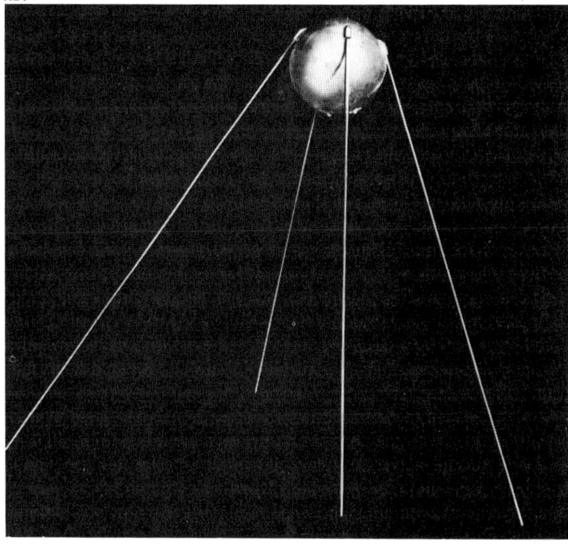

Sputnik I was the world's first artificial satellite. It was launched on 4 October 1957. This success encouraged a whole series of pioneering space ventures.

5

5 Richard Nixon and Leonid Brezhnev celebrated signing the first agreement on Strategic Arms Limitation (SALT) in 1972. It was designed to stabilize the Soviet-American nuclear relationship and reduce risks of war. Like agreements on trade and agriculture signed during the Nixon-Brezhnev summits, SALT has been hampered by suspicions following President Nixon's resignation from the US presidency.

7

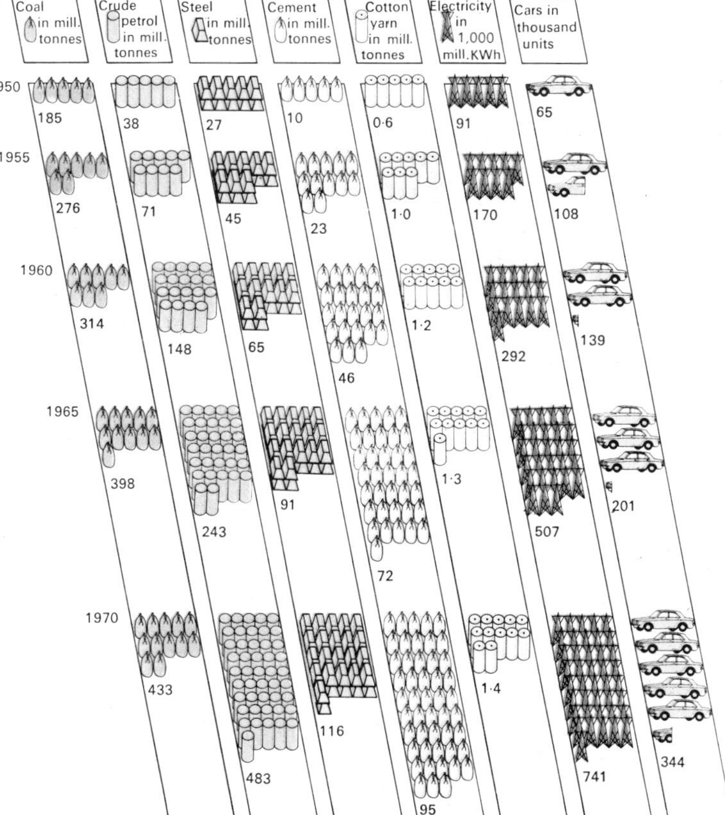

6 Gosudarstveni Universalni Magasin – GUM – is Moscow's biggest department store, selling anything from luxury fur coats to simple hairpins. Queues abound, but, as in shops elsewhere in the USSR, GUM is better stocked than before, reflecting a rise in general living standards.

6

7 Soviet industrial development since the war has been impressive, even allowing for statistical exaggerations. In that period output increased tenfold, more than doubling during the 1960s after reforms which gave more scope to individual initiative. Expansion of heavy industry is still stressed, but consumer production is growing in importance. Productivity per man however is still lower than in the West, hence the Soviet interest in Western technology.

8 International football matches draw great crowds in Moscow. Sport receives generous official encouragement, as part of the view that physical accomplishment makes for healthy, contented citizens and international prestige.

8

Eastern Europe since 1949

A successful coup made Czechoslovakia a communist country in February 1948 and extended the area of intensive Russian influence in Eastern Europe. Each country under communist control became a "people's democracy" – a one-party dictatorship closely modelled on that in the Soviet Union. The characteristic features of these regimes were: strict censorship of the press and control of all aspects of culture and religion; central economic planning; rapid and forced industrialization; at least partial collectivization of agriculture and in foreign policy submission to the line laid down in Moscow. Soviet control of Eastern Europe was guaranteed by the presence of Soviet troops in most of the satellite countries and numerous Soviet advisers and instructors.

After the defection of Yugoslavia (always the most independent of the satellite countries) from the Soviet bloc in 1948, purges took place in Albania, Bulgaria, Czechoslovakia, Hungary, Poland and Romania. These purges often culminated in show trials of officials accused of being sympathetic to the idea of the "separate roads to socialism" advocated in Yugoslavia [3]. Non-communists, too, especially members of the Churches, were subjected to persecution and harassment during that period [2].

Hungary – to encourage the others

After Stalin's death some of the most unpopular features of his policy towards Eastern Europe were modified by his successors, and East European leaders were allowed some degree of autonomy in their domestic policies. But in October 1956 Hungary openly revolted against its communist regime and repudiated its Soviet alliance [4]. At the same time in Poland the leadership of the party was restored to Wladyslaw Gomulka (1905–), who had been dismissed and imprisoned in 1948 for the alleged adoption of an independent line. After a show of indecision, Soviet tanks were used to crush the Hungarian uprising, but the Soviet Union stopped short of more permanent intervention in Poland. This was a sensible decision. Within a year Gomulka cancelled the liberal concessions that had been wrung from the regime by the intelligentsia in the autumn of 1956. However Poland kept its private agriculture while other East European countries went ahead with plans for full collectivization in the late 1950s.

No action was undertaken against Albania, which defected from the Soviet bloc in 1960 and promptly took China's side in the great Sino-Soviet quarrel that was just beginning. Romania opted for a more independent foreign policy in 1964, having for several years strenuously opposed Soviet plans for economic integration within Eastern Europe. But domestically both Albania and Romania remained one-party dictatorships.

The 1968 invasion of Czechoslovakia

Czechoslovakia, which had been the Soviet Union's model satellite for 20 years, provoked the most serious crisis in postwar Eastern Europe in 1968. Alexander Dubcek (1921–), who had become party leader and president in that year [Key], embarked on a course of energetic liberalization, of which the Soviet and some other East European leaders publicly disapproved. Censorship was relaxed, and a higher degree of local

CONNECTIONS

See also
306 Europe from confrontation to détente
240 The Soviet Union since 1945
234 The division of Europe
212 Socialism in the West

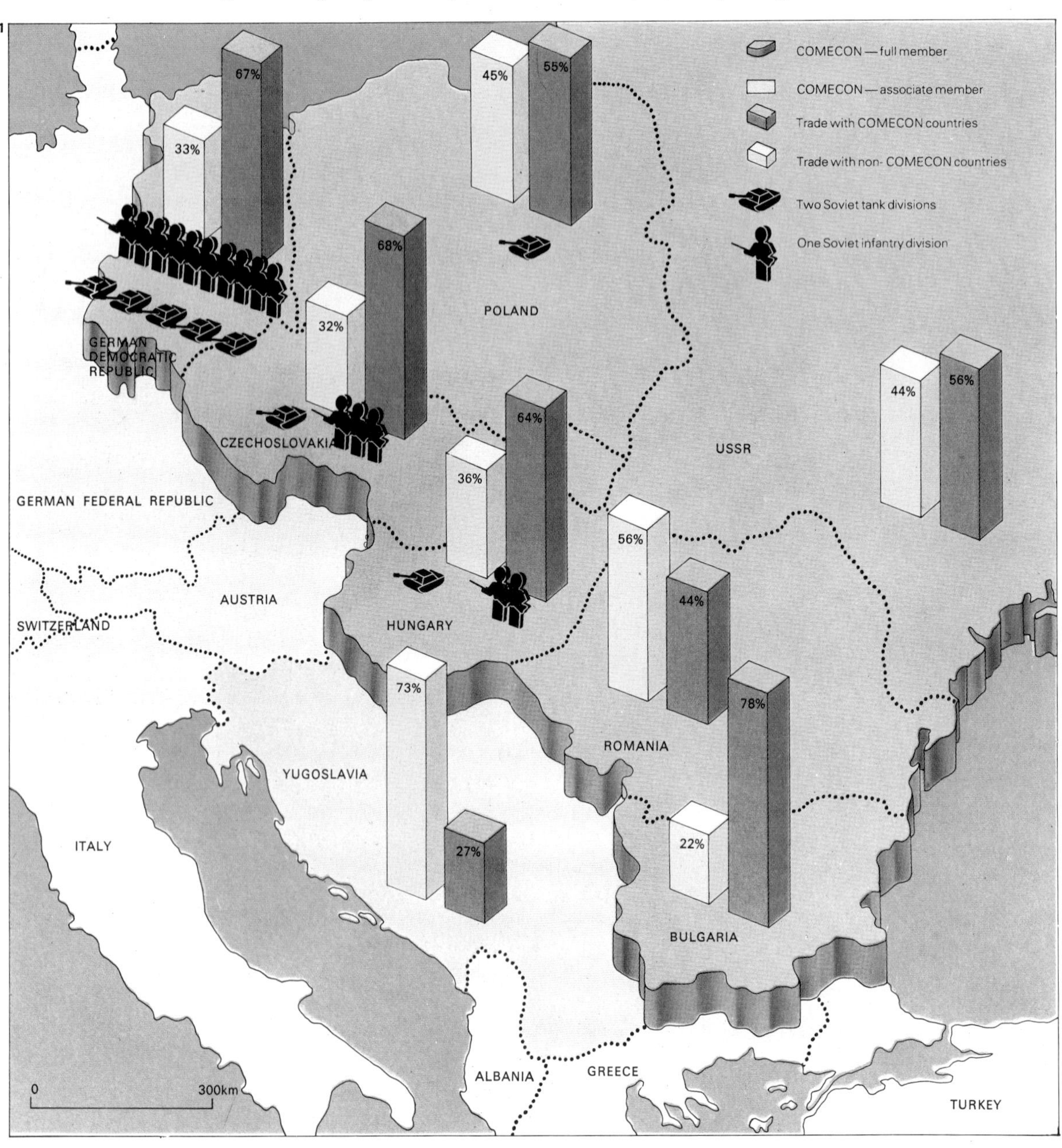

1 COMECON, the Council for Mutual Economic Assistance (which includes Cuba and Mongolia), was founded in 1949 as Stalin's answer to the Marshall Plan in Western Europe. Revitalized in 1958 by Nikita Khrushchev, the Soviet leader, to consolidate Soviet economic control of Eastern Europe, COMECON has now embarked on a policy of integration adopted at Bucharest in 1971 and further elaborated at Budapest in June 1975. In 1973 COMECON with 366 million people accounted for only 12% of world trade (the EEC, by comparison, with a population of 253 million, accounted for 40%). However, its trade with the West and the EEC in particular, is growing fast, with an increasing proportion generated by "joint ventures" between partners from Eastern and Western Europe. Higher costs of Western imports are now provoking more inter-COMECON joint ventures and greater investment in Soviet projects for the exploitation of natural resources. However, Eastern Europe still needs the West for its advanced technology. Yugoslavia, an observer in COMECON, conducts over 70% of its trade with the non-communist world, and Romania deals direct with the EEC. Albania trades with both East and West. To make a direct connection between the presence of Soviet troops and a country's trading pattern seems dubious.

autonomy was granted to the national minorities. In economic planning a move was made to reduce the high level of centralization, prices were allowed a closer relationship to market forces, and individual enterprises were given greater freedom. When Dubcek refused to bow to pressure from his allies, Warsaw Pact troops from the Soviet Union, Poland, East Germany, Hungary and Bulgaria marched in on 21 August 1968 [5]. Czechoslovak leaders were arrested and taken to Moscow, but when no replacements of any stature could be found they were allowed to stay in nominal power for a few months before being finally replaced in 1969.

Although Czechoslovakia's experiment was brutally suppressed, Hungary, under its leader Janos Kadar (1912–), was allowed to carry out a relatively successful series of reforms. Kadar's popular shift towards the consumer goods sector, was emulated elsewhere in Eastern Europe. Poland's new leader, Edward Gierek (1913–), who had replaced Gomulka after workers' riots in December 1970, made "Kadarization" one of the basic tenets of his policy. East Germany, too, embarked on its own version of "consumer revolution" in 1971 after the dismissal of its conservative leader, Walter Ulbricht (1893–1973).

The Soviet bloc closes ranks

Although agreements were reached which lowered some barriers between West Germany on one side and the Soviet Union, Poland and East Germany on the other in the 1970–72 period, there was a new ideological tightening up throughout Eastern Europe. This was due partly to Soviet fear of creeping liberalization and partly to a "backlash" among industrial workers and party officials against the material gains achieved from reform by the professional and managerial classes. New economic predicaments also helped the Soviet Union to turn COMECON's [1] focus eastwards once more. Western inflation had in the mid-1970s made imports from the West suddenly much more expensive; at the same time the Soviet Union raised the prices of oil and the other raw materials of which it holds a virtual monopoly of supply to Eastern Europe.

KEY

Alexander Dubcek (front, second right) kept in uneasy step with other Eastern European leaders at their meeting in Bratislava on 3 August 1968. Less than three weeks later, Warsaw Pact troops invaded Czechoslovakia. The pact was concluded between the Soviet Union and her satellites in May 1955. It forms the cornerstone of Soviet policy in Eastern Europe, bolstered by the presence there of 31 Soviet divisions.

2

2 Cardinal Jozsef Mindszenty (1892–1975), Primate of Hungary (centre), and a strong anti-communist, was imprisoned for life in 1949 after a dramatic show trial. Freed by the rebels in the 1956 rising, he remained in political asylum in the US embassy in Budapest until 1971 when he was ordered to Rome by the Pope. He was fervently opposed to Vatican attempts to come to terms with communist regimes in Eastern Europe.

3

3 Nikita Khrushchev (1894–1971) [*right*], Malenkov's successor as Soviet leader, went to Yugoslavia in May 1955 to repair the rift caused by Yugoslavia's assertion of independence in 1948. Khrushchev blamed the quarrel on Beria, the ex-chief of Russian police, executed in 1953. Josip Broz Tito (1892–) [*left*], the Yugoslav leader, insisted on formal Soviet recognition of Yugoslavia's ideological autonomy.

4

4 Stalin's statue was torn down in Budapest on 2 November 1956, a dramatic moment in the uprising against Soviet domination and the brutal Hungarian regime. Within two days 150,000 Soviet troops and 2,500 tanks were "pacifying" Hungary. The executions that followed the uprising soon gave way to the intelligent government of Janos Kadar. His policy combined better living standards with wider ideological freedom.

5

5 The Warsaw Pact troops who invaded Czechoslovakia in August 1968 met with no military resistance. However, the many spontaneous acts of obstruction such as raising roadblocks and setting fire to Soviet tanks were humiliating for the Soviet leaders who claimed that the intervention had been requested by Czechoslovak leaders. But supporters of the invasion were, in reality, few and had little encouragement.

6 Communism went on show in 1973 with the World Festival of Youth and Students, the largest propaganda rally since 1945. Held in East Berlin, it was a spectacular expression of East Germany's sense of achievement in the year of her world-wide recognition. However, despite the evidence of such displays, youth in Eastern Europe is also interested in Western culture and ideas and often dubious of Soviet bloc ideology and politics.

6

7 The Berlin Wall was built on 13 August 1961 to stop the continual exodus of large numbers of East Germans to the West. Between 1949, when Germany was divided, and 1961, more than 2,700,000 people escaped into West Berlin. Many East Germans still attempt to reach the West despite the dangers. However, this may change; East Germany now has higher living standards than any other communist nation and is the world's seventh largest industrial power.

7

China: the People's Republic

The Chinese People's Republic was established on 1 October 1949 [Key] by a mandate from a constituent assembly convened under the aegis of the Chinese Communist Party (CCP). The immediate task of the new government was to rehabilitate the war-ravaged economy inherited from Chiang Kai-shek's Nationalist administration after its forced withdrawal to the offshore province of Taiwan. A gradualist policy was adopted, characterized by the creation of a coalition of the various elements in Chinese society and the avoidance of violent class struggle. The communists did not baulk at suppressing their most intractable class enemies, but they were preoccupied with carrying out measures to ensure economic survival.

Major reforms of the 1950s

Mass support gave the new government the authority to take steps to conquer hyperinflation [1]. Land reform affecting over 80 per cent of the population was completed by early 1953. As a result the government gained control over surplus agricultural production; it also won the peasant backing it needed to weaken social institutions based on a kinship system dominated by elders [2]. This made it easier to set up new communist institutions in place of the old system. Another major reform was the implementation of the 1950 Marriage Law which greatly improved the status of women.

From 1953–7 China underwent a transition to socialism as commerce and industry were nationalized and agricultural institutions transformed. These changes were not accomplished without dissent but, as a 1957 rectification campaign showed, the power of the enlarged party machine considerably exceeded that of its critics. Meanwhile in foreign affairs China was aligned with the USSR, whose aid was crucial to industrialization during the first five-year plan (1953–7) and bitterly opposed to the United States, her major adversary in the Korean War (1950–53), proponent of the policy of "containment" and supporter of Taiwan.

Hoping to expand production rapidly by amalgamating collective farms into communes [3] and by adopting a backyard approach to industrialization [4], China began the Great Leap Forward in 1958 marking the implementation of a Chinese strategy of economic development and the rejection of the Soviet strategy employed in the preceding five years. As a result, an ideological dispute between China and the USSR gathered momentum, leading to a withdrawal of Soviet technicians and their blueprints in 1960. In the event the Great Leap Forward failed, owing to dissent, bad weather and an underestimation of the problems [5]. The outcome was an economic crisis and a forced retreat from Maoist principles.

The Cultural Revolution

The retreat was only temporary. Once economic recovery had been achieved in 1963 Mao Tse-tung (1893–1976), who had given up his post as head of state in 1959 to be replaced by Liu Shao-chi (1898–), resumed his efforts to realize socialism in China [6]. By now the ideological split between the USSR and China was being reflected within the CCP and the specifically Maoist attempts at running the economy had been openly criticized. Mao Tse-tung coun-

CONNECTIONS

See also
214 East Asia 1919–1945
144 The opening up of China
292 The wars of Indochina
254 Non-alignment and the Third World
298 The United Nations and its agencies

In other volumes
278 Man and Society

1 Queues outside banks in 1948 marked a collapse of confidence in China's currency and in the ability of the Nationalists to manage the economy. Inflation set off by the irresponsible issue of bank-notes was a problem during the Japanese war and it accelerated between 1945 and 1948 when the Shanghai price index rose 135,742 times, causing a hyperinflation. This the communist government inherited.

2 Burning of land title deeds and the public condemnation of landlords were common during the nationwide land reform campaign conducted by the Chinese government between mid-1950 and early 1953. The political and social impact of this campaign was as important as its economic effect. Socially, the destruction of the old system was underlined by the public humiliation of landlords and the venting of grievances by peasants led by communist cadres. Politically, the richer classes were isolated; economically, land redistribution among 300 million peasants stimulated their willingness to increase production.

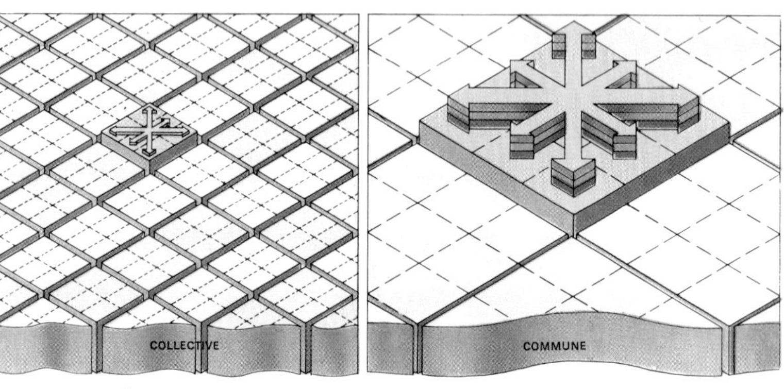

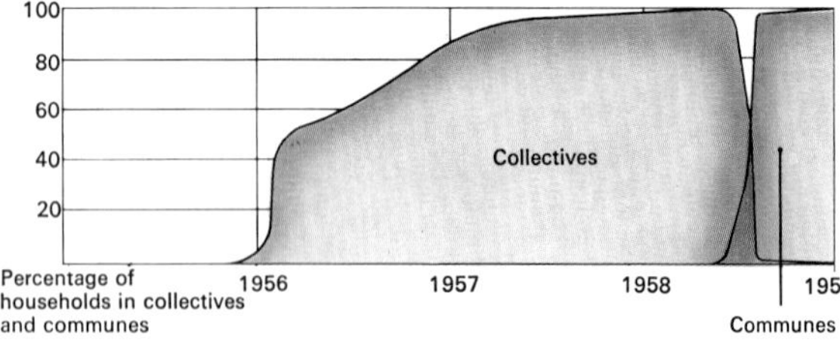

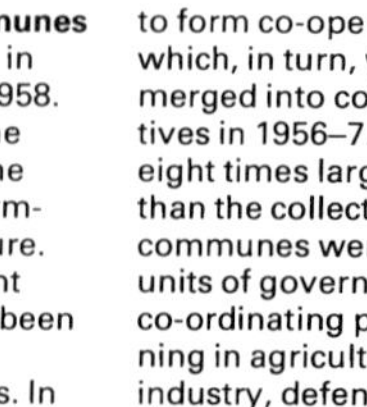

3 People's communes were introduced in the summer of 1958. This was to be the culmination of the socialist transformation of agriculture. In 1953–4 peasant households had been organized into mutual aid teams. In 1955 these merged to form co-operatives which, in turn, were merged into collectives in 1956–7. About eight times larger than the collectives, communes were also units of government co-ordinating planning in agriculture, industry, defence and education.

4 Backyard furnaces and foundries epitomized the Great Leap Forward, a drive launched in February 1958 to accelerate expansion of the Chinese economy. By mobilizing underemployed rural labour in small, labour-intensive industries it was intended to complement the production of urban-based, capital-intensive industries at little extra cost to investment funds. Called "walking on two legs", this strategy of economic development was widely promoted.

tered by launching a campaign to reverse a deteriorating ideological situation and a weakening in his personal influence. The campaign, the Cultural Revolution, aimed on the one hand at purging the CCP and on the other at ridding China of aspects of traditional culture incompatible with socialism. Party members were ousted and the state structure usurped by revolutionary committees in circumstances that sometimes led to violence. The key to Mao Tse-tung's success was his ability to mobilize support, especially from the young people, [7], coupled with the loyalty of the armed forces.

During the Cultural Revolution Mao Tse-tung presided over the rebuilding of the CCP and the mass organizations, a restoration of the state system, a restructuring of the education system and a reassessment of Chinese culture. The spilling over of the excesses of the Cultural Revolution into foreign affairs damaged China's international position for a while. Some 45 divisions of Soviet forces were deployed along the frontier, giving rise to armed clashes in 1969 [8]. China's foreign relations now became marked by alignment with the Third World, friendship with the medium-sized developed countries, trade and diplomacy with Japan and the USA and continuing confrontation with the USSR.

Admission to the United Nations

The success of China's new foreign policy was characterized by her admission to UN membership in 1971 and by a visit by the American president, Richard Nixon (1913–), in 1972 [9]. The eclipse of Lin Piao (1908–), the defence minister and Mao's heir apparent, who was reported killed during a flight to the USSR in 1971, suggests that an accommodation with capitalism at the expense of a reconciliation with the Soviet bloc was not unanimously approved. Nevertheless, China moved to the Fourth National People's Congress in January 1975 (the first for a decade), a new constitution and, for the first time since 1966, a fully manned state structure. Mao died in 1976 and was succeeded by Hua Kuo-feng. In the disturbances that followed, Mao's widow and some other prominent politicians were arrested and accused of treason.

KEY

Mao Tse-tung, as Chairman of the Chinese Communist Party and chairman elect of the government, stood in Tien An Men Square in Peking to proclaim the establishment of the People's Republic of China on 1 October 1949.

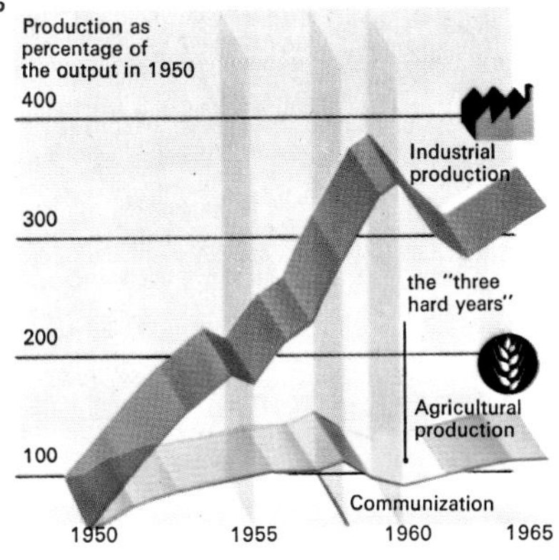

5 During the Great Leap Forward agricultural and industrial output dropped. Inadequate planning and accounting led to miscalculation of potential yields and failure to meet the targets set. Lack of experience and disorganization meant that many communes were ill-equipped and badly run. The worst weather for a century in 1959–60 led to economic crisis and the policies of the Great Leap Forward were shelved.

6 Exemplary production units singled out by Mao Tse-tung in 1964 were the Tachai agricultural brigade in Hsiyang county, Shansi province, and the Taching oilfield in Heilungkiang province. In Tachai [A] peasants transformed a poor environment and increased grain output without state aid or material incentives. In Taching [B] workers created a prototype agro-industrial community developed without foreign aid by reliance on their own technological innovations. Both show the importance attached to self-reliance, hard work and persistence in Chinese economic development after 1960. Then, as a result of frequent ideological differences, the Soviet Union withdrew her many technicians and cancelled all her aid programmes to the Chinese People's Republic.

7 The Little Red Book of quotations from Mao Tse-tung became the "bible" of the Cultural Revolution of 1966–8. It was studied on a nationwide basis as a pocket guide for action in any set of circumstances. But it was put to most use in the hands of young people, particularly Red Guards recruited from middle schools, universities and factories. As "successors to the revolution", they formed a main force in the campaign by Mao Tse-tung and his supporters against Liu Shao-ch'i, then head of state, and aspects of traditional culture standing in the way of Maoist policies.

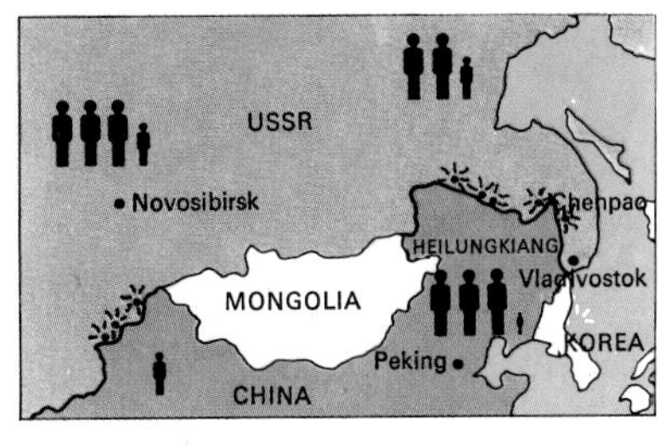

8 Border clashes between Chinese and Soviet forces on the Ussuri River frontier in Heilungkiang in March 1969 showed the extent to which Sino-Soviet relations had deteriorated in the course of the ideological disputes of the late 1950s. After the worst fighting over Chenpao or Damansky island, China claimed that the Soviet Union had provoked 4,189 incidents.

9 The visit of Richard Nixon, the US president, in 1972 marked a new era in China's foreign relations. Less hostile Sino-US attitudes had indirectly contributed to the admission of the People's Republic of China to the UN in 1971. It also led to better relations between China and Japan and increasing diplomatic isolation of the Nationalist government of Chiang Kai-shek in Taiwan.

Decolonization

Decolonization has been one of the greatest transforming processes in the world since 1945. A new word in the political vocabulary, it has achieved widespread usage and currency only since the middle 1950s as far-flung colonies have gradually achieved independence from their rulers.

Processes of decolonization

The term decolonization covers a wide range of processes by which power is transferred from the departing colonial authority to the newly independent nation. To date, transfer has usually been peaceful and by agreement – for example from Britain to Ceylon (now called Sri Lanka in 1948, Ghana in 1957 and Jamaica in 1962. In a few but important instances, strife has been an integral part of the process of decolonization but was not directly connected with the issue of independence – the Mau Mau emergency in Kenya, the *enosis* dispute in Cyprus, and British confrontation with Indonesia over the creation of the Malaysian Federation. In some of the best-known examples of decolonization, independence has been wrung by force from a reluctant colonial power – from The Netherlands in Indonesia in 1949, and from France in North Vietnam in 1954 and in Algeria in 1962 [6]. In the Congo in 1960, the Belgians granted independence to a territory that was wholly unprepared for it, and bloody chaos ensued [5]. But there can be two-way effects of decolonization – as in Portugal in 1974 when internal dissent and colonial unrest resulted in a revolution that hastened the independence of its colonies in 1975.

The process of decolonization, and the consequent emergence of new states, has resulted in major changes to the political map of the world [1]. In 1914 there were only eight sovereign states in the whole of Asia and Africa, and of these only Japan was regarded as a power of real account in world affairs; almost everywhere else throughout those continents the rule of dominating influence of the West Europeans, the Russians or the Americans prevailed. Only since World War II has the great retreat from and dismantling of the overseas empires of the West Europeans come about, first in Asia in the late 1940s, and then only slightly in North Africa in the early 1950s. After that, decolonization gathered pace, was in full flood between 1955 and 1965 [3], and eventually reached the Pacific and parts of the world that were once remote.

The quickening pace

Most of the principal overseas empires of the West European powers were already dissolving when the fifteenth session of the UN General Assembly began in September 1960 and an Anti-Colonialist Charter drawn up by 43 African and Asian countries was adopted without dissent. The British Empire was moving into a state of more or less voluntary liquidation: India [Key], Pakistan, Burma, Ceylon, Ghana, Malaya, Cyprus and Nigeria [4] had become independent. "Empire-into-Commonwealth" was an accomplished but continuing fact, although the wider problem of the role of the white man remained unresolved in the apartheid regime of South Africa and in Rhodesia. In his forthright "wind of change" speech to the South African Parliament in February 1960, the British prime minister, Harold

CONNECTIONS

See also
254 Non-alignment and the Third World
302 Underdevelopment and the world economy
292 The wars of Indochina
220 The Commonwealth
218 British foreign policy since 1914
134 The story of the West Indies
216 Indian nationalism
130 Imperialism in the 19th century
140 India in the 19th century
142 Africa in the 19th century

1

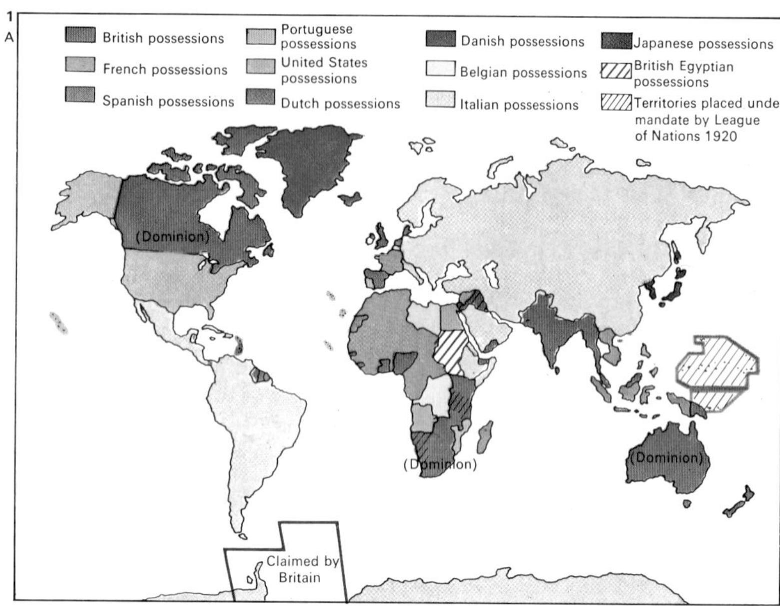

1 In 1926 there were more than 80 separate colonies and dependencies [A]. These comprised over 33% of the population and land area of the world. Seven West European countries (Britain, France, The Netherlands, Belgium, Portugal, Spain and Italy), whose total home population was about 200 million, controlled about 700 million people in overseas colonies. The British and French empires were by far the largest. Most of the new states of the post-1945 world [B] have come from these two empires. While the British Empire was truly worldwide, the French was predominantly in Africa and Indochina.

2

2 The election of U Thant of Burma as UN Secretary-General, after the death of Dag Hammarskjöld in the Congo in 1961, symbolized the growing number of voices and votes of new and non-aligned states in UN affairs, especially of Asian and African members. UN membership is valued by all newly independent states as an important symbol of their status.

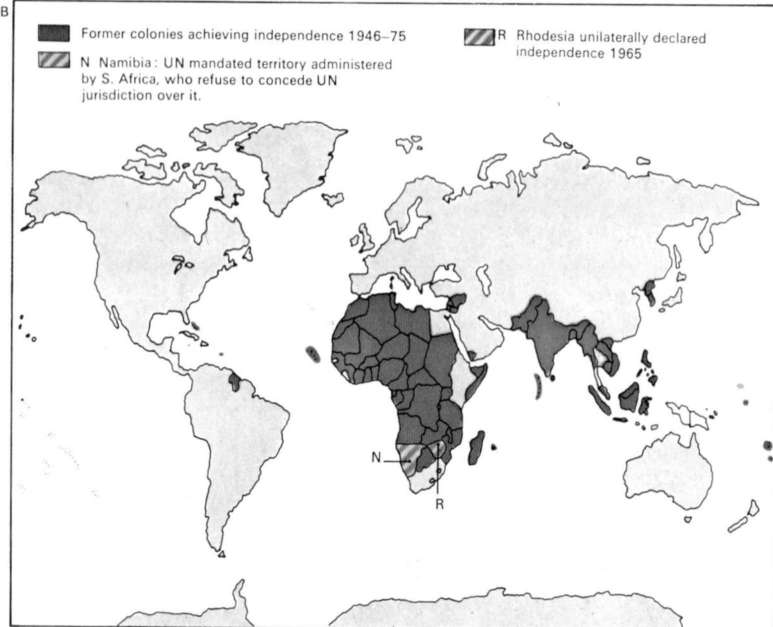

3

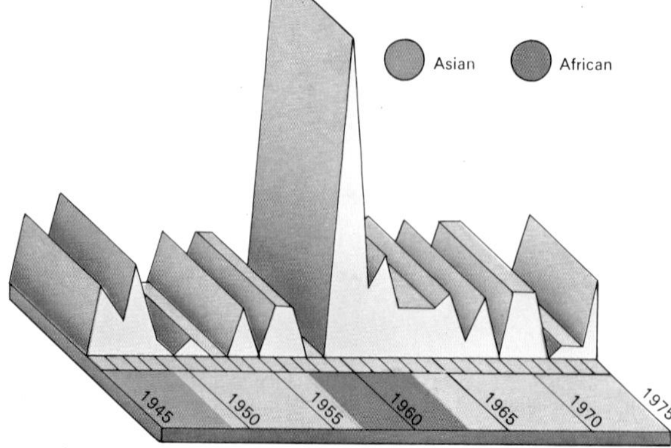

3 Decolonization had three main phases. First, from 1944 to 1949, it occurred in the southern flanks of Asia – Lebanon, Syria and Israel; then India, Pakistan, Burma, Ceylon; and the Philippines and Indonesia. From 1950 to 1956 little decolonization took place. Libya, Morocco and Tunisia became independent peaceably, and Algerians began the war for independence that ended in 1961. From 1956 to 1963 African decolonization got rapidly under way, with the Sudan in 1956, Ghana in 1957 and Guinea in 1958. In 1960 all of the French African colonies became independent, plus Nigeria and Belgian Congo.

Macmillan (1894–), had rightly predicted that the rate of decolonization was quickening. In the same year France's colonial presence was to shrink considerably in Africa and soon to disappear completely. So, too, was that of Belgium from the Congo. Of the West European powers, only Portugal continued to insist that its mission in its territories overseas was permanent, although the revolution of 1974 brought a sudden change of attitude.

Adjustment after decolonization

The whole period of decolonization, now virtually over, has created acute problems of adjustment for both former rulers and ruled. Some ex-imperial powers – notably Britain – have found the transition to lesser power status and a lower world standing acutely uncomfortable. Only since her decision to stay in the EEC, and with the Commonwealth discussions on world economic issues in June 1975, has Britain begun to find a new role as intermediary and honest broker between rich and poor, developed and developing countries. Most other colonial powers have domestic difficulties over decolonization.

The new states themselves have had to evolve political systems appropriate to their new situation and not necessarily those bequeathed by the outgoing authority [7]. Thus the abandonment of parliamentary constitutions in favour of one-party systems; the rejection of Russo-American models of development through industrialization in favour of the Chinese model of concentration on agriculture and self-sufficiency; and the adoption of a foreign policy independent of the decolonizing power can all be seen as a continuing the process of decolonization.

But if colonialism is almost dead "neo-colonialism" is alive. The United States, historically the greatest advocate of anti-colonialism, is also the country most often charged with "neo-colonialism". It may take the form of economic control through multinational corporations, military influence through arms aid and advisers, or even political "destabilization" as practised against the Marxist regime of President Allende in Chile. The Soviet Union is accused in similar terms, chiefly by China.

KEY

The inauguration of Earl Mountbatten as Viceroy of India in 1947 prefaced her independence from Britain later in the same year. This event symbolized the advent of the age of decolonization, carried out with a formal transfer of power.

4

4 Nigeria achieved independence peaceably from Britain in 1960. Power was handed over to a working federal parliament and government. But six years later, Nigeria suffered two military coups in one year and a bloody, but unsuccessful, attempt to create a new secessionist state of Biafra.

5

5 The Belgians' abrupt departure in 1960 from the Congo (now called Zaïre) led to bitter civil war, much bloodshed and to the attempted, but ultimately unsuccessful, secession of the copper-rich province of Katanga. The introduction of a UN peace-keeping force caused great controversy.

6

6 Algeria is one of the few countries since 1945 to have won independence by means of a successful war against France. This lasted from 1954 until 1962. De Gaulle, who had returned to power backed by the slogan "Algérie Française", conceded independence in July 1962. Algeria then began to play an active part in Arab League affairs and later on, as an oil-producing country with limited reserves, within OPEC. A number of important Afro-Asian and non-aligned conferences have been held in Algiers, especially the short-lived Afro-Asian meeting of October 1965 and the 1973 summit.

7

7 Most independence day ceremonies for the newly independent states may at first seem to involve only changes of personnel, who in style and outlook often resemble their predecessors. They may even wear wigs and carry maces. But parliaments, British or French style, are not always resilient institutions and often give way to military rule. For most of these new-state societies their sense of community goes deeper than their constitutionalism. Almost all Third World societies are pluralistic, with deep economic and racial differences which stand in the way of political stability and sustained economic progress.

Australia since 1918

Australia lost nearly 60,000 men killed during World War I. This gave authority to its representation at the Paris Peace Conference, where the prime minister, William Morris Hughes (1862–1952) [1], successfully defended the "White Australia" policy (a government policy that restricted the entry of non-Europeans into the country) and obtained for Australia a "C" class League of Nations mandate over the former German colony in northeastern New Guinea. The territory became a United Nations trusteeship after World War II, was administered jointly with Australia's colony of Papua, and both became independent as the joint state of Papua New Guinea in 1975.

Economic and political changes

Manufacturing industry in Australia, stimulated by World War I, received continuing protection through tariffs supported by a trade union movement that was growing in strength, ideology and militancy. But tariffs did not help the farmers. In 1922, the new Australian Country Party (whose basic policy is aimed at increasing the effectiveness of primary industries) won enough seats to depose Hughes and join the Nationalists in government. With two short exceptions, such coalitions with the Nationalists — later renamed (with a Labor rump) the United Australia Party, and later again the Liberal Party — have held office for 30 of the 44 years to 1977. Labor (which is committed to "democratic socialization of industry, production, distribution and exchange") came into office for one three-year term during the economic depression [2] and not again until 1941.

The wage-price spiral of the 1920s restricted the opportunities for industrial exports [3], so that the economy continued to be dependent on rural exports vulnerable to world price fluctuations. Government appeals to Britain for "men, money and markets" were partly satisfied by migrant settlers [4], preferred access to the London capital market and imperial trade preferences that were rationalized at a conference in Ottawa in 1932.

At the 1926 Imperial Conference, the British dominions, including Australia, were declared to be a state of equal status with Britain, and this was formalized in the Statue of Westminster, 1931.

Foreign policy expedients

Since early in the century, Australians and their governments had feared attack or absorption by an Asian power, especially Japan. Allied to Britain from 1902 until 1921, Japan had been a helpful if slightly ambiguous partner during World War I. Its expansion into Manchuria in 1931 fed Australian fears. But Japan had also become a major new export market for Australia [9], which thus encouraged rather than opposed Japan's aggression.

Australia's attitudes to foreign affairs during the 1930s tended to copy those of Britain: sanctions against, then appeasement of, Italy; eyes averted from Japanese aggression in China; and appeasement of Germany until 1939.

Australia entered the war against Germany one-and-a-quarter hours after Britain, and sent forces to the Middle East, Greece, the United Kingdom and South-East Asia.

CONNECTIONS

See also
126 Australia and New Zealand to 1918
190 World War I: Britain's role
230 World War II: Britain's role
220 The Commonwealth
250 New Zealand since 1918
218 British foreign policy since 1914

1 W. M. ("Billy") Hughes was born in London, but went to Australia at the age of 20. He became involved in union politics and served as prime minister from 1915 to 1923 as leader of a Nationalist Government. He was a notable wartime leader and an astringent and turbulent politician, but failed to win support for his policy of conscription for war service overseas. He was expelled from the Labor Party and founded the United Australia Party.

2 The world depression of 1929–33 seriously affected Australia, whose economy was almost entirely dependent on exports of primary products, and on British loans for industrial development. The Labor Government had to accept the orthodox economies of the day, such as wage restraint, public spending cuts and high unemployment. The economy did not recover fully until the late 1930s. This linocut was produced in 1933 by the group of Workers' Artists.

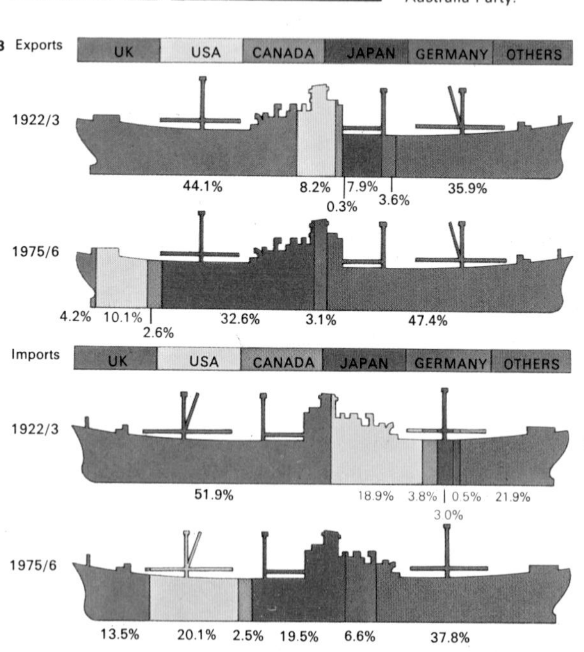

3 The traditional pattern of Australian trade, based on the export of wool, wheat, meat and minerals, and the import of consumer durables and industrial goods, began to change in the 1940s. There was also a shift away from the old markets and in particular from Britain, which had dominated trade earlier in the century. The main new markets were in Asia, and especially Japan, which in 1966 became Australia's largest customer. Britain's entry into the EEC in 1973 seriously affected the Australian export industries, especially that of dried and canned fruits, and dairy products.

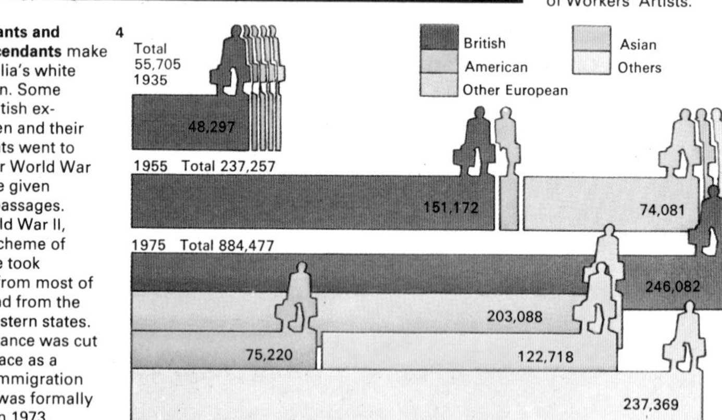

4 Immigrants and their descendants make up Australia's white population. Some 80,000 British ex-servicemen and their dependants went to settle after World War I and were given assisted passages. After World War II, another scheme of assistance took refugees from most of Europe and from the Middle Eastern states. The assistance was cut in 1974. Race as a factor in immigration selection was formally dropped in 1973.

5 Australia's mineral resources are among the world's greatest. Although mining began in the 19th century, its expansion since World War II has been fundamental to the country's economic growth. Most of the extractive industries are concentrated in Victoria and New South Wales. Many valuable minerals are found in Australia, including gold, silver, platinum, uranium, lead and copper, as well as iron ore and coal, which is here being exploited in an open-cast mine.

Australians played an important part in helping to clear North Africa of Axis forces; in providing aircrews for the RAF in Britain and Europe; in naval actions in many theatres; and in stopping the Japanese southward thrust in New Guinea. The nation was an essential granary and major supplier for Britain and for American forces.

Industry grew rapidly during the war. So did the power of the central government, which assumed sole control over income taxation. After the war, restricted imports (for a period), a high rate of investment, and overseas demand for primary products during and after the Korean War brought boom conditions. Rising living standards continued until the early 1970s, stimulated by discoveries of massive mineral deposits [5].

Postwar realignment

A conservative Liberal-Country parties government under Sir Robert Menzies (1894–   ) [7] was elected in late 1949, and remained in office for 23 years. During that time, Australia initiated the Colombo Plan for economic and technical co-operation in South and South-East Asia and became an ally of the United States under various security arrangements, including ANZUS (Australia, New Zealand, United States) and SEATO (South-East Asia Treaty Organization). It saw the need to encourage both Britain and the United States to remain committed to the security of South-East Asia, and sent forces to Malaya to help combat communist terrorists, and after the 1963 Indonesian "confrontation". Troops were also sent to Vietnam to fight under American command there [8].

By 1966 Japan had become Australia's most important customer, notably for minerals, the two economies becoming to a degree interdependent. The Labor Government of Gough Whitlam sought to weaken ties with the United States, and spent massively on education and a national health service. But it was hard hit by the recession of 1974, and Whitlam was removed from office in 1975. A conservative government was elected to cope with unemployment and inflation, which had begun to affect the Australian easy way of life [10].

KEY

Sydney Opera House was opened in 1974 and has come to represent the growing cultural identity of Australia in the modern age. Designed by the Danish architect Joern Utzon (1918–   ) in response to a competition in 1956, its cost increased far above original estimates because of engineering problems caused by its revolutionary design. Extra money to pay for the building was raised by lotteries. The final cost was more than A$100 million.

6

6 Isaac Isaacs (1855–1948) was the first Australian-born governor-general of the country on his appointment in 1931. The governor-general has been the representative of the Crown in Australia since Federation in 1901, and he normally takes a purely nominal constitutional role in the country's politics. The office became controversial in 1975 when the incumbent John Kerr (1914–   ) dismissed the Labor prime minister Gough Whitlam (1916–   ), although he had a working majority in Parliament. In the ensuing election it was claimed that the action was undemocratic, and led to demands for the abolition of the office of governor-general. But Labor was defeated at the polls.

7

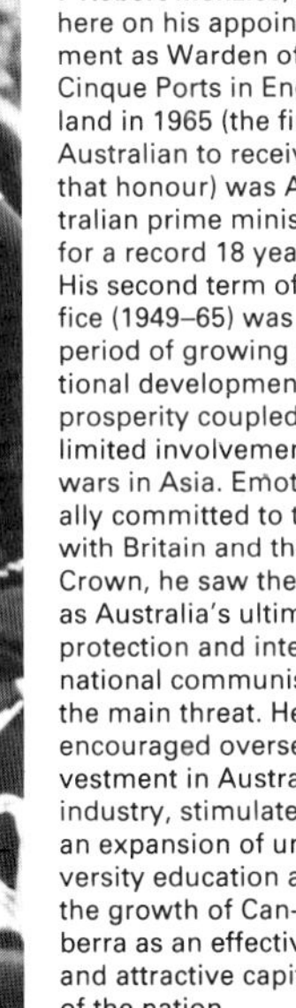

7 Robert Menzies, seen here on his appointment as Warden of the Cinque Ports in England in 1965 (the first Australian to receive that honour) was Australian prime minister for a record 18 years. His second term of office (1949–65) was a period of growing national development and prosperity coupled with limited involvement in wars in Asia. Emotionally committed to ties with Britain and the Crown, he saw the USA as Australia's ultimate protection and international communism as the main threat. He encouraged overseas investment in Australian industry, stimulated an expansion of university education and the growth of Canberra as an effective and attractive capital of the nation.

8

8 Australia's defence links with the USA prompted Menzies to respond to American and South Vietnamese appeals to help defend South Vietnam against communist insurgents and North Vietnamese attacks. A number of training advisers were sent in 1962 and 1964 and the first infantry battalion went in 1965, to be followed by two more battalions, supporting arms and services, air force and naval elements, under overall US command. The main Australian ground operations were in Phuoc Tuy province. Initially there was little opposition to the war, but as it dragged on, the issue of conscription became controversial and divided the nation.

9 Port Hedland, north Western Australia, is a centre for iron-ore export. In 1971, about 25 million tonnes of iron ore was exported to Japan. Many of the Australian extractive industries are financed by Japanese investment. But imports of Japanese manufactured goods are still low, and Australian investment in Japanese industry is still more substantial than the Japanese contribution to Australian industry.

9

10

10 Australia's open spaces, long hours of sunshine and high standard of living mean that life is healthy and often out-of-doors. The typical home is a bungalow on a quarter-acre plot of land, with a barbecue (shown here) in the back garden. Rising costs of beef and lamb may eventually limit the Australians' propensity for meat-eating, but in the late 1970s their calorie consumption was among the highest in the world. Sport, including skiing in the winter and swimming or cricket in the summer, is important to many Australians.

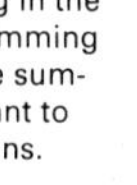

New Zealand since 1918

New Zealand in 1918 was still virtually an outpost of Great Britain. Although it had been an independent, self-governing Dominion since 1907, its economy depended entirely on the British market for agricultural exports, its defence policy relied on British naval protection, and its self-awareness was European if not exclusively British in character. Half a century later the Pacific Basin had become a focus of its economic, strategic and cultural attention.

The aftermath of World War I

The prosperity of the World War I years, when Britain bought all the wool, meat, cheese and butter that New Zealand could produce, continued for a brief period into the peace. Confident that "you couldn't go wrong farming", New Zealanders indulged in a bout of land speculation helped along by £23 million of government loans to ex-servicemen.

When export prices collapsed in 1921, so did the land boom. Many farmers, unable to meet mortgage repayments, sold out at giveaway prices; others left their farms derelict. Despite the reality of rural poverty, there was a positive outcome. The universal shortage of credit, and the unpreparedness of the government to meet the economic downturn, drove farmers to co-operate in the setting-up of marketing-boards, for meat in 1922, for dairy products in 1923.

Townsmen were even worse hit by the slump. The economy deteriorated further in the early 1930s, with the onset of worldwide depression. Unemployment became an epidemic; wages were savagely forced down, while diminished government revenue resulted in heavy cuts in public spending. Public discontent was expressed in riots in Auckland and Wellington [1].

The economy and the Labour Party

The political consequences were no less dramatic. In the 1935 general election, the Labour Party assumed office for the first time. Elected mainly on the votes of small farmers and town workers, Labour announced, and in large measure carried through, an extensive welfare programme [2]; higher wages and shorter hours for the town worker, pensions and benefits for the old, the widowed and the disabled, and a state rental housing scheme of subsidized accommodation [3]. Farmers were guaranteed a minimum price for however much they produced.

The first Labour government held office for 14 years. On the whole they were good years for the government and the economy. The world began to emerge from the Depression after 1935, and when another downturn threatened in 1938-9, World War II boosted demand for food exports. The war also fostered the extension of government economic controls.

In 1949 Labour lost power to the more free-enterprise inclined National Party, led by Sidney Holland (1893–1961). Since then, the two parties have alternated in office. National, led by Holland, Keith Holyoake (1904–), John Marshall (1912–), and Robert Muldoon (1921–), have held office in 1949–57, 1960–72, and since 1975. Labour, under Walter Nash (1882–1968), Norman Kirk (1923–74) and Wallace Rowling (1927–) have been in

CONNECTIONS

See also
126 Australia and New Zealand to 1918
124 Colonizing Oceania and Australasia
190 World War I: Britain's role
230 World War II: Britain's role
220 The Commonwealth
292 The wars of Indochina

1

2

1 The army and navy had to be called in when violence broke out in New Zealand cities during 1932, the low point of the Depression. Registered unemployment was 80,000 in a population of 1,500,000.

2 Labour's social security system, set up in 1938, included a medical and hospital service. All were guaranteed treatment, irrespective of ability to pay. The local branch of the British Medical Association objected to the scheme and *New Zealand Herald* cartoonist Minhinnick made this comment on Labour's answer. Principals are Prime Minister Michael Savage and Minister of Finance Walter Nash.

3 Since March 1937, when the state housing scheme came into effect, New Zealand has built more than 77,000 state houses or flats, to standards shown by these, which were built in 1938 in Wellington. The houses are allocated on a basis of need, formerly measured by income but since 1973 by a more complicated points system. The state also offers low-cost mortgages to young couples.

3

4

4 A "friendly invasion" by the United States Marines occurred after the long-standing New Zealand nightmare of a Japanese invasion seemed about to be realized in 1941. With the New Zealand Division serving in North Africa, only a handful of badly equipped reservists were available to repel an invasion. But the Battle of the Coral Sea on 7-8 May 1942 stopped the Japanese advance. The Americans, some of whom are seen here with a Maori girl at Rotorua, North Island, retained training camps in New Zealand until the end of 1944.

power in 1957–60, and 1972–5. Both parties are committed to full employment, high state spending and export promotion. Both recognize that welfare services are becoming increasingly expensive and difficult to finance out of general revenue. Both are aware of the social problems brought to inner city areas by an increasing Maori population, and by the immigration of Pacific Islanders. For both, the dominant economic problems have been to encourage local industry without over-protecting it, and to restrain the country's propensity to spend more on imports than it makes from the agricultural products that earn more than 80 per cent of its export income.

Fewer ties with Britain

World War II also brought home the lesson that New Zealand's defence could no longer be based on British protection. In 1939 as in 1914, New Zealand prepared for a faraway war. A division was sent to the Middle East, naval and air force units and men were dispatched to Europe. The home economy was dominated by the "Food for Britain" slogan. But the entry of Japan in 1941 posed a threat to the homeland. As the Japanese advanced southwards, New Zealand had to rely entirely on the power of the United States [4]. In 1951 New Zealand joined ANZUS, the defensive alliance with Australia and the United States. It sent small forces to Korea in 1950 and to support the Americans and Australians in Vietnam in 1965.

Pacific interest and commitment has been a continuing feature of the postwar period. In the early 1950s New Zealand started to co-operate with the Colombo Plan in supplying various forms of aid, chiefly agricultural, to South Asia. New Zealand troops served in the Malayan emergency of the 1950s and co-operation with Malaysia and Singapore continues through the Five Power Defence Arrangements (1971).

Since Britain entered the EEC in 1973 the need to diversify both markets and products has been starkly apparent. By 1976, although Britain was still New Zealand's largest single customer, almost half the trade was with the Pacific Basin, and the proportion was increasing [6,7].

KEY

New Zealand's tourist attractions, some of which are shown here, often have a dual economic value. Thermal regions, for example, not only attract tourists but also power geothermal electricity stations; total fish exports totalled more than $19 million in 1974; and even deer, culled as a pest and for sport, contribute to overseas income – venison worth nearly $7 million was exported in 1974. En route to becoming one of the world's main primary producers, as well as a Pacific playground, New Zealand followed a path similar in some respects to that of the United States in its frontier stage. But isolation and a small population have strongly influenced its development.

5

5 Rugby (here the Maori All Blacks play the Lions in 1950) has been called the religion of New Zealand. The All Blacks who toured Britain in 1905 lost only one game to Wales. New Zealanders still dispute the deciding try.

6

6 Kinleith pulp and paper mill produces 200 million tonnes of paper and 130 million tonnes of pulp each year. It is one of six such plants. Japan is the largest customer, taking almost half the output.

7 New Zealand milk powder unloaded in Brazil exemplifies the useful outlet that Latin America has become in recent years for dairy products and some meat. Intensive efforts by New Zealand to find new export markets such as these began even before Britain joined the EEC in 1973. Of NZ's 1967-8 dairy exports (476,000 tonnes), 301,500 went to the UK. In 1975-6 the totals were 458,000 and 149,000.

7

8

8 Air New Zealand, a development of Tasman Empire Airways Ltd, flies DC-10 aircraft on profitable routes to Los Angeles, Hong Kong and Singapore, but a third of its annual 3 million passenger kilometres (nearly 2 million passenger miles) is still flown across the Tasman Sea. The 1,200-mile flight is made by tourists, businessmen and migrant workers.

9

9 Kiri Te Kanawa (1948–) is, with Inia Te Wiata (1915–71) and Donald McIntyre (1934–), one of several New Zealand opera singers who have acquired an international reputation since the war. She is under contract to the Royal Opera House and has sung at the Paris Opera and the Metropolitan Opera, New York.

Southern Africa since 1910

When World War I broke out in 1914 Rhodesian police immediately occupied the Caprivi Strip in German South West Africa and Union forces immediately destroyed German coastal wireless stations. A pro-German rebellion in Transvaal was speedily crushed. On 9 July 1915 German South West Africa surrendered to Louis Botha (1862–1919). On 4 September 1916 Jan Christiaan Smuts (1870–1950) [3] took Dar es Salaam and most of German East Africa, operations which ended only in 1918.

Wealth and prosperity

In 1918 the Union of South Africa was the lodestone for all southern Africa, importing labour even from Mozambique. In gold [5] and diamonds [2] it had a wealth unique in Africa, and an era of prosperity seemed ahead. The collapse of the postwar boom impoverished White urban workers and Afrikaner farmers, and in 1924 brought to power a coalition of Nationalists and the Labour Party. In 1923 Rhodesian settlers obtained self-government, following a referendum in the previous year rejecting union with South Africa. In general a new African élite was emerging, teachers, preachers, clerks, some traders and some farmers. In the Union of South Africa the African National Congress was founded in 1913, with parallel organizations in Southern Rhodesia in 1934, Nyasaland in 1943, and Northern Rhodesia in 1949. The pass laws restricting the free movement of Africans and refusal to recognize their trade unions were bitterly felt African grievances.

Smuts, prime minister of South Africa 1919–24 and 1939–48, played a major role in World War II on the side of the Allies, thereby losing the support of the neutrals within South Africa. The French African territories were promised independence in 1944, and the independence of Burma, Ceylon, India and Pakistan taught their own lesson. Following Smuts's defeat at the polls in 1948, government attitudes towards race became more aggressive. The word apartheid, coined in 1929 to express separate white and African development, now took on a new meaning with the installation of Daniel Malan (1874–1959) as prime minister (1948–54) and leader of the Nationalist Party, which has remained in power since 1948. "Race" meant Afrikaner dominance, and the expansion of the economy after 1948 gave apartheid a specious seal of success. White South Africans enjoyed one of the highest standards of living in the world but the average monthly wage for a White was thirteen times that for a Black.

International tensions and UDI

In 1953 a Federation of Central Africa – Northern and Southern Rhodesia and Nyasaland – was brought into being by Britain in spite of considerable African opposition. That opposition brought the Federation to an end in 1963. Meanwhile the Union, under Hendrik Verwoerd (1901–66) as minister for native affairs, and then prime minister 1958–66, sought a solution to African antagonism by creating Bantustans, where Africans could eventually develop autonomous African states. At Cape Town on 3 February 1960 the British prime minister Harold MacMillan (1894–) made his "wind of change" speech, in

CONNECTIONS

See also
128 South Africa to 1910
142 Africa in the 19th century
190 World War I: Britain's role
220 The Commonwealth
218 British foreign policy since 1914

1 The fibre-producing sisal plant was originally a native of South America, but was smuggled into German East Africa in 1891. Only 41 plants survived the journey and from these all the sisal plantations of eastern and southern Africa, an important export crop, descend.

2 Diamonds were first discovered in South Africa near the Orange River in 1867, but these workings were soon surpassed in wealth by the dry diggings at Kimberley – the Great Hole – In 1871. The mines in South Africa and Namibia (shown here) are the wealthiest in the world, with an annual output of more than £60 million in value.

3 Jan Christiaan Smuts is one of the major political figures in the history of South Africa. He helped to formulate the 1910 constitution and was prime minister twice. He played leading roles in the League of Nations and the United Nations.

4 Fort Hare University was the first college to be opened to non-Whites (in 1916), and in 1969 it was restricted to members of the Xhosa tribe only. Other similar universities are those of Zululand, for Zulus only, the North, for Tsonga, Sotho and Venda only, one for Coloureds in Cape Town, and one for Indians in Durban. In addition South Africa has nine universities for Whites only. In 1972, 2.1 per cent of the White population and 0.2 per cent of the non-Whites managed to achieve university education.

5 Gold is the basis of South Africa's wealth. The precious metal was mined in Rhodesia from earliest times and its discovery at Witwatersrand in 1886 produced one of the world's great gold rushes. That goldfield, located in the Transvaal province of South Africa, is still the world's richest. By 1910, gold amounted to 59% of South Africa's exports. It has attracted enormous foreign investments and has encouraged the development of the railways as well as a number of manufacturing industries.

which he condemned apartheid and demanded that legitimate African aspirations to be recognized. On 21 March the Pan-African Congress demonstrated at Sharpeville, and a massacre ensued [7]

On 31 May 1961 the Republic of South Africa came into being, having withdrawn from the British Commonwealth. The Republic was also expelled from many international organizations that found South African presence distasteful. Despite strenuous diplomatic efforts, South Africa remained almost friendless: racial discrimination, police brutality, imprisonment without trial – all brought their consequences. In 1964 Northern Rhodesia and Nyasaland became independent as Zambia [6] and Malawi, while the white government in Rhodesia (formerly Southern Rhodesia) moved away from its previous attempts to provide for limited African political involvement. On 11 November 1965, having been refused independence without majority rule by Britain, Rhodesian prime minister Ian Smith (1919–) unilaterally declared independence (UDI). Britain declined to use force but joined with United Nations in sanctions [8], which have largely been evaded by South African aid. In the early 1970s guerrilla operations, at first scattered, escalated and posed a serious threat to the white regime. Britain continued to make efforts to conciliate the parties.

Developments among the smaller states

Britain gave independence to Basutoland and Bechuanaland as Lesotho and Botswana in 1966, and to Swaziland in 1968. Both Lesotho and Swaziland are heavily dependent upon the Republic. In 1976 the Republic purported to grant independence to Transkei, but under conditions such that no other country accorded it sovereign recognition. In 1967 the UN General Assembly had declared the continuation of the South African mandate over South West Africa unlawful, and had appointed an administrative council for it as Namibia. This action was ignored by the Republic, which gave it a parliament elected on a slender franchise together with the promise of independence at the end of 1977.

KEY

The Houses of Parliament, Cape Town, are the legislative centre of South Africa; the administrative capital is Pretoria. Cape Colony first enjoyed self-government under the constitution of 1872; this was superseded by the British South Africa Act, 1909, which established the Union of South Africa in 1910. The policy of apartheid made important changes to the constitution. Originally there was no restriction of race or colour for voters, but in 1956 the clauses of the South Africa Act protecting Coloured voters were abrogated and parliament has since been elected by Whites only.

6

6 Kenneth Kaunda (1924–), president of Zambia since that country's independence in 1964, was originally a schoolmaster. He became a district secretary of the African National Congress in 1950 and was twice gaoled for his political activities. Later he became prime minister of Northern Rhodesia. His book, *Humanism in Zambia and its Implementation* (1967), explains the theoretical and practical aspects of his moderate socialist policy.

7

7 The Sharpeville massacre, in March 1960, happened when the police opened fire on a demonstration against the discriminatory laws passed by the government, killing 67 and wounding 186 African demonstrators. The racial policies of the Afrikaner Nationalist Party, in power since 1948, stimulated African unrest. From 1952 onwards the African National Congress organized agitation against the legislation aimed at non-Whites.

8

8 Escorted convoys between Rhodesia and South Africa were organized because of guerrilla attacks by African nationalists, whose operations by 1976 were seriously threatening the security of the white regime in Rhodesia. Apart from being Rhodesia's only land link for trade, South Africa has played an important role in supporting the white regime while seeking a majority Rhodesian government with which it could co-operate.

10

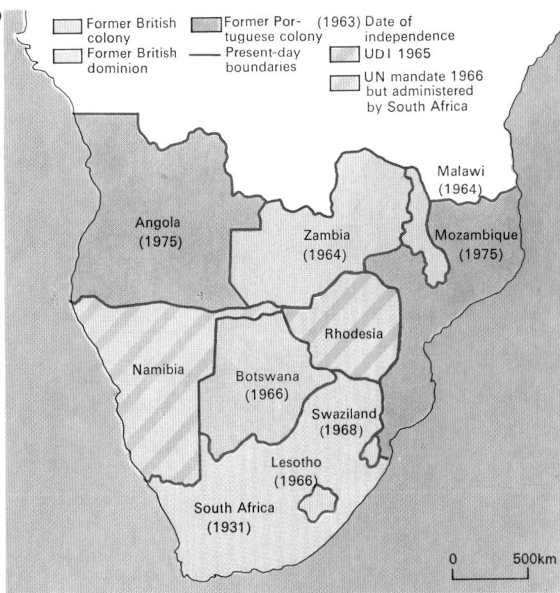

9

9 Rioting erupted in June 1976 at Soweto, a black township on the outskirts of Johannesburg, when thousands of black youths protested against teaching in Afrikaans as well as English in Bantu schools. Savagely repressed with 176 killed and 1,222 wounded, the unrest spread into an unprecedented wave of defiance that forced some reforms in townships and the rescinding of the language order. The riots showed a new African militancy.

10 Although most countries achieved independence peacefully in southern Africa, there remain a number of points of conflict within the current political map. In Rhodesia (Zimbabwe) UDI was only a culmination of the growing disagreements with Britain's resolve to transfer power to a majority government. Since then a number of diplomatic moves have failed and the conflict is increasingly moving towards the battlefield. South Africa claimed to have made Transkei independent in 1976, but this has not received international recognition. Guerrilla warfare is also being waged in South West Africa (Namibia) where the UN's decision to end South Africa's mandate, in 1966, was disregarded, although independence has been promised for 1977.

Non-alignment and the Third World

In Europe, Asia and North Africa in the early 1950s, the term "neutralist" was applied to countries that were outside the alliance systems of the Great Powers and wished to remain dissociated from the cold war struggle between the United States and the Soviet Union. Leaders such as Jawaharlal Nehru (1889–1964) of India, Gamal Abdel Nasser (1918–70) of Egypt and Josip Broz Tito (1892–) of Yugoslavia [Key] denied the need to enter alliances, to acquire nuclear weapons [3] or to allow foreign military bases to be set up in their respective countries.

Motives for non-alignment

A neutralist stance had been adopted by the United States itself during the nineteenth century. But the violation of the neutrality of several European countries in two world wars and the global scope of the power struggle that began after World War II led to a belief, particularly in the United States [5], that neutralism was a wishful attitude which failed to recognize that effective protection against "international communism" could be obtained only within the shelter of an alliance of the "Free World". However, for leaders of the militarily weak new nations of Africa and Asia a neutralist stance had three compelling advantages. It allowed them to assert an independence that would have been compromised by their military dependence on one of the Great Powers. It enabled them, by skilful diplomacy, to draw on aid from both the Western and Soviet blocs. And it gave them an opportunity to attempt objective moral leadership at a time when both power blocs were taking up rigid attitudes.

The neutralist, or "non-aligned" nations as they more accurately called themselves, emerged as a coherent force in world politics with the organization of the Bandung Conference in April 1955 in Indonesia, a country that played a leading role in the movement against colonialism [2]. The conference was dominated by Premier Chou En-lai (1898–1976) of China whose moderate attitudes at the conference did much to diminish Asian tensions. Further conferences were held in Belgrade in September 1961 [4]; Cairo, October 1964; Lusaka, September 1970; and Algiers, September 1973 [9]. The conferences steadily increased in the numbers of those attending and in importance.

The political label "Afro-Asian bloc" first gained general currency at Bandung. The more current term "Third World" or *Le Tiers Monde* [1], was coined in France in the mid-1950s to denote decolonized areas that wished to avoid conscription into American alliances or overseas base agreements. (They were collectively designated by some American strategists, including the later United States Secretary of State Henry Kissinger [1923–], as "the grey areas".) The voting power of this bloc at the United Nations made it a force that none of the Great Powers could afford to ignore.

Developments in the 1960s

Events of the later 1950s and 1960s led to significant shifts in the over-simplified tripartite division of the world into communist, Western and non-aligned blocs. The credibility of India's neutralist stance was reduced by her call for Western aid during her clash with China in 1962. Egypt became heavily dependent on Soviet military aid after the

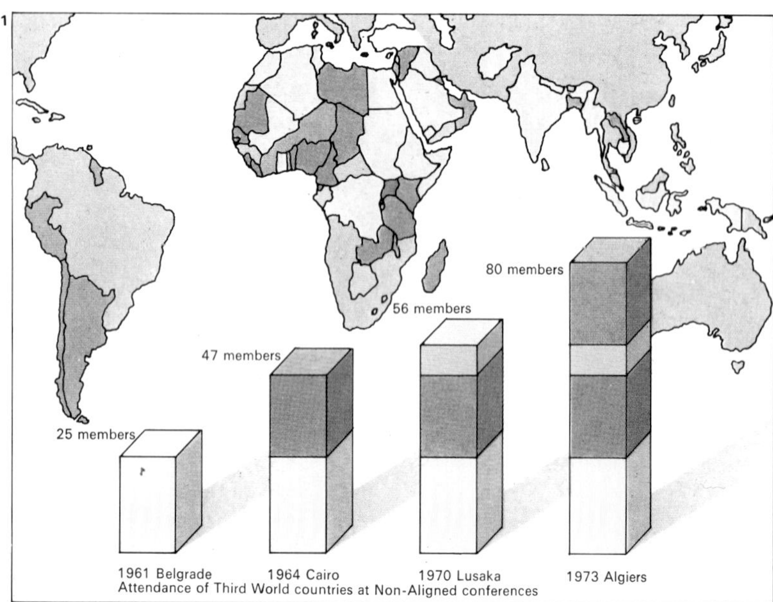

1 Membership of the Third World has grown with the spread of decolonization and now includes substantial parts of Latin America. China's leaders also claim membership. "Third World" is a general political label applied to newly independent, ex-colonial, poor or developing nations and peoples.

2 Demonstrations in Indonesia in the early 1960s against the establishment of Malaysia marked a phase of intense anti-colonialism under the feverish leadership of Achmad Sukarno (1901–70), who took Indonesia out of the UN in 1965 and proposed a rival organization of New Emerging Forces.

3 The mushroom cloud of China's first nuclear explosion in October 1964 while the Cairo non-aligned summit was meeting also marked the first entry to the "nuclear club" of a member of the Afro-Asian bloc. Nuclear testing, the spread of nuclear weapons and the possibility of nuclear blackmail by the Great Powers have been central and recurrent worries of non-aligned nations.

4 The first large meeting of the non-aligned nations at Belgrade in September 1961 drew representatives from 25 countries. Earlier, a number of smaller meetings had been called between Tito, Nehru, Nasser and some other leaders. Non-aligned nations had also conferred in some larger forums, in particular at the UN General Assembly late in 1960. The 1961 conference and subsequent meetings had to resolve frequent controversy about the admission of new members and whether they were genuinely non-aligned. But the number of nations attending grew steadily and the conferences provided the opportunity for broad discussions of topical world issues.

Arab-Israeli war of 1967 and, in the same year, the failure of a communist coup in Indonesia turned that country towards a more Western alignment. At the same time, Sino-Soviet tensions and dwindling of the cold war led to more subtle and complex international groupings [10].

Third World economic policies

In the mid-1970s relations between the Great Powers became less hostile, and Third World opposition to alliances and pressure against colonialism were subsidiary to economic concerns, particularly the wish to see the emergence of a new international economic order. Non-alignment continued to be a predominantly Afro-Asian movement, but it was the Arab and Latin American members who did most to infuse the non-aligned movement with new vitality.

The Arab nations led the way by seizing the initiative after November 1973 when OPEC, the Organization of Petroleum Exporting Countries, unilaterally quadrupled the price of oil and dealt a major blow to the existing worldwide distribution of wealth. The Latin Americans broadened the argument from oil to natural resources in mid-1975 when Cuba [7] proposed that all countries wishing to protect their natural resources should join the non-aligned. The most important issue on the agenda at the Lima conference of foreign ministers of non-aligned states in August 1975 – the statute on foreign investment, multinational companies and technology – was modelled closely on regulations established in the Andean Pact, Latin America's economic integration movement launched in 1968.

These moves were aimed at retaining control of national development and strategic resources. Foreign investment was viewed as acceptable only as long as it contributed to national goals. The non-aligned movement grew from being a negative reaction to the cold war into a positive policy to protect national resources and control foreign investment.

Non-aligned leaders intended to ensure that in future the rich, industrialized nations would no longer find it easy to negotiate with weak producers' associations.

KEY

Tito, Nehru and Nasser (pictured left to right at the Belgrade Conference, 1961) worked together as leaders and promoters of non-alignment from the mid-1950s. Nehru, spokesman for newly independent India, advanced non-alignment as a positive moral force and advocated non-nuclear "areas of peace". Tito represented independent Marxism resisting the pressures of Stalinist Russia. Nasser, leader of the new nationalist government in Egypt and of the larger Arab world, successfully played off cold war competitors with rival aid bids and rid Egypt of British military bases.

5 John Foster Dulles (1888–1959), US Secretary of State from 1953–9, was an unrelenting opponent of communism and the chief advocate of American strategy to contain China and the USSR by military alliances. Announcing in June 1956 that 42 nations were allied with the US, he achieved some notoriety when he said that "except in very exceptional circumstances, neutrality is an immoral and short-sighted conception".

6 Nuclear-free zones and zones of peace or neutrality are being proposed, debated and actively promoted in South-East Asia, southern Asia, the Indian Ocean and parts of Africa. A lead was given in 1967 by Mexico and some other nations when the Treaty for the Prohibition of Nuclear Weapons in Latin America (the Treaty of Tlateloco) was signed. Most other zones have yet to be ratified.

7 Strident anti-US attitudes emerged in Cuba after Fidel Castro (1927–) came to power early in 1959. A large Cuban delegation attended a turbulent 15th session of the UN General Assembly in September 1960. Cuba's role as a small nation defying a neighbouring superpower was further dramatized by an abortive US-backed invasion by Cuban exiles at the Bay of Pigs in April 1961 and a Soviet attempt to arm Cuba with nuclear missiles in 1962. Cuba has campaigned to make Havana a Third World capital linking Afro-Asia and Latin America.

8 OPEC, the Organization of Petroleum Exporting Countries, meeting at Geneva in January 1974, represented the most powerful cartel in the world – a position gained through the importance of oil in the world economy.

9 A World Food Conference sponsored by the UN in Rome in November 1974 and attended by 1,250 delegates from some 130 nations originated with the 1973 Algiers conference of non-aligned countries. The idea was adopted by Henry Kissinger, US Secretary of State, with Western backing.

10 Commonwealth prime ministers, seen at Kingston, Jamaica, in 1975, make up an international grouping that includes aligned and non-aligned, nuclear and non-nuclear, rich and poor nations. This voluntary association of former members of the British Empire engages in continuous consultation.

Latin America in the 20th century

The history of Latin America in the twentieth century is, above all, the story of attempts to break out of the economic, political and social patterns of the nineteenth century and of the resistance such attempts have encountered. Developments in Latin America have been increasingly affected by outside influences. The great Depression brought a collapse of world prices for Latin American exports and two world wars further stimulated industrialization and modernization by cutting the region off from traditional markets and sources of capital goods. There has been a rapid growth of the major cities such as Buenos Aires, Mexico City and São Paulo, swollen, in some instances, by immigration.

Dictatorships and the military in politics

Industrialization and modernization did not themselves bring fundamental political and social change to Latin America. Trade and industry were dominated by foreign enterprises, increasingly those of the United States. Nor did the growing middle classes in Latin America play the social role of their counterparts in the United States or Western Europe, and middle-class political parties seldom carried out essential reforms when they gained office. This situation encouraged the emergence of a new kind of dictator – one who sought the support of the urban workers. Such a dictator was General Juan Perón (1895–1974) of Argentina [3].

The military has remained a significant element in Latin American politics. Military intervention was given a considerable impetus from 1929 by the Depression which caused political convulsions in most Latin American countries. It was later encouraged by the cold war. Often faced by weak and ineffective civilian governments, the military has tended to regard itself as the true guardian of the national interest. Nationalism has always been strong in the Latin American military, and – although the latter has generally been conservative and, in recent decades, strongly anti-communist – this has sometimes been allied with radicalism, especially among younger army officers. As early as the 1920s a military president, Colonel Carlos Ibáñez (1877–1966) [2], carried out a programme of social reform in Chile. The most far-reaching of such programmes, however, has been that of the Peruvian military government which seized power in 1968. It began with the expropriation of a prominent United States-owned oil company and continued with the United States' interests as prime targets of Peruvian nationalism.

Antipathy towards the United States

Latin American nationalism has for a long time been directed mainly at the United States, which is by far the most important foreign presence in the region [Key]. The United States has usually exerted its influence in favour of stability and the status quo and against revolutionary changes which would threaten her interests. Fear of communism has often led her to support Latin American dictatorships. When, in 1961, President John F. Kennedy (1917–63) began the Alliance for Progress – an ambitious programme of economic and social development in Latin America involving substantial reforms and the promotion of democracy – it met with apathy and resistance. Latin Americans have since denounced "aid" as

CONNECTIONS

See also
86 Latin American independence
302 Underdevelopment and the world economy
254 Non-alignment and the Third World
302 The world's monetary system
138 The expansion of Christianity
130 Imperialism in the 19th century

In other volumes
148 The Physical Earth
216 The Natural World

1

1 The ideology of the Mexican revolution is symbolized in these huge murals by Rivera, Orozco and Siqueiros. The revolution was nationalist and the murals are a vivid expression of cultural nationalism. They depict great violence: the oppression of the Indians by the Spanish conquerors and the furious reaction of the Mexican peasants and workers. The Indians and their leaders are idealized in these murals, the oppressors grotesquely caricatured. In this picture Marx is exhorting the workers, while the Church and the capitalists are engrossed in wealth.

2

3

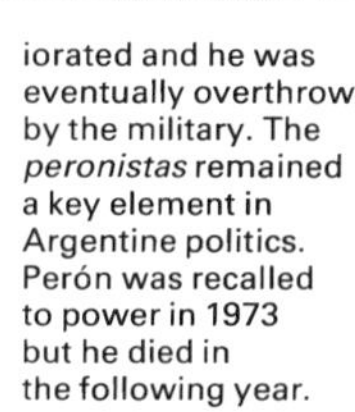

2 Colonel Carlos Ibáñez became President of Chile in 1927 and pursued policies combining nationalism and social reform. But they were undermined by the great Depression and he resigned.

3 General Juan Perón was President of Argentina from 1946–55. Assisted by his wife, Eva, he won over the urban masses with social benefits. After Eva's death in 1952 his position deteriorated and he was eventually overthrown by the military. The *peronistas* remained a key element in Argentine politics. Perón was recalled to power in 1973 but he died in the following year.

4

4 Fidel Castro, the charismatic leader of revolutionary Cuba, seen here addressing one of the countless gatherings at which he explains his policies, is probably the most widely known Latin American figure since Simón Bolívar. Although "Castroism" has not spread to other parts of the continent, Castro's success in defying the dominance of the US in the area has profoundly affected the latter's policies and prestige in Latin America.

5

5 Salvador Allende became the first freely elected Marxist head of state when he won the Chilean presidential election in 1970. Although faced with Congressional opposition and US hostility he embarked upon an ambitious socialist programme. Both his supporters and his opponents resorted to unconstitutional tactics. Economic chaos and violence culminated in his overthrow by a military coup and his violent death in 1973.

increasing their dependence upon the United States and serving the latter's interests more than their own.

Despite United States influence and the durability of traditional social structures there have been three authentic revolutions in Latin America during the twentieth century: in Mexico (1910), Bolivia (1952) and Cuba (1959). The Mexican revolution [1] brought about a new system of government, a sizeable redistribution of land and an improvement in the status of the Indians. It also asserted Mexican nationalism by taking over the foreign-owned oil industry in 1938. The Bolivian revolution, although less far-reaching, destroyed the privileges of the great landowners, nationalized the tin mines (Bolivia's main source of foreign exchange) and also raised the status of the Indians. The Cuban revolution has been the most radical, leading to the creation of an avowedly Marxist state aligned with the Soviet Union under the leadership of Fidel Castro (1927–) [4].

The Cuban example has not been followed elsewhere in Latin America, although there has been a marked increase in urban guerrilla violence in some countries, notably Argentina. The victory of Salvador Allende (1908–73) [5] in the Chilean elections of 1970 – even though he was ousted and killed three years later – was significant.

Third World co-operation

Meanwhile, the countries of Latin America have come to identify themselves with the developing countries of Asia and Africa [6] and to co-operate with them in endeavouring to obtain better trading terms from the richer industrialized nations. They have also tried to co-operate more closely with each other and to increase their trade outside the Western Hemisphere in order to lessen dependence upon the United States.

Brazil [7], traditionally more friendly towards the United States than the Spanish American countries, has for a long time entertained ambitions to be a great power. With its considerable economic progress from the mid-1960s onwards (under a military government), Brazil no longer sees itself as a developing nation. It could fulfil its ambitions by the end of this century.

KEY

The Pan-American Union building in Washington, DC, is the headquarters of the Organization of American States (OAS) which links Latin America with the United States.

6

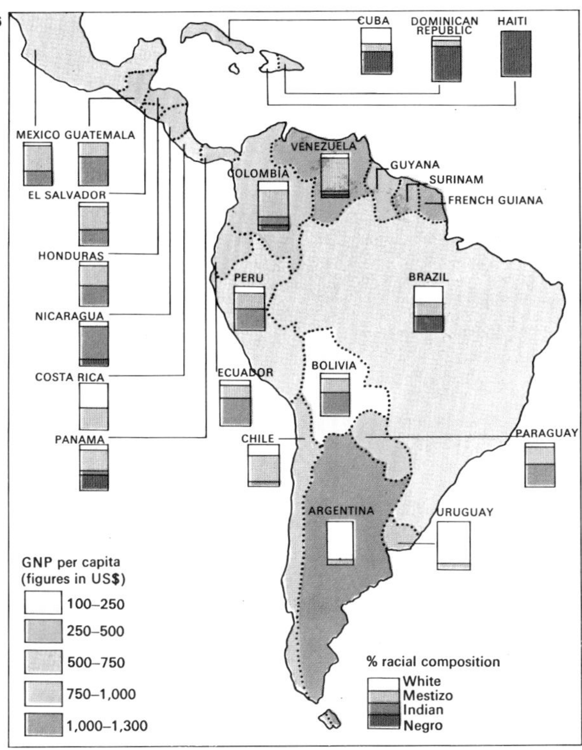

6 The identification of Latin America with the Third World of developing countries is illustrated by this map showing the average per capita income and racial composition of each country. The gap between rich and poor is often extreme.

7

7 The Ministry of Foreign Affairs building is in Brasília, the new capital of Brazil. Brasília symbolizes Brazil's ambition to become a power and to exploit the hitherto largely untapped wealth of the country's interior.

8

8 Shanty towns, such as this one in Rio de Janeiro, demonstrate the glaring disparity between rich and poor in Latin America, as well as the population drift into the burgeoning cities from the stagnant countryside areas.

9

9 The Transamazonian highway in Brazil exemplifies modernization in Latin America. When complete, the network of roads is designed to integrate the vast, unpopulated but richly endowed Amazon basin with the more developed coastal regions of Brazil.

10

10 General Alvarado (1910–), head of the Government of Peru, has pursued a policy of revolutionary nationalism since 1968, and actively fostered co-operation among Third World countries. In 1975 a conference of non-aligned nations was held in Lima.

Evolution of the Western democracies

World War II left Europe divided into two political camps. Except for Greece, where a communist-led revolt was raging, Eastern Europe was under Soviet occupation, including East Germany – although the Western Allies held part of Berlin. The first task was to put the war-torn countries of Europe on their feet. This was done by the USA through the multi-billion-dollar European Recovery Programme [3], also called Marshall Aid after its initiator, Secretary of State George Marshall (1880–1959) [2].

Breach with the Soviet Union

The Marshall Plan was designed to redevelop industries throughout Europe on an aid-sharing basis through the Office of European Economic Co-operation (OEEC). The Soviet Union was invited to join in but refused in July 1947, while vetoing the participation of other countries within its sphere of influence. These included Czechoslovakia, which, had just been incorporated in the Soviet bloc by the coup of 20–25 February 1948.

In addition to having complete power over East Germany, the Soviet government wished to exact heavy reparations from West Germany through the arrangements for Four-Power control. This meant continued dismantling of factories, thus preventing the recovery of West Germany that the other powers held to be vital for the economic recovery of Europe. By 1948 the breach between the Soviet Union and its former, Western allies was complete.

The Americans and British introduced a currency reform into West Germany that had striking success in bringing goods once more into the shops and restarting the wheels of industry. But when they extended the reform to their occupation zones in West Berlin the Russians imposed a blockade (June 1948). This was beaten by the organization of a gigantic airlift and the Russians called off the blockade in May 1949. Meanwhile the Western Allies effected the transition of West Germany to independent status as the German Federal Republic in May 1949.

A strong wave of idealism, strengthened by Churchill's call for a United States of Europe (Zürich 1947), brought into being the Council of Europe (1949), comprising most countries outside the Soviet bloc. But it disappointed many of its promoters because the British kept it a loose intergovernmental body with no real powers.

Steps to unity

Britain, with its predominant position in Western Europe in the immediate postwar years, had similarly been able to thwart American hopes that the OEEC could become a supranational body. Promoters of West European unity led by Jean Monnet (1888–) accordingly took another initiative. With the aid of Robert Schuman (1886–1963), the French Foreign Minister [4], supported by West Germany and Italy, they set out to bring together their countries and others in a supranational organization to administer their coal and steel industries jointly [5]. In May 1950 the Schuman Plan was launched, which led to the setting up in 1951 of the Coal and Steel Community of France, West Germany, Italy, Belgium, The Netherlands and Luxembourg. This act of statesmanship was made possible by the displacing of wartime resentments, through

CONNECTIONS

See also
236 Britain since 1945: 1
238 Britain since 1945: 2
306 Europe from confrontation to détente
300 The world's monetary system
234 The division of Europe
242 Eastern Europe since 1949
240 The Soviet Union since 1945
192 The Peace of Paris
208 The twenties and the Depression
226 Causes of World War II

In other volumes
278 Man and Society

1

1 General de Gaulle's entry into Paris at the head of Free French forces on 26 August 1944 marked the beginning of the end of World War II in the west. But the German armies resisted until May 1945. As Russian forces fought their way across eastern Europe and the Western Allies advanced through Italy and across the Rhine, the postwar political division of Europe began to take shape.

2

3

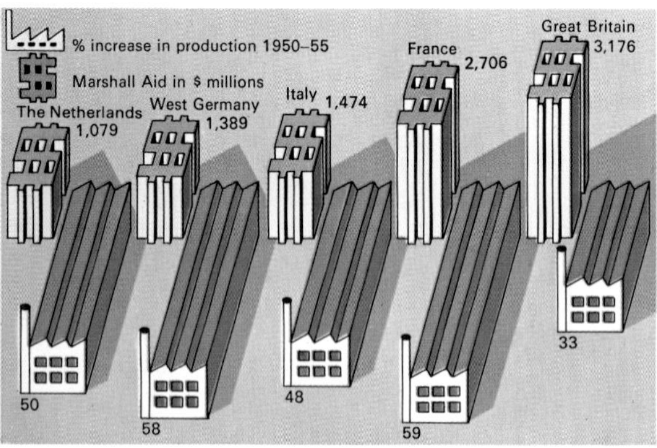

2 US Secretary of State Marshall [left], seen with the British Foreign Secretary, Ernest Bevin (1881–1951), initiated the aid scheme named after him in 1947–9 to restore a weakened Europe that might otherwise follow the path of communism.

3 The European recovery programme, set up to administer Marshall Aid, disbursed $13,150 million between 1948 and 1952 in addition to the $9,500 million already granted for Western Europe since the end of the war and $500 million worth of private food parcels. By mid-1951 industrial production was 42% higher than the prewar level, while agricultural output was 10% higher. Trade had more than doubled while in the early 1950s coal and steel production made massive advances.

4 Two Frenchmen initiated the scheme for the European Coal and Steel Community of France, West Germany, Italy, The Netherlands, Belgium and Luxembourg, established by the Treaty of Paris in 1951. Foreign Minister Robert Schuman [A] based this organization on proposals for pooling coal and steel output which were put forward by Jean Monnet [B], in charge of plans for French modernization.

B

4 A

5

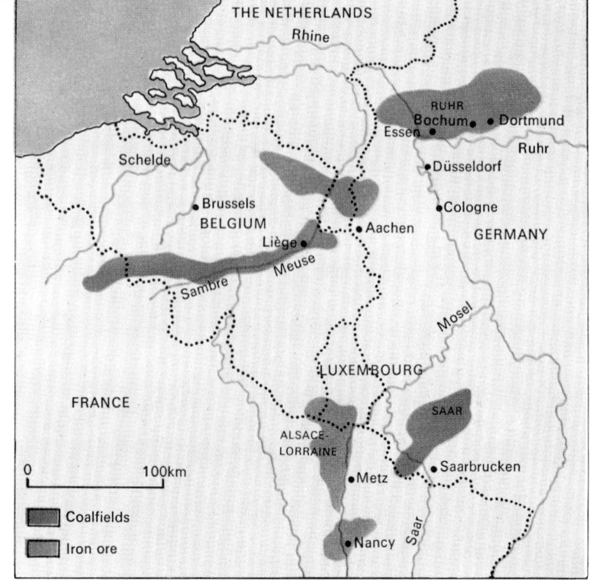

5 Coal seams cross frontiers in northern Europe and the ECSC countries saw that they could build up an efficient coal and steel industry only by devising a supranational system. In this way the coal and coke of the Ruhr could supply the steel industry of Lorraine while the Benelux countries (Belgium, Luxembourg, The Netherlands) and Italy could enjoy similar advantages. By combining these industries (vital for armaments) the risk of another war was reduced.

personal and political reconciliations [Key].

Belgium, The Netherlands and Luxembourg had set up a customs union (Benelux) at the end of the war. This served as a pilot scheme when these countries took the lead in a further step towards the unity of Western Europe, with the creation [6] of the European Economic Community and European Atomic Community. The immediate objectives of the EEC were to create conditions of fair trade first for manufactured goods then, by the Common Agricultural Policy, for agricultural products.

Although the EEC Commission, with powers of initiative and supervision, is a supranational body, main decision-making is in the hands of a Council of Ministers of member states. Attempts to increase the powers of the Commission as a decision-making body were thwarted by General de Gaulle [1] as President of France, and the Council continued as an inter-governmental body with every minister retaining the individual right of veto.

De Gaulle's position was particularly strong after he returned to power in 1958, settled France's colonial problem in Algeria and initiated the French Fifth Republic, whose constitution gave large powers to the president. He asserted the right of France to leadership of "European Europe" in opposition to American influence. In pursuit of this policy he blocked moves for America's close ally, Britain, to join the EEC. In 1966, he also took France out of the North Atlantic Treaty Organization (NATO). West Germany joined NATO in 1955.

Expansion of the EEC

After de Gaulle resigned in 1969, France's veto on British entry to the European Community was soon removed [7]. Denmark and Eire joined at the same time, in 1973. The countries of the European Free Trade Association [8], an industrial customs union that Britain had set up as a rival to the EEC in 1956, were given favoured relations with the Community. An EEC system of associated states that had begun with the ex-French colonies, was extended to a large number of African and Caribbean states by the Lomé Convention of 1975.

KEY

Reconciliation of France and Germany laid the foundation for a new political and economic structure within which the countries of Western Europe could be integrated. After more than 80 years of suspicion, tension and conflict, including three major wars, the two countries joined forces in the Schuman Plan (1950) leading to the European Coal and Steel Community. In January 1963 the West German Chancellor, Konrad Adenauer (1876–1967), and the French President, Charles de Gaulle (1890–1970) [left and right respectively], met to sign the Franco-German Treaty of Friendship.

6

6 The signing of the Treaty of Rome in March 1957 set up the European Economic Community after intensive negotiations under the chairmanship of Paul-Henri Spaak of Belgium (1899–1972). Six member states established a common market for industrial and later for agricultural products, along with schemes to "harmonize" regulations affecting, for instance, working conditions to ensure fair competition.

7

7 Consultations in May 1971 between Edward Heath (1916–), Prime Minister of Britain [left], and Georges Pompidou (1911–74), President of France, cleared the way for Britain's entry to the EEC, a step that successive British governments had tried to take since 1961. De Gaulle twice vetoed British entry but a changed French attitude enabled Britain, Denmark and Eire to become full EEC members in 1973.

8 The European Free Trade Association was set up under the leadership of Britain before she entered the EEC to offset advantages the "Six" were gaining. EFTA, a customs union for industrial goods, facilitated trade between its member-states, although trade growth was faster within the more powerful EEC. Some countries, such as Greece, were associated with the EEC pending full membership. EFTA continued after Britain entered the EEC, with special trading links to the EEC.

9 Riots in Paris in 1968 were led by students who, in both France and Germany, sought reforms of higher education and had other political aims. Sit-ins and growing violence developed into a general strike in France. De Gaulle's regime recovered after the army pledged support but was badly shaken. Promises of far-reaching educational reforms and generous wage settlements ended the strikes and unrest, although De Gaulle himself did not long remain in office as president.

8

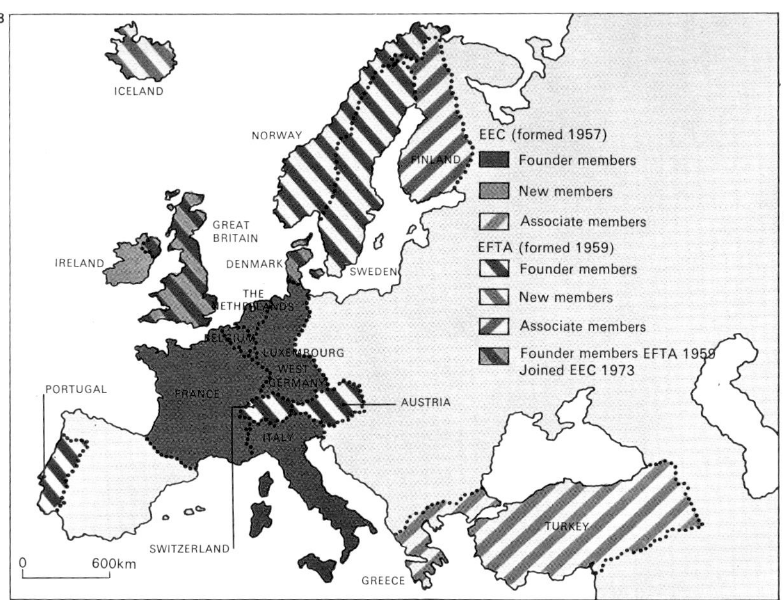

9

10 Arab representatives appeared unexpectedly at the first summit meeting of the enlarged EEC at Copenhagen in 1973. The summit closely followed the October Arab-Israeli war and consequent oil embargo. A steep increase in oil prices indicated a fundamental change in the relative positions of oil-producing and industrial nations, particularly affecting Europe. The Arabs arrived in Copenhagen in search of support against Israel.

10

Scotland in the 20th century

Two main political developments have occurred in Scotland in the twentieth century. One was the rise to power of the Labour Party, which was presaged by events in the nineteenth century and led by figures such as Keir Hardie (1856–1915). The other was the rise of the Scottish National Party (SNP) and the spread of Scottish nationalism, which has grown in response to economic and political developments both inside and outside Scotland during this century.

Nationalism and political changes

The displacement of the Liberal Party by the Labour Party in Scotland began before World War I with the work of Keir Hardie and several small socialist groups of which the Independent Labour Party (ILP) was the most important. The breakthrough came during and after the war, partly as a result of the great bitterness in labour relations on Clydeside [1, 2]. Shipyard and munitions workers there reacted angrily to the sweeping actions of the wartime coalition government, to wage controls and to the "dilution" of labour, as well as to alleged profiteering by manufacturers. In 1906 there were only two Scottish Labour MPs; by 1923 there were 35, and they were the largest party in Scotland – a position they have generally maintained since World War II even when the rest of Britain was returning to a Conservative administration.

The SNP, founded in 1928 as the National Party, sought Home Rule at first rather than independence. Ridden with factions and weak in membership, it made little impact until after 1962. Thereafter it grew fast, and Winifred Ewing's (1929–) victory at the Hamilton by-election in 1967 [Key], followed by the discovery of oil in the North Sea [10], increased support for the party [9] and made the prospect of an independent Scotland seem economically attractive. By 1974 the SNP had the allegiance of nearly a third of Scottish voters, 11 MPs, and a chance to displace Labour as the largest single political party in Scotland.

Both the Liberal and Labour parties have historically had a commitment to forms of Home Rule; the Liberals introduced unsuccessful bills in 1913 and 1914, and the ILP put it high on their programme, although a private member's bill in 1927 failed. Even the signing of almost two million names to a national "covenant" calling for a Scottish parliament within the framework of the United Kingdom in 1950 failed to move postwar governments to renewed action.

The success of the SNP produced new devolution proposals in the 1970s, but the failure of the Labour administration in 1977 to push through its original bill setting up a Scottish assembly with limited powers left the future fluid and uncertain.

Economic problems

Dissatisfaction with the Union can be related to the economic weaknesses of modern Scotland. In 1913 national income per head was probably only five per cent or less below the British average. Both in absolute and relative terms it had grown rapidly in the previous century with the differential between Scotland and England constantly narrowing. In the interwar years, however, severe depression in the heavy industries that dominated the Scottish economy (there were

CONNECTIONS

See also
164 Scotland in the 19th century
210 The British labour movement 1868–1930
236 Britain since 1945: 1
238 Britain since 1945: 2
284 Scotland culture since 1850

1 Women were introduced with other unskilled workers (dilution) to maintain the workforce numbers in the vital heavy industries of the Clyde during World War I as more and more men joined up. But dilution, with government attempts to direct labour, held down wages at a time of inflation, rising profits and rent increases, and placed a great strain on labour relations. Clydeside in particular was the scene of strikes and unrest.

2 Industrial unrest reached a peak in Glasgow shortly after World War I. In January 1919, munitions workers, threatened with unemployment, called a strike and a red flag was raised above the town hall in support of demands for a 40-hour week. The Lord Provost asked the demonstrators to return in two days for his answer; when they did so, the police attempted to clear the meeting and a riot ensued in which the strike leaders, including Emanuel Shinwell (1884-), a later minister, were arrested. The next day, tanks and troops were called in, but the strike had already collapsed.

3 John Wheatley (right) (1869–1930), a self-educated miner, was influential in bringing the Catholic vote to the Labour Party in Scotland despite the initial opposition of the clergy. Later, as minister of health in the first Labour Government in 1923, he introduced the first really effective Housing Act, designed to deal with the housing shortage. It increased rent subsidies and government finance to assist local authorities to build more council houses.

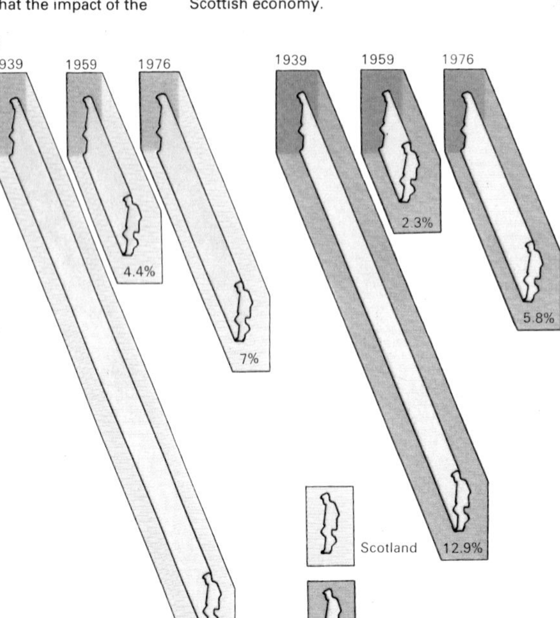

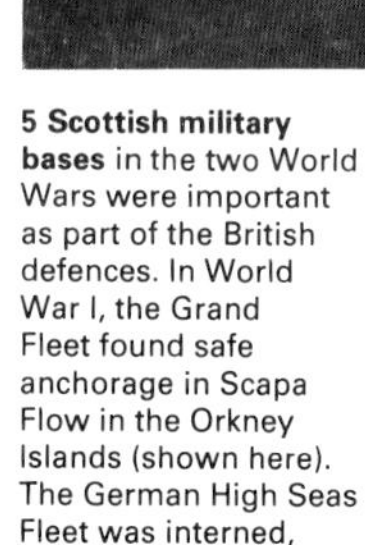

4 Scottish and UK levels of unemployment differed only marginally until the outbreak of World War I. But the greater dependence of Scotland on heavy industry meant that the impact of the depression was intensified and unemployment rose to exceed significantly the UK rate. This high rate of unemployment has tended to persist, despite efforts to diversify the Scottish economy.

5 Scottish military bases in the two World Wars were important as part of the British defences. In World War I, the Grand Fleet found safe anchorage in Scapa Flow in the Orkney Islands (shown here). The German High Seas Fleet was interned, and scuttled itself there in 1919. In World War II Scapa was again a naval base and the fortifications against submarine and air attack strengthened. In 1956 the base was abandoned and an important source of employment was lost to the region.

nearly 400,000 unemployed in 1932) caused the gap to widen to ten per cent and more [4]. Between 1921 and 1931 so many Scots emigrated that the population actually fell.

Since 1940 the economy has performed better, but well-paid employment has often been hard to come by for the Scot who stayed at home. Government regional policies aimed at producing new industries (such as motor vehicles at Linwood and Bathgate) have not cured the problems in the old industrial centres such as Clydeside and Dundee. As a result of the discovery and exploitation of North Sea oil in the 1970s, the gap between Scottish and English earnings is narrower again now than at any time since before World War I, but the prosperity is mainly in the north and rests only on this fragile base.

Although the modern Scot is much better off than his nineteenth-century predecessor, the annual rate of economic growth has not been as high as it was in the late nineteenth century. One consequence is that resources have never been sufficient to remove the stain of urban deprivation. Glasgow still has some of the worst slums in northern Europe, some of them now in modern council-built tenements.

Modern Scotland

The nature of the modern state has also added to Scottish frustrations. Since 1945 more Scottish firms have been directed by private capital operating from England, Europe and America, or have been nationalized and run by civil servants answerable to London. Despite the high calibre of the Scottish Office in Edinburgh, there has been a sense in which for the first time since Union in 1707 the Scots have begun to feel no longer in command of their own country.

Nevertheless modern Scotland is not completely introverted. The fame of John Logie Baird (1888–1946), the inventor of television, and of Alexander Fleming (1881–1955), the discoverer of modern antibiotics, is worldwide. In other fields the Edinburgh Festival (founded in 1947) has an international reputation; and the successful tourist trade has become an important earner of foreign currency.

KEY

Winifred Ewing's extraordinary victory at Hamilton in 1967 crowned nearly 40 years of struggle out of obscurity by the SNP. Although she later lost the seat, in 1970, by 1974 the SNP had gained the allegiance of almost a third of the Scottish voters and had 11 MPs at Westminster.

6

6 Cumbernauld is one of Scotland's most successful postwar towns. Designed in 1956 as an overspill town from Glasgow, it was intended to relieve some of the worst housing conditions and overcrowding in the city. Since then it has successfully attracted light industry and skilled workers, but Glasgow has been left with older, often declining, firms and fewer skills among its workforce.

7

7 The Scottish fishing industry, prosperous before 1914, was badly hit by foreign competition in the interwar years and has had mixed prosperity since. Over-fishing by foreign vessels close to the limits has also reduced the catch.

8 Hydroelectric power in the Highlands was first systematically developed in 1943. Among the most spectacular and successful schemes that were undertaken was this one on the River Awe, in Argyllshire.

8

9 In the late 1960s the Scottish National Party came to the fore in Scotland and united the disparate voices of Scottish discontent through a straight appeal to national self-interest, to "Put Scotland First".

9

10 The discovery of North Sea oil transformed British and Scottish politics in the 1970s. The SNP claimed the oil for Scotland, but the UK government, hard pressed by balance of payments problems and worried about British energy supplies, would not contemplate devolving control over it. Eighty per cent of the oil reserves are located off the Orkney and Shetland Isles, ironically areas that do not always consider themselves as being part of Scotland.

10

Wales in the 20th century

World War I introduced a number of crucial changes in the nature of the Welsh economy. In rural society the most significant change occurred in the pattern of land ownership. The massive estates that had dominated the countryside since Tudor times were put up for sale and bought by freehold farmers. In 1887, only ten per cent of the total cultivated surface of Wales was owned by peasant proprietors. By 1970, 61 per cent was in their hands.

Short-lived prosperity

Landlords had been prompted to sell by the boom years of World War I. But this prosperity proved both artificial and fleeting. The repeal of the Corn Production Act of 1917 meant that Welsh farmers no longer had an incentive to cultivate land. The development of motorized transport made milk production the most lucrative alternative. Mechanization, however, reduced the number of farm hands required, and they were forced to find alternative jobs either in the industrial south or in England.

Economic prosperity in industrial communities during the war years was no less artificial than in rural areas. Once the wartime demand for coal and steel contracted in 1923, the Welsh export market suffered a sharp decline. As oil became increasingly used by the navy, coal-mining areas were rapidly caught up in a deepening industrial recession. Reflecting the decline in the coal industry in South Wales, the number of miners employed fell from 265,000 in 1920 to 138,560 in 1933. South Wales had produced a record 57 million tonnes of coal in 1913. Yet on the eve of the nationalization of the coal industry in 1947 only a dozen mines remained in production. The decline of the iron, steel, tinplate and slate industries was no less disquieting. Stiff competition from foreign steelmakers with updated plant led to the closure of the Cyfarthfa and Dowlais ironworks, and unemployment descended "like the ashes of Vesuvius" on the industrial towns of South Wales.

By 1932, one-fifth of the working population of Wales was unemployed [4]. Shortages and restrictions created a bleak, disillusioned society which remained constantly under the strain of poverty and hardship [3]. For many, migration was the only outlet: the Rhondda Valley lost a fifth of its population between 1921 and 1939 [2] and a thousand people left Merthyr annually.

State assistance

In 1932 to meet the emergency South Wales was declared a "special area" by the government and a campaign was launched to modernize the traditional industries and to develop alternative industries. The most decisive development occurred when Richard Thomas and Company were persuaded in 1935 to open a strip mill at Ebbw Vale. Post-1945 developments were even more crucial. The contraction of the coal industry was offset by a huge expansion in steel production, particularly in the new plants established at Port Talbot [Key] and Llanwern. New tinplate works were established as old mills closed.

World War I also saw in sweeping political changes in Wales. With the decline of Nonconformity and the large estates, Liberalism lost its hold on the affections of

CONNECTIONS

See also
166 Wales 1536-1914
236 Britain since 1945: 1
238 Britain since 1945: 2

1 A World War I recruiting drive in Cardiff used children to win volunteers. Enthusiasm for the war remained high even after conscription was introduced in 1916, and more than 280,000 Welshmen served in the forces. Pacifists such as Keir Hardie (1856–1915), MP for Merthyr, were in a minority.

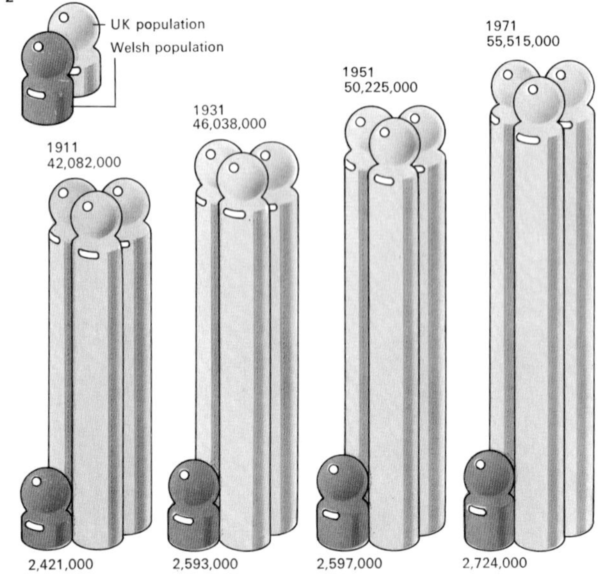

2 A decline in the Welsh population during the 1920s and 1930s reversed a growth trend that had been steady since the census of 1801. During the Depression years Wales lost all its natural population increase and a further 191,000 people, many of them to southeast England and the Midlands.

3 Soup kitchens such as this became an important means of supplementing the diet of many miners' families in South Wales during the Depression years when whole areas were progressively impoverished through unemployment. Malnutrition was common during periods of industrial action and unemployment, when long-term idleness created severe domestic problems. Fortunately, working-class communities were bound together by selfless effort, collective spirit and genuine compassion. These traditional values helped to alleviate the harsh social problems of the time. The Welsh poet Idris Davies wrote: "No manna fell on these communities and self-help was the only answer".

4 Miners from the Rhondda Valley and other Welsh mining areas marched to London to join a hunger demonstration by 200,000 people, from all parts of Britain, in Hyde Park on 8 November 1936. Many of the Welsh marchers had to sing on the streets for pennies to buy themselves food. The suffering and humiliation of the Depression years left deep scars on the hearts and minds of working-class people in Wales. Their sense of injustice led to a strengthened trade union movement and increased willingness among industrial workers to force action by demonstrations and strikes. In 1932 Welsh unemployment reached a quarter of a million.

the Welsh people. By contrast, the Labour Party emerged as the dominant political party in South Wales during the interwar years [5]. After 1945, socialism penetrated North Wales, and when the Labour Party won 32 seats out of 36 in the general election of 1966 it reached the peak of its dominance.

Welsh nationalism

Plaid Cymru (the Welsh Nationalist Party) was slower to achieve parliamentary success [6]. From the 1960s onwards, however, it extended its membership in both rural and industrial areas. In 1974 three Plaid members were elected to Parliament and the party has since established itself as the major rival to the Labour Party in Wales.

In 1964, the Labour Government established a Welsh Office in Cardiff, but because this body was granted little executive authority it scarcely began to fulfil the demands of the devolutionists, who called for the setting up of a representative assembly within Wales. The Kilbrandon Commission, established in 1968–9, came out in favour of a large measure of devolution for Wales.

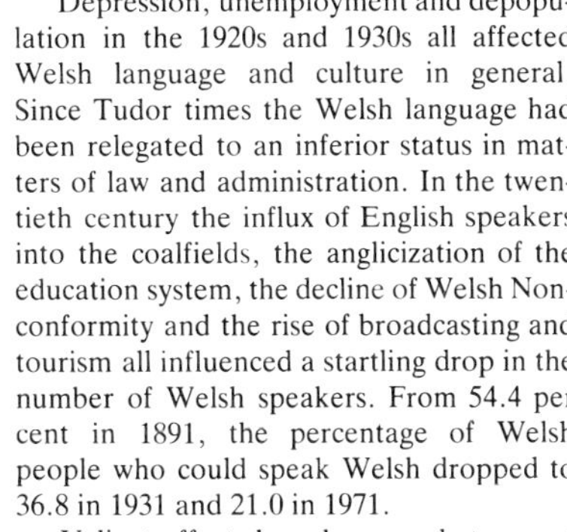

Depression, unemployment and depopulation in the 1920s and 1930s all affected Welsh language and culture in general. Since Tudor times the Welsh language had been relegated to an inferior status in matters of law and administration. In the twentieth century the influx of English speakers into the coalfields, the anglicization of the education system, the decline of Welsh Nonconformity and the rise of broadcasting and tourism all influenced a startling drop in the number of Welsh speakers. From 54.4 per cent in 1891, the percentage of Welsh people who could speak Welsh dropped to 36.8 in 1931 and 21.0 in 1971.

Valiant efforts have been made to arrest this decline [7]. *Urdd Gobaith Cymru* (the Welsh League of Youth), founded in 1922, fosters the language by inviting children to camps, sporting events and eisteddfodau; a growing number of schools teach in the vernacular at both primary and secondary level; and Welsh authors and publishers receive substantial grants. The Welsh Language Act (1967) has granted – in principle at least – Welsh equal validity with English.

KEY

The Port Talbot steelworks at West Glamorgan became a major factor in the Welsh economy after the massive Abbey Works and hot strip mill was built there in 1947, modernizing the existing plant. In the wake of the dramatic collapse of the coal industry, the expansion of steel production has brought changes in the industrial and social structure of Wales that are as far-reaching in many ways as the transformation that occurred during the first Industrial Revolution.

5

5 Aneurin Bevan (1897-1960), son of a Tredegar miner, entered Parliament in 1929 as Labour member for Ebbw Vale. He rapidly became the most stimulating socialist thinker of his day. A colourful personality and a brilliant spontaneous debater, he preached the gospel of democratic socialism with wit and passion. After editing the socialist *Tribune* (1942-5), he became Minister of Health and principal architect of the National Health scheme. Later, in opposition, he led a left-wing Labour group critical of the rearmament policies of the 1950s. Hugh Gaitskell defeated him for the party leadership in 1955.

6

6 Saunders Lewis (1893-   ), Welsh author, has been an inspiration to the nationalist movement as one of the founders, and later as president, of Plaid Cymru from 1925. The party, fired by Ireland's success in winning independence, made slow headway until after World War II. But its activities, at times explosive, were a major factor in achieving formal recognition for the Welsh language in schools and in such sensitive areas as broadcasting, which is now carried in two languages. The growing strength of the party at the polls has been accompanied by moves towards greater political autonomy.

7

7 Civil disobedience has been a tactic of *Cymdeithas yr Iaith Gymraeg* (the Welsh Language Society) since 1969. In 1971 members interrupted a High Court case in London to publicize their cause. The society, founded in 1962, aims to secure for the Welsh language equal status with English.

8

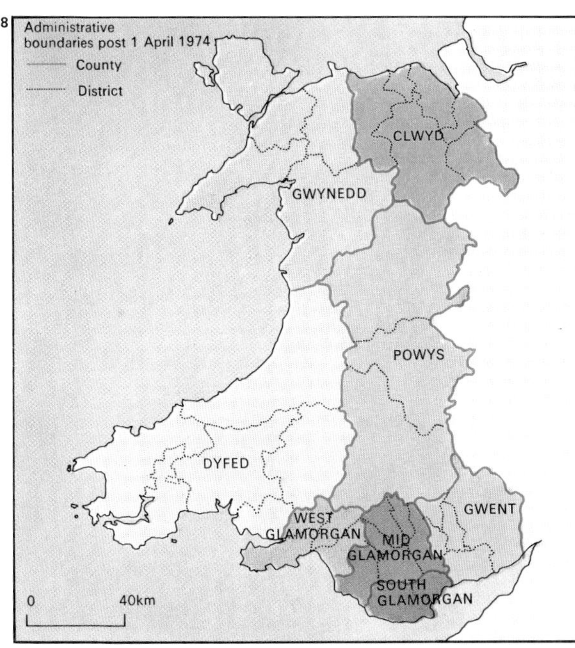

8 A new structure of local government administration was established for Wales in April 1974, dismantling a framework of shires that had lasted for more than 400 years. The 13 Welsh counties set up by the Tudors under the Act of Union in 1536 were abolished and in their place emerged eight units based broadly on ancient medieval divisions.

USA: the affluent society

The pervasive theme of American society since the end of World War II has been growth, bringing prosperity, innovation and, not least, growing pains. This growth has been most evident in the number of people living in the United States [1]. The population at the time of the 1940 census was 131 million; by 1970 the population was 203 million, an increase of 72 million. The population explosion had been fed more by the baby boom after the end of World War II, by the "second generation" baby boom of the late 1960s, and by people living longer, than by continued immigration from Europe. By 1972 the rate of immigration was about one-sixth what it had been before World War I and less than six per cent of the country's population was foreign born.

The increase in population has meant a vast expansion in the size of conurbations, although typically the city centres themselves have lost population. Those who remain in older urban areas are often black [6]. Among large American cities nine have populations that are 40 per cent or more black, including Washington, DC, which is more than two-thirds black. The growth in population has been greatest in the so-called "Rim States" along the American coast, from Florida through to Texas and California. In 1940 California had less than half the population of New York; by 1970 it had become the largest state in the Union.

The rise of the bureaucratic leviathan

The population explosion has been mirrored by an enormous growth in government. The number of public employees has trebled since the 1930s and more than doubled since 1945, and now constitutes nearly 20 per cent of the total workforce. The expansion of the American military is shown by the fact that there were 28 million ex-servicemen in America in 1975.

The growth in government is reflected in the creation of three new cabinet departments (Health, Education and Welfare; Housing and Urban Development; and Transportation), a response to the federal government's commitment to expand its capabilities for looking after its citizens, and mobilizing national resources.

Superficially, party politics has changed less than society as a whole. The presidency is contested by candidates of the Democratic and Republican parties, as it was a century ago. But the voting has been very unstable. Throughout most of the period, the Democratic Party has controlled both houses of the United States Congress [3]. Moreover, although the country claims to have a two-party system, in three postwar elections the president elect took less than half the vote, because of divisions within the two parties.

Expanding economy and prosperity

The American government has been able to expand activities at home and abroad because of the continuing growth of the nation's economy. In 1950 the gross national product was $284,000 million; by 1971 it had increased almost fourfold to $1,050,000 million. The growth in total national resources meant that, even without raising tax rates, the flow of money into the federal treasury increased massively. The amount of money left in the pockets of individual consumers also increased, although by a lesser

CONNECTIONS

See also
292 The wars of Indochina
300 The world's monetary system
288 American writing: into the 20th century
266 20th-century sociology and its influence
258 Evolution of the Western democracies
240 The Soviet Union since 1945
244 China: the People's Republic

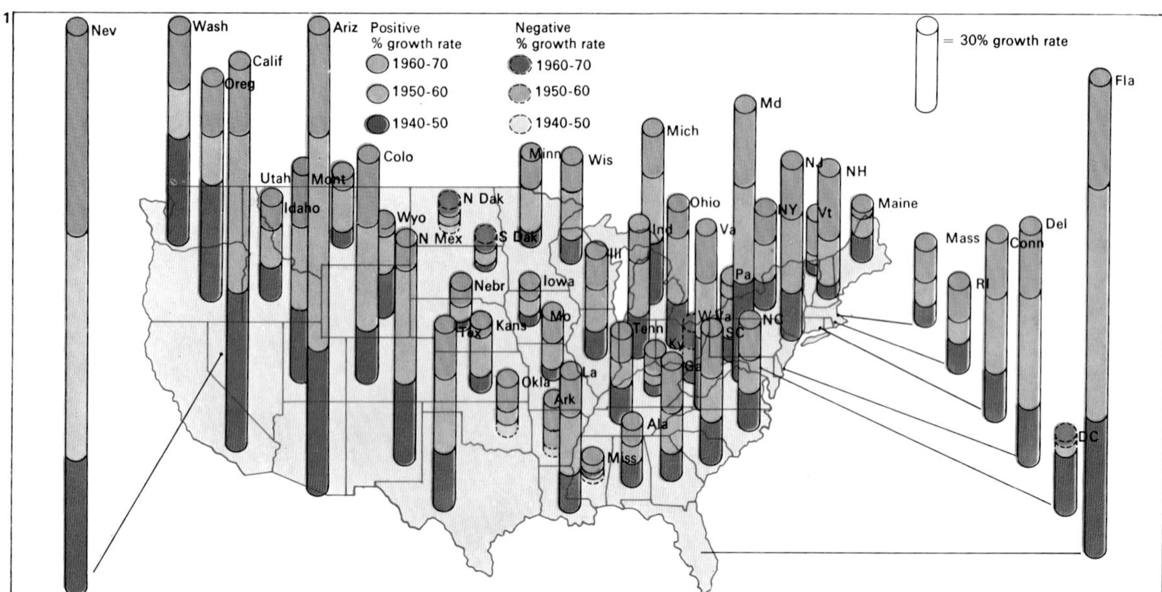

1 Rapid population growth in America after the war was due more to a marked increase in the birth-rate and life expectancy than to immigration; since 1945 total population has increased by over 50%. This in itself did not greatly affect population distribution across the continent. But there has been a significant movement of people to new centres of growth, north and south mixing in this internal migration. Florida and California, for example, were centres for this migration, as the diagram indicates. America's manpower and wealth provided the means of a world-wide "defence" effort postwar, but Vietnam showed that these alone are not enough.

2A 

B 

C 

D 

E 

F 

G 

2 Postwar presidents have been almost equally divided in party terms, Eisenhower [B], Nixon [E] and Ford [F] being Republicans, and Truman [A], Kennedy [C], Johnson [D] and Carter [G], Democrats. But all gave priority to foreign affairs. Truman found this compensated for domestic policy set-backs, but Johnson lost by his foreign policy the support that his domestic war on poverty had gained. Nixon found that his success abroad could not bury the Watergate affair. Of these men John F. Kennedy, the Harvard-educated son of a millionaire, came from the east coast; the others were all brought up in small towns, or came from unsophisticated farming regions.

rate, because a portion of the increase went to looking after the increased number of children and elderly and to employ the larger number of Americans of working age. The family income of Americans has risen steadily, even when allowance is made for the effects of creeping inflation. The real income of the average American family doubled from 1947 to 1971, when it exceeded $10,000 a year.

Higher earnings meant Americans could afford to buy more of everything. The great postwar housing boom caused a drop in the proportion of Americans living in substandard houses from nearly two in five in 1945 (many living in old farmhouses) to one in 20 in the early 1970s. The number of cars sold more than doubled from prewar years, totalling more than 8.5 million in 1970 [5]. Americans have also been investing more money in education. The proportion of young people receiving a high-school diploma (a secondary school leaving certificate) increased from one-half to three-quarters.

One of the biggest changes in American society in the postwar era occurred through the courts and the statute books, with the integration of blacks as full citizens in American society. A series of United States Supreme Court decisions culminated in 1954 in the declaration that segregation was unconstitutional. This led to major changes in education patterns throughout the country as subsequent court orders enjoined increasingly stringent methods of assuring a balance of blacks and whites in the schools.

The raising of black consciousness

In the 1960s blacks began to turn to the streets, protesting peacefully under leaders such as Martin Luther King (1929–68) [4], or rioting as an expression of frustration, as in the Watts area of Los Angeles, in Detroit, Newark and even in Washington, DC. Black family income, reflecting generations of discrimination, does not yet equal that of whites. Nonetheless it has been rising, both in absolute terms and as a proportion of white income, as more blacks receive better education and as the federal government enforces stricter practices for equal opportunity in most areas of employment.

KEY

The supermarket, with its variety and abundance of goods, symbolizes the affluence of postwar America. In the decade following World War II this wealth was highlighted by the austerity of a Europe recovering from conflict.

3

Congress
Presidency
1945 47 49 51 53 55 57 59 61 63 65 67 69 71 73 75
Truman Eisenhower Kennedy Johnson Nixon Ford Carter
49 53 57 61 65 69 73 77
Democrat controlled
Republican controlled

3 Since 1944 Republican presidents have generally faced a Congress held, almost continuously, by the Democrats. However, internal Democratic divisions have reduced the potential for conflict.

4 Martin Luther King organized the Montgomery, Alabama, bus boycott of 1955–6, the first great civil rights protest in the south. This nationwide spokesman for the black community was murdered in 1968.

5

Televisions 1950 1960 1970
Automobiles 1950 1960 1970
Washing machines 1950 1960 1970
New housing units started (in thousand units) 1950 1960 1970
1396 1296 1469
3364 4273 4094
4851 5708 7464
6666 6675 6547
Sales in thousands of units

5 The consumer goods boom in postwar USA began a "democracy of consumption": new homes, cars, washing machines and televisions became virtually the birthright of most Americans. Typical was the demand for television sets, first for black-and-white sets in the 1950s and later for colour sets, as technological advance made black-and-white television obsolescent. The boom in house construction brought mass production to the building industry, with economies of scale and standardization of product. A record of building well over a million houses a year meant that by the mid-1970s the number of homes built in the postwar era would have been able to provide a new house for almost every US family in 1939. Consumer durables also generated further costs – most notably the motor car. It consumed tracts of land for highways in and around cities and oil to fuel engines. Until the oil crisis of the 1970s resources to maintain this boom seemed boundless.

6 The growing black population left the rural south for the industrialized conurbations, moving to the car factories of Detroit and to Chicago and New York, traditional routes for blacks in search of work, and also to new growth areas such as Los Angeles and Houston. This influx provoked an outflow of white residents to the suburbs. The whites were partly attracted by suburban life and partly fearful of the urban ghettos. As a result of this movement, America's most important cities today often contain its greatest social, political and economic problems, generated by years of racial antagonism.

6 Black population as a percentage of total population

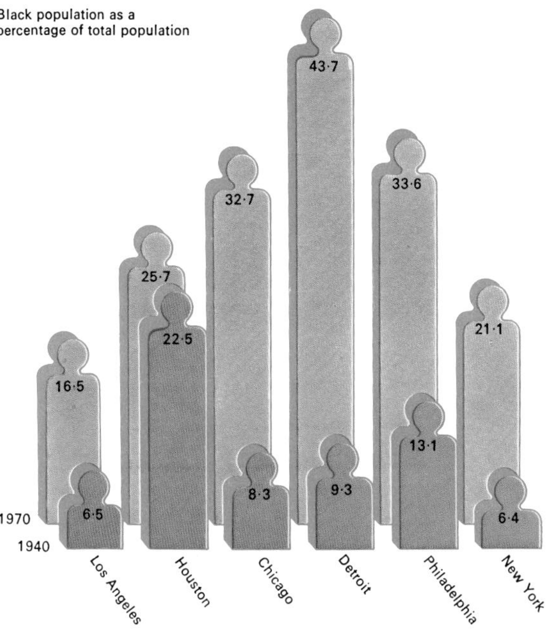

20th-century sociology and its influence

In the years following the end of World War II, sociology began to change from a theoretical system to a practical tool that could be used by government and industry. But it did not lose sight of its origins. It had begun from a desire to explain – and to counteract – the forces in industrialization that divided people, both economically and socially. Modern sociologists have continued to concentrate on ways of reducing inequalities and of increasing social integration.

The "good society"

The insecurity and disruption of the 1930s and 1940s had served to increase the concern of sociologists with the "good society". The good society was seen by some theorists as involving a high level of integration and stability, a common core of values and an emphasis on community. One school of thought that echoed these themes was that of structural functionalism which developed a picture of society as a self-regulating organism, in which all the various elements (institutions) perform necessary functions. Functionalism originated with Emile Durkheim and was developed in America by Talcott Parsons (1902–) and Robert Merton. Bronislaw Malinowski and Alfred Radcliffe-Brown founded British social anthropology with their studies in New Guinea, Africa and elsewhere of small-scale "primitive" cultures. The accessibility of the constituent elements of these small societies make an exhaustive study seem possible.

The conservative tenor of structural functionalism is apparent in its concentration on moral integration, in its emphasis upon existing social institutions and in its tendency to identify their functions with the interests of the more powerful groups in society.

Sociology and "social engineering"

Functionalism provided a theoretical basis for the widespread use of sociologists as "social engineers", dealing with particular problems for industry or government. A variety of different policies was drawn from functionalist analyses – while some, for example, stressed the need for different social levels, others advocated integration. Busing and comprehensive schooling are government policies adopted to promote integration and equality by bringing together privileged and underprivileged children at school [9]. Delinquency is another problem for which governments have increasingly employed sociologists. Functionalist analysis underlay the 1958 "Mobilization for Youth" programme in the United States, which hoped to narrow the gap between the goals desired by, and the actual opportunities offered to, the underprivileged – a gap that the originators of the programme believed to be a cause of delinquency [5]. In their postwar major rehousing and urban renewal schemes, governments of many industrialized nations have employed sociologists in planning and design [4] in an attempt to provide a solution to the concentration of social problems that seemed inherent in prewar slums everywhere.

The pioneering experiments at the Western Electric Company into the productivity and working conditions found in their factories in 1927 showed the great importance of "human factors" in raising productivity levels. The value of those findings led to

CONNECTIONS

See also
172 Masters of sociology
264 USA: the affluent society
156 The foundations of 20th-century science
212 Socialism in the West

In other volumes
204 Man and Society
248 Man and Society
262 Man and Society
264 Man and Society
274 Man and Society
278 Man and Society

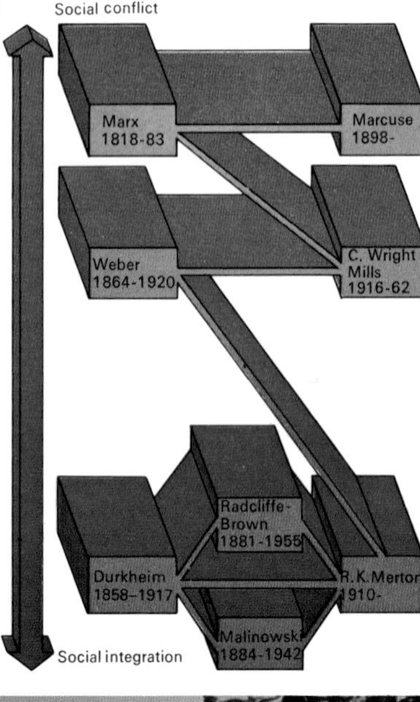

1 Sociological thought in the 20th century is, in many respects, as divided as that of the 19th century when many of its current disagreements began. There is no one sociological theory, but instead a number of different theories, some complementary and some conflicting. One of the most fundamental of these concerns the model of society with which the sociologist starts out. Some, such as Talcott Parsons, define society as a harmonious, self-regulating system. Others, like Marcuse, argue that society is not as harmonious as it may sometimes appear, but is deeply divided by economic inequalities which lead to both conflict and violent unrest.

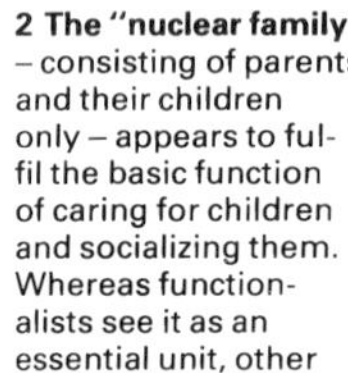

2 The "nuclear family" – consisting of parents and their children only – appears to fulfil the basic function of caring for children and socializing them. Whereas functionalists see it as an essential unit, other sociologists and social psychologists such as R. D. Laing attack it for being so tightly integrated that it may breed neurosis and repression, a point taken up strongly by the Women's Liberation Movement.

3 The Israeli kibbutz, one of the experiments in group living in Western society, demonstrates that, contrary to early functionalist thinking, the nuclear family is not the only possible structure that can care for children. In theory, children on the kibbutz are raised collectively, although some sociologists have pointed out that a strong sense of the family unit remains despite the communal features.

4 Governments have increasingly employed sociologists to assist in social planning. Postwar prosperity and recognition that slums were a focus of social problems, led to large-scale rehousing schemes. In many cases, however, "improvements" were carried out with little thought of what effect they would have on the people rehoused. Established communities with strong, supportive social systems were broken up in the move to well-designed but socially anonymous new towns. Various measures, from the way houses were grouped in small units to the siting of shops (shown here is the Postgate shopping centre, Scotland), have been tried to re-create a community feeling.

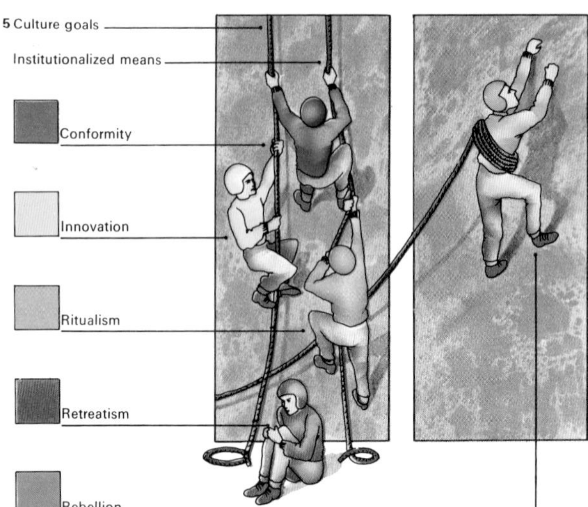

5 Individual adjustment within a structure of socially defined goals and means is shown on this diagram by Merton. With this model it is possible to analyse the behaviour of the delinquent (who may pursue a socially acceptable goal, but who does not follow morally prescribed means of attaining it) and of rebels who seek to change goals and means in society.

the large-scale employment of sociologists by management to work in such areas as marketing and industrial relations.

The legacy of Karl Marx

Structural functionalism was paralleled by Marxist theory. Whereas the structural functionalists stressed the notions of integration and co-operation, those inspired by Marx saw society as composed of conflicting classes divided by their differing economic positions.

Influenced by Marx, C. Wright Mills in his book *The Power Elite* pointed to a three-fold power concentration – the corporations, the military and the political – whose interests and actions were closely related [6]. But he argued that the power basis of this alliance could not be explained simply in Marxist, economic terms but required a wider analysis of social organization. Marxist analysis greatly influenced the Black Power movement, whose leaders were disillusioned with the philosophy of integration advocated by the Civil Rights movement, and who questioned whether integration was possible or even desirable. Following the race riots across the United States in 1968, black and white politicians and sociologists argued for increased aid and social legislation for the ghettos, proposals rejected by the Black Power movement as mere palliatives.

The Vietnam War and the rise of student protest also brought to the fore a well-developed but previously uninfluential school of sociology – the Frankfurt School. It emphasized the control of knowledge through the mass media. The media were seen to be the new opiate of the masses, in part explaining popular acceptance of what is, according to Marxists, an oppressive economic state.

The development of this theme by Herbert Marcuse [7] rose to prominence in the theoretical base of the growing student protest movement [10]. According to Marcuse, students along with marginal and dispossessed groups are the contemporary revolutionary agents, precisely because they are outside the hypnotic culture of consumer society. However, the complex and incisive work of the Frankfurt School has, as yet, had little influence.

KEY

The 20th century has been characterized for many by a widening gap between living standards and expectations (developed, for example, through advertising). Sociologists have viewed this gap in different ways. Some have seen it as a cause of unrest and social problems; others have attributed the apathy of the underprivileged towards improving their situation to the use of advertising and the creation of a "consumer dreamworld". This gap has also contributed to the use of sociology by governments who have increasingly intervened to reduce inequalities. In the commercial field, sociologists have developed techniques to maintain and exploit the gap.

6

6 The basis of power in American society, according to C. Wright Mills, greatly depends on the common social background of the political, military and business leaders. Educated similarly, attending the same social events, yet careful to maintain a popular image – here President Eisenhower opens the 1960 baseball season – they sustain a common outlook that obviates the need for a conscious conspiracy to preserve their rule.

7

7 Herbert Marcuse (1898–), Professor of Sociology at Berkeley University, shown here in discussion with students, provided a stimulating critique of modern society. His analysis of modern democracy as characterized by "repressive liberalism", in that freedom to disagree is more apparent than real, gave rise to a new approach to the study of social institutions. In the achievement of a truly liberated society, Marcuse allotted a central role to students. His work constitutes an important strand in the ideology of the student movement of the 1960s. Many of those involved in the student unrest of 1968 acknowledged Marcuse as their mentor.

8 Social science research has undergone a rapid expansion since the mid-1950s, as this diagram of US Federal support shows. But institutions like the Ford Foundation have provided the largest proportion of money in this area. The methods and findings of sociology have been applied to a wide variety of public and private fields, from military strategy to housing, and from marketing to industrial relations.

8

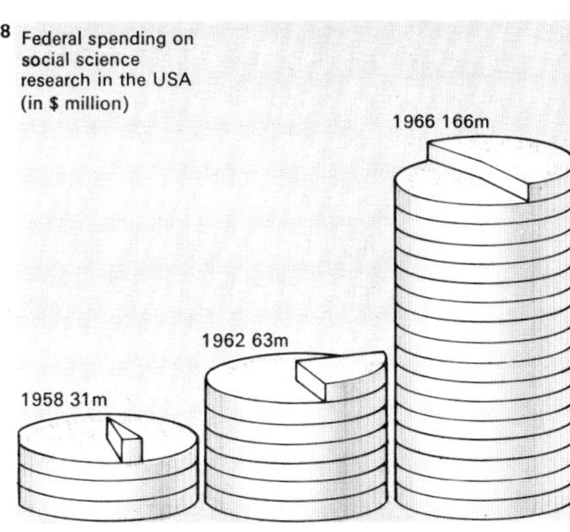

10 Social science students figured prominently in demonstrations against American involvement in Vietnam, such as this at Fort Dix, New Jersey, in 1970. Reaction to the war in Vietnam illustrates the paradox of sociology's influence on 20th-century political affairs. On the one hand sociology is charged with inciting conflict and change; on the other, it is accused of assisting in the maintenance of the existing social system. The involvement of radical social science students in opposing the war was more than equalled by the time, effort and particularly money spent on social scientific research designed to make the war more efficient.

10

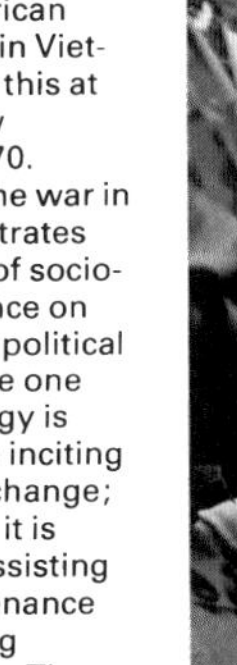

9

9 The policy of busing black, underprivileged children to white schools encompasses two key sociological ideas. The first is the belief that educational achievement is as much a matter of environment as of heredity (emphasizing the need to equalize opportunities in the classroom). The second is that of racial integration. Public discontent with this policy, typified in the Boston busing "war" shown here, points to the limitations of such attempts at social engineering.

Hollywood

During the silent film era, Hollywood had established at the centre of the film-making process a group of glamorous stars who personified the dreams and wishes of cinema audiences. Charles Chaplin, Mary Pickford, Douglas Fairbanks, Rudolph Valentino [1], Gloria Swanson and newer stars such as the haunting Greta Garbo [4] influenced the lifestyles of millions to whom the cinema represented escape from a drab world.

Big studio organization

After 1927, American control of the patents for sound equipment tightened Hollywood's grip on the world film industry, particularly when the European studios suffered financial reverses during the Depression. To meet the threat of falling audiences (a third of American cinemas had closed by 1933), Hollywood itself was reorganized by Wall Street financiers who gained control of the eight major production studios and set out to mass produce films by methods that would guarantee maximum profits. Individuality was subordinated to team productions in which dozens of scriptwriters might work on a single film. High-quality staging, costuming and photography and massive publicity machines projected the personalities of a new generation of screen idols, many recruited from the theatre. To exploit the particular talents of stars such as Joan Crawford, Jean Harlow or Clark Gable, formula films were devised with plots that varied only marginally. Slapstick and melodrama, the two most important genres of mass entertainment in the silent era, gave way to the farce of repartee (represented by the zany Marx Brothers), romantic dramas of society life, sex comedies and musical spectaculars, beginning with *Broadway Melody* (1929).

The straight transference of plays to movies, the shackling of cameras to clumsy soundproof booths and the restricted movement imposed by crude microphone equipment tended to rob films of their fluidity in the early days of sound. Back projection and huge studio lots were used to minimize the need for shooting on location and Hollywood was further removed from the realities of everyday life by a "code of decency" administered by the Hays Office. Movie moguls such as Louis B. Mayer (1885–1957), head of Metro-Goldwyn-Mayer (MGM), were able to impose a bland view of life on the entire output of their studios. Apart from the tough gangster films for which Warner Brothers became famous, the Hollywood movies of the 1930s were designed almost wholly to entertain.

Entertainment opiates

The optimistic gloss of Hollywood was reflected both in the choreographed Art Deco fantasies of Busby Berkeley [3] and in the rise of child stars such as Shirley Temple, Mickey Rooney [7] and Deanna Durbin. At the same time, the demand for entertainment was often met with a high degree of professional skill, revealed most clearly in the brilliant animated work of Walt Disney and in the verve and wit of comedies produced by Ernst Lubitsch at Paramount and George Cukor at MGM. In the Technicolor splendours of *Gone With The Wind* (1939), big-studio organization achieved its ultimate objective – a film that would remain popular (and profitable) for generations [5].

CONNECTIONS

See also
200 Origins of film
276 Cinema as art
208 The twenties and the Depression
288 American writing: into the 20th century

1 Rudolph Valentino, from the moment he appeared as a gaucho in *The Four Horsemen of the Apocalypse* (1921), became the romantic idol of millions of women. An Italian migrant and former tango dancer, his rise to super-stardom, intensified by his impact in *The Sheik*, personified an American dream of sudden fame and riches. An orgy of public grief followed his death (from peritonitis) in New York in 1926.

2 The picture palace was a place of escape and enchantment in the 1920s when this London cinema showed both live and filmed entertainment. Fantastic decorative flourishes were added inside and out. The impulse to dazzle audiences with fountains, marble pillars, gilded turrets, chandeliers and massed choirs reached a climax in 1927 when S. L. (Roxy) Rothafel opened a "Cathedral of the Motion Picture" in New York.

3 The musical was one of Hollywood's most enduring contributions to the popular art of cinema. In such films as *Footlight Parade* (1933), Busby Berkeley, a former dance director, broke away from a fixed camera angle to create stunning scenic effects with beautiful choreographic patterns of chorus girls or top-hatted men. The magical dancing of Fred Astaire and Ginger Rogers brought a more intimate style, while in the 1940s the musical tradition was again reshaped by the verve of Judy Garland and Gene Kelly. Perhaps the purest form of escapism, the musical began to lose ground only in the 1950s.

4 The phenomenon of stardom has never been demonstrated more hauntingly than by Greta Garbo. Her steady gaze into the camera (in the 1927 film *Love*) had a unique effect on both male and female audiences. Clare Boothe Luce described her as "a deer in the body of a woman living resentfully in the Hollywood zoo", and her performances in the films she made between 1926 and 1939 did indeed make her a legendary figure. She was born in Sweden in 1905, went to Hollywood as the protégée of the director Mauritz Stiller in 1925 and retired in 1941. Hollywood helped to create an unforgettable Garbo style by providing her with some of its better directors and cameramen.

The most successful attempt to emulate the Hollywood system was made in Britain where American backing enabled Alexander Korda to establish the world's second-largest film industry. Aided by the widespread introduction of colour at the end of the 1930s, the studio system in both countries survived World War II and box-office takings rose to a peak in 1946. During the next ten years, however, Hollywood was increasingly affected by rising production costs, labour disputes, anti-trust laws, foreign taxes and witch hunts for alleged communists. At the same time, the competing attraction of television halved audiences in a single decade.

To counter the challenge of television the studios tried to provide a more lifelike film image. Experiments with three-dimensional effects failed. But with Cinemascope (1953) the technique of film-making as mass entertainment moved into a significant third phase. Using versions of an anamorphic lens invented by a French optician, Henri Chrétien, nearly 40 years earlier, Hollywood began to mount multi-million dollar "blockbusters". Despite such notable epics as *Ben-Hur* (1959) and the emergence of new super-stars such as Marilyn Monroe (1926–62) [8], the big-studio system with its top-heavy executive structure began to break down in the late 1950s. Individual directors and stars began to regain control of production and make films whose themes would appeal to discriminating audiences. By the 1960s, more films were being shot increasingly on location. The Western, a distinctive Hollywood genre [6], was transplanted to Europe in "spaghetti Westerns".

Hollywood nostalgia

As mass entertainment, films remain most important in Asia, whose rising output has matched a production decline in the West. But the Hollywood era has been rediscovered in a nostalgic flood of old movies sold to television, which reveal the craftsmanship of the gangster films, comedies and musicals made in the 1930s and 1940s. At the same time, there are signs that a reorganized Hollywood industry will hold its place as a producer of big-scale films such as *The Godfather* [9].

KEY

A glittering première (of the 1930 Dietrich film *Morocco*) at Grauman's Chinese Theatre captured all the glamour of Hollywood in its heyday. "Strip off the phoney tinsel", said Oscar Levant, "and you'll find the real tinsel underneath."

5

5 *Gone With The Wind* (1939) has been seen by more people than any other film. Produced in the early days of colour, it ran nearly four hours, had a wilful heroine (Vivien Leigh), a he-man (Clark Gable), saintly supporting leads (Leslie Howard and Olivia de Havilland) and a story of high passions and turbulent events (the American Civil War). In true Hollywood style, it was directed by three men, chiefly by Victor Fleming.

6

6 Gary Cooper (1901–61), a famous cowboy hero, reached the climax of a long career when he played Marshal Will Kane in Fred Zinnemann's 1952 film *High Noon*. In the ritual of the Western showdown Hollywood found both an image of frontier virtues and a parable of moral conflict that could be restated endlessly. After James Cruze directed *The Covered Wagon* in 1923, the Western became the principal medium of the scenic epic. Great directors like John Ford and stars like Cooper, Henry Fonda, James Stewart and John Wayne made the cowboy a figure of courage, honesty and endurance.

7

7 The super-typical American family invented by Hollywood's dream factory appeared in an MGM series about the life of Andy Hardy. As played by the irrepressible Mickey Rooney, this small-town boy represented all the bounce and vigour of American youth without offending anybody's mother. The rose-tinted series was hugely popular and made Rooney himself the top star at the US box-office in 1939.

8

8 Marilyn Monroe was at the height of her fame as a sex symbol when she posed in a scene from *The Seven Year Itch*, (1955), one of two films she made for comedy director, Billy Wilder. But within seven years she was dead of a drug overdose. The warmest and most tragic of all Hollywood sirens, she appeared just as the film capital's ambivalent attitude to sex (long hedged by a "code of decency") was giving way to a less restrictive approach.

9

9 A scene from *The Godfather* (part 1) in which Al Pacino played the vengeful son of a Mafia chief, could have come straight from any of a dozen gangster films made during the 1930s, heyday of Edward G. Robinson and James Cagney. But *The Godfather* and its sequel are films of the 1970s, testimony to the enduring appeal of Hollywood's gangster idiom. Directed by Francis Ford Coppola, *The Godfather* had, by 1973, earned more than any other film in history.

Music from Stravinsky to Cage

The history of serious Western music in the twentieth century has been mostly one of experiment and innovation. Already in the first decades, existing conventions that had governed Western music for centuries were finally giving way under the intense search for new expression in sound.

Experiment and innovation

In fact, harmony (that body of classical rules governing the way sounds were put together, so determining key and to a large extent musical form) had been disintegrating quietly since the lush sounds of Richard Wagner's opera *Tristan and Isolde* (1865). The continuous stream of luxurious chromatic music cultivated by Wagner from then into the 1880s now bloomed in the music of Claude Debussy (1862–1918) into a colourful range of sound patterns.

In art the innovations of the French Impressionist painters (to whom Debussy has been compared musically) were overtaken in a reaction by the Expressionists and Cubists. Similarly in music Arnold Schoenberg (1874–1951) [4] and his pupils Anton Webern (1883–1945) and Alban Berg (1885–1935) [9] in Austria moved directly from Wagner's influence to the exploration of sounds in a more abstract sense.

About the same time, Igor Stravinsky (1882–1971) [2] was writing for the Paris-based Ballets Russes of impresario Serge Diaghilev (1872–1929) a series of vital, imaginative ballet scores – *The Firebird* (1910), *Petrushka* (1911), *The Rite of Spring* (1913). These were rich in asymmetrical, rhythms and orchestral colourings.

In Hungary, Béla Bartók (1881–1945) [5] was composing a vividly personal music strongly rhythmic and striking in its sophisticated use of modal and dissonant folk elements. In America, Charles Ives (1874–1954), was producing prolifically an original if uneven corpus of music that has come to be generally considered America's most individualistic and unconventional.

In Russia, the mystic and harmonically adventurous Alexander Scriabin (1872–1915) called for the projection of coloured lights in his "Prometheus: The Poem of Fire" (1909–11). And at the boundary of music, the short-lived Italian Futurists used the sounds of machine guns, aeroplanes and steam whistles in their new music of noise. The notion of music as "organized sound" was to be taken up in America by an expatriate Frenchman, Edgard Varese (1885–1965), in the 1920s, using percussion that included sirens and whistles.

Continuation of traditional music

Against the excesses of experimentation, composers such as Richard Strauss, Paul Hindemith, Dimitri Shostakovich, Edward Elgar, Aaron Copland, Zoltán Kodály and the group *Les Six* (Poulenc, Milhaud, Honegger, Durey, Tailleferre and Auric) in France continued in a more traditional vein derived from the nineteenth century, as others would for decades afterwards. Stravinsky himself, also in reaction, embarked on a Neoclassical period that lasted from *The Soldier's Tale* (1918) until the *Symphony in C* (1940), a time during which elements of formal restraint characterized his works. This trend was adopted by many composers at the time, as was the slight

CONNECTIONS

See also
208 The twenties and the Depression
264 USA: the affluent society
152 USA: reconstruction to World War I

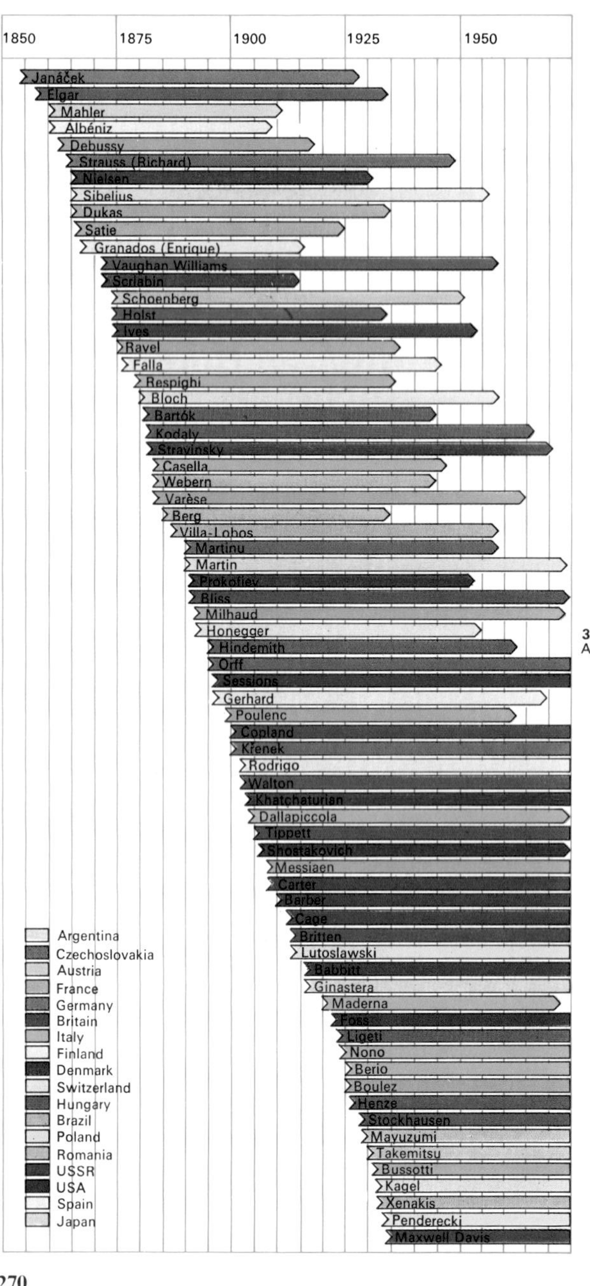

1 These 70 important composers from the rich, diverse 20th-century world of serious Western music, represent many styles. Symphonists in traditions established last century – Sibelius, Shostakovich, Nielsen or Vaughan Williams – are contemporary with the 12-note composition school of Schoenberg, Berg and Webern. Electronic music composers, Stockhausen, Milton Babbitt or Xenakis, contrast with those rooted in a more traditional nationalism like Casella, Falla or Khatchaturian. In recent years an international "modern" style, that depends on abstract notions of sound, has come to be recognized.

2
1750 1775 1800 1825 1850 1875 1900 1925 1950
Russian
Neo-classical
Serial
Russian influences
Teacher
Other early influences
Serial influences
Stravinsky
Glinka
Borodin
Mussorgsky
Tchaikovsky
Rimsky-Korsakov
Wagner
Debussy
Glazunov
Scriabin
Schoenberg
Webern
Berg

2 Stravinsky is one of the giants of 20th-century music, largely because his work shows an outstanding originality through his changes of style. The diagram shows which of his predecessors and contemporaries most influenced him and the stages through which he moved. Born in Russia he transformed his native harmony and rhythms in his early scores, especially for the Ballets Russes in Paris; after World War I a restrained Neoclassical quality informed his works; and from the 1950s until his death he found "serial" music a dynamic inspiration, as in his *Canticum sacrum* (1956).

3 A

B

3 The impact of recording in general on the appreciation and spread of music this century has been incalculable. From the first commercially successful 3-minute shellac discs made by the Italian tenor Enrico Caruso (in 1903 he received the first-ever gold disc for one million records sold of the aria "Vesti la giubba" from Leoncavallo's *Pagliacci*) to the 4-channel quadrophonic and video reproduction of the 1970s, a vast audience outside the concert hall has been given easy contact with every kind of music and performance through records and tapes. The illustration shows old and new styles of recording; Poland's noted pianist and prime minister (1918–20) Ignacy Paderewski (1860–1941) making an acoustic recording at his home in Switzerland in 1911, the sound being cut directly on to a wax disc [A]; and the New Philharmonia Orchestra and chorus under Raymond Leppard recording onto magnetic tapes [B].

influence of emergent jazz seen in music by composers such as Milhaud, Copland, Kurt Weill (1900–50), Walton, Křenek and George Gershwin (1898–1937).

Schoenberg's 12-note system

In reaction on a parallel plane Schoenberg, committed to "the emancipation of dissonance", produced in 1912 the classic *Sprechstimme* (speech-melody) work *Pierrot Lunaire.* This was for five musicians and a reciter who loops and slides through the poems (a composition called by Stravinsky "the mind and solar plexus of early twentieth-century music"). He eventually refined a 12-note method of composition this was to dominate the rest of his work. In this the 12 notes of the chromatic scale are arranged in rows or series (hence "serial music") that replaced traditional keys and harmony.

His pupil Alban Berg did not adopt the 12-note method until his *Violin Concerto* (1935), where he applied it undogmatically and romantically, while Anton Webern showed far-reaching insight into the extensions of music possible through it.

From 1945 a resurgence in "post-Webern" experimental music focused even more on the sounds themselves and the treatment of durations, dynamics, rests and colours, also in fixed series. The piano piece *Mode de valeurs et d'intensités* (1949) by Olivier Messiaen (1908–) [6] became a key work in the evolution of Pierre Boulez (1925–) and Karlheinz Stockhausen (1928–) [Key]. The latter's work was influenced by Pierre Schaeffer's Paris radio studio where, from 1948, *musique concrète* using tape recorders and natural sounds was being advanced [11]. Soon, Stockhausen composed the pioneer *Electronic Study I* (1953), the first piece composed wholly from electronic pure sine waves.

Experiments continued in the integration of "theatre of the absurd" methods into musical creation and performance (Cage [10], Mauricio Kagel, Cardew, and latterly Stockhausen), and in new means of determining the sounds by mathematics and computer (Yannis Xenakis [1922–]). By the 1970s, there were signs of a more disciplined settling-down of the experimental fervour.

KEY

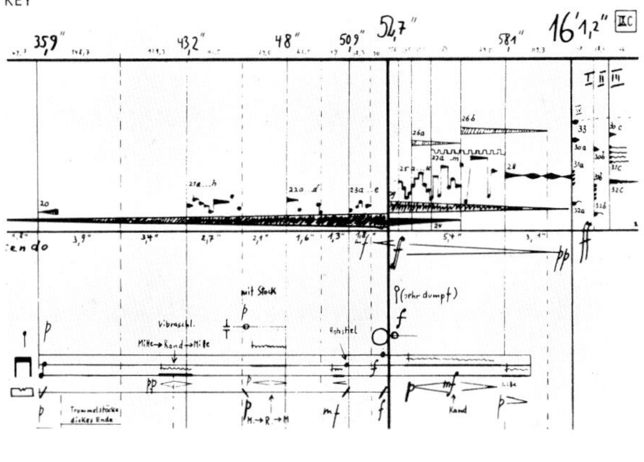

New notation has been a major innovation in music this century. Shown here is a page from the score of Karlheinz Stockhausen's *Kontakte* ("Contacts") for electronic sounds, piano and percussion (1960). The electronic sounds that issue from loudspeakers (indicated by Roman numerals I to IV) in the four corners of a hall, are described graphically above the thick line, while the live sounds made by the two performers are represented below. Time in seconds is given at the top to enable the players to co-ordinate precisely with the tape. The percussion instruments are shown by symbols. Composers also use graphs and drawings.

4

4 Arnold Schoenberg has been as much celebrated in 20th-century music for the dominating influence of his 12-note method of composition as for his own works. Yet his music, from the early Brahms and Wagner-influenced pieces like *Transfigured Night* (1899) to the late (1949) *Phantasy for Violin* with piano accompaniment, reveals a striking, adventurous imagination that is not confined by a rigid method.

5

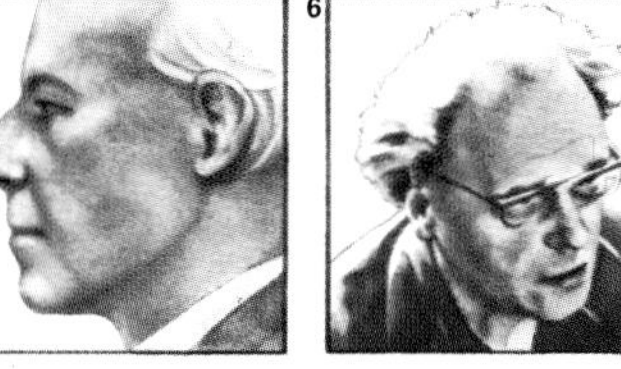

5 Béla Bartók is the most strikingly successful of the modern composers who found folk-music a vivid source of inspiration. Professor of piano at Budapest Academy for nearly 30 years, he began in 1905 to transcribe Hungarian folk-songs on field trips with his friend, the composer Kodály. By the end of his life he had noted and recorded about 8,000 tunes and his music drew imaginatively from their style.

6

6 Olivier Messiaen has been the most durable, imaginative and individually poetic French composer of the 20th century since Debussy. From 1931 organist at the Church of the Trinity in Paris, and a teacher at the Paris Conservatoire since 1942, he has written music characterized by unusual rhythmic series and influenced by oriental melody and plain chant, bird-song and religious themes.

7

7 Benjamin Britten (1913–76) had been for many years the central figure in the development of 20th-century British music. Turning from full Romantic expression he integrated new sounds and classical techniques (influenced at first by Stravinsky and Gustav Mahler) into the English choral and vocal tradition, always with a concern for directness of expression and melodic clarity.

8

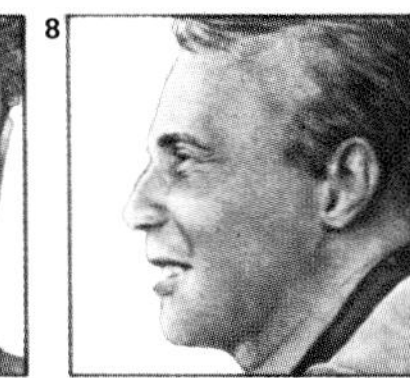

8 Hans Werner Henze (1926–), generally recognized as one of the most outstanding of the younger generation of composers, studied and worked in his native Germany before turning to composition full-time. He has produced opera, ballet, symphonic works, chamber music and music for voices, all of which demonstrate his chief virtue: the constant assimilation of contemporary styles in an original way.

9

9 Alban Berg and Anton Webern were Schoenberg's two most brilliant pupils. Each demonstrated and developed the influence of Schoenberg's ideas and method in his own way, although all three were collectively seen as the Viennese school of early 20th-century composition and were close friends. Webern's very precise music was to have the greater influence later in the century.

10

10 John Cage (1912–) has been a fearless and prolific American explorer of sounds and silences. His "absurd" experiments have had a stimulating influence on *avant-garde* painting, theatre and multi-media happenings. From early performances (1938) on a prepared piano (nuts, bolts, rubbers, etc. between the strings) he has prescribed chance music using several radios, silence – his *4'33"* for silent player(s) – and even funny stories to a piano background.

11

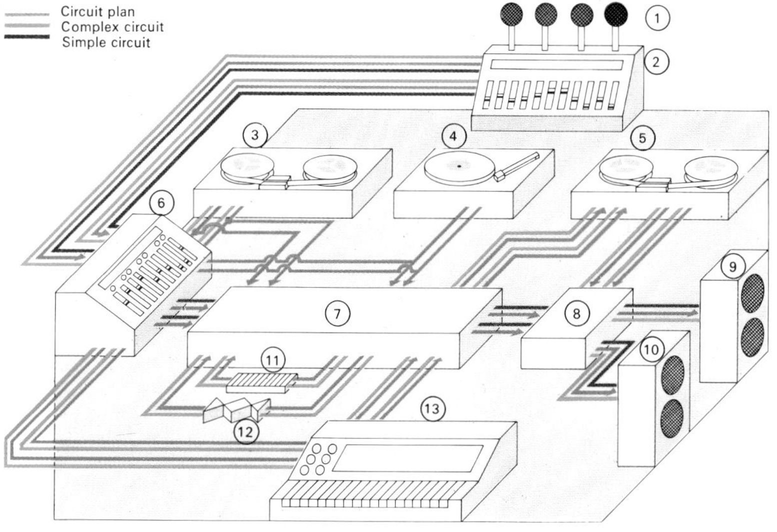

11 A small electronic music composition studio, based on a synthesizer [13], is shown here. The synthesizer's waves, together with signals from microphones and mixer [1, 2], tape decks [3, 5] and record player [4] are modified by the use of mixers [6], filters [11] and reverberation units [12]. All outputs go to a patch-board [7] and then through amplifier [8] and speakers [9, 10] as sounds.

Jazz and pop

Western popular music during the twentieth century has been dominated by the United States, and especially the new forms resulting from the interaction of differing African and European musical traditions of melody, harmony, rhythm and instrumentation. Black artists have played a crucial role, especially in the first half of the century. The folk-music of the slaves [2] had, by 1900, been transformed into a new kind of music – jazz.

The new sound was rhythmic, emotional and vital. It could be joyous or sad and could be played either by a full band or a soloist. Above all, it could be danced to.

The stages of jazz

Jazz went through four main periods – while the wider field of "popular" music tagged along behind. In the period from 1890 to 1917, jazz became popular among most black Americans. The first jazz style, known as ragtime (played on the piano), emerged from St Louis, with Scott Joplin (1868–1917) as its principal exponent. Then came a second jazz style, the classic blues, performed by professional entertainers such as "Ma" Rainey on the music hall and tent show circuit.

New Orleans was not the only town in the United States where jazz could be heard in this period, but it was certainly the most important [3]. In 1910 the city with its 89,000 black population had at least 30 bands. They were small units playing improvised pieces that had developed from parade marching tunes. White Americans called the twenties "the jazz age", but the "jazz" that dominated their dance music was highly diluted and often had little in common with black music, except the syncopation.

In the 1930s jazz suffered badly with the Depression and many musicians were forced to move to Europe. But in 1935, jazz suddenly leapt back in America – this time as "swing". Swing was big-band music and used large brass sections to provide a tidal wave of sound. It appealed particularly to young white audiences, and bandleaders such as Benny Goodman (1909–) and Glenn Miller (1904–44) achieved a popularity as great as the important black big-band leader, Duke Ellington (1899–1974) [4].

The public flocked to hear swing, but many black musicians began to react against the "composed" and "arranged" styles. They wanted to get back to small groups and the chance for greater improvisation.

The development of the blues

The result, in the 1940s, was "bop", a musically sophisticated product of young black musicians such as trumpeter Dizzy Gillespie (1917–) and saxophonist Charlie "Bird" Parker (1920–55) [5]. The blues had also been evolving. The "12-bar" style was not only the basis for much early New Orleans band music, but was used by guitar-soloists as the basis for powerful "folk-blues" or "country-blues", songs about their lives and problems. But as black workers left the farming lands of the south to move to the northern industrial cities, they took their blues with them and the music changed with the new environment. In the cities – particularly Chicago – rhythm and blues was formed. It was played on electric guitars with bass and drum backing and the sound was now harder and more driving.

Only one white American style flourished

CONNECTIONS

See also
274 Classical and modern ballet
268 Hollywood
278 Recent trends in the visual arts

1

1 The development of jazz and pop has been the result of the interaction of two musical forces: black music taken by the African slaves to the USA and white, originally European, folk-music. The USA was the cultural melting pot as black music developed through blues into the various jazz styles and then mixed with the urban rhythm and blues to produce rock. By the 1970s it was a free-for-all.

2 The African slaves brought with them songs that had rhythmic complexity and used certain musical patterns. The most characteristic was the "call and response" pattern. In its most primitive form it could be found in functional songs – work songs and "field hollers". Gangs working in the plantations eased the work with repetitive songs in which the lead singer was echoed by a chorus reply.

3 Louis Armstrong (left, foreground) (1900–71) was born in New Orleans and learned to play the cornet at reform school. Later he met the famed King Oliver (1885–1938) who became his teacher, and whom he replaced in Kid Ory's band. In 1927 he formed his own band. Armstrong became the best known exponent of "Dixieland" jazz, establishing the eminence of the virtuoso soloist.

4 Duke Ellington (left), whom many regard as the most important single talent jazz has produced, was a composer, songwriter, arranger and pianist. The most masterful exponent of big band jazz, he developed a unique style by working on the individual sounds of the first-rate instrumentalists in his band. He gave the blues its finest orchestral form, and wrote "composed jazz" that still left room for improvisation. The subtlety of his orchestration was unique.

5 Charlie Parker was as influential in the 1940s and 1950s as Armstrong had been earlier. Born in the slums of Kansas City, he played in big bands, then rebelled against their repetitive styles to become the leading revolutionary of "bop". His alto sax playing was complex and tortured, but for all his experimentation his roots were in early blues. An unhappy vagabond and drug addict, he has been called the "Rimbaud of modern jazz".

successfully against all this black competition – and that was country music [6], centred in Nashville, Tennessee. There were also a handful of extraordinary, itinerant white folk-singers, who travelled across America in the 1930s. Woodie Guthrie (1912–67) was the most important of them.

By the 1950s the big band jazz era had passed – leaving only ballad-singers and crooners such as Frank Sinatra (1915–) and Bing Crosby (1904–) – and the new "modern jazz" was popular only amongst a minority. The emergent postwar youth culture found a new style by mixing the smoother white country music styles with the energy and aggression of rhythm and blues – "rock 'n' roll" was born. The music was rough, noisy and sexual and its most popular exponents initially were whites. Elvis Presley [7] was its greatest vocalist, although it was black guitarist Chuck Berry who wrote the best rock 'n' roll songs.

By the late fifties rock 'n' roll was all-pervasive and split into several forms. Soloists playing acoustic guitars resurrected folk-songs, and then moved on to write new material, often in protest against social or political targets. Bob Dylan (1941–) [9] brought the new music to respectability by writing intelligent lyrics.

Contemporary rock music

The experiments of the sixties began with a "blues boom" – a mixing of rock 'n' roll with authentic black rhythm and blues styles. Guitarists such as Eric Clapton and Jimi Hendrix mastered the blues, then began to push the music forward in longer, semi-improvised pieces. The "underground" [10] – a youth rebellion against conformity, greatly influenced by drugs – further changed the music. "Acid rock" attempted musically to re-create drug experience through lengthy instrumentals and the use of elaborate lighting. The style started in San Francisco, with bands such as the Grateful Dead and Jefferson Airplane. In Britain, Paul McCartney (1942–) and John Lennon (1940–) of the Beatles [8], came under this influence as they progressed from simple, clever songs to the complexities of the "Sergeant Pepper" album.

KEY

Bessie Smith (*c.* 1898–1937), one of the greatest jazz-blues singers of all time, was born into poverty in Chattanooga, Tennessee. At the age of 11 she began touring the southern states with the Rabbit's Foot Minstrel Show, where she was greatly influenced by "Ma" Rainey. Bessie was extraordinarily popular from 1924 until 1927 when the Depression hit show business and the taste for the blues began to wane. A large, handsome, tragic woman who was alone for most of her life, she sang about the transitory nature of men, money and drink.

6 Jimmie Rodgers (1897–1933), an important country singer and guitarist, was the first man to be installed in Nashville's "Country Music Hall of Fame". The son of a Mississippi railwayman, he himself worked on the railway as a flagman, brakeman and baggage man, but left because of ill health and became an entertainer. He wrote his own songs, which incorporated yodelling with a blues influence.

7 Elvis Presley was born in 1935 in East Tupelo, Mississippi, and moved to Memphis as a cinema usher after leaving high school. He came to the attention of the local record company and became a show business phenomenon by being the first white artist to mix the wildness of black rhythm and blues with country music. He has survived because of his mastery of vocal technique.

8 The Beatles were for eight years – from 1962 to 1970 – the most successful group in the history of popular music. From playing in Hamburg and Liverpool clubs they became a legend, transforming rock 'n' roll with their fine melodies and harmonies. They had a truly progressive and experimental attitude to songwriting and record production which developed, if with uneasy passages, the more successful they became.

9 Bob Dylan was the leader of the "folk-rock" wave that swept America and the UK during the 1960s. His singing was first influenced by Woody Guthrie, in whose style he wrote protest classics such as "Blowin' in the Wind". He later moved to amplified blues styles and has remained a remarkable lyricist.

10 Pink Floyd were originally a London rhythm and blues band but they soon switched to mixed-media experiments and the use of elaborate light shows. By the late 1960s they were the leading British "underground" band. They have pioneered lengthy rock symphonic works using a mass of electronic equipment.

Classical and modern ballet

The Romantic movement, represented by the writings of Byron and the paintings of Delacroix, soon spread to ballet. Dancers abandoned masks and began to act the emotions required in the ballet, thereby effacing the distinction between dance and mime. The techniques of the ballet were expanded to express these moods and emotions.

The first Romantic ballets

La Sylphide, the first Romantic ballet, was first presented in 1832 with choreography by Filippo Taglioni. The role of the sylphide was created by his daughter, Marie Taglioni (1804–84) [2], the dancer most closely associated with the Romantic ballet. It was she who first wore the shortened skirt, still referred to as "ballerina" length.

La Sylphide was the first of many ballets featuring strange and mysterious creatures. Wilis, the spirits of girls who die before their wedding day, appeared in *Giselle*, the undoubted masterpiece of the Romantic era. It was first given in Paris in 1841 with Carlotta Grisi (1819–99) in the title role.

By the middle of the nineteenth century Romantic ballet became merely a vehicle for the ballerina's virtuosity and fell into decline. Dancers, choreographers and musicians turned to Russia, its state ballet school founded in 1735, its artistic tradition kept alive by men such as Marius Petipa (1819–1910), a Frenchman who went to St Petersburg to become principal dancer. He showed off the ballerina to advantage, using the *corps de ballet* (an ensemble of dancers who accompany the lead) usually as a decorative background. One act was often given over to a series of unrelated dances known as *divertissements*, the highlight being the *pas de deux* by the two principals.

The importance of Russia

The Sleeping Beauty [3], generally considered to be Petipa's masterwork, was given its first performance at St Petersburg in 1890 with an inspired score by Tchaikovsky. When Petipa fell sick, Lev Ivanov, his assistant ballet master, took over the choreography of *The Nutcracker* and with Petipa created *Swan Lake*, one of the greatest of all ballets.

The elderly Petipa was quick to appreciate the early work of the young Michel Fokine (1880–1942) [4]. However, Fokine rebelled against many of the traditions of the Petipa ballets. He abolished the antiquated mime and replaced the classical ballet skirts (tutus) with costumes appropriate to the period in which the ballet was set. So it was not surprising that the impresario, Sergei Diaghilev (1872–1929), chose this young rebel to be his ballet master and choreographer in western Europe.

The influence of Diaghilev

During the first season of Russian opera and ballet presented by Diaghilev in Paris in 1909, the Fokine works included *Les Sylphides*, with music by Chopin, and *Prince Igor* (music by Borodin). The company was a tremendous success and Tamara Karsavina (1885–), Anna Pavlova (1881–1931) and Vaslav Nijinsky (1890–1950) [Key] became famous overnight. Four years later Diaghilev broke away from the Russian Imperial Theatres and formed the Ballets Russes, which became one of the greatest ballet companies the world has ever known.

CONNECTIONS

See also
38 Early ballet
106 Music: the Romantic period
270 Music from Stravinsky to Cage
176 Cubism and Futurism

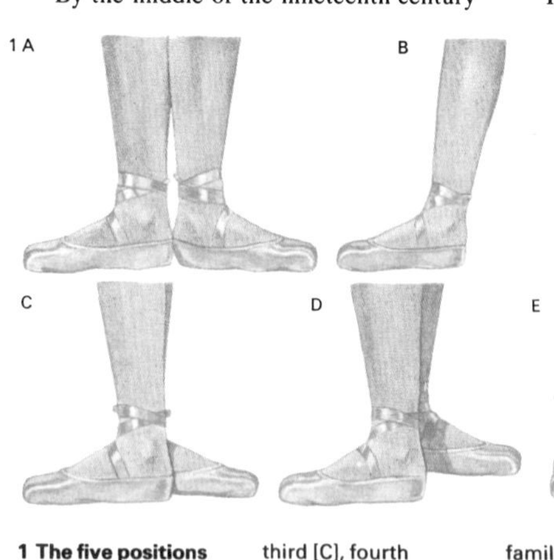

1 The five positions of the feet are the starting point for all ballet steps. In the first position [A], the heels touch; in the second [B] and the fourth [D] the feet are 30cm (12in) apart; in the third [C], fourth and fifth positions [E] the feet are parallel. Turnout is where the legs are rotated outwards from the hips. The arm movements are known as *portes de bras*. The two most familiar poses are the arabesque and the attitude. Steps can conveniently be classified as either jumps (eg the *jeté*), beats (the entrechat) or turns (the pirouette, a full turn on one foot).

2

4 Michel Fokine's *Scheherazade* (1910) first Romantic ballet, danced the part of the sylphide (fairy-like being) who falls in love with James, a Scotsman. Taglioni wore a bodice which left the shoulders bare, a knee-length muslin skirt, tights and pink satin point shoes, and this has become the accepted costume for the Romantic ballerina. *La Sylphide*, with different music by Lvenskjold and new choreography by August Bournonville, has been in the repertory of the Royal Danish Ballet in Copenhagen since 1836. The leading roles in the ballet are now most closely associated with Margarethe Schanne and Erik Bruhn.

3

3 *The Sleeping Beauty* was chosen by the Royal Ballet – at that time the Sadler's Wells Ballet – to reopen the Royal Opera House, Covent Garden after World War II. Margot Fonteyn and Robert Helpmann (1909–) dance in the last act, sometimes given on its own as *Aurora's Wedding*. London had seen nothing so splendid as the sets and costumes by Oliver Messel since the Bakst decor for the Diaghilev production at the Alhambra Theatre in 1921. Petipa's masterpiece (1890) is the cornerstone of the Royal Ballet repertory. The version most often mounted today is by the present company director, Kenneth MacMillan, with sets by Peter Farmer.

Diaghilev felt that ballet was part of a complex spectacle made up of poetry, literature, painting, music and choreography, and he tried to gather all these elements in the ballets created by his successive choreographers: Fokine, Nijinsky, Léonide Massine (1896–) Bronislava Nijinska (1891–1972), George Balanchine (1904–) and Serge Lifar (1905–). When he died in 1929 his company disbanded and dispersed, spreading his ideas throughout the Western world. Marie Rambert (1898–) came to London and formed what became the Ballet Rambert. At about the same time Ninette de Valois (1898–) formed the company that today is known as the Royal Ballet. De Valois's first ballerina was Alicia Markova (1910–) and when she left Margot Fonteyn (1919–) flowered into the *prima ballerina assoluta* of British ballet.

Serge Lifar stayed in Europe and became dancer, ballet master and choreographer at the Paris Opéra. George Balanchine [6] went to the United States and became director and choreographer of New York City Ballet. However, companies which already had a strong tradition of their own were not so affected by Diaghilev. Thus the Royal Danish Ballet in Copenhagen has continued to train dancers in the style of August Bournonville (1805–79), and in Russia the two major companies – the Bolshoi Ballet in Moscow and the Kirov (formerly the Maryinsky) Ballet in Leningrad – still present their post-revolutionary works with Soviet themes and spectacular dancing as the principal ingredients of the display.

There have been many developments in modern dance both in America and Europe since the pioneering work of Isadora Duncan (1878–1927), most notably from Martha Graham (1893–) [5]. Subject matter has become more realistic and the use of dance in the cinema has greatly expanded this genre; for example in *West Side Story* (1961) Jerome Robbins (1918–) devised modern ballet sequences for a highly successful film. With the current proliferation of dance companies throughout the world and a growing audience for both classical and modern ballet, there can be no doubt that ballet will continue to flourish for many years to come.

KEY

Vaslav Nijinsky was probably the most accomplished male dancer of this century. Fokine created several ballets for him, including *Petrushka* and *Scheherazade*, but Nijinsky also created several for himself. His performance, shown here, in his own *L'Après-midi d'un faune* (1912) caused a scandal and there was a riot at the première of his *La Sacre du Printemps* in 1913. Madness ended his career after only eight years.

4

4 Michael Fokine's *Scheherazade* (1910) had Ida Rubinstein as the Shah's favourite wife and Vaslav Nijinsky as her Negro slave. This was one of several Oriental ballets given by Diaghilev. Léon Bakst's brilliantly coloured decor had a great influence on fashion and interior design.

5

5 Martha Graham, seen here in her ballet *Hérodiade* produced in 1944, has created an entirely original style of dancing, a school of dance and a company in New York. The Graham dancer places much less emphasis on leaving the floor and executing the dance in mid-air than the classical dancer.

6

6 George Balanchine, artistic director of the New York City Ballet, has created numerous ballets for it since 1948. His company and the American Ballet Theatre are the two foremost classical companies in the city. Perhaps most notable have been the works he produced with Igor Stravinsky, including *Apollo Musagetes* (1928), choreographed for Diaghilev, and *Agon* (1957). *Agon* (shown here) is a plotless one-act ballet danced in black-and-white practice costume to a twelve-note musical score. This ballet, together with other works from the repertory, including *Dances at a Gathering*, choreographed by Jerome Robbins, is also performed by Britain's Royal Ballet.

7

7 *Marguerite and Armand* was the first ballet to be created for Margot Fonteyn and Rudolf Nureyev. Frederick Ashton used music by Franz Liszt and decor by Cecil Beaton to recreate in dance the familiar story of the Lady of the Camellias. Ashton's collaboration with Fonteyn during the formative years of the Royal Ballet produced masterworks such as *Ondine* and *Symphonic Variations*. Nureyev's partnership with Fonteyn resulted in a wonderful *pas de deux* in *Le Corsaire* and a memorable *Giselle*. He not only dances in a wide range of ballets and styles but also has produced several of the classics, including a sumptuous *Sleeping Beauty*.

Cinema as art

The cinema as a mirror of man's thoughts combines and extends the ancient arts of painting, music, dance, theatre, literature and architecture. It is the most persuasive and total medium of creative expression, ranging from a close reproduction of reality to the most extravagant fantasies.

Mainstream influences

At the outset of the sound era three main styles had emerged in cinematic art. The first was montage, a method of editing and reassembling isolated shots pioneered by D. W. Griffith and refined by Sergei Eisenstein. The second was *mise-en-scène*, in which the director attempted to present a view of life by planning longer narrative sequences as Erich von Stroheim (1885–1957) had done in 1923 in *Greed* [2]. The third was documentary, a journalistic approach which derived from the sensitive study of Eskimo life, *Nanook of the North* [1], by Robert Flaherty (1884–1951). The history of films as art is largely the story of how these three styles have been used and blended by directors seeking to express their individual vision.

The coming of sound in 1927 was resisted by many who felt that the unique art of the silent film would be debased and restricted thereby. One of the first to show that sound films could persuade, move and inspire was Lewis Milestone in a 1930 adaptation of Erich Maria Remarque's anti-war novel, *All Quiet On The Western Front.* With a few exceptions, such as Joseph von Sternberg's (1894–1969) *The Blue Angel* (1930), the pressures of commerce in America and of propaganda in Russia and Germany [3] hampered the use of film as a medium of personal expression during much of the 1930s.

More independent work emerged in France where major production studios had collapsed during the Depression. Jean Vigo, René Clair (1898–) and Marcel Pagnol (1895–) used film to satirize or reflect the mood of their country or to turn ordinary life into poetry. In *La Grande Illusion* (1937) and *La Règle du Jeu* (1939) Jean Renoir (1894–) [4] foreshadowed modern cinema by shaping his films round an idea rather than a well-made plot and by resisting fragmentation of the camera's view. A more startling enlargement of the camera's range of vision came with *Citizen Kane* (1941), Orson Welles's original recapitulation of a newspaper magnate's life [5], which used newsreel techniques developed in the late 1930s.

The realistic tradition

In the mid-1920s, the Soviet director Dziga Vertove had argued for a "cinema of actuality". The most successful attempt to find drama in the reality of working life emerged in Britain with the documentary work begun by John Grierson and continued by Basil Wright during the 1930s. In the postwar era, realism influenced feature films made by Italian directors such as Roberto Rossellini (1906–), Vittorio de Sica (1902–76) and Luchino Visconti (1906–), who improvised action on real locations, often using untrained actors [6]. The use of untrained actors became an article of faith for French director Robert Bresson (1907–). The end of the war also brought a resurgence of the poetic French tradition with outstanding work by Clair, Cocteau and Marcel Carné.

In the 1950s the growth of film societies

CONNECTIONS

See also
268 Hollywood
200 Origins of film
278 Recent trends in the visual arts

1 Eskimo life with its stark daily battle for survival inspired a film that is generally regarded as the starting-point of the documentary, *Nanook of the North* (1921). Its director was an explorer, Robert Flaherty, who set out not to make a scenic travelogue in the conventional style of the day but to show the humour and tenacity of an Eskimo hunter and his family. This creative treatment of reality led on to British documentaries of the 1930s which took the camera into the daily lives of working people, pioneering many techniques (such as synchronous sound interviews) that would later be used for in-depth television reporting.

2 *Greed*, a penetrating study of human behaviour set in San Francisco before 1914, was the product of an individualistic actor and director, Erich von Stroheim. His film was slashed to about a quarter of its original 10-hour length but remains a significant early example of the way in which the camera could reveal life by dwelling on a group of characters carefully manipulated by the director.

3 The use of film as a political weapon originated in Russia but was perfected by the Nazi regime in Germany during the 1930s when the Goebbels ministry brought a previously energetic film industry under almost complete control. Leni Riefenstahl in *Olympia*, her masterly film of the 1936 Olympics, was alone able to reconcile the demand for propaganda with creative use of the camera's possibilities.

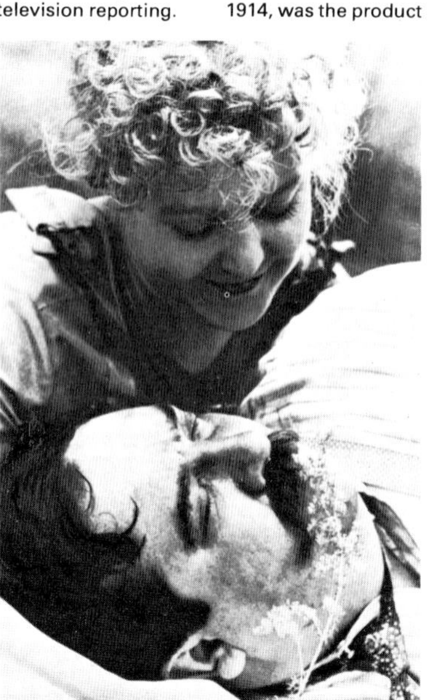

4 Lyrical photography and warm observation of ordinary people distinguished *Une Partie de Campagne*, Jean Renoir's 1936 film based on a story by Maupassant. Renoir, son of the painter Auguste, broke away from the vogue for montage editing in several films he made during the 1930s, preferring fluid use of his camera and composition of scenes in depth, a technique that was further developed in the postwar period.

5 Orson Welles starred in his own film *Citizen Kane*, a devastating study of the psychology of a business tycoon who bore a close resemblance to the newspaper owner William Randolph Hearst. Although he was only 25, Welles established himself immediately as a director of rare ability, combining a dynamic use of simulated newsreels, weird camera angles and original lighting and sound effects with photography of unusual depth.

and international film festivals introduced Western audiences to the artistry of Asian films. Akira Kurosawa's *Rashomon* [7] and Teinosuke Kinugasa's *Gate of Hell* won major prizes for Japan in 1951 and 1954 while Satyajit Ray's (1921–) story of Bengal life, *Pather Panchali*, was acclaimed at the Cannes film festival in 1956. The tormented personal statements of the Swedish director Ingmar Bergman (1918–) [8] found a wide audience. In Italy the potential of better lenses was exploited by Michelangelo Antonioni (1912–) [9], who used landscape to express the inner world of his characters, and by Federico Fellini (1920–), whose influential films reworked parts of his own life and fantasies.

The New Wave and modern cinema

The decline of commercial cinema in the 1950s left a large minority audience of discriminating filmgoers. Together with the availability of technically refined 16mm film and equipment, this led to the re-emergence of wholly individualistic *avant-garde* film, spearheaded by Maya Deren and Stan Brakhage in America. Although rarely surfacing in a commercialized way [10], the *avant-garde* has become an international art form with techniques that vary from fast-cut distorted images to half-hour takes of whatever passes before the lens.

In the commercial cinema a new group of film-makers, led by the New Wave directors in France, took up the idea of the author-director controlling a small team to realize his personal vision. The disruption of hardening conventions by directors such as Jean-Luc Godard (1930–) and François Truffaut (1932–) [Key] fed back into mainstream film-making everywhere. Even in Eastern Europe, where political control of the cinema was weakening a little, Milos Forman (1932–) and Ivan Passer pushed forward naturalism, and in Hungary Miklos Jancso choreographed intricate patterns of actors and camera movement in ultra-long takes.

Since 1960 the availability of silent, hand-held cameras and lightweight tape recorders has allowed the development of *cinéma vérité* – a style that is able to record more of the actuality of life than ever before.

KEY

François Truffaut gave an insight into the mechanics of modern film-making in *La Nuit Américaine* (1973) which showed him directing in the matter-of-fact style of the New Wave. This is a term that was used to describe a group of French directors led by Truffaut himself, Jean-Luc Godard, Alain Resnais and Claude Chabrol who, at the end of the 1950s, made films in an improvising, free-moving style which broke with many conventions of narrative films. Truffaut's *Jules et Jim* (1961) was among several of their successful films.

6

6 Neo-realism gained a world audience with the release in 1945 of *Rome, Open City*. In both this and a subsequent film, *Paisa*, Roberto Rossellini used a mixture of trained and untrained actors, real and staged action, to reconstruct the story of the Italian partisan movement with a startling sense of authenticity. The neo-realists, led by a writer, Cesare Zavattini, wanted to discard falsehood and take the camera into the streets and fields to film actual situations that would express the drama of life in post-war Italy. Until sentimentality crept in, the movement produced some fine work, notably in *Bicycle Thieves* (1948).

7 *Rashomon* established the reputation of Japanese film-makers when it won the Grand Prix at Venice in 1951. Its use of telephoto lenses was eagerly taken up by many Western directors.

7

8 In *The Seventh Seal* (1957), the black-draped figure of Death was one of many powerful images used by the Swedish director Ingmar Bergman to explore the meaning of human life, Bergman set this and other films in medieval Europe, making heavy use of symbolism and allegory and drawing intense contrasts between happiness and suffering. He has also made perceptive studies of contemporary life and particularly of marital relationships. Although enigmatic and oppressive, his films have won a worldwide intellectual following.

8

9

9 The brooding face of Monica Vitti was used by Antonioni in *L'Avventura* (1959) in audaciously slow-paced sequences that explored the thoughts of his characters and the emptiness of their lives.

10 Underground films reached the surface in *Flesh* (1969), Paul Morrisey's film that exploited in commercialized form the personalities and methods previously used by the pop artist Andy Warhol.

10

Recent trends in the visual arts

The visual arts since the mid-1950s are unprecedented in their variety. Never before has the definition of art included so many different kinds of activity. The diversity of today's art phenomena embraces such creations as Gilbert and George's "Singing Sculpture" [6], the miniature fish farm devised by the American artist Newton Harrison as a demonstration of a possible solution to world food problems, and Conrad Atkinson's written and visual record of social injustices and inequalities. All these are unified by a single factor – the art gallery, the context in which they appear.

It is paradoxical that the very system denounced by many contemporary *avant-garde* artists as a symptom of a corrupt society should provide a fertile arena for their activities. But art has developed to a point at which it queries its own existence and this is partly an aspect of the current self-doubts of Western civilization.

One can trace the dilemma of art in the mid-1970s back to problems confronted by the pioneers of abstract art in the early years of the century. Both Kasimir Malevich (1878–1935) and Wassily Kandinsky (1866–1944) discovered that, if abstraction was to develop into an independent means of expression, the formal elaboration of the image brought with it unwanted associations with the outside world. If abstraction was an impossibility how was painting to continue without a return to discredited illusionism?

Texture and op art

Among European artists the solution was generally to provide some additional interest to compensate the spectator for the lack of a subject. This could take the form of an exploration of texture. The outstanding exponent of this procedure is the Spaniard Antoni Tapies (1923–) [3]. Another solution is an offshoot of the constructivist tradition involving the creation of patterns that give the appearance of movement. Popularly known as op art, the best known examples are the "dizzying" compositions of Bridget Riley (1931–).

Because a mistrust of any kind of illusion is a characteristic of postwar artists, the tendency towards the presentation of real movement – "kinetic art" [4] – is not surprising. This was not a new idea; since the early 1930s Alexander Calder (1898–) had been producing delicate and colourful mobiles.

Action painting and flat colour

More radical views of what art could be about are suggested by developments in the USA. The action paintings of Jackson Pollock (1912–56) encouraged an unprecedented awareness of the painting as an object. Not only could the spectator participate vicariously in the excitement of painting them, but their "all-over" pattern resisted the kind of spatial interpretation that had actually been courted by earlier abstract painters. Painting had long since ceased to be regarded as a beautiful girl or an uplifting moral scene. Now its existence as a coloured flat surface seemed to stand up without recourse to the spiritual justifications sought by Kandinsky.

The influential American art critic Clement Greenberg (1909–) proposed that the essential development of modern art was towards each medium divesting itself of qual-

CONNECTIONS

See also
280 Art and architecture in 20th century Britain
284 Scottish culture since 1850
204 Abstract art
202 Dada, Surrealism and their legacy
176 Cubism and Futurism
174 Fauvism and Expressionism
270 Music from Stravinsky to Cage
272 Jazz and Pop

1

1 Richard Hamilton's "Just what is it that makes today's homes so different, so appealing?" Effectively launched British "Pop Art" in 1956. In contrast to the directness of Warhol, Hamilton creates a complex composition using collages to incorporate images from a wide variety of sources including romantic comics and advertising material. A figure from a body-building magazine is the dominant image.

2 In his "Veils" series of the 1950s, Morris Louis tried to establish a format permitting the purest possible experience of colour. In his later paintings colours were not merged but separated into stripes.

3

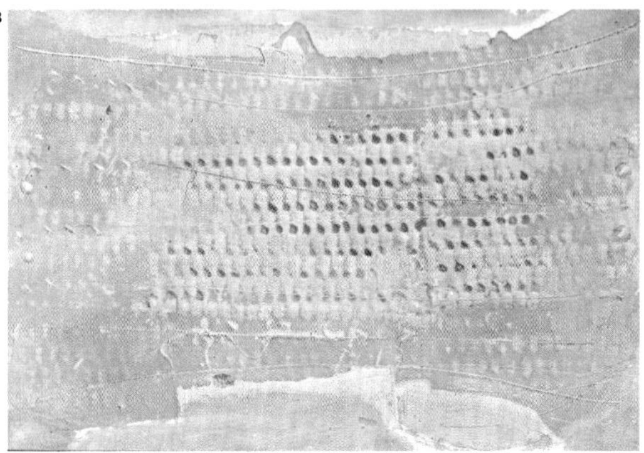

3 Many Spanish painters have specialized in texture painting. Here, in "Composition" (1958) by Antoni Tapies, sand and plaster are used in a way that evokes crumbling walls. The general air of desolation and delapidation is characteristic of much recent European painting and sculpture. It is a kind of anti-aestheticism in reaction to the over-refined, rather bloodless abstraction produced by fashionable painters in the immediate postwar years.

4

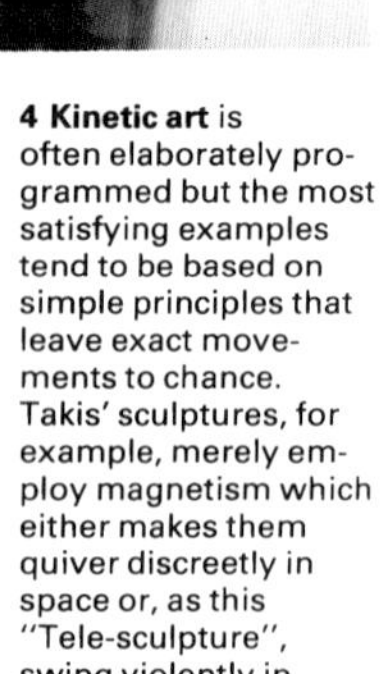

4 Kinetic art is often elaborately programmed but the most satisfying examples tend to be based on simple principles that leave exact movements to chance. Takis' sculptures, for example, merely employ magnetism which either makes them quiver discreetly in space or, as this "Tele-sculpture", swing violently in an arbitrary motion around the electrified coil. In other works Takis has added sound by causing magnetic vibrations against wires or gongs. When several of these sculptures are exhibited together and their movement amplified the result is undeniably powerful but also demonstrates much showmanship.

ities belonging to other forms. A painter particularly encouraged by Greenberg was Morris Louis (1912–62), whose respect for the flatness of painting was such that he even soaked his surfaces with colour rather than disturb their two dimensionality [2].

A neat way of producing a two-dimensional painting is to paint a flat subject as Jasper Johns (1930–) did when he painted straight renditions of targets and flags. Another possibility is to use an image already processed into two-dimensional form, such as the blown-up frames from cartoon strips of Roy Lichtenstein (1923–) or the standardized "Marilyn Six-Pack" [Key] of Andy Warhol (1930–). The commercial sources of these works has led to the term pop art which has also been applied to British artists, such as Richard Hamilton (1922–) [1], who define their stance towards their material.

Trends in sculpture

Sculpture, like painting, has become an art form primarily concerned with itself. In order to divest sculpture of its associations with the human figure, sculptors made use of such devices as brilliant synthetic colour, expanded it to an inhuman scale and fragmented its traditional monolithic form [5]. One extreme development known as "minimalism" reduces sculpture to such simple elements that one is forced into contemplation of the work's basic physical nature or into a total loss of interest in it.

This preoccupation of the traditional media with their own specific problems, generally branded as "formalism", has led many artists away from them. The much sought after freedom of the artist in capitalist societies is seen as a trap to involve the artist in the problems of art alone, a feeling enhanced by the suspicion that American Abstract Expressionism was a propaganda weapon in the cold war. The alternative, however, has tended to take the form of an attempt to escape from the commercialism of the art-world by avoiding the creation of conveniently saleable artworks. This explains the development of both performance art and earthworks, which are immovable from the site at which they are made.

KEY

Andy Warhol's "Marilyn Six-Pack" (1962) shows the uncompromising directness with which American Pop artists present their material. Warhol's subject-matter has always become a standardized image before it reaches him. Warhol comments on the way in which the media dilute experience by processing it. The repetition of the motif is a device also used by other painters and sculptors. With Warhol the result is that we are more aware of technical factors that differentiate each section rather than the content of the image. Other artists aim to break with tradition in composition, and attempt to create works in which all the parts function in an equal fashion.

5

5 "Bird in Arras IV" (1969) is by Tim Scott (1937–), one of a group of British sculptors who gained prominence in the mid-1960s using brightly coloured shapes in synthetic material. The group was primarily concerned with creating an abstract sculpture by completely removing any figure conventions and by avoiding mathematical methods of composition that leave little to the imagination.

6

6 Few artists have merged art and life as totally as Gilbert and George in their performance of "Singing Sculpture", in which they mimed mechanically for eight hours to Flanagan and Allen's "Underneath the Arches". Their art is the expression of elaborately created personae – a parody of respectability which is the limit in non-conformity. Perhaps the ultimate attempt to link art and reality is the American Alan Sonfist's bequest of his body to the New York Museum of Modern Art. He considers that the process of decay will be his final art-work. Some Austrian artists have already resorted to the disembowelling of animals.

7

7 Mathematical systems are the basis of the compositions of many of today's artists. The resulting combination of repetition and change can give rise to a special optical resonance of the kind deliberately cultivated by Victor de Vassarely (1908–), a Hungarian-born painter working in Paris, who is often regarded as the inventor of op art. Although totally abstract, his paintings create potent sensations of space, movement and volume.

8

8 Packaging is the one idea that dominates the work of Christo. This idea he has carried out with astonishing consistency and thoroughness, graduating from supermarket trolleys, through shop fronts, to part of the Australian coastline. Since such works cannot be exhibited in galleries, they are economically dependent for their very expensive realization on the sale of documentary "souvenirs", such as this drawing of a valley in Colorado which was covered by Christo with an enormous curtain. While the traditional arts have become totally immersed in their own specialized problems, an art such as Christo's perhaps demonstrates a means of escaping from some of the older limitations.

Art and architecture in 20th-century Britain

Twentieth-century British art and architecture is mainly a history of a varying relationship with European and American movements. For despite the presence of a number of internationally important artists, Britain did not emerge as an innovatory centre.

Painting and sculpture before 1940

There was little contact in the first decade of the century with the more advanced tendencies of European art. The British vanguard was represented by Philip Wilson Steer (1860–1942) and Walter Sickert (1860–1942), who practised what was essentially a local version of Impressionism, although Steer was also affected by the Aesthetic movement. In 1910 and 1912 the critic and painter Roger Fry (1866–1934) organized two Post-Impressionist exhibitions that brought to London paintings by Van Gogh, Cézanne, Matisse and the Cubists. The general public was outraged but the impact led to a remarkable, if short, burst of innovatory art between 1911 and 1914.

Although the central figure of the Camden Town Group, formed in early 1911, was Walter Sickert, its younger members, including Spencer Gore (1878–1914) and Harold Gilman (1876–1919), applied the monumental simplicity of Gauguin and the colouristic freedom of Van Gogh to scenes of everyday life. Fry himself founded the Omega Workshops, in which the most notable painters were Duncan Grant (1885–) and Vanessa Bell (1879–1961). Fry's theoretical opposition to narrative and illustrative art helped to lead Grant and Bell to take up abstract painting by 1914.

The most radical of all these movements was Vorticism. The Vorticists were a loosely knit group in which the painter and writer Wyndham Lewis (1884–1957), the painters David Bomberg (1890–1957) [1] and Henri Gaudier-Brzeska (1891–1915), and the sculptor Jacob Epstein (1880–1959) figured. Like the Italian Futurists, whose influence they scorned, they wanted to create a harsh, precise and mechanistic art for the new age.

The brilliant and superficial portraits of Augustus John (1878–1961) and the Neo-classicism of Mark Gertler (1892–1939) in the 1920s mark a return to conservatism. Although the erection of Epstein's sculpture "Rima" in Hyde Park in 1925 created a scandal, the strident primitivism of Epstein's work had little to do with the progress of art on the European continent. But in 1929 Paul Nash began a series of pictures which could not have been created without an appreciation of de Chirico and Metaphysical painting and of Surrealism. Henry Moore (1898–), following Constantin Brancusi took as his starting-point the substance of the material with which he was to work. Although the human figure remained his central theme, after 1934 Moore's work and the sculpture of Barbara Hepworth (1903–76) reflected both surrealistic and abstract aims. Her carvings are neither representational nor geometric, but evoke natural processes. The paintings of Ben Nicholson (1894–) pursued the pure Abstractionist ideas of Piet Mondrian and the De Stijl movement.

Developments since 1940

In the 1940s a new romantic trend emerged, perhaps encouraged by wartime isolation. The former Abstractionist John

CONNECTIONS

See also

1

1 David Bomberg's "In the Hold" (1913–14) is not a fully abstract painting: traces of the original drawings of figures in movement are still visible, such as the man bending over on the left. The way the picture is divided by a grid so that it fragments into multi-coloured facets is anti-illusionistic, however, and prophetic of later Abstractionism. The British Vorticists, led by Wyndham Lewis, in opposition to Italian Futurism, did not seek to express dynamic movement but believed the spirit of the age to be best suited by a harsh, static geometry. The Vorticists were in violent reaction against both Victorian taste and ideals and the atmospherics of Impressionist painting.

2

2 "Event on the Downs" (1934) by Paul Nash exemplifies a native tradition of visionary landscape being transformed and enriched by interaction with the *avant-garde* in Europe. There is a Surrealist incongruity in the confrontation of tree-stump and tennis-ball, but the point is not shock. By the implied movement of the ball rolling down the slope, Nash expresses the rise and fall of hill and vale. Nash's powerful placing of objects in a landscape reappears in his haunting pictures of German bombers brought down in England during World War II.

3

4

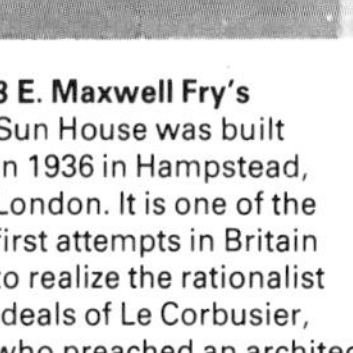

3 E. Maxwell Fry's Sun House was built in 1936 in Hampstead, London. It is one of the first attempts in Britain to realize the rationalist ideals of Le Corbusier, who preached an architecture based on strict engineering principles. The house is made of concrete, glass and steel used quite explicitly and uncompromisingly. No applied ornament clouds its visual purity.

4 "The Eclipse" (1950) by Victor Pasmore (1908–) is also called "Square Motif, Blue and Gold": it is transitional between a representational and an abstract stance. Pasmore was known in the 1940s for Whistlerian studies of the Thames, and his conversion to abstract painting caused a stir: non-figurative painting was still a controversial issue in the 1950s.

Piper (1903–) began painting evocative landscapes, Lawrence Lowry (1887–1976) and Graham Sutherland (1903–) exploited the potential of industrial scenes.

After 1945 many painters turned new to Abstractionism. Some, such as Kenneth Martin (1905–) were attracted to the Constructivist tradition. Others, such as William Scott (1913–), explored a more subjective and personal style. Figurative painting was vitalized by the disturbing private images of Francis Bacon (1910–), Lucian Freud (1922–) and Michael Ayrton (1921–75). The most successful sculptors of the 1950s such as Kenneth Armitage (1916–) were making powerfully expressive use of stylized figures.

The impact of American art at the end of the 1950s influenced younger artists towards experiments in pure form without content, most notably in the sculpture of Anthony Caro (1924–). Pop Art seemed particularly appropriate to the colourful, youth-orientated commercialism of the 1960s, in which the painter David Hockney (1937–) [6] figured. Recently, in Britain as elsewhere, artists have been moving away from the traditions of painting and sculpture towards performance, Land Art [8] and political discourse.

Modern British architecture

The Art Nouveau designs of Charles Rennie Mackintosh (1868–1928) for the Glasgow School of Art at the turn of the century had no successors, and early twentieth-century building was characterized by a neo-baroque eclecticism, practised most successfully by Edwin Lutyens (1869–1944). Functionalism first appeared in Britain in 1926, in a house in Northampton by the German Peter Behrens (1868–1940). Soon men such as E. Maxwell Fry (1899–) [3] and the partnership of Cornell, Ward and Lucas began to make full use of metal and reinforced concrete. Berthold Lubetkin (1901–) pioneered highrise residential blocks with Highpoint 1 in Highgate, London, in 1935. Since 1945 extensive government building has made the modern style almost official and its ethos of efficiency has made it equally attractive to commercial interests.

KEY

"Reclining Figure" is a stone sculpture of 1938 by Henry Moore. He had early reacted against the academic tradition, looking to primitive art as the more vital tradition. He dogmatically believed in direct carving but exhibited with both the Abstractionists and the Surrealists in the late 1930s, although he vigorously opposed these movements at various times. In this carving, fluid holes vary the monotony of the surface while its weathered texture lets it blend into the landscape.

5

5 The Royal Festival Hall was built for the Festival of Britain of 1951 by Leslie Martin (1908–) and Robert Matthew (1906–). It signalled acceptance of the new style. Contemporary influences from Europe and America are evident in the way the acoustics dominate the design of the interior of the auditorium.

6

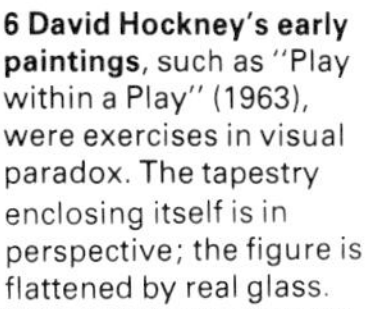

6 David Hockney's early paintings, such as "Play within a Play" (1963), were exercises in visual paradox. The tapestry enclosing itself is in perspective; the figure is flattened by real glass.

7

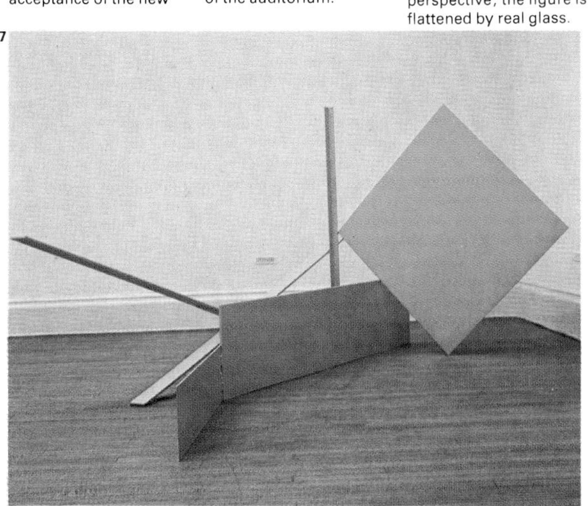

7 Anthony Caro made "Yellow Swing" from welded steel plates in 1965. His characteristic use of colour and refusal to provide any base for his sculpture challenged American art on its own terms and influenced his pupils at St Martin's School of Art, London. The aim of "Yellow Swing" is to create an active relationship with the ground and with the spectator. Caro later added textural interest to the formal interplay by using rusty scrap steel and visible welds. The rejection of both monumental function and domestic scale has made such sculpture dependent on a wealthy art world for support.

8

8 The Land Art movement in Britain is represented by environmental works such as "England" by Richard Long (1945–). Made in 1968 by pulling off the heads of daisies, "England" exists only in the form of documentary photographs. Long has not only taken art into the landscape, he has brought sticks and stones into the gallery.

Irish culture since 1850

The Irish Famine (1846–7) greatly accelerated the linguistic change which had been in progress in nineteenth-century Ireland. Landless labourers in the west were the chief victims and of these a large proportion were Irish-speaking. The Irish language, and such vernacular literature as was still produced, now became quite marginal. Where the language was still spoken, it became such a great social disadvantage that the practice of punishing children in school for speaking Irish was not only tolerated but approved.

Cultural revival

Later nineteenth-century Ireland, therefore, saw the process of anglicization brought virtually to completion. The older culture was celebrated sentimentally as something consigned to the past. The ballads of Tom Moore (1779–1852) remained popular, and nationalist sentiment ensured the popularity of such minor poets as Thomas Rolleston (1857–1920) and Samuel Ferguson (1810–86). The drawing-rooms of Dublin and Belfast accepted the change readily enough; but nineteenth-century Ireland was mainly rural and traditional: an underlying resentment at the destruction of cultural identity soon produced powerful reactions. The Famine was hardly over when language revival movements were being attempted in the provinces. Then, in 1884, in County Tipperary, one of the most important cultural institutions of modern Ireland was founded, designed to encourage ancient games such as hurling [3] and to discourage "foreign" games associated with the English. The Gaelic Athletic Association (GAA) checked anglicization by separating the Irish people, in their sports, from those associated with the British garrisons.

In 1893 a small group in Dublin founded *Conradh na Gaeilge* (the Gaelic League [1]), the most important of the linguistic revival organizations. This had remarkable success, especially in the establishment of nationwide classes to teach Irish.

Irish drama and literature

The literary revival – in English – was not a single or a unified movement. Many people belonging, at one level or another, to the half-world of Anglo-Ireland were attracted by the sentimental mistiness of the "Celtic twilight". Those with real talent, and some with genius, went on to make contact with the reality behind the mists.

William Butler Yeats (1865–1939) [2] found in the speech and memory of the peasant and in what he himself understood of Gaelic literature, the recollection of a clean, heroic world, far removed from the commercialism of his own time or the banalities of the popular press. George Bernard Shaw (1856–1950) found realism in Ireland and tilted at cant in England. John Millington Synge (1871–1909) appreciated the realism but heard closely the heroic echoes, James Stephens (1882–1950) perceived the fantasy of the old literature, and less significant writers, such as Æ, (George Russell) (1867–1935) hovered on the edge of Celtic whimsy. James Joyce (1882–1941) [5] is somewhat separate. He was rooted not in the world of Anglo-Ireland, but in the native tradition, which he did not regard as exotic. In his novels he wrote freely about the familiar angli-

CONNECTIONS

See also
26 Irish culture to 1850
162 Ireland from Union to Partition
100 Poetry and theatre in the 19th century
290 Emergent literatures of the 20th century
294 Ireland since Partition

1 The Gaelic League *(Conradh na Gaeilge),* a society founded in Dublin in 1893 for reviving the Irish language and publishing in Irish, included among its early members Stephen McKenna, William Gibson and Lord Ashbourne (1837–1913), seen here. Its language classes achieved great success and, although founded purely as a cultural movement, it became a nursery of nationalist revolutionaries. One of its founders, Eoin MacNeill, headed the Irish Volunteers in the 1916 rebellion.

2 Yeats was the leading poet of the literary revival. He went through many phases of development, in all of which he was passionately involved with his idea of Ireland. As an Anglo-Irishman, he was initially attracted by the exoticism of the ancient traditons and beliefs of Sligo and other parts. But he lived in several different worlds, one of which was London, the capital of an empire. As he developed, he drew much on other cultures: on Japan for theatrical modes, on India for spirituality, and on worlds of the past for a vision of aristocratic perfection beyond the reach of the philistinism he despised. Much of Yeats's best work belongs to the last ten years of his life.

3 Hurling, the national game of Ireland, is a tough, fast field-sport with a superficial resemblance to hockey. The All-Ireland championship final in Dublin is an annual highlight. It is one of several traditional Irish games, such as Gaelic football, that were cultivated by the Gaelic Athletic Association with a consciously revivalist and de-anglicizing purpose. Members of the GAA were prohibited from playing "foreign" games such as soccer, rugby and cricket. This gave the organization a strongly nationalist character, making it a fertile recruiting ground for revolution. Its countrywide organization makes the association a force that cannot be ignored today, especially in rural areas.

4 This illustration by Louis le Brocquy was one of those made for the major early Irish epic *Táin Bó Cuailgne,* in the translation by the poet Thomas Kinsella (1928–). The work depicts an heroic Iron Age world – a saga of war, heroes, gods and goddesses and magic, most of which was written in a terse prose style with short passages of rhetorical verse. It has attracted many writers, including Yeats and Synge, who based plays on it. Le Brocquy, who left Ireland for France, is one of the leading contemporary Irish painters. Most of his later work is wholly abstract, but in the *Táin* his pictures, all in stark black-and-white, catch the spirit of the epic.

cized, europeanized Ireland of his own time.

The literary movement, largely through the searchings of Yeats and Lady Gregory (1852–1932) for an Irish literary identity, produced the Abbey Theatre [Key], where realism vied uneasily with high stylization. This institution weathered the political storms and was the most important force of the literary revival to survive into the new Irish Free State. Sean O'Casey (1880–1964) [7] maintained its impetus throughout the 1920s. After this the Abbey went into a slow decline and for a long time the real life of the Dublin theatre was in the Gate Theatre, where the productions of Hilton Edwards and Míchéal MacLiammóir (1899–) brought the world to Dublin's stage.

Meantime, Irish governments for several decades pursued obscurantist policies, and the banning of books on "moral" grounds by officially sponsored boards became something of an international scandal. This policy, although it produced a dreary dearth in bookshops, did little harm to writing. The Cork writers Frank O'Connor (1903–66) and Sean O'Faoláin (1900–) sold their work largely outside the country. A little later, Samuel Beckett (1906–) [8] lived abroad, and Brendan Behan (1923–64), although he lived in Ireland, achieved his main successes overseas. The foolish banning perhaps hurt most severely poets such as Patrick Kavanagh who lived in Ireland without a significant world market.

Music and the visual arts

The visual arts have been feebler than the literary in modern Ireland. There were a few painters of modest distinction in the mineteenth century, such as William Orpen (1878–1931) and the portraitist John Yeats (1839–1922), the poet's father. John's son, Jack B. Yeats (1871–1957), developed from run-of-the-mill book illustration to a highly abstract Expressionist style. Since the 1920s and 1930s there have been numerous painters and sculptors working in standard international idioms. Ireland, in medieval times noted for its music, has greatly lagged behind in the nineteenth and twentieth centuries. Traditional music, however, has had a popular revival since the 1950s [6].

KEY

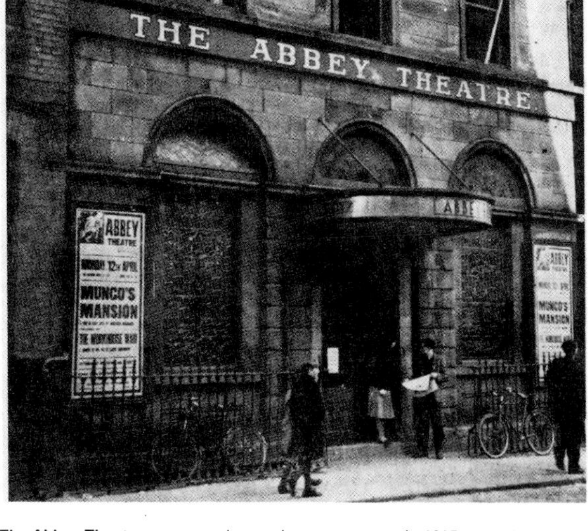

The Abbey Theatre, Dublin, was founded in 1904 by, among others, W. B. Yeats and Lady Gregory and staged many plays of the Irish literary revival. J. M. Synge's *Playboy of the Western World* in 1907 caused a riot among the audience for its portrayal of the Irish peasantry.

5

5 James Joyce, unlike most writers of the literary revival, came from a Catholic middle-class background. His first major work was *Dubliners* (1914); *Portrait of the Artist as a Young Man* (1916) is largely autobiographical, in which the central character finally decides to leave Ireland, like Joyce, to become a writer. From 1904 Joyce lived nearly all his life abroad. Between 1914 and 1922, Joyce worked on his major novel, *Ulysses*, an account of the events of one day in Dublin. The work, now regarded as one of the most important novels of the 20th century, was delayed by charges of obscenity.

6

6 Séan Ó Riada (1931–71) was an important figure in the postwar revival of traditional music in Ireland. In 1951 *Comhaltas Ceoltóirí Eireann* (Traditional Music Society of Ireland) was founded to sponsor traditional Irish music. It had remarkable success: its *Fleadh Ceoil* (music festival) attracts thousands of people. Ó Riada, a serious composer and performer who took a keen interest in all aspects of the tradition and his arrangements for his own group of players, *Ceoltóirí Cualann*, has created a widespread taste for this kind of music. His work was ended by his untimely death, at the age of 40.

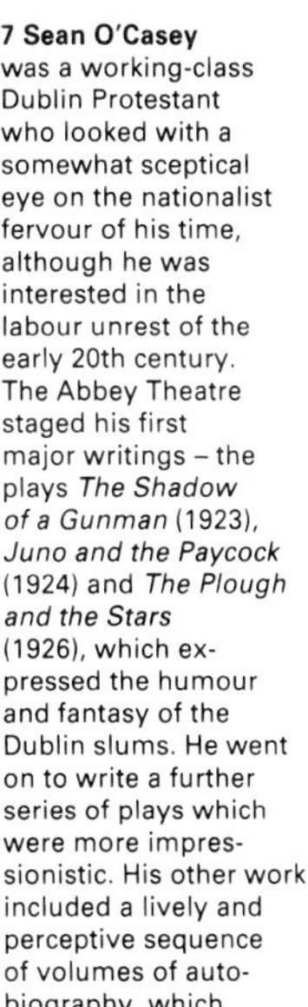

7 Sean O'Casey was a working-class Dublin Protestant who looked with a somewhat sceptical eye on the nationalist fervour of his time, although he was interested in the labour unrest of the early 20th century. The Abbey Theatre staged his first major writings – the plays *The Shadow of a Gunman* (1923), *Juno and the Paycock* (1924) and *The Plough and the Stars* (1926), which expressed the humour and fantasy of the Dublin slums. He went on to write a further series of plays which were more impressionistic. His other work included a lively and perceptive sequence of volumes of autobiography, which was published in one volume as *Autobiographies* (1963).

7

8

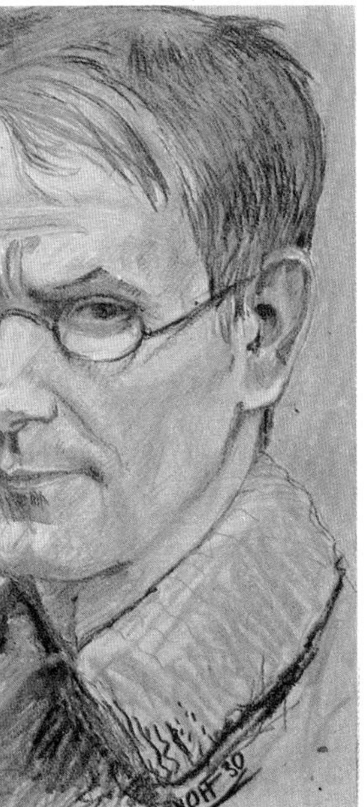

8 *Waiting for Godot* (1952) was the play with which Samuel Beckett first gained international fame. Like James Joyce, who influenced him strongly in his earlier work, Beckett has lived in exile for most of his life, living in France and writing some of his work originally in French. The word-play and humour in his novels and later in his plays show something of this influence, but his pessimistic view of life came out in his later work. It is seen in *Endgame* (1958) and *Krapp's Last Tape* (1958).

Scottish culture since 1850

The Scottish educational tradition has relied strongly on the written word and the communication of factual knowledge and this undoubtedly has contributed to the high esteem in which it is held today. Scottish religious tradition discouraged both music and the visual arts. Nineteenth-century society was sharply divided between middle and working class, but there was considerable opportunity to move from the working class to the professions by the educational ladder. On the other hand, there was much more limited scope for the development of artistic or musical talent.

Music and the visual arts

Important changes began in music when, in the mid-1800s, Edinburgh University set up a Chair of Music. The various churches came gradually to allow the use of organs and a structure of musical education and activity was slowly built up. The creation of the Edinburgh Festival in 1947 [Key], and later of a youth orchestra, gave the city more musical facilities than usual, while Glasgow founded the Scottish Opera [6]. In Thea Musgrave (1928–) Scotland has produced one of the most distinguished of modern composers, a musician of great and varied initiative, prepared to experiment boldly with new techniques and structure.

At the end of the century the rise of the architect Charles Rennie Mackintosh (1868–1928) for a time brought Scotland to the forefront of artistic innovation. Mackintosh combined the influence of Art Nouveau with a real feeling for building materials and motifs from seventeenth-century Scottish vernacular building. Before him, Scottish architecture had either used a classical style, as in the work of Alexander ("Greek") Thomson (1817–75), or exploited Gothic motifs, as in the work of David Bryce (1803–76), with little regard to appropriateness or physical comfort. Mackintosh's most famous work, the Glasgow School of Art [3, 4], became the first significant modern building in Europe that was widely influential, particularly in Germany and Austria. His simple geometric line and concern for the integral design of exterior and interior presaged later movements in modern architecture. The advances he made in architecture and design, however, were not followed.

Painting lapsed after the late nineteenth century, to revive again between the wars on a more permanent basis. After 1945, particularly in Glasgow, there evolved a flourishing group of painters who, by selling their work locally, cultivated the taste of the public for their individual styles. Perhaps the most significant of these was Joan Eardley (1921–63), whose feeling for the environment of poverty in Glasgow was particularly acute. Sculpture has done less well, and Scotland's most distinguished sculptor, Eduardo Luigi Paolozzi (1924–) [9], developed mainly under foreign surrealist influence and has done his work in England.

Literature, discovery and history

In the nineteenth century Scottish literature was aimed at the general English-speaking market. Robert Louis Stevenson (1850–94) [1] probably had more to say to the young writer than any other writer of the 1880s, but he had struggled so hard to free himself from the Edinburgh middle class ethos that only

CONNECTIONS

See also
30 Scottish culture to 1850
70 Georgian art and architecture
98 The novel and the press in the 19th century
178 Origins of modern architecture
280 Art and architecture in 20th-century Britain

1 Robert Louis Stevenson came to the forefront of Scottish letters in the late 1870s as an essayist, light versifyer and novelist. As a youth he was stricken with a severe illness that strengthened his desire to write. His work is conspicuous for its modernity and for its willingness to study sympathetically the semi-underworlds of various societies. Two of his most popular works are: *Kidnapped* (1886), an adventure story set in the aftermath of the Jacobite rebellion, and *The Strange Case of Dr Jekyll and Mr Hyde* (1886), an early, macabre treatment of schizophrenia. Perhaps his best known work is *Treasure Island* (1883), a tale about pirates and treasure.

2 Highland Games, so-called, are held in many parts of Scotland in summer. They involve piping contests, Highland dancing and various specialized sports such as tossing the caber. As displays, they are popular both with Scots and tourists, but they cannot be regarded as being deeply founded in traditional culture. They are largely a creation of the same 19th-century romantic movement that developed the artificial "clan" tartans, but the music is traditionally Highland. The two most famous of the Games are the Royal Highland Gathering or Braemar Gathering, held at Braemar, Aberdeenshire, and the Northern Meeting Highland Gathering, held at Inverness.

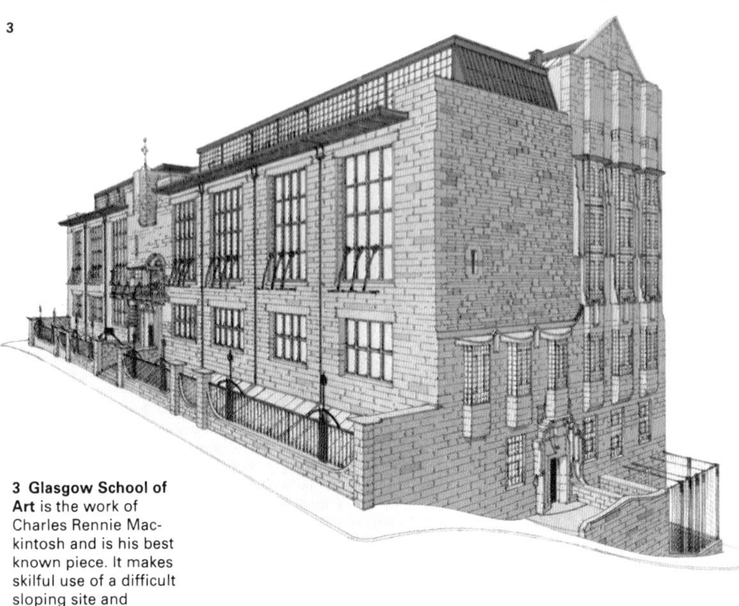

3 Glasgow School of Art is the work of Charles Rennie Mackintosh and is his best known piece. It makes skilful use of a difficult sloping site and handles stone, glass and ironwork boldly.

4 The Mackintosh Room, in the Glasgow School of Art, was designed by Charles Mackintosh and displays some of the best features of his work — his strong emphasis on verticality in the windows, his sense of space, and his original handling of the fireplace. It is free from the boneless tendency of Art Nouveau. Mackintosh stood for a reduction in the clutter of interiors and an emphasis on plain surfaces. His aesthetic inventiveness and perceptiveness strongly influenced European architectural design. The room was originally the Board Room of the School of Art, but now houses a collection of Mackintosh's furniture. His tables are generally quite sturdy but most of his chairs look obviously uncomfortable and appear likely to be broken in use.

at the end of his life did he attempt to use his insight into the Scottish past with his unfinished novel — *Weir of Hermiston.*

The contribution of the Scottish universities to discovery is shown in the work of James Clerk Maxwell (1831–79), the originator of modern physics. An interesting gap in Scottish intellectual life was the failure of the general public and the schools to treat Scottish history as a subject for serious thought and research. A generation of late nineteenth-century scholars did valuable work, but until recently they lacked successors and made little impression.

The problem of language

Between the wars Scottish literature became overtly national and began to grapple with the special problem of language. Edwin Muir (1887–1959) [7], who as an Orcadian had an awareness of national feeling, wrote "Scotsmen feel in one language and think in another". The problem was not only the issue of Scots or English, but the divergence within Scots idiom. This has been tackled with characteristic vigour and some success by Hugh McDiarmid (1892–) [5], who opened up a new range of possibilities in his handling of Scots words, and attempted to develop a synthetic dialect called Lallans. In spite of the handicap of this medium, his has been the dominant poetic voice in Scotland this century. The language problem lessened with the development of broadcasting because the modern child is brought up hearing a variety of personal dialects. A recent Orcadian poet, George MacKay Brown (1921–), seems to have no difficulty in using English as the language of feeling.

Broadcasting has increased the pressures on Scotland's other language, Gaelic. Nineteenth-century Gaeldom was in thrall to extreme Calvinist repression. The visual arts were dead, and literature, confined to songs and sermons (for poetry in Gaelic is synonymous with song) came under the religious disapproval of song. State schooling imposed English instruction, and was followed by the influence of radio. Surprisingly, in spite of these pressures, poetry has remained alive, and in Sorley Maclean (1911–) has today a distinguished voice.

KEY

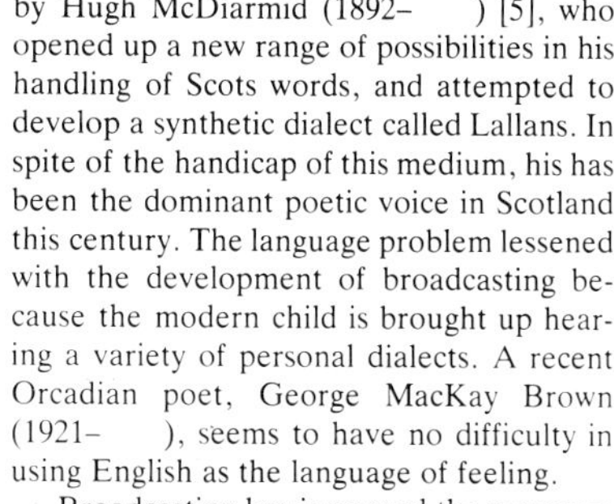

Edinburgh, a city endowed with abundant natural advantages for pageantry and display, has held the Edinburgh International Festival of Music and Drama – an annual three-week celebration of music, drama and art – since 1947. Artists and audiences come to the city from many parts of the world to perform and to enjoy a wide and varied offering of cultural entertainment. In addition to the traditional piping and military tattoo, the opera, concerts, ballet and theatre, a large body of fringe drama and art shows is increasingly making itself seen and heard. Although popular and successful, the festival has not made much use of Scottish creativity.

5

5 Hugh McDiarmid is the literary pseudonym of the communist, nationalist author and poet, C. M. Grieve. His has been one of the most influential voices in Scottish literature since the early 1920s, revealing the rhythmic possibilities of the Scots tongue. One of the founders of the Scottish Nationalist Party, he has always seen his writings as nationalistic. He received an honorary Edinburgh doctorate in 1957.

6 Glasgow Opera House is the home of the Scottish Opera. In the comparatively short time since it was formed, the company has built up an enviable reputation. It has provided Scottish audiences with the opportunity of seeing and hearing distinguished works performed by international artists in productions of significance. The opening season of Scottish Opera was performed in Glasgow in 1962.

6

7 Edwin Muir, poet and essayist, was more Orcadian than Scots. He writes in his moving autobiography of the traumatic effect of moving from life on an Orkney farm to the urban wilderness of a big city.

7

8

8 Cottages at Dunbar form part of a rebuilt area round the harbour. They were restored in 1952 by the Scots architect Basil Spence (1907–76), famed for his major works such as Coventry Cathedral (1951–62). Here he represents the movement to restore small-scale traditional buildings. The houses won an award in 1952.

9 The sculptor Eduardo Paolozzi was born in Leith but has done most of his work outside Scotland. In the 1950s his material was mainly rough cast bronze. With this he produced figures using machinery and débris, but with the emphasis on crude human forms rather than on mechanical aspects. The results resembled technological monsters.

9

The arts in Wales

It is generally agreed that what is now called the Welsh language was in existence by the second half of the sixth century. From that time on there developed a rich native literary tradition that, although subject to the eroding pressures of another language and culture, has over the years remained remarkably resilient and buoyant.

Poetry and literature

At the head of the Welsh poetic tradition stand the names of Taliesin and Aneirin, dating from the second half of the sixth century AD. Taliesin is more than half legendary; Aneirin wrote an epic called *Gododdin*, which celebrated the defeat of the Saxons at the Battle of Catraeth in *c*.603. Saga poetry of this type continued to be produced by the bards until the Edwardian conquest in the thirteenth century rendered court poetry obsolete. Nevertheless the twelfth and thirteenth centuries were the golden age of Welsh poetry. A rich variety of works were written or translated into Welsh, and a new metrical form based on the rhymed couplet, called the *cywydd*, became the vehicle of Dafydd ap Gwilym (*c*.1325-80) whose verse establishes him firmly among the outstanding poets of medieval Europe.

From the sixteenth century onwards the bardic craft went into decline and poetry became the province of the amateur. Welsh scholars were anxious to ensure that the Welsh language should become a fitting medium for the ideals of the Renaissance and, with the publication of the first Welsh Bible in 1588, a standard was set for literary expression in the vernacular.

Evangelical revivalism in the eighteenth century produced a flood of creative writing, especially hymns [3], which was not matched in the nineteenth century. But with the upsurge of national consciousness on the eve of the twentieth century, a renaissance of Welsh poetry occurred. A new generation of first-rate poets including T. Gwynn Jones (1871–1949) and R. Williams Parry (1884–1956), heavily influenced by lyrical romanticism, set new standards of diction and linguistic purity. Postwar nationalism brought Saunders Lewis (1893–) to the fore as the outstanding literary figure of his day, and since the outbreak of World War I Welsh literary contributions have increased enormously.

Literature written by Welshmen in the English tongue dates from the sixteenth century. Prior to the twentieth century, however, most Anglo-Welsh writing was uneven in quality. A new generation of creative poets and prose-writers emerged in the 1930s. Men such as Glyn Jones (1905–), Idris Davies (1936–) and Dylan Thomas (1914–53) [5] poured out a copious wave of literature which flooded English markets. This flourish, however, was not sustained in the 1950s, and it is significant that most modern Anglo-Welsh writers are now deeply involved in the intricate debate concerning the nature of national identity.

The musical tradition of Wales

It is a commonly held view that Wales is a musical nation. Giraldus Cambrensis (1146–1220) recounts that the Welsh were great lovers of music, but medieval Welsh music was discarded in its entirety in the wake of the Reformation. Ensuing musical

CONNECTIONS

See also
166 Wales 1536–1914
262 Wales in the 20th century

In other volumes
184 History and Culture 1

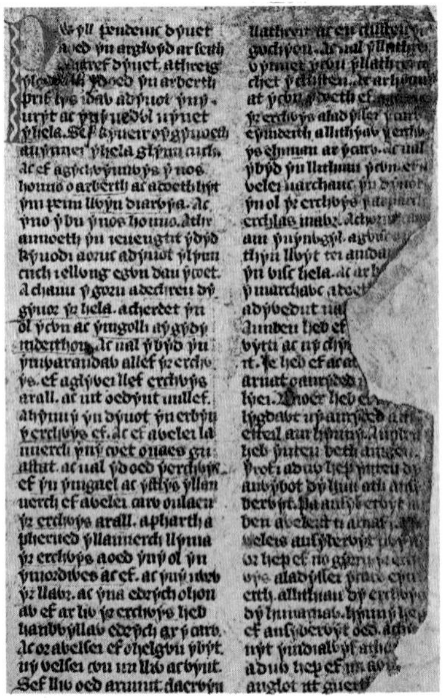

1 The 11 stories known as *The Mabinogion* form one of the most famous collections of saga literature of medieval Europe. The Mabinogi, written between the 11th and the 13th centuries, comprise the "Four Branches of Pwyll", "Branwen", "Manawydan" and "Math"; two brief works, "The Dream of Macsen Wledig" and "Lludd and Llefelys"; the most remarkable of early Arthurian tales, "Culhwch and Olwen"; the nostalgic "Dream of Rhonabwy"; and a trilogy of later Arthurian romances: "The Lady of the Fountain", "Peredur" and "Geraint son of Erbin". These tales are shot through with fascinating mythological themes and motifs.

2 John Parry (*c.* 1710–82), "The Blind Harpist", was harpist to Sir Watkins and his family of Wynnstay, Denbighshire. He was a diligent collector of Welsh harp airs and made the earliest known collection of exclusively Welsh melodies, in 1742. In his day, Parry was regarded as one of the best harpists in the kingdom. Although regrettably there is no musical source material extant before the 17th century, in the 18th century awareness of the value of musical tradition led musicians and musicologists such as Parry to collect melodies for instruments such as the triple harp, the *crwth* (bowed harp) and the *pibgorn* (single-reed pipe).

3 William Williams "Pantycelyn" (1717–91), was unquestionably the finest Welsh hymn writer of the age. A man of unusual attainments, he wrote more than a thousand hymns and a number of creative prose works that captured the unique flavour of Methodism, with its evangelical fervour. With the coming of Methodism in the 18th century, hymnody supplanted psalmody. The lyrical quality that distinguishes so many of Williams's hymns marks him as a herald of the Romantic movement.

4 "Llanberis Pass and Tŵr Padarn" was painted by Richard Wilson (1713–82) of Penegoes, Montgomeryshire, the most famous 18th-century Welsh landscape artist. He was one of a number of British artists among whom interest in the "wild" Welsh landscape increased rapidly from the 1740s onwards. Wilson moved to London in 1729 and studied under the portrait painter Thomas Wright. He soon established influential connections and became one of the founder members of the Royal Academy in 1768. Wilson was well known among his contemporaries for his "meditative love of nature".

5 Dylan Thomas is the only Anglo-Welsh poet of the 20th-century to have gained international fame. Born in Swansea, Thomas moved to London in 1936 where he earned a reputation not only as a poet, writer and broadcaster, but also as a spendthrift and a drunkard. His volume of *Collected Poems* (1952) sold phenomenally and his "play of voices" *Under Milk Wood* is one of the best known of all dramatic prose fantasies. Laid low by financial and moral pressures, he died aged only 39.

tastes were heavily influenced by the popular music of London, and harpists tended to ape the novel Italian instrumental and operatic music. At popular levels, the Welsh peasantry retained their folk-music traditions [2], particularly the art of singing *penillion* (stanzas), a highly specialized form of vocal music which was accompanied only by a harpist.

During the eighteenth century, hymn-singing became an intrinsic part of revivalist meetings and, in the nineteenth century, amateur music-making was stimulated by the tonic solfa movement, the singing festivals and the local eisteddfodau. Between the two world wars interest in instrumental music burgeoned as a new generation of Welsh composers, notably Daniel Jones (1912–), Arwel Hughes (1909–) and Grace Williams (1906–77), penned imaginative orchestral pieces. In modern times lively composers such as William Mathias (1934–) and Alun Hoddinott (1929–) have successfully introduced recent innovations, often showing strong Welsh-folk influences, to Welsh musical tastes. Even so, only now is Wales beginning to develop and sustain those institutions which are necessary to a truly professional musical culture.

Drama and the visual arts

Prior to the twentieth century Wales had virtually no tradition of professional drama and, before the advent of Saunders Lewis, doyen of Welsh dramatists, original Welsh plays were few. Similarly, the growth of an indigenous artistic tradition has been hampered by poverty, by the lack of true urban settlements and the dearth of Welsh institutions. The old puritan hostility to painting remained deeply rooted in Welsh society and talented artists such as Augustus John (1878–1961) and J. D. Innes (1887–1914) were forced to learn their trade in London.

In many ways, however, the prospects for the arts in Wales are bright. The Welsh Arts Council is responsible for allocating funds for literature, music, art and drama. This much needed financial assistance will place the arts in Wales on a more secure economic basis and will help to foster and enrich the cultural life of the Welsh people.

KEY

The Gorsedd ceremony of the Bards at the annual National Eisteddfod of Wales was first held in 1791 and subsequently became part of the eisteddfod. It was invented by the remarkable forger and scholar Iolo Morganwg (1747-1826), and is guild of bards headed by an arch-druid. After 1860 druidic pageantry became an essential part of the eisteddfod.

6

6 Ronald S. Thomas (1913–) is probably the most outstanding contemporary Welsh poet writing in English. A Cardiff-born clergyman who learned the Welsh language as an adult, he has steeped himself in the Welsh literary heritage. His poetry, which includes imagery from the Welsh rural, hill-farming communities in which he chooses to live and work, often focuses on the Welshman's seemingly vain struggle to protect his environment and traditions. Many of his most striking images are based on the natural world – the earth, trees, stars and wild animals – yet his view of the natural world is completely unromantic and unsentimental.

7

7 "Do not go gentle into that good night" was painted by Ceri Richards (1903–) in 1956. Henri Matisse and subsequently the Surrealists, especially Max Ernst, have visibly influenced his work, which reveals a highly individualistic style – combining a masterly command of colour, rhythm and line. Richards frequently paints visual interpretations of musical themes and poetry. This particular work is one of three based on a poem of the same title by Dylan Thomas. More abstract, yet no less lyrical and evocative, is Richards' interpretation of the piano music of Debussy in "Le Cathédrale Engloutie" (1961).

8

8 D. Gwenallt Jones (1899–1968), poet and social critic, articulates in his writings the predicament of Welsh society and culture in the 20th century. Gwenallt's hard-hitting and militant poetry expresses a distinct loathing of capitalism and materialism in socialist, nationalist and Christian terms. His verse vividly portrays the many hardships and sufferings that the Welsh rural and industrial communities went through during the economic slump of the interwar years.

9

9 The Welsh National Opera Company, founded in 1946 by an enthusiastic and forward-looking group of opera-lovers and businessmen, was established in order to present grand opera in Wales and to contribute to the musical and cultural life of the community. Since its foundation the company has opened its own opera school in Cardiff and this has proved an invaluable training ground for inexperienced singers The baritone Geraint Evans (1922–) and the soprano Gwyneth Jones (1936–), who has sung with the National Opera, are two Welsh opera singers who have won a deservedly high reputation in the world's leading opera houses.

American writing: into the 20th century

America's literature of the past two centuries has reflected her increasing self awareness. At and for some time after independence (1783) the literature was wholly colonial, but thereafter a persistent theme has been the conflict in writers between their American consciousness and their European heritage.

The early giants

The first three great figures were James Fenimore Cooper [1], Washington Irving (1783–1859) and Edgar Allan Poe (1809–49). Irving – traveller, biographer, writer of tales, essayist – hardly believed in his America and chose to live for 17 years in Europe. Poe, also a masterly short story writer, was a poet more lauded in France than in his home country.

Then came the era of "transcendentalism", led by Ralph Waldo Emerson (1803–82), who may be called the father of the first truly indigenous literature. Emerson and Henry David Thoreau (1817–62), the great recluse, believed that only in America could individuality co-exist with group harmony. Nathaniel Hawthorne saw more clearly than Emerson the difficulties involved in the achievement of a harmonious future; he was also perpetually haunted by the problem of evil, a theme of his friend Herman Melville (1819–91) [2]. With *Leaves of Grass* (1855), by Walt Whitman (1819–92) [3], the first completely native American poetry emerged, epitomizing a period often called the "American Renaissance". Contributing to the dominance of New England writers were the so-called "Brahmins" who included the poets Henry Wadsworth Longfellow (1807–82) and James Russell Lowell (1819–91). One poet of great originality – Emily Dickinson (1830–86) – lived apart from her contemporaries in total obscurity.

Realism and social criticism

Two of the most distinctive characteristics of American writing – humour and directness – came together in the stories and novels of Mark Twain (Samuel Langhorne Clemens). Twain (1835–1910) [4] was the indispensable bridge between romanticism and realism. The apostle of realism was the novelist William Dean Howells (1837–1920) and its most accomplished writer was his friend Henry James (1843–1916) [5]. The naturalists, who pushed realism to the limit in portraying the harsh side of American urban life, are epitomized by Theodore Dreiser (1871–1945) with his celebrated novel *An American Tragedy* (1925). A broader range of social criticism was provided by Sinclair Lewis (1885–1951), first American to win the Nobel prize for literature, Thomas Wolfe (1900–38), John Steinbeck (1902–68) and Nathanael West (1903–40) as well as by the influential experimentalist John Dos Passos (1896–1970). From Paris the guru of the "Lost Generation", Gertrude Stein (1874–1946), influenced many other experiments, from the lyrical prose of Sherwood Anderson (1876–1941) to the clipped realism of Ernest Hemingway (1899–1961) [9] and the "jazz age" despair of Scott Fitzgerald (1896–1940) [Key]. Among more regional writers the major novelist was the complex dissector of the south, William Faulkner (1897–1962) [8].

Social criticism has also been a strong theme in American drama, which flowered in

CONNECTIONS

See also
290 Emergent literatures of the 20th century
268 Hollywood
212 Socialism in the West
208 The twenties and the Depression
98 The novel and the press in the 19th century

1 In *The Last of the Mohicans* by James Fenimore Cooper (1789–1851) the author tries to portray fairly both the aspirations of white Americans and Indians. The hero is the ideal American.

2 Captain Ahab in Melville's *Moby Dick* (1851) hunted the great whale that had bitten off his leg. Melville was largely ignored until this century, but *Moby Dick*'s quality is now fully recognized.

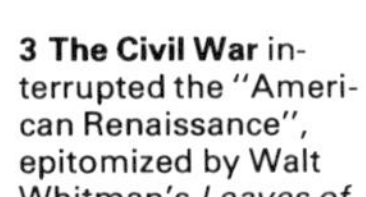

3 The Civil War interrupted the "American Renaissance", epitomized by Walt Whitman's *Leaves of Grass*. His devoted nursing of wounded soldiers reflects the democratic companionship of his poetry.

4 Mark Twain's gusto concealed a bleak pessimism. *The Adventures of Huckleberry Finn* (1884) is a crucial examination of the "American dream", and of the hypocrisies that undermine it.

5 Lamb House in Rye, Sussex, was the home of Henry James. He left the USA for England in 1877 and was the classic expatriate whose awareness remained American. His novels, universally acknowledged, study American–European interactions, then social life and finally the influence of morals on destiny.

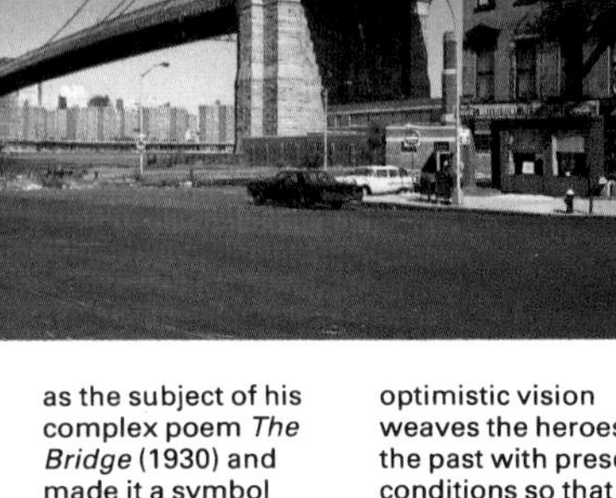

6 Brooklyn Bridge was built between 1869 and 1883 and spans New York's East River between Brooklyn and Manhattan. The poet Hart Crane used it as the subject of his complex poem *The Bridge* (1930) and made it a symbol of the migration across the continent from the Atlantic to the Pacific. Crane's optimistic vision weaves the heroes of the past with present conditions so that his poem shows an awareness of the problems of modern industrial society.

the 1920s, but only Arthur Miller (1915–) [10] ranks with those whose plays have developed more personal themes – Eugene O'Neill (1888–1953), Tennessee Williams (1914–) and Edward Albee (1928–).

Poetry and recent developments

Early this century a revival of poetry was led by Edwin Arlington Robinson (1869–1935) and Robert Frost (1874–1963). Their traditionalism has persisted in poets such as J. C. Ransom (1888–1974) and Allen Tate (1899–), while the symbolist Wallace Stevens (1879–1955) and Hart Crane (1899–1932) [6] were independent figures. The later group of Robert Lowell (1917–), Theodore Roethke (1908–63), Anne Sexton (1928–74), Sylvia Plath (1932–63) and John Berryman (1914–72) developed freer styles from traditionalist beginnings. Their highly personal style is often called "confessional".

Modernism derived from Ezra Pound (1885–1972) [7] and T. S. Eliot (1888–1965), who became an English citizen, drawing on European imagery for *The Waste Land* (1922), his influential lament for eroded spiritual values. A more consciously indigenous innovator was William Carlos Williams (1883–1963). An heir of Williams, as well as of Whitman and the outspoken novelist Henry Miller (1891–) [11] is Allen Ginsberg (1926–).

Ginsberg, the novelist Jack Kerouac (1922–69) and William Burroughs (1914–) lead the "Beat" generation that arose in the 1950s. Still more popular in the 1950s was *The Catcher in the Rye* (1951) by J. D. Salinger (1919–), a novel about an adolescent boy's rejection of adult "phoniness".

A most important development has been the emergence of black and Jewish literature. Black literature is epitomized in the works of James Baldwin (1924–) [12], Richard Wright (1908–60) and the more subdued Ralph Ellison (1914–); exponents of the new Jewish literature include Saul Bellow (1915–), Bernard Malamud (1914–), Philip Roth (1933–) and Norman Mailer (1923–). Mailer's work demonstrates vital engagement with some of the important issues that now concern American society.

KEY

F. Scott Fitzgerald (1896–1940) was an archetypal example of the expatriate American writer. He spent much of his time in Europe (here he is with his wife Zelda and daughter Scottie at Annecy in France in 1931), apparently divided between the "American dream", to which he was dedicated, and the urge to discover his cultural origins: *Tender is the Night* (1934) reflects his period in Europe. He was part of the "jazz age" yet he was a horrified critic of it, as *The Great Gatsby* shows. He returned to America to work as a scriptwriter in Hollywood (as did many other writers) but his disenchantment with its pressures and values led him to alcoholism.

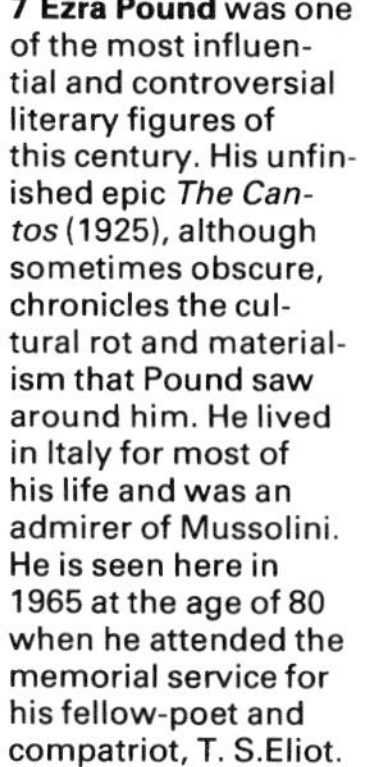

7 Ezra Pound was one of the most influential and controversial literary figures of this century. His unfinished epic *The Cantos* (1925), although sometimes obscure, chronicles the cultural rot and materialism that Pound saw around him. He lived in Italy for most of his life and was an admirer of Mussolini. He is seen here in 1965 at the age of 80 when he attended the memorial service for his fellow-poet and compatriot, T. S. Eliot.

7

8

8 William Faulkner spent most of his life in and around north Mississippi, a region he recreated in his fiction as Yoknapatawpha County. He was a regionalist who also elevated the anguish of the South to the status of universal myth. He was awarded the Nobel prize in 1949.

9

9 Ernest Hemingway wrote on the masculine frontiers of war, bull-fighting, big-game fishing and hunting – pursuits in which he found a code of courage and honour to set against the despairs of life. His famous laconic style was formed during his early years as a reporter and later as a war correspondent.

10

10 The dramatist Arthur Miller married Marilyn Monroe in 1956. This union of a national sex-symbol and a sensitive liberal intellectual failed – a theme of Miller's play *After the Fall* (1964).

11

11 The Beat generation regarded Henry Miller as a prophetic ancestor. He left a safe job in New York to become a Bohemian writer of candid sexual autobiography in Paris in the 1920s.

12

12 James Baldwin was born in poverty in New York's Harlem. He is internationally regarded as the most eloquent and savage indicter of racism and as the chief representative of the ever-growing black literary movement. His essays are powerfully intelligent, although sometimes marred by justified anger; but his most highly acclaimed writing is in his first three novels, *Go Tell It on the Mountain*, *Giovanni's Room* and *Another Country*.

13

13 Broadway and the area surrounding it in New York has been the theatrical heartland of America since the original Broadway Theatre opened (1847). The tension between pure entertainment and dramatic truth is symbolized by the nearby Radio City Music Hall in the Rockefeller Center, a showplace of the cinema industry that has attracted much dramatic talent. Writers such as Tennessee Williams have mastered both cinema and the live stage.

Emergent literatures of the 20th century

One of the chief features of twentieth-century writing has been experimentalism – a response to the uncertainty felt after the collapse of universal religious belief in the West. Another has been the emergence of new, distinctive literatures (or revival of old ones), often involving an element of political or racial protest that is mainly, but not wholly, confined to the West. After the Russian Revolution there was a decade of comparative freedom for Russian writers and much distinctive poetry and fiction was written; but by 1930 a censorship had been instituted that is still largely in force. *Dr Zhivago*, the major novel of Boris Pasternak (1890–1960), and most of the works of Alexander Solzhenitsyn (1918–), have never appeared in Russia, and many other excellent writers have been imprisoned or exiled.

The makers of myths

The general tendency, despite the persistence of realist methods, has been towards the use of mythological themes. While James Joyce's [1] *Ulysses*, for example, is in one aspect a novel of everyday, emphatically "ordinary" experience, the famous legend of Ulysses underlies it. Even Alain Robbe-Grillet (1922–), the French exponent of the atheistic "new novel", which seeks to destroy the traditional idea of the novel as a "story" as well as to demonstrate that the world is wholly indifferent to the hopes and aspirations of human beings, made use of this myth in his first novel *Les Gommes* (The Erasers) (1953), even if only to prove that it is irrelevant, irrational and illusory.

Although twentieth-century man has abandoned many traditional beliefs in the pursuit of new freedoms he has at the same time fallen back on the old, "irrational" myths, or on individual myth-making. Thus in his poetry Leopold Senghor [Key] has relied on ancient Negro virtues, more or less ignored by the West until the emergence of African nationalism. And in the 1920s the French surrealists, led by André Breton (1896–1966), relied on material supplied by dreams and by "automatic writing" produced not by the conscious but by the unconscious mind. There is consequently a kind of religious element in much modernist literature: not a reaffirmation of the old conventional dogmas or even, necessarily, an acknowledgment of the existence of God, but a search for values lying concealed beneath the only apparent rationalism of human behaviour. The Indian writer Rabindranath Tagore [6], much read in the West, was a humanist, but he sought to discover everything that was valid in traditional Hinduism.

Existentialists and reformers

Where religion eschewed in favour of pure atheism, where, as in the novels of the French philosopher Jean-Paul Sartre (1905–), the world is seen as an absurd place accidentally created, the emphasis is on an "existentialist" effort to discover a better and juster system: man is alone in the universe and must choose, of his own free will, to act in the interests of others. For Sartre this effort involves a return to and, when necessary, a modification of Marxist ideas. This is condemned as "revisionist" by the official French Communist Party, of which he has never been a member.

Latin American literature [3, 4, 5] has

CONNECTIONS

See also
288 American writing: into the 20th century
282 Irish culture since 1850
98 The novel and press in the 19th century
256 Latin America in the 20th century
254 Non-alignment and the Third World
302 Underdevelopment and the world economy

1 James Joyce (1882–1941), born in Dublin but absent from it after 1912, typified in his career the development of modernism. His short stories, *Dubliners* (1914), are realistic. So is the semi-autobiographical *A Portrait of the Artist as a Young Man* (1916). But what grew out of it, *Ulysses* (1922), is not; it is a mixture of realism, myth, interior monologue (the detailed tracing of internal thoughts), surrealist fantasy and deliberate pastiche. *Finnegans Wake* (1939), written in a complex language of the author's own creation, requires keys to be understood. Although inaccessible to most readers, it is a masterpiece of wit and vitality.

2 The copihue (*Lapargeria rosea*) is the national flower of Chile and a potent symbol of the beauties and mysteries of that country which Pablo Neruda (N.R. Reyes, 1904–73) wrote about so elegantly. The most loved of all Chilean poets, Neruda was a diplomat who died under mysterious circumstances a few days after Chile's civilian government was overthrown by the military. His poetry is typically modernist in its irreverence for political power and its linking of inner states with outward appearances. Neruda was much influenced by surrealism. He won the Nobel prize, as did another Chilean, Gabriela Mistral.

3 Gabriel García Marquéz (1928–), a Colombian novelist, is one of the most distinguished of Latin American writers. The themes of his novels, which include *No One Writes to the Colonel* (1961), are typical: his town of Macondo symbolizes Latin America yet is treated in a convincingly realistic way. Like Neruda in poetry and the Guatemalan Nobel prizewinner Miguel Asturias (1899–1974), he conveys a sense of the spiritually regenerative power of his country.

4 The Argentinian writer Jorge Luis Borges (1899–) is unusual in Latin American literature. Although he has written lyrically nostalgic poetry, he is primarily an intellectual writer (much influenced by English literature) of short fiction concerned with the invention of various metaphysical systems. His vision is sceptical but he pays warm tribute to man's ingenuity.

5 Mexico City is the cultural centre of one of the few Latin American countries to have achieved some measure of political reform. But Mexican literature shows that the objectives of successive left-wing revolutions have not been realized. The pioneer novels of Augustín Yañez (1904–) subtly examined the relations between Church and state. The leading communist novelist, José Revueltas (1914–), served a prison term. The leader of the younger generation of novelists is Carlos Fuentes (1928–), whose *Where the Air is Clear* (1958), showing Mexico City as deeply corrupt, was a bestseller. Among poets, Octavio Paz (1914–) has the highest international reputation.

come into full flower in the last half-century because it simultaneously asserts the need for political reform and acknowledges the mysteriousness of human existence. The still largely unexplored interior of the continent, with flora and fauna as yet unnamed by scientists, has been an irresistible symbol of the equally unexplored depths of the human mind – with its beauties and its horrors. Thus Pablo Neruda [2] could be both a communist activist and a celebrator of such mysteries.

The new African literatures show much variation. Amos Tutuola [10] reflects the complex mythological world of the African, showing its profundity; his fellow Nigerian, novelist Chinua Achebe (1930–), depicts the same kind of world in an entirely different way: he is a realist who nevertheless includes in his portraits of Nigerian life the powerful influence of purely tribal ways.

Moral and political protest

A feeling of protest at man's injustice to man naturally finds its most eloquent expression in the work of writers from countries living under extremist regimes: Solzhenitsyn in Russia and Athol Fugard [9] in South Africa.

Literary experimentalism has been applied to both form and subject-matter. New forms have sometimes been complex and demanding, but novelists such as François Mauriac (1885–1970) have written recognizable and coherent novels on subjects that are in every sense modern. Mauriac was a Roman Catholic, but the questions his novels pose are "existentialist" in the sense that they attack all conventional solutions of the problem of evil. Likewise, and more drastically, the works of his French compatriot André Gide (1869–1951) reveal a self-confessed bisexual who wants to discover, if possible, a basis for a viable human morality.

In England Graham Greene (1904–) is an avowed left-winger. In his novels there is an acute awareness of social and political problems and his Catholicism is explicitly keyed to the fight for justice. In sum, twentieth-century literature has responded to the challenge of its desperate context: it has combined imaginative genius with political concern and has been an articulate critic of social and political issues.

KEY

Leopold Senghor (1906–), President of Senegal, was, until he entered politics, the chief poet of the "Negritude" movement. This drew attention to Black culture's potential contribution to world literature by its raw vitality and original rhythms. His poem "Chaka" deals eloquently with the problem of reconciling political power with poetic practice.

6

6 The Bengali poet, dramatist and novelist, Rabindranath Tagore (1861–1941) created a literature nearer to spoken Bengali than had ever been written before. He played an important part in the liberation of India, but made himself unpopular by quarrelling with Mahatma Gandhi. As a writer his work includes the drama *The King of the Dark Chamber* (1910) and several novels, of which *Gora* (1908) is considered by some to be superior to Rudyard Kipling's picture of Indian life in *Kim*. But he is best remembered as a poet whose own translation of his *Gitanjali* (1909) was acclaimed by W. B. Yeats and others.

7

7 The Muslim counterpart of the Hindu Tagore was Mohammed Iqbal (1873–1938). Less creatively gifted than Tagore, he was more active politically. His philosophical poetry in Urdu and Persian embodied his belief that salvation lay in a socialistically revitalized Islam and his part in the creation of modern Pakistan is recognized there by a public holiday to mark the day of his death.

8

8 Yukio Mishima (1925–70), a leading Japanese novelist, created a sensation by committing suicide in protest against Japan's abandonment of right-wing discipline. His message, which went unheeded, was not obviously conveyed in his fiction. But he sharply portrayed most facets of Japanese life and was a keen analyst of the homosexual temperament and of obsessive psychological states.

9

9 Inter-racial problems are played out in this scene from *Statements after an Arrest under the Immorality Act* by Athol Fugard (1932–), a South African (half Afrikaans) whose drama has been banned by his government and who has been exiled. His plays reveal the absurdity of racial laws such as the illegality of inter-racial relationships. But they are imaginative and non-propagandist.

10

10 Amos Tutuola (1920–) is a Nigerian novelist of little education but great creative energy. Writing in a strange, entrancing and basic idiomatic English, he succeeded in *The Palm-Wine Drinkard* (1952) in conveying a sense of the spiritual validity of the complex legends and beliefs of his own countrymen. He mixes fantasy and magic with fact and legend to produce startlingly powerful allegories of the predicament of human beings who are ignorant of their unconscious motivations.

11

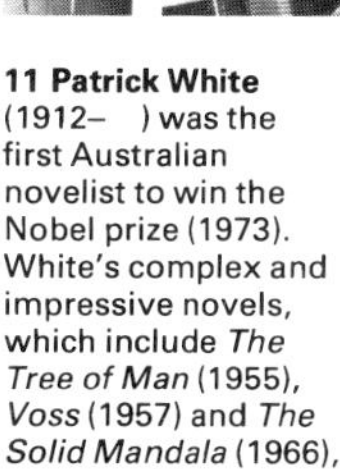

11 Patrick White (1912–) was the first Australian novelist to win the Nobel prize (1973). White's complex and impressive novels, which include *The Tree of Man* (1955), *Voss* (1957) and *The Solid Mandala* (1966), are rooted in Australia and yet are absolutely characteristic of modern fiction. He treats the innermost thoughts of his characters (by direct or symbolic means) as the ultimate form of realism; his books are complex but rewarding. His message is existentialist in its demand for utter truth to self, authenticity of being and relentless honesty. His heroes and heroines are often insane or dangerous and always unreasonable but they triumph over life by their courage.

The wars of Indochina

In the brief interlude between the Japanese surrender at the end of World War II and the arrival of Allied troops to enforce it in French Indochina, the communist-dominated Vit Minh movement [Key] seized power in Vietnam and proclaimed the country's independence on 2 September 1945. With British support, however, the French returned to Vietnam and as a result the Viet Minh were forced to try to negotiate independence. But their hopes were dashed at the Fontainebleau Conference [2] in 1946 and fighting broke out towards the end of that year [6].

The Geneva Agreements

In spite of heavy American financial support, the French were unable to defeat the Viet Minh backed by China and the Soviet Union and growing war weariness at home compelled them to seek a negotiated settlement. An international conference convened in Geneva in 1954 met in the shadow of the Viet Minh victory at Dien Bien Phu [4]. Vietnam was temporarily partitioned, and reunification elections were to be held in 1956.

After Geneva the communist regime in North Vietnam concentrated upon socialist reconstruction and instructed its followers in the south to restrict their activities to the political sphere. An anti-communist regime in the south had supported peaceful decolonization and did not sign the Geneva Agreements. By 1956, under the leadership of Ngo Dinh Diem (1901–63) [5], it had consolidated its authority, with American support, and felt strong enough to block reunification elections on the northern regime's terms and to move against communist supporters in the south. In January 1959, faced with the near destruction of its apparatus in the south, the Communist Party's central committee in Hanoi gave the order for armed struggle to begin.

By the autumn of 1961 President John F. Kennedy (1917–63) felt obliged to send large numbers of military advisers to South Vietnam. These did not turn the tide of insurgency and on 8 February 1965 President Lyndon B. Johnson (1908–73) ordered American bombing of North Vietnam to deter the movement of manpower and weaponry to the south. But the war on the ground [9] continued and the United States was forced to commit further aid and growing numbers of its own troops to the fighting from April 1965 onwards [8].

American withdrawal

In January 1968 the communists, who now included large numbers of North Vietnamese regular soldiers, launched the Tet or New Year offensive through South Vietnam. After some intensive fighting, it was beaten back, but it weakened America's will to fight. President Johnson announced, on 31 March, a cutback in the bombing of North Vietnam and his own withdrawal from the forthcoming presidential election campaign.

His successor, President Richard M. Nixon (1913–), pinned his hopes upon "Vietnamization". Although the United States continued to provide air and sea support for the South Vietnamese forces, US combat troops were gradually withdrawn.

Meanwhile, negotiations between the Americans and the North Vietnamese had begun in Paris in May 1968 and after the stalemate of a second major communist

CONNECTIONS

See also
246 Decolonization
264 USA: the affluent society
298 The United Nations and its agencies
244 China: the People's Republic
240 The Soviet Union since 1945
266 20th-century sociology and its influence

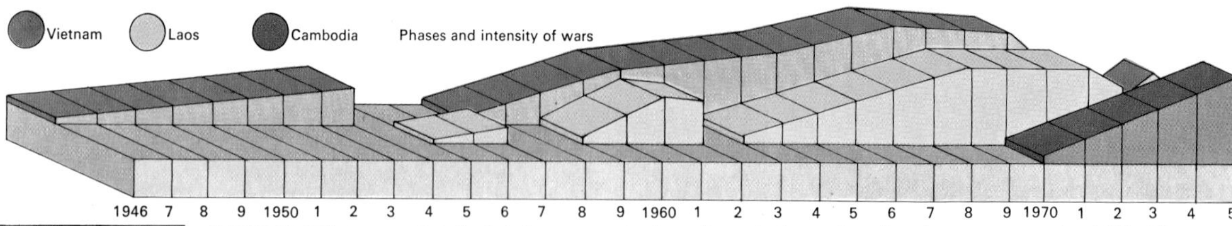

1 The intensity of war in Indochina increased over a 30-year period, spreading from Vietnam alone to Laos and finally to Cambodia, although fighting was on a smaller scale in those countries.

2 At the Fontainebleau Conference in July 1946 Ho Chi Minh insisted on the unity of Vietnam, which the French had divided into the colony of Cochin China in the south, Tonkin in the north and Annam in the centre. The conference broke down when France made Cochin China a separate republic.

3 Catholic influence (shown in this classroom and chapel at an orphanage in An Loc) was important in the educational system introduced to Vietnam by the French and was a factor in the anti-communism of many in the south. In 1939 about 1.6 million Vietnamese (about 8% of the population) were Catholic.

4 The raising of the Viet Minh flag on the French command post at Dien Bien Phu on 7 May 1954 marked the greatest military setback ever suffered by a European colonial power at the hands of local forces. This French fortress in north-west Vietnam fell to General Vo Nguyen Giap (1912–) after a 55-day siege.

5 Ngo Dinh Diem a Roman Catholic, was bitterly opposed to both French colonialism and communism. These traits initially won him US support when he became prime minister of South Vietnam in 1954. Gradually, however, nepotism and his authoritarian rule alienated the US. In 1963 the administration of President Kennedy connived at a coup by dissident South Vietnamese generals. Diem was assassinated in November 1963.

Vo Nguyen Giap

Communist general who sacrificed his troops to force France and the US from Vietnam

■ Vo Nguyen Giap, North Vietnamese general. Born: 25 August, 1911, in Quang Binh province, Vietnam. Died: 4 October, 2013, in Hanoi, aged 102.

VO NGUYEN Giap – who has died in Hanoi – was the North Viet-

camp

and

Vietn

A

with

join

surg

it ir

tha

tio

pi

vi

ga

sc

la

e

U

s

Don McCullin *Shell-shocked US Marine, The Battle of Hue 1968 (printed 2013).* ARTIST ROOMS National Galleries of Scotland and Tate. Presented by the artist 2014 © Don McCullin. Courtesy of Hamiltons Gallery, London

Nixon in

thdrawal

n.

died the

Mao Ze-

political

ism and

fare were

successful

as it did on succ…

These lessons were driven home during the Tet offensive of 1968, when North Vietnamese regulars and Viet Cong guerrillas attacked scores of military targets and provincial capitals throughout South Vietnam, only to be thrown back with staggering losses. Gen Giap had expected the offensive to set off uprisings and show the Vietnamese that the Americans were vulnerable.

Militarily, it was a failure. But the offensive came as opposition to the war was growing in the US, and the televised savagery of the fighting fuelled another wave of protests. President Lyndon B Johnson decided not to seek re-election, and with

Giap's victories … never be known. About 94,000 French troops died in the war to keep Vietnam, and the struggle for independence killed about 300,000 Vietnamese fighters. In the US war, about 2.5 million North and South Vietnamese died out of a total population of 32 million. America lost about 58,000 service members.

"Every minute, hundreds of thousands die on this earth," Gen Giap is said to have remarked after the war with France. "The life or death of a hundred, a thousand, tens of thousands of human beings, even our compatriots, means little."

Vo Nguyen Giap was born on 25 August, 1911, in An Xa in Quang Binh province, the southernmost part of what would later be North Vietnam. His father was an educated farmer and nationalist who encouraged his children to resist the French.

Mr Giap earned a degree in law and political economics in 1937 and then taught history at the Thanh Long School, a pri-

for privileged

noi. He also

Marx and was

sed by Mao's

ning political

egy to win a

hi Minh, the

tnamese Communist Party, chose Giap to lead the Viet Minh, the military wing of the Vietnam Independence League. In late 1953, the French established a stronghold in the northwest at Dien Bien Phu, near the border with Laos, garrisoned by 13,000 Vietnamese and North African colonial troops as well as the French Army's top troops and its elite Foreign Legion.

After an eight-week siege by Communist forces, the last French outposts were overrun on 7 May, 1954. The timing was a political masterstroke, coming on the very day that negotiators met in Geneva to discuss a settlement. Faced with the failure of their strategy, French negotiators gave up and agreed to withdraw. The country split into a Communist-ruled north and a non-Communist south.

In the late 1950s and early 60s, President Dwight D Eisenhower and later President John Kennedy looked on with rising anxiety as Communist forces stepped up their guerrilla war. By the time Kennedy was assassinated in Dallas in 1963, the US had more than 16,000 troops in South Vietnam.

Gen Westmoreland relied on superior weaponry to wage a war of attrition, in which he measured success by the number of enemy dead.

Though the Communists lost in any comparative "body count" of casualties, Gen Giap was quick to see that the indiscriminate bombing and massed firepower of the Americans caused heavy civilian casualties and alienated many Vietnamese from the government the Americans supported.

With the war in stalemate and Americans becoming less tolerant of accepting casualties, Gen Giap told a European interviewer, South Vietnam "is for the Americans a bottomless pit".

The US government began peace talks in Paris in May 1968. The next year, Nixon began withdrawing US troops.

Removed from direct command in 1973, Gen Giap remained minister of defence, overseeing the North's final victory when Saigon fell on 30 April, 1975. He also guided the invasion of Cambodia in 1979, which ousted the Khmer Rouge.

He was removed as minister of defence in 1980. In August 1991, he was ousted.

In his final years, Gen Giap was an avuncular host to visitors to his villa in Hanoi, where he read Western literature, enjoyed Beethoven and Liszt and became a convert to pursuing socialism through free-market reforms.

"In the past, our greatest challenge was the invasion of our nation by foreigners," he once said. "Now that Vietnam is independent and united, we can address our biggest challenge: poverty and economic backwardness."

5.10.2013

Clì, measan agus olannan ann an Souk El Had, Agadir; shuas, *Tagine* circe ann an Tafraoute

innte. Seadh, sia mìle! Dh'fhaodadh tu rud sam bith a cheannach an-seo bho bhratan-ùrlair àlainn gu measan allmharach gu criadhadaireachd chruthachail gu feòil cho ùr 's gum bi i fhathast beò nuair dh'iarras tu i. Seo an t-àite gus sgilean barganachaidh fheuchainn neo a gheurachadh!

Se deagh làrach a th' aig Agadir oir chanainn gur ann anns na sgìrean mun cuairt a tha na rudan as sònraichte, as tarraingiche ri fhaicinn. Fhuair sinn carbad air màl fad latha son mu £30 gus turas rathaid a ghabhail dhan sgìre bhrèagha bheanntaich air a bheil an Anti-Atlas. B'e seo fear de na tursan rathaid a bu shònraichte a ghabh mi riamh, leis na diofar sheallaidhean a chunnaic sinn air an t-slighe. Anns an fhàsach seo, chaidh sinn seachad air na mìltean de ghobhair, le balaich bheaga 'nam measg a' feuchainn an cumail còmhla. Chunnaic sinn seann ghearastain dhustach, air an togail bho chionn linntean air mullaichean chnoc creagach.

Stad sinn ann am baile beag iomallach, Tafraoute, gus beagan rùrachaidh a dheanamh 's greim bìdh fhaighinn. Tha biadh Morocco cliùiteach feadh an t-saoghail, ach 's dòcha gur ann son an stiubha *Tagine* a cliùitiche e. Tha mi toilichte innse nach e briseadh dùil a bh' anns an *Tagine* le feòil chirce a fhuair mi ann an taigh-bìdh cho mì-fhoirmeil 's nach robh fiùs ainm air.

A' suidhe an sin a' gabhail sùgh orainnseir abaich cho milis ri mil, a' ghrian gam bhlàthachadh 'na mo lèine-T thana 's mi coimhead duilleach nach robh ach gorm, cha robh agam ach aon smaoin – seo mar a nì thu Foghar!

Twitter @DavidGalavants

Dh'fhuirich Dàibhidh san taigh-òsta còig-rionnag *Rue Tikida Palace*, Agadir. Tha e spaideil, goireasach agus ri taobh na tràghad. Tha prìsean a' tòiseachadh mu £70 an urra son na h-oidhche, a' gabhail a-staigh biadh là 's oidhche, deoch, seirbheis rum.

...gu dìleas chì sinn àrdachadh ann an 2021 no 2031'

ìosal, tha Mgr MacAoidh den bheachd gum bu chòir sin a bhith ga thuigsinn an coimeas ri lughdachadh dhen t-sluagh fo aois 15 ann an Alba san fharsaingeachd. "Chaidh sluagh na h-Alba suas 4.6 ás a' cheud ged a thuit an àireamh chloinne san dùthaich 5.8 ás a' cheud. Ach aig a' cheart àm chaidh an àireamh de dhaoine òga le Gàidhlig suas. Tha sin a' ciallachadh gu bheil rud mór mór air tachairt ann an saoghal na Gàidhlig."

Chan ann a-mhàin air clann a tha am Bòrd a' coimhead ge-tà. Tha ro-innleachd aca gus barrachd inbhich, gu h-àraidh pàrantan, a thoirt a-steach air gnothaichean. "Bha treiseag ann far an robh daoine a' cur an cuid chloinne gu sgoiltean Gàidhlig gun a bhith a' dèanamh oidhirp aig an taigh. Tha sinne a' toirt air daoine a bhith mothachail air cho cudromach 's a tha an cànan san dachaigh agus a-nise tha sinn a' faicinn barrachd phàrantan ag iarraidh clasaichean Gàidhlig. Tha sin a' tachairt gu h-àraidh anns na bailtean móra agus chan e a-mhàin pàrantan chloinne a tha a' faighinn foghlam tro mheadhan na Gàidhlig.

"Tha an sgeama Gàidhlig sna bunsgoiltean a' toirt cothrom do chlann beagan còmhraidh agus amhrain ionnsachadh agus tha sin gu math cudromach cuid eachd. Chì pàrantan cho toilichte 's a tha an fheadhainn bheaga a' dèanamh sin agus bidh iad fhéin ag iarraidh Gàidhlig ionnsachadh. Cuideachd tha mìltean air clàradh air son cùrsaichean Ùlpan 's tha sinn a' toirt cothroman ionnsachaidh do luchd obrach nan comhairlean ionadail air feadh na dùthcha, rud nach robh ann roimhe seo. Agus tha sin uile a' dèanamh diofar."

Tha na figearan as ùire air son an fheadhainn uile gu léir aig a bheil comas air choreigin sa chànan, eadar bruidhinn, tuigsinn, sgrìobhadh agus leughadh, air dol sìos: 87,056 ann an 2011 an coimeas ri 92,400 ann an 2001, lughdachadh de 6 ás a' cheud. Nach eil sin fhéin a' riochdachadh dùbhlan don Bhòrd?

"Se an dùbhlan as motha an crìonadh eaconomaigeach, oir tha an t-airgead cho gann an coimeas ris na tha sinn ag iarraidh a dhèanamh. A dh'aindeoin sin, tha sinn a' toirt prìomhachais dha trì roinnean bunaiteach: an dachaigh, foghlam agus leasachadh a dh'ionnsaigh Gàidhlig a bhith ga bruidhinn mar chainnt choimhearsnachd."

Ge-tà, tha am Bòrd mothachail gu bheil àitean sna h-Eileanan Siar, 's dòcha gu h-àraidh ann an Leòdhas, far nach eil daoine fhathast air þlàthachadh ri foghlam tro mheadhan na Gàidhlig. Tha e coltach gu bheil baile Ghlaschu, aig a bheil deagh chliù son sgoiltean agus planaichean cànain, agus fiùs Dùn Éideann, le sgoil ùr Ghàidhlig, fada air thoiseach air Steòrnabhagh.

"Mean air mhean tha na h-eileanan deas air Leòdhas a' tighinn thairis gu foghlam Gàidhlig. Ach mun cuairt air Steòrnabhagh, far a bheil sluagh nas motha, sin far a bheil na h-àireamhan de sgoilearan Gàidhlig nas ìsle. Tha sinn ag obair leis a' Chomhairle feuch nach atharraich sinn cùisean anns na bliadhnaichean a tha romhainn.

"Rinneadh obair ionmholta ann an sgìre Shiabost mar-thà, agus théid barrachd dhaoine thrèanadh ann an obair leasachaidh choimhearsnachd. Tha an fhianais ag innse dhuinn gu bheil an crìonadh a bha cho domhainn air tighinn gu crìoch. Tha sinn a-nise sùileachadh air an ath chunntas ann an 2021. Ma dh'obraicheas a h-uile buidheann còmhla, gu dìleas agus gu dripeil a' cur air adhart adhbhar na Gàidhlig, chì sinn àrdachadh anns na figearan an ath thuras, no co-dhiù ann an 2031."

TUILLEADH FIOS

www.scotlandscensus.gov.uk

...n bhalach, 'm bàrr a' chroinn 's Leac an Lì 'nam shealladh . . .

an-diugh rathad mòr an rìgh seach frith-rathad. Ghluais teaghlach le dithis bheaga á Birmingham o chionn ghoirid.

Air an Tairbeart tha bùth iongantach, Buth Ailig, far am faigh croitear agus bean taighe gach nì air a bheil feum. Tha pailteas de bhiadh ann agus a bharrachd gheibhear snàthadan, sìoltachain, uèir stobach, cuibhrigean agus cluasagan. Bidh na seann daoine a thig air astar a' deanamh cèilidh agus gheibh iad cupa cofaidh.

Tha Somhairle MacLeòid a' còmhnaidh ann an Aird Àsaig. Chuir e seachad bliadhnachan ann an Glaschu ach bha aire air na beanntan, na stuadhan, na lochan agus an iasgach, agus cha b' fhada gus an do thill e dhachaigh. Chan eil miann idir aige a dhol air saor-làithean.

Ged a tha na seallaidhean àlainn agus Gàidhlig nan Tearach tlachdmhor, tha e duilich a chluinntinn gu bheil bodaich anns na bailtean, sgoiltean a' dùnadh agus Bùth Ailig an impis a dhol á bith. Mar a chanas e fhèin, "Seann aois agus cus de *red tape*."

Tha Mairead Rods ag aithris gu pongail.

TORMOD E DÒMHNALLACH

offensive in March – May 1972 and renewed American bombing raids upon Hanoi in December of that year, a peace agreement was signed on 27 January 1973.

Communist armed forces from North Vietnam were not obliged by the agreement to withdraw from the south [7] and further fighting began almost immediately as both sides jockeyed for position. The final collapse of the South Vietnamese goverment to communist forces [10] came on 30 April 1975.

Laos and Cambodia

Laotian nationalism split, in 1949, into pro-communist (Pathet Lao) and anti-communist sections and it was to the latter that the French conceded independence in 1953. With the United States striving to preserve an anti-communist government and the Viet Minh supporting the Pathet Lao, a full-scale civil war developed in 1960 and an international conference at Geneva in 1961–2 only temporarily defused the crisis. United States' bombing of North Vietnamese and Pathet Lao positions in Laos, controlling supply routes to South Vietnam, increased steadily after 1964. The January 1973 ceasefire in Vietnam was followed by one in Laos on 22 February and by mid-1975 the Pathet Lao had virtually taken over the country.

Cambodia also obtained its independence from France in 1953. Under its ruler, Prince Norodom Sihanouk (1922–), it managed to maintain a position of neutrality in the Indochina conflict for some years, but with the escalation of the war in Vietnam was forced to act as the main supply route for arms to the communists and to grant them virtual freedom of action in border areas. On 18 March 1970 Sihanouk was overthrown by a right-wing coup and Cambodia was plunged not only into its own civil war, but also into the wider Indochina conflict [1]. While America and South Vietnam attacked the communists in Cambodia, Sihanouk proclaimed a government-in-exile in Peking and allied himself with the left-wing Khmer Rouge rebels who had taken up arms against his own regime in 1967. The American bombing of Cambodia ended in August 1973 and the "Red Khmers" took the capital, Phnom Penh, on 18 April 1975.

KEY

Ho Chi Minh (1890–1969), principal figure in the Viet Minh struggle against the French and leader of North Vietnam after 1954, was born Nguyen That Thanh in north-central Vietnam. He left Vietnam in 1911 and was converted to communism in France after World War I. As a Comintern agent he founded the Indochinese Communist Party in 1930 but did not return to Vietnam from France until 1941, when he set up the Viet Minh front. In 1944 he organized the Viet Minh seizure of power in August 1945. Ho Chi Minh, whose adopted name means "he who enlightens", is shown (right) with the premier of Vietnam, Pham Van Dong (1902–).

6

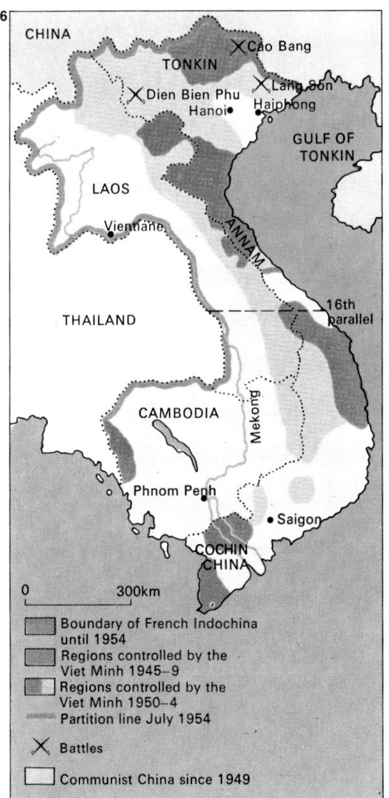

6 The war in Indochina between the Viet Minh and the French leading to the Geneva Agreements of July 1954 had two main phases. A French defeat at Cao Bang and the subsequent loss of Lang Son in October 1950 marked the onset of a more aggressive strategy by the Viet Minh, supported by aid from the newly established Chinese People's Republic. A French recovery followed, but only temporarily.

7

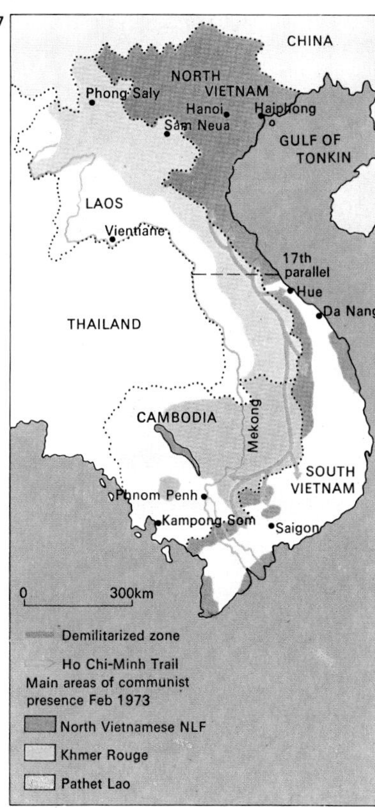

7 The ceasefire position early in 1973 left the main prizes of the long Indochina war still to be won. Communist forces held key border areas in South Vietnam, Laos and Cambodia along the Ho Chi Minh Trail carrying military supplies from North Vietnam. The peace agreement, designed chiefly to allow US withdrawal, called for a political settlement but both sides prepared for a military solution.

8

6·1 23 16·7 9·3

1966 1968 1970 1972

280 545 330 40

US military aid in dollars (thousands of millions)

US combined forces (totals in thousands)

8 US military aid to South Vietnam rose to a peak in 1968 when American combat troops totalled 545,000. Actual (incremental) US war expenditure that year was $23,000 million with $1,000 million more in aid. Spending fell as this effort produced only stalemate.

9

9 An American patrol in rough country epitomizes the problem faced by the US in Vietnam where sophisticated technology failed to win the war on the ground. Alongside the fighting an ultimately more important struggle for the allegiance of Vietnam's mainly peasant population was being won by the communists at village level.

10

10 As Saigon fell American helicopters evacuated their allies on 29 April 1975.

Ireland since Partition

By July 1921, the British government was at last ready to recognize Irish nationalism, and the credentials of Eamon de Valera [5] as spokesman for the Irish people. On 6 December the Irish delegates, led by Arthur Griffith (1872–1922) and Michael Collins (1890–1922), returned to Dublin with an agreement for a new Irish Dominion, within the Commonwealth, with the six counties of Northern Ireland separate [Key].

The Treaty embodying the terms was passed by the Dail on 7 January 1922, by only 64 votes to 57. Opponents included de Valera, who called it a betrayal. This rhetoric was soon to be backed with arms. The majority, however, led by Griffith, Collins, William Cosgrave [1] and Kevin O'Higgins [3], claimed that they had won the freedom to achieve their ultimate goals.

From Civil War to World War II

By the time the Civil War ended in victory for the pro-Treaty party in May 1923, Griffith was dead and Collins, his successor, had been murdered. Cosgrave became leader of the new state through its formative years. Anti-Treaty politicians continued to boycott Parliament until 1927. In December 1925, hopes in Dublin of incorporating Northern Ireland in the Free State were dashed when the Boundary Commission failed.

O'Higgins was assassinated in 1927. The ensuing tough, anti-terrorist legislation finally persuaded de Valera, who had formed the Fianna Fail Party a year earlier, to take his seat in Parliament. In 1932, with Labour support, his party won power. By then much had been gained internationally. Inside the developing Commonwealth, the Irish pursued full sovereign status for the dominions, and played a full part in the Imperial Conferences of 1926 and 1930, and in the preparation of the 1931 Statute of Westminster. At Geneva, where the Free State joined the League of Nations in 1923, the cause of small states was championed, and non-permanent membership of the League Council gained in 1930. De Valera did not like some of the commitments to Britain that he inherited, and Anglo-Irish relations for the next six years were marked by economic and constitutional disputes. During those years, de Valera gained an overall majority, unilaterally cancelled some of the Treaty terms, re-defined Irish nationality and, in 1937, adopted a new constitution, changing the country's name to Eire.

Differences with Britain were resolved in 1938, bringing much-needed financial and trade agreements – and Irish control of three naval installations. It was this that enabled de Valera to keep his country neutral in World War II. Neutrality was favoured by most Irishmen, although many volunteered to serve in the British forces.

Inflation and austerity

In 1948, electoral discontent brought to power a coalition government, led by John Costello (1891–1976) of Fine Gael (Cosgrave's former party). This administration lasted until 1951, during which time Eire withdrew from the Commonwealth and declared itself a republic (1949). Following the collapse of the coalition, de Valera headed a minority administration for three years, until Costello was again returned. But inflation and Costello's austerity programme

CONNECTIONS

See also
162 Ireland from Union to Partition
24 The English in Ireland
238 Britain since 1945:2
282 Irish culture since 1850

1 William Cosgrave (1880–1965), minister for local government in the first Dail, showed wisdom and steadiness when he unexpectedly became president in 1922. After defeat, he was leader of the opposition (1932–44).

2 The Shannon hydroelectric scheme, constructed between 1925 and 1929, was the first major venture by the new Irish government. The scheme provided countrywide electrification over a national grid devised in 1927. It set a pattern for state aid that was extended into many areas in need of development capital. It was agriculture, however, rather than heavy industry, to which money was diverted, notably towards the new beet sugar and the turf-processing sectors.

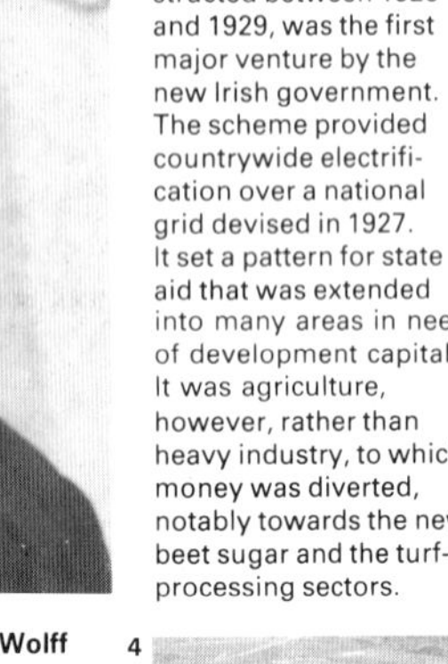

3 Kevin O'Higgins (1892-1927) was the dominant figure in the first Irish government, establishing law and order and serving in many cabinet posts and from 1923 as vice-president until he was assassinated.

4 Harland and Wolff shipyards, Belfast, founded in 1863, are Northern Ireland's main employers and exporters. Business has declined during the 20th century but the firm is still a barometer of the economy.

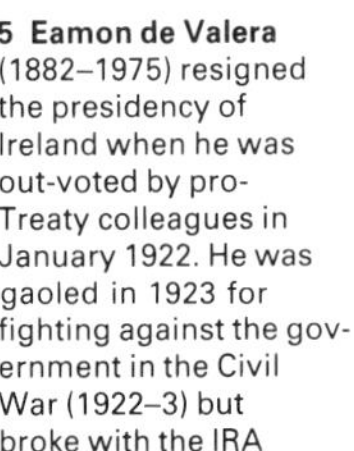

5 Eamon de Valera (1882–1975) resigned the presidency of Ireland when he was out-voted by pro-Treaty colleagues in January 1922. He was gaoled in 1923 for fighting against the government in the Civil War (1922–3) but broke with the IRA and became the dominant figure in Irish politics after reorganizing the Republican Party as Fianna Fail. The architect of Eire, he was a political leader from 1927 to 1973. Born in New York of a Spanish father and Irish mother, he went to Ireland as a child.

contributed to disenchantment at the polls, and de Valera's Fianna Fail were voted back in 1957 to begin 16 years of rule.

De Valera himself retired to become president in 1959, handing over to Sean Lemass (1899–1971), who was prime minister until 1966 when Jack Lynch [9] succeeded him. Lemass made new contacts with the Northern Ireland government in 1965, when he went to Belfast for talks with the Ulster premier, Captain Terence O'Neill (1914–   ), but it was Lynch who had to respond to armed conflict in the north.

In March 1973, only three months after taking Eire into the European Economic Community, Lynch's government was defeated at the polls by another combination of Fine Gael and Labour, led by William Cosgrave's son Liam.

The troubles of Northern Ireland

Northern Ireland had been a state, albeit subordinate to Westminster, since 1920. The first prime minister, James Craig (1871–1940), had hoped for unity between conflicting factions, but the Catholic minority stood aloof from the moulding of the state. As a result, the civil service, police, judiciary and educational system were very much tailored to Protestant needs.

Postwar Northern Ireland was transformed by the Welfare State. Social benefits so far outstripped those of Eire that the Catholics in Northern Ireland were at last ready to identify themselves with the state, but only on the basis of equal citizenship. Demands for equality were backed by the short-lived People's Democracy, and later by the Social Democratic and Labour Party of Gerry Fitt (1926–   ) and John Hume (1937–   ), the Alliance Party, and even by moderate Ulster Unionists.

When the civil rights movement [7] met Protestant resistance, violence erupted, enabling the Irish Republican Army (IRA) to revive the issue of nationalism. Fighting between Catholic and Protestant extremists brought British troops into the streets from 1969 [10], internment without trial (1971), and the end of the Northern Ireland Parliament at Stormont, scrapped in favour of direct rule from Westminster (1972).

Britain's partition of Ireland (1920) with a limited form of self-government for both north and south was accepted by the six counties of Ulster in 1921, when a parliament was established at Belfast. Southern Ireland used the 1920 Act to elect the Dail Eireann (outlawed since December 1919), but went on struggling for full independence. IRA (Irish Republican Army) forces were pitted against the British army, the Royal Irish Constabulary (supported by former soldiers nicknamed "Black and Tans") and Auxiliaries whose reprisals upset liberal opinion in Britain, encouraging Lloyd George's government to seek a settlement. A truce was established on 11 July 1921, and four months later negotiations began in earnest.

6 **A modern Dominican church** at Athy, Co Kildare, reflects the continuing importance of Catholicism in Irish life. In the republic in 1971, Roman Catholics made up 93.9% of the population, showing 2,795,666 adherents, as against 119,437 Protestants (4%) and 63,145 (2.1%) others. In Northern Ireland the groupings were: Protestant 811,272 (53.9%), Roman Catholic 477,919 (31.1%) and others 230,449 (15%). The strong contrast does not need underlining.

7 **Demonstrations in 1968** by the Civil Rights Association sought reforms in Northern Ireland's system of police, local government and housing to remove widespread discrimination in favour of Protestants. Predominantly Catholic, the movement was identified as a republican front by Protestant Unionists. Extremists under Ian Paisley (1926–   ) stirred anti-Catholic feeling and despite some reforms community distrust grew. After the disbanding of Stormont, Brian Faulkner (1921–77) failed in his efforts to rule with a Catholic-Protestant executive, established temporarily in 1974.

8 **Guinness's brewery in Dublin** is among the growth industries that have enabled the Irish Republic to reduce unemployment and emigration, two chronic features of the economy for more than a century.

9 **Jack Lynch** (1917–   ), was premier of the Irish Republic from 1966 to 1973, and re-elected in 1977. He was faced with the developing troubles in the north and the activities of the IRA, not only there but also in the south, which compromised his position in 1973.

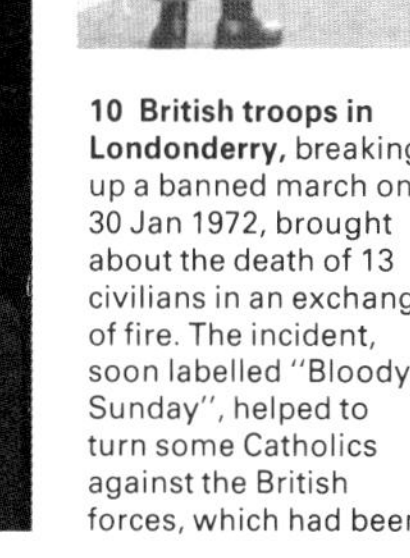
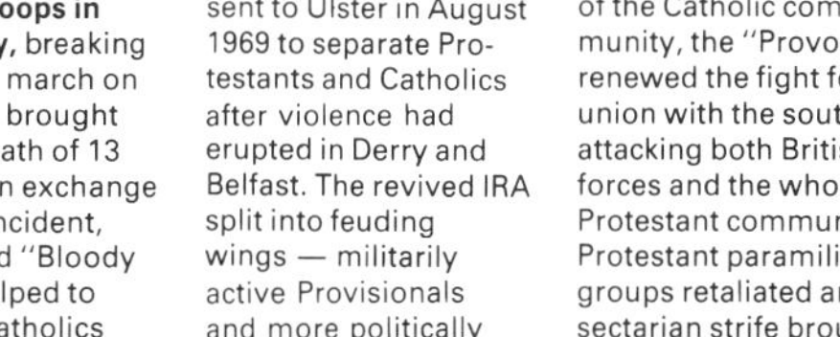

10 **British troops in Londonderry,** breaking up a banned march on 30 Jan 1972, brought about the death of 13 civilians in an exchange of fire. The incident, soon labelled "Bloody Sunday", helped to turn some Catholics against the British forces, which had been sent to Ulster in August 1969 to separate Protestants and Catholics after violence had erupted in Derry and Belfast. The revived IRA split into feuding wings — militarily active Provisionals and more politically oriented Officials. Posing as defenders of the Catholic community, the "Provos" renewed the fight for union with the south, attacking both British forces and the whole Protestant community. Protestant paramilitary groups retaliated and sectarian strife brought 1,600 deaths by January 1977, together with widespread injuries and property damage from bombs. A peace movement, founded in 1976 by Betty Williams, Mairead Corrigan and Ciaran McKeown, offered hopes of reconciliation, and strengthened the concern for a peaceful settlement.

The question of Israel

Zionism, the movement by Jews to set up a state in their ancient homestead of Palestine emerged late in the nineteenth century as a form of the nationalism then sweeping Europe. It represented an attempt to channel the Jewish sense of corporate existence into a secure political entity that would provide an answer to continuing persecution. Among Arabs at about the same time the nationalist concept began to fertilize a deep-rooted sense of separate identity lying dormant under Ottoman rule.

Origins of the conflict

Zionism and Arab separatism clashed from the beginning. In 1882 the first modern Jewish agricultural settlement was founded in Palestine, where Jews had been a minority for centuries. Muslim and Christian notables of Jerusalem urged the Ottoman administration to prevent further immigration. Nonetheless, the Zionist movement grew slowly and the Jewish population of Palestine gradually increased.

During World War I the defeat of the Turks was a vital military objective for the Allies. Britain secured the assistance of Hussein Ibn Ali (1854–1931), ruler of the Hejaz and guardian of Mecca, by pledging in vague terms to help Arab independence.

On 2 November 1917, Zionist hopes also seemed near fulfilment when the British foreign secretary, Arthur Balfour (1848–1930), declared: "His Majesty's government view with favour the establishment in Palestine of a national home for the Jewish people . . .". But meanwhile, in 1916, the Allies had agreed secretly to a postwar Middle East division of spoils that paid no immediate heed to Zionist hopes and reduced the Arab state to a Franco-British puppet. At the San Remo Conference in 1920 Palestine, which had been under direct British military rule since 1918, came under British mandate.

Over the next 25 years the situation in Palestine steadily worsened. The Jewish population increased [1] and so did Arab violence against the Jews, erupting in riots in 1921 and 1929. With the advent of racist persecution in Nazi Germany during the 1930s the Zionists felt that increased immigration was desperately necessary. On the other hand, new and more extremist Arab Palestinian leaders advocated halting immigration by force. Britain finally crushed an Arab revolt of 1936–9 [3], but its 1939 White Paper restricting Jewish immigration was a political victory for the Arabs.

During World War II, under the stress of Nazism, Zionism became a mass movement, lobbying the US government and public for support. After 1945 American Zionists shipped money and arms to the Haganah, a semi-underground Jewish army, and to more extremist guerrillas. In the face of British refusal to increase immigration, Jewish guerrilla violence and British counter-violence intensified. Finally Britain referred the problem to the UN, which in August 1947 recommended partition.

The birth of Israel

The British left on 14 May 1948; that same day the state of Israel was proclaimed [Key] and the armies of five Arab states attacked it. But armistices in 1949 left Israel holding most of the territory it had been granted,

CONNECTIONS

See also
298 The United Nations and its agencies
222 The rise of fascism
188 World War I
240 The Soviet Union since 1945

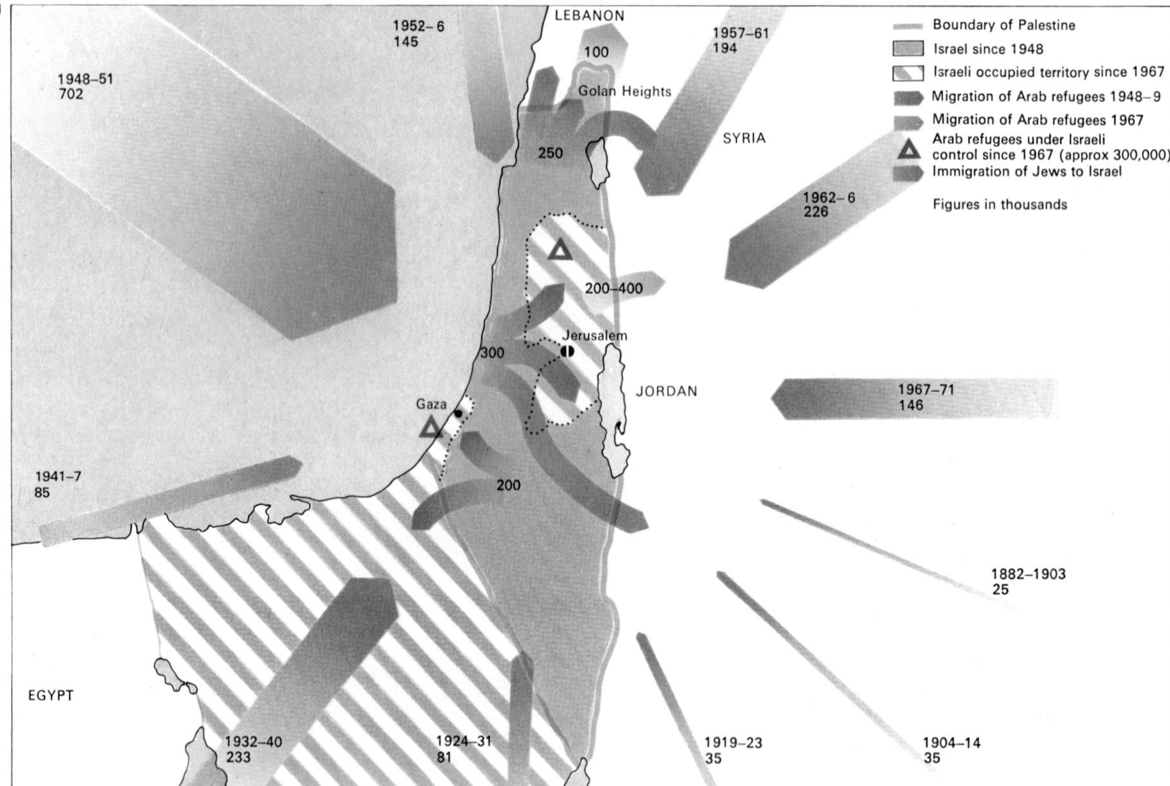

1 Migration to Palestine began in the late 19th century as groups of Jews sought freedom from persecution and reaffirmation of Jewish dignity through establishing settlements there. After the foundation of the World Zionist Organization by Theodor Herzl (1860–1904) in 1897, more Jews arrived buying land for collectives. During the Mandate, immigration fluctuated. Up to 1948 most arrivals were from Europe. After this, many Jews living in Arab countries migrated or fled to Israel. The Palestinians left their homes in two waves, the majority (more than half a million) in 1948 and a second group of between 200,000 and 400,000 during the war in 1967. Most were herded into UN refugee camps, only Jordan granting them citizenship. After 1967, 300,000 of them lived in refugee camps run by Israel.

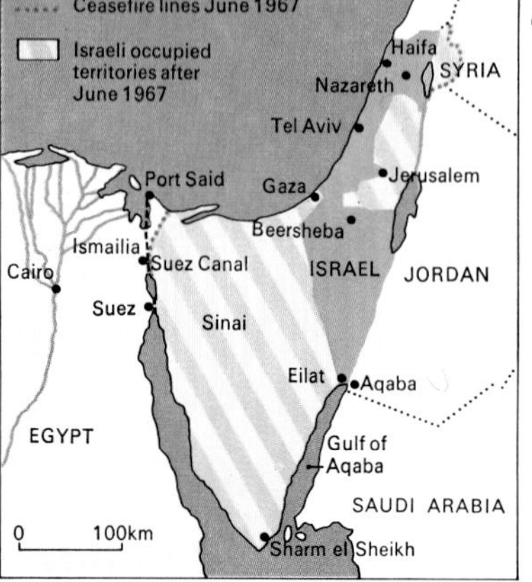

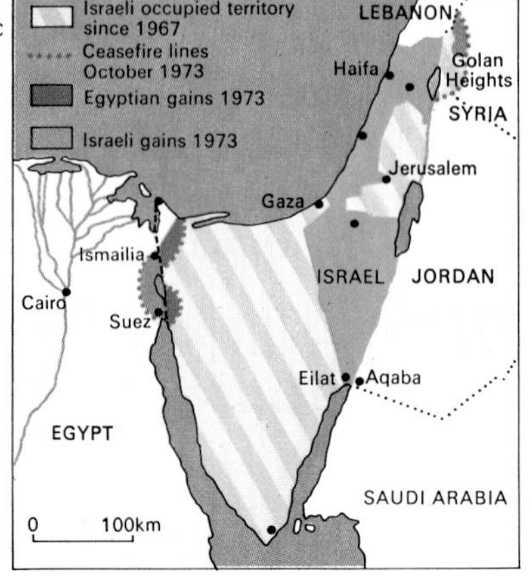

2 Israel's borders at the time of the 1949 armistice were wider than envisaged in the 1947 UN partition plan [A]. Arab Palestine had been largely incorporated into Jordan. After the 1967 war, Israel occupied East Jerusalem, the West Bank, the Golan Heights and the whole of Sinai [B]. The war of October 1973 and subsequent disengagement agreements returned some of the Golan to Syria and the Suez Canal and part of Sinai to Egypt [C].

together with some of the territory allotted for an Arab Palestinian state [2]. In the absence of this state, Jordan acquired the West Bank and Egypt the Gaza Strip. About 600,000 Palestinian Arabs lost their homes.

Israel was left surrounded by hostile neighbours and Arab humiliation and defeat demanded redress. Open war broke out on three further occasions. In 1956, with its shipping blocked by Egypt, Israel joined in an Anglo-French conspiracy to recapture the nationalized Suez Canal. In a lightning attack the Israelis occupied the Sinai Peninsula on the east bank of the Suez Canal. Pressure by the US and USSR forced Israel to withdraw from Sinai but a UN force was established in the Gaza Strip to act as a buffer.

When Egypt ordered the departure of the UN force in 1967 and on 22 May closed the Strait of Tiran, the Israelis seized the initiative on 5 June by a pre-emptive strike on the airfields of Egypt, Jordan, Syria and Iraq. After six days of fighting Israel held all Jerusalem [5] as well as the Suez Canal; the Jordanian army had been forced across the Jordan and the Syrian Golan Heights were occupied. This time Israel did not withdraw.

The lesson of the first strike was not lost on the deeply humiliated Arabs. On 6 October 1973 the forces of Egypt and Syria attacked simultaneously, Egyptian troops crossing the canal while Syrian troops advanced over the Golan plain. At the end of 16 days Egypt and Syria had gained a little territory and a great deal of prestige.

Distant hopes of peace

Stagnating in camps, the exiled Palestinians meanwhile had formed desperate guerrilla groups, which eventually united in 1969 under the umbrella of the Palestinian Liberation Organization. In October 1974 the PLO was recognized by all Arab countries as the sole representative of the Palestinians [6].

The realities of the Arab-Israel conflict have often been blurred by its being a focus of superpower rivalry, with the US supplying Israel and USSR arming the Arabs. In 1975, however, Egypt and Israel arrived at an interim peace agreement in which there were seeds of hope. Irreconcilable nationalist aims remain the core of the problem.

KEY

David Ben-Gurion (1886–1973), first prime minister of Israel, proclaimed the establishment of the Jewish state in the Museum of Modern Art in Tel Aviv on 14 May 1948, the day on which the last British high commissioner departed.

3

3 Arab revolts broke out in April 1936 against British rule in Palestine, partly as a result of declining prosperity but mainly because of mounting Jewish immigration. Spontaneous and horrifying attacks on Jews occurred throughout the country. At the same time Arab leaders called a six-month general strike in an effort to force the British to suspend Jewish immigration. At first directed against the Jews, the revolt later became anti-British, and eventually armed bands of unemployed also attacked Arabs who opposed them. The unrest ended in 1939.

4

4 Martial law was imposed in Tel Aviv in 1945. Jews saw the immigration limits in Britain's White Paper of 1939 as a betrayal, and reaction was muted only by the outbreak of war. Ben-Gurion said: "We shall fight with Britain as if there was no White Paper and we shall fight the White Paper as if there was no war". An unofficial Jewish army, the Haganah, had existed since the 1920s and in 1937 a more extreme group formed the Irgun (or Etzel). Allied in September 1945, these groups set out to change British policy by increasingly violent attacks on British troops. British military reaction was viewed as counter-violence.

5

5 Jerusalem, a city sacred to Judaism, Christianity and Islam, was visualized in all external partition plans for Palestine as an international city. In the 1948–9 war it was divided, with the east and Old City held by Jordan and the west by Israel. During the Six Day War of 1967 the city was forcibly reunited by the Israelis. New buildings encircling the whole city (in the distance here) are evidence of Israel's determination to retain control in its own hands.

6

A

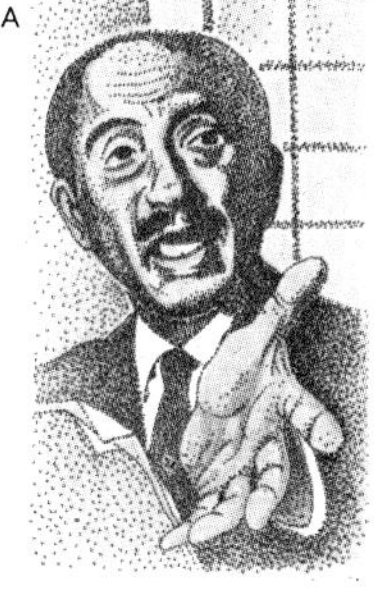

B

C

D

6 Arab opposition to Israel has taken different forms. Under Anwar Sadat (1918–) [A] Egypt, the main combatant, adopted a new and much-criticized course in 1975 by concluding an interim peace agreement with Israel. From a rigidly Islamic standpoint that rejected the idea of any part of the Islamic world under non-Muslim rule, Saudi Arabia used its enormous oil wealth [B] to help "front-line" Arab states like Syria maintain a bellicose attitude. While the PLO [C] aimed politically at a "secular democratic state" in the whole of Palestine, its militant extremist wings [D] captured world headlines with violent attacks inside and outside Israel.

The United Nations and its agencies

The name "United Nations" was devised by United Sates President Franklin D. Roosevelt (1882–1945) and was first used in the Declaration by the United Nations of 1 January 1942, when representatives of 26 nations pledged their governments to continue fighting together against the Axis powers. The new United Nations (UN) was effectively a drastically reorganized and updated version of the League of Nations.

The charter of the UN was drawn up by the representatives of 50 countries at the United Nations Conference on International Organization, which met at San Francisco from 25 April to 26 June 1945. The charter was signed on 26 June 1945 and the UN began officially on 24 October 1945 [1].

Peace and security

In theory, UN membership is open to all peace-loving states that accept the obligations of the charter. In fact, the principle of universality has been accepted, so that apart from Switzerland (with its rigid neutrality) all independent nations have joined or are doing so. By 1976 there were 144 members.

The UN is not a world government or suprastate. All member states are sovereign and equal. The charter provides that the UN shall not intervene in the internal affairs of any country, except when it is acting to maintain or restore international peace.

In the Security Council the five permanent members (France, UK, USA, USSR and the People's Republic of China) each have a veto. But conflicting outlooks – particularly the ideological cold war between the USSR and the West – have meant that one or other of the Great Powers has been able to frustrate the General Assembly's wishes, although the Uniting for Peace Resolution of 1950 gave the Assembly authority to recommend enforcement action over a veto.

The UN has been involved in more than 100 situations where peace has been at risk [2, 5]. For example, the Security Council played an important part in solving the dispute between The Netherlands and Indonesia over the latter's independence in 1949; it prevented a threatening situation from escalating into outright hostilities when foreign troops intervened in the Lebanon and Jordan in 1958; it contributed towards the peaceful transition of colonies to independence through organizing plebiscites and referenda, and on numerous occasions the secretary-general of the UN [4] has used quiet diplomacy to prevent conflicts over issues that could have become explosive.

The preamble to the UN charter determines to "reaffirm faith in fundamental human rights, in the dignity and worth of the human person, in the equal rights of men and women". Major steps to this end have been the 1946 Convention on the Political Rights of Women, the 1948 Universal Declaration of Human Rights, the 1951 Convention on Genocide and the 1965 Convention on the Elimination of Racial Discrimination.

Economic and social work

More than 80 per cent of the UN's funds are devoted to helping poorer countries develop their own human and economic resources [9].

Under the supervision of the Economic and Social Council, there are seven functional commissions that make studies, issue reports or draft international treaties relating

CONNECTIONS

See also
302 Underdevelopment and the world economy
254 Non-alignment and the Third World
246 Decolonization
192 The Peace of Paris
296 The question of Israel

In other volumes
106 Man and Society
108 Man and Society
316 Man and Society

1

1 Joseph Paul Boncour (1873–1972) signs the United Nations Charter for France at the first meeting of the organization in San Francisco in 1945. Since the first 50 members appended their signatures to the charter the membership has grown to almost treble that original number. As they have joined, the very many emergent nations have gradually weakened the Great Powers' 20-year domination of the UN.

2

2 UN troops cross the Han River in Korea as they move to meet the North Korean invaders of South Korea in 1950. It was the UN's first military intervention in a war – but almost by default. The USSR, at that time boycotting the Security Council, was unable to veto a recommendation that the UN should go to the aid of South Korea. Sixteen nations responded to the call to arms, but in the event it was overwhelmingly the US that provided the men, equipment and overall command to drive the North Koreans back across the dividing line of the 38th Parallel.

3

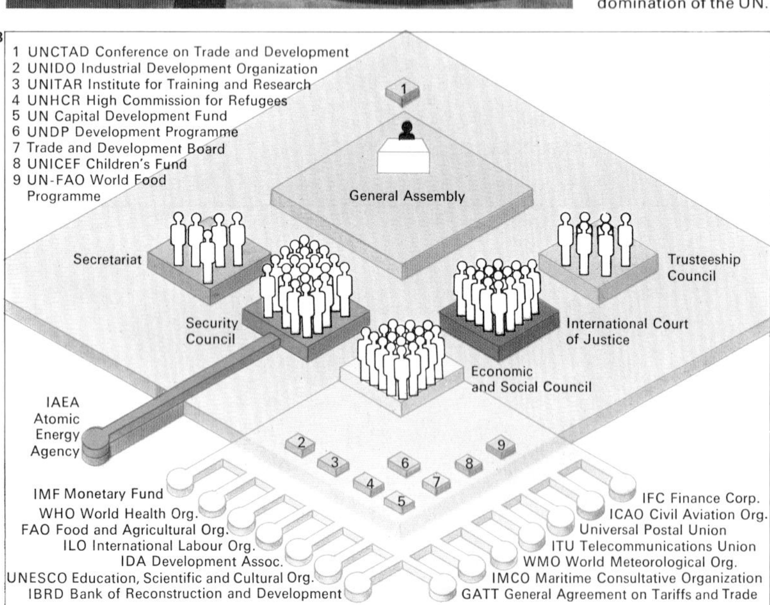

3 The "political" aspect of the UN is dominated by the General Assembly and the Security Council, but apart from these there are four other bodies. The Economic and Social Council (ECOSOC), under the supervision of the General Assembly, co-ordinates the UN's economic and social work and that of 14 of its specialized agencies. The Trusteeship Council was established to supervise the affairs of 11 trusteeship territories, of which all but one (the Pacific Islands) have now achieved independence. The International Court of Justice is the principal judicial organ and all UN members are parties to its statutes and can refer cases to it. It consists of 15 judges elected by the General Assembly and Security Council voting independently. The judges serve an initial term of nine years. Lastly, the Secretariat services all the other organs and administers the programmes and policies laid down by them.

4 A

B

C

D

4 The chief administrator of the UN is the secretary-general, a man proposed by the Security Council and elected by the Assembly. Since 1946 there have been four; Trygve Lie (1896–1968) [A] of Norway, (1946–53); Dag Hammarskjöld (1905–61) [B] of Sweden, whose term ended tragically in an air crash in N. Rhodesia; the Burmese, U Thant (1909–) [C], who retired in 1971; and Kurt Waldheim (1918–) [D] of Austria.

to subjects such as human rights and control of narcotic drugs. There are also five regional economic commissions – one each for Africa, Western Asia, Asia and the Pacific, Europe and Latin America. Increased stress on direct operational field activities is reflected in the stepped-up pace of the United Nations Development Programme, a voluntarily financed operation carried out by the UN and 15 related agencies.

The emergence of a new majority

Until the 1960s the balance of power within the UN General Assembly lay with the Western Alliance, partly because of the composition of the Security Council, but as colonial territories acquired independence in the 1960s so new states with traditions and interests very different from those of the US and the European liberal democracies joined the UN. The influence of these new states became manifest in the General Assembly, where an increasing emphasis was placed on the evils of colonialism and apartheid and on the need for economic development. The numerical majority of present members are from Africa, Asia, Latin America and the Middle East. By 1970 it was apparent that the balance of power in the Assembly had positively shifted to a non-aligned group, which did not necessarily support either side in the East/West ideological battle [7]. The states of the Western Alliance found themselves in a minority as resolutions favouring the non-aligned group was passed, often with Eastern European backing.

The full effects of this change, however, were not felt until 1974, when the special session of the Assembly adopted a declaration and a programme of action on the establishment of a new international economic order. In the declaration, UN members solemnly proclaimed their determination to work urgently for "the establishment of a new international economic order based on equity, sovereign equality, interdependence, common interest and co-operation among all States, irrespective of their economic and social systems, which shall ensure steadily accelerating economic and social development in peace and justice for present and future generations".

KEY

The "Parliament of the World", the UN General Assembly, has its permanent home in Manhattan, New York. It can discuss and make recommendations on any subject mentioned in the UN charter except when the Security Council is discussing it, but it has no power of enforcement. It elects members to the other UN agencies, appoints the secretary-general and fixes and allocates the budget.

5 Potential "powder keg" situations throughout the world have seen the presence of UN peace-keeping forces since the organization moved to back South Korea when it was invaded in 1950. They have been used to separate forces in the Middle East, to control armed conflict, and keep internal order in the Congo after its independence (1960–64) and in Cyprus, (1964 onwards) where clashes between Greek and Turkish communities erupted into an invasion of the island by Turkey in the course of 1974. Non-combatant observers have been in Indonesia, Korea, Lebanon, Jammu and Kashmir, West Iran and the Yemen.

5

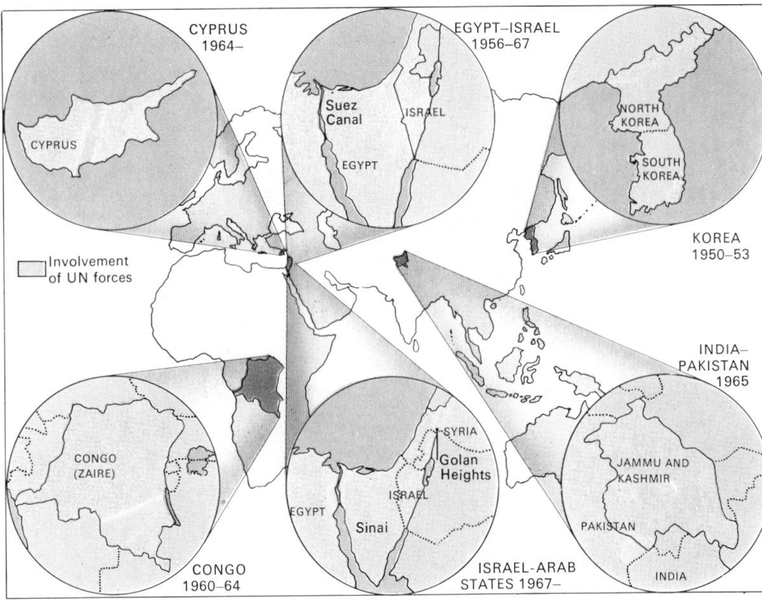

6

6 The giant monuments of Abu Simbel were saved from the waters of Lake Nasser by UN agencies, in particular by UNESCO. As yet the UN's cultural work, like its exercises in international diplomacy, has been less impressive than its continuing battle against disease and famine through the work of the World Health Organization and the Food and Agriculture Organization.

7

7 The UN membership consists of sovereign states that accept the obligations contained in the UN Charter. From time to time non-self-governing territories have been allowed to put their case to the committees of the Assembly but a precedent was set in 1975 when the head of the Palestine Liberation Organization, Yasser Arafat, was allowed to address the Assembly.

8 After the Arab-Israeli War resumed in October 1973 two ceasefire resolutions sponsored by the Soviet Union and the US were adopted by the Security Council. But the fighting continued and it was the eight non-aligned members of the Security Council who then proposed the dispatch of this non-combatant observer force, whose function was to supervise the ceasefire conditions.

8

9

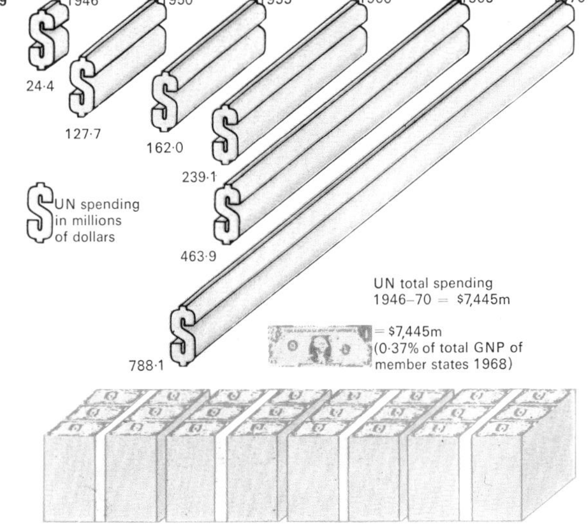

9 UN members contribute according to their yearly product, the USA paying most. The total UN expenditure since 1946 is only a fraction of the annual wealth produced by its members.

The world's monetary system

The establishment of a new and more stable international monetary system was one of the most important tasks for world leaders as World War II drew to a close. At the Bretton Woods Conference in 1944 negotiators had bitter memories of the 1930s when the breakdown of the gold standard [1] as a semi-automatic system of adjusting imbalances in trade and payments between nations was followed by a period of unstable exchange rates, restrictive trade practices and deep economic slump in most major countries. It was the aim of the conference to devise a monetary system that would encourage international co-operation and end instability.

The Bretton Woods system

The essential features of the new system were: stable, or fixed, exchange rates; the creation of a new central organization, the International Monetary Fund (IMF), to oversee the new arrangements and assist countries in balance-of-payments difficulties [Key, 2]; and assistance, through the newly established World Bank (International Bank for Reconstruction and Development), to poor countries. Stable exchange rates required each IMF member to report to the Fund the value of its currency (in terms of gold). Since all currencies were thus "priced" in terms of a single denominator, gold, this also established rates of exchange between them. These rates were to be regarded as essentially fixed and a major change in the value of a currency was permitted only when a country was suffering from "fundamental disequilibrium" in its balance of payments. To correct a "fundamental" surplus (exports greater than imports) a country would revalue (making its exports more expensive and its imports cheaper); to adjust a deficit it would devalue.

The US dollar, and to a lesser degree the British pound sterling, came to play a central role in the new system. Sterling had long had an important position as a major trading or "hard" currency [4]. The dollar's pre-eminence was largely a postwar phenomenon and reflected the economic and political strength of the United States in a world in which most other leading countries were still ravaged by the results of the war. Together with the fact that the US Treasury undertook to convert foreign holdings of dollars into gold at a fixed price of $35 per ounce (thus making the dollar "as good as gold") this prompted other countries to accumulate holdings of dollar balances on which they could earn interest. The dollar and sterling thus acted as key "reserve currencies", supplementing gold. The Bretton Woods system became fully operational only in 1958 when, after a prolonged period of postwar reconstruction, all major currencies became freely convertible one for another.

Pressure on sterling

The crucial requirement for the smooth functioning of the Bretton Woods system was the willingness of countries to hold the two reserve currencies. In general they did so until 1964, after which a series of currency crises progressively undermined the fixed exchange rate system. Pressure centred initially on sterling. International confidence was eroded by Britain's chronic economic problems at home and overseas. There was heavy selling of sterling by international

CONNECTIONS

See also
302 Underdevelopment and the world economy
298 The United Nations and its agencies
264 USA: the affluent society

In other volumes
278 Man and Society
314 Man and Society

1 Under the gold standard imbalances in trade are settled by transfers of gold between countries. If the value of exports and imports balances [A] a country neither loses nor gains gold. The value of money circulating in a country is directly tied to its stock of gold [B]. When a deficit arises because imports are greater than exports [C] an outflow of gold takes place to settle the difference [D]. This reduces the volume of money at home, depressing wages and prices [E]. Goods for export are cheaper, more are sold, and equilibrium is restored with a smaller gold stock [F].

2 The resources of the International Monetary Fund come from quotas subscribed by its members [1–5], 25% in gold [yellow] and the rest in their own currency. Any member in balance-of-payments difficulties can borrow from the Fund the currency of other members up to a top limit of 200% of its own quota. Country 1 is borrowing 150% [6] while country 2 draws the full 200% [7]. So that appropriate balance of currencies is maintained repayments [8, 9] must be made within five years in the currencies of members whose money has been borrowed from the Fund.

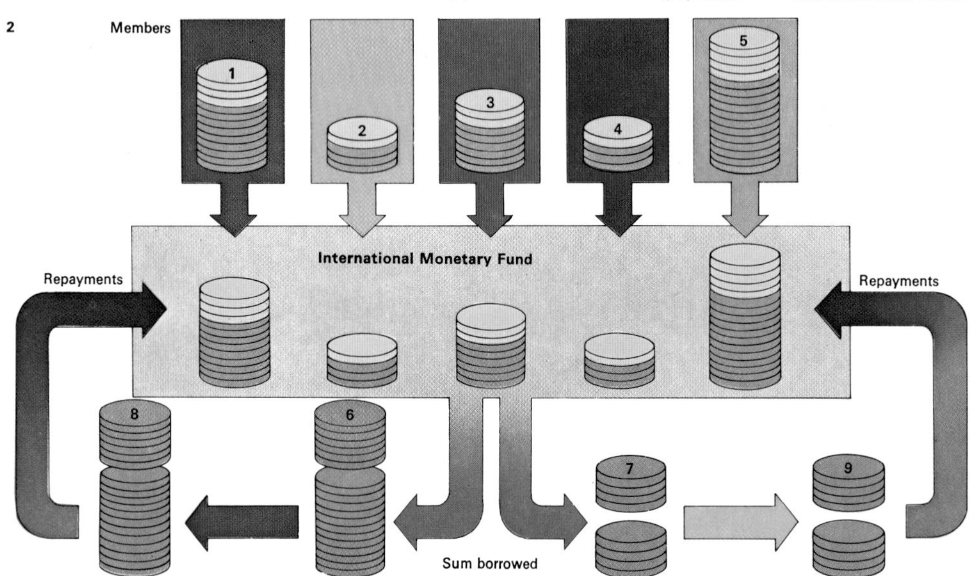

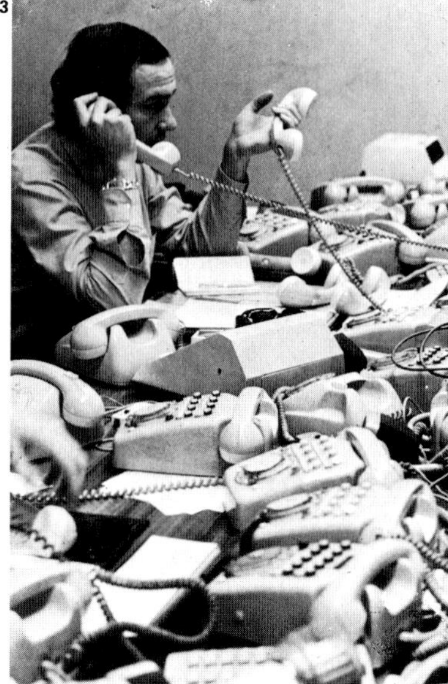

3 The flow of world money is very fast indeed. The foreign exchange rooms of bankers such as Samuel Montagu turn over millions of pounds a day.

holders on many occasions, facilitated by the gradual build-up of large quantities of easily transferable or "hot" money in the Eurodollar market [8]. Selling could be stemmed only at the cost of the Bank of England's running down its own holdings of foreign currency in order to buy up sterling in the exchange markets and thus prevent the exchange rate falling below its agreed value. Even the provision of additional funds to the Bank of England by other central banks and by the IMF (through loans and by boosting total world reserves through the creation of a new reserve asset, the Special Drawing Right [7]) could not succeed in saving sterling, and in November 1967 the pound was devalued by 14.3 per cent.

The crisis of confidence soon spread to the dollar, took the form of persistent demand by holders of dollars for their conversion into gold and resulted in a serious drain on US gold reserves. In August 1971 President Nixon took steps to check this outflow and shocked the world by announcing the ending of the longstanding US commitment to sell gold for dollars. President Nixon's surprise package prompted new international negotiations and resulted in the Smithsonian Agreement of December 1971. This provided for a substantial revaluation of all major currencies against the dollar and was intended to produce a more realistic dollar exchange rate.

Floating exchange rates

The Smithsonian Agreement failed to restore confidence [6] and renewed pressure against sterling early in 1972 culminated in a decision in June to allow the pound to "float" and find its own value in the foreign exchange markets. Early in 1973 Italy, Switzerland, Japan and eventually all the major European currencies had to follow suit and allow their currencies to float against the dollar.

This system of generalized floating [5] still prevails. The authorities, however, do not let the markets freely determine the rate, but intervene occasionally to serve national interests. Despite repeated attempts, both within the IMF and outside to reach agreement on a more stable monetary system, negotiations remain deadlocked.

KEY

The International Monetary Fund has its headquarters in Washington. Set up in 1945 to stabilize exchange rates and help finance world trade, it draws its membership from all the major non-communist countries of the world.

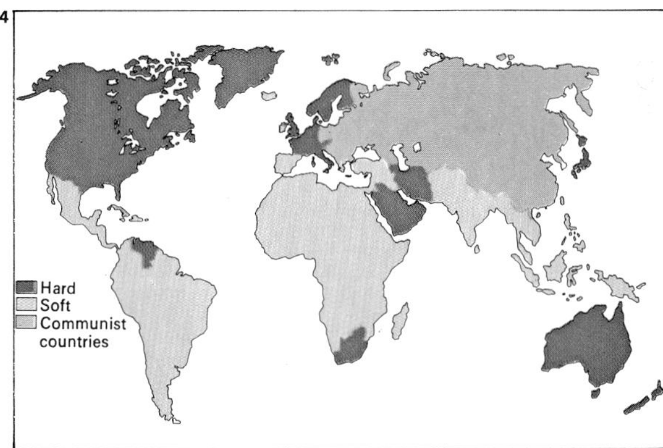

4 World currencies can be roughly split into "hard" or "soft", but in some areas these categories are changing, notably as a result of oil revenues. Hard currencies were once those convertible at a fixed rate and much used for trade. Soft currencies included those in limited use or not convertible. With the breakdown of fixed rates the terms now have a more general meaning of strong and weak currencies.

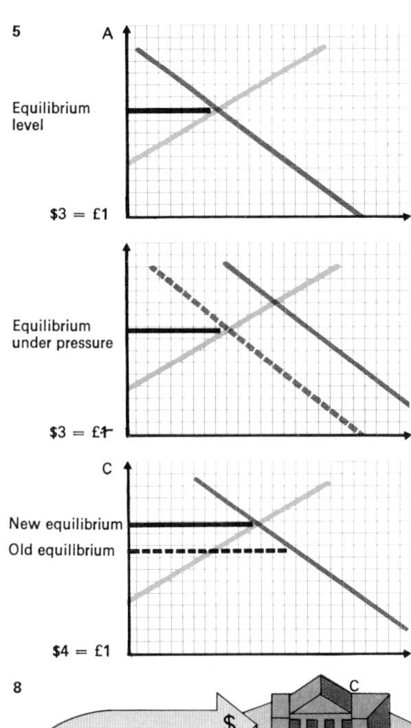

5 A floating rate of exchange finds its level according to supply and demand in the world's money markets at any given point in time. Assuming that a home currency is $ and the foreign currency £, the exchange rate settles at a level that will equate demand [red] and supply [blue] [A]. If demand for imports increases, in the absence of countervailing measures, demand for foreign currency will exceed supply at the old exchange rate [B]. The price of foreign currency therefore rises or, in other words, the exchange rate of the home currency depreciates [C]. The world has had a system of floating or fluctuating exchange rates since 1972.

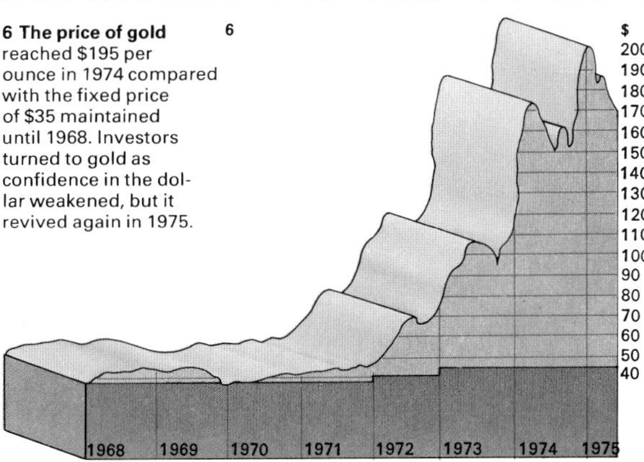

6 The price of gold reached $195 per ounce in 1974 compared with the fixed price of $35 maintained until 1968. Investors turned to gold as confidence in the dollar weakened, but it revived again in 1975.

8 A Eurodollar is created when a dollar passes to a holder outside the United States and, instead of being converted to another currency or deposited within the United States, it is deposited with a bank outside the United States. There are other "Euro" currencies, such as Eurosterling. The term signifies that the currency concerned is deposited outside its country of origin. Once a European bank [A] has received a Eurocurrency deposit from, for instance, a French exporter it can lend it in turn to other banks in need of funds [B, C, D] and it may finally be borrowed by a British businessman who wants to finance investment. The Eurocurrency market emerged in the late 1950s and constitutes a vast international pool of highly mobile money sometimes used for currency speculation. Latest estimates put its size at $200,000 million.

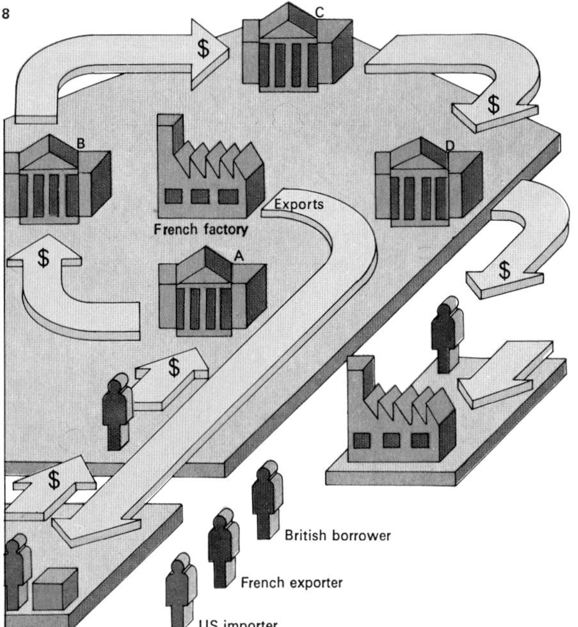

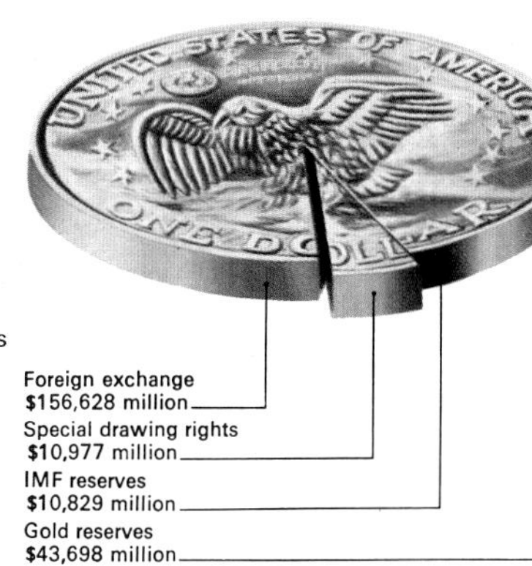

7 Special Drawing Rights (SDRs), introduced in 1970, were created by IMF to increase the volume of resources for financing world trade. They have two main advantages. First, they are a stable, internationally acceptable form of exchange. Second, they enable the IMF to make transferable loans to those countries that need additional foreign reserves to finance trade deficits. In this way they act as a convenient international system of debits and credits.

Underdevelopment and the world economy

The decades following World War II have been characterized by a marked division between a small group of mostly industrialized nations where general living standards and prosperity have risen quite rapidly, and the overwhelming majority of nations where poverty remains acute [Key, 1]. In the former group are found the highly industrialized countries of North America, Australia and New Zealand, and most of Europe and Japan, and in the latter the extensive regions of underdevelopment in South America, Asia and Africa, although Brazil has shown a very marked rise in gross national product (GNP) in this period.

Patterns of trade

Simultaneously with this steadily widening gap in material standards, dozens of new nations have been created in the process of decolonization. But self-government has not brought economic freedom. The pattern of trade established during the colonial period means that the new nations are still frequently dependent on the old metropolitan countries. Their economic role remains largely one of supplying agricultural goods and industrial raw materials [3], serving as markets for the surplus manufactures of the industrialized nations and acting as a reservoir of cheap labour. Finally, much of the trade and industry of the ex-colonies is in the hands of international companies based in the rich countries and whose profits do not accrue where they are created.

As a consequence of these traditional ties, the less developed countries have also suffered the booms and recessions of the industrialized world. There have been sharp swings in demand for the primary products sold by the poorer nations leading to violent fluctuations in commodity prices and therefore in their foreign earnings [2]. This (together with the inevitable unpredictability of agricultural production) makes planning a development programme almost impossible because unpredictable export earnings force planners to curb necessary imports of machinery and capital equipment.

Although the rich countries provide some overseas aid [6], the flow of funds is inadequate and few of the less developed countries have what economists call "self-sustained growth" – that is, profit levels are not high enough to finance expansion on the scale desired. Indeed there is much argument about whether the conditions that led to industrialization and economic take-off throughout the 1800s in western Europe and North America still exist and whether it is even feasible for the less developed countries to copy the industrialized West.

The developing nations and cartels

If the governments of the Third World nations are to eradicate poverty and maintain social and political stability it would nevertheless seem that they have no alternative but to take their peoples down the road to industrialization [4] in the hope of finding a formula for self-sustained economic growth. This means mechanizing industry and agriculture and has led to demands that the existing industrialized nations should provide the requisite funds. For example, it has been suggested that they should lower the present customs duties and quotas they impose on some of the industrialized goods

CONNECTIONS

See also
300 The world's monetary system
298 The United Nations and its agencies
254 Non-alignment and the Third World
246 Decolonization
256 Latin America in the 20th century
264 USA: the affluent society

In other volumes
156 The Physical Earth
176 The Physical Earth
248 The Natural World
108 Man and Society
316 Man and Society

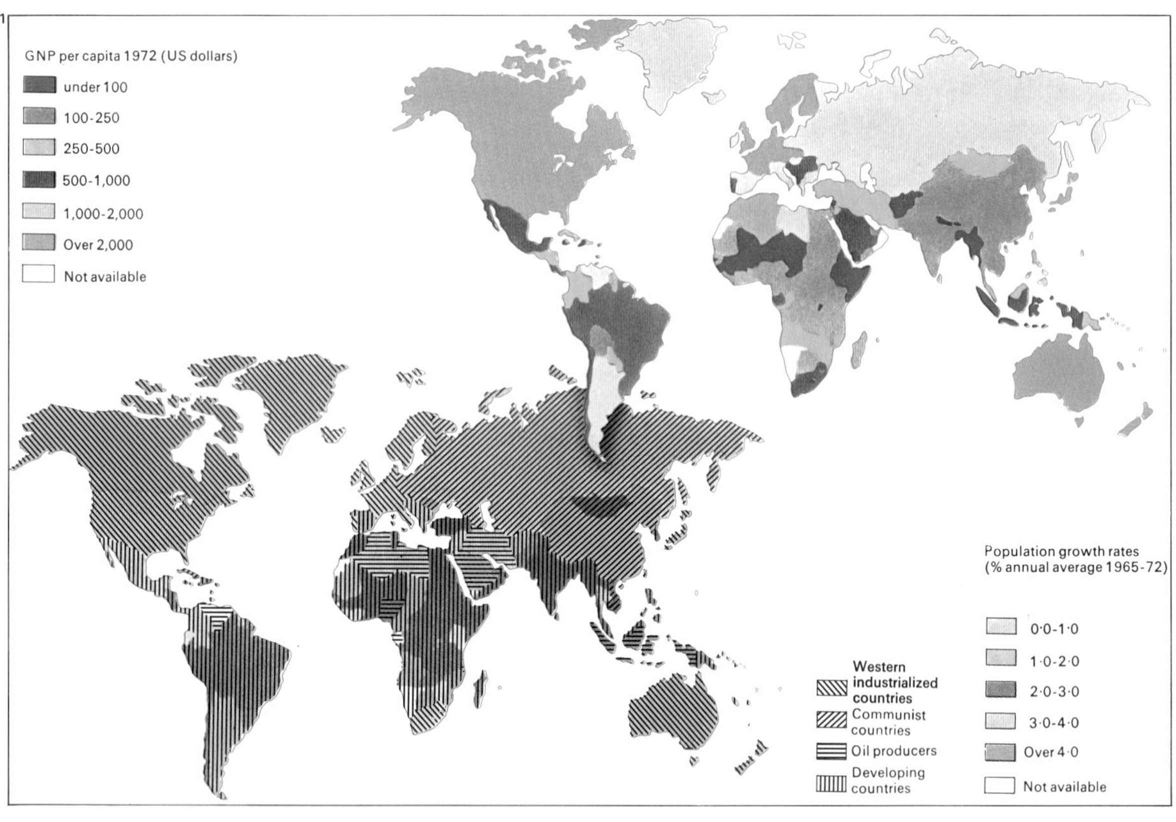

1 Some 600 million people live in countries which in 1970 had per capita incomes of between $2,000 and $3,000 a year; another 2,000 million live in countries where per capita income is estimated at less than $200. In countries where small labour-intensive landholdings predominate rural population increase is often stimulated beyond the ability of the land to support it, encouraging migration to the towns where the urban labour market cannot support it either. Indeed the poorest countries are usually those in which the population growth rate is highest: there were over 538 million people in India by 1970 and the numbers have been swelling at the rate of 2.3% a year. Mexico's annual rate of increase in the 1960s was 3.5%. By contrast population in many western European countries is rising by less than 1% per annum.

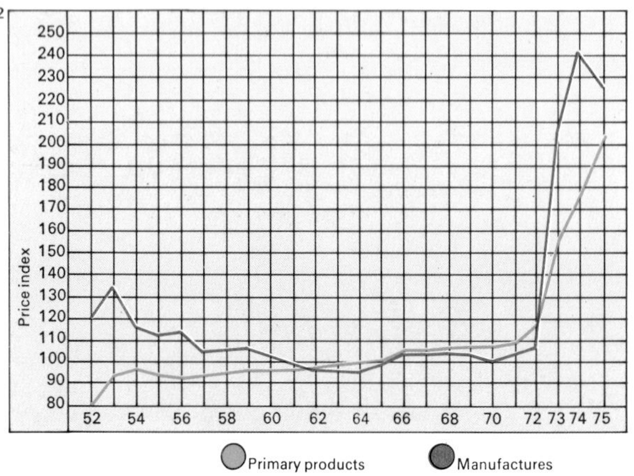

2 World export prices between 1950 (the peak of the commodity boom of the Korean War) and 1970 moved first of all in favour of the products of the less developed countries but after a period of relative strength this advantage was lost. Some economists blame the weak economic performance of the less developed countries on a marked deterioration in their "terms of trade" – the fall in the price they get for exports relative to the cost of their imports.

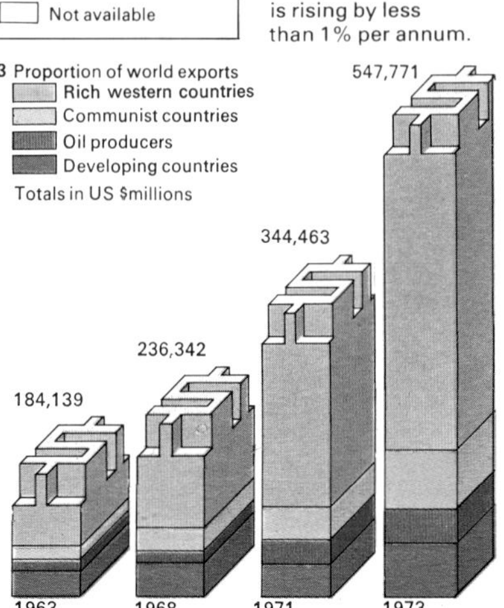

3 Export figures for the 1960s and early 1970s show that developing countries accounted for a relatively small proportion of world trade. This began to change in 1974, but only as a result of higher oil prices. The exports of most developing countries are still agricultural products like coffee or sugar and raw materials for industry like rubber or tin. Only about 25% of their exports – frequently textiles – are manufactured goods.

they import from the Third World – such as textiles – which at present often encounter high tariffs because they compete with the industries of the industrialized nations. For these reasons political tensions have been increasing between rich and poor countries.

Some less developed countries have also attempted to achieve higher prices for their primary products by banding together in associations. One of the most successful of these has been the Organization of Petroleum Exporting Countries (OPEC). Because of its near monopoly in the export of oil, OPEC succeeded in getting a fivefold increase in the oil price during 1973 and 1974. Other groups of commodity producers have not been as successful and higher oil prices and the resultant higher price of manufactured goods have hurt developing countries, such as India, which do not possess oil. In spite of this, the example set by OPEC has proved an inspiration for other producers of raw materials, although many economists argue that such associations, or "cartels" as they are called, cannot last for long because supply and demand will eventually drive the price back down to a sustainable level.

The desire of developing nations for changes in the world trading system has also led to political initiatives such as the United Nations special conference on raw materials in 1974 which adopted a programme for a "New International Economic Order". A resolution to this effect was approved by most countries despite opposition from many of the richer nations. However, it is generally recognized that a new economic order can be established only if the industrial countries are prepared to meet a far-reaching list of demands from the Third World nations.

Future prospects

The success of OPEC, the possible growth of more such cartels and the fear of political upheaval in the Third World should existing levels of poverty persist have produced statements of willingness on the part of the industrialized group to make at least some concessions. But despite growing concern about Third World problems and sincere efforts by certain countries, the general level of aid has been dropping since the 1960s [6].

KEY

Undernourishment, disease and bad housing loom over 80% of the world's 4,000 million people, in stark contrast to the affluence of a few nations.

4

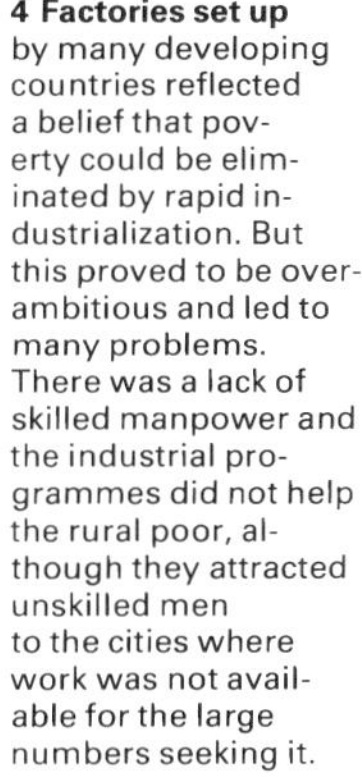

4 Factories set up by many developing countries reflected a belief that poverty could be eliminated by rapid industrialization. But this proved to be over-ambitious and led to many problems. There was a lack of skilled manpower and the industrial programmes did not help the rural poor, although they attracted unskilled men to the cities where work was not available for the large numbers seeking it.

5

5 Rice-planting in India and elsewhere in the 1960s raised the hopes that some less developed countires could become more self-sufficient in food by a "Green Revolution". This was the introduction of new, high-yielding rice and wheat plants which could greatly increase harvests. Although modestly successful in some areas, the costs involved have proved formidable for peasants borrowing at high local interest rates.

6 The rich countries provide foreign aid both in goods and in funds. Here [A] a US helicopter lands supplies. But aid is inadequate when set against the real food shortages and lack of jobs. Some 80% of the money invested in developing nations comes from their own limited resources. Official development assistance from 17 of the world's richest nations [B] in 1974 was just 0.33% of their combined GNP, much less than the figure of 0.53% that was given in 1960.

6 A

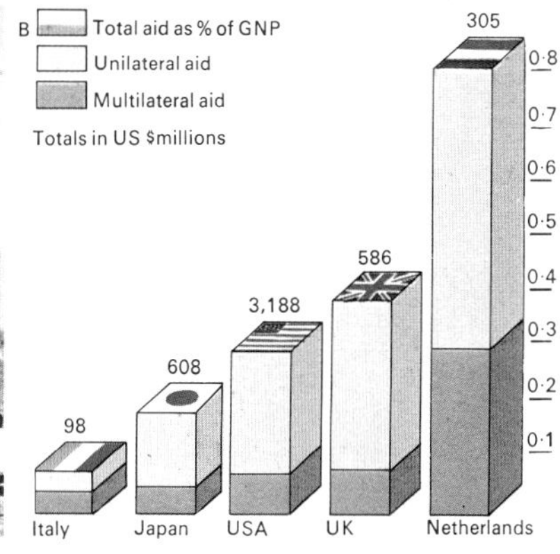

7

7 Shanty towns have grown on the fringes of urban areas such as Bombay, shown here, because poverty in poor countries is less severe there than in the countryside. But migration puts a great strain on services and facilities that are already stretched.

8

8 Aid is not spent on welfare alone. Some of it is committed to prestige projects such as the Organization for African Unity building in Addis Ababa.

Modern Christianity and the New Beliefs

Developments in the life of the Christian Churches during the latter half of the twentieth century have been faster and more far-reaching than at any stage since the Protestant Reformation of the sixteenth century. The main features have been the ecumenical movement (for the reunion of the Churches), the Churches' deeper commitment to the service of the secular world and the cause of world justice, and the dialogue with "unbelief", notably Marxism. Two outstanding events have been the foundation of the World Council of Churches, to which the Roman Catholic Church does not belong, and the Second Vatican Council (1962–5).

The work of the World Council

The World Council of Churches [1], formed in 1948, today includes 271 Churches working in 90 countries. It is neither a church nor a union of churches, but a forum for the joint study of theology and ecumenism and of Christian insights into the socio-economic and political problems of society; it also organizes relief and other social services for the deprived regions of the world. The principal Churches represented are Anglican, Baptist, Congregationalist, Lutheran, Methodist, Moravian, Old Catholic, Orthodox, Presbyterian and Reformed; the Society of Friends is also a member.

There has not yet been a fusion of major Churches on a global basis, but there have been hundreds of unions of Christian groupings on a local level.

At first the Roman Catholic Church, although sympathetic, stood apart from the World Council, but soon it began to send observers to World Council meetings and eventually to have permanent links with it in the fields of social theology and action. A new ecumenical climate, fostered by Pope Pius XII (pontificate 1939–58) on the Roman Catholic side, received dramatic impetus from the pontificate of Pope John XXIII (1958–63) and the visit to Pope Paul VI (1897–) in 1966 by the Archbishop of Canterbury, Dr Michael Ramsey (1904–). The joint theological commission they set up has already reached a degree of unanimity over, for instance, the central doctrine of the Eucharist, for which few Christians would have dared to hope a decade before. Two of the major hurdles yet to be surmounted are the questions of the teaching authority in the Church and the position of the Papacy.

Effects of the Second Vatican Council

The Second Vatican Council of the Roman Catholic Church [2], summoned by Pope John, attended by observers from other Churches and completed under Pope Paul, to some extent narrowed the theological gaps among the Churches in regard to revelation (the Bible and tradition), authority (the collegial authority of the bishops), the nature of the Church and the recognition that all Christians are united in Christ by baptism. The Council opened many doors to dialogue between Christianity and the great non-Christian religions, and also between Christians and unbelievers. The Church's commitment to the service of the world was reinforced by Pope Paul's many journeys overseas, especially his visits to the United Nations Organization and to the developing countries of the Far East and Latin America. His first visit to Jerusalem, where he met the

CONNECTIONS

See also
138 The expansion of Christianity
172 Masters of sociology
266 20th-century sociology and its influence

In other volumes
224 Man and Society
218 Man and Society

1 The World Council of Churches held its first General Assembly in Amsterdam in 1948. The WCC includes the main Christian Churches apart from the Roman Catholic Church. It is not an amalgamation of Churches but a forum for theological discussion intended to lead to ultimate Christian reunion. It is also concerned with applying Christian teaching to the problems of world justice. It gives aid to development projects in deprived regions but has been criticized for helping freedom fighter movements and thus "fostering violence".

2 The Second Vatican Council was opened in 1962 in St Peter's, Rome, by Pope John. It brought together nearly 3,000 bishops and other Roman Catholic Church leaders whose purpose was to renew the spirit of the Church from within. The Council, in its theological statements, narrowed the gap between itself and the other Christian communions. It committed itself to being "The Church of the Poor", and opened the way to dialogue with non-Christian religions and also with the communists, thus ending the postwar period of direct confrontation with the communist powers.

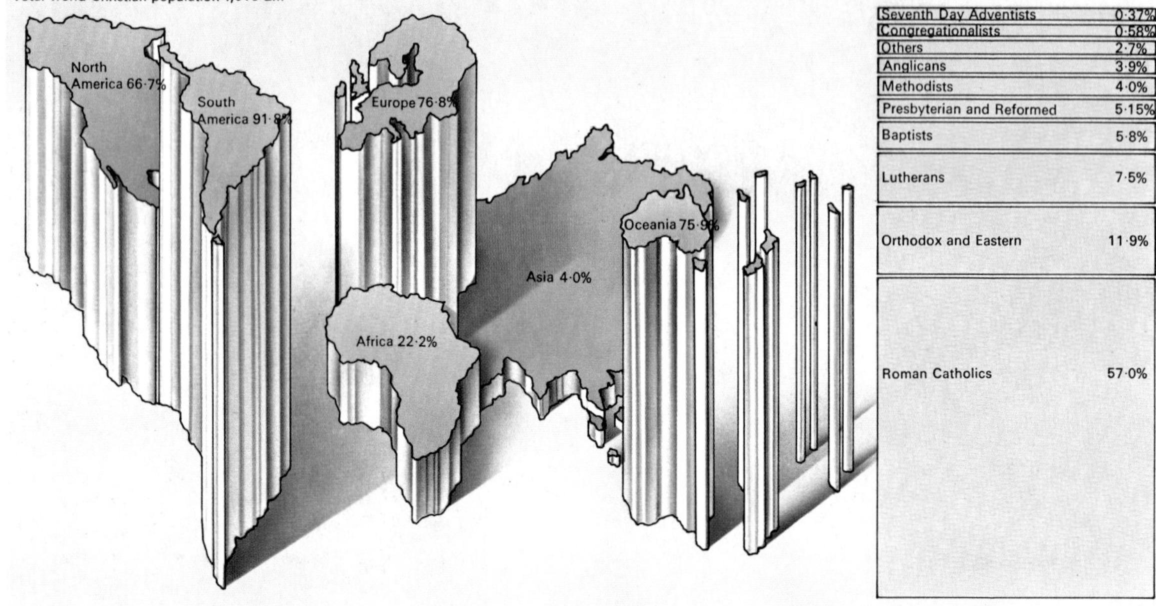

3 The distribution of the world's Christians is historically determined. Through the Roman Empire Christianity spread throughout Europe, to be transmitted worldwide by European emigration and by colonial and missionary activity. The Catholic Church still has by far the largest Christian congregation, claiming almost 60 per cent of the estimated total world Christian population. World Christianity divides into three main streams: the Catholic (Orthodox, Anglican and Roman), the Protestant (Calvinist and Lutheran) and Free Church (Congregationalist, Baptist and Methodist). Map figures show the estimated percentage of Christians within the population of each continent.

Orthodox Ecumenical Patriarch Athenagoras in 1964 [Key], was seen as the first great step towards healing the breach, nearly a thousand years old, between Rome and the Orthodox Churches of the East. Pope Paul's pontificate is also notable for a series of conversations with the communist powers.

Revolt against tradition

The 1960s was a period of intense interest in the concept of the "Death of God" theology identified with Protestant thinkers such as Paul Tillich (1886–1965) [4] and Dietrich Bonhoeffer (1906–45) ("religionless Christianity"), and popularized in *Honest to God* (1963) by an Anglican bishop, Dr John Robinson. Broadly speaking this line of thought rejected the traditional "analogous" way of talking about God. He was not "a person" somewhere "out there" but the transcendent "ground of being", and manifested to the world in the life of Christ.

Eventually the "Death of God" theology faded and was replaced by the more positive concept of the "Theology of Hope", which owes much to the thought of the Jesuit scientist Pierre Teilhard de Chardin (1881–1955). It has been called a "this world" theology, which in Latin America has provoked what is now called the "theology of liberation" [7].

One of its first practical exponents was the Colombian priest Camilo Torres who, despairing of converting the rich oppressors of the poor, joined the local guerrillas in 1965, was soon afterwards killed by the police and became known as the Christian counterpart of Che Guevara (1928–67). The corresponding witness of the Anglican Church has been most notable in South Africa [8], beginning with the championship of the African people by Father, now Bishop, Trevor Huddleston (1913–).

Finally, the last generation has witnessed the rise of movements in the Christian communions that lay less stress on intellectual religious experience and more on emotional fervour and "discernment". A convert-making revivalism in the 1950s and 1960s, such as that of the American evangelist Billy Graham (1918–), has been succeeded by what is known as "Pentecostalism" or "Charismatic Renewal" [9].

KEY

The meeting in Jerusalem in 1964 between Pope Paul VI and the Orthodox Ecumenical Patriarch Athenagoras was the first of its kind since the Eastern Church broke away from the Papacy nearly 1,000 years ago.

4

4 Paul Tillich, the great German Protestant theologian, rejected traditional ideas of God and called Him instead "the ground of our being", a theme taken up by Bishop John Robinson.

5

5 Rudolf Bultmann (1884–), another prominent German theologian, became famous for demythologizing the New Testament, stressing Christ as a spiritual rather than an historical figure.

6

6 A freedom march of black demonstrators in Dallas, Texas, is led by a white priest. One of the Church's most obvious contributions to the new social order in the postwar period has been its active opposition to all forms of discrimination and its struggle to defend the rights of the black people of North America and oppressed peoples in various parts of the African continent.

7

7 Archbishop Helder Camara of Recife, Brazil, is one of the religious leaders of the battle for social revolution in Latin America. The Archbishop's methods have remained non-voilent, unlike those of Father Camilo Torres, the priest turned guerrilla. The Church's struggle for the underprivileged in the sub-continent has taken many forms: the constitutional struggle through the Christian Democrat parties and the dialogue between Christians and Marxists, social action via the Church's co-operatives and credit union, housing and educational programmes and the proclamation of the theology of liberation. Many priests in Latin America have suffered heavily for their actions.

8

8 Students in Cape Town demonstrated outside the cathedral in 1972. The meeting was called to support the principle of racial equality in education. The students obtained the permission of the Anglican dean to hold their meeting on Church property because street demonstrations were banned, but this did not protect them from brutal intervention by the local police.

9

9 The Children of Jesus, swooning in ecstasy, are one of many spontaneous prayer groups seeking knowledge of God and Christ through emotional experience as distinct from reasoned theology. Some of the movements appear to be extremist and grounded in emotional instability. Others, such as the Pentecostalists, who talk about "Charismatic Renewal", are a more convincing blend of quasi-mysticism and "service of the brethren".

Europe from confrontation to détente

Between 1955 and 1975 Europe moved from the cold war to the beginnings of co-operation. In a military sense the confrontation continued because the countries of both the North Atlantic Treaty Organization (NATO) and the Warsaw Pact built up their armed strength and deployed nuclear weapons [1]. But gradually the confrontation came to be accepted as a guarantor of stability in the relations between the Eastern and Western powers.

The meaning of détente

Détente, however, took many years to develop. Originally, it appeared that there were two reasons for optimism. The first was the denunciation of Stalin's methods by Nikita Khrushchev (1894–1971) in 1956. De-Stalinization seemed to promise greater liberalism in Eastern Europe and an improvement in East-West relations. The agreement on a neutral and independent Austria through the Austrian State Treaty appeared to confirm this [Key].

The second ground for hope lay, paradoxically, in German rearmament. When West Germany joined NATO in 1955, the Soviet response was to organize its allies, including East Germany, in the Warsaw Pact. While this reaction appeared threatening, the Soviet Government clearly expected that each superpower would now recognize the final division of Germany and that this would provide the basis for peaceful co-existence. Both hopes were speedily disappointed. When Soviet control in Hungary was threatened [2], Soviet tanks soon demonstrated the limits of the new liberalism. At the same time the Western powers refused to recognize East Germany. In response, Khruschev tried to make them do so by creating a series of crises over Berlin [3]. These crises, which continued from 1958 to the building of the Berlin Wall in 1961, appeared at times to threaten a third world war and helped to accelerate the arms race.

The fear of nuclear war

From 1957 onwards, nuclear missiles were introduced into the arms race (and tactical nuclear weapons into Europe). Crises between the superpowers became increasingly dangerous and it was the most intense of these crises, over the emplacement of Soviet missiles in Cuba in 1962 [5, 7], which induced the superpowers to reconsider their relations and move towards détente.

At their level the two superpowers agreed – tacitly at least – to respect each other's spheres of influence and this implied an acceptance of the alliances as they stood. But at the European level there were certain attempts to change the existing system. In Eastern Europe such attempts arose from a desire to win greater independence from the Soviet Union. In Western Europe they arose from a sense of growing economic power and partly from a wish to see greater liberalization in the East. The most articulate spokesman of this Western European approach was President Charles de Gaulle (1890–1970) [9], who went so far as to take France out of NATO (but not the alliance) in 1966 in an attempt to create a more flexible political system in Europe.

However, from 1963 onwards the two superpowers developed an increasingly close understanding, based on the attempt not only

CONNECTIONS

See also
234 The division of Europe
258 Evolution of the Western democracies
242 Eastern Europe since 1949
298 The United Nations and its agencies
240 The Soviet Union since 1945
218 British foreign policy since 1914
238 Britain since 1945: 2

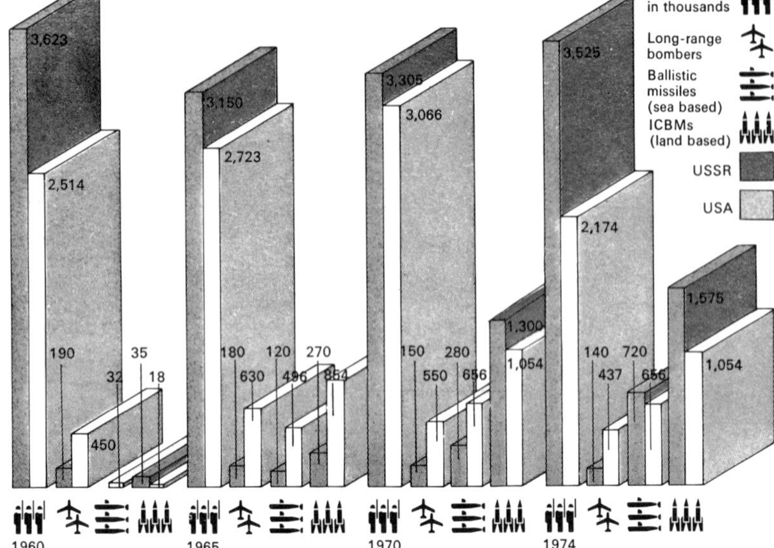

1 The USSR maintains a much larger army than the US, and now has a greater number of land and sea-based strategic missiles, although the number of nuclear warheads held by each side is uncertain.

2 In 1956 there was a popular revolt against communism in Hungary. The rebel government, headed by Imre Nagy (1895–1958), demanded that the Russian troops leave. Instead, more tanks arrived in November and during the next two weeks thousands of "freedom fighters" were killed by Soviet troops. Despite de-Stalinization in Russia, the Hungarians were not allowed to break up the Eastern bloc.

3 Instead of leading to better relations, the scaling down of the policies of Joseph Stalin (often referred to as de-Stalinization) and peaceful co-existence proved to be but a prelude to crisis. Through pressure on Berlin, Nikita Khrushchev tried to force the West to acknowledge the division of Germany. But the two superpowers also tried to manage the crisis through a common understanding. Although at this 1959 meeting Eisenhower and Khrushchev [left] failed to resolve the crisis it set a precedent for later consultations and suggested that the powers recognized that their interest in avoiding war was more important than victory from a conflict.

4 A new crisis arose when an American intelligence aircraft was shot down in Russia in May 1960. The pilot, Gary Powers, was captured. At the Paris conference in May Khrushchev demanded that Eisenhower apologize for the incident; when the US President refused, Khrushchev left the conference, which then broke up. He also withdrew his offer to Eisenhower to visit the Soviet Union. Eisenhower had previously accepted responsibility for the incident.

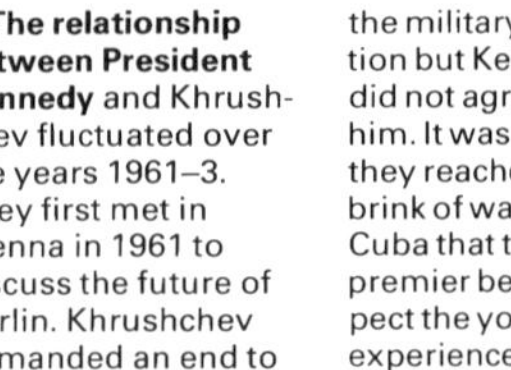

5 The relationship between President Kennedy and Khrushchev fluctuated over the years 1961–3. They first met in Vienna in 1961 to discuss the future of Berlin. Khrushchev demanded an end to the military occupation but Kennedy did not agree with him. It was not until they reached the brink of war over Cuba that the Soviet premier began to respect the young, inexperienced Kennedy.

to avoid nuclear war but also to control the arms race that might produce it. Their agreements began in 1963 with the renunciation of nuclear tests in the atmosphere or space; they continued through the attempt to halt the spread of nuclear weapons (which might have made other conflicts more dangerous) in the non-proliferation treaty of 1968, and they culminated in a whole series of talks and agreements designed to control the dangerous new weaponry that each was capable of developing – the Strategic Arms Limitation Talks (SALT) [8].

This understanding on controlling the arms race also helped to provide the basis for other agreements, most notably the Berlin Agreement of 1972, that reduced conflict; in addition a series of economic agreements designed to create a positive interest in détente were reached.

Problems in Eastern Europe

But this process of increased understanding was not smooth. The period of relative Soviet tolerance ended in 1968 when the Soviet Union and members of the Warsaw Pact invaded Czechoslovakia to destroy the programme for democratic government of Alexander Dubcek (1921–).

Thereafter it was the West German Chancellor, Willy Brandt (1913–) [10], who restored European détente at about the same time as the two superpowers began the SALT talks. Brandt's *Ostpolitik* established political and economic agreements between West Germany, the Soviet Union and Poland, and subsequently East Germany. It was this last agreement that led to the recognition of East Germany by all the Western powers. Since his *Ostpolitik* was also instrumental in bringing about the Berlin Agreement, it laid to rest two of the major causes of tension of the entire cold war.

Détente was by then firmly established, and became the basis of US foreign policy under Henry Kissinger (1923–). At the same time the Conference on Security and Co-operation in Europe, culminating at Helsinki in 1975, established the guidelines for agreement over a range of issues. It was still unclear how far détente could lead to real co-operation, but the foundations had been laid.

KEY

The independence of Austria was restored by treaty in 1955, as the Allies had agreed it would be after Stalin's death. The Soviet Foreign Minister, Molotov (1890–) is seen here signing the treaty in Vienna. However, the Warsaw Pact had been set up the day before, enabling Soviet troops to remain in neighbouring Hungary. Whether this was to be the beginning of the end of the cold war was not known. Until an agreement on Germany was reached, the outcome was uncertain.

6

7

6 By 1961 the refugee flood from East Berlin threatened East Germany itself. The Soviet Government hesitated to start a new crisis by sealing the city off, but finally began the wall in August.

7 In October 1962 the US discovered that Russia had set up missile bases in Cuba, installed by Khrushchev as a counter-balance to US superiority. President Kennedy demanded at the Security Council the removal of the missiles and isolated Cuba with naval forces. Khrushchev offered to withdraw if he was allowed Turkish bases. This resulted in stalemate which saw the USSR back down after a week.

8 Nuclear arms: US-USSR agreements

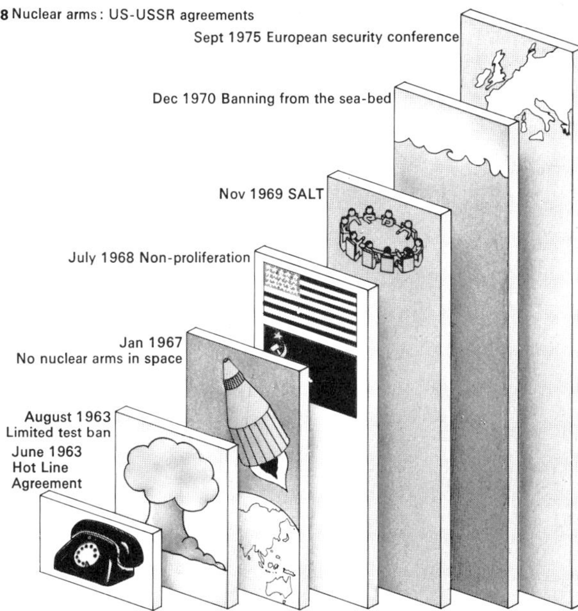

9

10

8 After the Cuban crisis, East and West tried to come to agreement on control of the arms race and on forms of co-operation which gave each side an interest in maintaining détente. They substituted agreement for threat.

9 Charles de Gaulle wanted the European powers to develop their own interests irrespective of the superpowers. He found a natural ally in Romania, which was trying to establish its independence from the Soviet bloc.

10 Willy Brandt was German Chancellor from 1969–74 when he harmonized the attitudes of détente that developed in Europe during the 1960s. He established a new relationship with the USSR and improved relations with the East European countries. His record as an opponent of Nazism and his willingness to offer amends for German atrocities won him trust abroad. His *Ostpolitik* led to Western recognition of East Germany and ended the postwar German problem.

Rulers of Britain

The first native ruler of all England after the departure of the Romans emerged in the early ninth century. In 828 Egbert, King of Wessex from 802, became "Bretwalda", overlord of the kings in England. His ancestor Cerdic founded Wessex in 519; his successors consolidated their position until they became sole kings of England. Of the subsequent monarchs only Sweyn, Canute I, Harold I, Canute II, Harold II and William I were not descended from him.

Egbert was the first of his line to have a religious coronation, a custom that has survived to modern times. Only since the Act of Settlement of 1701 has succession been by strict primogeniture, female heirs succeeding in the absence of males. Before that the law of succession was undefined. Certain conventions are discernible: descent from Cerdic; seniority by birth; designation of his successor by the reigning monarch; recognition by the great men of the realm; and the consent of the Archbishop of Canterbury, in the name of the Church, to the religious coronation. Also admissible was the right of conquest, recognized in the cases of Sweyn, William the Conqueror and Henry VII.

A claimant could also succeed even if he had not been designated by his predecessor nor had seniority by birth – William II was designated successor by his father, passing over his elder brother, Robert of Normandy. William II died suddenly and the youngest brother, Henry I, simply seized the throne, the nobility and the archbishop consenting. Stephen and John acted similarly. Henry IV and Henry VII were usurpers. Henry VIII vainly designated the heirs of his younger sister Mary, Queen of France and Duchess of Suffolk, to succeed him, failing heirs to his son Edward VI.

Parliament summoned James VI of Scotland to the throne in 1603, sanctioned the rule of Oliver Cromwell and his son, restored Charles II, and deemed James II (who had fled) to have abdicated. It welcomed William III and Mary II to the throne, regulated the succession by the Act of Settlement – bringing the Hanoverians to the throne – and likewise accepted Edward VIII's abdication in 1936.

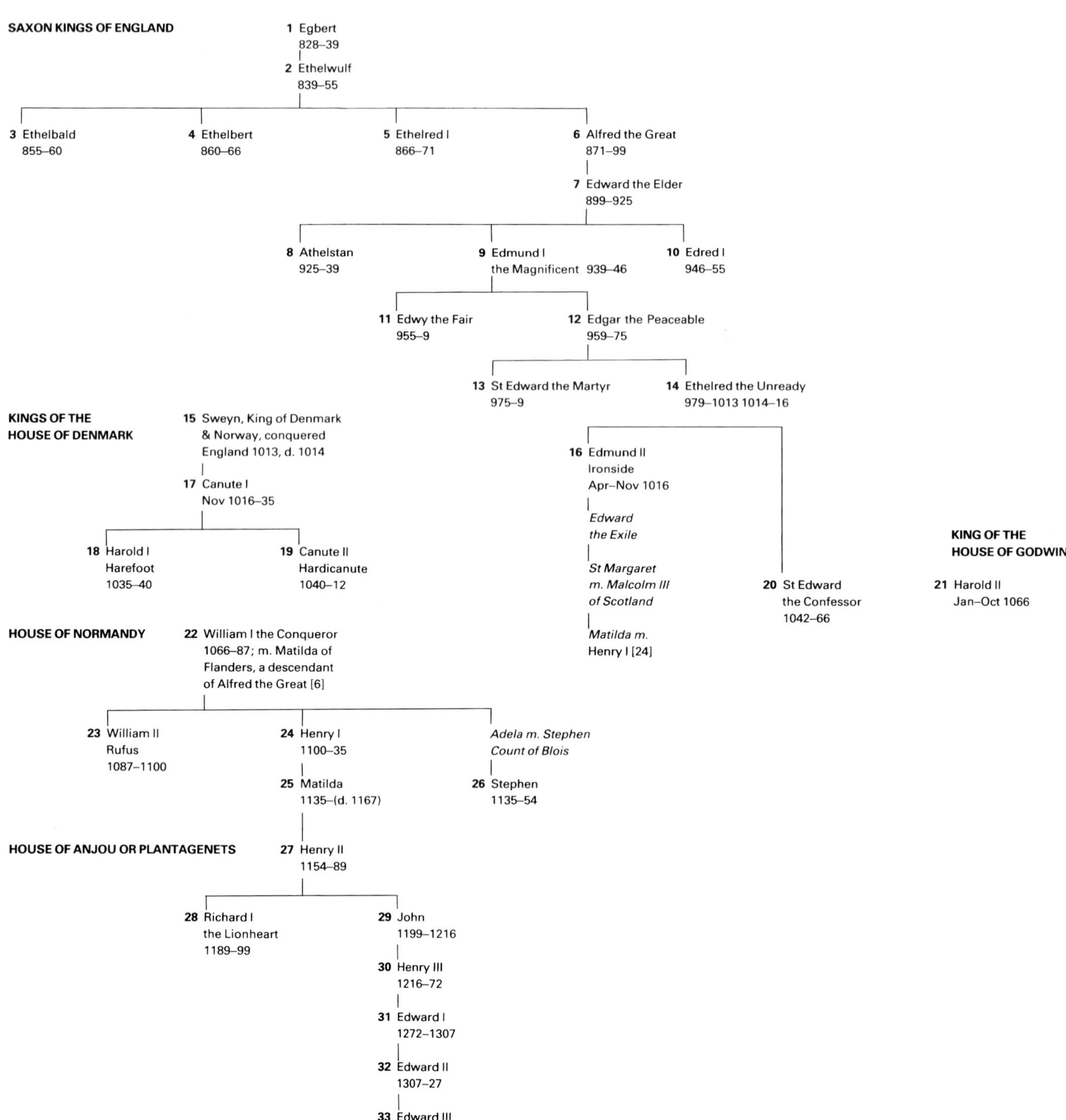

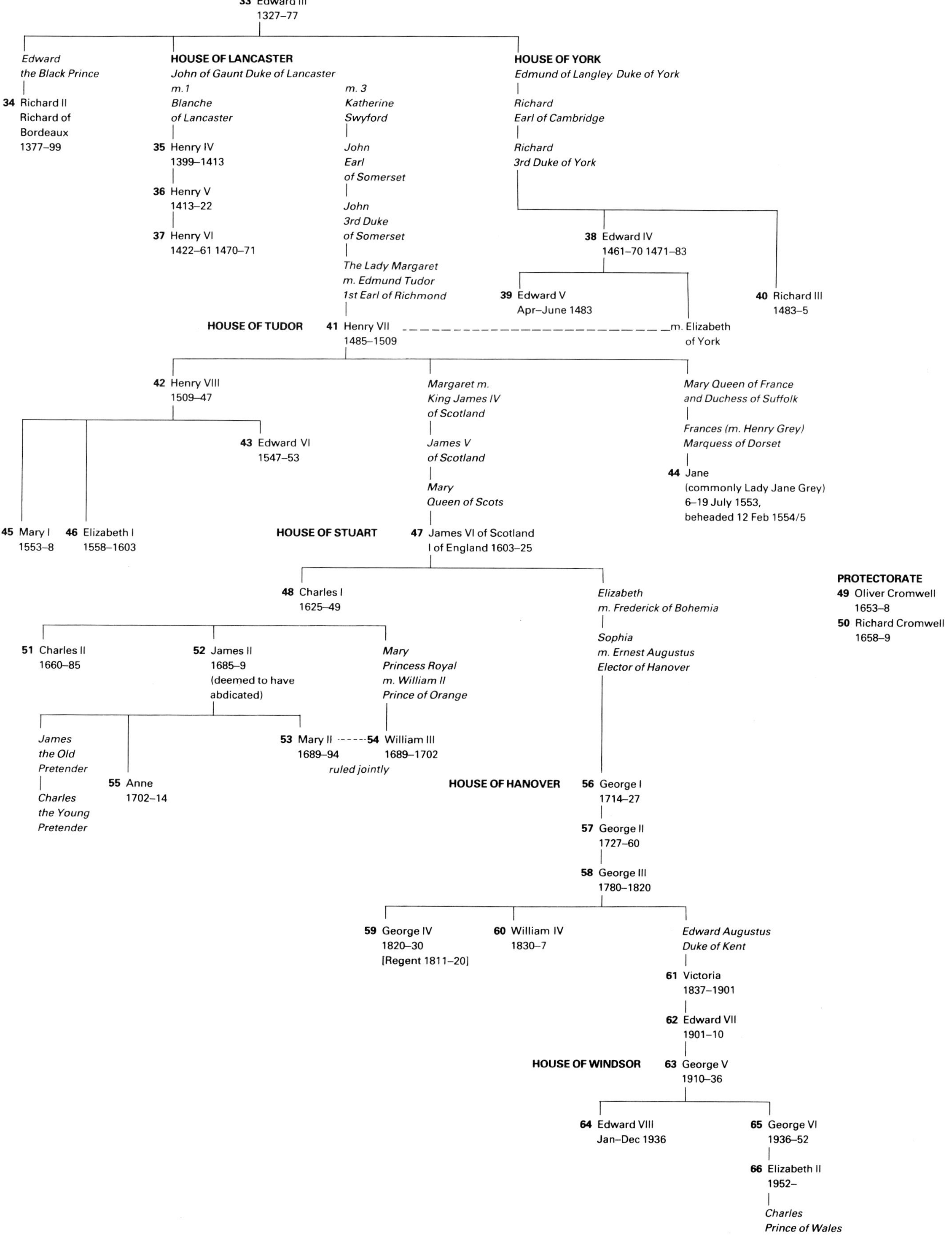

33 Edward III
1327–77
Edward
the Black Prince
34 Richard II
Richard of
Bordeaux
1377–99
HOUSE OF LANCASTER
John of Gaunt Duke of Lancaster
m. 1
Blanche
of Lancaster
35 Henry IV
1399–1413
36 Henry V
1413–22
37 Henry VI
1422–61 1470–71
m. 3
Katherine
Swyford
John
Earl
of Somerset
John
3rd Duke
of Somerset
The Lady Margaret
m. Edmund Tudor
1st Earl of Richmond
HOUSE OF YORK
Edmund of Langley Duke of York
Richard
Earl of Cambridge
Richard
3rd Duke of York
38 Edward IV
1461–70 1471–83
39 Edward V
Apr–June 1483
40 Richard III
1483–5
HOUSE OF TUDOR
41 Henry VII
1485–1509
m. Elizabeth
of York
42 Henry VIII
1509–47
43 Edward VI
1547–53
45 Mary I
1553–8
46 Elizabeth I
1558–1603
Margaret m.
King James IV
of Scotland
James V
of Scotland
Mary
Queen of Scots
Mary Queen of France
and Duchess of Suffolk
Frances (m. Henry Grey)
Marquess of Dorset
44 Jane
(commonly Lady Jane Grey)
6–19 July 1553,
beheaded 12 Feb 1554/5
HOUSE OF STUART
47 James VI of Scotland
I of England 1603–25
48 Charles I
1625–49
Elizabeth
m. Frederick of Bohemia
Sophia
m. Ernest Augustus
Elector of Hanover
PROTECTORATE
49 Oliver Cromwell
1653–8
50 Richard Cromwell
1658–9
51 Charles II
1660–85
52 James II
1685–9
(deemed to have
abdicated)
Mary
Princess Royal
m. William II
Prince of Orange
James
the Old
Pretender
Charles
the Young
Pretender
55 Anne
1702–14
53 Mary II
1689–94
54 William III
1689–1702
ruled jointly
HOUSE OF HANOVER
56 George I
1714–27
57 George II
1727–60
58 George III
1780–1820
59 George IV
1820–30
[Regent 1811–20]
60 William IV
1830–7
Edward Augustus
Duke of Kent
61 Victoria
1837–1901
62 Edward VII
1901–10
HOUSE OF WINDSOR
63 George V
1910–36
64 Edward VIII
Jan–Dec 1936
65 George VI
1936–52
66 Elizabeth II
1952–
Charles
Prince of Wales

Time Chart

In history as it comes to be written, there is usually some Spirit of the Age which historians can define, but the shape of things is seldom so clear to those who live them. To most thoughtful men it has generally seemed that theirs was an Age of Confusion. **Joseph Wood Krutch** "The Loss of Confidence", *The Measure of Man* (1954)

That men do not learn very much from the lessons of history is the most important of all the lessons that history has to teach. **Aldous Huxley,** "A case of voluntary ignorance", *Collected Essays* (1960)

Eternity is a terrible thought. I mean, where's it going to end? **Tom Stoppard,** *Rosencrantz and Guildenstern Are Dead* (1967)

1720-1760 Reason and the Enlightenment

Principal events

A series of dynastic and trade wars overtook Europe, adding to the growing conflict between centralized monarchical authority, the nobility and the newly strong mercantile class. In France the supremacy of the monarch over the nobles broke down, producing political stalemate. In Prussia, Russia and Portugal, however, the liberal ideas of the Enlightenment were harnessed to the growth of royal absolutism and industrial reform.

The English moved inland in India into the vacuum of the collapsing Mogul Empire, prized both for the value of its produce and the quality of its culture. Here they competed successfully with the French despite the unwillingness of the English government to take on imperial responsibility. At the same time the American colonies, whose economies were beginning to grow, were becoming impatient with Britain's rigid mercantilist policies.

1720-4

The English South Sea Company and the French Mississippi scheme, which had both aimed to restore royal finances, collapsed in 1720.
The Pragmatic Sanction, 1713, establishing the indivisibility of Austria-Hungary, was accepted in 1720.
In North America, Spain occupied Texas, 1720-2, to prevent a French invasion, and the Piedmont region was colonized by Swiss, Germans and Scots.

1724-8

Peter the Great of Russia died, 1725, having encouraged industrial growth, centralized the administration and subdued the nobility.
The ministry of Fleury, 1653-1743, in France began, 1726, introducing a period of peace and economic growth which led to a strengthening of the middle classes.
The Russian border with China was fixed in 1727.

1728-32

The Anglo-Spanish War, 1727-8, forced Spain to end her siege and confirm England's possession of Gibraltar, 1729.
Anna, *r.* 1730-40, Empress of Russia, founded the Corps of Cadets to encourage the nobles' participation in administration.
By the Treaty of Vienna, 1731, the Holy Roman Empire dissolved the Ostend East India Company, England's colonial trading rival in cotton, spices and saltpetre.

1732-6

England prohibited trade between her American and West Indian colonies by the Molasses Act of 1733.
War over the succession, 1733-5, weakened Poland.
Georgia, the last of the English colonies, was founded in 1733.
The French Compagnie des Indes was firmly established in India by 1735.
Class distinctions between the merchant and military groups in Japan became blurred during a long period of economic decline.

National events

In this period of Whig oligarchy, the constitutional roles of cabinet and prime minister were evolving. Commercial growth was encouraged by the government's fiscal policy and, with the increasing population and new agricultural techniques, prepared the ground for the Industrial Revolution.

Robert Walpole, 1676-1745, became Britain's first prime minister, 1721. His policy of keeping a permanent national debt brought prosperity.

George II, *r.* 1727-60, assigned to power the former opposition to George I. He was much influenced by his wife, **Caroline of Ansbach**.

Agricultural enclosures were beginning to increase the production of food so that it could support the growing urban population.

Walpole's attempt to introduce excise duty on wine and tobacco, 1733, brought great unrest, but the Tories were too weak to overthrow him.

Gulliver's Travels

Dutch East Indiaman

Marble bust of Voltaire

Carl Linnaeus

House of Menander, Pompeii

Religion and philosophy

The influence of English empiricist ideas on the philosophical tradition stemming from Descartes led to the great intellectual development known as the French Enlightenment. Montesquieu, Voltaire, Rousseau and other "philosophes" who contributed to the French *Encyclopédie* believed in the power of reason and knowledge to liberate man from restrictive political and religious systems.
On religious questions these thinkers tended towards deism or even atheism, and accepted a materialist conception of the universe. In politics they were liberals. Montesquieu sought to classify social systems and analyse their function. The Physiocrats laid the foundation of scientific economics. Others such as Condillac elaborated the basic ideas of materialist philosophy.
In Britain David Hume showed how empiricism could lead to an extreme scepticism.

Christian Wolff, 1679-1754, a follower of Leibniz, made rationalist anti-traditional philosophy popular in Germany. Puritan Pietists engineered his expulsion from the University of Halle in 1723, but he later became its chancellor.
Ba'al Shem Tov, *c.* 1700-60, founded **Hasidism** in Poland. This vibrant orthodox movement within Judaism stressed the joy of religious practice and expression, and rejected academic formalism and élitism.

Giambattista Vico, 1688-1744, an important forerunner of the modern social scientists, outlined his ideas in *Universal Law*, 1720-1, and elaborated them later in his masterpiece, *The New Science*, 1725. Vico held that societies pass from a bestial stage through a patrician stage ruled by an hereditary elite, to a stage where men are equal. He warned that man was never wholly rid of his bestial aspect and might always regress into barbarism.

Voltaire, 1694-1778, returned to France in 1729 after a two-year visit to England. His *Lettres philosophiques*, 1734, advocating the empiricism of **Isaac Newton** and **John Locke** and the merits of the English political system, had a great influence on the French Enlightenment. Voltaire was a deist, and an active liberal fighting for the exercise of tolerance in both religion and politics.

In his ***Treatise on Human Nature***, written from 1734 to 1737, **David Hume**, 1711-76, argued from empiricist presuppositions that knowledge was unattainable. He said that since connections were unobservable our belief in them was irrational. Hume held that the basis of moral judgement was man's subjective reaction of approval or disapproval of the effects actions have to himself and others.

Literature

European literature was dominated by the critical spirit of the Age of Reason expressed in the work of essayists and satirists such as Pope and Swift in Britain or Voltaire and Montesquieu in France, where polemical writing was in the ascendant. The same desire to grasp social reality, found expression in the English novel, whether in the vein of a new realism, with Defoe or Fielding, in the psychological studies of Laurence Sterne or the picaresque novels of Smollett. In Italy, Goldoni's comedies began a parallel move in the theatre away from stock characterization and towards a greater realism.
The basis of modern Russian poetry was established by Lomonosov.

Daniel Defoe, 1660-1731, a prolific writer and one of the founders of modern journalism, turned to fiction (disguised as fact) and revealed a powerful imagination. His novels, including *Robinson Crusoe*, 1719-20, and *Moll Flanders*, 1722, are noted for their highly realistic descriptions.

Jonathan Swift, 1667-1745, poet, polemicist and churchman, published *Gulliver's Travels*, 1726, a highly imaginative satire on mankind. Swift wrote brilliantly abusive essays.
The Beggar's Opera by John Gay, 1685-1732, was first played in 1728. It uses elements of Italian opera and traditional songs to create a new style of political satire.

German ideas of the Aufklärung (Enlightenment) were summed up in *Critische Dichtkunst*, 1730, a critical work by the playwright **J. C. Gottsched**, 1700-66. He argued that literature must imitate classical models and that its purpose is didactic. But the lyrical poetry of **F. G. Klopstock**, 1724-1803, who took Greek verse as his model, anticipated Romanticism.

The Italian **Scipione Maffei**, 1675-1755, published his erudite study of the history of Verona, *Verona Illustrata*, in 1732.
Montesquieu, 1689-1755, a leading French thinker and satirist, wrote his *Considerations of the Grandeur of the Romans and their Decadence*, 1734, an outstanding piece of socio-political analysis.

Art and architecture

Late manifestations of the more emotional baroque and rococo forms were seen in Austria and Germany as well as in European colonial architecture in the mid-18th century. But as concepts of "good taste" emerged during the Enlightenment, combined with a more exact and careful study of the aesthetics of classical art, the exuberance of the early 18th century became restrained within realist, or neoclassical modes. Interest in fantasy shifted from the Baroque to chinoiserie or rococo "Gothick", in the search for new stylistic forms.
Native Indian art was in decline and European styles were introduced to India by the advancing colonialists. The impact of European expansion in the cultural sphere was also found in China.
In Japan colour printing techniques were developed and the art reached a new peak in the work of Utamaro.

Easter Island was discovered by the Dutch, 1722. Archaeologists have since been baffled by the significance of, and building methods used to erect, the megalithic statues found there.
Austrian art reached its peak in the architecture of palaces, churches and monasteries, especially those of **Lukas von Hildebrandt**, 1688-1745. Known as "Austrian Baroque", this style with its florid shapes and lavish decoration paved the way for late German Baroque.

Indian art was in decline with the collapse of the Mogul Empire and European architectural styles began to be introduced in colonial towns, including Bombay where **St Thomas Cathedral** was built.
Catholic Bavaria accepted Italian baroque forms which in the later work of **Balthasar Neumann**, 1687-1753, took on an almost rococo lightness. His church at Vierzehnheiligen, 1743-72, is richly painted in pink, gold and white.

Palladianism, a revival of interest in the restrained classicism of Vitruvius, Palladio and his English follower Inigo Jones, marked an English reaction against Baroque. It was pioneered by **Colen Campbell**, *d.* 1729, and taken up by **Lord Burlington**, 1694-1753, who encouraged **William Kent**, *d.* 1748, and **Isaac Ware**, *d.* 1766.
Giuseppe Castiglione, 1698-1768, settled in China *c.* 1730, and was the first Western painter to be appreciated there.

Venice took the lead in Italian art with the painting of **Giovanni Tiepolo**, 1696-1770, and **Antonio Canaletto**, 1697-1768. **Servandoni**, 1695-1766, began work on the façade of San Sulpice in Paris, 1732. It relied on antique architecture and heralded a reaction against Rococo.

Music

Italian influence on European music waned, except in opera and song. The French evolved instruments and musical theory, but the Germans and Austrians, patronized by their princes, made the most use of these developments and ushered in the classical age of music. In the work of Joseph Haydn, 1732-1809, the symphony found a champion to establish its form.

Traité de l'harmonie, 1722, by **Jean Philippe Rameau**, 1683-1764, provided the foundation of harmonic thought for two centuries, with its clear statement of the function of tonality.

Light opera emerged in Germany, where **Reinhard Keiser**, 1674-1739, wrote operas with catchy tunes. He wrote a comic opera in 1726 which used spoken dialogue rather than recitative.

Virtuoso players such as **Antonio Vivaldi**, *c.* 1675-1741, advanced the techniques of their instruments and led to a distinction between music for professional and amateur players.

Religious cantata and oratorio were developed on a grand scale by **Bach** and **Georg Friedrich Handel**, 1685-1759, to embrace all musical techniques but without the use of operatic staging.

Science and technology

Great technological innovation were created and stimulated by the Industrial Revolution. In England the textile industry, with its need for large-scale bleaching and dyeing processes, gave a boost to practical chemistry and to machine technology. The flying shuttle produced the large quantities of cloth that demanded bleaching, and new methods were invented to provide the great amounts of acid employed in the process. Similarly, the need to transport more raw materials and finished products by sea than ever before encouraged navigational innovation. An early form of the sextant and the first accurate chronometer were invented.
Meanwhile pure scientific research continued in the form of discoveries, particularly in plant physiology and growth. Early work on electricity was performed at Leyden University and in America, providing the basis of later experiments into the nature of electric currents and their potential.

Smallpox inoculations were first administered in the New World during an epidemic in 1721, when **Zabdiel Boylston**, 1679-1766, inoculated 240 persons, of whom all but six survived.

The chronometer was developed from 1726 by John Harrison, 1693-1776, an Englishman, to aid navigation, as longitude could be determined only by time. He invented the compensating pendulum, so that his chronometers would keep perfect time in whatever climate they were used.
Plant physiology was founded by the publication of *Vegetable Staticks*, in 1727, by **Stephen Hales**, 1677-1761. Measuring plant growth and sap production, Hales realized that air is necessary for plants to grow.

Stellar aberration, a change in the position of stars caused by the Earth's motion, was detected in 1729 by an Englishman, **James Bradley**, 1693-1762. This was the first absolute confirmation of Copernicus' theory that the Earth moves around the Sun.
Cobalt was discovered in 1730 by George Brandt, 1694-1768, a Swedish chemist.
The reflecting quadrant, a forerunner of the sextant, aided navigation. It was invented in 1730 by John Hadley, 1682-1744.

Systema Naturae was published by Carl Linnaeus, 1707-78, a Swedish botanist, in 1735. He defined the differences between species and formed the idea of classifying plants and animals into species and genera, classes and orders.
The flying shuttle was invented in England in 1733 by John Kay, 1704-64.
Rubber was found in South America by Charles Marie de la Condamine, 1701-74, while on an expedition to measure the curvature of the Earth, 1735.

1736-40

Russia and Austria clashed with the Turks over their Polish policy. The Russians captured Azov but by the Treaty of Belgrade, 1739, were prevented from keeping a fortified Black Sea base there.
Commercial rivalry in America between England and Spain brought an end to a period of peace for England, with the War of Jenkins' Ear, 1739-41. The war resulted from a dispute over trading rights in the Spanish colonies.

1740-4

Frederick II the Great of Prussia, *r.* 1740-86, introduced religious toleration and agricultural reform, consolidated royal authority and reformed the army. In 1740 he occupied Silesia, thus striking the first blow in the War of the Austrian Succession, 1740-8.
Elizabeth of Russia, *r.* 1741-62, gave new authority to the Senate.
The Marathas took Bengal, 1742-4, and disturbed English trade in Bombay.

1744-8

Frederick II began the Second Silesian War, 1744-5. France and Prussia defeated the Austrians and their allies at the battle of Fontenoy, 1745.
In North America, English forces took Louisburg, 1745, and made new conquests from the French in the West Indies.
In India, the Frenchman **Joseph Dupleix**, 1697-1763, took Madras, 1746. However, all these conquests were restored by the **Treaty of Aix-la-Chapelle**, 1748.

1748-52

Louis XV, *r.* 1715-74, met united opposition from the nobility and clergy in France when he tried to introduce new taxes on their wealth to pay for his war expenses, 1751.
Robert Clive, 1725-74, seized Arcot, 1751, in search of personal power and booty, and thus established English authority over southern India, ousting the French opposition.
The Chinese invaded Tibet, 1751, following a growth in Chinese population and wealth.

1752-6

Sebastião Pombal, 1699-1782, introduced Enlightenment ideas to Portugal, 1751-77, ruthlessly attacking clerical and noble privileges and stimulating industrial growth.
Dupleix was recalled to France in 1754, leaving India to the British. Delhi was sacked by Afghan invaders, 1756-7.
Moscow university was founded in 1755 to promote education among the Russian nobility.
Lisbon was destroyed by an earthquake in 1755.

1756-60

In the Seven Years War, 1756-63, Austria was at first defeated by Frederick II.
Clive won control of Bengal at Plassey, 1757.
The Marathas occupied the Punjab in 1758.
Pombal expelled the Jesuits from Portugal in 1759.
Most of Canada came under British control after the surrender of **Montreal**, 1760. This ended the need for British garrisons to defend the American colonies.

The death of Queen Caroline, 1683-1737, weakened Walpole's authority, which relied in part on favour at court.

Commitment to Austria in the war of Austrian Succession led to the fall of Walpole, 1742.

The last Jacobite rebellion in Scotland was destroyed at Culloden in 1746.

Robert Clive's military and commercial activities in India, although unpopular at home, stimulated a further expansion of overseas trade.

Government by a regular cabinet comprising the heads of the main administrative departments was regularly adopted.

The Seven Years War greatly expanded the empire and brought a new commercial confidence in spite of political instability.

Frederick II of Prussia reviewing troops

Jean Jacques Rousseau

Robert Bakewell's improved sheep

Sextant by J. Bird

John Wesley, 1703-91, an Anglican minister, founded the Methodist movement in England. After a spiritual experience in 1738 Wesley began evangelical open-air preaching and drew up a set of "Rules" for his followers, who formed "bands" – groups for mutual encouragement and for teaching and prayer. They believed in a personal relationship with God and were noted for their good works. The Methodists finally broke with the Church of England in 1795.

The puritanical Wahhabi movement within Islam was founded by **Muhammed ibn 'Abd al-Wahhab**, 1703-92. He advocated a return to the original principles of Islam, and condemned as polytheistic the decoration of mosques and the cult of saints, which he saw as intervening in the personal and direct relationship between the faithful and God. In 1744 the powerful Saudi family in central Arabia adopted the principles of the Wahhabi sect.

In *The Spirit of Laws*, published 1748, the French social theorist **Charles Montesquieu**, 1689-1755, examined the relationships between a society's laws and its other characteristics such as religion and economic organization, drawing on an immense range of information about other cultures. He elaborated a study of types of governmental systems and analysed the prerequisites of their proper functioning.

The first volume of the French *Encyclopédie* appeared, 1751. Edited by **Denis Diderot**, 1713-84, and completed in 1772, this was a monument to the "philosophes" of the French Enlightenment and aimed to advance reason, knowledge and liberty. The contributors, who included **Etienne Condillac**, 1715-80, the Lockean philosopher, were deists or atheists who held liberal political views and a materialist conception of the universe.

Jean Jacques Rousseau, 1712-78, published his *Discourse on Inequality* in 1755. In this work, and in *The Social Contract*, 1762, he argued that in a natural state men were equal and that it was only society that creates inequality and misery. He argued that the injustices of society could be minimized if citizens resigned their rights to a government that acted on the "general will".

The Physiocrats of the 18th century were the first scientific school of economics. They regarded agriculture rather than manufacturing as the source of wealth, and advocated the doctrine of *laissez-faire*, or free trade, against the complex trade regulations then in force. The most important Physiocrat was **Francois Quesnay**, 1694-1774, whose *Tableau economique*, 1758, was the first work to attempt an analysis of the workings of an entire economy.

Voltaire (F. M. Arouet), 1694-1778, wit, poet, dramatist and epitome of the Enlightenment in his scorn for prejudice and distrust of accepted ideas, wrote the philosophical poems *Le Mondain*, 1736, and *Discours sur l'Homme*, 1738. Stressing the value of experience, he later satirized ideas of human perfectability in *Candide*, 1759, a tale of innocence abused.

The crowning achievement of Augustan poetry in England, *The Dunciad* of **Alexander Pope**, appeared in its final version, 1743. This was a mock heroic attacking the betrayal of literature by hack writers, using elements of Homer, Virgil, Dante and Milton and defending a role for the poet as a conserver of the values of society.

Italy's greatest comic dramatist, Carlo Goldoni 1707-93, wrote *The Servant of Two Masters*, 1745. A skilful and prolific craftsman, he substituted a script and more realistic treatment of character and situation for commedia del' arte, the traditional Italian comic form in which actors playing stock roles improvised upon an outline scenario.

The English novel, developed by **Samuel Richardson**, 1689-1761, in *Pamela*, 1740-1, flowered in the masterpiece *Tom Jones*, 1749, by **Henry Fielding**, 1707-54. **Laurence Sterne**, 1713-68, mastered a vein of black humour in *Tristram Shandy*; Tobias Smollet, 1721-71, the picaresque tradition in *Roderick Random*, 1748.

The father of modern Russian literature, Mikhail Lomonosov, *c.* 1711-65, published his *Grammar*, 1755. Poet and linguist, he set up verse rules and three styles of literary diction that opened up new possibilities in Russian literature.
Romanticism was foreshadowed in France by the **Abbé Prévost**, 1697-1763, the prolific author of *Manon Lescaut*, 1731.

Realism in Chinese fiction, exemplified in the satirical novel *Unofficial History of Scholars* by **Wu Ching-tse**, 1701-54, was further developed by **Tsao Chan**, *c.* 1719-63, in *The Dream of a Red Chamber*. In this novel, the grandeur and decline of a Chinese family was described with convincing detail and a new sense of humanity.

French art was divided between the officially accepted art in the rococo vein, like the frivolous, mildly erotic work of **François Boucher**, 1703-70, who had adopted much of Tiepolo's technique, and the more solid realistic genre scenes of **Jean Chardin**, 1699-1779, which reflected a contemporary taste for northern painting, especially 17th century Dutch masters.
Herculaneum was discovered in 1738.

Colour printing was developed in Japan, *c.* 1742, with outstanding results by **Kitagawa Utamaro**, 1753-1806, who was one of the greatest exponents of the ukiyo-e school of painting. This "floating world" art form was famous for its depiction of sensuous women.
Spanish colonial architecture was executed in a baroque style, especially in Mexico. The collision with existing cultures introduced new motifs like the Puebla tiles on the Church of San Francisco, Acatepec.

Chinoiserie, a taste for Chinese art and design, became popular in Europe in the 1740s. In England, **William Hogarth**, 1697-1764, attacked the social abuses of his time in his engravings and paintings. He often followed a narrative of events in a series of paintings as in "Marriage à la Mode", 1745.

The "Gothick rococo" became a fashion in England with the remodelling of Strawberry Hill House, 1749, by **Horace Walpole**, 1717-97. The library fireplace combined motifs from medieval tombs in Westminster and Canterbury Cathedrals.
British and French artists such as **Joshua Reynolds**, 1723-92, and **Jacques-Germain Soufflot**, 1713-80, would revolutionize art and architecture after studying art in Rome *c.* 1750.
Pompeii was found in 1748.

A torrent of publications heralded a change in taste in European art, foreshadowing **Neoclassicism**, which would be based on a detailed study of ancient Greek and Roman art. The archaeological discoveries engraved by **Piranesi**, 1720-78, in *Antichita Romana*, 1757, and such dissertations on taste as *Dialogue on Taste*, 1754, by **Allan Ramsay**, 1713-84, resulted in an ability to distinguish different phases in antiquity.

Russian architecture was based largely on French developments, baroque forms with rococo decoration, producing the splendour of the Winter Palace, 1754-64, in St Petersburg by **Bartolomeo Rastrelli**, 1700-71.
A positive reaction against Rococo in France was seen in the fleeting fashion of **"Le Gout Grec"** and also in a more significant dependence on antique precedents in the design of Sainte Geneviève (now the Panthéon) by **Soufflot** in Paris.

Contrapuntal writing reached a masterful zenith under Bach, with music of great power and intricacy, as in the "Kyrie" from his *Mass in B minor*, 1738.

Equal temperament was worked out in Germany. It made modulation to distant keys possible, as in the *Well-tempered Klavier*, 1722-44, by **J. S. Bach**, 1685-1750.

The symphony orchestra gained the basis of its present form at Mannheim court under **Johann Stamitz**, 1717-57, who trained his players to produce controlled extremes of loud and soft.

American settlers began making a distinctive music with easily carried instruments. Barn dances were held as buildings were completed and hymn-singing meetings were held in homes.

The symphony in the hands of Haydn developed greatly from 1750 to 1760, advancing its instrumentation and the form of its contrasting movements, usually four in number.

Sonata form was advanced by **C. P. E. Bach**, 1714-88, who made imaginative use of key relationships and conflicts in the development sections of first movements of his symphonies.

Daniel Bernoulli, 1700-82, a Frenchman, related fluid flow to pressure in 1738.

Anders Celsius, 1701-44, a Swede, devised the Celsius scale of temperature, *c.* 1744, with 0° as the freezing-point of water and 100° as the boiling-point.
The crucible method of making steel by heating scrap iron was found in England, 1740, by Benjamin Huntsman, 1704-76.
Mikhail Lomonosov, 1711-65, working in Russia, rejected the phlogiston theory and suggested the law of the conservation of mass.

Traité de Dynamique, 1743, by **Jean d'Alembert**, 1717-83, solved problems in mechanics.
The Leyden jar, developed at the University of Leyden, 1745, was able to store a large charge of static electricity. It was used in the first investigations into the nature of electricity.
John Roebuck, 1718-94, a British inventor, developed a process for manufacturing sulphuric acid, used to bleach textiles, on a large scale in 1746.

Benjamin Franklin, 1706-90, working in America, flew a kite in a thunderstorm, 1752, to prove that lightning is electrical and from his results developed a lightning conductor.
Selective breeding, pioneered by **Robert Bakewell**, 1725-95, in England, improved livestock, while the experimental farming of **Viscount Townshend**, 1674-1738, improved crop rotation.
Georges Buffon, 1707-88, published the first volume of his massive *Histoire Naturelle*, 1749-88.

Immanuel Kant, 1724-1804, in Germany, published his views on the formation of the solar system in 1755, anticipating the work of Laplace. He also suggested that galaxies of stars exist and that the tides slow the rotation of the Earth. Both of these ideas were verified much later.
René Réaumur, 1683-1757, proved that digestion is a chemical process and invented an 80 degree thermometer scale.

Carbon dioxide was discovered in 1756 by Joseph Black, 1728-99, a British chemist.
The sextant of John Bird (1758) made navigational observations far more accurate.
Lomonosov was the first man to observe atmosphere on the planet of Venus, 1761.
John Dollond, 1706-61, produced the first achromatic lenses in 1757 in England.

1760-1800 Revolution in America and France

Principal events

The old order in Europe was fundamentally shaken by three major revolutions – in America, France and England – which changed the political and economic basis of Western society and would ultimately transform the world. The American War of Independence represented the overthrow of the old colonial and trading system and installed the ideas of liberty and democracy as the ideals of the United States. The French Revolution of 1789 swept away the privileges of the outdated *ancien régime* and established a new idea of popular right, which would be carried by Napoleon's conquests to stir the rest of Europe to revolt.

In England the Industrial Revolution began in earnest in the 1780s, providing the basis for a fundamental transformation of Western and ultimately global society by accelerating urbanization and creating new sources of wealth, new social classes and democratic demands.

1760-4

Prussia increased its military power after 1760 and an inconclusive settlement to the Seven Years War followed. **The Treaty of Paris**, 1763, confirmed English supremacy in Canada and India. **The War** left French government finances in a precarious state despite expanding trade. **The Pontiac Conspiracy**, an American Indian revolt, was suppressed by the English in Canada, 1763-6.

1764-8

The Sugar Act and Stamp Act, 1764-5, by which Britain aimed to recover revenue from the American colonies, aroused local opposition. **England ruled Bengal and Bihar** by 1765, maintaining a puppet Mogul emperor. **Ali Bey**, *r.* 1768-73, declared Egyptian independence from Turkish rule, 1766. **Catherine II** of Russia, *r.* 1762-96, consulted a convention of all social classes to reform Russian law, 1767.

1768-72

The American colonies began their westward expansion, settling Tennessee in 1769. **French trade with India** increased after the French East India Company lost its monopoly, 1769. Opposition to absolutism in France increased among intellectuals. **James Cook**, 1728-79, began the exploration of Australia in the *Endeavour*, 1768-71. In the **Boston Massacre**, 1770, British troops fought with American colonists.

1772-6

After Pugachev's revolt, a large peasant and cossack uprising, 1773-5, Catherine II reformed Russian provincial administration. **The Regulating Act** established an English governor-general in India, 1773. **Warren Hastings**, 1732-1818, reformed the Bengal administration. **Demands by the American colonists** that they be represented in the English Parliament led to the **American War of Independence**, 1775-83.

National events

The Industrial Revolution introduced factory-based machine production and resulted in the growth of a wealthy industrialist class and large new towns in the north without parliamentary representation. Radical societies for electoral reform grew up, some interested in French Jacobinism.

Overseas trade doubled between 1720 and 1760. Canals such as the Bridgwater, 1761, facilitated the movement of heavy goods around Britain.

John Wilkes, 1727-97, was thrice expelled by the Commons after winning election, 1768. He championed free reporting of Parliament.

Wilkes's elections and expulsions finally led to the establishment of freer elections. Political stability was restored by Lord North, 1770-82.

The East India Company was regulated, 1772-3. In 1775 the **American War of Independence** began.

Thomas Paine

American Revolution

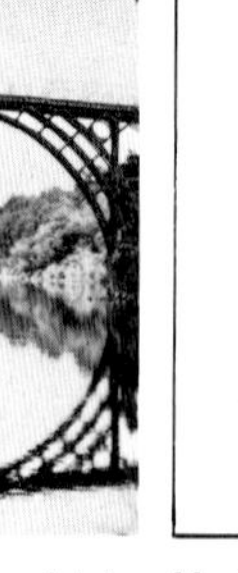

Iron bridge, Coalbrookdale

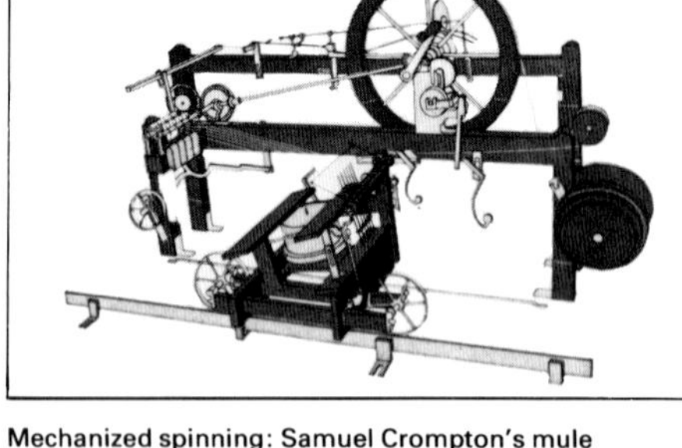

Mechanized spinning: Samuel Crompton's mule

Montgolfier's balloon

Religion and philosophy

The question of the existence of God became subordinate for many European thinkers to questions of social organization.

In America the revolution was associated with ideas of democracy, liberty and equality which in turn inspired the French Revolution.

In Britain new economic thinking reflected the emergence of the industrial system. Adam Smith laid the foundation of modern economics, fostering the liberal doctrine of the free market and the absence of state encroachment on individual freedom. Bentham argued that desire for utility, avoidance of pain and pursuit of pleasure motivated behaviour. The Scottish Enlightenment advanced social thought with Ferguson's and Monboddo's work on social development and man's origins.

Kant, however, laid the basis for German idealism, with his opposition to pure empiricism, claiming that such concepts as time were innate.

The Scottish School of Common Sense Philosophy was begun by **Thomas Reid**, 1710-96, who argued in *An Inquiry into the Human Mind on the Principles of Common Sense*, 1764, that Hume's scepticism about attaining true knowledge was against common sense. **Dugald Stewart**, 1753-1828, sought rejection of fruitless metaphysical speculation and the creation of scientific philosophy.

Adam Ferguson, 1723-1816, an early British sociologist, put forward the theory that man's unceasing desire to control nature was the cause of social development, in his *Essay on the History of Civil Society*, 1766. **The Judaic religion** was interpreted by **Moses Mendelssohn**, 1729-86, in terms of the metaphysics of Leibniz, paving the way for a synthesis of Judaism and modern philosophical and scientific thought, later to develop into **Reform Judaism**.

Johann Herder, 1744-1803, German poet and thinker, was among the first modern thinkers to question the limits of reason. He emphasized the immediacy and therefore the power of feeling – ideas that were later to become the essence of the Romantic movement. **Paul d'Holbach**, 1723-1803, the most ardent materialist of the French Encyclopédists, wrote in his *System of Nature*, 1770, that man's life is determined from birth.

Lord Monboddo, 1714-99, a British anthropologist, believed man's present social state evolved from a previous animal one. This conflicted with the view then current that man was unique. He began publication of his work on language in 1773. **The "Shakers"**, a group of puritanical nonconformists led by **Ann Lee**, 1736-84, began their first colony in America in 1774. They believed total sexual abstinence was the basis of man's spiritual salvation.

Literature

Forerunners of Romanticism emerged in Germany, France and Britain. The emphasis on unity and order in literary style and the sceptical and rational attitudes of mind that marked the Enlightenment were beginning to give way to increasing respect for human instincts and emotions, sincerity of feeling and freedom and naturalism of style. This transition, initiated by Rousseau in France, was carried on in Germany by the *Sturm und Drang* movement whose greatest voice, Goethe, combined passion with discipline. The work of the British poets Gray, Cowper, Burns and Blake exemplified the transition from classicism to romanticism in English poetic style. Samuel Johnson's work advanced literary criticism.

Jean Jacques Rousseau, 1712-78, whose concept of the "noble savage" deeply influenced romanticism, published *La Houvelle Héloïse*, 1761, a novel advocating simple relationships in a natural setting. The *Encyclopédie*, edited 1751-72 by **Denis Diderot**, 1713-84, and **Jean Le Rond d'Alembert**, 1717-83, expressed the scepticism of the Enlightenment.

Gothic themes involving the supernatural and the *crime de passion* appeared in the ultra-romantic novel *The Castle of Otranto*, 1764, by **Horace Walpole**, 1717-97. **Thomas Percy**, 1729-1811, published *Reliques of Ancient English Poetry*, 1765. **Karl Bellman**, 1740-95, a Swedish poet, began his *82 Epistles* in 1765.

Thomas Gray's *Poems by Mr Gray*, 1768, included the "Elegy Written in a Country Churchyard". Gray, 1716-71, treated themes of history and death in a sensitive, meditative manner. **Gotthold Lessing**, 1729-81, used Shakespearian models influentially in *Laokoon*, 1766.

Sturm und Drang (Storm and Stress), a German literary movement stressing subjectivity and contemporary unease, found a genius in **Johann Wolfgang von Goethe**, 1749-1832. His novel *The Sufferings of Young Werther*, 1774, began a cult of the hero ruled by the heart rather than head. Romantic pessimism was exemplified in the poems of **Novalis**, 1772-1801.

Art and architecture

The arts in Europe, and particularly in France, reflected the critical spirit of the Enlightenment by returning to an austere style based on moral and aesthetic theories. Antiquarian and archaelogical investigation had transformed ideas on cultural development so that the various styles of Greek and Roman antiquity, the Middle Ages and the Renaissance, could now be distinguished. Neoclassicism, which developed towards the end of the 18th century, incorporated this knowledge, adopting Greek and Roman ideals of beauty and ethics derived from antique sculpture, architecture, painting and literature. This historical concern was also to lead to acceptance of eclecticism and the concept of a modern style.

The European colonial presence in Asia tended to paralyse the development of indigenous artistic styles, but native traditions survived in areas remote from foreign influence.

Robert Adam with his brother James introduced a new eclectic style of architecture to town and country houses in Britain, like **Syon House**, 1762-9, in which they combined elements of English Palladianism with details of Roman architecture and Renaissance palaces. **Neoclassical painting** was developed in Rome, under the impetus of the German archaeologist **Johann Joachim Wickelmann**, 1717-68, by his follower **Anton Raffael Mengs**, 1728-79.

Soufflot's church of Ste Geneviève in Paris progressed. The design combined Greek post and lintel systems and attempted to achieve the lightness of Gothic architecture. **The Royal Academy of Art**, London, was founded in 1768 under royal patronage. The first President, **Joshua Reynolds**, in "13 Discourses", 1769-1790, promoted the "Grand Manner" in English painting.

An empirical, scientific attitude to art in England was shown by **George Stubbs**, 1724-1806, in the *Anatomy of the Horse*, published 1766, and in "The Experiment with the Air Pump", 1768, by **Joseph Wright** of Derby, 1734-97. **French Neoclassic architecture** was governed by the severe unadorned classicism seen in the works of **Jacques Gondouin**, 1737-1818, of which the **Ecole de Médecine**, Paris, 1769, is a fine example.

Reynold's supremacy in English portraiture was challenged in 1774 when **Thomas Gainsborough**, 1727-88, moved to London. His "William Henry, Duke of Gloucester", *c.* 1775, was deliberately glamorous and richly coloured. His later paintings introduced a more lyrical note to English portraiture. **Indian artists** in the late 18th and early 19th centuries were dominated by European techniques. An exception was the Patua paintings of east India.

Music

The classical age of European music was dominated by Joseph Haydn, 1732-1809, and Wolfgang Mozart, 1756-91. Composers pursued variety within movements, building bigger structures by manipulating musical themes and utilizing key relationships and contrasts of instrumental sound, appealing equally to the heads and hearts of their educated audiences.

The symphony and sonata grew in complexity under the hand of Haydn from *c.* 1760. The first movement had contrasting themes worked over in a development section.

Counterpoint declined in importance and the continuo disappeared. Contrapuntal forms such as fugue continued to be used but usually as part of a movement in a larger work.

Christoph Gluck, 1714-87, reformed opera in Paris, stressing the balance between the musical and the dramatic elements. He expressed his ideals in a preface to *Alceste*, 1769.

String quartets were written in large numbers. They were an ideal vehicle for the development of classical designs and allowed the composer to hear his work immediately.

Science and technology

In Britain the Industrial Revolution began to transform the face of the nation. James Watt produced the first rotary engine, which could be used to power factories anywhere in the country, while the spinning jenny and the water frame furthered mechanization of the textile industry. Agricultural improvements, including more efficient crop rotation and selective breeding, increased the amount of food and provided a surplus for the towns. Developments in hygiene and medicine, such as the water closet, vaccination and the widespread use of soap, would form the basis for substantial improvements in urban living conditions, many of which, however, were not realized until the nineteenth century.

Science was linked with liberty in revolutionary France as many academies of science were founded after 1789, while American technology worked against freedom – the success of the cotton gin helping to prolong slavery in the South.

Joseph Black, 1728-99, a Scottish chemist, defined the difference between heat and temperature, and discovered specific and latent heat, 1760-3. His basic work on heat enabled his friend **James Watt** to build a steam engine. **The spinning-jenny** was invented in England, 1764, by **James Hargreaves.** It could spin several threads at once.

The Lunar Society, an informal society of technologists, was founded in England *c.* 1765. **Neurology** was established with the work of Swiss physiologist **Albrecht von Haller**, 1708-77. Haller located nerves and showed that nerve impulses stimulate muscles. **Henry Cavendish**, 1731-1810, a British scientist, discovered hydrogen in 1766. He also made fundamental, unpublished discoveries in electricity. In 1798 he calculated the Earth's mass.

The water frame was invented in 1768 by **Richard Arkwright**, 1732-92. Powered by water, it spun cotton into a strong thread. **James Watt**, 1736-1819, patented his steam engine in 1769. This engine used a separate cylinder for condensing steam and worked quickly and efficiently. Watt's engine was the first to produce rotary motion. **Luigi Galvani**, 1737-98, an Italian, found in 1771 that two metals in contact with a frog's leg cause it to twitch. Unwittingly he had produced current electricity.

Oxygen was discovered, 1772, by the Swede **Carl Scheele**, 1742-86. He withheld his findings until after the independent discovery by **Joseph Priestley**, 1733-1804, in 1774. Scheele was also involved in the discovery of chlorine, 1774, tungsten, 1781, and other elements. **Daniel Rutherford**, 1749-1819, discovered **nitrogen** in 1772.

1776-80

The American colonies declared their independence, 1776, and allied with France, 1778, and Spain, 1779. The English overran the southern states 1778, but were weakened by a French blockade of shipping. **The French government** was ruined by this war in spite of the continued financial reforms of **Jacques Necker**, 1732-1804. **Pombal**, 1699-1782, completed the reorganization of the administration in Portuguese Brazil, 1777.

1780-4

American independence was assured by the British surrender at Yorktown, 1781, and formally recognized at the 1783 **Treaty of Paris**.
A sudden growth in the English cotton industry after 1780 marked the beginning of the English **Industrial Revolution**. **Russia** occupied the Crimea in 1783.
Hastings made an effective peace with the Marathas, 1784.

1784-8

The United States began trading with China, 1784, but suffered post-war depression through loss of contact with the West Indies, 1784-7. **The American Constitution** was signed in Philadelphia, 1787. **The aristocratic parliaments** in France blocked proposals for financial reform, 1787. **The founding of *The Times* newspaper** in England, 1788, accompanied the growth of an informed middle class in Europe.

1788-92

England established convict settlements in Australia, 1788. **Louis XVI**, *r.* 1774-92, was forced to summon the estates-general in 1789 because of the financial crisis. **The French Revolution** began when a group of middle-class radicals took over the administration with the help of the Paris mob and tried to set up a constitutional monarchy, 1789. **George Washington**, 1732-99, became the first president of the United States, 1789.

1792-6

France was declared a republic, 1792. Louis was executed, 1793, and during the ensuing terror, 1793-4, many of the nobility were also guillotined as a result of the fear of a counter-revolution backed by Austrian forces. **The French overran Holland** and established the Batavian Republic in 1795. **Revolutionary ideas** led to the freeing of slaves in the French West Indies, arousing hostility among the European powers.

1796-1800

By the **Treaty of Campo Formio**, 1797, Austria ceded Belgium to France. **Napoleon**, 1769-1821, defeated Austria, 1796, but his plans to invade England, 1798, failed and he was prevented by **Horatio Nelson**, 1758-1805, from cutting England off from India at the Battle of the Nile, 1798. In 1799 he overthrew the moderate **Directory** and established a dictatorship, 1799-1804.

Relaxation of anti-Catholic laws reflected the confidence of the government but resulted in the destructive **Gordon riots** in London in 1780.

William Pitt the Younger, 1759-1806, formed his first ministry. Further colonial expansion became necessary to replace the American colonies.

The economic boom based on coal mining and cotton production began, 1786, bringing the development of new towns in the Midlands and North.

Pitt survived the 1788 crisis of George III's temporary insanity. **Until the execution of Louis XVI**, British opinion generally backed the Revolution.

The London Corresponding Society was founded among the artisan class to campaign for electoral reform, 1792, but was suppressed by Pitt, 1796.

Payments under the **Speenhamland** system of Poor Law became common after 1795. **The Combination Acts**, 1799, banned trades union activity.

"The Oath of the Horatii" by David

James Watt's rotary steam engine

French Revolution: the execution of the king

Founder of modern economics, **Adam Smith**, 1723-90, argued that although manufacturers do not intend to satisfy the general good, they are led to do so by the "invisible hand" of the competitive market. When a producer satisfied his self-interest by selling goods for which there is a demand, he also satisfies a general social need. Smith's *Wealth of Nations* was published in 1776.

Immanuel Kant, 1724-1804, in his *Critique of Pure Reason*, 1781, wrote that although knowledge cannot transcend experience, the concepts that organize perception are innate to the human mind and prior to experience. In *Metaphysics and Morals*, 1785, he argued that man's idea of morality is *a priori* and that people act morally when the maxim on which they act is one which they can desire all men to follow.

Liberalism, the belief that the state should not encroach on individual freedom, was proposed by **Jeremy Bentham**, 1748-1832, in *A Fragment on Government*, 1776. He argued in his *Principles*, 1784, for utilitarianism, the theory that the happiness of the majority of individuals was the greatest good. This was to be achieved by allowing each individual the freedom to maximize his useful achievement by avoiding pain and pursuing pleasure.

In France, 1789-90, Church lands were nationalized and religious orders suppressed. **Edmund Burke**, 1729-97, in *Reflections on the Revolution in France*, 1790, argued that the replacement of practical politics by utopianism had led to extremism. **Tom Paine**, 1737-1809, in America, wrote *The Rights of Man*, 1791, to oppose *Burke's Reflections*. Paine believed revolution could be avoided only if the causes of the discontent were eradicated.

Equal opportunities for women to develop their talents were demanded by **Mary Wollstonecraft**, 1759-97, in her *Vindication of the Rights of Women*, 1792. Her husband, **William Godwin**, 1756-1836, published *Enquiry Concerning Political Justice*, 1793. A radical, he argued that government power over citizens inevitably bred corruption. **The Cult of Reason** and later the **Cult of the Supreme Being** were substituted for Christianity in France, 1793-4.

The English Evangelical Movement had emerged within the Church of England, influenced by Methodism. Its followers believed in the certainty of salvation, emphasizing evangelism and social walfare. **Reverend Thomas Malthus**, 1766-1834, published his *Essay on the Principle of Population*, 1798, rejecting the possibility of infinite improvements in human conditions on the grounds that population expands more rapidly than the available food supply.

Comedy of manners, revived in England by **Oliver Goldsmith**, c. 1730-74, in *She Stoops to Conquer*, 1773, reached a peak in *The School for Scandal*, 1777, by the Irish wit **R. B. Sheridan**, 1751-1816. **Italian patriotism** was stirred by **Vittorio Alfieri**, 1749-1803, whose 19 verse tragedies in the classical mode opposed tyranny.

The influential French novel *Les Liaisons Dangereuses*, 1782, by **P. A. F. Choderlos de Laclos**, 1741-1803, had a savage tone in contrast to the vogue for high moral sentiment established by Rousseau. The privileges of the upper class were satirized by **Pierre de Beaumarchais**, 1732-99, in *The Barber of Seville*, 1775, and *The Marriage of Figaro*.

Scottish folk traditions found a passionate and lyrical voice in the poems of **Robert Burns**, 1759-96, whose *Kilmarnock Edition*, 1786, established him as a skilled writer of songs, satires and narratives. A deep love of nature, a major theme in romanticism, is found in the blank verse of "The Task", 1785, by the English poet **William Cowper**, 1731-1800.

A new tradition of candid biography was begun by **James Boswell**, 1740-95, in his *Life of Johnson*, 1791, bringing to life his friend **Samuel Johnson**, 1709-84. Johnson, a brilliant conversationalist, editor, poet and critic, dominated English literature after 1750 with his *Dictionary*, 1755, the Gothic novel *Rasselas*, 1759, and *Lives of the Poets*, 1779-81.

William Blake, 1757-1827, one of the most powerful, imaginative artists in English literature, published the lyrical *Songs of Experience*, 1794, complementing *Songs of Innocence*, 1789. Poet, painter, engraver and, above all, visionary, he issued prophetic warnings of the danger of industrialization and materialism in *The Marriage of Heaven and Hell*.

The Romantic movement in England began with *Lyrical Ballads*, 1798, by **William Wordsworth**, 1770-1850, and **Samuel Taylor Coleridge**, 1772-1834. **The novels of Jean Paul**, 1763-1825, including *Hesperus*, 1795, combined the idealism of Fichte with the romantic sentimentality of *Sturm und Drang*.

Classicism in Russia under Catherine the Great was led by foreign artists such as the French sculptor **Etienne-Maurice Falconet**, 1716-91, who executed the equestrian statue of Peter the Great, 1769, and Scottish architect **Charles Cameron**, c. 1740-1812, who went to Russia in 1779. Cameron decorated several apartments in the palace of Tsarkoe Selo (now Puskino) near Leningrad for Catherine.

Neoclassical painting was firmly established in France with "Oath of the Horatii", 1784, by Jacques-Louis David, 1748-1825, in which the subordination of colour to drawing enforces the theme of heroic self-sacrifice as exemplified by ancient Rome. "The Nightmare" by **Henry Fuseli**, 1741-1825, and the works of **William Blake**, 1757-1827, reveal an emphasis on the bizarre and supernatural in contrast with Academic aims.

The Academy on Fine Arts in Mexico City, founded in 1785, was staffed primarily by Spanish-trained artists who were largely instrumental in introducing Neoclassicism to Mexico. **English caricature** was developed by James Gillray, 1757-1815, in "A New Way to Pay the National Debt", 1786 and by **Thomas Rowlandson**, 1756-1827, who illustrated Smollet, Goldsmith, Sterne and Swift, and produced "Imitations of Modern Drawings", 1784-88.

The "Style Troubadour" originated in France when anecdotal scenes from the lives of wise kings of early French history were used by anti-royalists to accentuate the incompetence of Louis XVI. **Classicism in English architecture** was exemplified in the work of John Soane, 1753-1837, whose austere and original manipulation of the antique was seen in his **Bank of England Stock Office**, 1792, with its top-lit vaulted hall.

Painting in Revolutionary France was used as a political weapon. David's "The Death of Marat", 1793, combines classicism with an element of realism to deify a revolutionary hero and muster republican support. **John Flaxman**, 1755-1826, published his illustrations to Homer's *Iliad* and *Odyssey* in 1793. With their simple outline figures they immediately became a major model for Neoclassical painters and influenced later generations.

The success of Napoleon's Italian campaign 1796-7, galvanized French art with the public display in the Louvre of looted art treasures. This enhanced the image of Napoleon as a national hero. **The Capitol, Richmond, Va**, 1789-98, built by Thomas Jefferson, 1743-1826, was based on the Maison Carrée, Nîmes, and brought Neoclassical architecture to the United States.

African music as described by Western observers probably resembled music heard today, in which groups of instruments, such as marimbas and drums, freely explore areas of sonority.

Mozart was one of the first great composers who tried to live independently without the support of a patron, but he died a pauper. His work in opera and other fields of music shows the effects of his original thought. While others were content with the stock characters of Italian opera, he created such works of individual genius as *Don Giovanni*, produced in Prague, 1787.

Domenico Cimarosa, 1749-1801, "the Italian Mozart", composed his most celebrated opera, *Il matrimonio segreto* in Vienna in 1792.

Niccolò Paganini, the Italian virtuoso violinist, 1782-1840, made his debut in Genoa in 1793, playing his own variations on "La Carmagnole".

The violin was taken to India by British rulers. Indian musicians absorbed it into their music, utilizing its subtleties of intonation and tone colour.

A practical water closet was patented in 1778 in England by **Joseph Bramah**, 1748-1814. Bramah's many inventions introduced practical techniques that founded the engineering industry. **The spinning mule** was invented in England in 1779 by **Samuel Crompton**, 1753-1827. It was able to spin high quality thread on many spindles at once. **Cheap soap** resulted from the work c. 1780 of **Nicholas Leblanc**, 1742-1806, in France. He patented his process of producing soda from salt in 1791.

Uranus was discovered by **William Herschel**, 1738-1822, in 1781. It was the first planet to be discovered that was not known to ancient civilizations. **The first manned flight** took place in 1783 in a hot air balloon made and flown by the French **Montgolfier** brothers Joseph, 1740-1810, and Jacques, 1745-99. **James Watt** invented the double-acting engine in 1784.

Chlorine was first used for bleaching cloth in 1785 by **Claude Berthollet**, 1748-1822, a French chemist. **The threshing machine** was patented, 1788, by a Scotsman, **Andrew Meikle**, 1719-1811. **Jacques Charles**, 1746-1823, a French physicist, formulated **Charles's law** c. 1787, that at constant pressure the volume of a gas is related to its absolute temperature. **The power loom**, invented in 1785 by **Edmund Cartwright**, 1743-1823, mechanized weaving.

Theory of the Earth, 1788, by **James Hutton**, 1726-97, began modern geology by viewing all geological change as continuous. *Traité Elémentaire de Chimie*, by **Antoine Lavoisier**, 1743-94, written in 1789, founded modern chemistry with its insistence on measurement and standard nomenclature. Lavoisier stated the law of the conservation of mass; he also defined chemical reaction. In 1790 **Watt** applied his fly-ball governor to control the speed of a steam engine.

Coal gas was first produced in 1792 by **William Murdock**, 1754-1839, a British inventor. **The cotton gin**, a device used to strip cotton from bolls, was invented, 1793, by **Eli Whitney**, 1765-1825, revitalizing cotton growing in the United States. **The metric system** was adopted in France in 1795. **Scientific institutes** abounded in revolutionary France, as did prizes for scientific developments. Institutes included the Jardin des Plantes and the Ecole Polytechnique.

Vaccination, discovered in 1796 by Edward Jenner, 1749-1823, led to the eradication of smallpox. **The nature of heat**, or kinetic energy, was discovered in 1798 by the American, **Count Rumford**, who noticed that the boring of cannons produced heat and reasoned that heat is a form of motion and not a fluid. **The battery** was invented by **Count Volta**, 1745-1827, in 1800. He developed Galvani's observations into a practical idea for an electricity supply.

1800-1825 The rise of industrial power

Principal events

Inspired by a vision of himself as head of a European empire, Napoleon Bonaparte overran most of Europe but was unable to maintain his conquests. With his final defeat at Waterloo in 1815, the *ancien régime* was restored to France. His conquests, however, sparked off a multitude of constitutional and nationalist demands throughout Europe, while his occupation of Spain encouraged the Latin American countries to grasp their independence. They remained, however, unable to reorganize themselves economically or to free themselves politically from European influence.

In England, the Industrial Revolution caused the emergence of a new wealthy class and social tensions gave a greater urgency to demands for parliamentary reform, while her naval strength and leadership of the final coalition against Napoleon left her the dominant trading power in the world.

1800-2

Napoleon established the prefecture as the main instrument of local government, subject to central control. He improved education and made a compromise with the Church, 1800-1.

His aggressive nationalist campaigns led to victory over the empire and Austria, conquest in Italy, 1800, and temporary peace with England in 1802.

With the murder of Paul I, **Alexander I**, *r.* 1801-25, became Tsar of Russia.

1802-5

Nationalist feeling brought a Serbian uprising against the Ottoman rule in 1804.

Napoleon assumed the title of emperor in 1804.

Britain resumed the war against him in 1803 and was joined by **Russia, Austria** and **Sweden** in 1805. Russia was defeated at **Austerlitz**, 1805, but Britain's naval victory at **Trafalgar**, 1805, resulted in a crippling blockade of French shipping.

Napoleon proclaimed himself **King of Italy**, 1805.

1805-7

After defeating Prussia, 1806, Napoleon allied with Russia, 1807, and set up the **Continental System** (which Russia was forced to leave for economic reasons in 1810) to exclude British trade from Europe.

The Holy Roman Empire came to an end when **Francis II**, *r.* 1792-1835, who was also Emperor of Austria, renounced the title, 1806.

The slave trade was abolished in the British Empire, 1807, although slavery continued in the colonies.

1807-8

A nationalist revolt broke out in Spain when Joseph Bonaparte, 1768-1844, assumed the throne, 1808. Britain exploited this to attack Napoleon in the Peninsular War, 1808-14.

Austria and Prussia reformed their army and taxation systems to improve military capacity.

Archduke Charles of Austria appealed to the Germans to oppose Napoleon but was defeated at **Wagram**, 1809.

France assumed control of Swedish foreign affairs.

National events

Stimulated by the Napoleonic Wars, the Industrial Revolution created large, overcrowded towns centred on factories in which whole families were employed. Social and economic discontent ensued, resulting in widespread agitation for electoral reform, fostered by the growing middle classes.

1800-2

Robert Owen, 1771-1858, set up a co-operative cotton mill at New Lanark, 1800.

Britain and Ireland were constitutionally united, 1801.

1802-5

A Factory Act, 1802, tried ineffectually to limit children's working hours in factories.

Nelson destroyed the Franco-Spanish fleet at Trafalgar, 1805.

1805-7

England declared a blockade of French ports.

Napoleon's Continental System, 1806, raised food prices and depressed the textile industry.

1807-8

Parliament, based on the gentry's authority and agricultural constituencies, was growing increasingly out of touch with the interests of the new industries.

The town of New Lanark

Napoleon at Eylau, 1807

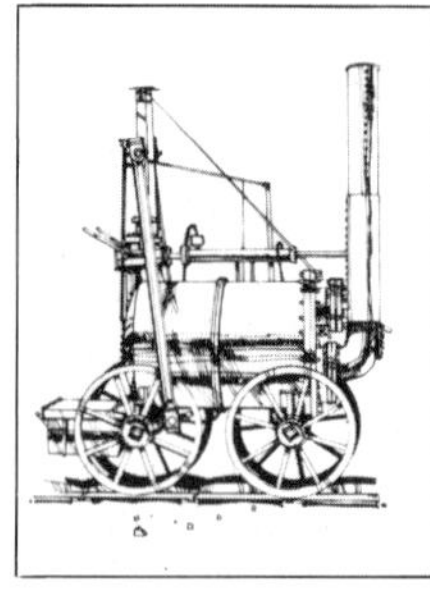

Trevithick's steam engine

"The 3rd of May, 1808" by Goya

Religion and philosophy

Classical economic theory was developed and systematized in the work of Say in France and Ricardo in England, the latter influencing both sides of the debate about *laissez-faire* doctrines, which dominated social thought in the early 19th century. At the same time reaction to the social evils of industrial capitalism ranged from Sismondi's warning of class antagonisms to the social experiments of Robert Owen and the Utopianism of Charles Fourier.

The major philosophical school of the period, German idealism, emerged in a country as yet relatively sheltered from the major social upheavals of the time. In particular, Hegel, who would greatly influence the young Karl Marx, argued that historical progress was identical with the advancement of human consciousness, while Schopenhauer and Schelling emphasized man's darker, irrational impulses and prepared the way for Freud and existentialism.

1800-2

Friedrich Schelling, 1775-1854, published his *Transcendental Idealism*, 1800, in Germany, grounding his idealism on external nature. His philosophy of man stressed the force of irrationalism which, as the source of all evil, could dominate the intellect, wherein lay the power for good.

William Paley, 1743-1805, an Anglican, advanced the idea in *Natural Theology*, 1802, that the design evident in the world implied the work of a creator.

1802-5

Jean Baptiste Say published his *Traité d'économie politique* in 1803, putting forward his "law of markets", which states that supply creates its own demand with the consequence that depression is the result of over-production in some markets and under-production in others, an imbalance that would automatically correct itself.

The *Code Napoléon* of 1804 nationalized French law and established the principle of equal citizenship.

1805-7

G. W. F. Hegel, 1770-1831, the German idealist philosopher, published his first great work, *Phenomenology of Mind*, in 1807. In this and later works Hegel expressed the view that reality is essentially a whole (which he called the Absolute) comprising both mind (subject) and matter (object). The physical world would cease to be alien and objective when, with the attainment of total comprehension, object and subject merged into the Absolute.

1807-8

Charles Fourier, 1772-1837, writing in France (*The Social Destiny of Man*, 1808) and **Robert Owen** in England (*A New View of Society*, 1813) advocated social reconstruction on the basis of workers' co-operatives. Owen organized mills at New Lanark, 1800-24, on principles of welfare and justice but his American experimental community at New Harmony, 1825, was short-lived, as were the communes based on the even more Utopian ideas of Fourier.

Literature

English Romanticism reached its peak with the work of the poets Wordsworth and Byron, who explored the quest for harmony with nature and stressed the independence of genius from social convention. The historical novel developed by Scott linked the interest in the past with an implicit concern for national identity – a trend echoed in the German concentration on folk-tales and mythology. In other respects, however, German Romanticism, as displayed by Goethe and Schiller, involved a less violent break with 18th-century humanism.

Chateaubriand and Madame de Staël tried to introduce the ideas of the movement to France, but met little success as classicism still reigned until the work of the poets in the 1820s.

1800-2

A more subjective emphasis developed in German literature with the two unfinished novels of **Novalis**, 1772-1801, published in 1798-1800. Both were *Bildungsromans* novels on the education of the hero and the development of character and temperament, like Goethe's *Wilhelm Meister*.

William Wordsworth wrote his autobiographical poem *The Prelude* in 1799-1805.

1802-5

Le Génie du Christianisme, 1802, by **Francois René de Chateaubriand**, 1768-1848, introduced to French literature a mystical Christianity which had a great influence on the French romantic writers.

The need for political freedom in Germany was the subject of the play *William Tell*, 1804, by the poet and dramatist **Johann Schiller**, 1759-1805.

1805-7

The German romantic poets Brentano, 1778-1842, and **von Arnim**, 1781-1831, collected folk-poems, in *Des Knaben Wunderhorn*, 1805-8.

Heinrich Kleist, 1771-1811, published *Penthesilea*, 1808.

1807-8

The humanism of Johann Goethe, 1749-1832, was expressed in his play *Faust, Part 1*, 1808.

The poetry of the Italian Ugo Foscolo, 1778-1827, extolled the past and the value of art as a permanent shrine to virtue.

The lyric dramatic poetry of **Adam Oehlenschläger**, 1779-1850, was popular in Denmark.

Art and architecture

European art in the early 19th century saw a reaction to Neo-classicism and the beginnings of Romanticism, involving a shift from formal rules to an emphasis on the subjective – feelings, impressions, imagination – and a preference for fantasy, excess and the poetic. The influence of Romanticism also inspired an interest in historical and foreign styles of architecture, while in painting it produced the freer technique and the more expressive use of colour found in the work of Delacroix. The choice of subject-matter also changed to include contemporary and historical scenes which reflected the nationalist ideals of the time – a tendency heightened in France by the exotic career and lavish commissions of Napoleon. While Neoclassicism came to be rejected by artists in Western Europe its influence spread into Russia and the New World, where it dominated architecture.

1800-2

The term "picturesque" was introduced in England at the end of the 18th century as an aesthetic category, characterized by irregularity, variety and roughness of texture; it had a decisive effect on landscape painting and architecture.

British industrial architecture had developed over the 18th century to include iron frame constructions which were fire-proof and functional.

1802-5

The Greek revival in European architecture, *c.* 1760-1830, was now at its height. Its essence was the exact reproduction of Greek models, which were admired for their simplicity and associated with the beginnings of civilization. It was predominantly used for public buildings like the theatre at Besançon 1784, by **Claude-Nicolas Ledoux**, 1736-1806, although domestic forms, sometimes in the style of a Greek temple, were popular.

1805-7

Napoleon commissioned portraits and commemorative scenes to enhance his imperial image. Canvases presented an aura of magnificence with allusions to the Roman Empire as in "Napoleon as Emperor", 1806, by **Jean-August-Dominique Ingres**, 1780-1867, and "Napoleon at Eylau", 1808, by **Antoine Gros**, 1771-1835.

Napoleonic architecture like the Paris Bourse, 1807, by **A. T. Brongniart**, 1739-1813, was similarly inspired.

1807-8

Davidian ideals were questioned in France by his pupils. Some, like **Anne-Louis Girodet**, 1767-1824, in his "Entombment of Atala", 1808, used unusual light effects and mystical subjects.

The investigation of the symbolic, mystical and religious aspects of landscape took place primarily in Germany, with the work of **Caspar David Friedrich**, 1774-1840, whose "The Cross in the Mountains", 1808, was painted as an altarpiece.

Music

Romanticism in European music began to replace classicism as personal expression in the arts took precedence over ideals of formal balance. But the first romantic composers, such as Ludwig van Beethoven, 1770-1827, and Carl von Weber, 1786-1826, were trained in classical techniques and brought restraint to bear on the new, sensuous style of music.

1800-2

The piano repertory was rapidly extended by **Muzio Clementi**, 1752-1832, and Beethoven, as the instrument gained a greater range of notes and sound quality.

1802-5

Short forms for the piano were established by such composers as **John Field**, 1782-1837, who wrote nocturnes – one-movement lyric pieces, later made popular by **Frédéric Chopin**, 1810-49.

1805-7

Beethoven straddled the classical and romantic eras, extending the range of sonata form and composing works, such as the *Coriolanus Overture*, 1807, directly inspired by literary ideas.

1807-8

Programme music, interpreting the events and moods of a specific story, emerged as a feature of romantic music, using evocative sounds as in Beethoven's *Pastoral Symphony*, 1808.

Science and technology

Progress in technology and science in Europe divided between Britain and France. France became the centre of pure science while Britain forged ahead in industrial science. Although automation was invented in France, its potential was not fully exploited. Similarly, atomic theory was first proposed in England but was refined in Europe.

The most important technological innovation was that of powered transport in England and the USA, which opened new areas of industrial expansion; in the same period gas lighting transformed city life. Many scientific discoveries, too, would have subsequent importance. Electrical science developed with the discovery of electromagnetism and would stimulate enquiry into the nature of matter as well as producing new sources of energy, while modern chemistry developed under the influence of Gay-Lussac and Avogadro. The study of fossils raised new questions about the age and origins of life.

1800-2

Automation using punched cards to control the production of silk fabric was invented in France, 1801, by **Joseph Marie Jacquard**, 1752-1834.

Ultra-violet light in the Sun's spectrum was discovered, 1801, by **Johann Ritter**, 1776-1810.

The interference of light was shown, in 1801, by **Thomas Young**, 1773-1829, restoring the wave theory first put forward by Christiaan Huygens. Young also studied elasticity, giving his name to the tensile modulus, or scale of elasticity.

1802-5

Screw-cutting machines and lathes were developed at the engineering works of the Englishman **Henry Maudslay**, 1771-1831, who invented the screw micrometer and schooled many fine engineers.

Jean Baptiste Lamarck, 1744-1829, a French naturalist, coined the term "biology" in 1802. In his *Philosophie Zoologique*, 1809, he held that evolution occurred.

The first railway locomotive was built in England by **Richard Trevithick**, 1771-1833, and first ran in 1804.

1805-7

Gas lighting was introduced in European cities *c.* 1806.

The Clermont, built by **Robert Fulton**, 1765-1815, an American engineer, inaugurated the first regular steamboat service along the Hudson River in 1807, although a short-lived service had been run *c.* 1790.

The Geological Society of London was founded in 1807.

1807-8

Potassium and sodium were discovered, 1806-7, by **Humphry Davy**, 1778-1829, an English chemist.

Jean Fourier, 1768-1830, a French mathematician, discovered that a complex wave is the sum of several simple waves.

The atomic theory, stating that the same elements have the same atoms and that a compound is made up of atoms of elements combined in fixed proportions, was propounded in 1808 by **John Dalton**, 1766-1844.

1808-12

Napoleon's empire reached its greatest extent, 1812, when with Austria and Prussia he invaded Russia. French forces took Moscow, 1812, but were unable to sustain the Russian winter and were forced to retreat with serious losses.
Paraguay and Venezuela became independent from Spain, 1811, marking the final collapse of Spanish imperial authority.
Napoleon married Marie Louise of Austria in his search for an heir, 1810.

1812-14

The Duke of Wellington, 1769-1852, led the Allied forces into Paris, 1814. Napoleon abdicated and was exiled to Elba.
The monarchy was restored in Louis XVIII, r. 1814-24.
The Congress of Vienna, 1814-15, restored monarchs to the Austrian and Prussian thrones and the kingdom of the Netherlands was founded as a buffer against France.
Britain won definitive control of the **Cape of Good Hope**, 1814, on the route to India.

1814-17

Napoleon returned from Elba and Austria, Britain, Prussia and Russia formed a new alliance. Louis fled from Paris to return only after Napoleon's defeat at **Waterloo** and exile to St Helena, 1815.
The Holy Alliance aimed to crush the spread of radicalism in Austria, Prussia and Russia.
Prince Metternich, 1773-1859, crushed similar aspirations in the German states.
Ferdinand I, r. 1815-25, regained the Italian throne.

1817-20

The American and Canadian border was fixed in 1818.
The first immigrants settled the Australian grasslands, 1817-18.
A revolt in Naples against Ferdinand I was crushed in 1821 and Ferdinand returned with a more liberal constitution.
The British founded Singapore to rival Dutch Malacca as the centre for Far Eastern trade.
Only Nepal, the Sikh and Sind states, and Afghanistan were independent of British rule in India by 1818.

1820-22

Ferdinand VII of Spain, r. 1814-33, was captured by liberal rebels in 1820.
The nationalist Greek war for independence from the Ottoman Empire began, 1821. The Great Powers intervened in 1827 and established a Greek kingdom in 1832.
Spain lost Mexico and Peru, 1821, while Brazil became independent from Portugal, 1822.
Opium trade between India and China flourished in the reign of **Hsüang Tung**, 1821-50.

1822-25

A Spanish liberal revolt was crushed with French help, 1823.
The Monroe Doctrine, 1823, asserted that the American continent could no longer be an arena for European colonial activity.
Britain surpassed other European countries in her industrial and trading position, but agitation began for more *laissez-faire* trading policies.
The Anglo-Burmese wars began in 1824, following Burmese aggression.

Unemployed domestic weavers destroyed new machines in the **Luddite riots**, 1811-16, fearing the impact of mechanization on their craft.

The expansion of small country banks reached its peak, 1814, facilitating the transfer of agricultural wealth into industrial investment.

The Corn Laws, 1815, protected landowners by excluding foreign corn and keeping the price of bread high.

A Manchester reform meeting ended in the massacre of **Peterloo** in which soldiers fired into the crowd, killing several demonstrators, 1819.

The Cato Street conspiracy, which aimed to overthrow the government, was detected, 1820, and the conspirators hanged.
The post-war slump ended.

A liberal Conservative ministry reduced duties on imports and repealed the **Combination Acts**, 1824.

The Brighton Pavilion by Nash

Lord Byron

The Peterloo massacre

Hegel's *Science of Logic* was published in 1812-16. In it he developed a dialectical method of reasoning – opposing a thesis with its antithesis to establish a synthesis – with the aim of revealing the nature of the Absolute. Hegel believed history to be a dialectical progression towards the Absolute and considered the Prussian state to be the culmination of this dialectical progression.

The conservative tradition in France found exponents in **Joseph de Maistre**, 1753-1821, and **Louis Bonald**, 1754-1840, who insisted on the supremacy of Christianity and the absolute rule of the Church and pope. De Maistre published his *Essay on Constitutions* in 1814, in which he used his facility for logical argument to oppose the progress of science, liberalism and the empirical methods of the "philosophes", especially Voltaire.

The rising science of economics was systematized by **David Ricardo**, 1772-1823, in his *Principles of Political Economy and Taxation*, 1817, which aimed to set out the laws governing the division of the social product among the classes in society. The systematization of this work made him the leading exponent of the classical school in England and he was influential both among *laissez-faire* economists and their opponents, such as Robert Owen and, later, Karl Marx.

Jean Charles Sismondi, an early theorist of economic crisis, warned in his *New Principles of Political Economy*, 1819, against the effects of unregulated industrialism, predicting acute social conflict. In Germany, **Artur Schopenhauer**, 1788-1860, published *The World as Will and Idea* in 1819. His emphasis on man's will and irrational impulses prepared the way for the departure from the 18th-century idea of rational progress.

Thomas Erskine, 1788-1870, a Scottish theologian, in *Internal Evidence for the Truth of Christian Religion*, 1820, held that the meaning of Christianity lay in its conformity with man's spiritual and ethical needs.
Friedrich Schleiermacher, 1768-1834, a German theologian and founder of modern Protestant theology, in *The Christian Faith*, 1821-2, saw religious feeling as a sense of absolute dependence and sin as a desire for independence.

The anti-union Combination Acts were repealed in Britain, 1824, after a campaign led by **Francis Place**, 1771-1854, and **Joseph Hume**, 1777-1855. Place advocated birth control as a means by which the working class could limit their numbers to improve wages.
Leopold Ranke, 1795-1886, a German historian, wrote *History of the Latin and Teutonic Nations, 1494-1514*, in 1824, seeking a history based on scientific methods.

Madame de Staël, 1766-1817, wrote *De L'Allemagne* in 1810.
The works of Esaias Tegner, 1782-1846, convey his belief in an ideal world and his version of Swedish national glory, as described in his poem *Svea*, 1811.
The English middle classes were given perceptive scrutiny by **Jane Austen**, 1775-1817, in *Pride and Prejudice*, 1813 and *Sense and Sensibility*, 1811.

The English Romantic poet Percy Bysshe Shelley, 1792-1822, wrote his first major poem, *Queen Mab*, 1811-13.
The first historical novel was the medieval romance *Waverley*, 1814, by **Sir Walter Scott**, 1771-1832. His evocations of the past had extensive influence, notably with **Manzoni** in Italy.
Brothers Grimm made their collection of folk-tales, 1812-15.

Grotesque themes were studied by the German composer and author **Ernst Hoffmann**, 1776-1822, in his tales and in the novel *The Devil's Elixir*, 1813-15.
The psychological analysis of a broken love affair is the subject of *Adolphe*, 1816, by **Benjamin Constant**, 1767-1830 – an ardent liberal and a political journalist, closely alligned with the French Romantics.

The spirit of the Romantic movement was epitomized in the notorious life of **Lord Byron** 1788-1824, and the tormented homeless Byronic hero became popular in European fiction. In contrast, his *Don Juan* (first two cantos, 1819), was a biting, unromanticized social satire.
William Hazlitt, 1778-1830, commented on contemporary English life in astute essays.

Alphonse Lamartine, 1790-1869, moved from the private anguish of *Meditations*, 1820, to the mystical lyricism of *Harmonies Poétiques*, 1830.
A search for eternal perfection is the theme of *Prometheus Unbound*, 1820, by the English poet Shelley. His friend **John Keats**, 1795-1821, wrote superlative odes, including the *Ode to a Nightingale*, 1819.

The Confessions of an English Opium Eater, 1822, was the autobiography of **Thomas De Quincey**, 1785-1859.
Thomas Peacock, 1785-1866, skilfully used his satirical novels to attack English views, obsessions and political dogmas.

The pastiches of Indian, Chinese and Egyptian styles seen in the Prince Regent's Brighton Pavilion, from 1810, by the English architect John Nash, 1752-1835, reflected the world-wide process of exploration and colonization by the European powers.
A quasi-religious order, the Nazarenes, was founded in Vienna, 1809, by **Friedrich Overbeck**, 1789-1869 and **Franz Pforr**, 1788-1812. They moved to Rome in 1810.

The work of Francisco de Goya, 1746-1828, combined objective reportage and a sense of personal horror at the Napoleonic Wars in "The Disasters of War", 1810-14. He later developed the new medium of lithography.
In Russia, Alexander I commissioned classical buildings for St Petersburg and encouraged French Neoclassicism in all branches of the arts, particularly portraiture.

In France, Théodore Géricault, 1791-1824, took contemporary events as the vehicle for his themes and used large canvases to strengthen their impact in, for example, "The Raft of the 'Medusa'", 1817.
German Neoclassical architecture first developed after the fall of Napoleon, with the emphasis on purely Grecian forms seen in the Neue Wache, Berlin, 1816, by **Karl Friedrich Schinkel**, 1781-1841.

Neo-Renaissance architecture was developed in Germany at the same time by **Leo von Klenze**, 1784-1864. His Palais Leuchtenberg, Munich, 1816, is the first German example, although the style existed earlier in France.
In 1816 Dom João VI of Brazil invited French architects, painters and sculptors to Rio de Janeiro to "civilize" Creole taste. Their neoclassical style dominated the arts for the next hundred years.

English Romantic portraiture with its emphasis on drama and psychological investigation was exemplified in the forceful works of **Thomas Lawrence**, 1769-1830, whose portraits of the "Heads of State of Europe" in the Waterloo Chamber, Windsor Castle, 1818-20, celebrate the triumph over Napoleon.

Romantic tendencies in art were crystallized in the work of **Eugène Delacroix**, 1798-1863, where truth was no longer merely factual but a glimpse into man's soul. Delacroix set out to achieve this with the use of expressive colour and an emphasis on mood and the poetic. His work marked a decisive shift in French painting away from the importance of form and stressed violence and drama, as in his "Massacre at Chios", 1824.

Orchestral concerts became popular in London, Paris and Vienna as the middle classes began to support music. The patron and his salon declined in influence.

A greater variety of orchestral instruments allowed richer tone contrasts. There were few large orchestras so many orchestral works were published as piano transcriptions.

Lieder, a German form of lyric song, was raised by **Franz Schubert**, 1797-1828, to new heights as his piano parts provided a counterpart to the voice, echoing the mood of the lyric.

Opera centred on Paris. The light lyricism of *The Barber of Seville*, 1816, by **Gioacchino Rossini**, 1792-1868, gave way to the romanticism of **Carl von Weber's** *Der Freischütz*, 1821.

Conductors were required to marshal the expanded orchestra, as it could no longer be led by an instrumentalist. In 1820, **Ludwig Spohr**, 1784-1859, introduced the conductor's baton.

The symphony found new depths of expression, especially with **Beethoven**, as a personal creation requiring a large orchestra. His *Ninth Symphony*, 1824, is an example.

Joseph Gay-Lussac, 1778-1850, a French chemist, announced in 1808 that gases combine in certain proportions by volume and suggested that the proportions are linked to the formula of the compound formed. Gay-Lussac's work also led to the correct atomic weights of elements. **Amadeo Avogadro**, 1776-1856, an Italian, argued in 1811 that equal volumes of gases at the same temperature and pressure contain equal numbers of molecules.

Georges Cuvier, 1769-1832, a French anatomist, broadened Linnaeus' classification to include phyla and fossils, thus founding palaeontology. The hardness of materials was classified in 1812 by a German, Friedrich Mohs, 1773-1839.
Chemical symbols as used today were introduced in 1814 by Jöns Berzelius, 1779-1848, a Swede, who later made a correct list of atomic weights.
Dark lines in the Sun's spectrum were identified, 1814, by Joseph von Fraunhofer, 1787-1826.

The safety lamp was invented by **Humphry Davy** in 1815, to prevent explosions in mines. The optical experiments of **Jean Biot**, 1774-1862, after 1815, led to the founding of polarimetry.
The first geological map, of England and Wales, was published, 1815, by **William Smith**, 1769-1839.
The single wire telegraph was invented in 1816.
Photography was born out of the experiments in 1816 of **Nicéphore Niepce**, 1765-1833.

Electromagnetism was found, 1820, by **Hans Oersted**, 1777-1851, who noticed that a compass needle was deflected by a wire carrying a current. He was also the first to prepare aluminium, 1825.
Thomas Seebeck, 1770-1831, a Russo-German physicist invented the thermocouple, an instrument for measuring temperature as electricity, 1821.

André Ampère, 1775-1836, studied the effects of electric currents in motion, founding and naming the science of electrodynamics by 1822. He also invented the solenoid.

The electromagnet, the first machine to use electricity, was made by **William Sturgeon**, 1783-1850, an English physicist.
Sadi Carnot, 1796-1832, published *On the Motive Power of Fire*, 1824, in which he showed that only a fraction of the heat produced by burning fuel in an engine is converted into motion, which depended only on the temperature difference in the engine. This was the basis of modern thermodynamics.

1825-1850 Liberalism and nationalism

Principal events

The spread of industrialism from England to north Europe brought the rise of a solid middle class advocating liberal and nationalist ideas, as well as a new urban radicalism focused by regular economic booms and slumps. In spite of attempts to suppress them, these ideas spread throughout Europe, culminating in the nationalist and radical revolts of 1848. At the time this was a failure but the ideals of 1848 would be realized later as Italy and Germany achieved unification and the old empires collapsed.

The United States expanded vigorously westwards, her population and industry increasing, while European colonialism was most active in Asia. The impact of British culture was felt in India for the first time and the process of penetration of China began in earnest with the end of the Opium Wars which forced China to open her ports to foreign trade.

1825-28

Following a liberal **Decembrist revolt**, 1825, **Nicholas I**, *r.* 1825-55, introduced repressive measures. **Charles X**, *r.* 1824-30, alienated bourgeois support in France by restoring to nobles land lost in the Revolution. **The British in India** ended their titular subservience to Mogul rule, 1827. **The Javanese rebelled** against Dutch rule but were put down.

1828-1830

Lord Bentinck, 1774-1839, built new canals and roads in India and prohibited the customs of suttee and thuggee. **Turkey recognized Greece's independence** after British and Russian intervention, 1829. **Uruguay** was established as an independent buffer state between Argentina and Brazil, 1828. **The Workingmen's Party** was established in the US, pledged to social reform. **Charles X** faced rising opposition in France after 1825.

1830-33

The French liberal opposition expelled Charles X, replacing him with **Louis-Philippe**, pledged to rule with middle-class support, 1830. This sparked off a liberal and nationalist revolt in Belgium, which became independent of the Netherlands. **Nationalist risings** in Italy and Germany were unsuccessful. **The French** established their authority in Algiers, 1830. **The Young Italy** group was launched by **Giuseppe Mazzini**, 1805-72, in Italy in 1831.

1833-5

Slavery was abolished throughout the British Empire, 1833. **The German customs union**, completed in 1834, became the focus of German nationalism. **Regional opposition** to the liberal Spanish regime led to the **Carlist Wars**, 1834-9, in support of **Don Carlos**. **Louis-Philippe** renounced his radical support and introduced strict censorship, 1835. **British trade with China** increased after the East India Company's monopoly ended.

National events

The social problems that came with industrialization reached their peak in the 1840s, manifesting themselves in the Chartist movement. Parliamentary reform in 1832 recognized the new importance of the industrial towns and was followed by a series of attempts to regulate conditions of work.

The first public railway between Stockton and Darlington was opened, 1825, leading to a boom in competitive railway building after 1830.

The growth of an Irish nationalist movement brought the repeal of anti-Catholic laws, 1829. **The "new" police force** was set up in London in 1829.

The Reform Act, 1832, introduced new industrial constituencies. The supremacy of the landowners was nevertheless maintained in Parliament.

The Poor Law Amendment Act, 1834, set up workhouses for the poor. **The Tolpuddle Martyrs** were transported, 1834, for trades-union activity.

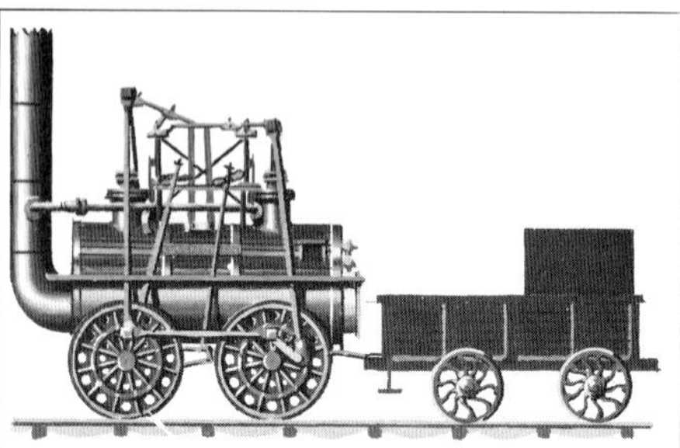

Stephenson's *Locomotion*, 1825

Simon Bolivar

Delacroix: "1830 Revolution"

"View of Mount Fuji" by Hokusai

Religion and philosophy

Several European thinkers, principally in France, advocated the application of the observational, or "positive", methods used in the natural sciences to the study of social phenomena.

The growth of the industrial system stimulated new and radical thought. Saint-Simon and, after him, Comte, argued that industrial society should be governed by a new "priesthood" trained in the positivist method, while the existing forms of social organization were criticized by utopian thinkers who looked forward to ideal societies free from inequality and injustice.

Marx and Engels, who were influenced by French utopianism as well as German idealism and British political economy argued that in order to end the inequality and injustice of existing society there must be a working-class revolution.

A number of Adventist sects prophesying the return of Christ emerged during these years, especially in the USA.

Henri de Saint-Simon, 1760-1825, in *New Christianity*, 1825, sought to combine the ideals of Christianity with science and a belief in industrialism to form a new religion of socialism based on a science of society. He advocated management of society by experts in the new social science, to aid welfare and progress. His theories were further developed by his one-time secretary, **Auguste Comte**, 1798-1857.

François Guizot, 1787-1874, published his *History of Civilisation in France*, 1829, a study of social institutions using empirical data and a historical approach. **James Mill**, 1773-1836, in his *Analysis of the Human Mind*, 1829, developed Bentham's pleasure-pain doctrine in the field of psychology. **John Darby**, 1800-82, set up the **Plymouth Brethren** sect in Ireland, 1830, emphasizing the second coming of Jesus Christ.

Mormonism was founded in New York State by Joseph Smith, 1805-44, in 1830 and based on the *Book of Mormon*. Opposition to their enclosed community life caused them to settle in Utah, 1847. **Modern Adventism** was founded in the USA by **William Miller**, 1781-1849, who began in 1831 to prophesy the imminent end of the world. The failure of his predictions produced many breakaway sects, including the **Seventh Day Adventists.**

The Oxford Movement within Anglicanism began, 1833, under the leadership of **John Keble**, 1792-1866, and Cardinal **John Newman**, 1801-90. They sought to revive the ideals of the medieval Church and reintroduced elaborate rituals. **Félicité Lamennais**, 1782-1854, a French Catholic priest, argued in *Thoughts of a Believer*, 1834, for the separation of Church and state and attacked the papacy for its interference in politics.

Literature

While Romanticism spread to Russia with Lermontov and Gogol, its brooding introversion began to break down in Western Europe and the political implications of its rebellion were explored. The Jung Deutschland group insisted on the political role of literature and in France the realistic depiction of the past or of contemporary society became important new themes. The social and nationalist commitment of writers found expression in the revolutions of 1848, in which authors such as Hugo and Lamartine played an important role.

In England novelists explored social relationships. Dickens concentrated on the evil results of industrialization with a wealth of characterization equalled only by Balzac.

James Fenimore Cooper, 1789-1851, gained international fame with his historical novel *The last of the Mohicans*, 1826, in which he portrayed the American Indians. *I Promessi Sposi*, 1825-27, by the Italian **Alessandro Manzoni**, 1785-1873, is a love story set in 17th century Italy with delicate Christian idealism.

The Cenacle group of poets was set up in Paris, 1827, by **Victor Hugo**, 1802-1885, and **Sainte-Beuve**, 1804-1869. Hugo's *Odes et Ballades*, 1826, celebrated the theme of liberty. The romantic poetry of **Alfred de Musset**, 1810-57, and the stoical pessimism of **Alfred de Vigny**, 1797-1863, reflected French disillusionment in the post-Napoleonic era.

Social realism in the French novel began with the work of **Stendhal (Henri Beyle)**, 1783-1842. A soldier and critic, he wrote *Le Rouge et Le Noir*, 1830. **George Sand**, 1804-76, produced idealized novels of peasant life, such as *Mauprat*, 1837, and **Prosper Merimée**, 1803-70, brought archaeological exactness to the romantic novel.

The first great modern Russian writer, **Alexander Pushkin**, 1799-1837, broke down the archaic stiffness of the written style, and introduced realistic language. He used peasant folk songs in the poem *The Dead Princess*, 1834, and medieval history as the source for the novel *Eugene Onegin*, 1823-31, and the tragic drama *Boris Godunov*.

Art and architecture

Naturalism in painting – the devotion to truth to nature – received a special impetus from work in Britian where scientific advances and the Industrial Revolution affected art.

British landscapists such as John Constable studied natural effects in a scientific manner rather than composing classical panoramas, and French artists emulated his innovations.

The interest in history and in different historical styles continued throughout Europe and America. A large number of paintings of historical scenes were produced but it is in architecture that the range of interest in different styles was clearest. Italian models remained a source for the style of secular public buildings but the Gothic revival received a new impetus.

England's new wealthy middle classes began to impose their taste on painting and architecture, while rapid urbanization generated the need for new solutions in town planning.

The American ornithologist and painter **John James Audubon**, 1785-1851, published *The Birds of America*, from original drawings in England between 1827 and 1838. **Italian Renaissance architecture** inspired the Travellers' Club, London, by **Charles Barry**, 1795-1860, an example of the neo-Renaissance style. **In France oil paints** were sold in tubes for the first time *c.* 1830, which made painting out of doors much easier.

A craze in France for things English known as *anglomanie* developed in the post-Napoleonic era. Watercolour painting became popular in Paris where **Richard Parkes Bonington**, 1802-28, worked after 1820. British oil painting techniques were studied by French artists, especially after the exhibition of Constable's work in the Paris Salon of 1824, and **Delacroix** illustrated Byron's poem in his "The Execution of Doge Marino Faliero".

The School of Architecture, Berlin, 1832-5, by **Karl Friedrich Schinkel** was based on strict units of measurement and the design used iron frame structures. It had vestiges of classical detail but came close to a "modern", simple and functional style. **English industrial towns** such as Manchester developed the back-to-back house as a solution to the housing needs generated by the Industrial Revolution.

In England, the taste and buying power of the new, wealthy middle classes resulted in an increasing demand for genre and small-scale historical or romantic scenes with emphasis on narrative, as in "Giving a Bite", 1834, by **William Mulready**, 1786-1863. **Japanese colour printing** reached its height in "The Thirty-six views of Mount Fuji", 1834-5, by **Hokusai** and the poetic "Views of Kyoto", 1834, by **Ando Hiroshige**, 1797-1858.

Music

Romanticism evolved further in Europe, where composers came to consider music a kind of poetry that penetrated to the heart, ennobling the soul and stimulating the imagination. It was at times extravagant and replete with epic works and cult figures, stimulating growth in orchestration techniques and producing such new forms as the symphonic poem.

Paganini, arrived in Vienna in 1828 and set out to refute rumours that he had been in prison between 1801 and 1804.

The mouth organ, the Chinese sheng, came to Vienna in 1829. **Felix Mendelssohn**, 1809-47, revived Bach's music in 1829 after a century of eclipse, giving it the status it enjoys today.

Hector Berlioz, 1803-69, made inspired use of the orchestra to express the extra-musical allusions of programme music, as in his *Symphonie fantastique*, 1830, with its elaborate story line.

Frederic Chopin, 1810-49, a Polish exile working mostly in Paris from 1831, wrote dazzling short pieces for piano in such forms as nocturne, mazurka and polonaise, often based on dances.

Science and technology

The railway system which now grew up in Britain, providing cheap transport for labour and raw materials, proved a vital precondition for the expansion of any industrial society. The process of industrialization, however, brought more and more people into the cities and led to a severe worsening of the conditions of working people. In Bristol, England, the death rate doubled between 1831 and 1841, though advances in medicine and public health began to improve matters from 1840 onwards. The discovery of asepsis was particularly important in lowering the child mortality rate.

In pure science the discovery of alternative geometries to that of Euclid prompted new enquiries into formerly accepted theories, clearing the way for Mach and Einstein.

The discovery of Brownian motion finally established the existence of unobservable particles, in this case molecules; an important step towards the eventual acceptance of atomic theory.

The first public steam railway opened in 1825 between Stockton and Darlington in England. The first train was drawn by *Locomotion No. 1*, built by the British engineer **George Stephenson**, 1781-1848. **Robert Brown**, 1773-1858, a British botanist, observed the random motion of particles, **Brownian motion**, in 1827, thus proving that molecules existed. **Ohm's law**, relating current, voltage and resistance, was laid down in 1827 by German physicist **Georg Ohm**, 1787-1854.

The polarimeter, which analyses the passage of polarized light through matter, was developed in 1828, by British physicist **William Nicol**, 1768-1851. **Embryology** was founded, 1828, by the German biologist **Karl Baer**, 1792-1876. He studied germ layers and the growth and development of the embryo. **Organic chemistry** began with the synthesis of urea by German chemist **Friedrich Wöhler**, 1800-82, in 1828. He showed that organic substances do not always come from living things.

Non-Euclidean geometry, was developed *c.* 1829-30 by **Nikolai Lobachevski**, 1793-1856. **Screw threads** were standardized, 1830, by the Briton **Joseph Whitworth**, 1803-87, making mass production of screws viable. ***Principles of Geology***, 1833, by **Charles Lyell**, 1797-1875, showed rocks evolve slowly. **Electromagnetic induction** was discovered, 1831, by a British physicist, **Michael Faraday**, 1791-1867. This discovery led to the electric generator.

Karl Gauss, 1777-1855, a German mathematician, devised a set of units for magnetism, presented in 1833. **Charles Babbage**, 1791-1871, a British mathematician, developed the principles of the mechanical computer in the 1830s. **The *Royal William*,** a Canadian vessel, was the first steamship to cross the Atlantic wholly under its own power, 1833. With the building of the ***Great Western***, in 1838, by **I. K. Brunel**, 1806-59, transatlantic services became regular.

1835-7

The Boers of South Africa began the **Great Trek**, 1835, to find new territory free from British rule.
British attempts to unite her colonies of **Upper and Lower Canada** led to revolt, 1837.
Victoria, *r.* 1837-1901, succeeded to the British throne.
Britain attempted to intervene in Persia, 1834-8, to forestall Russian influence, but was repelled by **Muhammed Shah**, *r.* 1835-48.

1837-40

A working-class radical Chartist movement developed in England, stimulated by European economic depression.
China seized opium imports from India, provoking the **Opium Wars** with Britain, 1839.
British industrialists' attempts to introduce free trade led to the foundation of the **Anti-Corn Law League**, 1839.
Ieyoshi, *r.* 1838-53, the Japanese shōgun, opposed mounting pressure for Occidental trade.

1840-44

Britain annexed **Natal**, 1843.
Upper and Lower Canada were united in 1840 and given responsible government.
The Straits Convention, 1841, between Russia, Britain, France, Austria and Prussia, closed the Dardanelles and Bosphorus to foreign shipping.
Frederick William IV of Prussia, *r.* 1840-61, encouraged German nationalism.
The Treaty of Nanking, 1842, opened Chinese ports to British trade.

1844-5

The persecution of Christians in Confucianist French Indo-china led to French military involvement, 1840-50.
Sanitary reform and slum clearance was introduced in England in the 1840s.
The Anglo-Sikh wars broke out in India, 1845-8.
Texas was annexed by the United States in 1845.
Utopian socialism became popular among intellectuals and the working classes.

1845-8

A potato famine in Ireland resulted in mass emigration to the United States, 1846-8.
The US invaded Mexico and occupied the capital in 1847, defeating Mexico after conflict over Texas, 1846-8.
Liberal hopes in Prussia were raised when the **Landtag** was called, 1847, by the king, who asked for funds to build railways.
The accession of the liberal **Pope Pius IX**, *r.* 1846-78, raised nationalist hopes in Italy.

1848-50

Britain annexed the Punjab and won Sikh loyalty, 1845-8.
An outburst of urban radicalism brought the expulsion of **Louis-Philippe** in France and the establishment of a republic, 1848.
Metternich resigned in Austria.
Hungary declared herself independent and German and Italian nationalist movements emerged.
By 1849 Austria had defeated the Italian revolt and racial disputes split Hungary.
The gold rush in California opened up the West, 1848-50.

The abolition of slavery in the colonies led to increases in the price of their products.
A great railway building boom began, 1836.

The radical Chartist movement grew up in the North with six demands including universal suffrage.
The penny post was introduced.

The Anti-Corn Law League led by the industrialists **Cobden**, 1804-65, and **Bright**, 1811-89, became the focus of extra-Parliamentary agitation.

Robert Peel, 1788-1850, gave the Bank of England a monopoly on printing money, 1844, and established its authority over other banks.

The Corn Laws were repealed, 1846.
Competition from food imports led to **agricultural improvements**.

The Public Health Act, 1848, attempted to introduce regular sewerage and clean water. The first attacks on slums began.

Charles Dickens

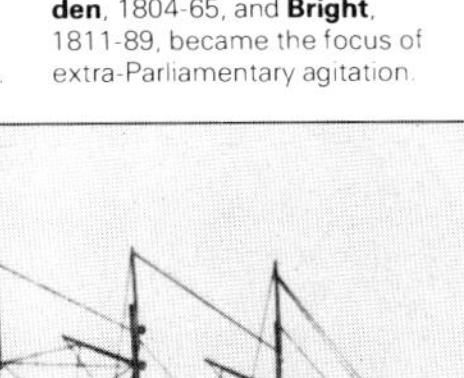
Brunel's *Great Britain*

Giuseppe Garibaldi

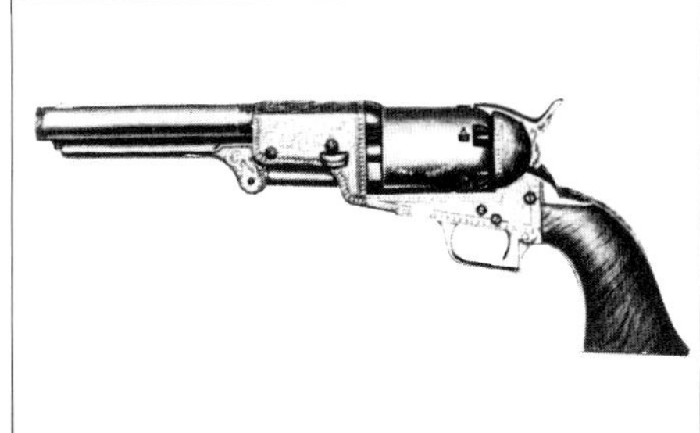
Colt Dragoon revolver, 1848

Probability theory and statistical methods applied to social phenomena by the Belgian **Adolphe Quetelet**, 1796-1874, led to the discovery that the frequency of suicide in a society was constant and therefore predictable.
Ralph Emerson, 1803-82, an American Unitarian, developed in *Nature*, 1836, a belief called **Transcendentalism**, proposing that spiritual exploration of one's soul in communion with nature led to the highest wisdom.

The terms "sociology" and **"positivism"** were coined by the Frenchman **Auguste Comte**, 1798-1857. His *Course of Positive Philosophy*, 1830-42, argued for the use of positivist methods in social studies. He stated that man's thought had passed through the theological and metaphysical stages and reached the positivist stage. He divided sociology into statics, the study of the interdependence of social institutions, and dynamics, the study of change.

Ludwig Feuerbach, 1804-72, in *The Essence of Christianity*, 1841, advocated a humanistic atheism. His critique of Hegel and his claim that God is nothing other than a projection of human nature influenced Marx.
Sören Kierkegaard, 1813-55, a Danish religious philosopher, published *Either/Or*, 1843. Regarded as the founder of existentialism, he argued against rationalism and emphasized the need to make choices between ethical or aesthetic alternatives.

The Babi movement, which developed into the **Bahai** religion, was founded in Persia, 1844, by **Ali Muhammed**, 1819-50. It drew inspiration from diverse sources and emphasized the unity of mankind.
Alexis de Tocqueville, 1805-59, the first sociologist to examine the impact of democracy on non-political institutions, held that democratic emphasis on equality might suppress individuality and lead to total conformity – a tyranny of the majority.

Pierre Proudhon, 1809-65, a French philosopher, held that "property is theft". His book *The Philosophy of Misery*, 1846, formulated anarchism as a political theory. Together with his fellow Frenchmen **Louis Blanc**, 1811-82, and **Louis Blanqui**, 1805-81, he called for an end to the capitalistic exploitation of labour and for a revolutionary change in the existing social and economic orders. These men participated in the revolt of 1848.

The Christadelphians, a pacifistic millenial adventist sect, was founded in 1848 in the US by **John Thomas**, 1805-71.
Karl Marx, 1818-83, and **Friedrich Engels**, 1820-95, wrote the *Communist Manifesto*, 1848, predicting the revolutionary overthrow of capitalism and its replacement by socialism. Marx saw history as a class struggle for ownership of society's material resources.

Heinrich Heine, 1797-1856, a German poet and a member of the **Jung Deutschland** group which described romantic idealism and wanted all literature to have a political role, wrote *Die Romantische Schule* in Paris, 1836.
Giacomo Leopardi, 1798-1837, used the romantic theme of man's helplessness in the face of nature, in *La Ginestra*, 1836.

The prolific French novelist Honoré de Balzac, 1799-1850, wrote over 90 interconnecting novels and stories, 1829-43, set in Paris and the provinces in the 1820s, which he called *La Comédie Humaine*.

Mikhail Lermontov, 1814-41, stands at the beginning of Russia's golden age of prose. Pushkin's successor as poet, playwright and novelist, he was directly influenced by Western literature, especially the work and the flamboyant romanticism of Lord Byron. His novel, *A Hero of Our Times*, on the tragic theme of a man without a purpose, appeared in 1840.

The fantastical and tormented aspects of Russian fiction stem from **Nikolai Gogol**, 1809-52. His novel *Dead Souls*, 1842, a moral satire on bureaucracy, centred on the career of a man who deals in dead serfs.
Adam Mickiewicz, 1798-1855, Poland's national poet, used themes of Lithuanian folklore in his vibrant epic works.

Powerful imagination and insight marked the works of the English novelists the **Brontë sisters. Emily**, 1818-48, wrote *Wuthering Heights*, 1847, an account of ferocious passions set against a raw elemental background. *Jane Eyre* by **Charlotte**, 1816-55, was published in the same year and *The Tenant of Wildfell Hall* by **Ann**, 1820-49 in 1848.

The English social novelist Charles Dickens, 1812-70, published one of his greatest novels, *Dombey and Son*, 1846-8, dealing with the Victorian family.
W. M. Thackeray, 1811-63, wrote his caustic attack on the false face of polite society, *Vanity Fair*, 1847-8, a satirical novel which is set on the eve of the Napoleonic wars.

American primitive painting continued in the work of the Quaker painter **Edward Hicks**, 1780-1849, whose naive landscape with animals, "The Peaceable Kingdom", uses simple forms, flat colours and two dimensions.
French romantic sculpture stressed dramatic movement and fleeting gestures, poses and expressions and was epitomized by the "Marseillaise" on the Arc de Triomphe, 1833-6, by **Francois Rude**, 1784-1855.

The Hudson River School of American landscape painters produced picturesque and romantic views of the eastern states, exemplified by **Thomas Cole**, 1801-48 and **Asher Durand**, 1796-1886.
An increased range of oil colours was introduced, including mauves, violets, bright greens and intense yellows.

The Gothic Revival in Europe which had begun *c.* 1750 and was based on serious investigation of medieval sources, prompted the decision to build the **Houses of Parliament**, London, 1836-68, in a Gothic style, and the similar exploration of national Gothic styles in the rest of Europe.
Political cartoons reached a high level of sophistication in England in the work of **George Cruikshank**, 1792-1878, and in France with **Honoré Daumier**.

Joseph Turner, 1775-1851, in England was one of the first painters to celebrate contemporary technology. His "Rain, Steam and Speed", 1844, proclaims the power of the machine in a diffuse, misty style suggesting the speed of the train.
The Barbizon School of French landscapists, including **Jean François Millet**, 1814-75, settled near Fontainebleau.
John Ruskin, 1819-1900, emphasized truth to nature in "Modern Painters".

Photography (daguerreotypes) and the style of early photographic portraits, where the subject had to maintain a still pose during the long exposure period, influenced such painters as **Ingres**, particularly in his portrait of "La Comtesse d'Haussonville", 1845.
Historical architecture spread in the US, as in Europe. The picturesque, Neo-Norman Smithsonian Institute in Washington, 1846, was built by **James Renwick, Jr**, 1818-85.

The Pre-Raphaelite Brotherhood was founded in England, 1848. Its painters sought a return to the purity and moral seriousness of 15th-century Italian art and rejected contemporary academic styles. Their work was characterized by great detail and the use of brilliant colours on a white ground, as in "The Awakening Conscience", 1852, by **William Holman Hunt**, 1827-1910, and "Christ in the House of his Parents", 1849, by **John Everett Millais**, 1829-96.

Robert Schumann, 1810-56, was an arch-Romantic, especially in his evocative piano music like *Carnaval*, 1834-5. He founded the *avant-garde* publication *Neue Zeitschrift für Musik* in 1834.

Mikhail Glinka, 1804-57, in such works as *A Life for the Tsar*, 1836, heralded the rise of a Russian national school, using folk music and inspired by the Napoleonic Wars.

Vienna, in Ferdinand I's reign, 1835-48, was the centre of popular dance music, exporting the Viennese waltz and works of the older **Johann Strauss**, 1804-49, to the rest of Europe.

Brass bands, especially in Germany, produced a popular music of their own. The cornet provided a high voice, and became a virtuoso solo instrument.

Franz Liszt, 1811-86, a great piano virtuoso who toured the capitals of Europe, supported the work of young musicians at the court of Weimar, 1848-61.

The symphonic poem or tone poem was established by Liszt while at Weimar, the first appearing in 1848-9. The form expressed a poem, story or idea in terms of orchestral music.

The electric telegraph, was patented in Britain by **Charles Wheatstone**, 1802-75, in 1837.
The Colt pistol, the first repeating firearm, was patented, 1835-6.
Robert Brown, 1773-1858, first discovered and named the nucleus of a living cell.

The Morse code was invented, 1837, by **Samuel Morse**.
Stellar parallax was detected in 1838 by **Friedrich Bessel**, 1784-1846, showing that the stars lie at immense distances from the Earth.
Vulcanization, was discovered in 1839 by the American inventor **Charles Goodyear**, 1800-60.
Theodor Schwann, 1810-82, and **Matthias Schleiden**, 1804-81, both German biologists, in 1839 first described the anatomy of animal and plant cells and the parts they play.

Jean-Louis Agassiz, 1807-73, a Swiss naturalist, studied glaciers and showed, 1840, that an Ice Age had once occurred, producing glacial action that had helped to shape the land masses.
Anaesthesia had its beginnings in 1842, when American surgeon **Crawford Long**, 1815-78, operated on an etherized patient.
The *Great Britain*, the first iron-hulled steamer powered by a screw propeller, crossed the Atlantic in 1843.
Artificial fertilizer was first prepared *c.* 1840.

Nitroglycerine was discovered 1846 by the Italian chemist **Ascanio Sobrero**.
Neptune was discovered, 1846, by the German astronomer, **Johann Galle**, 1812-1910.

The mechanical equivalent of heat was first measured and the principle of the conservation of energy put forward by **Julius von Mayer**, 1814-78, a German physicist, in 1842. The British physicist **James Joule**, 1818-89, and the German **Hermann Helmholtz**, 1821-94, established both more thoroughly *c.* 1847.
The rotary printing press was invented in 1847 by the American **Richard Hoe**, 1812-86.

Asepsis was demanded, 1847, by **Ignaz Semmelweiss**, 1818-65, a Hungarian physician. He showed that childbed fever could be prevented if hospital doctors cleaned their hands regularly.
The St. Lawrence Seaway was first opened in 1848.
Reinforced concrete was invented, 1849, by a French engineer, **Joseph Monier**, 1823-1906.
A telegraphic cable connection across the English Channel was laid in 1850.

1850-1875 Darwin and Marx

Principal events

The development of ruthlessly pragmatic political planning epitomized by the ministry of Bismarck in Prussia brought about the national unification of Italy and Germany, where the idealism of the 1848 revolutions had failed. Industrial expansion went hand in hand with cynical foreign policies which, with the death of liberal ideals, contributed to the growth of international tensions.

British imperial power was at its peak after the defeat of the Indian mutiny. Britain's economic supremacy, backed up by military strength, made her unchallengeable throughout the world.

The victory of the North in the American Civil War ended slavery and prepared the way for American industrial and political expansion, while the European powers extended their domination in South-East Asia and Japan set out to transform herself into a modern industrial society.

1850-3

Napoleon III, *r.* 1852-70, restored the French Empire after the upheavals of 1848.
The rights of national minorities were suppressed in the Austro-Hungarian Empire.
California became a free state in 1850 and a compromise was reached on slavery in the United States in 1850.
The vastly destructive Taiping Rebellion broke out in China in 1850.
The English in India subdued Burma, 1852-3.

1853-5

Russia's defeat in the Crimean War, 1854-6, checked her ambitions in eastern Europe.
The New York-Chicago rail link was completed, 1853.
The discovery by David Livingstone, 1813-73, of the Victoria Falls, 1855, sparked off European exploration of the African interior.
Europe experienced an economic boom, 1852-6.
Camillo Cavour, 1810-61, began the industrialization of Piedmont in northern Italy.

1855-8

After the Crimean War, the Black Sea was declared neutral and the virtual independence of Serbia recognized by the major European powers, 1856.
The Indian Mutiny, a series of mutinies by Sepoy troops and scattered popular uprisings against British rule, was ruthlessly suppressed, 1857-8.
John Brown, 1800-59, militantly pursued the anti-slavery cause in the southern USA.
China was finally opened up by the treaties of Tientsin, 1858.

1858-60

By the Government of India Act, 1858, British rule in India was transferred to the Crown. Administrative reforms and the building of railways with British investment followed.
Piedmont drove the Austrians from northern Italy with French assistance, 1859.
British troops raided Peking, 1860.
The Suez Canal construction was begun by de Lesseps, 1805-94.
Russia expanded at China's expense, 1858-60.

National events

The rising prosperity of the mid-Victorian system, based on a rapid expansion of population, industry and trade, resulted in higher wages and the birth of a strong labour movement. Electoral reform and governmental interest in social and economic planning changed the face of British politics.

The Great Exhibition, 1851, symbolized growing prosperity.

Lord Palmerston, 1784-1865, became prime minister in 1855.
The Northcote-Trevelyan Report, 1854, introduced Civil Service reform.

Palmerston's first ministry, 1855-8, reflected Britain's confidence in international politics and stability at home.

William Gladstone, 1809-98, signed a commercial treaty with France, 1860, hoping that her trade would lead to international harmony.

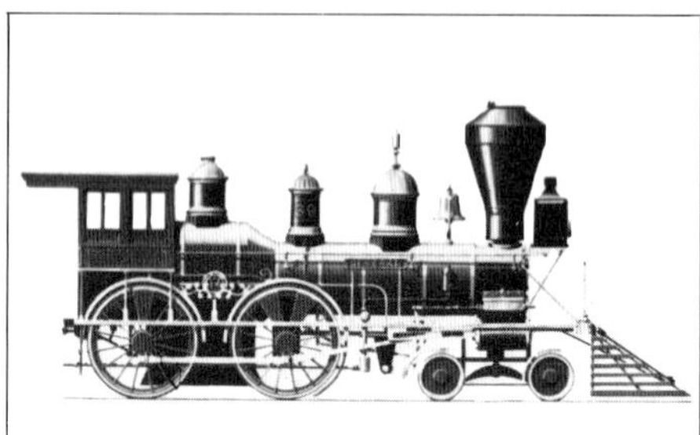

Western and Atlantic Railroad: the *General*, 1885

Bessemer steel producing process

Indian Mutiny: massacre at Delhi

Religion and philosophy

The spread of industrialization provoked a major reassessment of moral, social and political thought. Many Christians campaigned to relieve the worst aspects of urban poverty and sociology developed tools to describe the changes in social relationships, while many political and moral philosophies grew up which rejected urbanization and capitalism, laying stress on personal withdrawal or social revolution. The latter was advocated in particular by Karl Marx whose work provided a radical attack on capitalist economics and stated that the victory of the proletariat over the bourgeoisie was an historical inevitability.

Darwin's theory of evolution proved as influential as the ideas of Marx, challenging many of the basic tenets of Christian belief and forcing the importance of scientific thought to the fore. Many attempts were made to apply his ideas to the political and cultural fields.

The Taiping Rebellion in China, 1850, under the leadership of **Hung Hsiu-Chuan**, 1814-64, was a radical religious movement influenced by Protestant Christian teaching which demanded the communalization of property, equal redistribution of land and equality between men and women.
T. B. Macaulay, 1800-59, an English historian, published a *History of England* between 1848 and 1861, stressing the victory of liberalism since 1688.

Frederick Denison Maurice, 1805-72, the major theologian of 19th-century Anglicanism, published his *Theological Essays*, 1853. He helped to found the **Christian Socialist movement** in England, which tried to apply Christian ideals in an industrial society, and joined the co-operative movement.
Henry Thoreau, 1817-62, an American, described his attempt to practise **Transcendentalism** by living a simple life in the midst of nature, in *Walden*, 1854.

Frederic Le Play, 1806-82, a French sociologist, developed a technique of collating field data that influenced methods of statistical sampling in *European Workers*, 1855. He defined three basic family types, related to general social conditions.
Hippolyte Taine, 1828-93, a French historian, argued in *The French Philosophers* for the use of a scientific (positivist) method in the study of culture, 1857. He had a great influence on later 19th-century thought.

Charles Darwin, 1809-82, published *Origin of Species*, 1859, arousing opposition from the English Church because he contradicted Genesis and seemed to render the role of God in the creation superfluous. By implying that man stood at the pinnacle of evolutionary development, Darwin's theory reinforced contemporary ideas of the inevitability of social progress. After defeat in the debate, the Church ceased to intervene in science.

Literature

As the realist novel produced the powerful and candid tragedy of *Madame Bovary* by Flaubert, new literary styles were also emerging in French poetry. Baudelaire's attempt to explore his inner self would lead to the symbolist movement to which reality beyond the poet's own imagination was irrelevant.

In Russia, Dostoevsky and Tolstoy were writing and the moral, psychological and political issues they explored recurred in the literature of the English Victorian novelists from Charles Dickens to George Eliot.

American literature reached maturity with the poetry of Whitman – a distinct contrast with contemporary European styles – and the strong prose of Melville's epic novel *Moby Dick*.

The Victorian Alfred Lord Tennyson, 1809-92, showed in his elegiac poem *In Memoriam*, 1850, verbal grace and awareness of the moral dilemmas of the age.
The American novelist, Nathaniel Hawthorne, 1806-64, wrote *The Scarlet Letter*, 1850, and **Herman Melville** produced his allegorical *Moby Dick*, 1851, with an American delight in vigorous prose.

The fantastic and dream-inspired poems of *Les Chimeres*, 1854, by **Gérard de Nerval**, 1808-55, would influence later symbolists.
The *Poems* of **Matthew Arnold**, 1822-88 were an intellectual expression of Victorian unease, as was the poetry of **Robert Browning**, 1812-89, whose dramatic monologues are full of human insight.

Walt Whitman, 1819-92, published his autobiographical *Leaves of Grass* in 1855.
Modern poetry began with *Les Fleurs du Mal*, 1857, by the Frenchman **Charles Baudelaire**, 1821-67.
Theophile Gautier, 1811-72, proclaimed the concept of art for art's sake, 1857.
Gustave Flaubert, 1821-80, published *Madame Bovary*, 1856.

Adalbert Stifter, 1805-68, wrote of the harmony of man and nature in *Der Nachsommer*.
Ivan Goncharov, 1812-91 wrote *Oblomov*, 1859, mocking the bankruptcy of intellectualism.

Art and architecture

The reaction against academic precepts, which ruled the bulk of official painting in Europe, began in earnest and took the form of the assertion that the subject matter of everyday life was worthy of art.

The Impressionists broke new ground with their revolutionary techniques for representing light and colour, best seen in the works of Monet; while Realist painters like Courbet stated that art must have a social and political purpose. In the same period English art saw a distinct reaction against the aesthetics and values of industrial society with the work of the Pre-Raphaelites and the Arts and Crafts movement.

Town planning became a priority in the European capitals and the use of cast iron revolutionized municipal architecture.

Trading contact with the East and the Meiji restoration in Japan brought an interpenetration of Eastern and Western art.

Realism reached its height in the works of **Gustave Courbet**, 1819-77, such as "The Stonebreakers", 1850.
Jean Francois Millet, 1814-75, created an ennobling picture of peasants in his canvas "The Sower", 1850.
The Pre-Raphaelite Brotherhood, were defended by **John Ruskin**, 1851.
Prefabricated units of iron and glass were used by **Joseph Paxton**, 1803-65, in the Crystal Palace, London, 1851.

Honoré Daumier, 1808-79, painted children in "La Ronde", 1855.

Baron Haussmann, 1809-91, introduced squares, parks and boulevards into Paris, 1851-68. His wide streets were designed to create impressive views and to ensure easy policing of the city centre.
The New Building of the Louvre, 1852-7, and the Paris Opera, 1861-74, illustrate the powerful Neo-Renaissance and Neo-Baroque forms which were key to the civic architecture of the Second Empire.

Landscapes by **Jean Baptiste Camille Corot**, 1796-1875, such as "The Valley", 1855-60, anticipated the Impressionists with their use of muted colours and soft outlines.
The Arts and Crafts movement in England led by **William Morris**, 1834-96, and **Philip Webb**, 1831-1915, originated in a dissatisfaction with manufactured goods and a respect for the medieval craftsman. It produced wallpapers, furniture, tapestries and carpets usually by hand.

Music

Romantic ideals were reinforced in 1858 by Darwin's theory of evolution through the survival of the fittest, and confirmed the common view that man verges on perfection. The odd conclusion that art, like life, evolves from lower forms to higher led to an increasing distinction between serious and popular music, a distinction that remains prevalent in the West even today.

Musical forms still developed themes, much as the classics had done, but the work of composers such as **Liszt** gave music a mystical aura and set it apart from normal experience.

Late Romantic composers, such as **Richard Wagner**, 1813-83, and **Anton Bruckner**, 1824-96, aimed for grandeur by using large forces and a rubato (freer) beat to savour the sound.

Use of rubato tended to slow down performances and to produce rhythmic problems in large ensembles so that performances became dull and turgid.

Grand opera carried the Romantic ideal to a peak in the works of **Giuseppe Verdi**, 1813-1901, in such works as *Un Ballo in maschera*, 1859, evoked specific characters and moods.

Science and technology

Science and technology became more closely tied to the needs of industry in this period, especially in Germany where chemists produced dyes and explosives and in the United States where engineers enjoyed a high social status. The abolition of slavery in America and the rise of trade unionism in Europe both raised labour costs and so stimulated mechanization, while the Crimean War provided an incentive for the development of new and better kinds of steel.

While Darwin's theory of evolution, backed by Mendel's researches into genetics, was the most popular scientific breakthrough, chemistry and astronomy both advanced dramatically with the development of spectroscopy in Germany and the application of the Doppler Effect. The former permitted many new elements to be discovered, while the latter, through the measurement of red-shift, produced more accurate estimates of the size of the known universe.

Entropy, following from the second law of thermodynamics, was conceived, 1850, by the German **Rudolf Clausius**, 1822-88.
The sewing machine was developed, 1851, by **Isaac Singer**, 1811-75, an American.
Physiology advanced with the work, *c.* 1851, of the French scientist **Claude Bernard**, 1813-78.
The rotation of the Earth was demonstrated conclusively in 1851 by the French physicist **Jean Foucault**, 1819-68.

Symbolic logic was founded by **George Boole**, 1815-64, a British mathematician, with his *Laws of Thought*, 1854.
Agriculture gained a scientific basis with the publication in the 1840s and 1850s of work by the German **Justus von Liebig**, 1803-73, showing how plants use vital elements in cycles.

The first synthetic plastic material, later named celluloid, was patented in 1855 by the British chemist Alexander Parkes, 1813-90.
Mauve, the first artificial dye, was derived from aniline, 1856, by the British chemist **William Perkin**, 1838-1907. Production of the new synthetic substances stimulated the growth of modern organic chemistry.
Steel was produced cheaply in the converter patented in 1856 by the British metallurgist **Henry Bessemer**, 1813-98.

The principles of molecular structure, were discovered by **Friedrich Kekulé von Stradonitz**, 1829-96, in 1858.
The theory of evolution was put forward in 1858-9 by the British naturalists **Charles Darwin**, 1809-82, and **Alfred Wallace**, 1823-1913.
The first oil well was struck near Titusville, Pennsylvania, in 1858.
Atomic weights and chemical formulae were standardized by the Italian **Stanislao Cannizzaro**, 1826-1910, in 1858.

1860-3

Abraham Lincoln, 1809-65, was elected president of the US on an anti-slavery platform.
The Confederate states took Fort Sumter, 1861, and this sparked off the Civil War.
Giuseppe Garibaldi, 1807-82, liberated southern Italy from Neopolitan rule and gave it to Piedmont, uniting Italy, 1861.
Alexander II of Russia, *r.* 1855-81, emancipated the serfs in 1861.
French troops occupied Mexico City, 1863, and set up a puppet emperor, Archduke Maximilian.

1863-5

An allied Western expedition participated in a violent civil war in Japan, 1864.
Rome remained under papal rule as a virtual protectorate of France.
The northern American states defeated the South, 1865.
Slavery was abolished and Lincoln assassinated, 1865.
Karl Marx, 1818-83, presided over the First International in London 1864.
Christianity was spread in China by missionaries.

1865-8

After the defeat of Austria by Prussia in the war of 1866, the North German Confederation was set up.
American objections and opposition at home led to French withdrawal from Mexico, 1866.
The Dominion of Canada was established by the British North America Act, 1867.
The Dual Monarchy of Austria-Hungary was established, 1867 – an unpopular compromise on the nationalist question.

1868-70

Napoleon III instituted liberal reforms to quell the growing opposition, but his empire collapsed in the face of a Prussian invasion, 1870.
By 1870 the railway systems of France, England and Belgium were virtually complete, stimulating heavy industry.
The victory of Mutsuhito, *r.* 1868-1912, in Japan led to a policy of industrialization and an end to the shōguns.
Negro suffrage was enforced in the United States, 1870.

1870-3

A revolutionary commune set up in Paris, 1870-1, was ruthlessly suppressed. It rejected the authority of the French government after the surrender to Prussia.
Prussia's seizure of Alsace-Lorraine completed German unification and a German Empire was proclaimed under **Wilhelm I**, *r.* 1861-88.
An American attempt to open up Korea failed, 1871-3, but Britain forced the Treaty of Pangkor on Malaya.

1874-5

The conservative Third Republic was set up in France after the defeat of the Commune, in spite of repeated attempts to restore the monarchy.
Britain bought a decisive share in the **Suez Canal**, 1875, thus acquiring a quick route to India.
French power in Indochina, extended by Napoleon III, was confirmed, 1874.
The revelation of corruption in the Grant administration in the US resulted in financial panic and economic depression.

The Companies Act, 1862, introduced limited liability companies.
The first underground railway was built in London, 1860-3.

The demand for electoral reform among liberals and radicals was blocked by Palmerston's conservatism until his death in 1865.

Benjamin Disraeli, 1804-81, introduced household suffrage, 1867.
The Trades Union Congress was founded in 1868.

Gladstone became committed to disestablishing the Anglican Church in Ireland, where a nationalist movement was growing in strength, 1868-74.

The Education Act, 1870, made primary education open to all.
Depression after a financial crisis ended the peak of British economic supremacy.

The Public Health Act codified sanitary law and the Artisans' Dwelling Act tried to improve slum conditions, 1875.

Charles Darwin lampooned

Confederates in the American Civil War

Karl Marx

Wilhelm I of Prussia acclaimed German emperor

Ferdinand Lassalle, 1825-64, urged German workers to seek universal suffrage, 1862. His programme formed the basis of the German Social Democratic Party established in 1869 in opposition to the revolutionary Marxist International, and foreshadowed later social democratic and parliamentary movements.
Ernest Renan, 1823-92, wrote *The Life of Jesus*, 1863, in a positivist framework, stressing historical detail.

John Stuart Mill, 1806-73, an Englishman, attempted to create a political theory uniting the conditions of industrialism with basic tenets of human freedom. He argued in *Utilitarianism*, 1863, that what gives pleasure to man is good, while in *On Liberty*, 1859, he insisted that a man must be free to act as he wishes without disturbing the freedom of others. His political critique of English society brought him close to a socialist position.

Karl Marx, 1818-83, published Volume I of *Capital*, 1867, providing a theoretical analysis of the workings of capitalism. He argued that industrial profits were made by exploiting the workers, but claimed that capitalism would plunge into chaos through its inner contradictions.
William Booth, 1829-1912, shocked at the extent of poverty and degradation in London, began his Evangelical ministry in 1865 and later founded the **Salvation Army.**

Papal infallibility was asserted by the first Vatican Council, 1869-70, which ruled that the pope or an ecumenical council of bishops was immune from error when pronouncing on matters of faith or morals. The pope's increased prestige resulting from this ruling partly compensated for the loss of the Papal States to Italy.
The classic statement of the case for women's suffrage was presented by **J. S. Mill** in *The Subjection of Women*, 1869.

Bakunin, 1814-76, a Russian anarchist, stressed the need for violent revolution to overthrow the state and allow the essential goodness of man to develop. His rejection of centralization and subordination to authority in favour of the free spirit of revolt led to his expulsion from the Internationale, 1872, after conflicting with Marx.
Johann von Döllinger, the German theologian, 1799-1890, was excommunicated, 1871, for opposing the idea of infallibility.

An upsurge of religious revivalism associated with the Temperance Movement, which saw alcohol as the cause of working-class degradation, was started in America and Britain by **D. L. Moody**, 1837-99.
Wilhelm Wundt, 1832-1920, a German who published *Principles of Physiological Psychology*, 1873-4, established experimental psychology. He sought to investigate by introspection the immediate experiences of consciousness.

An impressionistic realist style based on detailed social observation was developed in France by the brothers **Edmond and Jules de Goncourt**, 1822-96 and 1830-70, in their *Journals*.
The realist novel in Russia produced a varied account of the conflicts in Russian society.

The debate between Slavophiles and Westernizers was described in *Fathers and Sons* by **Ivan Turgenev**, 1818-83, an ardent Westernizer.
Count Leo Tolstoy, 1828-1910, the greatest exponent of the Russian realist novel, dealt in his epic *War and Peace*, 1865-72, with the Napoleonic Wars, combining a panoramic vision with acute analysis of character.

The Parnassians, a group of French poets including **Charles Marie Leconte de Lisle**, 1818-94, and later **Paul Verlaine**, 1844-96, rejected the emotionalism and loose forms of the Romantics and wrote strictly disciplined, detached verse following **Theophile Gautier**. Their name derived from the journal *Le Parnasse Contemporain*, first published in 1866.

The great novels of Fyodor Dostoevsky, 1821-81, including *Crime and Punishment* 1866, *The Brothers Karamazov* 1879-80, and *The Idiot* 1868, study totalitarianism, the conflict between atheism and compassionate Christianity, and good and evil in man.
The English novelist Charles Dickens continued his prodigious output of social novels.

George Eliot, the pseudonym of Mary Anne Evans, 1819-80, described the conflicts within English provincial society in *Middlemarch*, 1871-2.
The early French symbolist poets, **Arthur Rimbaud**, 1854-91, and **Paul Verlaine**, 1844-96, aimed to devise a truly poetic visionary language; Rimbaud wrote *Une Saison en Enfer* in 1873.

Jules Laforgue, 1860-87, a symbolist, was one of the first French poets to use free verse.

The Salon de Refusés which exhibited some of the 4000 canvases rejected by the official Salon, was established in Paris in 1863. It included works by **Pissarro**, **Cezanne**, **Whistler**, and **Edouard Manet**, 1832-1883, whose "Déjeuner sur L'Herbe", of a naked woman enjoying a picnic with friends, created a sensation.
Gustave Doré, 1832-83, established his reputation with illustrations for Dante's *Inferno*, 1861.

Eugène-Emmanuel Viollet-le-Duc, 1814-79, published his *Dictionary of French Architecture from the 11th to the 16th Centuries*, 1858-75.
Japanese draughtsmanship, flat expanses of colour and subject matter, strongly influenced the Impressionists and later Van Gogh and Gauguin. Their impact on Parisian artists can be seen in **Manet's** "Portrait of Emile Zola", 1868, and in **Whistler's** "Princess of the Land of Porcelain", 1864.

Japanese prints were being collected in Paris in the 1860s and were exhibited at the **Exposition Universelle**, 1867.
The effects of light out of doors on the surface of an object or figure was first captured and faithfully recorded by **Claude Monet**, 1840-1926, in his painting "Women in the Garden", 1866.

Mural painting in France was revived by **Pierre Puvis de Chavannes**, 1824-98, whose monumental style and subdued colours seen in "Ludus Pro Patria", 1865-9, inspired the Symbolists of the 1880s.

Monet's painting "Impression, Sunrise", 1872, exhibited in 1874 with works by **Pierre Auguste Renoir**, 1841-1919, **Alfred Sisley**, 1839-99, **Edgar Degas**, 1834-1917, **Pissarro** and **Cézanne**, gave its name to the Impressionist movement with which these painters were associated. **Impressionism** abandoned traditional linear representation, aiming to capture the fleeting effects of light and colour by using small dashes and strokes of colour.

The later Pre-Raphaelite style in England, best represented by the work of **Edward Burne-Jones**, 1833-98, like the "Briar Rose" series, 1871-90, later influenced the Symbolist movement.
London's first garden suburb, Bedford Park at Turnham Green, was designed by the Victorian architect and associate of William Morris, **Richard Norman Shaw**, 1831-1912, in 1875.

The opera *Faust*, by Charles Francois Gounod, 1818–1893, was produced in London, Dublin and New York in 1863.

Negro spirituals emerged in the US as blacks took up the singing school tradition of colonial America, but mixed with it a rhythmic work-song style.

Light opera centred on Paris and Vienna, with the theatrical humour of **Jacques Offenbach**, 1819-80, and the lavish settings of the younger **Johann Strauss**, 1825-99.

Wagner revolutionized opera, using a continuously moving harmonic structure over which leitmotive (short themes) identified dramatic elements. He believed opera should combine all the arts.

César Franck, 1822-90, a Belgian working in the classical tradition, attracted little attention in his lifetime but influenced an important group of younger French composers.

Johannes Brahms, 1833-97, carried on Beethoven's tradition especially in the symphony. Brahms's classical control of the emotional impulses in his music gives it a rich dramatic quality.

The open hearth process for the production of steel was developed in France, following the invention of the regenerative furnace by **William Siemens**, 1823-83, and **Frederick Siemens**, 1826-1904.
A submarine telegraphic cable was laid across the Atlantic from Ireland to Newfoundland, 1858-66.
Colloids were distinguished in 1861 by the British chemist **Thomas Graham**, 1805-69. He also discovered **osmosis**.

The first underground railway opened in London in 1863.
The Massachusetts Institute of Technology was founded, 1865.

Dynamite was invented in 1866 by the Swedish inventor **Alfred Nobel**, 1833-96, who founded the Nobel Prize.
Antiseptic surgery was introduced by the British physician Lord Lister, 1827-1912, by 1867. Following Pasteur's research into the nature of disease, he used carbolic acid to disinfect the operating theatre.
Genetics was founded with the publication, 1865, of the experiments of the Austrian botanist **Gregor Mendel**, 1822-84.

Bacteriology was founded with the work of the French chemist **Louis Pasteur**, 1822-95, in the 1860s. Pasteur discovered that micro-organisms cause fermentation and disease, and used sterilization to kill bacteria.
Light was shown to be an electromagnetic radiation by the British physicist **James Clerk-Maxwell**, 1831-79. He predicted other such radiations.

The Periodic table of the elements was devised by the Russian chemist **Dmitry Mendelyev**, 1834-1907, in 1869.
The typewriter was developed commercially after the 1867 invention of American **Christopher Sholes**, 1819-90.
The Challenger expedition of 1872-6 founded oceanography.
Intermolecular forces were calculated by **Johannes van der Waals**, 1837-1923, in 1873. He accurately described the behaviour of real gases, using mathematical equations.

1875-1900 The age of imperialism

Principal events

Domination of the world outside the Americas lay with a few European states. Among them Britain was still the greatest imperial and industrial power but Germany now increasingly challenged this position. The US also grew in strength and by 1900 overtook Britain in the production of basic industrial materials.

The emergence of a group of fixed alliances in Europe served to polarize foreign affairs and the Balkans, in particular, presented an inflammatory arena for international conflict.

Improvements in communications, however, and the quest for new bases of economic and political power shifted the focus of rivalries between the states to Africa and Oceania. Britain greatly extended her empire but the other European states, the US and a newly modernized Japan also joined in the scramble. By 1899 all Asia was in the hands of Europe and China was in thrall to the West.

1875-8

Britain bought the Khedive's Suez Canal shares, 1875, and annexed South Africa, 1877. **The Slav nationalist forces** in the Balkans erupted against Turkey and the Bulgarian massacres, 1875, aroused a public outcry in Britain. Russia supported the insurgents, hoping to win new authority in the Balkans. **The Satsuma rebellion**, 1877, led by conservative forces in Japan, failed to halt the tide of reform and new ideas.

1878-80

After Russia's defeat of Turkey, 1878, Britain and Austria-Hungary intervened to check Russian ambitions and the Powers met at the Congress of Berlin, 1878, to decide the future of the Balkans. **Germany and Austria-Hungary** formed a Dual Alliance, 1879. **In Afghanistan** Britain sought to secure her position in India against Russian expansion, 1878-80. **Chile** began her successful war against Bolivia and Peru, 1879.

1880-3

British imperial expansion in Africa was checked by defeat in the Transvaal, 1881, but British occupation forces were installed in Egypt in 1882. **Under Bismarck**, 1815-98, Germany aimed to build a solid European power structure, signing the **Three Emperors' Alliance** with Russia and Austria, 1881, and a similar alliance with Italy, 1882. At home, Bismarck introduced sickness benefits to help weaken the growing appeal of socialism, 1883.

1883-5

Britain consolidated her position on the Afghan border and in Egypt but was defeated by native forces in the Sudan, 1885. **France** took Indochina, 1884. **The Treaty of Berlin**, 1884, defined the rights of 14 European powers in Africa. This helped stop the scramble for colonies which could have led to a major war. **Eastern Rumelia's union** with Bulgaria, 1885, provoked war with Serbia and Austria acted to save Serbia from invasion.

National events

Britain's economic supremacy was challenged by other powers. The extended franchise gave a more popular ring to politics and saw the foundation of an Independent Labour Party, 1893. The traditional parties changed the emphasis of their policies as imperialism and Ireland became the major political issues.

Legislation in 1875 legalized peaceful picketing and freed trade disputes from the law of conspiracy. **Victoria** took the title Empress of India in 1876.

Charles Parnell, 1846-91, mobilized the Irish nationalists and aimed to reform the Irish Land Law.

The Irish problem was accentuated by terrorism in Ireland. **Textile output** in England began to decline after 1880.

Virtual universal male suffrage, led to an intensive party activity at the constituency level.

Maxim machine gun

Paris Exhibition 1889: the Machine Hall

Queen Victoria

Religion and philosophy

Growing interest in the attempt to link social theory to biological evolutionism gave rise to more subtle sociological and anthropological studies in the English-speaking world. Drawing on the experience of colonial administration, men such as Tylor, Spencer and Frazer developed the notion of a natural progression between "primitive" and "advanced" societies. Meanwhile in Vienna, Freud began to formulate profoundly influential ideas on the subconscious and the nature of man.

In philosophy the absolute idealism of Hegel found its first supporters in England with Bradley, while in the United States pragmatic thinkers such as William James argued that the truth of an idea depends on its social function.

The ideology of anti-semitism grew up in the wake of heightened nationalist sentiment, while an evolutionary type of socialism grew more popular than its revolutionary counterpart.

Hinduism witnessed the rise of various reform movements in the 19th century under the impact of Western thought. Most important was that led by **Ramakhrishna**, 1836-86, an extreme ascetic in the Vedanta tradition, who believed that all religions were essentially identical. **The Theosophical Society**, which set out to foster the transmission of Eastern thought to the West, was founded in New York in 1875 by **Helen Blavatsky**, 1831-91.

The Jehovah's Witnesses, an evangelical movement believing in the second coming of Christ, were founded by **Charles Russell**, 1852-1916, in the United States. **Christian Science** was founded in Boston by **Mary Baker Eddy**, 1821-1910, who rejected medicine and saw prayer as the only cure for illness. **Heinrich von Treitschke**, 1834-96, fostered German nationalism in his *History of Germany in the 19th Century*, 1879-94.

A theory of social evolution was developed by the Englishman **Edward Tylor**, 1832-1917, in his *Anthropology*, 1881. Through studying primitive religion, he concluded that many existing social customs were "survivals" from earlier stages of development. Similar ideas were defended in America by **Lewis Morgan**, 1818-81, who developed the study of kinship systems.

In Russia, **Peter Kropotkin**, 1842-1921, was the leading theorist of the Anarchist movement. In *Words of a Rebel*, 1884, he emphasized non-violence and argued that co-operation rather than conflict was the basis of evolutionary progress. **The Fabian Society**, founded in Britain, 1884, advocated a gradual evolution towards socialism. **The Zionist Movement** held its first conference in Prussia in 1884 as anti-semitism grew.

Literature

The pessimistic application of theories of evolution is found in Zola's naturalistic novels which stressed the limitations on man's actions stemming from his inherited characteristics and the environment and portrayed the most sordid aspects of French lower-class life. English literature exchanged the exuberance of Dickens for the critical mood of Hardy.

Nationalism still acted as a vital cultural stimulus, creating a school of national regeneration in Spain in reaction to the political weakness highlighted by the war with Cuba, and in Italy celebrating unification. In both, writers turned to their national classics for models. The first self-conscious Latin American school grew up asserting independence from European traditions.

English literature after 1875 saw a reaction to the confidence of the high Victorian era, reflected in the decadent poetry of **Algernon Charles Swinburne**, 1837-1909. His sensual *Poems and Ballads* show traces of symbolist influences. **Gerard Manley Hopkins**, 1844-89, described the tensions of his religious vision in lyrical, experimental poetry.

Realism in the theatre was pioneered by the Norwegian **Henrik Ibsen**, 1828-1906, who dramatized social issues using ordinary conversation, as in *A Doll's House*, 1879. In England **G. B. Shaw**, 1856-1950, also attacked social complacency in plays enlivened by vivid characterization, satire and wit such as *Mrs. Warren's Profession* and *The Devil's Disciple*.

Native American humour was found in *The Adventures of Huckleberry Finn* by **Mark Twain**, 1835-1910. **The meeting of the New World with the Old** was explored by **Henry James**, 1843-1916, in *Portrait of a Lady*, 1881. He probed subtleties of character, temperament and motive, as in *Washington Square*, and the dazzling *The Golden Bowl*.

Naturalism in literature was inaugurated in France by **Emile Zola**, 1840-1902, who explored deterministic notions of the relation between heredity and environment and the casualties of urban society in the novels *Les Rougon Macquart*, 1871-93. **Guy de Maupassant**, 1850-93, followed him in his novels and short stories such as *Boule de Suif*.

Art and architecture

A self-conscious and revolutionary avant-garde emerged in European art at the end of the 19th century. In France Van Gogh, Gauguin and Cézanne, the major innovators of this time, developed their different styles out of their Impressionist origins. The symbolists rejected the Impressionist vision, turning instead to the past and to the exotic imagery of the later English Pre-Raphaelites in which Art Nouveau, an essentially decorative style and the first non-historical style to win wide acceptance, also had roots. Beginning in Belgium and England Art Nouveau owed its original character to a semi-abstract use of natural forms and had far-reaching effects in architecture and the applied arts.

Construction in metal became even more popular after the Paris exhibitions of 1878 and 1889, encouraged by the substitution of steel for iron, which also made possible the development of the skyscraper in the US.

A parallel to the Impressionist idea of forms dissolved in light appeared in the work of the greatest 19th-century sculptor **Auguste Rodin**, 1840-1917, who produced his first free-standing figure, "Age of Bronze", in 1877.

Ballet girls, working girls, and cabaret artists were the subject matter for such works of **Edgar Degas**, as "Scènes de Ballet", 1879. He worked in a great variety of media and was influenced by the action photographs of dancers and racing horses taken by **Muybridge**. **A Slavic revival in Russia** reached its peak in the '70s and '80s based on a careful documentation of national cultural history.

Official painting in England was represented by **Lawrence Alma-Tadema**, 1836-1912, and **Frederick Leighton**, 1830-96, who painted pseudo-classical scenes in a realistic though sentimental manner. **A move from Impressionism** was evident in the works of **Paul Cézanne**, 1839-1906, who achieved an almost abstract quality in such paintings as "L'Estaque", *c.* 1882-5, with its emphasis on form, colour, planes and light.

Neo-Impressionism or **Pointillism** was developed by **Georges Seurat**, 1859-91, **Paul Signac** 1863-1935, and **Camille Pissarro**, 1830-1903. A reaction against the spontaneity of Impressionism, it created the optical effect of light by means of dots of colour which were fused by the eye into continuous tones. **The Berlin Reichstag**, 1884-94, and the **Victor Emmanuele II** monument in Rome, 1885-1911, used antique forms to create a sense of civic grandeur.

Music

Romanticism began to decline as nationalism and impressionism became more important ideals in music. Meanwhile the future of American and European popular music was formed in the United States into the increasing appreciation of the rhythmic genius of Negro folk musicians and an awareness of the potential of the newly developed gramophone.

National qualities appeared in serious music both out of patriotism, heard in **Bedřich Smetana's** Czech *Má Vlast*, 1874-5, and of exoticism, in **Emmanuel Chabrier's** *España*, 1883.

Art songs were composed all over Europe, after decades of domination by German lieder writers. The form was finely worked by French composers like **Henri Duparc**, 1848-1933.

English light opera, notably the deft, tuneful creations of **W. S. Gilbert**, 1836-1911, and **Arthur Sullivan**, 1842-1900, became a craze throughout Europe, the US and Australia.

The origins of jazz and blues are found in the work songs that united poor blacks as they toiled in fields, and in gospel songs that united them in church.

Science and technology

Germany now took the lead in the science-based industries as a result of the emphasis on science and technology in education and a political system that gave power to industry. She possessed a flourishing heavy industry, became the centre of early motor-car development and led the field in medicine, now a preventative as well as a curative science, with the discovery of antibodies and of new drugs. Koch's work on tuberculosis was the most important advance. As a result of these technical discoveries combined with the widespread building of new hospitals, mortality rates dropped throughout western Europe. Other technological achievements that would alter society were the inventions of the telephone and phonograph.

Classical physics failed to explain discoveries made in radioactivity and the problem posed by the Michelson-Morley experiment, and entered a time of uncertainty that would only be resolved by Einstein's theory of relativity.

The telephone was invented in 1876 by the American inventor **Alexander Bell**, 1847-1922. **Bacteria were identified** by methods of growing and staining cultures, developed from 1876 onwards by the German bacteriologist **Robert Koch**, 1843-1910. He found the bacteria that cause tuberculosis, anthrax and cholera. **The phonograph was invented** in 1877 by the American inventor **Thomas Edison**, 1847-1931.

In 1879 **Edison** patented his incandescent light bulb. But in 1878 **Joseph Swan**, a British physicist, had patented the **first successful filament electric lamp**. In 1879-80 both Swan and Edison independently produced a practical light bulb. **Piezoelectricity**, electricity produced by the compression of certain types of crystal, was discovered, 1880, by a Frenchman **Pierre Curie**, 1859-1906.

The ether was proved not to exist by the experiments of the American physicists **Albert Michelson**, 1852-1931, and **Edward Morley**, 1838-1923, from 1881. This result led to the **theory of relativity**. **The electric tram** first ran in Berlin in 1881. **Cell division** was described in 1882 by the German anatomist **Walther Flemming**, 1843-1905. Following Koch's work, **Pasteur** used attenuated bacteria to confer immunity to anthrax, 1881, and against rabies, 1885.

H. C. Maxim, 1840-1916, invented the **Maxim machine gun** in England, 1884. **The steam turbine** was made in 1884 by the British engineer **Charles Parsons**, 1854-1931. **Motor transport** was founded in 1885 with the invention of the motor car by **Karl Benz**, 1844-1929, a German engineer. In the same year another German engineer, **Gottlieb Daimler**, 1834-1900, patented a gasoline engine which he used initially to power a motorcycle.

1885-8
The Canadian Pacific Railway was completed, 1885. **All American Indians** were confined to reservations by 1887. **The American Federation of Labor** was set up in 1886. **Germany** signed a Reinsurance Treaty with Russia, 1887, to minimize the danger of war between them in the Balkans. **Britain, Italy and Austria-Hungary** agreed to maintain the status quo in the Mediterranean and the Near East, 1887.

1888-90
The partition of Africa neared completion with Britain dominating the centre and south. **In Japan, Emperor Meiji**, *r.* 1868-1912, granted a Western-style constitution. In France the war minister **Georges Boulanger**, 1837-91, attempted to seize power. **The US overtook Britain** in steel production by 1890. **The Social-Democratic Party**, the most popular in Germany, was legalized, 1890.

1890-4
In Germany, **Kaiser Wilhelm II**, *r.* 1888-1918 dismissed Bismarck in 1890 and let the treaty with Russia lapse. **The European alliance blocs** took shape. **The Triple Alliance** of Germany, Austria-Hungary and Italy was renewed in 1891. Russia and France made a **Dual Alliance**, 1891. **Brazil** adopted a federal republican constitution, 1891. **In the US**, the **Populist Party** grew out of agrarian protest at currency deflation.

1894-6
In the victorious war against China, 1894-5, Japan gained Formosa and a free hand in Korea. The Powers' scramble for diplomatic and trading concessions in China began. **Sergei Witte**, 1849-1915, reformed Russian finances and stimulated industrialization and eastward expansion, 1890s. **The Dreyfus case**, 1894-9, revealed deep splits in French society between the liberal radicals and the Church and army.

1896-8
After the abortive Jameson raid, 1895-6, Britain faced a crisis with the Boer republics. **On the Nile**, Anglo-French tensions were relaxed after the conquest of the Sudan by **Lord Kitchener**, 1850-1916, and his meeting with the French at Fashoda, 1898. **France** aimed to consolidate the Saharan empire with a sphere of influence in Morocco. **Russia** threatened China, where the Powers gained territory and concessions, 1897-8.

1898-1900
In China, 1898, reactionary forces acted to stop the Westernizing **Hundred Days of Reform**, and an anti-foreign **"Boxer" rebellion**, 1900, brought disruption and resulted in foreign intervention and further impositions on Chinese sovereignty. **Faced with depression** the US raised its tariff wall, 1897, and joined in expansion overseas, fighting Spain, 1898, over **Cuba** and securing **Puerto Rico** and the **Philippines.**

Conservatives and Liberal Unionists opposed Gladstone's Irish Home Rule Bill, 1886. **"New unionism"** began to mobilize the working masses, 1888.

Intellectual socialist parties began to develop in the 1880s. **The Irish nationalists** were split by Parnell's divorce scandal, 1890.

The Independent Labour Party was founded, 1893. **Gladstone's** Liberals returned to power but the Lords rejected their second **Home Rule Bill**.

Joseph Chamberlain, 1836-1914, began a policy of colonial expansion in southern Africa. **The Land Act**, 1896, extended tenants' rights in Ireland.

Victoria's Diamond Jubilee Year 1897, marked the peak of British imperialist ambitions. **Chamberlain** opposed home rule.

Britain was hard pressed in the **Boer War**, 1899-1902. **The Labour Representation Committee** was founded in 1900.

Bell at his telephone, 1892

Benz Velo motor car, 1896

Art nouveau: Horta interior

Leopard from Benin

Sigmund Freud

Friedrich Nietzsche, 1844-1900, a German, vehemently rejected Christianity, science and conformist moralities in *Beyond Good and Evil*, 1886. **Edouard Drumont**, 1844-1917, popularized anti-semitism in *La France Juive*, 1886. **Ferdinand Tönnies**, 1855-1936, the German sociologist, published *Community and Association*, 1887, distinguishing "communities", involving moral consensus, from "associations", based on self-interest.

James Frazer, 1854-1941, a British anthropologist, surveyed a great range of beliefs and customs in *The Golden Bough*, 1890. He claimed that there was a natural progression from magical, through religious to scientific belief systems.

The Neoclassical school of economics was established by the Englishman **Alfred Marshall**, 1842-1924, in his *Principles of Economics*, 1890. He united the "classical" view that prices are determined by costs with the "marginalist" view of **W. S. Jevons**, 1835-82, that prices depend on the interaction of supply and demand. **K'ang Yü-wei**, 1858-1927, advocated social equality in China, and argued that Confucius had supported historical progress.

Herbert Spencer, 1820-1903, constructed and popularized a **comprehensive evolutionary theory** which saw all things as progressing from simplicity to complexity. His sociology, based on an analogy between societies and organisms, used the idea of the survival of the fittest. **F. H. Bradley**, 1846-1924, was an English proponent of Hegel. He argued that an idea's truth depended on its coherence with the set of ideas comprising the Absolute.

Philosophical pragmatism was expounded by the American psychologist **William James**, 1842-1910, in *The Will to Believe*, 1897. He argued that a belief was true if its acceptance aided the solution of practical problems. **Emile Durkheim**, 1858-1917, stirred French opinion with his sociological study *Suicide*, 1897, which attempted to link positivist social ideas with an interest in morality, which he saw as the basis of society.

Sigmund Freud, 1856-1939, elaborated the main tenets of psychoanalytic theory in *The Interpretation of Dreams*, 1900, developed in the course of clinical experience in Vienna. He divided the mind between the ego, id and superego, and argued that psychological disorders stemmed from the repression of sexual urges in early life. His emphasis on the subconscious influenced later irrationalism. **H. S. Chamberlain**, 1855-1927, spread racialist ideas.

Symbolist poetry developed in France with the *Poésies*, 1887, of **Stéphane Mallarmé**, 1842-98. He sought for an ideal world, but one of the intellect and not the emotions. **Maurice Maeterlinck**, 1862-1949, in Belgium, wrote symbolist plays. In Sweden **August Strindberg**, 1849-1912, wrote plays such as *Miss Julie* in a naturalist vein.

Italian nationalist ideas were voiced by **Giosuè Carducci**, 1835-1907. A scholar and anti-romantic, his *Odi Barbara* is a patriotic vision of Italy's glorious past and future destiny. **Verismo**, Italian realism, was developed by **Giovanni Verga**, 1840-1922, whose novels combined a perceptive study of the Sicilian class structure with the personal cares of *Mastro Don Gesualdo*.

In his novels of English rural life such as *Tess of the D'Urbervilles*, 1891, **Thomas Hardy**, 1840-1928, expressed a pessimistic view of life in which man was swamped by cosmic ironies. **The English Decadent movement** is epitomized by *The Picture of Dorian Gray*, 1891, by Oscar Wilde, 1854-1900, who was inspired by *A Rebours* by **J. K. Huysmans**.

Knut Hamsun, 1859-1952, a Norwegian, condemned an over-emphasis on social issues. **The stories and poems of Rudyard Kipling**, 1865-1936, examined the relationship between British and Indian culture. **H. G. Wells** pioneered science fiction in *The Time Machine*.

The height of the **Modernismo** movement in South America was realized in *Prosas Profanas* by the Nicaraguan poet **Ruben Dario**, 1867-1916. **The Generation of '98**, a group of Spanish intellectuals, set out to counteract Spanish apathy and revitalize Spanish culture. The group included the poet, philosopher and novelist **Miguel Unamuno**, 1864-1936.

Russian realist drama reached its peak with the work of **Anton Chekhov**, 1860-1904, which had a huge effect on European drama, notably *Uncle Vanya* and *The Seagull*. Chekhov was also a masterly short story writer, able to convey pessimistic themes with a humorous twist. **The Greek poet Cavafy**, 1863-1933, wrote about the ironies of man's existence.

Vincent van Gogh, 1853-90, a painter obsessed with the problems of expression, painted tormented landscapes and portraits using heightened colour and a frenzied and turbulent style seen in "The Sower", 1888. **Impressionist ideas** were taken to England by **Wilson Steer**, 1860-1942, and **Walter Sickert**, 1860-1942, who set up the **New English Art Club**, 1886, in protest against pseudo-classical styles.

Symbolist art which developed after 1886 sought an escape from the present and appealed to the imagination and the senses. In France its chief exponents were **Gustave Moreau**, 1826-98, whose paintings have a rich, jewel-like quality, and **Odilon Redon**, 1840-1916, who used his fantastical imagination in the illustration of Baudelaire's *Fleurs du Mal*, 1890. **The Eiffel Tower** was constructed in 1889, demonstrating contemporary engineering skills.

Art Nouveau, an international decorative style using flat, flowing and tendril-like forms, was popularized in England by the graphics of **Aubrey Beardsley**, 1872-98, for the magazine *Studio*, from 1893, and the buildings of **Victor Horta**, 1861-1947, in Belgium. **Gothic forms and wild extravagant decoration** characterized the buildings of the Spanish architect **Antoni Gaudi**, 1856-1926. His church of **Sagrada Familia** in Barcelona, from 1893, is related to Art Nouveau.

Synthetism, a style characterized by the use of strong flat colours organized into well-defined areas, was developed by **Paul Gauguin**, 1848-1903, in his paintings executed in Tahiti. He influenced the **Nabis**, a group of French painters led by **Paul Serusier**, 1865-1927, who rejected the naturalism of the Impressionists. **A large collection of Benin art**, the first African art well known in Europe, was brought back by a punitive expedition in 1897.

The Vienna Sezession, set up to promote the Austrian form of Art Nouveau, 1897, was concerned mainly with interior design. **Gustav Klimt**, 1862-1918, was its leading exponent. **The English Vernacular style** of domestic architecture pioneered by **C. A. Voysey**, 1857-1941, emphasized natural materials and solid construction and developed many of the designs which became part of a modern (non-historical) European style.

The sources of German Expressionism can be seen in the works of the Norwegian painter **Edvard Munch**, 1863-1944, who concentrated on the expression of intense emotion in his "Frieze of Life" c. 1890-1900. **The Chicago school of architecture** pioneered a modern American public style, creating the skyscraper. The Schlesinger-Mayer store by **Louis Sullivan**, 1856-1924, begun in 1899, rose to nine floors and had detailing suggestive of Art Nouveau.

Russian music gained its national qualities – lyricism, vitality and colourful orchestration – in the work of **Peter Tchaikovsky**, 1840-93 and **Nikolai Rimsky-Korsakov**, 1844-1908.

Symphonic traditions continued in Europe with vast works by **Anton Bruckner**, 1824-96, and **Gustav Mahler**, 1860-1911, who introduced folk elements.

National styles developed in the works of composers like **Jean Sibelius**, 1865-1957, in Finland and **Isaac Albéniz**, 1860-1909, in Spain.

Claude-Achille Debussy, 1862-1918, brought Impressionism to music. His intricate tone colour and continuous form utilized exquisite harmony and a whole tone scale.

The American John Sousa, 1854-1932, wrote superb **marches** for marine bands, including *The Stars and Stripes Forever*, 1897.

Ragtime was played in the US in the 1890s by black pianists, notably by **Scott Joplin**, 1868-1917. The bright syncopation of the style overlaid European dance and march forms.

Aluminium could be produced economically from 1886 by the electrolytic process developed, almost at the same time, by an American chemist, **Charles Hall**, 1863-1914, and a French chemist, **Paul Héroult**, 1863-1914. **Radio waves were produced** about 1887 by **Heinrich Hertz**, 1857-94, a German physicist. **Edison** set up a **research laboratory** at West Orange, 1887, with teams of inventors working together systematically.

The pneumatic tyre was invented in 1888 by the British vet **John Dunlop**, 1840-1921. **Photographic film** and paper was developed, 1884-8, by **George Eastman**, 1854-1932, an American inventor. Eastman's work made the cinema, as well as still photography, a possibility.

Diphtheria antitoxin was isolated in 1892 by the German biologist **Paul Ehrlich**, 1854-1915.

The diesel engine was invented by the German engineer **Rudolf Diesel**, 1858-1913, and demonstrated in 1896. French inventors **Auguste Lumière**, 1862-1954, and his brother **Louis**, 1864-1948, developed good cine equipment. The first public showing of their films took place in 1895. **X-rays** were discovered, 1895, by the German physicist **Wilhelm Röentgen**, 1845-1923. **Radioactivity was discovered** in 1896, by the French physicist **Antoine Becquerel**, 1852-1908.

The theoretical basis for space travel was provided in the work of **Konstantin Tsiolkovsky**, 1857-1935. In 1898 he proposed a liquid fuel rocket. The Polish-French chemist **Marie Curie**, 1867-1934, working with her husband **Pierre**, discovered polonium and radium in 1898. **The electron was discovered** in 1897 by the British physicist **J. J. Thomson**, 1856-1940. **Malaria** was shown to be transmitted by the mosquito by the British physician **Ronald Ross**, 1857-1932, in 1897.

Viruses were discovered, 1898, by the Dutch scientist **Martinus Beijerinck**, 1851-1931, while investigating the cause of tobacco mosaic disease. **Radioactivity** was found to include two different kinds of rays, alpha rays and beta rays, in 1899 by a British physicist, **Ernest Rutherford**, 1871-1937. In 1913 he used the rays to **penetrate the atom**. **Aspirin** was first marketed as a small drug by **Bayer AG**, a German pharmaceutical firm, in 1899.

1900-1925 Europe plunges into war

Principal events

World War I, arising from political and economic competition among the European Powers, dominated the period. In it, Europe suffered great losses in man-power and economic strength, while the United States and Japan won new political prestige. The need for organization on an unprecedented scale brought social and political upheaval in many countries. The old empires disappeared, leaving many new nationally based states, an embittered and dismembered Germany and a communist Russia. Fear of socialism grew stronger and was linked with economic discontent to stimulate fascism in Italy.

The new location of power outside Europe and the rise of nationalism in India and China marked the transition from an international order based firmly on Europe to a world arena of politics, which would lead to widespread decolonization after World War II.

1900-3

The US Steel Corporation integrated the industry, 1901.
As France and Italy made an entente in 1902, weakening Italy's links with Germany, **Britain** emerged successfully from the **Boer War** and formed an alliance with **Japan**. This countered the Russian presence in **Manchuria** and encouraged Japan's expansionist ambitions on the Asia mainland.
The United States extended its influence over **Panama**, 1903, winning canal-building rights.

1903-5

The Entente Cordiale, 1904, of France and Britain settled the two powers' outstanding colonial disputes, especially in Egypt.
Japan firmly established her military power, defeating Russia in 1905 in Manchuria.
In defeat Russia was convulsed, 1905-6, with a revolution against tsarist autocracy by the industrial and intellectual classes.
Intervention by Wilhelm II, r. 1888-1918, in Morocco, 1905, alarmed France which claimed supremacy there.

1905-8

The Powers met at Algeçiras in 1906 to settle the Moroccan question in favour of France.
Tsarist rule in Russia was re-imposed with only minor constitutional reforms.
Russia and Japan reached agreement over China in 1907.
Britain, France and Spain agreed to oppose German naval expansion in the Mediterranean.
An Anglo-Russian Convention covered Persia, Afghanistan and Tibet and brought Britain into Europe's power blocs.

1908-10

Increasing Anglo-German competition was expressed in a race to build warships.
With Russian agreement, Austria annexed Bosnia-Hercegovina in 1908.
Nationalist unrest in Catalonia disrupted Spain, 1909.
The Powers intervened to prevent a Serbo-Austrian war.
India secured constitutional reforms in 1909.
The former Boer republics helped form a new dominion, the **Union of South Africa**, 1910.

National events

The progressive forces of the new Liberalism, which stressed social reform, checked the Labour Party advance. But the Irish Question, constitutional crises and wartime pressure split the party. The power of the Lords was much reduced and Labour inherited the second party role, 1923.

The Education Act, 1902, put the responsibility for education in the hands of local authorities.

Joseph Chamberlain, 1836-1914, campaigned for protective tariffs favouring imperial exports. The Tories split on **Free Trade.**

Henry Campbell-Bannerman, 1836-1908, led the Liberals in their 1905 election landslide, introducing social reforms, army and navy reorganization.

A constitutional crisis ensued after the Lord's rejected the **"Peoples' Budget"**, 1909, which provided for social expenditure financed from income tax.

Wright Brother's flight

Model T Ford

Suffragettes, 1911

World War I: German skeleton in the trenches

Religion and philosophy

The philosophies of Bergson, Croce, Dilthey and Husserl, stressing intuition and immediate sympathy as the basic method of understanding, contributed to the development of a concept of the human sciences distinct from the natural sciences.

Under their influence Max Weber investigated the motives as well as the causes of human action, notably the effect of religion on man's supposedly "rational" economic behaviour. Russell and Wittgenstein, however, still took science and mathematics as the paradigm of knowledge in their work on the logical structure of language.

Psychoanalytic theory continued to explore the nature of the unconscious but two of Freud's colleagues, Adler and Jung, criticized his insistence on the sexual basis of neuroses.

The Russian Revolution accentuated the socialist split between violence and the peaceful battle for working-class rights.

The Pentecostal Movement began in America c. 1902.
Vilfredo Pareto, 1848-1923, an Italian and a positivist wrote *The Socialist Systems*, 1902, a refutation of Marxist economics and sociology. He accepted the existence of class conflict but saw the process Marx described as a series of progressive revolutions as no more than the successive replacement of ruling elites by each other.

Max Weber, 1864-1920, a German historian and sociologist, was concerned to combat the influential Marxist school of historical materialism. He opposed a simplistic belief in economic determinism and stressed the causal role of ideas in history.
In France, Maurice Barrès, 1862-1923, and **Charles Maurras**, 1868-1952, argued influentially for cultural unity, the supremacy of the state and the primacy of the national interest.

Henri Bergson, 1859-1941, in *Creative Evolution*, 1907, stressed the importance of change through a creative life-force, in opposition to the static scientific view of nature. His view that intuition was superior to scientific or intellectual perception was echoed by **William Dilthey**, 1833-1911, in Germany, to support an ethical relativism.
The Modernist Movement in Catholicism was condemned by Pope Pius X in 1907.

Georges Sorel, 1847-1922, in *Reflections on Violence*, 1908, celebrated the use of violence and rejected all bourgeois values. He influenced Mussolini.
The German historian **Friedrich Meinecke** (1862-1954) looked for a meaning within the historical process itself, but sought to avoid cultural relativism. His *Cosmopolitanism and the Nation State*, 1908, acknowledged the significance of the unification of Germany but regretted the death of the culture that preceded it.

Literature

The need for new forms of self-expression able to encompass a growing awareness of the unconscious gave rise to many strong and individualistic movements in European literature.
The surrealists evolved out of the symbolists, and their attempt to "trap" the subconscious in a spontaneous literary form broke down all restrictions of style.
In the English-speaking world a more formal school grew up with the modernist poets Pound and Eliot.
The German Expressionists were among the first to voice a lack of faith in society. Their prophecies were realized with World War I, the image of which haunted later writers.
Japan came into contact with Western realist and naturalist schools.

The Celtic literary Renaissance was a cultural reflection of Irish independence. Ancient legends were revived by **W. B. Yeats** 1865-1939, **J. M. Synge**, 1871-1909, and **Sean O'Casey**, 1884-1964.
James Joyce was self-consciously Irish but hostile to the Celtic Renaissance.

Impressionism, a German literary style which set out to describe complex emotional states by using symbolic imagery, is used in *Stundenbuch*, 1905, by **Rainer Maria Rilke**, 1875-1926.
Revolution and man's capacity to withstand extremities dominate such novels of **Joseph Conrad**, 1857-1924, as *Nostromo*, 1904.

With the publication of *Kormichye*, 1907, **Ivan Ivanovich**, 1866-1949, was acclaimed leader of the Russian Symbolist movement, which also included the poet **Alexander Blok**, 1880-1921.
Stefan George, 1868-1933, was influenced by Nietzsche in his desire to ennoble German culture with his esoteric poetry, such as *The Seventh Ring*, 1907.

Modern Japanese fiction began after 1905 with a powerful naturalist school including **Shimazaki Toson**, 1872-1943, and **Tayama Katai**. **Mori Ogai**, 1862-1922, reacted against their obsession with squalor in *Vita Sexualis*, 1909.
The American **Jack London**, 1876-1916, was stirred by social injustices to write popular tales and political tracts.

Art and architecture

Traditional forms and concepts of art were dramatically broken down between 1900 and 1925 as a variety of alternative aesthetic principles developed. In particular Cubism attempted to break away from the conventions of perspective that had ruled European art since the Renaissance, while Dadaism and Russian Constructivism aimed to destroy the distinction between art and life.

In architecture, too, definitive new styles emerged in the US and Europe with the publication of Frank Lloyd Wright's early designs and the establishment of the Bauhaus, both emphasizing assymetry and plain surfaces.

The cinema transformed the whole scope of the visual arts, developing from the early popular experiments of 1900 to the politically motivated films of Eisenstein in Russia (where the Revolution stimulated artistic innovation in many fields), the dramas of Griffith and the popular comedies of Chaplin.

A reaction in architecture against the highly decorated surfaces of Art Nouveau produced the simple rectangular forms seen in the **Convalescent Home**, Vienna, 1903, by **Josef Hoffman**, 1870-1956. In France this trend is found in the garage, rue Ponthieu, Paris, 1905, by **Auguste Perret**, 1874-1954.
The Intimiste painters Edouard Vuillard, 1868-1940, and **Pierre Bonnard**, 1867-1947, used Impressionist techniques in domestic scenes after 1900.

Cubism, rejecting traditional methods of portraying reality began in 1907 with **Picasso's** "Demoiselles d'Avignon", 1907. Other Cubists include **Juan Gris**, 1887-1927, **Robert Delaunay**, 1885-1941, **Fernand Léger**, 1881-1955, and **Georges Braque**, 1882-1963.
In Paris **Pablo Picasso**, 1881-1973, began his Blue period, 1901-4, and his Rose period, 1905, producing lyrical, conventionally representational paintings.

The Fauvist period in painting, 1905-8, was characterized by the use of flat patterns and intense, unnatural colours. "Open Window, Collioure", 1905, by **Henri Matisse**, 1869-1954, is typical, as are the works from this period of **André Derain**, 1880-1954, **Maurice Vlaminck**, 1876-1958 and **Raoul Dufy**, 1877-1953.

In Germany the Expressionist painting of the **Die Brücke group** 1905-13, distorted reality to produce a personal view of the world, depicting intense and painful emotions after the style of **Edvard Munch**. They were chiefly represented by **Emil Nolde**, 1867-1956.
The Italian Futurists produced work and manifestos that extolled the technological energy of modern life, from 1909.

Music

The Romantic tradition lingered on into the early twentieth century. Popular music began to make its mark, and many serious composers sought a radical break with the past while others turned to folk music for their inspiration. The radicalism in the arts that followed World War I produced a variety of new musical techniques as well as altering aesthetic principles.

Richard Strauss, 1864-1949, continued the romantic tradition with operas and tone poems in a grand Wagnerian style.

Giacomo Puccini, 1858-1924, brought Italian grand opera to a grand finale with such dramatic and melodic operas as *La Bohème*, 1896, *Tosca*, 1900, and *Madame Butterfly*, 1904.

Blues grew steadily more popular in the early 20th century in the United States. Their cross-rhythms and varying intonation brought great expression to the simple *aab* 12-bar form.

New Orleans became the cradle of **jazz** as ragtime bands, using instruments left over from the Civil War, took up improvisation and developed into small traditional jazz bands.

Science and technology

Einstein's theories of relativity and Planck's Quantum theory revealed a new picture of the ultimate workings of nature. Although Newton's theories still proved accurate enough for most predictions, Einstein held that there was no absolute motion, motion in respect of empty space. His relativity principle stated that motion must always appear as the relative motion of one object with respect to another. It related time, mass and length to velocity and mass to energy, and provided a theoretical basis for the development of nuclear physics.

Although World War I stimulated research and technology in Europe, the impetus for scientific advance shifted to America. The invention of the electronic valve which allowed the development of the radio transmitter to proceed, and the development of powered flight speeded up intercontinental communications and the introduction of mass-produced cars revolutionized private transport.

Blood groups were first distinguished in c. 1900.
Gamma rays were discovered in radioactivity, 1900, by **Paul Villard**, 1860-1934.
The quantum theory that energy consists of indivisible units, was proposed by **Max Planck**, 1858-1947, in 1900.
Guglielmo Marconi, 1874-1937, was the first to transmit radio signals across the Atlantic, in 1901.
The fingerprint system was introduced in Britain, 1901.

The first sustained flight by a power-driven aeroplane was made by **Wilbur** and **Orville Wright** in the USA, 1903.
Detroit became the centre of the motor car industry, 1903.
The first electronic valve was made in 1904.
The special theory of relativity was published in 1905 by **Albert Einstein**, 1879-1955.

Mass production of cars began in the US with the Model T Ford in 1908.
The third law of thermodynamics, that absolute zero cannot be attained, was put forward in 1906.
Emil Fischer 1852-1919, showed in 1907 that proteins are composed of amino acids – a vital step in molecular biology.
The first helicopter flew, 1907.
The cloud chamber, used in detecting the paths of atomic particles, was perfected 1906 by **Charles Wilson**, 1869-1959.

Ammonia was synthesized in 1908, enabling Germany to produce the first high explosives.
Chromosomes were established as the carriers of heredity, 1909.
Bakelite, a synthetic polymer used for making electric plugs was invented in 1909. Its success stimulated the development of plastics.
Combine harvesters were common in the US by 1910.
Louis Blériot made his first flight in 1907 and crossed the English Channel, 1909.

1910-13

In Mexico, Porfirio Diaz, *r.* 1877-1911, was overthrown and the US intervened by occupying Vera Cruz.
The Triple Entente powers of France, Russia and Britain made military and naval agreements. After the 1911 Agadir incident when the Germans sent a gunboat to frighten the French, they countered German ambitions in Morocco.
A nationalist republic was set up in China in 1911 under **Sun Yat-sen**, 1866-1925.

1913-15

Austria's Archduke Franz Ferdinand, 1863-1914, was assassinated in June 1914 by Serbian nationalists, setting off events leading to world war.
By 1915 Germany and Austria, with Turkey and Bulgaria, were fighting against the Entente allies, with Italy and Japan.
Military operations extended from the main "front" in France and Belgium to the Russian plains, the Balkans, the Middle East and the German colonies in Africa.

1915-18

Germany started a blockade of British shipping. The main naval battle at Jutland, 1916, was inconclusive. Germany's unrestricted submarine warfare (leading to the sinking of the ***Lusitania*** in 1915) provoked the US to enter the war, 1917.
The strain of war brought revolution to Russia in 1917. The tsar abdicated and **Bolshevik** forces led by **Lenin**, 1870-1924, won power and withdrew Russia from the war, after defeating the liberal government.

1918-20

Britain, France and the US defeated Germany in 1918. In the **Versailles Treaty**, 1919, inspired by the democratic ideals of **Woodrow Wilson**, 1856-1924, the US president, new ethnic Balkan states were established and Turkey was partitioned. **War guilt** and indemnity were assigned to Germany, where an abortive revolution disrupted the new republic, 1919.
Wilson's League of Nations was inaugurated in 1920 but without US participation.

1920-22

In Russia, reactionary forces with allied aid tried unsuccessfully to defeat the Bolsheviks.
Germany, struggling against economic chaos after the loss of the major industrial centres, secured Soviet friendship in the **Treaty of Rapallo**, 1922.
Japan, which had been granted Germany's rights in China by the Versailles Treaty, made peace with China, 1922.
Benito Mussolini, 1883-1945, established fascist power in Italy in 1922.

1922-5

Germany countered France's occupation of the Ruhr, 1923-5, with passive resistance and suffered massive inflation, which destroyed the economic strength of the middle classes.
The Dawes plan, 1924, eased the repayment schedule for German war reparations.
Mustafa Kemal, 1881-1938, president of the new Turkish republic, began modernization of Turkish society, 1923.
The American economy boomed until 1929.

Herbert Asquith, 1852-1928, secured passage of the **Parliament Act**, 1911, which limited the power of the Lords. **National insurance** was introduced.

Industrial disorder, suffragettes and Ulster opposition to **Irish Home Rule** inflamed society. The world crisis postponed civil war in Ireland.

The 1916 Easter Rebellion boosted the republican Sinn Fein in Ireland. **David Lloyd George**, 1863-1945, maximized the war effort.

Women over 30 gained the vote in 1918 in recognition of the suffragette campaign, 1903-14, and the importance of the role played by women in the war effort.

The 1920 Home Rule Act partitioned Ireland. Republicans rejected the 1921 Free State settlement but met defeat in a civil war, 1919-21.

Labour became the official opposition in 1923 and in 1924 first exercised minority rule under **Ramsay MacDonald**, 1866-1937.

Albert Einstein

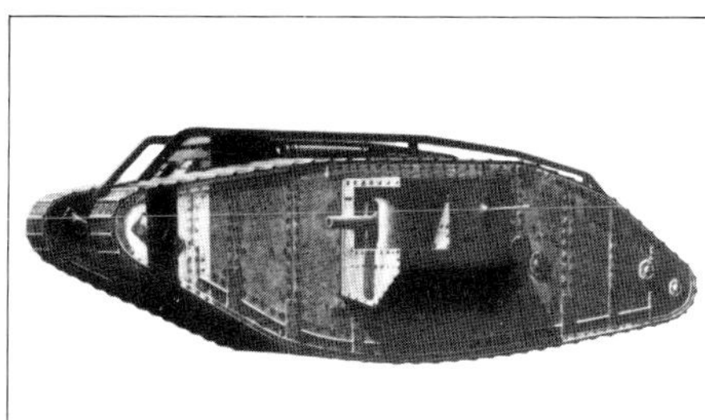

World War I tank

Russian Revolution: street scene in Petrograd, 1917

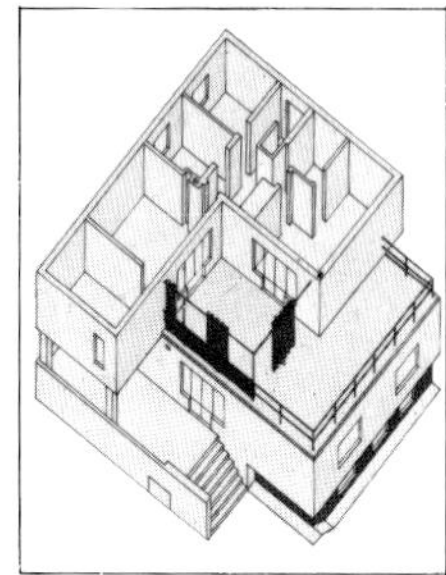

Bauhaus: house by Gropius

Sophisticated physics produced scientific theorists such as the Austrian **Ernst Mach**, 1838-1916, and the Frenchman **Henri Poincaré**, 1854-1912, who argued that unobservable entities like atoms should be regarded only as useful postulates about material nature.
Phenomenology was founded by the German philosopher **Edmund Husserl**, 1859-1938, who argued that true knowledge stemmed from the imaginative analysis of direct experience.

Bertrand Russell, 1872-1970, the English philosopher and mathematician, applied empiricist principles to language, which he claimed to be constructed solely from sensory ideas and logic. In *Principia Mathematica*, 1910-13, he attempted, with **A. N Whitehead**, 1861-1947, to derive mathematics from the axioms of logic.
Opposition to the war led **Rosa Luxemburg**, 1871-1919 to found the left-wing Spartacist party in Germany.

V. I. Lenin, 1870-1924, the Russian politician, argued in *The State and Revolution*, 1917, for a party of professional revolutionaries.
Freud's emphasis on the sexual basis of psychiatric disorders led to the defection of two of his followers, **Carl Jung**, 1875-1961, and **Alfred Adler**, 1870-1937. Jung developed a theory of the collective unconscious, while Adler tried to derive a psychology from man's tendency to strive for perfection.

The British anthropologists Malinowski 1884-1942 developed **functionalism** 1914-18 in anthropology, studying social phenomena in terms of their function within an integrated social structure, in opposition to evolutionist anthropology.
Oswald Spengler, 1880-1936, published *The Decline of the West*, 1918, claiming that Western civilization had ceased to be "creative" and had become concerned only with materialism.

The *Tractatus Logico-Philosophicus* of **Ludwig Wittgenstein**, 1889-1951, an Austrian living in London, was published in 1921; it argued that philosophy was an analytic, not a speculative, subject.
Aimee Semple McPherson, 1890-1944, built the Angelus Temple in Los Angeles in 1922 and preached the religion of the foursquare gospel.

The Hungarian **Gyorgy Lukacs**, 1885-1971, a Marxist influenced by Hegel's idealism, wrote of the role of creative awareness in the development or revolutionary consciousness, in *History and Class-consciousness*, 1923.
Benedetto Croce, 1866-1952, a historian who argued that the past could be understood only when seen in relation to current problems, became the spokesman for the opposition to Fascism in Italy after 1923.

The German Expressionists described visions of the collapse of society. **George Heym**, 1887-1912, prophesied a great war in *Umbra Vitae*, 1912, as did **George Trakl**, 1887-1914, in his poem *Sebastian in Traum*, 1914.
Guillaume Apollinaire, 1880-1918, dominated the surrealist and avant garde movements in Paris from 1913 until his death.

Franz Kafka, 1883-1924, described man's spiritual bereavement in symbolic terms in *The Trial*, 1914-15.
Ezra Pound, 1885-1972, worked on his *Cantos* from 1914 until his death. His allusive erudite style influenced many English poets.

The "literary revolution" in China in 1917, used the vernacular language in literature.
The English war poets voiced their horror of mass warfare. **Rupert Brooke**, 1887-1915, and **Wilfred Owen**, 1893-1918, died while on duty. **Robert Graves**, 1895-   , and **Siegfried Sassoon**, 1886-1967, also wrote about the period, both in prose and verse.

The Hindu writer **Rabindranath Tagore**, 1861-1941, translated the mystical *Gitanjali*.
Mohammed Iqbal, 1873-1938, wrote in Urdu and Persian and voiced a growing resentment against the West, in India.
André Gide, 1869-1951, kept his *Journals* from 1889-1949.
Hermann Hesse, 1877-1962, wrote *Demian*, 1919.

T. S. Eliot, 1888-1965, whose *The Waste Land*, 1922, is a dense and highly literary meditation on the situation of modern man.
Luigi Pirandello, 1867-1936, reflected the spiritual confusion of the post-war years in his play *Six Characters in Search of an Author*, 1921.

The Stream-of-consciousness technique was used by **Marcel Proust**, 1871-1922, to evoke the past in the long series of novels *A La Recherche du Temps Perdu*, 1913-27, and by **James Joyce**, 1882-1941, in *Ulysses*, 1922. This and *Finnegans Wake*, 1939, are highly experimental, original and questioning works.

Analytical Cubism, 1910-12, concentrated on pure form, excluding interest in colour. **Synthetic Cubism**, 1912-14, involved the construction of an image often by means of collage, such as the "Bottle of Anis del Mono", 1914, by **Juan Gris**.
Dar Blaue Reiter group of Expressionist painters, 1911-14, used colour and abstract forms to convey spiritual realities and included **Wassily Kandinsky**, 1866-1944, and **Paul Klee**, 1879-1940.

Russian Constructivism, 1913-mid-20s, was initiated by **Vladimir Tatlin**, 1885-1953, and exploited the concept of Synthetic Cubism. Its emphasis was on abstract structures made of a variety of materials. Tatlin's "Constructions", 1913-14, were made of wood, metal and glass.
The first long feature films included the Italian *Cabiria*, 1914, and *The Birth of a Nation*, 1915, directed by **D. W. Griffith**, 1875-1948, a drama about the American Civil War.

The Dadaist movement developed in Zurich in 1916 in the work of **Jean (Hans) Arp**, 1887-1966, and **Tristan Tzara**, 1896-1963. It was deliberately "anti-art" and aimed to outrage and scandalize a complacent society. Its chief exponent was **Marcel Duchamp**, 1887-1968, whose "Fountain", 1917, consisted of a urinal.
The films of **Charlie Chaplin**, 1889-   , including *The Tramp*, 1915, won international acclaim.

The de Stijl group founded in Holland by **Theo van Doesburg**, 1883-1931, and **Piet Mondrian**, 1872-1944, developed their art and architecture based on spatial relationships. They used straight lines, right-angles and primary colours.
The Bauhaus school of architecture, design and craftsmanship was founded in Germany in 1919 by **Walter Gropius**, 1883–1969. It attempted to reconcile art and design with industrial techniques.

In France **Fernard Léger's** paintings reflected contemporary interest in machinery. His "Three Women", 1921, reduces figures to machine like forms and uses metallic colours.
Frank Lloyd Wright, 1869-1959, the greatest and most influential of American architects, designed the Imperial Hotel, Tokyo, 1919-22, using an entirely new anti-earthquake construction.

Architects in Europe such as **Walter Gropius, Mies van der Rohe**, 1886-1969, and **Le Corbusier**, 1887-1965, convinced of the need for streamlined, functional buildings, used the new media of concrete and glass to achieve a modern style epitomized in Gropius' design for the **Bauhaus**, 1925-6.
The Russian Revolution stimulated experimental cinema, led by **Sergei Eisenstein**, 1898-1948, whose *Battleship Potemkin*, 1925, preached socialism.

The Ballets Russes of **Sergei Diaghilev**, 1872-1929, commissioned major works, such as **Igor Stravinsky's** *Petrushka*, 1911, and **Maurice Ravel's** *Daphnis and Chloe*, 1912.

The Rite of Spring, 1913, by **Stravinsky**, 1882-1971, gave new emphasis to the role of rhythm in serious music, using irregular metre and highly varied motifs.

Charles Ives, 1874-1954, became the first truly original American composer, working in several keys and rhythms at once in many works, such as his *Concord Sonata*, 1909-15.

Harmony reached a peak of complexity with **Stravinsky**, who worked in several keys simultaneously, and then split asunder in the key-less music of **Arnold Schoenberg** and **Béla Bartók**.

Bartók, 1881-1945, created a style marked by extreme dissonance and elegant melody, particularly in his six quartets, 1907-39. He collected and studied Hungarian folk music.

Twelve-note or serial music, created in 1924 by **Schoenberg**, 1874-1951, was based on an arbitrarily ordered series or row using the 12 notes of the chromatic (half tone) scale.

Electrical superconductivity was discovered in 1911.
Nuclear theory, that the atom contains a central nucleus, was announced by **Lord Rutherford**, 1871-1937, in 1911, in England.
Vitamins were recognized as essential to health in 1906; their classification in 1911 stimulated dietary studies.
Continental drift, the theory that the continents shift, was first proposed in 1912.
Cellophane was first manufactured in 1912.

The proton was recognized as the nucleus of the hydrogen atom by Lord Rutherford, in 1913.
Niels Bohr, 1885-1962, showed, 1913, how changes in the electron orbits of the atom produce energy.
The Geiger counter was used to measure radioactivity, 1913.
Atomic numbers were determined by an X-ray method discovered in 1914.
The life cycle of stars was determined by work done in 1914.
Stainless steel was made in Germany from 1914.

The general theory of relativity was published by Einstein in 1915.
Tractors, introduced by Ford in 1915, used the diesel engine.
World War I stimulated technological advance on both sides, particularly in weaponry and transport. **The diesel engine** was used in tanks and the aircraft industry expanded.
German development of **synthetic rubber and of cellulose** as a substitute for cotton led the search for new artificial fabrics.
Gas was used as a weapon.

The first transatlantic flight was made, 1919, by the British aviators **Alcock and Brown**. The flight lasted almost $16\frac{1}{2}$ hours.
The first mass spectrograph was developed in 1919.
The first commercial aeroplane service, between London and Paris, was set up 1919.

Diesel locomotives and railcars came into use *c.* 1920. The growing use of internal combustion engines led to a decline in the supremacy of coal as the major industrial fuel after 1910.
Radio broadcasting on a regular basis began in the United States in 1920.
Insulin, a hormone, was isolated in 1922 and first used in the treatment of diabetes.
The teleprinter was developed in 1921, greatly speeding the transmission of long-distance information.

Radioactive tracers, used for the determination of many biological reactions, were developed in 1923.
Electrons were shown to behave as waves as well as particles, in 1922-4. This discovery made possible the invention of the electron microscope, in 1932.
External spiral galaxies were found by Edwin Hubble, 1923.
Clarence Birdseye 1886-1956, experimented with quick-frozen foods commercially, 1924.

1925-50 From depression to recovery

Principal events

The legacy of mistrust and depression following World War I brought a worldwide economic crisis at the end of the 1920s. The stronger industrial powers survived with the aid of new economic and social policies but in Germany, where the obligation to pay war debts exacerbated the effects of national defeat, the Nazi regime took power whose militarist ambitions in Europe would help to precipitate World War II.

In the USSR a policy of forced industrialization was pursued under Stalin, destroying many of the ideals of the Revolution, while the basis for a communist China was laid after a long civil war. India won her independence, but only at the cost of partition.

World War II left Europe shattered and weak and Germany divided, with the capitalist and socialist blocs locked in a continuing, though ostensibly peaceful, struggle for power.

1925-8

Chiang Kai-shek, 1887-1975, ousted the left-wing from the Kuomintang (nationalist party) in China. He captured Peking and unified the country, 1928, against Japanese expansion.
Germany joined the League of Nations, 1926, which hoped to bring peace by disarmament.
Fascist rule in Italy became increasingly authoritarian.
Joseph Stalin, 1879-1953, began forcible industrialization in Russia, 1928, after Lenin's death, 1924.

1928-30

The Kellogg-Briand Pact was signed by 23 powers in 1928 to outlaw war.
The last allied forces left the Rhineland, 1929.
Leon Trotsky, 1879-1940, was exiled from Russia, 1929.
The Wall Street Crash, 1929, led to business depression in America, causing economic recession throughout Europe and a rise in left-wing activity.
Gandhi, 1869-1948, began a civil disobedience campaign against British rule in India.

1930-33

The Round Table Conferences on India failed to satisfy nationalist demands, 1930-1.
The Hoover moratorium on war debts helped Europe to survive the depression, 1931, but the economic slump in Germany brought fighting between left- and right-wing groups.
Japan occupied Manchuria, 1931, after fears that trading contacts with China would be cut.
A republic was set up in Spain, 1931, dominated by liberals and socialists.

1933-5

Japan left the League of Nations, 1933, after condemnation of her action in Manchuria.
Adolf Hitler, 1889-1945, elected German chancellor, set up a Nazi dictatorship, 1933.
Franklin D. Roosevelt, 1882-1945, introduced a New Deal of social and economic reforms in the US to end the slump, 1933.
Stalin began a massive purge of Russian party officials, 1935.
Civil war in China between the left-wing and nationalists led to **the Long March**, 1934-5.

National events

Amidst economic depression, a national government was set up in 1931. It faced continuing social distress, imperial decay and major European commitments. After the upheaval of war, Labour's promise in 1945 of social transformation as outlined in the Beveridge report was realized.

The 1926 general strike came in reaction to massive unemployment. Action by government and mainly middle-class volunteers left a legacy of bitterness.

A second minority Labour government, 1929-31, failed to cure the rising unemployment and economic depression owing to the financial crisis.

Ramsay MacDonald split the Labour Party in 1931 over economic measures. He joined the Conservatives in a **national government**, 1931-35.

Legislation in 1934 introduced an unpopular **means test** for those on national assistance. There were over two million unemployed.

Gandhi in Calcutta, 1925

The Depression: soup kitchen in Chicago, 1930

Spanish Civil War poster

Victims of Hitler's concentration camps

Religion and philosophy

Political thought was dominated by conflict between the democratic ideal and its opponents on the left and right. Marxist political theory developed divergent trends as the Russian and Chinese revolutions took their course, but its influence in the West declined as supporters of liberalism rallied to oppose fascism, with its ideological roots in 19th century irrationalism. A new democratic philosophy, sustained by Keynes' economic theories of consumer prosperity, became linked with attempts to control political violence on a worldwide scale, marked by the founding of the United Nations.

The Christian Church came face to face with growing secularization in the industrialized countries and the need to find a new approach to the problems of an emergent Third World.

Philosophy remained split between those primarily studying human consciousness, and those who used a scientific model to understand reality.

The American J. B. Watson 1878-1958, developed behaviourist psychology in *Behaviourism*, 1925, seeking to explain behaviour wholly in terms of responses to external stimuli.
In *Mein Kampf*, 1925-7, **Hitler** drew upon the ideas of **Gobineau**, 1816-82, who argued that the development of a civilization depends on racial superiority and purity, and requires military aggression. Hitler condemned democracy as based on invalid egalitarianism.

Existentialism was developed in Germany from Husserl's phenomenological ideas, by **Martin Heidegger**, 1889-1976, who was appointed professor of philosophy at Freiburg in 1928. He argued that authentic human existence consists in not being subordinated to the external world.
The word apartheid was first used to describe racial segregation in South Africa in 1929.

J. M. Keynes, 1883-1946, overthrew the neoclassical orthodoxy in economics with two books, *Treatise on Money*, 1930, and *The General Theory*, 1936. He stated that market forces which lowered wage rates would not cure economic depressions; production and investment would only increase if spending by consumers, business and government went up. His theory influenced the New Deal and economic planning in the West, until the 1970s.

Leon Trotsky, 1879-1940, a Russian Marxist, argued for permanent revolution in his *History of the Russian Revolution*, 1932-3. He claimed that socialism in Russia could not survive unless revolutions also took place in more advanced countries, and opposed Stalin's doctrine of socialism in one country.
Gandhi organized *satyagraha* (truth force) campaigns to foster Indian nationalism by non-violence and emphasized the values of village life.

Literature

The insistent excavation of personal experience which had begun with the Romantics and reached a peak with the stream-of-consciousness writings of Proust and Joyce found a new exponent in Virginia Woolf and the more consciously Freudian Surrealists. Much European writing of the interwar period, however, reflected a need to grasp social issues of the time. Some, such as Camus, accepted the fact of social commitment while admitting the ultimate meaninglessness of existence. Others like Brecht developed new artistic forms to embody their political vision with a lesser emphasis on the individual. In the Third World, too, where writers were inspired by the ideal of national independence, a new, more confident literature emerged.

A major writer who met the requirements of socialist realism was **Mikhail Sholokhov**, 1905- in his *Tales from the Don*, 1925.
The Bloomsbury Group in London included the novelist **E. M. Forster**, 1879-1970, and **Virginia Woolf**, 1882-1941, who used a personal style of imagery in *To the Lighthouse*, 1927.
Highly personal experience was used in the **English Realist novel**.

D. H. Lawrence, 1885-1930, challenged the taboos of class and sex in novels such as *Lady Chatterley's Lover*. **John Cowper Powys**, 1872-1963, studied man in his environment, while **Malcolm Lowry**, 1909-57, wrote of his devastating experiences in Mexico.
A group of left-wing poets in London in the 20s included **W. H. Auden** 1907-73, and **Stephen Spender**, 1909-

A forerunner of the **Theatre of the Absurd**, **Luigi Pirandello**, 1867-1936, explored the theme of mutual incomprehension.
American writing was richly varied, ranging from the southern novels of **William Faulkner**, 1897-1962, whose *Light in August* appeared, 1932 to *The Grapes of Wrath* by **John Steinbeck**, 1902-68, treating the hardship of the depression.

"The Lost Generation", a group of Americans in Paris in the 1920s and 1930s, included **Scott Fitzgerald**, 1896-1940, whose *Tender is the Night* was an elegy for the American Dream; the masculine **Ernest Hemingway**, 1899-1961, **Gertrude Stein**, 1874-1946, an influential experimentalist; and the less typical **Henry Miller**, 1891- who shattered sexual taboos.

Art and architecture

In Europe before World War II there was increasing integration between art forms. Furniture design, painting and architecture were developed by the de Stijl and Bauhaus groups. Formal developments in painting also affected architecture. By 1932 the new International Style had come into existence. The first Surrealist manifesto in 1924, with its emphasis on exploration of the unconscious, represented the culmination of the avant-garde movement in art, which linked radical artistic and political ideas.

Many of the artistic movements of the postwar period found expression in the cinema, but the depression caused the collapse of the film industries of many European countries and introduced a period of Hollywood supremacy based on large studio organizations, which had the effect of suppressing much individual talent, and leading to the development of styles suited to a mass market.

Expressionist techniques were used by **Chaim Soutine**, 1893-1943, in "Page Boy at Maxim's", 1927, and by **Marc Chagall**, 1889- , in "Russian Wedding", 1925.
Expressionist cinema was developed by **Fritz Lang**, 1890-1976, in his vision of the future *Metropolis*, 1926, while **Dali** explored surrealist cinema in *Le Chien Andalou*, 1928.
The Jazz Singer, 1927, was the first talking picture.

Surrealism, founded in Paris, explored the reality of the subconscious. Its leading exponent was **Salvador Dali**, 1904- , whose "Illuminated Pleasures", 1929, shows objects taken out of context and replaced in fantastic juxtapositions. Other important artists were **René Magritte**, 1898-1967, **Giorgio de Chirico**, 1888- , **Joan Miró**, 1893- , and **Max Ernst**, 1891-1976. **Ernst** and **André Masson**, 1896- , practised automatism, a free-brush style.

The International Style in architecture, 1932, recognized a new and independent style that had emerged in the twenties. This was typified in the Villa Savoye, 1928-31, by **Le Corbusier**, with its white rectangular exterior and horizontal windows. The individual style of the French painter **Georges Rouault**, 1871-1958, whose religious works achieve a stained glass quality, can be seen in his "Christ mocked by Soldiers", 1932.

In Germany anti-Nazi artistic expressions by artists such as **Otto Dix**, 1891-1969, **George Grosz**, 1893-1959, **Max Beckmann**, 1884-1950, and **Oskar Kokoschka**, 1886- resulted in either suppression or exile for the artists concerned.
Socialist Realism was officially adopted in the USSR under Stalin in 1934, using an explicit, academic style in order to convey clearly the message of the dignity of the working classes.

Music

Serious music split into several mutually exclusive schools, most of which could attract few listeners or performers in spite of the spread of the radio and gramophone. However, these did help to broaden the audience for popular music which, in various jazz forms and "musicals", flourished widely.

An English school, including **Frederick Delius**, 1862-1934, **Gustav Holst**, 1874-1934, and **Vaughan Williams**, 1872-1958, produced pastoral music after **Edward Elgar**, 1857-1934.

Louis Armstrong, 1900-71, created a solo style in jazz with his innovative trumpet improvisations of 1925-30.
Duke Ellington, 1899-1974, began an orchestral style in jazz.

Ionization, 1931, by **Edgard Varèse**, 1885-1965, written solely for percussion instruments, showed that a piece of serious music could be constructed successfully using rhythm only.

The Neoclassic movement reinterpreted classical form in modern sound. Initiated by **Stravinsky**, the style attracted **Sergei Prokofiev**, 1891-1953, and **Paul Hindemith**, 1895-1963.

Science and technology

Economic depression and war hindered some areas of science while advancing others. In the West, steelmaking, engineering and agricultural production fell during the thirties, but falling prices stimulated consumer industries and aviation, radio, the car industry and artificial fibres continued to develop. The USSR, too, was industrializing fast.

With the rise of Hitler, many nuclear physicists fled to America, where their research ensured that Germany's supremacy in physics was lost and that the Nazis would not be the first to possess nuclear weapons.

World War II made great use of science, both to destroy and to save lives. Electronics, radar, nuclear technology, jet aviation and antibiotics were all products of the war.

In Britain important work was done in astronomy, exploring the implications of Einstein's theories to produce conflicting concepts of the origin of the universe.

Modern sound recording began with the introduction of electric recording in 1925.
Liquid fuel rockets were first tested in America in 1926.
The big bang theory of the origin of the universe was first put forward in 1927, by **Abbé Lemaître**, 1894-1966.
Wave mechanics, was founded by **Erwin Schrödinger**, in 1928.
The Heisenberg uncertainty principle, that every observation has a degree of probability, was proposed in 1927.

John Logie Baird invented a high-speed mechanical scanning system, 1928, which led to the development of television.
The anti-bacterial activity of *Penicillium* mould was discovered, 1928, by **Alexander Fleming**, 1881-1955, but it was not made stable enough for medical use until 1943.
The distance of galaxies was related to their speed of recession as measured by the red shift, 1929, by **Edwin Hubble** 1889-1953.

The cyclotron and other circular particle accelerators were developed from a working model made in America c. 1930.
Wallace Carothers, 1896-1937, invented **nylon** in 1931.
Radio astronomy began in 1931 with the detection of radio signals from outer space.
Deuterium, heavy hydrogen, was discovered in 1931.
The first nuclear reaction using an accelerator was activated in 1932.
Neutrons were discovered, 1932.

Sky-scraper building in the US was interrupted by the depression of the 1930s.
The first radio-isotopes were prepared by **Frédéric Joliot-Curie**, 1900-58, and his wife **Irène**, 1897-1956, in 1934.
Experiments made by **Robert Watson-Watt**, 1892-1973, after 1935 led to the invention of **radar**.
The meson, a sub-atomic particle, was predicted in 1935.

1935-7

The governmental reforms of 1935 in India again fell short of nationalist demands. **Mussolini invaded Abyssinia** in 1935 to satisfy fascist imperial ambitions. The League failed to intervene effectively. **Hitler** re-militarized the Rhineland, 1936, and Mussolini proclaimed the **Rome-Berlin axis**. **A right-wing coup** after the Popular Front won the elections led to civil war in Spain, 1936-9. **The Japanese** began their attack of China in 1937.

Edward VIII, *r.* 1936, abdicated to marry a divorcee. **A tariff war** and a revised constitution for **Eire** weakened Anglo-Irish relations.

Messerschmitt-262

The Vienna Circle, a group of philosophers who met there, 1922-36, including **Moritz Schlick** 1882-1936 and **Rudolf Carnap**, formulated logical positivism, an empiricist philosophy of language according to which only statements that could be verified by observation were meaningful. **Pope Pius XI**, *r.* 1922-39, condemned fascism, 1931, and communism, 1937. He adopted a friendly attitude to Protestant liberalism, although opposing *laissez-faire* social policies.

Spanish folk traditions and modern cruelty were studied by the dramatist **Federico Garcia Lorca**, 1898-1936. In Germany an aesthetic and idealist style was used by **Thomas Mann**, 1875-1955. His *Joseph and his Brothers*, 1933-43, explores the theme of exile. **A sense of cultural collapse** inspired **Robert Musil**, 1880-1942, to write *The Man Without Qualities*.

Ben Nicholson, 1894-   , one of Britain's leading abstract artists, achieved worldwide recognition in the Cubist and Abstract Art exhibition, New York, 1936. The rectangular, textured "White Relief", 1935, is typical of his style at this time. **Frank Lloyd Wright**, the American architect, produced outstanding buildings, including his famous **Falling Water**, Bear Run, Pa. 1936-7.

Musical theatre reached a peak of sophistication in the United States, with lavish shows and beautiful songs, notably by **George Gershwin**, 1898-1937 in *Porgy and Bess*, 1935.

The first television service was opened in Britain in 1936. **New industries** were developed to escape the depression. In Britain and the US, the new interest in consumer expenditure, combined with the completion of the electricity supply, led to the growth of consumer durable industries, while in Germany road-building was encouraged. **The citric acid cycle**, which occurs in bodily energy production, was found in 1937.

1937-40

Germany annexed Austria, 1938. Europe's powers met at **Munich**, 1938, to discuss German claims in Czechoslovakia, but failed to restrain Hitler. **German threats** to take Danzig in Poland resulted in Anglo-French intervention. **A European war** began in Sept 1939. **Francisco Franco**, 1892-1975, became dictator of Spain after defeating the republicans. **Germany took France** in 1940. **Japan** had conquered most of eastern China by 1939.

Neville Chamberlain, 1869-1940, aimed to avoid war at Munich, 1938. He abandoned appeasement after Hitler took Prague in 1939.

Mao Tse-tung, 1893-1976, adapted Marxism-Leninism to Chinese conditions, and argued that the peasantry, as well as the industrial proletariat, could succeed in making a socialist revolution. Mao later maintained that socialism could only be reached by a permanent revolution to prevent the development of privilege. His studies in guerrilla warfare were important to his political success and influenced later Third World revolutionaries.

Experimental epic theatre was pioneered by the German Marxist **Bertold Brecht**, 1898-1956. **Many foreign writers** fought in the Spanish Civil War. **George Orwell**, 1903-50, described it in *Homage to Catalonia*, 1939. **Important English novelists**, dealing with traditional themes were **Graham Greene**, 1904- **Aldous Huxley**, 1894-1963, and **Evelyn Waugh**, 1903-66.

One of **Picasso's finest paintings**, "Guernica", 1937, was prompted by the destruction of this Basque town by German bombers during the Spanish Civil War. **Hollywood** won an international supremacy in film-making during the depression, using enormous casts in lavish productions such as *Gone with the Wind*, 1939.

Serial music developed further with the work of **Alban Berg**, 1885-1935, and **Anton Webern**, 1883-1945, eventually submitting all musical elements to mathematical procedures.

The Graf Zeppelin (LZ 130) was built in 1938, the largest airship to be made. It ran on a regular commercial transatlantic service. **The Volkswagen "Beetle"**, designed by **Porsche** to Hitler's requirements, was built 1938. **Nuclear fission**, developed as a source of energy in the US, was first achieved in 1939. **Einstein told the US president** of the possibility of making an **atomic bomb**, 1939, to pre-empt German research. **Food-dehydration** by vacuum-contact drying was developed.

1940-43

Germany waged a lightning war in the West, but failed to invade or destroy Britain, which fought at sea and in the air, and opposed Italy in north Africa. **In June 1941 Germany invaded Russia** and drove the Red Army back to Moscow. **Japanese aggression** in the Pacific, culminating in the attack on **Pearl Harbor**, Hawaii, 1941, brought the US into the war. **Hitler** began the systematic genocide of the Jews, 1941.

British troops were repulsed in Europe. **Winston Churchill**, 1874-1965, led a coalition government, uniting Britain against German air attacks, 1940.

Women munitions workers

Phenomenology was developed in France by **Maurice Merleau-Ponty**, 1908-61, in *The Structure of Behaviour*, 1942. Closely associated with him was **Jean-Paul Sartre**, 1905-   , who in *Being and Nothingness*, 1943, advanced the Existentialist claim that authentic existence requires the individual exercise of free choice between alternative possibilities, without reference to accepted social roles. **Oxfam** was founded, 1942, to combat Third World poverty.

The *Makioka Sisters*, 1943-8, an account of a Japanese family by **Tanikaki Junichiro**, 1886-1965, owes much to Western realism. **Serious native American drama** was created by **Eugene O'Neill**, 1888-1953 and **Tennessee Williams**, 1911-   , who explored the frustrations of urban society. The plays of **Arthur Miller**, 1915-   , deal with individual moral and political responsibility.

American artists turned increasingly to the depiction of provincial life in a realistic style. "Nighthawks", 1942, by **Edward Hopper**, 1882-1967, records with formal precision the isolation of a city at night. **Hollywood cinema** escaped the limitations of its genres (westerns, gangster films and love stories) with *Citizen Kane*, 1941, directed by **Orson Welles**, 1915-   , and *The Grapes of Wrath*, 1940, by **John Ford**, 1895-1973.

Glen Miller, 1904-44, leader of the US Army Air Force Band in Europe, entertained troops with his distinctive, "big band" saxophone sound.

Plutonium, the first artificial element, was made, 1940. **The first jet-powered aircraft** flew in 1941, using an engine made by **Frank Whittle**, 1907- **The first nuclear reactor** was built in 1942 in Chicago. **Penicillin**, the first antibiotic, was produced on a large scale from *Penicillium* mould in 1943. The German development of the **V2 rocket-bomb**, 1942, provided the basis for future rocket development. The war also brought improvements in electronics and medical equipment.

1943-5

In 1943 Russia stopped the Germans at Stalingrad and Anglo-American forces took north Africa and invaded Italy. **Guerrilla action**, especially in Yugoslavia and France, weakened Nazi control. **The invasion of Normandy** by Britain and US in June 1944 opened a "second front" and Allied forces from east and west met on the Elbe in April 1945. **The Allies** agreed on Soviet and western spheres of influence at Yalta in 1945.

With conscription British military forces totalled 4.5 million in 1944. **Labour** won a landslide victory in 1945.

Churchill, Roosevelt and Stalin at Yalta

Karl Popper, 1902-   , an Austrian living in England, argued that scientific theories must be open to falsification, so that scientific progress required a community in which accepted ideas were subject to criticism. In *The Open Society and its Enemies*, 1944, he went on to attack the belief that there are general laws in history, which he saw as leading to totalitarian politics.

Salvatore Quasimodo, 1901-68, opposed fascism in Italy in lyrical symbolist poetry. **The "negritude" movement**, calling for black cultural identity, was initiated by **Leopold Senghor**. **Latin American literature** flourished with the "poetry for simple people" of the Chilean **Pablo Neruda**, 1904-73, and the complex stories of the Argentinian **Jorge Luis Borges.**

Official war artists in Britain such as **Graham Sutherland**, 1903-   , and **John Piper**, 1903-   , recorded the devastating effects of the bombings. **Mies van der Rohe**, in the US from 1938, designed during the war years the campus of the Illinois Institute of Technology, using cubic simplicity and perfect precision in details.

The swing era, 1935-45, dominated the popular music interest in the US, featuring such big bands as **Benny Goodman's**, playing highly arranged jazz with an energetic beat.

Large diameter pipelines were introduced in the US, facilitating the distribution of oil, 1943. **DNA** was shown to carry hereditary characteristics, 1944. **The kidney machine** was developed in 1944. **IBM** produced a mechanical calculating machine, 1944. **DDT** was discovered in 1939 and introduced as an insecticide, 1944, as synthetic fertilizers became available, leading to an increase in agricultural yields.

1945-7

America dropped two atomic bombs on Japan and ended the war in the Pacific, 1945. **The United Nations** was formed in 1945. **The Truman doctrine**, 1947, promised aid to non-communist countries, particularly Turkey and Greece. **Britain granted independence to India**, 1947, which divided through religious conflict. **The Chinese communists** were aided by Japan's defeat, controlling Manchuria by 1947.

Clement Attlee, 1883-1967, led Labour's radical programme of strict economy combined with nationalization, welfare and decolonization, 1945-50.

T. Adorno, 1903-   , and **M. Horkheimer**, 1895-   , of the **Frankfurt School of Sociology**, argued in *Dialectics of Enlightenment*, 1947, that true knowledge could only be achieved by a social revolution which would liberate man from the idea that nature is independent of, and external to, him. **Dietrich Bonhoeffer**, 1906-45, a German Protestant, argued that God is dead and sought a conception of Christianity relevant to a secular society.

In ***Deaths and Entrances*** 1946, the exuberant imagery of the Welsh poet **Dylan Thomas** 1914-53, is at its best. **Russia's history from 1900-30** was the subject of the humanistic novel *Doctor Zhivago*, by **Boris Pasternak**, 1890-1960. **Italy's** leading novelists, **Cesare Pavese**, 1908-50 and **Alberto Moravia**, 1907-   , both condemned the estrangement of the modern age.

Emaciated single figures on wire frames characterized the work of the Swiss sculptor **Albert Giacometti**, 1901-66, such as "Man Pointing", 1947. The British sculptor, **Henry Moore**, 1898-   , used his material to express natural forms in terms of stone or wood. "Three Standing Figures", 1947-8, is characteristic of his work. **Italian neo-realist cinema** relied on simple stories and untrained actors, as in *Rome, Open City*, 1945, by Roberto Rossellini, 1906-77.

Be bop, a complex form of jazz featuring virtuoso improvisation, emerged in 1945 as a reaction to the widely popular swing style. Its principal creator was **Charlie Parker**, 1920-55.

The first nuclear bombs were made in the US in 1945 and tested at Alamogordo, New Mexico, in 1945. **Britain's first atomic power station** was built, 1947. **The sound barrier** was broken by the Bell XI rocket-propelled American aircraft, 1947. **Radiocarbon dating**, a method of accurately finding the ages of archaeological discoveries, was perfected in 1947.

1947-50

The USSR blockaded Berlin, 1948-9, to isolate it from the west. **Zionists** declared Israel's independence, 1948. **Indonesia** threw off Dutch rule. **Mao Tse-tung**, 1893-1976, set up the People's Republic of China, 1949. **The North Atlantic Treaty Organization** provided for mutual assistance against aggression among the Western powers. **The socialist coup** in Czechoslovakia, 1948, extended Soviet control of eastern Europe.

The National Health Service came into operation in 1948, and rationing was relaxed. More curbs were imposed on the power of the Lords.

Nuclear bomb test

Martin Buber, 1878-1965, a Jewish thinker influenced by the mysticism of the Hasidic tradition in Judaism, advocated a direct, personal relationship of man with God, and praised the new *kibbutzim* in Israel as almost ideal socialist communities, in *Paths to Utopia*, 1947. **The World Council of Churches** first met in 1948. **The welfare state** uniting private enterprise with state responsibility, took shape with the British National Health Service, 1948.

Jean-Paul Sartre, 1905-   , gave existential philosophy a literary form in his war trilogy, *Les Chemins de la Liberté*, 1945-9. The existential dilemma also masks the feminist novels of **Simone de Beauvoir**, 1908-   . **Albert Camus**, 1933-60, formulated his theories of **The Absurd** in his novels and essays, notably *The Outsider*, and *The Plague*, 1947.

Abstract Expressionism developed in the US after 1945, expressed in the drip paintings of **Jackson Pollock**, 1912-56, in the "black and white" paintings of **Willem de Kooning**, 1904-   , and in the blurred expanses of rich colours in the work of **Mark Rothko**, 1903-70. The Unité d'Habitation, 1946-52, by **Le Corbusier**, a huge block of 337 two-storey apartments, was the first building to use rough cast concrete.

Radio and gramophone disseminated music to all developed countries, spreading new forms and styles so widely that national schools could no longer emerge.

A Jaguar sports car, capable of 193km/h (120mph) was put into production in 1948. **The "steady-state" theory** of the universe was proposed by **H. Bondi and T. Gold**, 1948. **The transistor** was invented in 1948. Its invention made possible the miniaturization of electronic equipment and, with micro-circuitry, the computer. **A United States step rocket** sent a vehicle to a height of more than 390km in 1949. **The World Health Organization** was set up in 1948.

1950-1975 The modern world

Principal events

The division of the world into two major power blocs after World War II was confused by a Sino-Soviet ideological split, and after a series of dangerous incidents between Russia and America in the 1950s and early 1960s the Cold War gave way to a period of official détente.

In spite of continuing imperialism by the major powers, whether militarily or by economic intervention, Third World liberation from European control has accelerated, changing the composition of the United Nations as the newly independent African and Asian states have joined and forcing industrialized countries to pay a higher price for raw materials.

Economic planning has become increasingly world-wide with the rise of development economics and the attempt to control currency exchange rates. In the 1970s serious inflation has spread to all the industrialized countries.

1950-2

War between North and South Korea, which had its roots in the Communist triumph in China, produced UN intervention.

The Arab League powers formed a security pact and began a blockade of Israel, 1950.

No agreement on Germany's future was reached, but peace was made with Japan, 1951.

The US strengthened defence links with Japan and Formosa.

Six European powers joined a single **Coal and Steel Commission**, 1952.

1952-5

Geneva conference, 1955, divided Vietnam into North and South after a Communist victory at **Dien Bien Phu** had forced the French forces to withdraw.

Opposition to British and French imperialism brought terrorist campaigns in Algeria, Kenya, Cyprus and Malaya.

The USSR opposed the reunification of Germany, 1954, and the **Warsaw Pact** united the Soviet satellites in reaction to West Germany's incorporation in NATO, 1955.

1955-8

The Soviet leader, Nikita Khrushchev, 1894-1971, denounced Stalinist principles, 1956; a Sino-Soviet split resulted. Soviet troops invaded **Hungary**, to crush a nationalist rising, 1956.

President Nasser, 1918-70, of Egypt nationalized the Suez Canal Company, 1956, provoking Britain, France and Israel to military intervention.

The Treaty of Rome, 1957, established the **Common Market** in western Europe.

1958-60

Discontent in France over the Algerian war brought **Charles de Gaulle**, 1890-1970, to power in 1958.

China's Great Leap Forward, an economic push in agriculture and industry, 1958, ended in economic chaos, after the withdrawal of Soviet aid, 1960.

World opinion was aroused by the Sharpeville massacre, 1960.

Fidel Castro, 1927- , a Marxist, controlled **Cuba**, 1959.

The Belgian Congo's independence, 1960, led to anarchy.

National events

Britain granted independence to most of her empire and aligned herself with the US and Europe, but after the prosperity of the 1950s and 1960s governments have faced increasing violence in Northern Ireland while mounting inflation and worsening industrial relations prompted a questioning of the Welfare State.

Hugh Gaitskell, 1906-63, imposed health service charges in 1951 to meet defence costs. **Labour** lost the October election.

Britain exploded her first atomic bomb, 1952. **Churchill's ministry**, 1951-5, de-nationalized road transport and steel. **Food rationing** ended, 1954.

Britain lost prestige in the Suez crisis, 1956, but exploded hydrogen bombs, 1957. **The Campaign for Nuclear Disarmament** began, 1958.

Cyprus was given her independence, 1960. **Britain** agreed to join the European Free Trade Association, 1960.

Medivac in the Korean War

Le Corbusier: design

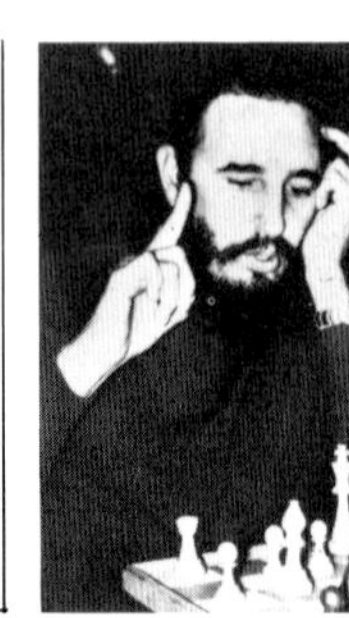

Fidel Castro

The Berlin Wall

The Beatles, 1963

Nyerere of Tanzania

Religion and philosophy

American sociology has been dominant in the West since World War II, expanding the use of surveys and other observational techniques into a major tool of government policy and developing in the work of Talcott Parsons a complex schema for the understanding of whole societies. Many of the general trends of thought seen in the industrialized countries also originated in the United States, whether in the work of theorists such as Marcuse, in the radical opposition to the Vietnam War or in the hippy movement, with its rejection of political activism and search for increased personal awareness and communal living.

In the same period Third World theorists have produced an analysis of the processes and effects of colonialism and the means of eradicating it.

The Christian churches have tried to overcome some of their differences, and in the Third World become linked with progressive social policies.

Franz Fanon, 1925-61, a West Indian, analysed the psychological and social repression of the black man in *Black Skin, White Masks*, 1952. He advocated an independent and socialist Third World.

Talcott Parsons, 1902- , an American, developed structural functional sociology in *The Social System*, setting out to construct a general model for societies, showing the interdependence of their institutions and emphasizing shared values.

Joseph McCarthy, 1908-57, led an American witch-hunt against liberals and Marxists as a result of Cold War tension.

The Oxford School of Ordinary Language Philosophy, including **Gilbert Ryle**, 1900- , and **J. L. Austin**, 1911-60, followed Wittgenstein's later ideas. In *Dilemmas*, 1954, Ryle tried to show that problems in philosophy derive from conceptual confusion and would be dissolved if we kept to the normal meaning of words.

Noam Chomsky, 1928- , an American, revolutionized linguistics by analysing the structure of language. He showed in *Syntactic Structures*, 1957, that grammatical speech depends upon a system of rules too complex to be learned by example.

Paul Tillich, 1886-1965, a Protestant, sought to fuse traditional religious values with a modern emphasis on individual responsibility in the *Dynamics of Faith*, 1956.

Structuralism, the attempt to find basic patterns or "structures" for a scientific study of man, was developed by the Frenchman **Claude Levi-Strauss**, 1908- in his *Structural Anthropology*, 1958. **Michel Foucault**, 1926- has applied this method to the history of ideas.

Jean-Paul Sartre, 1905- tried to link **Existentialism** and **Marxism** in the *Critique of Dialectical Reason*, 1960.

Literature

The rise of a worldwide reading public and of cheap and widely distributed books has allowed the writer greater freedom of experimentation. Increasingly confessional novels reflecting a sense of the isolation of the individual, and the use of a journalistic approach to deal with contemporary events have challenged the very concept of fiction, which has traditionally required a distance between the author and his subject. At the same time the beat writers, in seeking to celebrate the spontaneous, have questioned artistic form.

However, traditional literary forms remain the main vehicle for Third World writers, who have set out to portray the conflicts aroused in the individual by the colonization.

The Theatre of the Absurd, which saw man as a helpless creature in a meaningless universe, was explored by **Samuel Beckett**, 1906- , **Eugene Ionesco**, 1912- , and **Jean Genet**, 1910-.

The "new novel", without form or plot, was developed in the work of **Alain Robbe-Grillet**, 1922- and **Nathalie Sarraute**, 1900-.

Black American writers gained status with *The Invisible Man*, by **Ralph Ellison**, 1914- , and the writings of **James Baldwin**, 1924- . The Swiss dramatists **Max Frisch**, 1911- , and **Friedrich Dürrenmatt**, 1921- , and the Frenchman **Jean Anouilh**, 1910- , share a preoccupation with the tragi-comic and grotesque aspects of life.

English drama was active in the 1950s. Disillusionment with contemporary Britain was vented by **John Osborne**, 1929- in his play *Look Back in Anger*, 1956. **Harold Pinter**, 1930- wrote *The Birthday Party*, 1958, which he followed with more "comedies of menace". **Arnold Wesker**, 1932- , wrote socialist plays including *Roots*, 1958.

The work of **Jack Kerouac**, 1922-69, epitomizes the outlook of the American beat generation. Its writers include **William Burroughs**, 1914- , and **Henry Miller**, 1891- , and the poets **Allen Ginsberg**, 1926- , and **Lawrence Ferlinghetti**, 1919- , all of whom sought "spontaneous living" and the means to express it.

Art and architecture

Although America still dominates the visual arts, the increasingly international nature of the market has brought a new uniformity of style most clearly seen in architecture, where monumental concrete styles are found throughout the world.

In painting, attempts to explore the fundamentals of visual language have produced an ever-simplified abstract style and the breakdown of traditional distinctions between the disciplines and even between art and life; while Pop-art has incorporated into art the mass-produced images of consumer society.

The emergence in many parts of the world, including South America, India and eastern Europe, of the art film, aiming more at expression than at profit, has challenged the domination of Hollywood and forced the adoption of new formal styles and greater individual freedom in American commercial cinema, as well as a more critical view of modern society.

Skyscraper building in the US revived after World War I, with a new reliance on glass and steel, seen in Lever House, New York, 1952.

Le Corbusier designed **Chandigarh**, the new capital of the Punjab, 1950, in rough cast concrete.

The growth of film festivals after 1945 led to a less commercial cinema, and brought the work of the Japanese **Kurosawa**, 1910- , and the Indian **Satyajit Ray**, 1921- , to the West.

Pier Luigi Nervi, 1891- , regarded as the most brilliant concrete designer of his age, helped design the UNESCO building in Paris, 1954-8.

The International Style in architecture can be seen, post-war, at its most elegant in the Rødovre Town Hall, Copenhagen, begun in 1955 by **Arne Jacobsen**, 1902-71.

"Brutalism" in architecture was a term coined, 1953, for a functional style which, for example, let electric ducts be clearly seen.

Pop-art emerged in London in 1956 in the works of **Richard Hamilton**, 1922- , **Peter Blake**, 1932- , and **Eduardo Paolozzi**, 1924- , using motifs from commercial art.

Hard-edge painting with large, clearly defined areas of bright colour, was conceived in New York, 1958, and explored by **Ellsworth Kelly**, 1923-.

Kinetic art made use of light and movement for its effects as in *Mobile*, 1958, by **Alexander Calder**, 1898-.

Brazilian architecture centred on the building of a new capital, **Brasilia**. Its cathedral, 1959, by **Oscar Niemeyer**, 1907- , uses graceful curved concrete structures.

The New Wave of French cinema emerged in 1959 in reaction to the clichés of Hollywood. *400 Blows*, 1959 by **François Truffaut**, 1932- , and *Breathless*, by **Jean-Luc Godard**, 1930- , introduced stylistic innovations on a low budget.

Music

New elements have appeared in Western music, stemming from new ways of producing sound and of organizing the music. The open texture of Eastern music began to make its mark in the West as Western music, in turn, reached the East. Rock music began simply in the 1950s and soon became highly creative.

Traditional methods lived on in the operas of **Benjamin Britten**, 1913-76, and the symphonies of **Dimitri Shostakovitch**, 1906-75, who created personal styles of music by conventional means.

John Cage, 1912- , pioneered a music in which the score is a set of directions delineating a musical process, giving much freedom to performers. His notorious *4' 33"*, 1954, is all silence.

Musique concrète widened musical horizons after World War II in France, involving a collage of sounds, both musical and natural, processed into a recording.

Rock music promoted a strong eight-note beat over a static harmony in popular music. At first played on guitars by groups with a lead singer, it later became far more complex.

Science and technology

Scientific institutions set up by governments or industries have taken over from the individual experimenter, as the scale on which scientific research is conducted has mushroomed. The growth in prosperity in industrialized countries since 1945 has been accompanied by a boom in technologically sophisticated goods available to the general public; in particular, electronic equipment has been improved by miniaturization.

Much scientific research has been related to the rival arms and space programmes of the USSR and the US. But since the completion of the American Apollo Moon programme, the emphasis in the US has shifted to the ecological problems which man must solve if he is to have a future on Earth. The hunt for new energy resources has been stimulated by a rise in oil prices, and new foods have been developed to help cope with expanding population. Small-scale, technological innovations have benefited Third World economies.

Magnetic recording developed during the 1950s. Modern sound and video recording as well as computer operations depend on storing electrical signals in the form of magnetic patterns according to principles discovered by the Danish inventor **Valdemar Poulsen**, 1869-1942, in 1898.

Soya-bean farming increased, c. 1950, following a growing demand for vegetable oil during World War II.

The first hydrogen bomb was tested by the United States in 1952.

The structure of DNA was found in 1953, leading to closer understanding of protein synthesis in the body and the inheritance of characteristics by the next generation.

Polio vaccine was developed, 1953-5.

Oral contraception followed from the investigations in the 1950s into the role that sex hormones play in reproduction.

The link between smoking and lung cancer was first proposed in 1952.

The neutrino, a fundamental particle predicted in 1931, was detected in 1956.

Nuclear power was first generated on a viable industrial scale in Britain from 1956.

Britain introduced a **Clean Air Act**, 1956, after 4,000 died in a London smog, 1952.

International Geophysical Year, an international venture to investigate the Earth, took place, 1957-8.

The first artificial satellite, Sputnik 1, was launched by the USSR in 1957.

Explorer 1, the first American satellite, was launched, 1958, and detected radiation belts above the Earth.

Stereophonic records first became available in 1958.

The hovercraft was demonstrated in 1959.

Computers entered into commercial use, 1955, and were common by 1960.

The kidney machine was developed in the US c. 1960.

1960-3

South Africa left the British Commonwealth, 1961, after Britain accepted the trend of decolonization in Africa, 1960. **The Russians** built the Berlin Wall, 1961. After **Joyhn F. Kennedy**, 1917–63, the US president, intervened in the Cuban civil war, 1961. Soviet missile supplies to Cuba provoked a world crisis, 1962. Kennedy supported the **Civil Rights Movement**, encouraging the march on Washington, 1963. **Algeria** won her independence.

1963-5

The Nuclear Test Ban Treaty was signed by the US, USSR and Britain, 1963, but China exploded her first bomb in 1964. **After Kennedy's assassination**, **Lyndon B. Johnson**, 1908-73, passed civil rights bills and built up US forces in Vietnam to oppose the Communist rebels. **Britain** granted independence to **Kenya**, 1963, and **Malawi**, 1964. But Rhodesia declared her own independence under white rule.

1965-8

The Chinese Cultural Revolution aimed to weaken the bureaucracy and stimulate more public participation, 1966-8. **Growing American military activity** in Vietnam failed to bring victory. **Biafran** secessionist claims led to civil war in Nigeria. **Israel** defeated the Arab states in the 1967 **Six Days War** and extended her frontiers. **France** left NATO, 1966, to protest against American strength in Europe.

1968-70

Student revolt in France, 1968, was reflected throughout Europe. After referendum defeat, 1969, de Gaulle resigned. **Soviet troops** invaded Czechoslovakia to end liberal reforms. **Richard Nixon**, 1913- , resumed bombing North Vietnam, after peace talks and troop withdrawals, 1970. **Tanzania and Zambia** secured Chinese support for a railway linking the copperbelt to the sea.

1970-3

Massive balance of payments deficits forced a devaluation of the US currency, 1971. **China joined the UN**, 1971. **Bangladesh** was set up, 1971, after a civil war in Pakistan. **The EEC expanded** to include Britain, Ireland and Denmark. **Nixon** visited China, 1972, and secured rapprochement with the Soviet Union. **Salvador Allende**, 1908-73, a Marxist, was elected president of Chile, 1970, but was killed after a right-wing coup.

1973-6

US troops left Vietnam, 1973. **The Arabs** fought well in the October War, then forced up world oil prices to put pressure on Israel's western allies. **The Watergate scandal** in the US over White House corruption forced Nixon to resign, 1974. **A coup in Portugal**, 1974, led to developing revolution and the breakup of Portugal's colonial empire in Africa. **Communist forces** took control of South Vietnam and Cambodia, 1975.

Harold Macmillan, 1894- accepted decolonization and racial equality within the expanded Commonwealth. **Labour** debated nationalization.

Following Tory leadership disputes and a Labour victory in the election of 1964, the steel industry was nationalized, 1965.

Under Harold Wilson, 1916- , British policies on racial equality were rejected by the white minority of Southern Rhodesia, which claimed independence, 1965.

The Labour government faced continued economic problems in spite of devaluation, 1967. **Immigration** restrictions were introduced, 1968.

Edward Heath, 1916- , faced growing violence in Northern Ireland and a rising rate of inflation. He negotiated entry to the EEC, 1973.

Heath introduced a **3-day week** after a miners strike, 1973. **Labour** returned to office, 1974. Britain's first **referendum**, 1975, approved entry to the EEC.

Riots in Washington at the death of Martin Luther King

Ho Chi Minh

Appollo astronaut

Bangladesh famine victims

Prince Fahd of Saudi Arabia

R. D. Laing, 1927- , studied schizophrenia in a personal rather than clinical way in *The Divided Self*, 1960, and developed a humanistic school of anti-psychology. **The Ecumenical Movement** for Christian unity began in 1961-2 when the Eastern Orthodox and Catholic churches met with Protestants at the World Council of Churches, while the Vatican Council, 1962, tried to reconcile differences within Catholicism.

Herbert Marcuse, 1898- , associated with the "Frankfurt" School of Sociology, argued in *One Dimensional Man*, 1964, that in modern industrial society there is a process of "repressive tolerance" which diverts the creative impulses in man by satisfying his material needs. **Julius Nyerere**, president of Tanzania since 1964, set out to weaken Western influence by political non-alignment, to develop a village-based socialism, and to foster African nationalism.

The American Civil Rights Movement against racial intolerance of Negroes was led by **Martin Luther King**, 1929-68, who believed in the use of moral forces. In the mid-60s, however, black leaders such as **Eldridge Cleaver**, turned to violence. **The "flower-power" movement**, originating with American students in 1967, seeking awareness with the aid of mind-expanding drugs. A US counter-culture grew up, based on communes and anarchism.

The radical student movement of 1968, originating in America and Europe in opposition to the Vietnam War, stressed individual liberation from the constraints of capitalism, influenced by Third World revolutionaries such as **Che Guevara**, 1928-67, and the writings of **Marcuse**. **Pope Paul VI**, *r.* 1963- condemned the use of artificial methods of birth control, 1968, arousing widespread criticism.

Western religious groups stressing personal awareness included the "**Jesus Freaks**" and the **Divine Light Mission**, which has Hindu elements. **The Conservation movement** has argued that continued industrial growth is incompatible with the preservation of the natural world and its resources. *The Limits to Growth*, 1973, predicted the imminent disappearance of natural resources.

The rapidly growing population of developing countries was the subject of a campaign, 1973-4, organized by groups such as the **International Planned Parenthood Federation**, concerned to introduce birth control programmes to the Third World. The United Nations called 1975 **International Women's Year**, to encourage the participation of women in industry and world affairs.

The damaging effects of Western civilization in African culture are examined in the novels of **Chinua Achebe**, 1930- , and the plays of **Wole Soyinka**, 1934- , both Nigerians. The West Indian novelist **V. S. Naipaul**, 1932- , wrote of poverty in Trinidad with delicate irony in *A House for Mr Biswas*, 1961.

Postwar German society was explored in the writings of **Günter Grass**, 1927- and **Heinrich Böll**, 1917- **The American novel** flourished with many Jewish writers becoming prominent, among them **Saul Bellow**, 1915- , **Philip Roth**, 1933- and **Norman Mailer**, 1923- who also satirized politics in a journalistic style.

South American literature reached the West with translations of established writers. The Columbian **Gabriel García Marquéz**, 1928- , described the history of a family in a tropical town in *One Hundred Years of Solitude*, 1967. **Mexico's dual heritage** of savagery and civilization is the theme of the surrealistic poetry of **Octavio Paz**, 1914-

Change of Skin by the Mexican **Carlos Fuentes**, 1928- , is an "open novel" describing the fluctuations of experience. **Criticism of the Soviet regime** in the *Gulag Archipelago*, led to the exile of **Alexander Solzhenitsyn**, 1918- **The Japanese postwar generation** is described in the novels of **Yukio Mishima**, 1925-70.

Science fiction has become increasingly popular, notably in the works of American writers **Kurt Vonnegut Jr**, 1922- , **Isaac Asimov**, 1920- and **Ray Bradbury**, 1920- **Carlos Casteneda** published his *Journey to Ixtlan*, the last of a series of accounts of his meetings with a Mexican shaman.

Traditional English drama is represented by **Tom Stoppard**, 1937- , and **David Storey**, 1933- , while in America **Edward Albee**, 1928- , has made subtle use of conventional theatrical forms. **Experimental theatre groups** seek to work with the audience. **Formal experimentation** in the theatre has produced the almost silent plays of **Samuel Beckett**.

Distorted human forms confined within a claustrophobic space characterize the work of the Briton **Francis Bacon**, 1910- seen in his "Red Figure", 1962. **A move towards formalism** is seen in the dramatic use of curved concrete at the TWA buildings, Kennedy Airport, 1961, by **Eero Saarinen**, 1910-61. **Japanese architecture** united traditional forms with the new materials of steel and concrete in the work of **Tange**, 1913-

Two exponents of **Op-art**, who studied the effect of optical illusions juxtaposing colours and forms, were the Hungarian **Victor Vasarely**, 1908- , and the Briton **Bridget Riley**, 1931- **Pop-art in America** in the 1960s took images from cartoon comics as in *Whaam*, 1963, by **Roy Lichtenstein**, 1923- , and from commercial advertising in the work of **Andy Warhol**, 1930- **The Chinese sculpture** "Rent Collection Yard", 1965, depicted the miseries of the empire.

The "Happening", the creation of an environment simulating the effects of hallucinatory drugs, often with rock music and shifting patterns of colour, was pioneered in the US, *c.* 1965. **Realism** in British painting was exemplified in the works of **Lucian Freud** 1922- , whose portraits and townscapes show detailed draughtsmanship, and in the figure paintings of **David Hockney**, 1937- , like "Peter getting out of Nick's pool", 1966.

Land Art and **"Arte Povera"**, emerged in 1969 as an avant-garde movement which was concerned with art as assemblages of simple elements such as earth and rocks. **A politically committed documentary style** of film-making arose in Britain, seen in *Kes*, 1969, by the directors **Tony Garnett**, 1936- , and **Kenneth Loach**, 1936-

An exhibition in London and Paris, 1973, of **Chinese art** treasures, including archaeological discoveries made during the Cultural Revolution, restored cultural contacts between China and the West. **The Hong Kong film industry**, specializing in kung-fu and karate films, won immense popularity in the West after 1973 with films such as *Fist of Fury*, 1973, starring **Bruce Lee**, 1940-73.

Conceptual art, practised by **Barry Flanagan**, 1944- , and **Keith Arnott**, 1931- , in England, aimed to communicate through concepts rather than visual images. **An underground movement** of abstract artists in the USSR attempted unsuccessfully to hold an open-air exhibition in Moscow, 1974. **The epic disaster film** became popular in Hollywood, including *Earthquake*, 1974, and *The Towering Inferno*, 1974.

Graphic notation of symbols to portray sound became widespread in the 1960s as composers sought new sounds and effects from electronic and conventional instruments.

Simplicity and space marked the experimental music of the 1960s. Composers like **Terry Riley**, 1935- use simple repeated phrases that overlap in ever-changing patterns, as in his *In C*, 1964.

The tape recorder, invented in 1942, made all kinds of artificial sound reproduction possible. It was used creatively in popular music, as in the song "Sergeant Pepper", 1967, by the **Beatles**.

Poet-musicians became popular in the late 1960s, singing their own often highly individual compositions. Most influential was **Bob Dylan**, 1941- .

The synthesizer became a readily accessible instrument with the development of microelectronics. Its wide range of sounds may well spur future musical advances.

Electronic music permitted the composers to create and superimpose new sounds. Often contemplative, its sonority, forms and intonation give it Eastern qualities.

The bathyscape *Trieste* descended 11km (7miles) to the deepest part of the ocean, 1960. **The laser** was invented in 1960, and used for precision cutting and optical surgery. **Tiros I**, the first weather satellite, was placed in orbit by the United States in 1960. **Manned space flight** began in 1961 with a one-orbit mission by the Russian cosmonaut **Yuri Gagarin** 1934-68. **Telstar** the first communications satellite, was launched by the US in 1962.

Syncom, the first communications satellite which is constantly available for use, was put into orbit by the US in 1964. **Radiation** at a wavelength of 7 centimetres was first detected from space in 1965, providing support for the big bang theory. **The development of integrated circuits** in the 1960s brought new possibilities of miniaturization, stimulating the rise of the electronics industry in the US and making electronic equipment common in the West.

Plate tectonics developed as a theory to explain continental drift from 1965. **Mariner 4**, an American space probe, flew past Mars in 1965 and sent back the first pictures of another planet. **The first heart transplant** was performed in 1966. **Research into plant genetics** and soil fertility led to the **Green Revolution** in many Third World countries, 1966-70. **The Rance estuary power station** in France, harnessing tidal energy, was set up 1967.

DDT was banned in the US, 1969, following concern about its harmful side-effects. **The first Moon landing** was made in 1969 by members of the US Apollo 11 space mission. Space research has facilitated **invisible light astronomy**, and assisted **meteorology**. Spin-offs with industrial or domestic use include aluminium foil, and teflon, convenient for cooking utensils.

Earth resource satellites were first launched by the US in 1971 to detect and map the world's resources. **Germ warfare** was banned by international convention, 1972. **The rise in oil prices**, 1973, and fear over the limited nature of mineral fuel supplies has stimulated further research into alternative sources of energy. **Tidal, solar and geophysical energy** are being investigated.

The end of the US Moon programme released scientific resources for environmental research. Legislation in the US, 1974, spurred manufacturers to reduce lead pollution from the automobile. **The American Viking space probes** landed on Mars and sent back photographs 1976. **A Russian space probe** took pictures of Venus in 1975. **The "space-race"** ended when the Russian Soyuz spacecraft linked up with an Apollo in 1975.

INDEX

How to use the index The principles of using this index are explained at the beginning of the index to History and Culture Volume 1.

Picture Credits

16–17 [Key] Cooper Bridgeman; [4] Mansell Collection; [6] Henry E. Huntington Library & Art Gallery; [7] Cooper Bridgeman/Victoria Art Gallery, Bath. **18–19** [Key] Mansell Collection; [1] Cooper Bridgeman Library; [2] Sun Alliance & London Insurance Group; [3] Mansell Collection; [4] Radio Times Hulton Picture Library; [5] Eileen Tweedy; [6] Mansell Collection; [7] Fotomas Index; [8] Mansell Collection; [9] Fotomas Index. **20–1** [Key] County Records Office, Bedford; [1] Museum of English Rural Life, University of Reading; [3] Museum of English Rural Life, University of Reading; [5] Museum of English Rural Life, University of Reading; [6] Museum of English Rural Life, University of Reading; [7] Museum of English Rural Life, University of Reading; [8] Mansell Collection; [9] John Webb/by kind permission of the Earl of Leicester. **22–3** [Key] Source unknown; [1] Source unknown; [2] Trustees of the National Portrait Gallery; [3] Bibliothèque Nationale, Paris; [4] Mansell Collection; [5] Germanisches National Museum, Nuremburg; [6] Mansell Collection; [7] Lauros Giraudon; [8A] J. E. Bulloz. **24–5** [Key] National Gallery of Ireland; [1] Master & Fellows, Pembroke College, Cambridge; [2] Cooper Bridgeman Library; [3] C. M. Dixon; [4] Public Records Office of Northern Ireland; [5] Fotomas Index; [6] Radio Times Hulton Picture Library; [8] National Gallery of Ireland; [9] Mansell Collection; [10] Eileen Tweedy. **26–7** [Key] [1] C. M. Dixon; [2] Trinity College Library, Dublin; [3] National Museum of Ireland; [4] Commissioners of Public Works in Ireland; [5] C. M. Dixon; [6] National Portrait Gallery; [7] National Museum of Ireland; [8] National Museum of Ireland; [9] National Portrait Gallery; [10] National Portrait Gallery; [11] Dr. Peter Harbison. **28–9** [Key] Scottish Tourist Board; [1] Mary Evans Picture Library; [2] Mansell Collection; [4] Eileen Tweedy/BM; [5] Radio Times Hulton Picture Library; [6] MB photograph; [7] Mary Evans Picture Library; [8] Studio Swain, Glasgow/Mitchell Library, Glasgow; [9] Mansell Collection. **30–1** [Key] National Portrait Gallery; [1] C. M. Dixon; [2] National Museum of Antiquities of Scotland; [3] Woodmansterne; [4] Scottish Tourist Board; [5] Woodmansterne; [6] Studio Swain, Glasgow; [7] Tom Scott/Scottish National Portrait Gallery; [8] Woodmansterne/© National Trust for Scotland; [9] Picturepoint. **32–3** [Key] Radio Times Hulton Picture Library; [1] National Maritime Museum; [2] By kind permission of the President & Council of the Royal College of Surgeons of England; [3] Rijksmuseum van Naturlijke Historie; [4] Akademie der Wissenschaften der DDR/George Rainbird Ltd; [5] Trustees of the British Museum/Ray Gardner; [6A] India Office Library; [6B] Trustees of the British Museum/Ray Gardner; [7] Source unknown; [8] By permission of the Royal College of Physicians of London; [9] Radio Times Hulton Picture Library. **34–5** [Key] Crown Copyright Reserved; [1] Snark International; [2] By permission of Mrs Barbara Edwards/Carter Nash Cameron; [3] By courtesy of the Feoffes of Chetham's Library, Manchester; [4] Snark International; [5A] British Library/Ray Gardner; [5B] British Library/Ray Gardner; [6] Mansell Collection; [7] Trustees of the Tate Gallery/John Webb. **36–7** [Key] F. Lugt Collection, Holland; [1] Lauros Giraudon/Chateau de Versailles; [2] Giraudon/Louvre; [3] Michael Holford; [4] Bavaria Verlag; [5] Reproduced by permission of the Trustees of the Wallace Collection/John R. Freeman; [6] Telarci Giraudon/Louvre. **38–9** [Key] Giraudon; [1] British Museum/John R. Freeman; [2] British Museum/John R. Freeman; [3] Victoria & Albert Museum; [4] Mansell Collection; [5] Reproduced by permission of the Trustees of the Wallace Collection/John R. Freeman; [6] Angelo Hornak/V & A; [7] Victoria & Albert Museum; [8] British Museum/John R. Freeman. **40–1** [Key] Mansell Collection; [1] Scala; [2] Germanisches Nationalmuseum; [3] Bildarchiv Preussischer Kulturbesitz, Berlin; [4] Bildarchiv Preussischer Kulturbesitz, Berlin; [5] Ullstein Bilderdienst. **42–3** [Key] Royal Academy of Arts; [1] Reproduced by permission of the Trustees of the Wallace Collection/John R. Freeman; [2] Cooper Bridgeman/Towneley Hall Museum & Art Gallery, Burnley; [4] Trustees of the Tate Gallery/John Webb; [5] Photo Meyer, Wien; [6] Giraudon/Musées Royaux des Beaux-Arts; [7] Lauros Giraudon/Louvre. **44–5** [2] Lauros Giraudon/Musée de la Marine, Paris; [4] Radio Times Hulton Picture Library; [5] Angelo Hornak/V & A; [6] Trustees of the British Museum/John R. Freeman; [7] Ampliaciones y Reproducciones Mas, Barcelona; [8] Mansell Collection. **46–7** [Key] Bristol City Art Gallery; [1] Fotomas Index; [2] Picturepoint; [3] National Maritime Museum; [4] Picturepoint; [5] Eileen Tweedy/India Office Library; [6] Mansell Collection; [7] National Maritime Museum; [8] Mansell Collection; [9] Josiah Wedgwood & Sons Ltd. **48–9** [Key] India Office Library & Records; [1] Ann and Bury Peerless; [3] Ann and Bury Peerless; [4] Ann and Bury Peerless; [5] India Office Library & Records; [7] India Office Library & Records; [8] India Office Library & Records; [9] India Office Library & Records; [10] India Office Library & Records. **50–1** [Key] Michael Holford/V & A; [1] Ann and Bury Peerless; [2] Victoria & Albert Museum; [3] By kind permission of Air India; [4] Michael Holford; [5] India Office Library & Records/Ann and Bury Peerless; [6] By kind permission of Air India; [7] Michael Holford/V & A; [8] Victoria & Albert Museum. **52–3** [Key] William MacQuitty; [2] Trustees of the British Museum; [3] William MacQuitty; [4] William MacQuitty; [5] Radio Times Hulton Picture Library; [7] Trustees of the British Museum; [8] Cooper Bridgeman/National Maritime Museum. **54–5** [Key] Museum of Fine Arts, Boston; [1] Percival David Foundation; [2] Courtesy of the Smithsonian Institution, Freer Gallery of Art, Washington DC; [3] Nelson Gallery-Atkins Museum, Kansas City, Missouri (Nelson Fund); [4] National Palace Museum, Taiwan; [5] Trustees of the British Museum; [6] William MacQuitty; [7] Aubrey Singer, BBC/Robert Harding Associates; [8] Trustees of the British Museum; [9] Trustees of the British Museum. **56–7** [Key] International Society for Educational Information, Tokyo; [1] Bradley Smith; [2] Bradley Smith; [3] International Society for Educational Information, Tokyo; [4] Bradley Smith; [5] Bradley Smith; [6] Weidenfeld & Nicolson; [7] International Society of Educational Information, Tokyo; [8] Orion Press. **58–9** [Key] Werner Forman Archive; [2] Mansell Collection; [3] Mansell Collection; [4] Mansell Collection; [6] Mary Evans Picture Library; [7] Mansell Collection. **60–1** [Key] Trustees of the British Museum; [1] D. E. F. Russell/Robert Harding Associates; [2] John Picton; [3] Prof. Thurston-Shaw; [4] John Picton; [5] Trustees of the British Museum/Museum of Mankind/Angelo Hornak; [6] Trustees of the British Museum/Museum of Mankind; [7] Trustees of the British Museum/Museum of Mankind; [8] Photograph copyright 1976 By the Barnes Foundation; [9] Archivo e Studio; [10] From *Nuba Personal Art* by James C. Faris/Duckworth, 1972. **62–3** [6] By kind permission of the City of Bristol Archives/David Strickland; [7] Mansell Collection; [8] Bettmann Archive; [9] Bettmann Archive/Smithsonian Institution. **64–5** [3] Mary Evans Picture Library; [4] Metropolitan Museum of Art, New York, Gift of Mrs Russell Sage, 1910; [6] Copyright Yale University Art Gallery; [7] Mansell Collection. **66–7** [2] Ronan Picture Library; [5] A. F. Kersting; [6] Ronan Picture Library; [7] Derby Borough Councils & Art Gallery; [8] Josiah Wedgwood & Co; [9] Cooper Bridgeman/Lord Mountbatten. **68–9** [Key] Mansell Collection; [1] National Portrait Gallery; [2] National Portrait Gallery; [3] National Portrait Gallery; [4] Museum of London; [5] Radio Times Hulton Picture Library; [6] Eileen Tweedy; [7] Fotomas Index; [8] Radio Times Hulton Picture Library. **70–1** [Key] Trustees of the Croome Estate/Photo by Don J. Adams; [1] Iveagh Bequest, London; [2] National Gallery; [3] Aerofilms; [4] Cooper Bridgeman Library; [5] National Portrait Gallery; [6] Fotomas Index; [7] Angelo Hornak; [8] J. Bethell/National Trust. **72–3** [Key] The British Library; [1] Angelo Hornak; [2] Bavaria Verlag; [3] John R. Freeman; [4] Trustees of the Tate Gallery/John Webb; [5] Trustees of the British Museum/John R. Freeman; [6] Giraudon; [7] Archiv für Kunst und Geschichte, Berlin; [8] Mansell Collection: [9] Crown Copyright Reserved. **74–5** [Key] Giraudon; [4] Roger Viollet; [5] Giraudon/B. N. Estempes, Paris; [7] Giraudon/Chateau Versailles; [8] Réunion des Musées Nationales. **76–7** [Key] Roger Viollet; [2] Cooper Bridgeman/Coram Foundation; [3] Giraudon/Chateau Versailles; [5] Garanger Giraudon/Musée de l'Armée, Prague; [7] Angelo Hornak/V & A; [8] National Army Museum. **78–9** [Key] Picturepoint; [1] Radio Times Hulton Picture Library; [2] Radio Times Hulton Picture Library; [4] Robert Harding Associates; [5] Weidenfeld & Nicolson Archives/National Maritime Museum; [6] Fotomas Index; [7] Radio Times Hulton Picture Library; [8] Eileen Tweedy. **80–1** [Key] Trustees of the British Museum; [1] Courtesy of the Detroit Institute of Arts, Gift of Mr & Mrs Bert L. Smokler & Mr & Mrs Lawrence A. Fleischman; [2] Courtesy of Birmingham City Museum & Art Gallery; [3] Museo del Prado; [4] Kunst Dias Blauel/Stadelsches Kunstinstitut; [5] Telarci Giraudon/Louvre; [6] Lauros Giraudon/Louvre; [7] Telarci Giraudon/Louvre. **82–3** [Key] John R. Freeman; [2] Museen der Stadt, Vienna; [3] Crown Copyright Reserved; [4] Museen der Stadt, Vienna; [5] Trustees of the National Portrait Gallery; [6] Lauros Giraudon/Louvre. **84–5** [Key] Turkish Tourist & Information Office; [1] J. E. Bulloz; [3A] Bildarchiv der Österreichischen Nationalbibliothek; [3B] Bildarchiv der Österreichischen Nationalbibliothek; [4] William Allan; [5] National Army Museum; [6] Radio Times Hulton Picture Library; [7] Bildarchiv der Österreichischen Nationalbibliothek. **86–7** [Key] Bettmann Archive; [3] Mansell Collection; [4] Douglas Botting; [5] Mike Andrews; [6] Bettmann Archive; [7] Bettmann Archive; [8] Bettmann Archive. **88–9** [Key] Mary Evans Picture Library; [4] Mansell Collection; [8] By permission of the Science Museum, London; [9A] By permission of Stanley Gibbons Ltd/Angelo Hornak; [10] Mansell Collection. **90–1** [Key] Fotomas Index; [2] Mansell Collection; [3] Weidenfeld & Nicolson Archives; [4] Aerofilms; [5] Radio Times Hulton Picture Library; [6] David Strickland; [8] Michael Holford. **92–3** [Key] Freemans; [1] National Museum of Wales; [3] Fotomas Index; [6] Radio Times Hulton Picture Library; [7] Picturepoint; [8] Spectrum Colour Library; [9] Picturepoint. **94–5** [Key] Fotomas Index/BM; [1] Radio Times Hulton Picture Library; [2] Fotomas Index/BM; [3] Radio Times Hulton Picture Library; [4] National Portrait Gallery; [5] Fotomas Index; [6] Eileen Tweedy; [8] Radio Times Hulton Picture Library; [9] Radio Times Hulton Picture Library; [10] Radio Times Hulton Picture Library. **96–7** [Key] Weidenfeld & Nicolson Archives/National Monuments Record; [1] Radio Times Hulton Picture Library; [2] National Portrait Gallery; [3] Radio Times Hulton Picture Library; [4] Radio Times Hulton Picture Library; [5] Radio Times Hulton Picture Library; [7] Weidenfeld & Nicolson Archives; [8] National Portrait Gallery; [9] Aerofilms; [10] Radio Times Hulton Picture Library. **98–9** [Key] Mansell Collection; [1] Trustees of the Tate Gallery/John Webb; [2] J. Whitaker/Sphere Books; [3] Angelo Hornak/Scolar Press; [4A] John Frost Newspaper Collection/Angelo Hornak; [4B] Radio Times Hulton Picture Library; [5] Lauros Giraudon © S.P.A.D.E.M., Paris, 1976; [6] Roger Viollet; [7] Sally Chappell/V & A; [8] Giraudon/Louvre; [9] Society of Cultural Relations with the USSR. **100–101** [Key] University College, Oxford/Angelo Hornak; [1] Trustees of the British Museum; [2] John Massey Stewart; [3] Roger Viollet; [4] Scala; [5] Snark International; [6] Mander & Mitchenson Theatre Collection; [7] Society for Cultural Relations with the USSR; [8] Mander & Mitchenson Theatre Collection; [9] Radio Times Hulton

Picture Library. **102–3** [Key] Ashmolean Museum, Oxford; [1] Whitworth Art Gallery, University of Manchester; [2] Cooper Bridgeman/V & A; [3] Ralph Kleinhempel, Hamburg; [4] Cooper Bridgeman/Staatliche Kunstsammlungen, Dresden; [5] Gulbenkian Museum, Lisbon; [6] Trustees of the Tate Gallery; [7] Laing Art Gallery, Newcastle upon Tyne; [8] Trustees of the Tate Gallery. **106–7** [2] Spectrum Colour Library; [3] K. M. Andrew; [4] Novosti Press Agency; [5] Bildarchiv Preussischer Kulturbesitz, Staatsbibliothek, Berlin; [6] Bildarchiv Preussischer Kulturbesitz, Staatsbibliothek, Berlin; [7] Bildarchiv Preussischer Kulturbesitz, Staatsbibliothek, Berlin; [8] Bildarchiv Preussischer Kulturbesitz, Staatsbibliothek, Berlin; [9] Mary Evans Picture Library. **108–9** [Key] Mansell Collection; [3] Roger Viollet/Bibliothèque Nationale; [4] Ullstein Bilderdienst; [5] Scala; [7] Snark International. **110–11** [Key] Mansell Collection; [1] Bildarchiv Preussischer Kulturbesitz, Staatsbibliothek, Berlin; [3] John R. Freeman; [4] Out of copyright; [5] Mauro Pucciarelli; [6] Istituto per la Storia del Risorgimento, Rome. **112–13** [Key] Mansell Collection; [1] Punch Publications Ltd; [2] Robert Harding Associates; [3] Picturepoint; [4] J. Bethell/National Trust/Weidenfeld & Nicolson Archives; [5] Geoff Goode; [6] Mansell Collection; [7] National Portrait Gallery; [9] Cooper Bridgeman Library/the Guildhall Library & Art Gallery. **114–15** [Key] Radio Times Hulton Picture Library; [2] Fotomas Index; [4] Mansell Collection; [5] Radio Times Hulton Picture Library; [6] Radio Times Hulton Picture Library; [7] Radio Times Hulton Picture Library; [8] Radio Times Hulton Picture Library; [9] Robert Harding Associates/London Museum. **116–17** [Key] Helmut Gernsheim Collection; [1] Angelo Hornak/Wellington Museum; [2] Lauros Giraudon/Musée Fabre, Montpellier/© A.D.A.G.P. Paris, 1976; [3] Giraudon/Louvre; [4] Cooper Bridgeman; [5] Cooper Bridgeman/Sir Colin Anderson; [6] Trustees of the Tate Gallery; [7] Courtesy of Jefferson Memorial College, Thomas Jefferson University; [8] Walters Art Gallery, Baltimore; [9] Royal Academy of Arts, London; [10] Trustees of the Tate Gallery/John Webb. **118–19** [Key] J. E. Bulloz; [1] Giraudon/Musée Marmottan, Paris/© S.P.A.D.E.M. Paris, 1976; [2] Museum of Fine Arts, Boston, Arthur Gordon Thompkins Residuary Fund/© S.P.A.D.E.M. Paris, 1976; [3] Service de Documentation Photographique de la Réunion des Musées Nationaux/Musée Jeu de Paume/© S.P.A.D.E.M., Paris, 1976; [4] Service de Documentation Photographique de la Réunion des Musées Nationaux/Musé Jeu de Paume/© S.P.A.D.E.M., Paris, 1976; [5] Service de Documentation Photographique de la Réunion des Musées Nationaux/Musée Jeu de Paume; [6] K. X. Rousseu; [7] Service de Documentation Photographique de la Réunion des Musées Nationaux/Musée Jeu de Paume; [8] Service de Documentation Photographique de la Réunion des Musées Nationaux/Musée Jeu de Paume/© S.P.A.D.E.M., Paris, 1976; [9] Giraudon/Musée Jeu de Paume/© S.P.A.D.E.M., Paris, 1976; [10] Philadelphia Museum of Art/John G. Johnson Collection. **120–1** [Key] Brenda Houston Rogers; [1] Österreichische Nationalbibliothek; [2] Roger Viollet; [3] Novosti/National Russian Museum; [4] Mander & Mitchenson Theatre Collection; [5] Brenda Houston Rogers; [7] Gunter Englert; [8] Roger Viollet; [9] Stuart Robinson. **122–3** [Key] J. E. Bulloz; [1] Royal Academy of Arts; [2] A. F. Kersting; [4] From *Entretiens sur L'Architecture*, 1863; [5] From the *Pelican History of Art* by B. Abbott; [7] Mauro Pucciarelli; [8] The Museum of London; [9] Mauro Pucciarelli; [10] James Austin. **124–5** [Key] Geoff Goode/Trustees of the British Museum/Museum of Mankind; [2] Mansell Collection; [3] Trustees of the National Portrait Gallery; [4] Mary Evans Picture Library; [5] Trustees of the National Portrait Gallery; [8] Radio Times Hulton Picture Library; [9] Rijksmuseum, Amsterdam; [10] British Museum/Photo © Aldus Books, London. **126–7** [Key] The Art Gallery of New South Wales; [1] National Portrait Gallery; [2] National Portrait Gallery; [3] Waitangi National Trust; [4] Fotomas Index; [5] Weidenfeld & Nicolson Archives; [6] Fotomas Index/BM; [7] De Maus Collection/Alexander Turnbull Library; [8] Aukland Institute & Museum; [9] Australian Information Service; [10] Radio Times Hulton Picture Library. **128–9** [Key] Mansell Collection; [1] Popperfoto; [2] Mansell Collection; [4] Mansell Collection; [5] Alan Hutchison; [6] Cooper Bridgeman Library/National Army Museum; [7] Mary Evans Picture Library; [8] Radio Times Hulton Picture Library. **130–1** [Key] Kim Sayer; [3] Mary Evans Picture Library; [5] Mansell Collection; [6] Bildarchiv Preussischer Kulturbesitz, Berlin. **132–3** [Key] Mansell Collection; [1] Radio Times Hulton Picture Library; [2] Mary Evans Picture Library; [3] Maurice Rickards/photo by Angelo Hornak/courtesy Canadian Pacific; [4] Punch Publications Ltd; [5] Fotomas Index; [6] Royal Commonwealth Society; [7] Punch Publications Ltd. **134–5** [Key] Eileen Tweedy/BM; [1] Mansell Collection; [2] Mary Evans Picture Library; [3] Fotomas Index/National Maritime Museum; [5] Radio Times Hulton Picture Library; [7] Popperfoto; [8] ZEFA; [9] Popperfoto. **136–7** [Key] Popperfoto; [2] Mary Evans Picture Library; [3] Eileen Tweedy; [5] Mansell Collection; [6] Weidenfeld & Nicolson Archives; [9] "Reprinted with permission of The Toronto Star". **138–9** [Key] Werner Forman Archive/Anspach Collection, NY; [2] Romano Cagnoni; [3] Charles Perry Weimer; [5A] Anthony Atmore; [5B] Anthony Atmore; [6] Popperfoto; [7] Ann and Bury Peerless. **140–1** [Key] Punch Publications Ltd; [2] National Army Museum; [3] Rick Strange/Robert Harding Associates; [4] National Army Museum; [5] Mansell Collection; [6] National Army Museum; [7] Trustees of the National Portrait Gallery; [8] India Office Library & Records; [9] National Army Museum. **142–3** [Key] Source unknown; [2] Mary Evans Picture Library; [3] Radio Times Hulton Picture Library; [4] Sybil Sassoon/Robert Harding Associates; [5] Radio Times Hulton Picture Library; [6] From *Tribal Innovators* by I. Schapera; [7] Angelo Hornak/From *Life, Scenery & Customs of Sierra Leone & Gambia* by T. E. Poole; [8] Radio Times Hulton Picture Library. **144–5** [Key] Roger Viollet; [3] Radio Times Hulton Picture Library; [4] Library Archives of Weidenfeld & Nicolson; [8] V & A/Weidenfeld & Nicolson; [9] Mansell Collection. **146–7** [Key] Newcastle upon Tyne Public Libraries; [1] International Society for Educational Information, Tokyo; [3] Memorial Picture Gallery, Meiji Shrine, Tokyo; [4] Novosti Press Agency; [5] International Society for Educational Information, Tokyo; [6] Bradley Smith; [8] International Society for Educational Information, Tokyo. **148–9** All pictures from Western Americana. **150–1** [5] American History Picture Library; [9] Mansell Collection. **152–3** [1] Mansell Collection; [2] Western Americana; [3] Brown Brothers; [4] Angelo Hornak; [7] Brown Brothers. **154–5** [Key] Ronan Picture Library; [1] Royal Geographical Society; [2] Bradley Smith; [3] Tokyo National Museum; [4] National Maritime Museum; [6] Radio Times Hulton Picture Library; [7] Radio Times Hulton Picture Library. **156–7** [Key] Cavendish Laboratory/University of Cambridge; [1] The Royal Institution/Cooper Bridgeman; [2] Ronan Picture Library; [3] C. E. Abranson; [4] Mansell Collection; [5] The Royal Institution of Great Britain; [6] Ronan Picture Library; [7] Mansell Collection; [8] Popperfoto; [9] Mansell Collection; [10] Bavaria Verlag. **158–9** [Key] By kind permission of Marks & Spencer Ltd; [2] Radio Times Hulton Picture Library; [4] Radio Times Hulton Picture Library; [5] Angelo Hornak; [6] Mansell Collection. **160–1** [Key] Radio Times Hulton Picture Library; [1] National Portrait Gallery; [3] Radio Times Hulton Picture Library; [4] Punch Publications Ltd.; [5] Radio Times Hulton Picture Library; [6] Fotomas Index; [7] Radio Times Hulton Picture Library; [9] Fotomas Index. **162–3** [1] National Portrait Gallery; [4] Radio Times Hulton Picture Library; [5] Radio Times Hulton Picture Library; [6] Mansell Collection; [7] Radio Times Hulton Picture Library; [8] Mansell Collection; [9] Radio Times Hulton Picture Library. **164–5** [Key] Radio Times Hulton Picture Library; [2] Mary Evans Picture Library; [4] Woodmansterne; [5] Popperfoto; [7] National Portrait Gallery; [8] Radio Times Hulton Picture Library; [9] Weidenfeld & Nicolson Archives/Labour Party Photo Library. **166–7** [1] Eileen Tweedy/National Library of Wales; [2] Welsh Folk Museum/National Museum of Wales; [3] Eileen Tweedy/National Library of Wales; [4] National Museum of Wales; [5] Radio Times Hulton Picture Library; [6] Eileen Tweedy/National Library of Wales; [7] Eileen Tweedy/National Library of Wales; [8] National Portrait Gallery. **168–9** [Key] Novosti Press Agency; [1] Radio Times Hulton Picture Library; [2] Radio Times Hulton Picture Library; [3] John R. Freeman; (4) Novosti Press Agency; [5] Courtesy of Cultural Relations with the USSR; [6] Novosti Press Agency; [8] Radio Times Hulton Picture Library; [9] Novosti Press Agency; [10] Novosti Press Agency **170–1** [Key] Mansell Collection; [1] Punch Publications Ltd; [3] From the John Gorman Collection; [4] Radio Times Hulton Picture Library; [5] Giraudon; [6] British Red Cross Society; [7] Radio Times Hulton Picture Library. **172–3** [Key] Mansell Collection; [2] Manchester Public Libraries; [3A] Radio Times Hulton Picture Library; [3B] Radio Times Hulton Picture Library; [3C] Mansell Collection; [5A] Giraudon; [5B] Angelo Hornak/V & A. **174–5** [Key] Cooper Bridgeman/Rasmus Meyer Collection, Bildegalerie, Bergen; [1] Hans Hinz, Kunst Museum, Basle; [2] Trustees of Tate Gallery/© A.D.A.G.P., Paris, 1976; [3] Harry N. Abrams Inc/The Hermitage, Leningrad/© S.P.A.D.E.M. Paris, 1976; [4] Trustees of the Tate Gallery/© S.P.A.D.E.M. 1976/John Webb; [5] Hans Hinz, Basle; [6] Hans Hinz, Basle/© S.P.A.D.E.M., Paris, 1976; [7] Kunstsammlung Nordrhein Westfalen/© S.P.A.D.E.M., Paris, 1976; [8] By courtesy of Marlborough Fine Art (London) Ltd. **176–7** [Key] Trustees of the Tate Gallery/© A.D.A.G.P., Paris, 1976; [1] Collection Alex Maguy/© A.D.A.G.P., Paris, 1976; [2] Giraudon/Kunst Museum Basle; [3] Charles Uht/By permission of Nelson A. Rockefeller/© S.P.A.D.E.M., Paris, 1976; [4] Collection the Museum of Modern Art, New York, Bequest of Alma Erickson Levene/© A.D.A.G.P., Paris, 1976; [5] Philadelphia Museum of Art, The A. E. Gallatin Collection/Alfred J. Wyatt/© S.P.A.D.E.M., Paris, 1976; [6] Musée Leger/ S.P.A.D.E.M., Paris, 1976; [7] Collection, the Museum of Modern Art, New York. Acquired through the Lillie P. Bliss Bequest/© S.P.A.D.E.M., Paris, 1976; [8] Private collection, Milan; [9] Collection, the Museum of Modern Art, New York. Acquired through the Lillie P. Bliss Bequest. **178–9** [Key] Architectural Association; [1] A. F. Kersting; [2] Christabel Valkenberg; [3] Architectural Association; [4A] Architectural Association; [4B] Architectural Association; [5] Bill Engdahl/Hedrich Blessing; [6B] Architectural Association. **180–1** [Key] Ronan Picture Library; [1] Mary Evans Picture Library; [3] Ullstein Bilderdienst; [5] Österreichische Nationalbibliothek; [6] Source unknown; [7] Snark International; [8] Radio Times Hulton Picture Library. **182–3** [Key] National Portrait Gallery; [1] Photographie Giraudon; [2] Mary Evans Picture Library; [3] Radio Times Hulton Picture Library; [4] Mansell Collection; [6] Mansell Collection; [7] Punch Publications Ltd; [8] Popperfoto; [9] London School of Economics. **184–5** [Key] Source unknown; [1] Mansell Collection; [2] Cyril & Methodius National Library, Sofia, Bulgaria; [3] Mansell Collection; [4] Archiv für Kunst & Geschichte; [6] IBA, Zurich; [8] Radio Times Hulton Picture Library. **186–7** [Key] Roger Viollet; [1] Radio Times Hulton Picture Library; [2] Ullstein Bilderdienst; [7] Source unknown; [8] Ullstein Bilderdienst. **188–9** [Key] Imperial War Museum; [7] Robert Hunt Library; [10] Image Press. **190–1** [1] Imperial War Museum; [3] Heeresgeschichtliches Museum, Vienna; [5] Image Press; [6] Chaz Bowyer; [7] Ullstein Bilderdienst; [9] Image Press. **192–3** [Key] Roger Viollet; [2] Roger Viollet; [5] Radio Times Hulton Picture Library; [8] Mansell Collection; [9] Ullstein Bilderdienst. **194–5** [Key] Mansell Collection; [1] Radio Times Hulton Picture Library; [2] Radio Times Hulton Picture Library; [3] Eileen Tweedy/IWM; [4] Camera Press; [5] Wilfred Owen Estate; [8] Mansell Collection; [9] Radio Times Hulton Picture Library; [10] Radio Times Hulton Picture Library. **196–7** [Key] Novosti Press Agency; [1] Novosti Press Agency; [2] Mansell Collection; [3] Bettmann Archive; [4] Novosti Press Agency; [5] Novosti Press Agency; [6] Radio Times Hulton Picture Library; [7] Novosti Press Agency; [8] Snark International; [9] Geoff Goode/Courtesy The School of Slavonic Studies; [11] Radio Times Hulton Picture Library. **198–9** [Key] Camera Press; [1] Radio Times Hulton Picture Library; [2] Radio Times Hulton Picture Library; [4] Radio Times Hulton Picture Library;

[5] Popperfoto; [8] Radio Times Hulton Picture Library; [9] Life © Time Inc. 1976/Colorific; [10] Radio Times Hulton Picture Library; [11] Robert Hunt Library. **200–201** [Key] Kobal Collection; [1] By Courtesy of Madame Malthete Melies/© S.P.A.D.E.M., Paris, 1976; [2A] Out of copyright; [2B] Out of copyright; [3] Itala Films; [4] Kobal Collection; [5] Kobal Collection; [6] Kobal Collection; [7] Out of copyright; [8] Kobal Collection. **202–3** [Key] D. Bellon/Images et Textes/© A.D.A.G.P., Paris, 1976; [1] Gimpel Fils, London/© A.D.A.G.P., Paris, 1976; [2] Kunstsammlung Nordrhein Westfalen/© S.P.A.D.E.M., Paris, 1976; [3] Collection of the Museum of Modern Art, New York, Purchase; [4] Harry N. Abrams Inc/Permission of Nelson A. Rockefeller; [5] Collection of the Museum of Modern Art, New York. Acquired through an Anonymous Fund, The Mr & Mrs Joseph Slifka & Armand G. Erpf Funds, and by gift from the artist/© A.D.A.G.P., Paris, 1976; [6] Museum Boymans-Van Beuningen Rotterdam/© A.D.A.G.P., Paris, 1976; [7] Arts Council of Great Britain/© S.P.A.D.E.M., Paris, 1976; [8] Collection of the Museum of Modern Art, New York. The A. Conger Goodyear Fund; [9] National Museum, Moderna Museet, Stockholm. **204–5** [Key] Collection of the Museum of Modern Art, New York. Gift of Mr & Mrs David M. Solinger; [1] Kunstsammlung Nordrhein Westfalen, Düsseldorf/© A.D.A.G.P., Paris, 1976; [2] Stedelijk Museum, Amsterdam; [3] Angelo Hornak; [4] Haags Gemeentemuseum; [5] Trustees of the Tate Gallery; [6] Colourphoto Hans Hinz, Basle/Kunstmuseum, Basle/© A.D.A.G.P., Paris, 1976; [7] By kind permission of Henry Moore/Trustees of the Tate Gallery/John Webb; [8] Metropolitan Museum of Art, New York, George A. Heard Fund, 1957. **206–7** [Key] James Austin; [1] Angelo Hornak; [2] Cadbury Brown/Architectural Association; [4] Roger Whitehouse/Architectural Association; [6] James Stirling/Architectural Association; [7] Adrian Atkinson/Architectural Association; [8] Geoffrey Munro/Architectural Association; [9] James Stirling/Architectural Association. **208–9** [Key] Bettmann Archive; [4] Woolf, Laing, Christie & Partners; [5] Bettmann Archive; [6] Radio Times Hulton Picture Library; [7] Picturepoint; [9] Bettmann Archive. **210–11** [Key] London School of Economics; [1] Radio Times Hulton Picture Library; [2] Radio Times Hulton Picture Library; [3] Radio Times Hulton Picture Library; [5] Radio Times Hulton Picture Library; [6] Radio Times Hulton Picture Library; [7] Syndication International; [8] Radio Times Hulton Picture Library. **212–13** [Key] TUC Archive; [1] Mansell Collection; [2] Radio Times Hulton Picture Library; [3] Radio Times Hulton Picture Library; [4] Roger Viollet; [5] Roger Viollet; [6] Bettmann Archive; [8] Radio Times Hulton Picture Library; [9] Maurice Rickards. **214–15** [Key] Camera Press; [1] Source unknown; [3] Radio Times Hulton Picture Library; [4] Keystone Press; [5] Camera Press; [7] Keystone Press; [9] Robert Hunt Library. **216–17** [Key] Radio Times Hulton Picture Library; [1] Radio Times Hulton Picture Library; [2] National Army Museum; [3] Popperfoto; [4] Camera Press; [5] Mansell Collection; [6] Geoslides; [7] National Army Museum; [8] Robert Hunt Library; [10] Keystone Press. **218–19** [Key] Popperfoto; [1] Popperfoto; [2] ZEFA; [3] National Portrait Gallery; [4] Popperfoto; [5] Popperfoto; [6] Popperfoto; [7] Black Star Publishing Ltd; [8] Keystone Press Agency Ltd; [9] Popperfoto; [10] Camera Press. **220–1** [Key] Daily Mirror; [2] Trustees of the National Portrait Gallery; [3] Radio Times Hulton Picture Library; [5] Keystone Press; [6] Camera Press; [7] Royal Commonwealth Institute; [8] Associated Press. **222–3** [Key] Internationale Bildagentur; [2] Ullstein Bilderdienst; [2] Ullstein Bilderdienst; [3] Ullstein Bilderdienst; [5] Ullstein Bilderdienst; [6] International Bilderagentur; [8] Snark International. **224–5** [Key] Popperfoto; [1] Radio Times Hulton Picture Library; [2] Radio Times Hulton Picture Library; [3] Radio Times Hulton Picture Library; [4] British Airways; [5] Radio Times Hulton Picture Library; [6] Graham Newell; [7] Butlins; [8] Radio Times Hulton Picture Library; [9] Radio Times Hulton Picture Library; [10] Radio Times Hulton Picture Library. **226–7** [Key] Keystone Press; [1] Keystone Press; [2] Robert Hunt Library/Associated Press; [4] Archiv für Kunst und Geschichte; [7] Robert Hunt Library; [8] Cartoon by David Low by arrangement with the Trustees of the London *Evening Standard*; [9] Robert Hunt Library. **228–9** [2A] Robert Hunt Library; [2B] Robert Hunt Library; [2C] Imperial War Museum; [4] Robert Hunt Library; [6] Robert Hunt Library; [8] Robert Hunt Library/Imperial War Museum. **230–1** [1] Documentation Française; [2] Image Press; [3] Image Press; [4] Novosti Press Agency; [5] Popperfoto; [6] Imperial War Museum; [7] Official USAF Photo; [8] Imperial War Museum; [9] Popperfoto; [10] Australian War Memorial. **232–3** [Key] Radio Times Hulton Picture Library; [1] Radio Times Hulton Picture Library; [2] Radio Times Hulton Picture Library; [3] Imperial War Museum; [4] Fox Photos; [5] Roger Viollet; [6] Roger Viollet; [7] Angelo Hornak/Imperial War Museum; [8] Novosti Press Agency; [9] Orion Press/Camera Press; [10] Snark International; [11] IBA, Zurich; [12] Snark International. **234–5** [Key] Novosti Press Agency; [1] Robert Hunt Library; [2] Radio Times Hulton Picture Library; [3] IBA, Zurich; [4] Associated Press; [6] Keystone Press Agency; [7] IBA, Zurich; [8] Popperfoto; [9] Popperfoto. **236–7** [Key] Popperfoto; [1] Popperfoto; [2] Punch Publications Ltd; [3] Radio Times Hulton Picture Library; [5] Graham Nash/Conservative Party Archives; [6] Fox Photos; [7] Radio Times Hulton Picture Library; [8] Popperfoto; [9] Popperfoto. **238–9** [Key] Picturepoint; [1] Popperfoto; [2] David Strickland; [3] Popperfoto; [5] Camera Press; [6] Camera Press; [7] Rolls Royce Motors Ltd; [8] Camera Press; [9] Picturepoint. **240–1** [Key] Novosti Press Agency; [1] Popperfoto; [2] Popperfoto; [3] John F. Freeman; [5] Popperfoto; [8] Colorsport. **242–3** [Key] Camera Press; [2] Popperfoto; [3] Camera Press; [4] Associated Press; [5] Popperfoto; [6] Ginette Laborde, Paris; [7] Picturepoint. **244–5** [Key] Eastfoto; [1] Henri Cartier Bresson/John Hillelson Agency; [2] Eastfoto; [4] Camera Press; [6A] Sally and Richard Greenhill; [6B] Camera Press: [7] Camera Press; [9] Magnum Distribution. **246–7** [Key] Popperfoto; [2] Camera Press; [4] Popperfoto; [5] Camera Press; [6] Camera Press; [7] Australian News & Information Bureau. **248–9** [Key] Picturepoint; [1] Australian News & Information Bureau; [2] Anne-Marie Ehrlich; [5] Australian News & Information Bureau; [6] Popperfoto; [7] Popperfoto; [8] Camera Press; [9] Australian News & Information Bureau; [10] Australian News & Information Bureau. **250–1** [1] Weekly News; [2] New Zealand Herald; [3] Housing Corporation of New Zealand/Alexander Turnbull Library; [4] New Zealand Herald; [5] Weekly News; [6] New Zealand Forest Products Ltd, Auckland; [7] New Zealand Dairy Board; [8] Air New Zealand; [9] Frank Wallis. **252–3** [Key] Popperfoto; [1] Popperfoto; [2] De Beers Consolidated Mines Ltd; [3] National Portrait Gallery; [4] South Africa House; [5] Camera Press; [6] Camera Press; [7] Popperfoto; [8] Popperfoto; [9] Popperfoto. **254–5** [Key] Source unknown; [2] Camera Press; [3] Camera Press; [4] Associated Press; [5] Popperfoto; [7] Camera Press; [8] Keystone Press; [9] United Nations; [10] Agency for Public Information/Errol Harvey. **256–7** [Key] General Secretariat, Organization of American States; [1] National Palace of Mexico; [2] Radio Times Hulton Picture Library; [3] Camera Press; [4] Robert Cundy/Robert Harding Associates; [5] Associated Press; [7] G. A. Mather/Robert Harding Associates; [8] Ramano Cagnoni; [9] Douglas Botting; [10] Associated Press. **258–9** [Key] Associated Press; [1] Roger Viollet; [2] Popperfoto; [4A] Roger Viollet; [4B] Roger Viollet; [6] Associated Press; [7] Associated Press; [9] Camera Press; [10] Associated Press. **260–1** [Key] Popperfoto; [1] The Archives, Glasgow University; [2] Radio Times Hulton Picture Library; [3] Popperfoto; [5] Camera Press; [6] Picturepoint; [7] Picturepoint; [8] Picturepoint; [9] Camera Press; [10] Picturepoint. **262–3** [Key] Picturepoint; [1] Mansell Collection; [3] Radio Times Hulton Picture Library; [4] MB photograph; [5] Camera Press; [6] Welsh Arts Council; [7] Camera Press. **264–5** [Key] Shostal; [2A] Camera Press; [2B] Camera Press; [2C] Camera Press; [2D] Camera Press; [2E] Camera Press; [2F] Camera Press; [2G] Camera Press; [4] Associated Press. **266–7** [Key] Margaret Murray; [2] Geoff Goode; [3] Kibbutz Representatives; [4] Scottish New Towns Development Board; [6] Keystone Press; [7] Ted Lau, Fortune © Time Inc 1976/Colorific; [9] Camera Press; [10] Camera Press. **268–9** [Key] Copyright unknown; [1] Kobal Collection; [2] Kobal Collection; [3] Kobal Collection; [4] Kobal Collection; [5] Kobal Collection; [6] Kobal Collection; [7] Kobal Collection; [8] Kobal Collection; [9] Kobal Collection. **270–1** [Key] Universal [Edition] London Ltd; [3A] EMI; [3B] Photo Macdomnic, by courtesy of EMI; [4] Österreichische Nationalbibliothek; [9] Copyright unknown; [10] James Klosty/Edition Peters. **272–3** [Key] Max Jones; [2] American History Picture Library; [3] Max Jones; [4] Max Jones; [5] Redferns/pic by Herman Leonard (Charles Stewart Collection); [6] Max Jones; [7] Redferns/Stephen Morley; [8] Redferns/David Redfern; [9] CBS Records/photograph by Don Hunstein; [10] Jill Furmanovsky/Hipgnosis. **274–5** [Key] Mander & Mitchenson Theatre Collection; [2] Angelo Hornak/V & A; [3] Frank Sharman; [4] Musée des Arts Décoratifs, Paris; [5] Mander & Mitchenson Theatre Collection; [6] Fred Fehl; [7] Anthony Crickmay. **276–7** [Key] Columbia Warner Pictures/B.F.I.; [1] Robert J. Flaherty/B.F.I.; [2] Anycon Co/B.F.I.; [3] Ian Cameron; [4] Societé du Cinéma du Panthéon/B.F.I.; [5] RKO General Inc/B.F.I.; [6] Contemporary Films/B.F.I.; [7] Toho International/B.F.I.; [8] Aktiebolaget Svensk Filmindustri/B.F.I.; [9] Rizzoli Films SPA/B.F.I.; [10] Vaughan Films/B.F.I. **278–9** [Key] Leo Castelli, New York; [1] Tübingen Kunsthalle, Germany; [2] Kasmin Ltd; [3] Gallerie Reckermann, Cologne/Germot Langer; [4] Museum of Modern Art, New York. Gift of D. & J. de Menil; [5] Waddington Galleries; [6] Nigel Greenwood/Douglas Thompson; [7] Denise René Gallery, Paris; [8] Trustees of the Tate Gallery/John Webb. **280–1** [Key] Trustees of the Tate Gallery; [1] John Webb/Trustees of the Tate Gallery; [2] Cooper Bridgeman Library; [3] David Strickland; [4] John Webb/Trustees of the Tate Gallery; [5] Woodmansterne; [6] Kasmin Ltd; [7] John Goldblatt/© Floyd Picture Library; [8] Lisson Gallery. **282–3** [Key] Mander & Mitchenson Theatre Collection; [1] National Gallery of Ireland; [2] National Portrait Gallery; [3] Irish Tourist Board; [4] From *The Tain*/illustration by Louis le Brocquy/published by The Dolmen Press; [5] National Portrait Gallery; [6] Irish Tourist Board; [7] Picturepoint; [8] Andrew Wiard/Report. **284–5** [Key] Scottish Tourist Board; [1] National Portrait Gallery; [2] Scottish Tourist Board; [4] Woodmansterne; [5] Studio Swain/Mitchell Library, Glasgow; [6] Scottish Opera House, Glasgow; [7] Studio Swain/Mitchell Library, Glasgow; [8] Sir Basil Spence Partnership; [9] Trustees of the Tate Gallery. **286–7** [Key] The Wales Tourist Board; [1] Eileen Tweedy/National Library of Wales; [2] Eileen Tweedy/National Library of Wales; [3] Eileen Tweedy/National Library of Wales; [4] Eileen Tweedy/National Library of Wales; [5] National Portrait Gallery; [6] Jane Bown; [7] Trustees of The Tate Gallery; [8] Welsh Arts Council; [9] Welsh National Opera & Drama Co. **288–9** [Key] Mansell Collection; [1] Trustees of the British Museum/Ray Gardner; [2] Trustees of the British Museum/Ray Gardner; [3] Source unknown; [5] Sheila Orme/The National Trust; [6] SEFA; [7] Topix; [8] Henri Cartier Bresson/John Hillelson Agency; [9] Kobal Collection; [10] Alskog Inc; [11] Eva Sereny/Sygma/John Hillelson Agency; [12] Photo Baker, London; [13] Ian Whyles/Robert Harding Associates. **290–1** [Key] Camera Press; [1] The Beinecke Rare Book & Manuscript, Yale University; [2] John Watson; [3] Photo Baker, London; [4] Photo Baker, London; [5] Alan Durand/Robert Harding Associates; [6] Popperfoto; [7] Source unknown; [8] Associated Press; [9] Douglas H. Jeffrey; [10] Kindly loaned by Amos Tutuola; [11] Camera Press. **292–3** [Key] Roman Cagnoni; [2] Popperfoto; [3] Camera Press; [4] Camera Press; [5] Camera Press; [9] Camera Press; [10] Popperfoto. **294–5** [Key] Mander and Mitchenson Theatre Collection; [1] Mansell Collection; [2] Tony O'Malley Pictures Ltd; [3] Green Studio, Dublin; [4] Harland & Wolff Ltd; [5] Camera Press; [6] Camera Press; [7] Camera Press; [8] Arthur Guiness Son & Co (Dublin) Ltd; [9] Camera Press; [10] Popperfoto. **296–7** [Key] Institute of Contemporary History/Weiner Library/Geoff Goode; [3] Central Zionist Archives; [4] Central Press; [5] Camera Press. **298–9** [Key] United Nations; [1] Associated Press; [2] Associated Press; [4A]

United Nations; [4B] United Nations; [4C] United Nations; [4D] United Nations; [6] Camera Press; [7] Keystone Press; [8] Camera Press. **300–301** [Key] Terry Kirk/Financial Times; [3] Régis Bossu/Copyright SYGMA/Magnum. **302–3** [Key] Camera Press; [4] Paolo Koch; [5] C. Gascoigne/Robert Harding Associates; [6] Picturepoint; [7] Margaret Murray/Ikon Productions; [8] Peter Fraenkel. **304–5** [Key] Popperfoto; [1] World Council of Churches; [2] Camera Press; [4] Camera Press; [5] Associated Press; [6] Shostal; [7] Camera Press; [8] Associated Press; [9] Chris Steele Perkins. **306–7** [Key] Associated Press; [2] Associated Press; [3] Elliot Erwitt/Magnum; [4] Associated Press; [5] Associated Press; [6] Associated Press; [7] Neal Boenzi/The New York Times/John Hillelson Agency; [9] Camera Press; [10] Popperfoto.

Time Chart
310–11 Mrs Barbara Edwards/Carter Nash Cameron; Michael Holford/V & A; Mansell Collection; Scala; Mansell Collection; Mansell Collection; Ronan Picture Library. **312–13** Michael Holford; Michael Holford/Louvre; Giraudon. **314–15** Mansell Collection; Mansell Collection; Cooper Bridgeman; Mansell Collection; Mansell Collection. **316–17** Mansell Collection; Cooper Bridgeman; Cooper Bridgeman; Mansell Collection. **318–19** Ronan Picture Library; Mansell Collection; Ronan Picture Library; Robert Hunt Library; Mansell Collection; Bildarchiv Preussischer Kulturbesitz. **320–1** Mansell Collection; Mansell Collection; Bildarchiv Foto Marburg. **322–3** Mansell Collection; Mansell Collection; Imperial War Museum; Mansell Collection; Mansell Collection. **324–5** Radio Times Hulton Picture Library; Popperfoto; Maurice Rickards; Robert Hunt Library; Novosti Press Agency; Novosti Press Agency; Camera Press. **326–7** Camera Press; Camera Press; Camera Press; Keystone Press; Camera Press; Camera Press; Camera Press; Camera Press.

Artwork Credits

Art Editors
Angela Downing; George Glaze; James Marks; Mel Peterson; Ruth Prentice; Bob Scott

Visualizers
David Aston; Javed Badar; Allison Blythe; Angela Braithwaite; Alan Brown; Michael Burke; Alistair Campbell; Terry Collins; Mary Ellis; Judith Escreet; Albert Jackson; Barry Jackson; Ted Kindsey; Kevin Maddison; Erika Mathlow; Paul Mundon; Peter Nielson; Patrick O'Callaghan; John Ridgeway; Peter Saag; Malcolme Smythe; John Stanyon; John Stewart; Justin Todd; Linda Wheeler

Artists
Stephen Adams; Geoffrey Alger; Terry Allen; Jeremy Alsford; Frederick Andenson; John Arnold; Peter Arnold; David Ashby; Michael Badrock; William Baker; John Barber; Norman Barber; Arthur Barvoso; John Batchelor; John Bavosi; David Baxter; Stephen Bernette; John Blagovitch; Michael Blore; Christopher Blow; Roger Bourne; Alistair Bowtell; Robert Brett; Gordon Briggs; Linda Broad; Lee Brooks; Rupert Brown; Marilyn Bruce; Anthony Bryant; Paul Buckle; Sergio Burelli; Dino Bussetti; Patricia Casey; Giovanni Casselli; Nigel Chapman; Chensie Chen; David Chisholm; David Cockcroft; Michael Codd; Michael Cole; Terry Collins; Peter Connelly; Roy Coombs; David Cox; Patrick Cox; Brian Cracker; Gordon Cramp; Gino D'Achille; Terrence Daley; John Davies; Gordon C. Davis; David Day; Graham Dean; Brian Delf; Kevin Diaper; Madeleine Dinkel; Hugh Dixon; Paul Draper; David Dupe; Howard Dyke; Jennifer Eachus; Bill Easter; Peter Edwards; Michael Ellis; Jennifer Embleton; Ronald Embleton; Ian Evans; Ann Evens; Lyn Evens; Peter Fitzjohn; Eugene Flurey; Alexander Forbes; David Carl Forbes; Chris Fosey; John Francis; Linda Francis; Sally Frend; Brian Froud; Gay Galfworthy; Ian Garrard; Jean George; Victoria Goaman; David Godfrey; Miriam Golochoy; Anthea Gray; Harold Green; Penelope Greensmith; Vanna Haggerty; Nicholas Hall; Horgrave Hans; David Hardy; Douglas Harker; Richard Hartwell; Jill Havergale; Peter Hayman; Ron Haywood; Peter Henville; Trevor Hill; Garry Hinks; Peter Hutton; Faith Jacques; Robin Jacques; Lancelot Jones; Anthony Joyce; Pierre Junod; Patrick Kaley; Sarah Kensington; Don Kidman; Harold King; Martin Lambourne; Ivan Lapper; Gordon Lawson; Malcolm Lee-Andrews; Peter Levaffeur; Richard Lewington; Brian Lewis; Ken Lewis; Richard Lewis; Kenneth Lilly; Michael Little; David Lock; Garry Long; John Vernon Lord; Vanessa Luff; John Mac; Lesley MacIntyre; Thomas McArthur; Michael McGuinness; Ed McKenzie; Alan Male; Ben Manchipp; Neville Mardell; Olive Marony; Bob Martin; Gordon Miles; Sean Milne; Peter Mortar; Robert Morton; Trevor Muse; Anthony Nelthorpe; Michael Neugebauer; William Nickless; Eric Norman; Peter North; Michael O'Rourke; Richard Orr; Nigel Osborne; Patrick Oxenham; John Painter; David Palmer; Geoffrey Parr; Allan Penny; David Penny; Charles Pickard; John Pinder; Maurice Pledger; Judith Legh Pope; Michael Pope; Andrew Popkiewicz; Brian Price-Thomas; Josephine Rankin; Collin Rattray; Charles Raymond; Alan Rees; Ellsie Rigley; John Ringnall; Christine Robbins; Ellie Robertson; James Robins; John Ronayne; Collin Rose; Peter Sarson; Michael Saunders; Ann Savage; Dennis Scott; Edward Scott-Jones; Rodney Shackell; Chris Simmonds; Gwendolyn Simson; Cathleen Smith; Les Smith; Stanley Smith; Michael Soundels; Wolf Spoel; Ronald Steiner; Ralph Stobart; Celia Stothard; Peter Sumpter; Rod Sutterby; Allan Suttie; Tony Swift; Michael Terry; John Thirsk; Eric Thomas; George Thompson; Kenneth Thompson; David Thorpe; Harry Titcombe; Peter Town; Michael Trangenza; Joyce Tuhill; Glenn Tutssel; Carol Vaucher; Edward Wade; Geoffrey Wadsley; Mary Waldron; Michael Walker; Dick Ward; Brian Watson; David Watson; Peter Weavers; David Wilkinson; Ted Williams; John Wilson; Roy Wiltshire; Terrence Wingworth; Anne Winterbotham; Albany Wiseman; Vanessa Wiseman; John Wood; Michael Woods; Owen Woods; Sidney Woods; Raymond Woodward; Harold Wright; Julia Wright

Studios
Add Make-up; Alard Design; Anyart; Arka Graphics; Artec; Art Liaison; Art Workshop; Bateson Graphics; Broadway Artists; Dateline Graphics; David Cox Associates; David Levin Photographic; Eric Jewel Associates; George Miller Associates; Gilcrist Studios; Hatton Studio; Jackson Day; Lock Pettersen Ltd; Mitchell Beazley Studio; Negs Photographic; Paul Hemus Associates; Product Support Graphics; Q.E.D. [Campbell Kindsley]; Stobart and Sutterby; Studio Briggs; Technical Graphics; The Diagram Group; Tri Art; Typographics; Venner Artists

Agents
Artist Partners; Freelance Presentations; Garden Studio; Linden Artists; N.E. Middletons; Portman Artists; Saxon Artists; Thompson Artists

The quotation from Aldous Huxley's essay *A Case of Voluntary Ignorance* on page 311 is by permission of Mrs Laura Huxley and Chatto and Windus Ltd. The quotation from Tom Stoppard's play *Rosencrantz and Guildenstern Are Dead* on page 311 is by permission of the publishers, Faber and Faber Ltd. The map on page 167 is based on information in *An Historical Atlas of Wales*, by William Rees, published by Faber and Faber. The quotation from Aldous Huxley's essay "A Case of Voluntary Ignorance" in *Collected Essays* (1960) is by permission of Mrs Laura Huxley and the publishers, Chatto & Windus Ltd. The map on page 128 is based on information in *South African Atlas*, by G. S. P. Freeman-Grenville, published by Rex Collings.